University Physics
Semester 2

PEARSON
Custom Publishing

We work with leading authors to develop the strongest educational materials bringing cutting-edge thinking and best learning practice to a global market.

Under a range of well-known imprints, including Financial Times/Prentice Hall, Addison Wesley and Longman, we craft high quality print and electronic publications which help readers to understand and apply their content, whether studying or at work.

Pearson Custom Publishing enables our customers to access a wide and expanding range of market-leading content from world-renowned authors and develop their own tailor-made book. You choose the content that meets your needs and Pearson Custom Publishing produces a high-quality printed book.

To find out more about custom publishing, visit www.pearsoncustom.co.uk

A Pearson Custom Publication

University Physics
Semester 2

Compiled from:

Essential University Physics
Volume 1
by Richard Wolfson

Sears and Zemansky's
University Physics with Modern Physics
Twelfth Edition
by Hugh Young, Roger Freedman and A.Lewis Ford

PEARSON
Custom
Publishing

Pearson Education Limited
Edinburgh Gate
Harlow
Essex CM20 2JE

And associated companies throughout the world

Visit us on the World Wide Web at:
www.pearsoned.co.uk

First published 2009

Taken from:

Essential University Physics Volume 1
by Richard Wolfson

ISBN 978 0 80 533829 4

Sears and Zemansky's University Physics with Modern Physics Twelfth Edition
by Hugh Young, Roger Freedman and A.Lewis Ford

ISBN 978 0 32 150121 9

ISBN 978 1 84776 273 3

Printed and bound in Great Britain by Ashford Colour Press, Gosport, Hampshire.

Contents

PART THREE

PHYS 1011:

Waves, Light and Quanta

PERIODIC MOTION

13

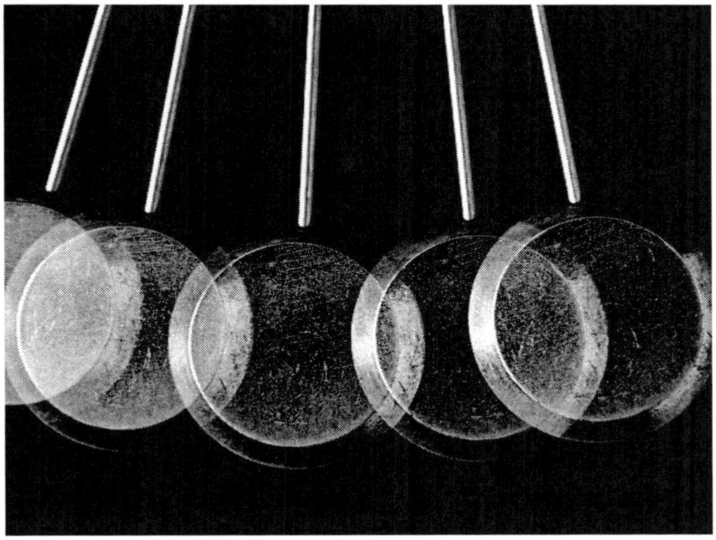

? Suppose you doubled the mass of a clock's pendulum (including the rod and the weight at the end) while keeping its dimensions the same. Would the clock run fast or slow?

LEARNING GOALS

By studying this chapter, you will learn:

- How to describe oscillations in terms of amplitude, period, frequency, and angular frequency.
- How to do calculations with simple harmonic motion, an important type of oscillation.
- How to use energy concepts to analyze simple harmonic motion.
- How to apply the ideas of simple harmonic motion to different physical situations.
- How to analyze the motions of a simple pendulum.
- What a physical pendulum is, and how to calculate the properties of its motion.
- What determines how rapidly an oscillation dies out.
- How a driving force applied to an oscillator at the right frequency can cause a very large response, or resonance.

Many kinds of motion repeat themselves over and over: the vibration of a quartz crystal in a watch, the swinging pendulum of a grandfather clock, the sound vibrations produced by a clarinet or an organ pipe, and the back-and-forth motion of the pistons in a car engine. This kind of motion, called **periodic motion** or **oscillation,** is the subject of this chapter. Understanding periodic motion will be essential for our later study of waves, sound, alternating electric currents, and light.

A body that undergoes periodic motion always has a stable equilibrium position. When it is moved away from this position and released, a force or torque comes into play to pull it back toward equilibrium. But by the time it gets there, it has picked up some kinetic energy, so it overshoots, stopping somewhere on the other side, and is again pulled back toward equilibrium. Picture a ball rolling back and forth in a round bowl or a pendulum that swings back and forth past its straight-down position.

In this chapter we will concentrate on two simple examples of systems that can undergo periodic motions: spring-mass systems and pendulums. We will also study why oscillations often tend to die out with time and why some oscillations can build up to greater and greater displacements from equilibrium when periodically varying forces act.

13.1 Describing Oscillation

Figure 13.1 shows one of the simplest systems that can have periodic motion. A body with mass m rests on a frictionless horizontal guide system, such as a linear air track, so it can move only along the x-axis. The body is attached to a spring of negligible mass that can be either stretched or compressed. The left end of the spring is held fixed and the right end is attached to the body. The spring force is the only horizontal force acting on the body; the vertical normal and gravitational forces always add to zero.

13.1 A system that can have periodic motion.

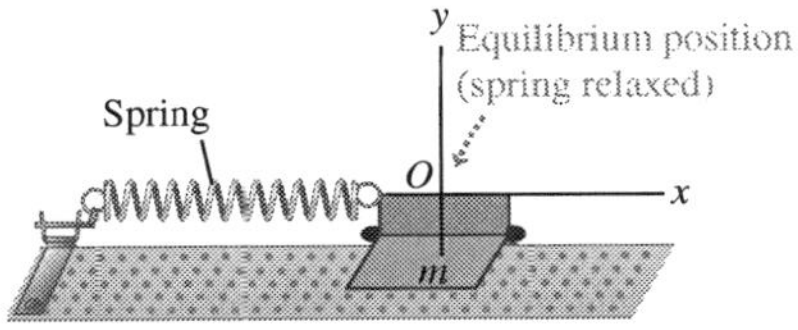

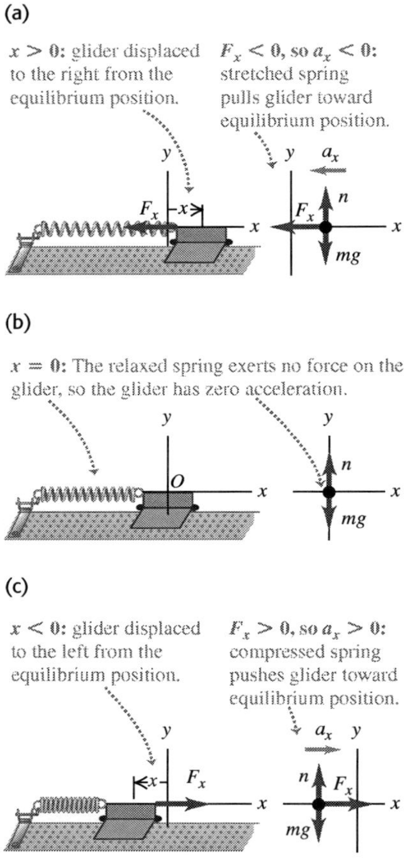

13.2 Model for periodic motion. When the body is displaced from its equilibrium position at $x = 0$, the spring exerts a restoring force back toward the equilibrium position.

It's simplest to define our coordinate system so that the origin O is at the equilibrium position, where the spring is neither stretched nor compressed. Then x is the x-component of the **displacement** of the body from equilibrium and is also the change in the length of the spring. The x-component of the force that the spring exerts on the body is F_x, and the x-component of acceleration a_x is given by $a_x = F_x/m$.

Figure 13.2 shows the body for three different displacements of the spring. Whenever the body is displaced from its equilibrium position, the spring force tends to restore it to the equilibrium position. We call a force with this character a **restoring force.** Oscillation can occur only when there is a restoring force tending to return the system to equilibrium.

Let's analyze how oscillation occurs in this system. If we displace the body to the right to $x = A$ and then let go, the net force and the acceleration are to the left (Fig. 13.2a). The speed increases as the body approaches the equilibrium position O. When the body is at O, the net force acting on it is zero (Fig. 13.2b), but because of its motion it *overshoots* the equilibrium position. On the other side of the equilibrium position the body is still moving to the left, but the net force and the acceleration are to the right (Fig. 13.2c); hence the speed decreases until the body comes to a stop. We will show later that with an ideal spring, the stopping point is at $x = -A$. The body then accelerates to the right, overshoots equilibrium again, and stops at the starting point $x = A$, ready to repeat the whole process. The body is oscillating! If there is no friction or other force to remove mechanical energy from the system, this motion repeats forever; the restoring force perpetually draws the body back toward the equilibrium position, only to have the body overshoot time after time.

In different situations the force may depend on the displacement x from equilibrium in different ways. But oscillation *always* occurs if the force is a *restoring* force that tends to return the system to equilibrium.

Amplitude, Period, Frequency, and Angular Frequency

Here are some terms that we'll use in discussing periodic motions of all kinds:

The **amplitude** of the motion, denoted by A, is the maximum magnitude of displacement from equilibrium—that is, the maximum value of $|x|$. It is always positive. If the spring in Fig. 13.2 is an ideal one, the total overall range of the motion is $2A$. The SI unit of A is the meter. A complete vibration, or **cycle,** is one complete round trip—say, from A to $-A$ and back to A, or from O to A, back through O to $-A$, and back to O. Note that motion from one side to the other (say, $-A$ to A) is a half-cycle, not a whole cycle.

The **period,** T, is the time for one cycle. It is always positive. The SI unit is the second, but it is sometimes expressed as "seconds per cycle."

The **frequency,** f, is the number of cycles in a unit of time. It is always positive. The SI unit of frequency is the hertz:

$$1 \text{ hertz} = 1 \text{ Hz} = 1 \text{ cycle/s} = 1 \text{ s}^{-1}$$

This unit is named in honor of the German physicist Heinrich Hertz (1857–1894), a pioneer in investigating electromagnetic waves.

The **angular frequency,** ω, is 2π times the frequency:

$$\omega = 2\pi f$$

We'll learn shortly why ω is a useful quantity. It represents the rate of change of an angular quantity (not necessarily related to a rotational motion) that is always measured in radians, so its units are rad/s. Since f is in cycle/s, we may regard the number 2π as having units rad/cycle.

From the definitions of period T and frequency f we see that each is the reciprocal of the other:

$$f = \frac{1}{T} \qquad T = \frac{1}{f} \qquad \text{(relationships between frequency and period)} \qquad (13.1)$$

Also, from the definition of ω,

$$\omega = 2\pi f = \frac{2\pi}{T} \quad \text{(angular frequency)} \tag{13.2}$$

Example 13.1 Period, frequency, and angular frequency

An ultrasonic transducer (a kind of loudspeaker) used for medical diagnosis oscillates at a frequency of 6.7 MHz $= 6.7 \times 10^6$ Hz. How much time does each oscillation take, and what is the angular frequency?

SOLUTION

IDENTIFY: Our target variables are the period T and the angular frequency ω.

SET UP: We are given the frequency f, so we can find these variables using Eqs. (13.1) and (13.2).

EXECUTE: From Eqs. (13.1) and (13.2),

$$T = \frac{1}{f} = \frac{1}{6.7 \times 10^6 \text{ Hz}} = 1.5 \times 10^{-7} \text{ s} = 0.15 \ \mu\text{s}$$

$$\begin{aligned}\omega &= 2\pi f = 2\pi(6.7 \times 10^6 \text{ Hz}) \\ &= (2\pi \text{ rad/cycle})(6.7 \times 10^6 \text{ cycle/s}) \\ &= 4.2 \times 10^7 \text{ rad/s}\end{aligned}$$

EVALUATE: This is a very rapid vibration, with large f and ω and small T. A slow vibration has small f and ω and large T.

Test Your Understanding of Section 13.1 A body like that shown in Fig. 13.2 oscillates back and forth. For each of the following values of the body's x-velocity v_x and x-acceleration a_x, state whether its displacement x is positive, negative, or zero. (a) $v_x > 0$ and $a_x > 0$; (b) $v_x > 0$ and $a_x < 0$; (c) $v_x < 0$ and $a_x > 0$; (d) $v_x < 0$ and $a_x < 0$; (e) $v_x = 0$ and $a_x < 0$; (f) $v_x > 0$ and $a_x = 0$.

13.2 Simple Harmonic Motion

The simplest kind of oscillation occurs when the restoring force F_x is *directly proportional* to the displacement from equilibrium x. This happens if the spring in Figs. 13.1 and 13.2 is an ideal one that obeys Hooke's law. The constant of proportionality between F_x and x is the force constant k. (You may want to review Hooke's law and the definition of the force constant in Section 6.3.) On either side of the equilibrium position, F_x and x always have opposite signs. In Section 6.3 we represented the force acting *on* a stretched ideal spring as $F_x = kx$. The x-component of force the spring exerts *on the body* is the negative of this, so the x-component of force F_x on the body is

$$F_x = -kx \quad \text{(restoring force exerted by an ideal spring)} \tag{13.3}$$

This equation gives the correct magnitude and sign of the force, whether x is positive, negative, or zero (Fig. 13.3). The force constant k is always positive and has units of N/m (a useful alternative set of units is kg/s^2). We are assuming that there is no friction, so Eq. (13.3) gives the *net* force on the body.

When the restoring force is directly proportional to the displacement from equilibrium, as given by Eq. (13.3), the oscillation is called **simple harmonic motion,** *abbreviated* **SHM.** The acceleration $a_x = d^2x/dt^2 = F_x/m$ of a body in SHM is given by

$$a_x = \frac{d^2x}{dt^2} = -\frac{k}{m}x \quad \text{(simple harmonic motion)} \tag{13.4}$$

The minus sign means the acceleration and displacement always have opposite signs. This acceleration is *not* constant, so don't even think of using the constant-acceleration equations from Chapter 2. We'll see shortly how to solve this

13.3 An idealized spring exerts a restoring force that obeys Hooke's law, $F_x = -kx$. Oscillation with such a restoring force is called simple harmonic motion.

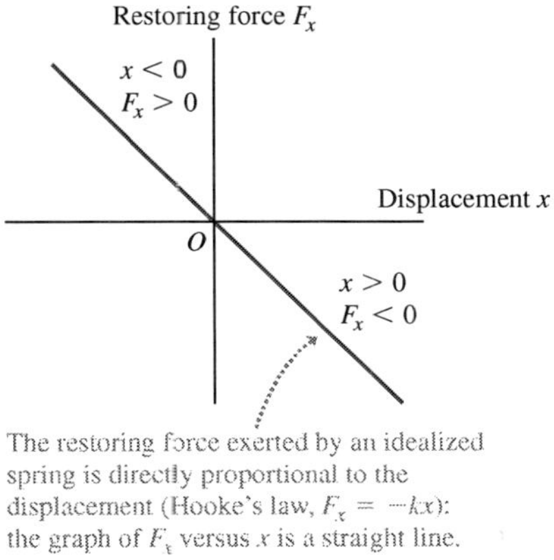

13.4 In most real oscillations Hooke's law applies provided the body doesn't move too far from equilibrium. In such a case small-amplitude oscillations are approximately simple harmonic.

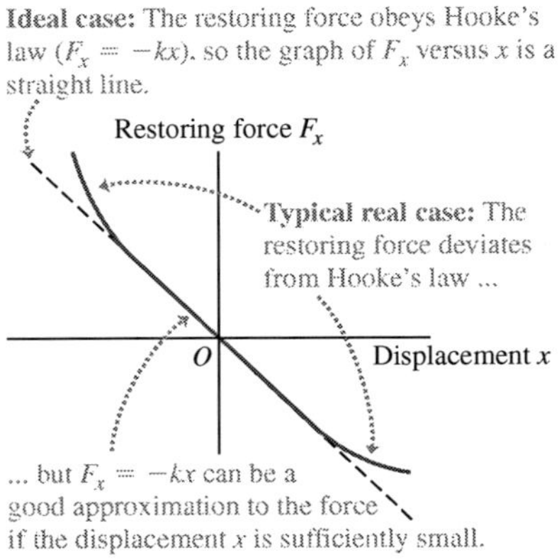

equation to find the displacement x as a function of time. A body that undergoes simple harmonic motion is called a **harmonic oscillator.**

Why is simple harmonic motion important? Keep in mind that not all periodic motions are simple harmonic; in periodic motion in general, the restoring force depends on displacement in a more complicated way than in Eq. (13.3). But in many systems the restoring force is *approximately* proportional to displacement if the displacement is sufficiently small (Fig. 13.4). That is, if the amplitude is small enough, the oscillations of such systems are approximately simple harmonic and therefore approximately described by Eq. (13.4). Thus we can use SHM as an approximate model for many different periodic motions, such as the vibration of the quartz crystal in a watch, the motion of a tuning fork, the electric current in an alternating-current circuit, and the oscillations of atoms in molecules and solids.

Circular Motion and the Equations of SHM

To explore the properties of simple harmonic motion, we must express the displacement x of the oscillating body as a function of time, $x(t)$. The second derivative of this function, d^2x/dt^2, must be equal to $(-k/m)$ times the function itself, as required by Eq. (13.4). As we mentioned, the formulas for constant acceleration from Section 2.4 are no help because the acceleration changes constantly as the displacement x changes. Instead, we'll find $x(t)$ by noticing a striking similarity between SHM and another form of motion that we've already studied in detail.

Figure 13.5a shows a top view of a horizontal disk of radius A with a ball attached to its rim at point Q. The disk rotates with constant angular speed ω (measured in rad/s), so the ball moves in uniform circular motion. A horizontal light beam shines on the rotating disk and casts a shadow of the ball on a screen. The shadow at point P oscillates back and forth as the ball moves in a circle. We then arrange a body attached to an ideal spring, like the combination shown in Figs. 13.1 and 13.2, so that the body oscillates parallel to the shadow. We will prove that the motion of the body and the motion of the ball's shadow are *identical* if the amplitude of the body's oscillation is equal to the disk radius A, and if the angular frequency $2\pi f$ of the oscillating body is equal to the angular speed ω of the rotating disk. That is, *simple harmonic motion is the projection of uniform circular motion onto a diameter.*

We can verify this remarkable statement by finding the acceleration of the shadow at P and comparing it to the acceleration of a body undergoing SHM, given

13.5 (a) Relating uniform circular motion and simple harmonic motion. (b) The ball's shadow moves exactly like a body oscillating on an ideal spring.

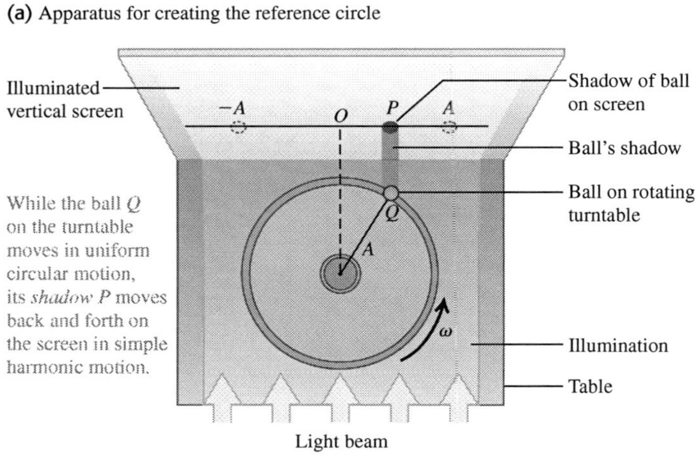

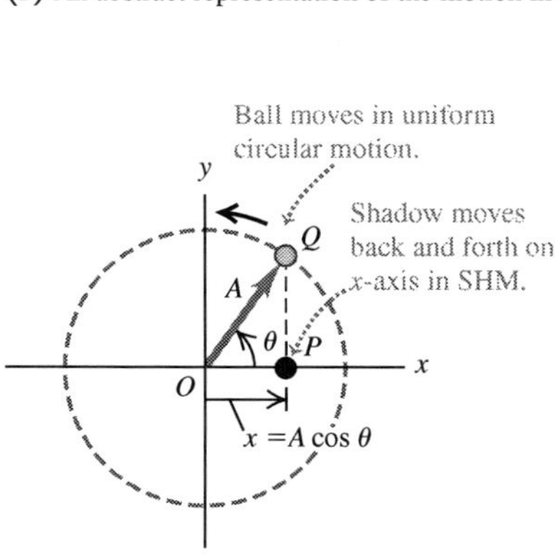

by Eq. (13.4). The circle in which the ball moves so that its projection matches the motion of the oscillating body is called the **reference circle;** we will call the point Q the *reference point.* We take the reference circle to lie in the xy-plane, with the origin O at the center of the circle (Fig. 13.5b). At time t the vector OQ from the origin to the reference point Q makes an angle θ with the positive x-axis. As the point Q moves around the reference circle with constant angular speed ω, the vector OQ rotates with the same angular speed. Such a rotating vector is called a **phasor.** (This term was in use long before the invention of the Star Trek stun gun with a similar name. The phasor method for analyzing oscillations is useful in many areas of physics. We'll use phasors when we study alternating-current circuits in Chapter 31 and the interference of light in Chapters 35 and 36.)

The x-component of the phasor at time t is just the x-coordinate of the point Q:

$$x = A\cos\theta \tag{13.5}$$

This is also the x-coordinate of the shadow P, which is the *projection* of Q onto the x-axis. Hence the x-velocity of the shadow P along the x-axis is equal to the x-component of the velocity vector of the reference point Q (Fig. 13.6a), and the x-acceleration of P is equal to the x-component of the acceleration vector of Q (Fig. 13.6b). Since point Q is in uniform circular motion, its acceleration vector $\vec{a}_Q$ is always directed toward O. Furthermore, the magnitude of $\vec{a}_Q$ is constant and given by the angular speed squared times the radius of the circle (see Section 9.3):

$$a_Q = \omega^2 A \tag{13.6}$$

Figure 13.6b shows that the x-component of $\vec{a}_Q$ is $a_x = -a_Q\cos\theta$. Combining this with Eqs. (13.5) and (13.6), we get that the acceleration of point P is

$$a_x = -a_Q\cos\theta = -\omega^2 A\cos\theta \quad \text{or} \tag{13.7}$$

$$a_x = -\omega^2 x \tag{13.8}$$

The acceleration of the point P is directly proportional to the displacement x and always has the opposite sign. These are precisely the hallmarks of simple harmonic motion.

Equation (13.8) is *exactly* the same as Eq. (13.4) for the acceleration of a harmonic oscillator, provided that the angular speed ω of the reference point Q is related to the force constant k and mass m of the oscillating body by

$$\omega^2 = \frac{k}{m} \quad \text{or} \quad \omega = \sqrt{\frac{k}{m}} \tag{13.9}$$

We have been using the same symbol ω for the angular *speed* of the reference point Q and the angular *frequency* of the oscillating point P. The reason is that these quantities are equal! If point Q makes one complete revolution in time T, then point P goes through one complete cycle of oscillation in the same time; hence T is the period of the oscillation. During time T the point Q moves through 2π radians, so its angular speed is $\omega = 2\pi/T$. But this is just the same as Eq. (13.2) for the angular frequency of the point P, which verifies our statement about the two interpretations of ω. This is why we introduced angular frequency in Section 13.1; this quantity makes the connection between oscillation and circular motion. So we reinterpret Eq. (13.9) as an expression for the angular frequency of simple harmonic motion for a body of mass m, acted on by a restoring force with force constant k:

$$\omega = \sqrt{\frac{k}{m}} \quad \text{(simple harmonic motion)} \tag{13.10}$$

When you start a body oscillating in SHM, the value of ω is not yours to choose; it is predetermined by the values of k and m. The units of k are N/m or kg/s^2, so

13.6 The **(a)** x-velocity and **(b)** x-acceleration of the ball's shadow P (See Fig. 13.5) are the x-components of the velocity and acceleration vectors, respectively, of the ball Q.

(a) Using the reference circle to determine the x-velocity of point P

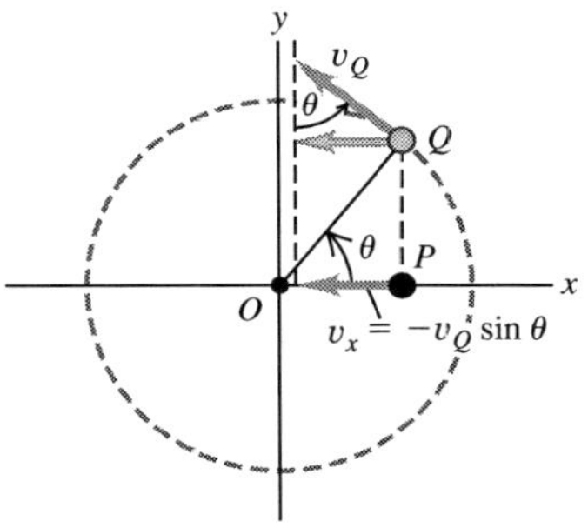

(b) Using the reference circle to determine the x-acceleration of point P

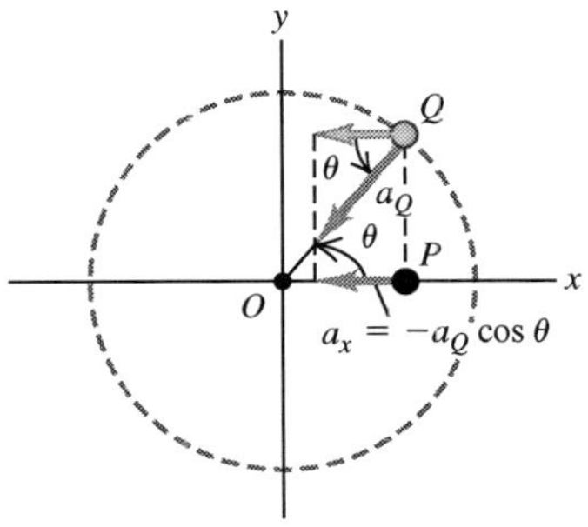

k/m is in $(\text{kg/s}^2)/\text{kg} = \text{s}^{-2}$. When we take the square root in Eq. (13.10), we get s^{-1}, or more properly rad/s because this is an *angular* frequency (recall that a radian is not a true unit).

According to Eqs. (13.1) and (13.2), the frequency f and period T are

$$f = \frac{\omega}{2\pi} = \frac{1}{2\pi}\sqrt{\frac{k}{m}} \quad \text{(simple harmonic motion)} \tag{13.11}$$

$$T = \frac{1}{f} = \frac{2\pi}{\omega} = 2\pi\sqrt{\frac{m}{k}} \quad \text{(simple harmonic motion)} \tag{13.12}$$

We see from Eq. (13.12) that a larger mass m, with its greater inertia, will have less acceleration, move more slowly, and take a longer time for a complete cycle (Fig. 13.7). In contrast, a stiffer spring (one with a larger force constant k) exerts a greater force at a given deformation x, causing greater acceleration, higher speeds, and a shorter time T per cycle.

13.7 The greater the mass m in a tuning fork's tines, the lower the frequency of oscillation $f = (1/2\pi)\sqrt{k/m}$ and the lower the pitch of the sound that the tuning fork produces.

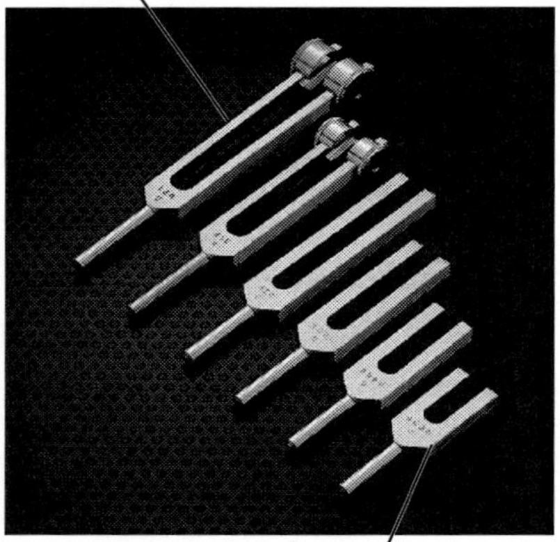

CAUTION Don't confuse frequency and angular frequency You can run into trouble if you don't make the distinction between frequency f and angular frequency $\omega = 2\pi f$. Frequency tells you how many cycles of oscillation occur per second, while angular frequency tells you how many radians per second this corresponds to on the reference circle. In solving problems, pay careful attention to whether the goal is to find f or ω.

Period and Amplitude in SHM

Equations (13.11) and (13.12) show that the period and frequency of simple harmonic motion are completely determined by the mass m and the force constant k. *In simple harmonic motion the period and frequency do not depend on the amplitude A.* For given values of m and k, the time of one complete oscillation is the same whether the amplitude is large or small. Equation (13.3) shows why we should expect this. Larger A means that the body reaches larger values of $|x|$ and is subjected to larger restoring forces. This increases the average speed of the body over a complete cycle; this exactly compensates for having to travel a larger distance, so the same total time is involved.

The oscillations of a tuning fork are essentially simple harmonic motion, which means that it always vibrates with the same frequency, independent of amplitude. This is why a tuning fork can be used as a standard for musical pitch. If it were not for this characteristic of simple harmonic motion, it would be impossible to make familiar types of mechanical and electronic clocks run accurately or to play most musical instruments in tune. If you encounter an oscillating body with a period that *does* depend on the amplitude, the oscillation is *not* simple harmonic motion.

Example 13.2 Angular frequency, frequency, and period in SHM

A spring is mounted horizontally, with its left end held stationary. By attaching a spring balance to the free end and pulling toward the right (Fig. 13.8a), we determine that the stretching force is proportional to the displacement and that a force of 6.0 N causes a displacement of 0.030 m. We remove the spring balance and attach a 0.50-kg glider to the end, pull it a distance of 0.020 m along a frictionless air track, release it, and watch it oscillate (Fig. 13.8b). (a) Find the force constant of the spring. (b) Find the angular frequency, frequency, and period of the oscillation.

SOLUTION

IDENTIFY: Because the spring force (equal in magnitude to the stretching force) is proportional to the displacement, the motion is simple harmonic.

SET UP: We find the value of the force constant k using Hooke's law, Eq. (13.3), and the values of ω, f, and T using Eqs. (13.10), (13.11), and (13.12), respectively.

EXECUTE: (a) When $x = 0.030$ m, the force the spring exerts on the spring balance is $F_x = -6.0$ N. From Eq. (13.3),

$$k = -\frac{F_x}{x} = -\frac{-6.0\ \text{N}}{0.030\ \text{m}} = 200\ \text{N/m} = 200\ \text{kg/s}^2$$

13.8 (a) The force exerted *on* the spring (shown by the vector F) has x-component $F_x = +6.0$ N. The force exerted *by* the spring has x-component $F_x = -6.0$ N. (b) A glider is attached to the same spring and allowed to oscillate.

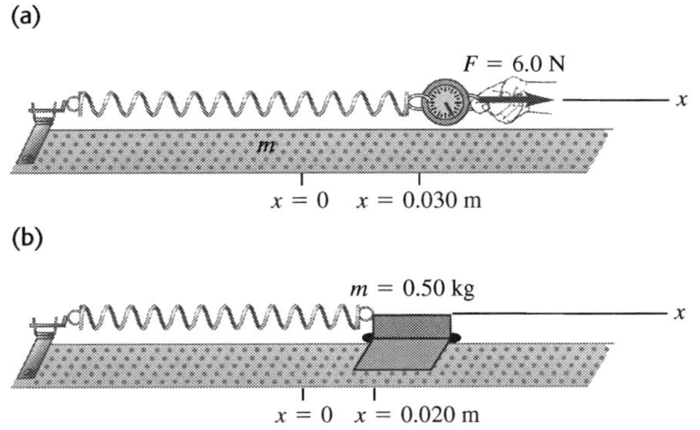

(b) Using $m = 0.50$ kg in Eq. (13.10), we find

$$\omega = \sqrt{\frac{k}{m}} = \sqrt{\frac{200\ \text{kg/s}^2}{0.50\ \text{kg}}} = 20\ \text{rad/s}$$

The frequency f is

$$f = \frac{\omega}{2\pi} = \frac{20\ \text{rad/s}}{2\pi\ \text{rad/cycle}} = 3.2\ \text{cycle/s} = 3.2\ \text{Hz}$$

The period T is the reciprocal of the frequency f:

$$T = \frac{1}{f} = \frac{1}{3.2\ \text{cycle/s}} = 0.31\ \text{s}$$

A period is usually stated in "seconds" rather than "seconds per cycle."

EVALUATE: The amplitude of the oscillation is 0.020 m, the distance to the right that we pulled the glider attached to the spring before releasing it. We didn't need to use this information to find the angular frequency, frequency, or period, because in SHM none of these quantities depend on the amplitude.

Displacement, Velocity, and Acceleration in SHM

ActivPhysics ONLINE

9.1 Position Graphs and Equations
9.2 Describing Vibrational Motion
9.5 Ape Drops Tarzan

We still need to find the displacement x as a function of time for a harmonic oscillator. Equation (13.4) for a body in simple harmonic motion along the x-axis is identical to Eq. (13.8) for the x-coordinate of the reference point in uniform circular motion with constant angular speed $\omega = \sqrt{k/m}$. It follows that Eq. (13.5), $x = A\cos\theta$, describes the coordinate x for both of these situations. If at $t = 0$ the phasor OQ makes an angle ϕ (the Greek letter phi) with the positive x-axis, then at any later time t this angle is $\theta = \omega t + \phi$. We substitute this into Eq. (13.5) to obtain

$$x = A\cos(\omega t + \phi) \qquad \text{(displacement in SHM)} \qquad (13.13)$$

where $\omega = \sqrt{k/m}$. Figure 13.9 shows a graph of Eq. (13.13) for the particular case $\phi = 0$. The displacement x is a periodic function of time, as expected for SHM. We could also have written Eq. (13.13) in terms of a sine function rather than a cosine by using the identity $\cos\alpha = \sin(\alpha + \pi/2)$. *In simple harmonic motion the position is a periodic, sinusoidal function of time.* There are many other periodic functions, but none so smooth and simple as a sine or cosine function.

13.9 Graph of x versus t [See Eq. (13.13)] for simple harmonic motion. The case shown has $\phi = 0$.

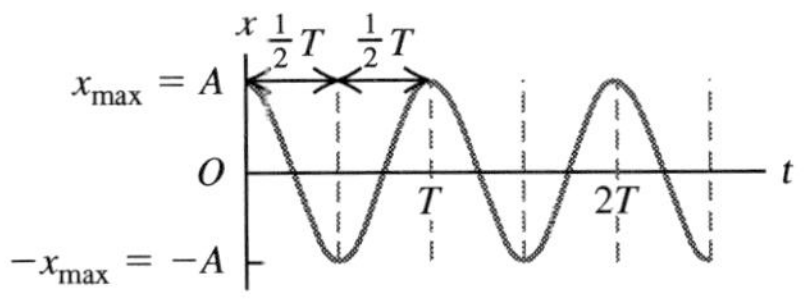

The value of the cosine function is always between -1 and 1, so in Eq. (13.13), x is always between $-A$ and A. This confirms that A is the amplitude of the motion.

The period T is the time for one complete cycle of oscillation, as Fig. 13.9 shows. The cosine function repeats itself whenever the quantity in parentheses in Eq. (13.13) increases by 2π radians. Thus, if we start at time $t = 0$, the time T to complete one cycle is given by

$$\omega T = \sqrt{\frac{k}{m}}T = 2\pi \qquad \text{or} \qquad T = 2\pi\sqrt{\frac{m}{k}}$$

13.10 Variations of simple harmonic motion. All cases shown have $\phi = 0$ [see Eq. (13.13)].

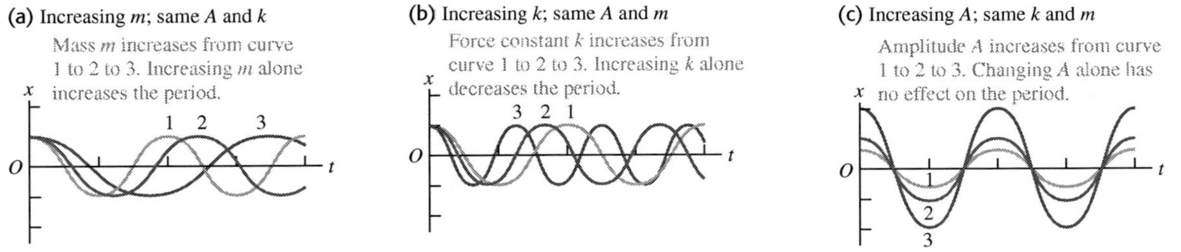

13.11 Variations of SHM: displacement versus time for the same harmonic oscillator with different phase angles ϕ.

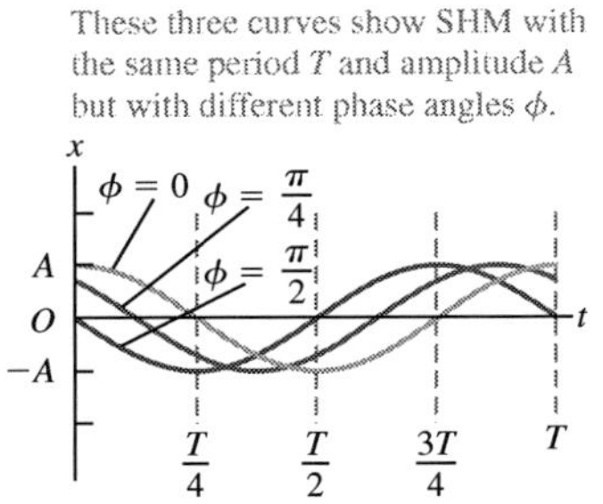

13.12 Graphs of (a) x versus t, (b) v_x versus t, and (c) a_x versus t for a body in SHM. For the motion depicted in these graphs, $\phi = \pi/3$.

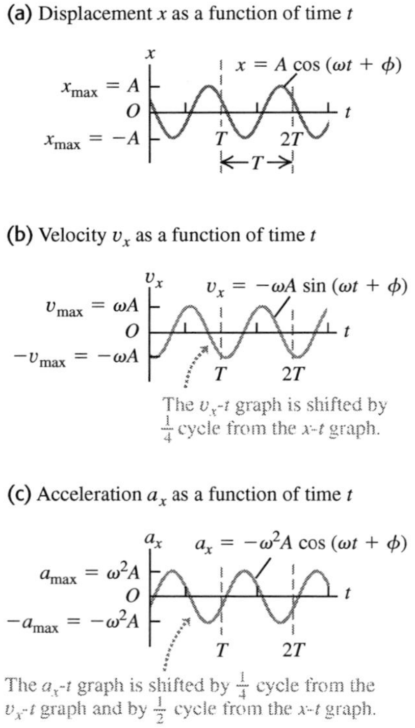

which is just Eq. (13.12). Changing either m or k changes the period of oscillation, as shown in Figs 13.10a and 13.10b. The period does not depend on the amplitude A (Fig. 13.10c).

The constant ϕ in Eq. (13.13) is called the **phase angle.** It tells us at what point in the cycle the motion was at $t = 0$ (equivalent to where around the circle the point Q was at $t = 0$). We denote the position at $t = 0$ by x_0. Putting $t = 0$ and $x = x_0$ in Eq. (13.13), we get

$$x_0 = A\cos\phi \qquad (13.14)$$

If $\phi = 0$, then $x_0 = A\cos 0 = A$, and the body starts at its maximum positive displacement. If $\phi = \pi$, then $x_0 = A\cos\pi = -A$, and the particle starts at its maximum *negative* displacement. If $\phi = \pi/2$, then $x_0 = A\cos(\pi/2) = 0$, and the particle is initially at the origin. Figure 13.11 shows the displacement x versus time for three different phase angles.

We find the velocity v_x and acceleration a_x as functions of time for a harmonic oscillator by taking derivatives of Eq. (13.13) with respect to time:

$$v_x = \frac{dx}{dt} = -\omega A\sin(\omega t + \phi) \qquad \text{(velocity in SHM)} \qquad (13.15)$$

$$a_x = \frac{dv_x}{dt} = \frac{d^2x}{dt^2} = -\omega^2 A\cos(\omega t + \phi) \qquad \text{(acceleration in SHM)} \qquad (13.16)$$

The velocity v_x oscillates between $v_{max} = +\omega A$ and $-v_{max} = -\omega A$, and the acceleration a_x oscillates between $a_{max} = +\omega^2 A$ and $-a_{max} = -\omega^2 A$ (Fig. 13.12). Comparing Eq. (13.16) with Eq. (13.13) and recalling that $\omega^2 = k/m$ from Eq. (13.9), we see that

$$a_x = -\omega^2 x = -\frac{k}{m}x$$

which is just Eq. (13.4) for simple harmonic motion. This confirms that Eq. (13.13) for x as a function of time is correct.

We actually derived Eq. (13.16) earlier in a geometrical way by taking the x-component of the acceleration vector of the reference point Q. This was done in Fig. 13.6b and Eq. (13.7) (recall that $\theta = \omega t + \phi$). In the same way, we could have derived Eq. (13.15) by taking the x-component of the velocity vector of Q, as shown in Fig. 13.6b. We'll leave the details for you to work out (see Problem 13.85).

Note that the sinusoidal graph of displacement versus time (Fig. 13.12a) is shifted by one-quarter period from the graph of velocity versus time (Fig. 13.12b) and by one-half period from the graph of acceleration versus time (Fig. 13.12c). Figure 13.13 shows why this is so. When the body is passing through the equilibrium position so that the displacement is zero, the velocity equals either v_{max} or $-v_{max}$ (depending on which way the body is moving) and the acceleration is zero. When the body is at either its maximum positive displacement, $x = +A$, or its maximum negative displacement, $x = -A$, the velocity is zero and the body is instantaneously at rest. At these points, the restoring force $F_x = -kx$ and the acceleration of the body have their maximum magnitudes. At $x = +A$ the acceleration is negative and equal to $-a_{max}$. At $x = -A$ the acceleration is positive: $a_x = +a_{max}$.

If we are given the initial position x_0 and initial velocity v_{0x} for the oscillating body, we can determine the amplitude A and the phase angle ϕ. Here's how to do it. The initial velocity v_{0x} is the velocity at time $t = 0$; putting $v_x = v_{0x}$ and $t = 0$ in Eq. (13.15), we find

$$v_{0x} = -\omega A \sin\phi \tag{13.17}$$

To find ϕ, we divide Eq. (13.17) by Eq. (13.14). This eliminates A and gives an equation that we can solve for ϕ:

$$\frac{v_{0x}}{x_0} = \frac{-\omega A \sin\phi}{A\cos\phi} = -\omega\tan\phi$$

$$\phi = \arctan\left(-\frac{v_{0x}}{\omega x_0}\right) \quad \text{(phase angle in SHM)} \tag{13.18}$$

It is also easy to find the amplitude A if we are given x_0 and v_{0x}. We'll sketch the derivation, and you can fill in the details. Square Eq. (13.14); then divide Eq. (13.17) by ω, square it, and add to the square of Eq. (13.14). The right side will be $A^2(\sin^2\phi + \cos^2\phi)$, which is equal to A^2. The final result is

$$A = \sqrt{x_0^2 + \frac{v_{0x}^2}{\omega^2}} \quad \text{(amplitude in SHM)} \tag{13.19}$$

Note that when the body has both an initial displacement x_0 and a nonzero initial velocity v_{0x}, the amplitude A is *not* equal to the initial displacement. That's reasonable; if you start the body at a positive x_0 but give it a positive velocity v_{0x}, it will go *farther* than x_0 before it turns and comes back.

13.13 How x-velocity v_x and x-acceleration a_x vary during one cycle of SHM.

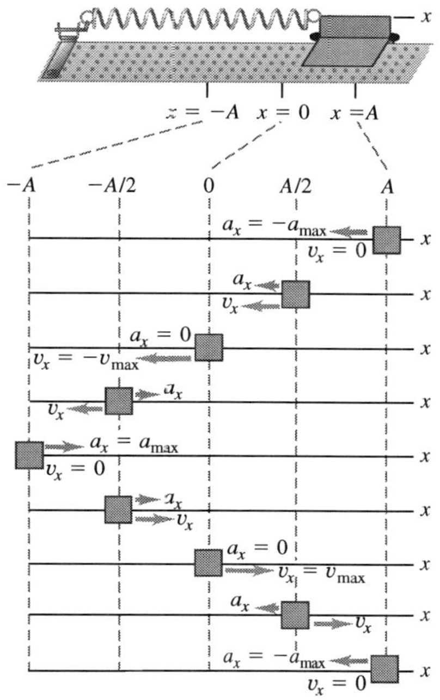

Problem-Solving Strategy 13.1 Simple Harmonic Motion I: Describing Motion

IDENTIFY *the relevant concepts:* An oscillating system undergoes simple harmonic motion (SHM) *only* if the restoring force is directly proportional to the displacement. Be certain that this is the case for the problem at hand before attempting to use any of the results of this section. As always, identify the target variables.

SET UP *the problem* using the following steps:

1. Identify the known and unknown quantities, and determine which are the target variables.
2. It's useful to distinguish between two kinds of quantities. *Basic properties* of the system include the mass m and the force constant k as well as quantities derived from m and k, such as the period T, frequency f, and angular frequency ω. *Properties of the motion* describe how the system behaves when it is set into motion in a particular way. They include the amplitude A, maximum velocity v_{max}, and phase angle ϕ as well as the values of x, v_x, and a_x at a particular time.
3. If necessary, define an x-axis as in Fig. 13.13, with the equilibrium position at $x = 0$.

EXECUTE *the solution* as follows:

1. Use the equations given in Sections 13.1 and 13.2 to solve for the target variables.
2. If you need to calculate the phase angle, be certain to express it in radians. The quantity ωt in Eq. (13.13) is naturally in radians, so ϕ must be as well.
3. If you need to find the values of x, v_x, and a_x at various times, use Eqs. (13.13), (13.15), and (13.16), respectively. If the initial position x_0 and initial velocity v_{0x} are both given, you can determine the phase angle and amplitude from Eqs. (13.18) and (13.19). If the body is given an initial positive displacement x_0 but zero initial velocity $(v_{0x} = 0)$, then the amplitude is $A = x_0$ and the phase angle is $\phi = 0$. If it has an initial positive velocity v_{0x} but no initial displacement $(x_0 = 0)$, the amplitude is $A = v_{0x}/\omega$ and the phase angle is $\phi = -\pi/2$.

EVALUATE *your answer:* Check your results to make sure they're consistent. As an example, suppose you've used the initial position and velocity to find general expressions for x and v_x at time t. If you substitute $t = 0$ into these expressions, you should get back the correct values of x_0 and v_{0x}.

Example 13.3 Describing SHM

Let's return to the system of mass and horizontal spring we considered in Example 13.2, with $k = 200$ N/m and $m = 0.50$ kg. This time we give the body an initial displacement of $+0.015$ m and an initial velocity of $+0.40$ m/s. (a) Find the period, amplitude, and phase angle of the motion. (b) Write equations for the displacement, velocity, and acceleration as functions of time.

SOLUTION

IDENTIFY: As in Example 13.2, the oscillations are SHM and we may use the expressions developed in this section.

SET UP: We are given the values of k, m, x_0, and v_{0x}. From them, we calculate the target variables T, A, and ϕ and the expressions for x, v_x, and a_x as functions of time.

EXECUTE: (a) The period is the same as in Example 13.2, $T = 0.31$ s. In simple harmonic motion the period does not depend on the amplitude, only on the values of k and m. In Example 13.2 we found that $\omega = 20$ rad/s. So from Eq. (13.19),

$$A = \sqrt{x_0^2 + \frac{v_{0x}^2}{\omega^2}}$$

$$= \sqrt{(0.015\ \text{m})^2 + \frac{(0.40\ \text{m/s})^2}{(20\ \text{rad/s})^2}}$$

$$= 0.025\ \text{m}$$

To find the phase angle ϕ, we use Eq. (13.18):

$$\phi = \arctan\left(-\frac{v_{0x}}{\omega x_0}\right)$$

$$= \arctan\left(-\frac{0.40\ \text{m/s}}{(20\ \text{rad/s})(0.015\ \text{m})}\right) = -53° = -0.93\ \text{rad}$$

(b) The displacement, velocity, and acceleration at any time are given by Eqs. (13.13), (13.15), and (13.16), respectively. Substituting the values, we get

$$x = (0.025\ \text{m})\cos[(20\ \text{rad/s})t - 0.93\ \text{rad}]$$

$$v_x = -(0.50\ \text{m/s})\sin[(20\ \text{rad/s})t - 0.93\ \text{rad}]$$

$$a_x = -(10\ \text{m/s}^2)\cos[(20\ \text{rad/s})t - 0.93\ \text{rad}]$$

The velocity varies sinusoidally between -0.50 m/s and $+0.50$ m/s, and the acceleration varies sinusoidally between $-10\ \text{m/s}^2$ and $+10\ \text{m/s}^2$.

EVALUATE: You can check the results for x and v_x as functions of time by substituting $t = 0$ and evaluating the result. You should get $x = x_0 = 0.015$ m and $v_x = v_{0x} = 0.40$ m/s. Do you?

Test Your Understanding of Section 13.2 A glider is attached to a spring as shown in Fig. 13.13. If the glider is moved to $x = 0.10$ m and released from rest at time $t = 0$, it will oscillate with amplitude $A = 0.10$ m and phase angle $\phi = 0$. (a) Suppose instead that at $t = 0$ the glider is at $x = 0.10$ m and is moving to the right in Fig. 13.13. In this situation is the amplitude greater than, less than, or equal to 0.10 m? Is the phase angle greater than, less than, or equal to zero? (b) Suppose instead that at $t = 0$ the glider is at $x = 0.10$ m and is moving to the left in Fig. 13.13. In this situation is the amplitude greater than, less than, or equal to 0.10 m? Is the phase angle greater than, less than, or equal to zero?

13.3 Energy in Simple Harmonic Motion

ActivPhysics ONLINE

9.3 Vibrational Energy
9.4 Two Ways to Weigh Young Tarzan
9.6 Releasing a Vibrating Skier I
9.7 Releasing a Vibrating Skier II
9.8 One- and Two-Spring Vibrating Systems
9.9 Vibro-Ride

We can learn even more about simple harmonic motion by using energy considerations. Take another look at the body oscillating on the end of a spring in Figs. 13.2 and 13.13. We've already noted that the spring force is the only horizontal force on the body. The force exerted by an ideal spring is a conservative force, and the vertical forces do no work, so the total mechanical energy of the system is *conserved*. We also assume that the mass of the spring itself is negligible.

The kinetic energy of the body is $K = \frac{1}{2}mv^2$ and the potential energy of the spring is $U = \frac{1}{2}kx^2$, just as in Section 7.2. (You'll find it helpful to review that section.) There are no nonconservative forces that do work, so the total mechanical energy $E = K + U$ is conserved:

$$E = \frac{1}{2}mv_x^2 + \frac{1}{2}kx^2 = \text{constant} \qquad (13.20)$$

(Since the motion is one-dimensional, $v^2 = v_x^2$.)

The total mechanical energy E is also directly related to the amplitude A of the motion. When the body reaches the point $x = A$, its maximum displacement from equilibrium, it momentarily stops as it turns back toward the equilibrium position. That is, when $x = A$ (or $-A$), $v_x = 0$. At this point the energy is entirely potential, and $E = \frac{1}{2}kA^2$. Because E is constant, it is equal to $\frac{1}{2}kA^2$ at any other point. Combining this expression with Eq. (13.20), we get

$$E = \frac{1}{2}mv_x^2 + \frac{1}{2}kx^2 = \frac{1}{2}kA^2 = \text{constant} \quad \text{(total mechanical energy in SHM)} \tag{13.21}$$

We can verify this equation by substituting x and v_x from Eqs. (13.13) and (13.15) and using $\omega^2 = k/m$ from Eq. (13.9):

$$\begin{aligned} E &= \frac{1}{2}mv_x^2 + \frac{1}{2}kx^2 = \frac{1}{2}m[-\omega A\sin(\omega t + \phi)]^2 + \frac{1}{2}k[A\cos(\omega t + \phi)]^2 \\ &= \frac{1}{2}kA^2\sin^2(\omega t + \phi) + \frac{1}{2}kA^2\cos^2(\omega t + \phi) \\ &= \frac{1}{2}kA^2 \end{aligned}$$

(Recall that $\sin^2\alpha + \cos^2\alpha = 1$.) Hence our expressions for displacement and velocity in SHM are consistent with energy conservation, as they must be.

We can use Eq. (13.21) to solve for the velocity v_x of the body at a given displacement x:

$$v_x = \pm\sqrt{\frac{k}{m}}\sqrt{A^2 - x^2} \tag{13.22}$$

The $\pm$ sign means that at a given value of x the body can be moving in either direction. For example, when $x = \pm A/2$,

$$v_x = \pm\sqrt{\frac{k}{m}}\sqrt{A^2 - \left(\pm\frac{A}{2}\right)^2} = \pm\sqrt{\frac{3}{4}}\sqrt{\frac{k}{m}}A$$

Equation (13.22) also shows that the *maximum* speed v_{max} occurs at $x = 0$. Using Eq. (13.10), $\omega = \sqrt{k/m}$, we find that

$$v_{max} = \sqrt{\frac{k}{m}}A = \omega A \tag{13.23}$$

This agrees with Eq. (13.15), which showed that v_x oscillates between $-\omega A$ and $+\omega A$.

Interpreting E, K, and U in SHM

Figure 13.14 shows the energy quantities E, K, and U at $x = 0$, $x = \pm A/2$, and $x = \pm A$. Figure 13.15 is a graphical display of Eq. (13.21); energy (kinetic, potential, and total) is plotted vertically and the coordinate x is plotted horizontally. The

13.14 Graphs of E, K, and U versus displacement in SHM. The velocity of the body is *not* constant, so these images of the body at equally spaced positions are *not* equally spaced in time.

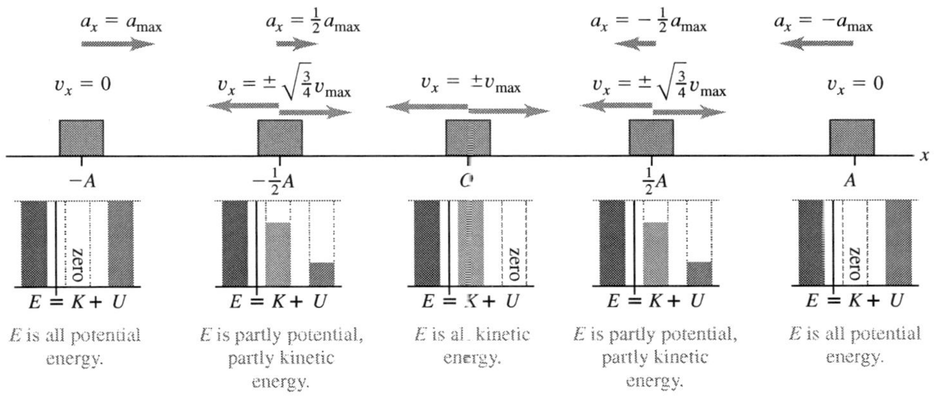

13.15 Kinetic energy K, potential energy U, and total mechanical energy E as functions of position for SHM. At each value of x the sum of the values of K and U equals the constant value of E. Can you show that the energy is half kinetic and half potential at $x = \pm\sqrt{\frac{1}{2}}A$?

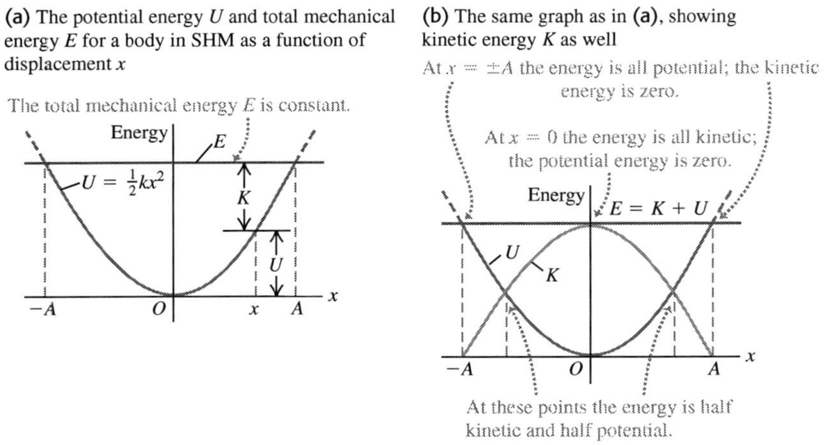

parabolic curve in Fig. 13.15a represents the potential energy $U = \frac{1}{2}kx^2$. The horizontal line represents the total mechanical energy E, which is constant and does not vary with x. At any value of x between $-A$ and A, the vertical distance from the x-axis to the parabola is U; since $E = K + U$, the remaining vertical distance up to the horizontal line is K. Figure 13.15b shows both K and U as functions of x. The horizontal line for E intersects the potential-energy curve at $x = -A$ and $x = A$, so at these points the energy is entirely potential, the kinetic energy is zero, and the body comes momentarily to rest before reversing direction. As the body oscillates between $-A$ and A, the energy is continuously transformed from potential to kinetic and back again.

Figure 13.15a shows the connection between the amplitude A and the corresponding total mechanical energy $E = \frac{1}{2}kA^2$. If we tried to make x greater than A (or less than $-A$), U would be greater than E, and K would have to be negative. But K can never be negative, so x can't be greater than A or less than $-A$.

Problem-Solving Strategy 13.2 Simple Harmonic Motion II: Energy

The energy equation, Eq. (13.21), is a useful alternative relationship between velocity and position, especially when energy quantities are also required. If the problem involves a relationship among position, velocity, and acceleration without reference to time, it is usually easier to use Eq. (13.4) (from Newton's second law) or Eq. (13.21) (from energy conservation) than to use the general expressions for x, v_x, and a_x as functions of time [Eqs. (13.13), (13.15), and (13.16), respectively]. Because the energy equation involves x^2 and v_x^2, it cannot tell you the sign of x or of v_x; you have to infer the sign from the situation. For instance, if the body is moving from the equilibrium position toward the point of greatest positive displacement, then x is positive and v_x is positive.

Example 13.4 Velocity, acceleration, and energy in SHM

In the oscillation described in Example 13.2, $k = 200$ N/m, $m = 0.50$ kg, and the oscillating mass is released from rest at $x = 0.020$ m. (a) Find the maximum and minimum velocities attained by the oscillating body. (b) Compute the maximum acceleration. (c) Determine the velocity and acceleration when the body has moved halfway to the center from its original position. (d) Find the total energy, potential energy, and kinetic energy at this position.

SOLUTION

IDENTIFY: The problem refers to the motion at various *positions* in the motion, not at specified *times*. This is a hint that we can use the energy relationships found in this section to solve for the target variables.

SET UP: Figure 13.13 shows the choice of x-axis. The maximum displacement from equilibrium is $A = 0.020$ m. For any position x we use Eqs. (13.22) and (13.4) to find the velocity v_x and acceleration a_x, respectively. Given the velocity and position, we use Eq. (13.21) to find the energy quantities K, U, and E.

EXECUTE: (a) The velocity v_x at any displacement x is given by Eq. (13.22):

$$v_x = \pm\sqrt{\frac{k}{m}}\sqrt{A^2 - x^2}$$

The maximum velocity occurs when the body is moving to the right through the equilibrium position, where $x = 0$:

$$v_x = v_{max} = \sqrt{\frac{k}{m}}A = \sqrt{\frac{200\ \text{N/m}}{0.50\ \text{kg}}}(0.020\ \text{m}) = 0.40\ \text{m/s}$$

The minimum (i.e., most negative) velocity occurs when the body is moving to the left through $x = 0$; its value is $-v_{max} = -0.40\ \text{m/s}$.

(b) From Eq. (13.4),

$$a_x = -\frac{k}{m}x$$

The maximum (most positive) acceleration occurs at the most negative value of x, $x = -A$; therefore

$$a_{max} = -\frac{k}{m}(-A) = -\frac{200\ \text{N/m}}{0.50\ \text{kg}}(-0.020\ \text{m}) = 8.0\ \text{m/s}^2$$

The minimum (most negative) acceleration is $-8.0\ \text{m/s}^2$, occurring at $x = +A = +0.020$ m.

(c) At a point halfway to the center from the initial position, $x = A/2 = 0.010$ m. From Eq. (13.22),

$$v_x = -\sqrt{\frac{200\ \text{N/m}}{0.50\ \text{kg}}}\sqrt{(0.020\ \text{m})^2 - (0.010\ \text{m})^2} = -0.35\ \text{m/s}$$

We choose the negative square root because the body is moving from $x = A$ toward $x = 0$. From Eq. (13.4),

$$a_x = -\frac{200\ \text{N/m}}{0.50\ \text{kg}}(0.010\ \text{m}) = -4.0\ \text{m/s}^2$$

At this point the velocity and the acceleration have the same sign, so the speed is increasing. The conditions at $x = 0$, $\pm A/2$, and $\pm A$ are shown in Fig. 13.14.

(d) The total energy has the same value at all points during the motion:

$$E = \frac{1}{2}kA^2 = \frac{1}{2}(200\ \text{N/m})(0.020\ \text{m})^2 = 0.040\ \text{J}$$

The potential energy is

$$U = \frac{1}{2}kx^2 = \frac{1}{2}(200\ \text{N/m})(0.010\ \text{m})^2 = 0.010\ \text{J}$$

and the kinetic energy is

$$K = \frac{1}{2}mv_x^2 = \frac{1}{2}(0.50\ \text{kg})(-0.35\ \text{m/s})^2 = 0.030\ \text{J}$$

EVALUATE: At the point $x = A/2$, the energy is one-fourth potential energy and three-fourths kinetic energy. You can check this result by inspecting Fig. 13.15b.

Example 13.5 Energy and momentum in SHM

A block with mass M attached to a horizontal spring with force constant k is moving with simple harmonic motion having amplitude A_1. At the instant when the block passes through its equilibrium position, a lump of putty with mass m is dropped vertically onto the block from a very small height and sticks to it. (a) Find the new amplitude and period. (b) Repeat part (a) for the case in which the putty is dropped on the block when it is at one end of its path.

SOLUTION

IDENTIFY: The problem involves the motion at a given position, not a given time, so we can use energy methods. Before the putty lands on the block, the mechanical energy of the oscillating block and spring is constant. When the putty lands on the block, it's a completely inelastic collision (see Section 8.3); the horizontal component of momentum is conserved, but kinetic energy decreases. Once the collision ends, the mechanical energy remains constant at its new value.

SET UP: Figure 13.16 shows our sketches. In each part we consider what happens before, during, and after the collision. We find the amplitude A_2 after the collision from the final energy of the system, and we find the period T_2 after the collision using the relationship between period and mass.

EXECUTE: (a) Before the collision the total mechanical energy of the block and spring is $E_1 = \frac{1}{2}kA_1^2$. Since the block is at the equilibrium position, $U = 0$, and the energy is purely kinetic (Fig. 13.16a). If we let v_1 be the speed of the block at the equilibrium position, we have

$$E_1 = \frac{1}{2}Mv_1^2 = \frac{1}{2}kA_1^2 \quad \text{so} \quad v_1 = \sqrt{\frac{k}{M}}A_1$$

During the collision the x-component of momentum of the system of block and putty is conserved. (Why?) Just before the collision

13.16 Our sketches for this problem.

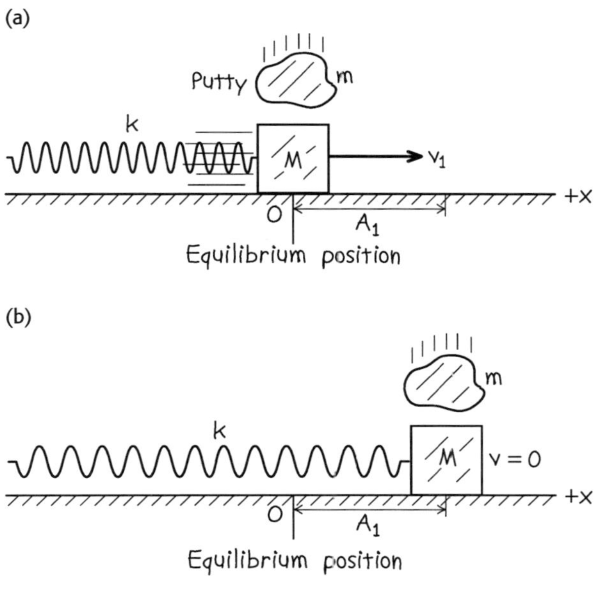

this component is the sum of Mv_1 (for the block) and zero (for the putty). Just after the collision the block and putty move together with speed v_2, and their combined x-component of momentum is $(M + m)v_2$. From conservation of momentum,

$$Mv_1 + 0 = (M + m)v_2 \quad \text{so} \quad v_2 = \frac{M}{M + m}v_1$$

Continued

The collision lasts a very short time, so just after the collision the block and putty are still at the equilibrium position. The energy is still purely kinetic but is *less* than before the collision:

$$E_2 = \frac{1}{2}(M + m)v_2^2 = \frac{1}{2}\frac{M^2}{M + m}v_1^2 = \frac{M}{M + m}\left(\frac{1}{2}Mv_1^2\right)$$
$$= \left(\frac{M}{M + m}\right)E_1$$

Since E_2 equals $\frac{1}{2}kA_2^2$, where A_2 is the amplitude after the collision, we have

$$\frac{1}{2}kA_2^2 = \left(\frac{M}{M + m}\right)\frac{1}{2}kA_1^2$$
$$A_2 = A_1\sqrt{\frac{M}{M + m}}$$

The larger the putty mass m, the smaller the final amplitude.

Finding the period of oscillation after the collision is the easy part. Using Eq. (13.12), we have

$$T_2 = 2\pi\sqrt{\frac{M + m}{k}}$$

(b) When the putty drops, the block is instantaneously at rest (Fig. 13.16b). The x-component of momentum is zero both before and after the collision: The block has zero kinetic energy just before the collision, and the block and putty have zero kinetic energy just after the collision. The energy is all potential energy stored in the spring, so adding the extra mass of the putty has *no effect* on the mechanical energy. That is,

$$E_2 = E_1 = \frac{1}{2}kA_1^2$$

and the amplitude after the collision is unchanged $(A_2 = A_1)$. The period still changes when the putty is added, though; its value doesn't depend on how the mass is added, only on what the total mass is. So T_2 is the same as we found in part (a), $T_2 = 2\pi\sqrt{(M + m)/k}$.

EVALUATE: Why is energy lost in part (a) but not in part (b)? The difference is that in part (a) the putty slides against the moving block during the collision, and energy is dissipated by kinetic friction.

Test Your Understanding of Section 13.3 (a) To double the total energy for a mass-spring system oscillating in SHM, by what factor must the amplitude increase? (i) 4; (ii) 2; (iii) $\sqrt{2} = 1.414$; (iv) $\sqrt[4]{2} = 1.189$. (b) By what factor will the frequency change due to this amplitude increase? (i) 4; (ii) 2; (iii) $\sqrt{2} = 1.414$; (iv) $\sqrt[4]{2} = 1.189$; (v) it does not change.

13.4 Applications of Simple Harmonic Motion

So far, we've looked at a grand total of *one* situation in which simple harmonic motion (SHM) occurs: a body attached to an ideal horizontal spring. But SHM can occur in any system in which there is a restoring force that is directly proportional to the displacement from equilibrium, as given by Eq. (13.3), $F_x = -kx$. The restoring force will originate in different ways in different situations, so the force constant k has to be found for each case by examining the net force on the system. Once this is done, it's straightforward to find the angular frequency ω, frequency f, and period T; we just substitute the value of k into Eqs. (13.10), (13.11), and (13.12), respectively. Let's use these ideas to examine several examples of simple harmonic motion.

Vertical SHM

Suppose we hang a spring with force constant k (Fig. 13.17a) and suspend from it a body with mass m. Oscillations will now be vertical; will they still be SHM? In Fig. 13.17b the body hangs at rest, in equilibrium. In this position the spring is stretched an amount Δl just great enough that the spring's upward vertical force $k\,\Delta l$ on the body balances its weight mg:

$$k\,\Delta l = mg$$

Take $x = 0$ to be this equilibrium position and take the positive x-direction to be upward. When the body is a distance x *above* its equilibrium position (Fig. 13.17c), the extension of the spring is $\Delta l - x$. The upward force it exerts on the body is then $k(\Delta l - x)$, and the net x-component of force on the body is

$$F_{\text{net}} = k(\Delta l - x) + (-mg) = -kx$$

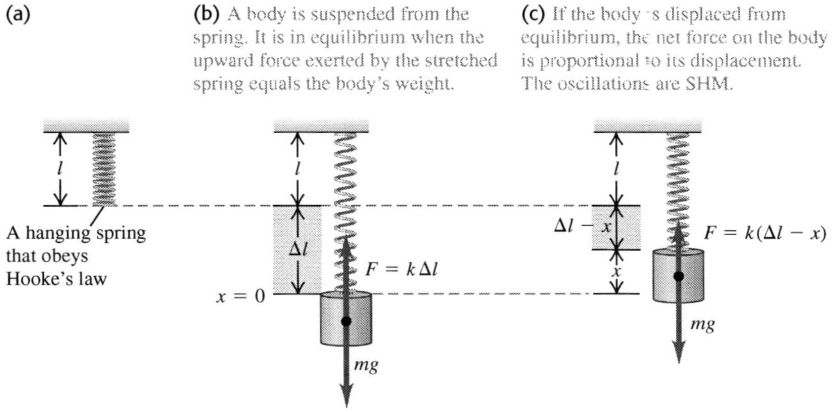

13.17 A body attached to a hanging spring.

13.18 If the weight mg compresses the spring a distance Δl, the force constant is $k = mg/\Delta l$ and the angular frequency for vertical SHM is $\omega = \sqrt{k/m}$—the same as if the body were suspended from the spring (See Fig. 13.17).

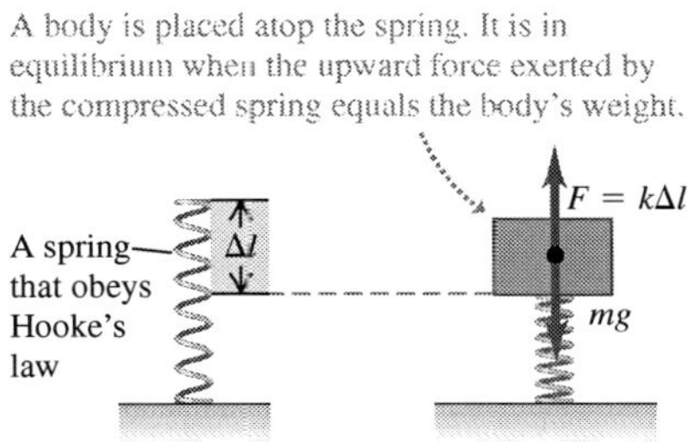

that is, a net downward force of magnitude kx. Similarly, when the body is *below* the equilibrium position, there is a net upward force with magnitude kx. In either case there is a restoring force with magnitude kx. If the body is set in vertical motion, it oscillates in SHM with the same angular frequency as though it were horizontal, $\omega = \sqrt{k/m}$. So vertical SHM doesn't differ in any essential way from horizontal SHM. The only real change is that the equilibrium position $x = 0$ no longer corresponds to the point at which the spring is unstretched. The same ideas hold if a body with weight mg is placed atop a compressible spring (Fig. 13.18) and compresses it a distance Δl.

Example 13.6 Vertical SHM in an old car

The shock absorbers in an old car with mass 1000 kg are completely worn out. When a 980-N person climbs slowly into the car to its center of gravity, the car sinks 2.8 cm. When the car, with the person aboard, hits a bump, the car starts oscillating up and down in SHM. Model the car and person as a single body on a single spring, and find the period and frequency of the oscillation.

SOLUTION

IDENTIFY: The situation is like that shown in Fig. 13.18.

SET UP: The compression of the spring when the extra weight is added tells us the force constant, which we can use to find the period and frequency (the target variables).

EXECUTE: When the force increases by 980 N, the spring compresses an additional 0.028 m, and the coordinate x of the car changes by -0.028 m. Hence the effective force constant (including the effect of the entire suspension) is

$$k = -\frac{F_x}{x} = -\frac{980\ \text{N}}{-0.028\ \text{m}} = 3.5 \times 10^4\ \text{kg/s}^2$$

The person's mass is $w/g = (980\ \text{N})/(9.8\ \text{m/s}^2) = 100$ kg. The *total* oscillating mass is $m = 1000\ \text{kg} + 100\ \text{kg} = 1100$ kg. The period T is

$$T = 2\pi\sqrt{\frac{m}{k}} = 2\pi\sqrt{\frac{1100\ \text{kg}}{3.5 \times 10^4\ \text{kg/s}^2}} = 1.11\ \text{s}$$

and the frequency is

$$f = \frac{1}{T} = \frac{1}{1.11\ \text{s}} = 0.90\ \text{Hz}$$

EVALUATE: A persistent oscillation with a period of about 1 second makes for a very unpleasant ride. The purpose of shock absorbers is to make such oscillations die out (see Section 13.7).

Angular SHM

A mechanical watch keeps time based on the oscillations of a balance wheel (Fig. 13.19). The wheel has a moment of inertia I about its axis. A coil spring exerts a restoring torque τ_z that is proportional to the angular displacement θ from the equilibrium position. We write $\tau_z = -\kappa\theta$, where κ (the Greek letter kappa) is a constant called the *torsion constant.* Using the rotational analog of

13.19 The balance wheel of a mechanical watch. The spring exerts a restoring torque that is proportional to the angular displacement θ, so the motion is angular SHM.

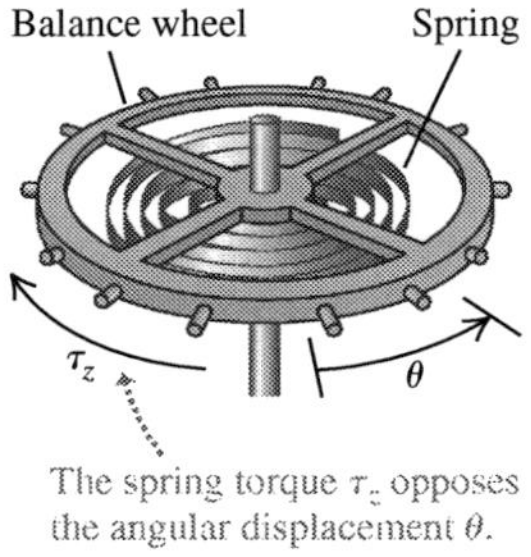

Newton's second law for a rigid body, $\Sigma\tau_z = I\alpha_z = I\,d^2\theta/dt^2$, we can find the equation of motion:

$$-\kappa\theta = I\alpha \quad \text{or} \quad \frac{d^2\theta}{dt^2} = -\frac{\kappa}{I}\theta$$

The form of this equation is exactly the same as Eq. (13.4) for the acceleration in simple harmonic motion, with x replaced by θ and k/m replaced by κ/I. So we are dealing with a form of *angular* simple harmonic motion. The angular frequency ω and frequency f are given by Eqs. (13.10) and (13.11), respectively, with the same replacement:

$$\omega = \sqrt{\frac{\kappa}{I}} \quad \text{and} \quad f = \frac{1}{2\pi}\sqrt{\frac{\kappa}{I}} \quad \text{(angular SHM)} \tag{13.24}$$

The motion is described by the function

$$\theta = \Theta\cos(\omega t + \phi)$$

where Θ (the Greek letter theta) plays the role of an angular amplitude.

It's a good thing that the motion of a balance wheel *is* simple harmonic. If it weren't, the frequency might depend on the amplitude, and the watch would run too fast or too slow as the spring ran down.

*Vibrations of Molecules

The following discussion of the vibrations of molecules uses the binomial theorem. If you aren't familiar with this theorem, you should read about it in the appropriate section of a math textbook.

When two atoms are separated from each other by a few atomic diameters, they can exert attractive forces on each other. But if the atoms are so close to each other that their electron shells overlap, the forces between the atoms are repulsive. Between these limits, there can be an equilibrium separation distance at which two atoms form a *molecule.* If these atoms are displaced slightly from equilibrium, they will oscillate.

As an example, we'll consider one type of interaction between atoms called the *van der Waals interaction.* Our immediate task here is to study oscillations, so we won't go into the details of how this interaction arises. Let the center of one atom be at the origin and let the center of the other atom be a distance r away (Fig. 13.20a); the equilibrium distance between centers is $r = R_0$. Experiment shows that the van der Waals interaction can be described by the potential-energy function

$$U = U_0\left[\left(\frac{R_0}{r}\right)^{12} - 2\left(\frac{R_0}{r}\right)^6\right] \tag{13.25}$$

13.20 (a) Two atoms with centers separated by r. (b) Potential energy U in the van der Waals interaction as a function of r. (c) Force F_r on the right-hand atom as a function of r.

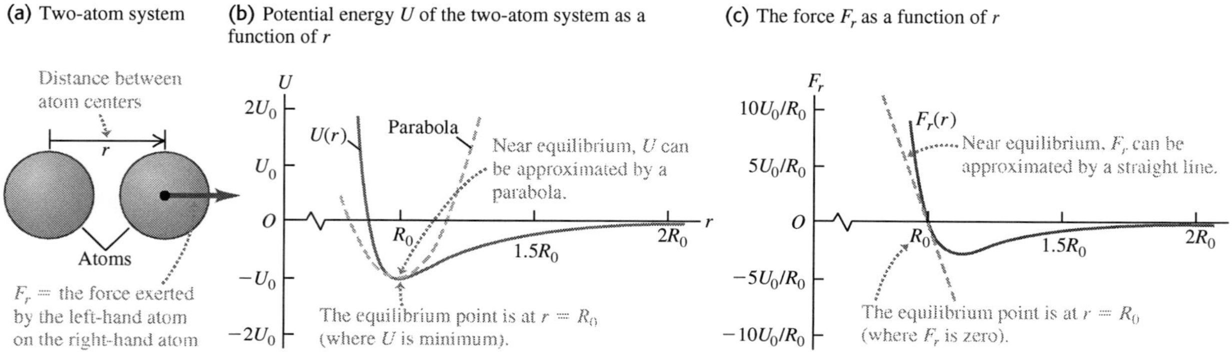

where U_0 is a positive constant with units of joules. When the two atoms are very far apart, $U = 0$; when they are separated by the equilibrium distance $r = R_0$, $U = -U_0$. The force on the second atom is the negative derivative of Eq. (13.25):

$$F_r = -\frac{dU}{dr} = U_0\left[\frac{12R_0^{12}}{r^{13}} - 2\frac{6R_0^6}{r^7}\right] = 12\frac{U_0}{R_0}\left[\left(\frac{R_0}{r}\right)^{13} - \left(\frac{R_0}{r}\right)^7\right] \quad (13.26)$$

The potential energy and force are plotted in Figs. 13.20b and 13.20c, respectively. The force is positive for $r < R_0$ and negative for $r > R_0$, so it is a *restoring* force.

Let's examine the restoring force F_r in Eq. (13.26). We let x represent the displacement from equilibrium:

$$x = r - R_0 \quad \text{so} \quad r = R_0 + x$$

In terms of x, the force F_r in Eq. (13.26) becomes

$$\begin{aligned} F_r &= 12\frac{U_0}{R_0}\left[\left(\frac{R_0}{R_0+x}\right)^{13} - \left(\frac{R_0}{R_0+x}\right)^7\right] \\ &= 12\frac{U_0}{R_0}\left[\frac{1}{(1+x/R_0)^{13}} - \frac{1}{(1+x/R_0)^7}\right] \end{aligned} \quad (13.27)$$

This looks nothing like Hooke's law, $F_x = -kx$, so we might be tempted to conclude that molecular oscillations cannot be SHM. But let us restrict ourselves to *small-amplitude* oscillations so that the absolute value of the displacement x is small in comparison to R_0 and the absolute value of the ratio x/R_0 is much less than 1. We can then simplify Eq. (13.27) by using the *binomial theorem:*

$$(1 + u)^n = 1 + nu + \frac{n(n-1)}{2!}u^2 + \frac{n(n-1)(n-2)}{3!}u^3 + \cdots \quad (13.28)$$

If $|u|$ is much less than 1, each successive term in Eq. (13.28) is much smaller than the one it follows, and we can safely approximate $(1 + u)^n$ by just the first two terms. In Eq. (13.27), u is replaced by x/R_0 and n equals -13 or -7, so

$$\frac{1}{(1 + x/R_0)^{13}} = (1 + x/R_0)^{-13} \approx 1 + (-13)\frac{x}{R_0}$$

$$\frac{1}{(1 + x/R_0)^{7}} = (1 + x/R_0)^{-7} \approx 1 + (-7)\frac{x}{R_0}$$

$$F_r \approx 12\frac{U_0}{R_0}\left[\left(1 + (-13)\frac{x}{R_0}\right) - \left(1 + (-7)\frac{x}{R_0}\right)\right] = -\left(\frac{72U_0}{R_0^2}\right)x \quad (13.29)$$

This is just Hooke's law, with force constant $k = 72U_0/R_0^2$. (Note that k has the correct units, J/m^2 or N/m.) So oscillations of molecules bound by the van der Waals interaction can be simple harmonic motion, provided that the amplitude is small in comparison to R_0 so that the approximation $|x/R_0| \ll 1$ used in the derivation of Eq. (13.29) is valid.

You can also show that the potential energy U in Eq. (13.25) can be written as $U \approx \frac{1}{2}kx^2 + C$, where $C = -U_0$ and k is again equal to $72U_0/R_0^2$. Adding a constant to the potential energy has no effect on the physics, so the system of two atoms is fundamentally no different from a mass attached to a horizontal spring for which $U = \frac{1}{2}kx^2$. The proof is left to you (see Exercise 13.39).

Example 13.7 Molecular vibration

Two argon atoms can form a weakly bound molecule, Ar_2, held together by a van der Waals interaction with $U_0 = 1.68 \times 10^{-21}$ J and $R_0 = 3.82 \times 10^{-10}$ m. Find the frequency for small oscillations of one of the atoms about its equilibrium position.

SOLUTION

IDENTIFY: This is just the situation shown in Fig. 13.20.

Continued

SET UP: Because the oscillations are small, we can use Eq. (13.11) to obtain the frequency of simple harmonic motion. The force constant is given by Eq. (13.29).

EXECUTE: The force constant is

$$k = \frac{72U_0}{R_0^2} = \frac{72(1.68 \times 10^{-21}\ \text{J})}{(3.82 \times 10^{-10}\ \text{m})^2} = 0.829\ \text{J/m}^2 = 0.829\ \text{N/m}$$

This is comparable to the force constant of a loose, floppy toy spring like a Slinky™.

From the periodic table of the elements (see Appendix D), the average atomic mass of argon is

$$(39.948\ \text{u})(1.66 \times 10^{-27}\ \text{kg/1 u}) = 6.63 \times 10^{-26}\ \text{kg}.$$

If one of the argon atoms is fixed and the other atom oscillates, the frequency of oscillation is

$$f = \frac{1}{2\pi}\sqrt{\frac{k}{m}} = \frac{1}{2\pi}\sqrt{\frac{0.829\ \text{N/m}}{6.63 \times 10^{-26}\ \text{kg}}} = 5.63 \times 10^{11}\ \text{Hz}$$

The oscillating mass is very small, so even a floppy spring causes very rapid oscillations.

EVALUATE: Our answer for f isn't quite right. If there is no net external force acting on the molecule, the center of mass of the molecule (located halfway between the two atoms) doesn't accelerate. To ensure this, both atoms must oscillate with the same amplitude in opposite directions. It turns out that we can account for this by replacing m with $m/2$ in the expression for f. (See Problem 13.86.) This makes f larger by a factor of $\sqrt{2}$, so $f = \sqrt{2}(5.63 \times 10^{11}\ \text{Hz}) = 7.96 \times 10^{11}\ \text{Hz}$. An additional complication is that on the atomic scale we must use quantum mechanics, not Newtonian mechanics, to describe motion; happily, the frequency has the same value in quantum mechanics.

Test Your Understanding of Section 13.4 A block attached to a hanging ideal spring oscillates up and down with a period of 10 s on earth. If you take the block and spring to Mars, where the acceleration due to gravity is only about 40% as large as on earth, what will be the new period of oscillation? (i) 10 s; (ii) more than 10 s; (iii) less than 10 s.

13.21 The dynamics of a simple pendulum.

(a) A real pendulum

(b) An idealized simple pendulum

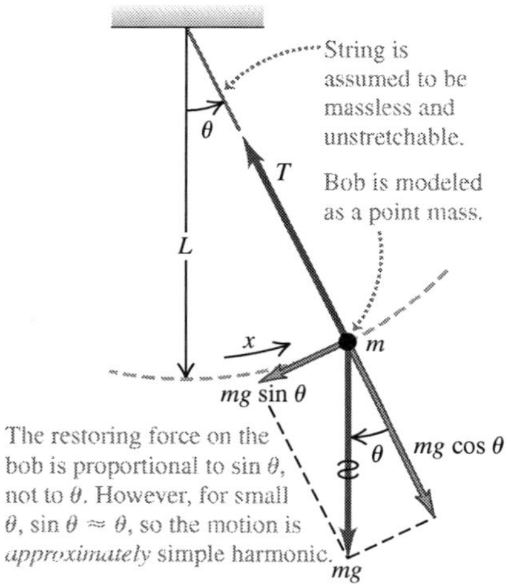

13.5 The Simple Pendulum

A **simple pendulum** is an idealized model consisting of a point mass suspended by a massless, unstretchable string. When the point mass is pulled to one side of its straight-down equilibrium position and released, it oscillates about the equilibrium position. Familiar situations such as a wrecking ball on a crane's cable or a person on a swing (Fig. 13.21a) can be modeled as simple pendulums.

The path of the point mass (sometimes called a pendulum bob) is not a straight line but the arc of a circle with radius L equal to the length of the string (Fig. 13.21b). We use as our coordinate the distance x measured along the arc. If the motion is simple harmonic, the restoring force must be directly proportional to x or (because $x = L\theta$) to θ. Is it?

In Fig. 13.21b we represent the forces on the mass in terms of tangential and radial components. The restoring force F_θ is the tangential component of the net force:

$$F_\theta = -mg\sin\theta \tag{13.30}$$

The restoring force is provided by gravity; the tension T merely acts to make the point mass move in an arc. The restoring force is proportional *not* to θ but to $\sin\theta$, so the motion is *not* simple harmonic. However, if the angle θ is *small*, $\sin\theta$ is very nearly equal to θ in radians (Fig. 13.22). For example, when $\theta = 0.1$ rad (about 6°), $\sin\theta = 0.0998$, a difference of only 0.2%. With this approximation, Eq. (13.30) becomes

$$F_\theta = -mg\theta = -mg\frac{x}{L} \quad \text{or}$$

$$F_\theta = -\frac{mg}{L}x \tag{13.31}$$

The restoring force is then proportional to the coordinate for small displacements, and the force constant is $k = mg/L$. From Eq. (13.10) the angular frequency ω of a simple pendulum with small amplitude is

$$\omega = \sqrt{\frac{k}{m}} = \sqrt{\frac{mg/L}{m}} = \sqrt{\frac{g}{L}} \quad \text{(simple pendulum, small amplitude)} \qquad (13.32)$$

The corresponding frequency and period relationships are

$$f = \frac{\omega}{2\pi} = \frac{1}{2\pi}\sqrt{\frac{g}{L}} \quad \text{(simple pendulum, small amplitude)} \qquad (13.33)$$

$$T = \frac{2\pi}{\omega} = \frac{1}{f} = 2\pi\sqrt{\frac{L}{g}} \quad \text{(simple pendulum, small amplitude)} \qquad (13.34)$$

? Note that these expressions do not involve the *mass* of the particle. This is because the restoring force, a component of the particle's weight, is proportional to m. Thus the mass appears on *both* sides of $\Sigma\vec{F} = m\vec{a}$ and cancels out. (This is the same physics that explains why bodies of different masses fall with the same acceleration in a vacuum.) For small oscillations, the period of a pendulum for a given value of g is determined entirely by its length.

The dependence on L and g in Eqs. (13.32) through (13.34) is just what we should expect. A long pendulum has a longer period than a shorter one. Increasing g increases the restoring force, causing the frequency to increase and the period to decrease.

We emphasize again that the motion of a pendulum is only *approximately* simple harmonic. When the amplitude is not small, the departures from simple harmonic motion can be substantial. But how small is "small"? The period can be expressed by an infinite series; when the maximum angular displacement is Θ, the period T is given by

$$T = 2\pi\sqrt{\frac{L}{g}}\left(1 + \frac{1^2}{2^2}\sin^2\frac{\Theta}{2} + \frac{1^2 \cdot 3^2}{2^2 \cdot 4^2}\sin^4\frac{\Theta}{2} + \cdots\right) \qquad (13.35)$$

We can compute the period to any desired degree of precision by taking enough terms in the series. We invite you to check that when $\Theta = 15°$ (on either side of the central position), the true period is longer than that given by the approximate Eq. (13.34) by less than 0.5%.

The usefulness of the pendulum as a timekeeper depends on the period being *very nearly* independent of amplitude, provided that the amplitude is small. Thus, as a pendulum clock runs down and the amplitude of the swings decreases a little, the clock still keeps very nearly correct time.

13.22 For small angular displacements θ, the restoring force $F_\theta = -mg\sin\theta$ on a simple pendulum is approximately equal to $-mg\theta$; that is, it is approximately proportional to the displacement θ. Hence for small angles the oscillations are simple harmonic.

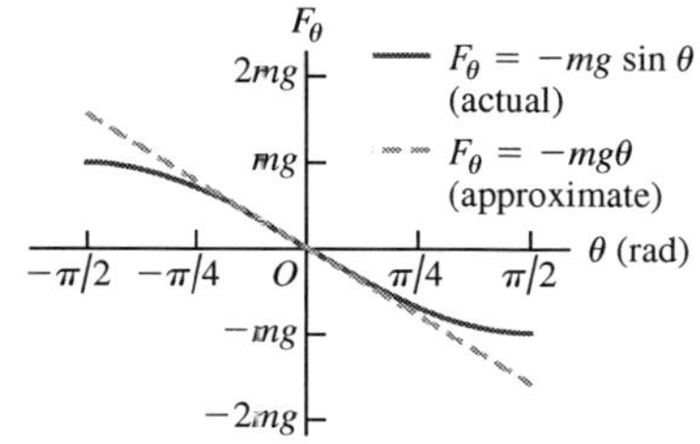

ActivPhysics Online

9.10 Pendulum Frequency
9.11 Risky Pendulum Walk
9.12 Physical Pendulum

Example 13.8 A simple pendulum

Find the period and frequency of a simple pendulum 1.000 m long at a location where $g = 9.800 \text{ m/s}^2$.

SOLUTION

IDENTIFY: Since this is a simple pendulum, we can use the ideas of this section.

SET UP: We use Eq. (13.34) to determine the period T of the pendulum from its length, and Eq. (13.1) to find the frequency f from T.

EXECUTE: From Eqs. (13.34) and (13.1),

$$T = 2\pi\sqrt{\frac{L}{g}} = 2\pi\sqrt{\frac{1.000 \text{ m}}{9.800 \text{ m/s}^2}} = 2.007 \text{ s}$$

$$f = \frac{1}{T} = \frac{1}{2.007\text{s}} = 0.4983 \text{ Hz}$$

EVALUATE: The period is almost exactly 2 s. In fact, when the metric system was first established, the second was defined as half the period of a 1-meter pendulum. This wasn't a very good standard for time, however, because the value of g varies from place to place. We discussed more modern time standards in Section 1.3.

Test Your Understanding of Section 13.5 When a body oscillating on a horizontal spring passes through its equilibrium position, its acceleration is zero (see Fig. 13.2b). When the bob of an oscillating simple pendulum passes through its equilibrium position, is its acceleration zero?

13.23 Dynamics of a physical pendulum.

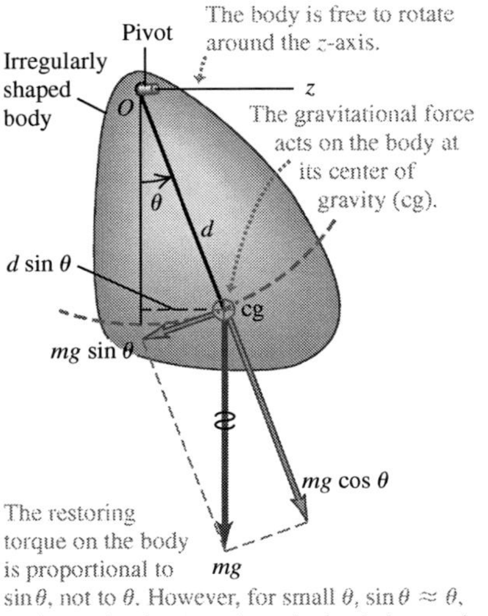

13.6 The Physical Pendulum

A **physical pendulum** is any *real* pendulum that uses an extended body, as contrasted to the idealized model of the *simple* pendulum with all the mass concentrated at a single point. For small oscillations, analyzing the motion of a real, physical pendulum is almost as easy as for a simple pendulum. Figure 13.23 shows a body of irregular shape pivoted so that it can turn without friction about an axis through point O. In the equilibrium position the center of gravity is directly below the pivot; in the position shown in the figure, the body is displaced from equilibrium by an angle θ, which we use as a coordinate for the system. The distance from O to the center of gravity is d, the moment of inertia of the body about the axis of rotation through O is I, and the total mass is m. When the body is displaced as shown, the weight mg causes a restoring torque

$$\tau_z = -(mg)(d\sin\theta) \tag{13.36}$$

The negative sign shows that the restoring torque is clockwise when the displacement is counterclockwise, and vice versa.

When the body is released, it oscillates about its equilibrium position. The motion is not simple harmonic because the torque τ_z is proportional to $\sin\theta$ rather than to θ itself. However, if θ is small, we can approximate $\sin\theta$ by θ in radians, just as we did in analyzing the simple pendulum. Then the motion is *approximately* simple harmonic. With this approximation,

$$\tau_z = -(mgd)\theta$$

The equation of motion is $\sum\tau_z = I\alpha_z$, so

$$-(mgd)\theta = I\alpha_z = I\frac{d^2\theta}{dt^2}$$

$$\frac{d^2\theta}{dt^2} = -\frac{mgd}{I}\theta \tag{13.37}$$

Comparing this with Eq. (13.4), we see that the role of (k/m) for the spring-mass system is played here by the quantity (mgd/I). Thus the angular frequency is

$$\omega = \sqrt{\frac{mgd}{I}} \quad \text{(physical pendulum, small amplitude)} \tag{13.38}$$

The frequency f is $1/2\pi$ times this, and the period T is

$$T = 2\pi\sqrt{\frac{I}{mgd}} \quad \text{(physical pendulum, small amplitude)} \tag{13.39}$$

Equation (13.39) is the basis of a common method for experimentally determining the moment of inertia of a body with a complicated shape. First locate the center of gravity of the body by balancing. Then suspend the body so that it is free to oscillate about an axis, and measure the period T of small-amplitude oscillations. Finally, use Eq. (13.39) to calculate the moment of inertia I of the body about this axis from T, the body's mass m, and the distance d from the axis to the center of gravity (see Exercise 13.49). Biomechanics researchers use this method to find the moments of inertia of an animal's limbs. This information is important for analyzing how an animal walks, as we'll see in the second of the two following examples.

Example 13.9 Physical pendulum versus simple pendulum

Suppose the body in Fig. 13.23 is a uniform rod with length L, pivoted at one end. Find the period of its motion.

SOLUTION

IDENTIFY: Our target variable is the oscillation period of a rod, which acts as a physical pendulum. We need to know the rod's moment of inertia to do this.

SET UP: We use Table 9.2 (Section 9.4) to find the moment of inertia of the rod, and then substitute this value into Eq. (13.39) to determine the period of oscillation.

EXECUTE: From Table 9.2, the moment of inertia of a uniform rod about an axis through one end is $I = \frac{1}{3}ML^2$. The distance from the pivot to the center of gravity is $d = L/2$. From Eq. (13.39),

$$T = 2\pi\sqrt{\frac{\frac{1}{3}ML^2}{MgL/2}} = 2\pi\sqrt{\frac{2L}{3g}}$$

EVALUATE: If the rod is a meter stick $(L = 1.00\text{ m})$ and $g = 9.80\text{ m/s}^2$, then

$$T = 2\pi\sqrt{\frac{2(1.00\text{ m})}{3(9.80\text{ m/s}^2)}} = 1.64\text{ s}$$

The period is smaller by a factor of $\sqrt{2/3} = 0.816$ than the period of a simple pendulum with the same length, calculated in Example 13.8. The cg of the rod is half as far from the pivot as the cg of the simple pendulum, which means the torque is half as great. By itself that would give the rod a period $\sqrt{2}$ times greater than the simple pendulum. But the rod's moment of inertia around one end, $I = \frac{1}{3}ML^2$, is one-third that of the simple pendulum, which by itself would make the rod's period $\sqrt{1/3}$ that of the simple pendulum. The moment of inertia factor is more important in this case, which is why the rod has a shorter period than the simple pendulum.

Example 13.10 *Tyrannosaurus rex* and the physical pendulum

All walking animals, including humans, have a natural walking pace—that is, a number of steps per minute that is more comfortable than a faster or slower pace. Suppose this natural pace corresponds to the oscillation of the leg as a physical pendulum. (a) How does the natural walking pace depend on the length L of the leg, measured from hip to foot? Treat the leg as a uniform rod pivoted at the hip joint. (b) Fossil evidence shows that *Tyrannosaurus rex,* a two-legged dinosaur that lived about 65 million years ago at the end of the Cretaceous period, had a leg length $L = 3.1$ m and a stride length $S = 4.0$ m (the distance from one footprint to the next print of the same foot; Fig. 13.24). Estimate the walking speed of *T. rex.*

13.24 The walking speed of *Tyrannosaurus rex* can be estimated from leg length L and stride length S.

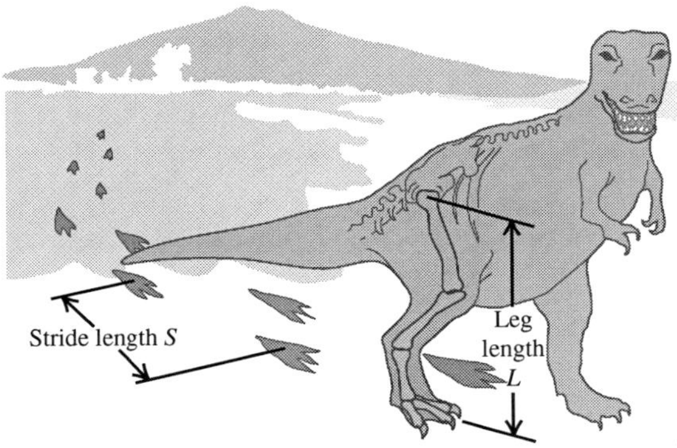

SOLUTION

IDENTIFY: Our target variables are (a) the relationship between the walking pace and the leg length and (b) the walking speed of *T. rex.*

SET UP: We treat the leg as a physical pendulum, with a period of oscillation as found in Example 13.9. The shorter the period, the faster the walking pace. We can find the walking speed from the period and the stride length.

EXECUTE: (a) From Example 13.9 the period of oscillation of the leg is $T = 2\pi\sqrt{2L/3g}$, which is proportional to $\sqrt{L}$. Each period (a complete back-and-forth swing of the leg) corresponds to *two* steps, so the walking pace in steps per unit time is just twice the oscillation frequency $f = 1/T$. Hence the walking pace is proportional to $1/\sqrt{L}$. Animals with short legs (small values of L), such as mice or Chihuahuas, have rapid walking paces; humans, giraffes, and other animals with long legs (large values of L) walk at slower paces.

(b) According to our model for the natural walking pace, the elapsed time for one stride of a walking *Tyrannosaurus rex* is

$$T = 2\pi\sqrt{\frac{2L}{3g}} = 2\pi\sqrt{\frac{2(3.1\text{ m})}{3(9.8\text{ m/s}^2)}} = 2.9\text{ s}$$

The distance moved during this time is the stride length S, so the walking speed is

$$v = \frac{S}{T} = \frac{4.0\text{ m}}{2.9\text{ s}} = 1.4\text{ m/s} = 5.0\text{ km/h} = 3.1\text{ mi/h}$$

This is about the same as a typical human walking speed!

Continued

EVALUATE: Our estimate must be somewhat in error because a uniform rod isn't a very good model for a leg. The legs of many animals, including *T. rex* as well as humans, are tapered; there is a lot more mass between the knee and the hip than between the knee and the foot. Thus the center of mass is less than $L/2$ from the hip; a reasonable guess would be about $L/4$. The moment of inertia is therefore *considerably* less than $ML^2/3$, probably somewhere around $ML^2/15$. Try these numbers out with the analysis of Example 13.9; you'll get a shorter oscillation period and an even faster walking speed for *T. rex*.

Test Your Understanding of Section 13.6 The center of gravity of a simple pendulum of mass m and length L is located at the position of the pendulum bob, a distance L from the pivot point. The center of gravity of a uniform rod of the same mass m and length $2L$ pivoted at one end is also a distance L from the pivot point. How does the period of this uniform rod compare to the period of the simple pendulum? (i) The rod has a longer period; (ii) the rod has a shorter period; (iii) the rod has the same period.

13.7 Damped Oscillations

The idealized oscillating systems we have discussed so far are frictionless. There are no nonconservative forces, the total mechanical energy is constant, and a system set into motion continues oscillating forever with no decrease in amplitude.

Real-world systems always have some dissipative forces, however, and oscillations die out with time unless we replace the dissipated mechanical energy (Fig. 13.25). A mechanical pendulum clock continues to run because potential energy stored in the spring or a hanging weight system replaces the mechanical energy lost due to friction in the pivot and the gears. But eventually the spring runs down or the weights reach the bottom of their travel. Then no more energy is available, and the pendulum swings decrease in amplitude and stop.

13.25 A swinging bell left to itself will eventually stop oscillating due to damping forces (air resistance and friction at the point of suspension).

The decrease in amplitude caused by dissipative forces is called **damping,** and the corresponding motion is called **damped oscillation.** The simplest case to analyze in detail is a simple harmonic oscillator with a frictional damping force that is directly proportional to the *velocity* of the oscillating body. This behavior occurs in friction involving viscous fluid flow, such as in shock absorbers or sliding between oil-lubricated surfaces. We then have an additional force on the body due to friction, $F_x = -bv_x$, where $v_x = dx/dt$ is the velocity and b is a constant that describes the strength of the damping force. The negative sign shows that the force is always opposite in direction to the velocity. The *net* force on the body is then

$$\sum F_x = -kx - bv_x \tag{13.40}$$

and Newton's second law for the system is

$$-kx - bv_x = ma_x \qquad \text{or} \qquad -kx - b\frac{dx}{dt} = m\frac{d^2x}{dt^2} \tag{13.41}$$

Equation (13.41) is a differential equation for x; it would be the same as Eq. (13.4), the equation for the acceleration in SHM, except for the added term $-b\,dx/dt$. Solving this equation is a straightforward problem in differential equations, but we won't go into the details here. If the damping force is relatively small, the motion is described by

$$x = Ae^{-(b/2m)t}\cos(\omega' t + \phi) \quad \text{(oscillator with little damping)} \tag{13.42}$$

The angular frequency of oscillation ω' is given by

$$\omega' = \sqrt{\frac{k}{m} - \frac{b^2}{4m^2}} \quad \text{(oscillator with little damping)} \tag{13.43}$$

You can verify that Eq. (13.42) is a solution of Eq. (13.41) by calculating the first and second derivatives of x, substituting them into Eq. (13.41), and checking whether the left and right sides are equal. This is a straightforward but slightly tedious procedure.

The motion described by Eq. (13.42) differs from the undamped case in two ways. First, the amplitude $Ae^{-(b/2m)t}$ is not constant but decreases with time because of the decreasing exponential factor $e^{-(b/2m)t}$. Figure 13.26 is a graph of Eq. (13.42) for the case $\phi = 0$; it shows that the larger the value of b, the more quickly the amplitude decreases.

Second, the angular frequency ω', given by Eq. (13.43), is no longer equal to $\omega = \sqrt{k/m}$ but is somewhat smaller. It becomes zero when b becomes so large that

$$\frac{k}{m} - \frac{b^2}{4m^2} = 0 \quad \text{or} \quad b = 2\sqrt{km} \tag{13.44}$$

When Eq. (13.44) is satisfied, the condition is called **critical damping.** The system no longer oscillates but returns to its equilibrium position without oscillation when it is displaced and released.

If b is greater than $2\sqrt{km}$, the condition is called **overdamping.** Again there is no oscillation, but the system returns to equilibrium more slowly than with critical damping. For the overdamped case the solutions of Eq. (13.41) have the form

$$x = C_1 e^{-a_1 t} + C_2 e^{-a_2 t}$$

where C_1 and C_2 are constants that depend on the initial conditions and a_1 and a_2 are constants determined by m, k, and b.

When b is less than the critical value, as in Eq. (13.42), the condition is called **underdamping.** The system oscillates with steadily decreasing amplitude.

In a vibrating tuning fork or guitar string, it is usually desirable to have as little damping as possible. By contrast, damping plays a beneficial role in the oscillations of an automobile's suspension system. The shock absorbers provide a velocity-dependent damping force so that when the car goes over a bump, it doesn't continue bouncing forever (Fig. 13.27). For optimal passenger comfort, the system should be critically damped or slightly underdamped. Too much damping would be counterproductive; if the suspension is overdamped and the car hits a second bump just after the first one, the springs in the suspension will still be compressed somewhat from the first bump and will not be able to fully absorb the impact.

13.26 Graph of displacement versus time for an oscillator with little damping [see Eq. (13.42)] and with phase angle $\phi = 0$. The curves are for two values of the damping constant b.

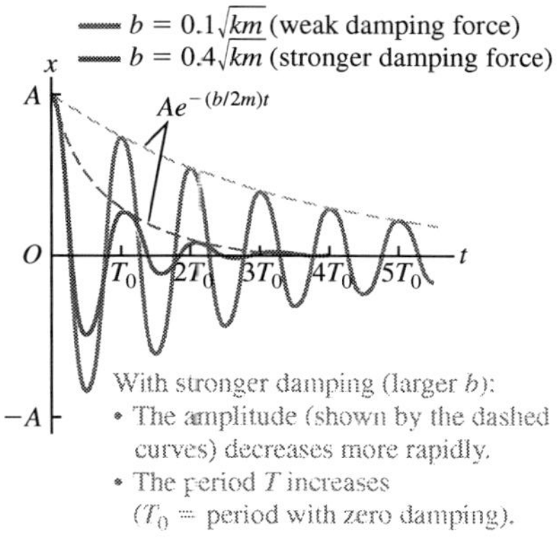

13.27 An automobile shock absorber. The viscous fluid causes a damping force that depends on the relative velocity of the two ends of the unit.

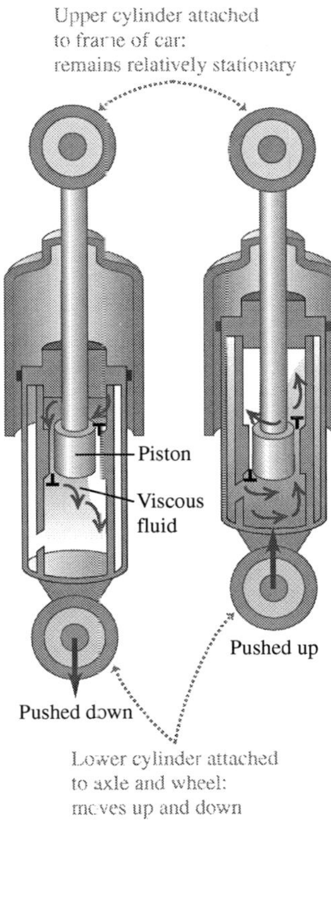

Energy in Damped Oscillations

In damped oscillations the damping force is nonconservative; the mechanical energy of the system is not constant but decreases continuously, approaching zero after a long time. To derive an expression for the rate of change of energy, we first write an expression for the total mechanical energy E at any instant:

$$E = \frac{1}{2}mv_x^2 + \frac{1}{2}kx^2$$

To find the rate of change of this quantity, we take its time derivative:

$$\frac{dE}{dt} = mv_x\frac{dv_x}{dt} + kx\frac{dx}{dt}$$

But $dv_x/dt = a_x$ and $dx/dt = v_x$, so

$$\frac{dE}{dt} = v_x(ma_x + kx)$$

From Eq. (13.41), $ma_x + kx = -bdx/dt = -bv_x$, so

$$\frac{dE}{dt} = v_x(-bv_x) = -bv_x^2 \quad \text{(damped oscillations)} \tag{13.45}$$

The right side of Eq. (13.45) is **negative** whenever the oscillating body is in motion, whether the x-velocity v_x is positive or negative. This shows that as the body moves, the energy decreases, though not at a uniform rate. The term $-bv_x^2 = (-bv_x)v_x$ (force times velocity) is the rate at which the damping force does (negative) work on the system (that is, the damping *power*). This equals the rate of change of the total mechanical energy of the system.

Similar behavior occurs in electric circuits containing inductance, capacitance, and resistance. There is a natural frequency of oscillation, and the resistance plays the role of the damping constant b. We will study these circuits in detail in Chapters 30 and 31.

Test Your Understanding of Section 13.7 An airplane is flying in a straight line at a constant altitude. If a wind gust strikes and raises the nose of the airplane, the nose will bob up and down until the airplane eventually returns to its original attitude. Are these oscillations (i) undamped, (ii) underdamped, (iii) critically damped, or (iv) overdamped?

13.8 Forced Oscillations and Resonance

A damped oscillator left to itself will eventually stop moving altogether. But we can maintain a constant-amplitude oscillation by applying a force that varies with time in a periodic or cyclic way, with a definite period and frequency. As an example, consider your cousin Throckmorton on a playground swing. You can keep him swinging with constant amplitude by giving him a little push once each cycle. We call this additional force a **driving force.**

Damped Oscillation with a Periodic Driving Force

If we apply a periodically varying driving force with angular frequency ω_d to a damped harmonic oscillator, the motion that results is called a **forced oscillation** or a *driven oscillation.* It is different from the motion that occurs when the system is simply displaced from equilibrium and then left alone, in which case the system oscillates with a **natural angular frequency** ω' determined by m, k, and b, as in Eq. (13.43). In a forced oscillation, however, the angular frequency with which the mass oscillates is equal to the driving angular frequency ω_d. This does *not* have to be equal to the angular frequency ω' with which the system would oscillate without a driving force. If you grab the ropes of Throckmorton's swing, you can force the swing to oscillate with any frequency you like.

Suppose we force the oscillator to vibrate with an angular frequency ω_d that is nearly *equal* to the angular frequency ω' it would have with no driving force. What happens? The oscillator is naturally disposed to oscillate at $\omega = \omega'$, so we expect the amplitude of the resulting oscillation to be larger than when the two frequencies are very different. Detailed analysis and experiment shows that this is just what happens. The easiest case to analyze is a *sinusoidally* varying force—say, $F(t) = F_{max} \cos \omega_d t$. If we vary the frequency ω_d of the driving force, the amplitude of the resulting forced oscillation varies in an interesting way (Fig. 13.28). When there is very little damping (small b), the amplitude goes through a sharp peak as the driving angular frequency ω_d nears the natural oscillation angular frequency ω'. When the damping is increased (larger b), the peak becomes broader and smaller in height and shifts toward lower frequencies.

We could work out an expression that shows how the amplitude A of the forced oscillation depends on the frequency of a sinusoidal driving force, with

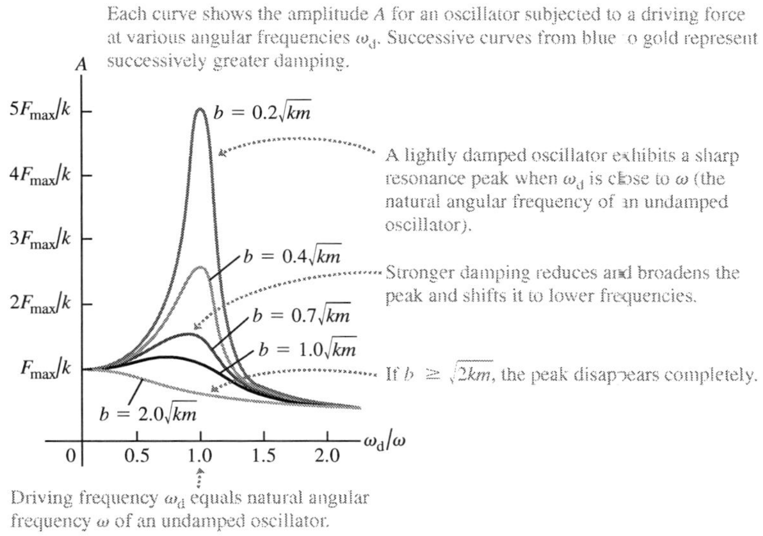

13.28 Graph of the amplitude A of forced oscillation as a function of the angular frequency ω_d of the driving force. The horizontal axis shows the ratio of ω_d to the angular frequency $\omega = \sqrt{k/m}$ of an undamped oscillator. Each curve has a different value of the damping constant b.

maximum value F_{max}. That would involve more differential equations than we're ready for, but here is the result:

$$A = \frac{F_{max}}{\sqrt{(k - m\omega_d^2)^2 + b^2\omega_d^2}} \quad \text{(amplitude of a driven oscillator)} \quad (13.46)$$

When $k - m\omega_d^2 = 0$, the first term under the radical is zero, so A has a maximum near $\omega_d = \sqrt{k/m}$. The height of the curve at this point is proportional to $1/b$; the less damping, the higher the peak. At the low-frequency extreme, when $\omega_d = 0$, we get $A = F_{max}/k$. This corresponds to a *constant* force F_{max} and a constant displacement $A = F_{max}/k$ from equilibrium, as we might expect.

Resonance and Its Consequences

The fact that there is an amplitude peak at driving frequencies close to the natural frequency of the system is called **resonance.** Physics is full of examples of resonance; building up the oscillations of a child on a swing by pushing with a frequency equal to the swing's natural frequency is one. A vibrating rattle in a car that occurs only at a certain engine speed or wheel-rotation speed is an all-too-familiar example. Inexpensive loudspeakers often have an annoying boom or buzz when a musical note happens to coincide with the resonant frequency of the speaker cone or the speaker housing. In Chapter 16 we will study other examples of resonance that involve sound. Resonance also occurs in electric circuits, as we will see in Chapter 31; a tuned circuit in a radio or television receiver responds strongly to waves having frequencies near its resonant frequency, and this fact is used to select a particular station and reject the others.

Resonance in mechanical systems can be destructive. A company of soldiers once destroyed a bridge by marching across it in step; the frequency of their steps was close to a natural vibration frequency of the bridge, and the resulting oscillation had large enough amplitude to tear the bridge apart. Ever since, marching soldiers have been ordered to break step before crossing a bridge. Some years ago, vibrations of the engines of a particular airplane had just the right frequency to resonate with the natural frequencies of its wings. Large oscillations built up, and occasionally the wings fell off.

Nearly everyone has seen the film of the collapse of the Tacoma Narrows suspension bridge in 1940 (Fig. 13.29). This is usually cited as an example of resonance driven by the wind, but there's some doubt whether it should be called that.

13.29 The Tacoma Narrows Bridge collapsed four months and six days after it was opened for traffic. The main span was 2800 ft long and 39 ft wide, with 8-ft-high steel stiffening girders on both sides. The maximum amplitude of the torsional vibrations was 35°; the frequency was about 0.2 Hz.

The wind didn't have to vary *periodically* with a frequency close to a natural frequency of the bridge. The airflow past the bridge was turbulent, and vortices were formed in the air with a regular frequency that depended on the flow speed. It is conceivable that this frequency may have coincided with a natural frequency of the bridge. But the cause may well have been something more subtle called a *self-excited oscillation,* in which the aerodynamic forces caused by a *steady* wind blowing on the bridge tended to displace it farther from equilibrium at times when it was already moving away from equilibrium. It is as though we had a damping force such as the $-bv_x$ term in Eq. (13.40) but with the sign reversed. Instead of draining mechanical energy away from the system, this anti-damping force pumps energy into the system, building up the oscillations to destructive amplitudes. The approximate differential equation is Eq. (13.41) with the sign of the b term reversed, and the oscillating solution is Eq. (13.42) with a *positive* sign in the exponent. You can see that we're headed for trouble. Engineers have learned how to stabilize suspension bridges, both structurally and aerodynamically, to prevent such disasters.

Test Your Understanding of Section 13.8 When driven at a frequency near its natural frequency, an oscillator with very little damping has a much greater response than the same oscillator with more damping. When driven at a frequency that is much higher or lower than the natural frequency, which oscillator will have the greater response: (i) the one with very little damping or (ii) the one with more damping?

CHAPTER 13 SUMMARY

Periodic motion: Periodic motion is motion that repeats itself in a definite cycle. It occurs whenever a body has a stable equilibrium position and a restoring force that acts when it is displaced from equilibrium. Period T is the time for one cycle. Frequency f is the number of cycles per unit time. Angular frequency ω is 2π times the frequency. (See Example 13.1.)

$$f = \frac{1}{T} \qquad T = \frac{1}{f} \tag{13.1}$$

$$\omega = 2\pi f = \frac{2\pi}{T} \tag{13.2}$$

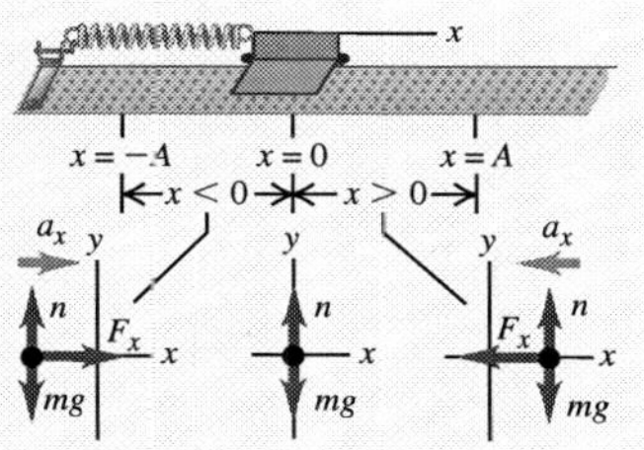

Simple harmonic motion: If the restoring force F_x in periodic motion is directly proportional to the displacement x, the motion is called simple harmonic motion (SHM). In many cases this condition is satisfied if the displacement from equilibrium is small. The angular frequency, frequency, and period in SHM do not depend on the amplitude, but only on the mass m and force constant k. The displacement, velocity, and acceleration in SHM are sinusoidal functions of time; the amplitude A and phase angle ϕ of the oscillation are determined by the initial position and velocity of the body. (See Examples 13.2, 13.3, 13.6, and 13.7.)

$$F_x = -kx \tag{13.3}$$

$$a_x = \frac{F_x}{m} = -\frac{k}{m}x \tag{13.4}$$

$$\omega = \sqrt{\frac{k}{m}} \tag{13.10}$$

$$f = \frac{\omega}{2\pi} = \frac{1}{2\pi}\sqrt{\frac{k}{m}} \tag{13.11}$$

$$T = \frac{1}{f} = 2\pi\sqrt{\frac{m}{k}} \tag{13.12}$$

$$x = A\cos(\omega t + \phi) \tag{13.13}$$

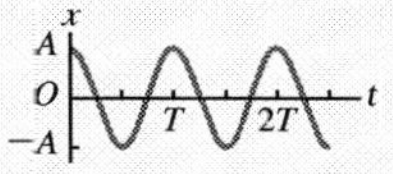

Energy in simple harmonic motion: Energy is conserved in SHM. The total energy can be expressed in terms of the force constant k and amplitude A. (See Examples 13.4 and 13.5.)

$$E = \frac{1}{2}mv_x^2 + \frac{1}{2}kx^2 = \frac{1}{2}kA^2 = \text{constant} \tag{13.21}$$

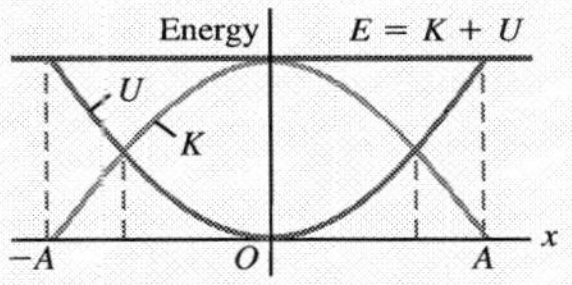

Angular simple harmonic motion: In angular SHM, the frequency and angular frequency are related to the moment of inertia I and the torsion constant κ.

$$\omega = \sqrt{\frac{\kappa}{I}} \quad \text{and} \quad f = \frac{1}{2\pi}\sqrt{\frac{\kappa}{I}} \tag{13.24}$$

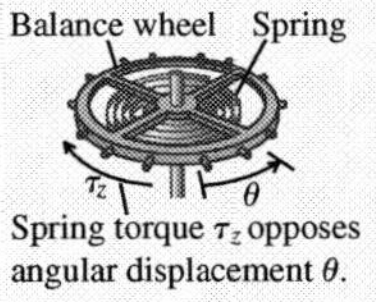

Simple pendulum: A simple pendulum consists of a point mass m at the end of a massless string of length L. Its motion is approximately simple harmonic for sufficiently small amplitude; the angular frequency, frequency, and period then depend only on g and L, not on the mass or amplitude. (See Example 13.8.)

$$\omega = \sqrt{\frac{g}{L}} \tag{13.32}$$

$$f = \frac{\omega}{2\pi} = \frac{1}{2\pi}\sqrt{\frac{g}{L}} \tag{13.33}$$

$$T = \frac{2\pi}{\omega} = \frac{1}{f} = 2\pi\sqrt{\frac{L}{g}} \tag{13.34}$$

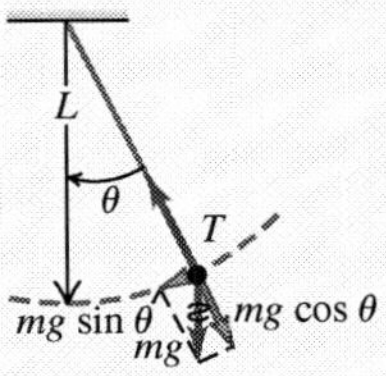

Physical pendulum: A physical pendulum is any body suspended from an axis of rotation. The angular frequency and period for small-amplitude oscillations are independent of amplitude, but depend on the mass m, distance d from the axis of rotation to the center of gravity, and moment of inertia I about the axis. (See Examples 13.9 and 13.10.)

$$\omega = \sqrt{\frac{mgd}{I}} \qquad (13.38)$$

$$T = 2\pi\sqrt{\frac{I}{mgd}} \qquad (13.39)$$

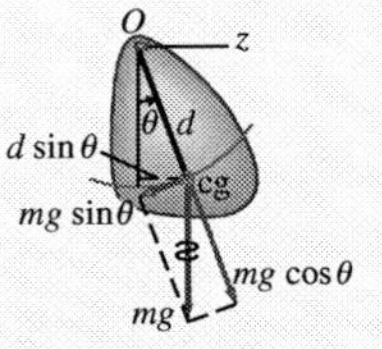

Damped oscillations: When a force $F_x = -bv_x$ proportional to velocity is added to a simple harmonic oscillator, the motion is called a damped oscillation. If $b < 2\sqrt{km}$ (called underdamping), the system oscillates with a decaying amplitude and an angular frequency ω' that is lower than it would be without damping. If $b = 2\sqrt{km}$ (called critical damping) or $b > 2\sqrt{km}$ (called overdamping), when the system is displaced it returns to equilibrium without oscillating.

$$x = Ae^{-(b/2m)t}\cos\omega' t \qquad (13.42)$$

$$\omega' = \sqrt{\frac{k}{m} - \frac{b^2}{4m^2}} \qquad (13.43)$$

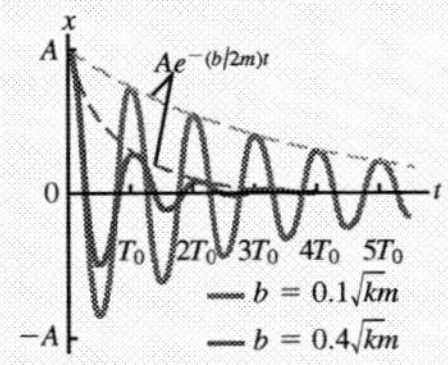

Driven oscillations and resonance: When a sinusoidally varying driving force is added to a damped harmonic oscillator, the resulting motion is called a forced oscillation. The amplitude is a function of the driving frequency ω_d and reaches a peak at a driving frequency close to the natural frequency of the system. This behavior is called resonance.

$$A = \frac{F_{max}}{\sqrt{(k - m\omega_d^2)^2 + b^2\omega_d^2}} \qquad (13.46)$$

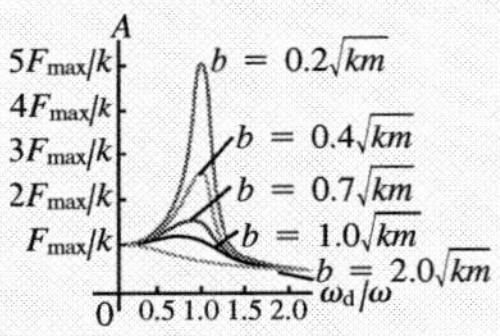

Key Terms

periodic motion (oscillation), *419*
displacement, *420*
restoring force, *420*
amplitude, *420*
cycle, *420*
period, *420*
frequency, *420*
angular frequency, *420*
simple harmonic motion (SHM), *421*
harmonic oscillator, *422*
reference circle, *423*
phasor, *423*
phase angle, *426*
simple pendulum, *436*
physical pendulum, *438*
damping, *440*
damped oscillation, *440*
critical damping, *441*
overdamping, *441*
underdamping, *441*
driving force, *442*
forced oscillation, *442*
natural angular frequency, *442*
resonance, *443*

Answer to Chapter Opening Question ?

Neither—the clock would still keep time correctly. If its rod has negligible mass, then the pendulum is a simple pendulum and its period is independent of the mass [see Eq. (13.34)]. If the rod's mass is included, the pendulum is a physical pendulum. Doubling its mass m also doubles its moment of inertia I, so the ratio I/m is unchanged and the period $T = 2\pi\sqrt{I/mgd}$ [Eq. (13.39)] remains the same.

Answers to Test Your Understanding Questions

13.1 Answers: (a) $x < 0$, (b) $x > 0$, (c) $x < 0$, (d) $x > 0$, (e) $x = 0$, (f) $x > 0$ Figure 13.2 shows that the net x-component of force F_x and the x-acceleration a_x are both positive when $x < 0$ (so the body is displaced to the left and the spring is compressed), while F_x and a_x are both negative when $x > 0$ (so the body is displaced to the right and the spring is stretched). Hence x and a_x always have *opposite* signs. This is true whether the object is moving to the right $(v_x > 0)$, to the left $(v_x < 0)$, or not at all $(v_x = 0)$, since the force exerted by the spring depends only on whether it is compressed or stretched and by what distance. This explains the answers to (a) through (e). If the acceleration is zero as in (f), the net force must also be zero and so the spring must be relaxed; hence $x = 0$.

13.2 Answers: (a) $A > 0.10$ m, $\phi < 0$; (b) $A > 0.10$ m, $\phi > 0$ In both situations the initial $(t = 0)$ x-velocity v_{0x} is nonzero, so from Eq. (13.19) the amplitude $A = \sqrt{x_0^2 + (v_{0x}^2/\omega^2)}$ is greater than the initial x-coordinate $x_0 = 0.10$ m. From Eq. (13.18) the phase angle is $\phi = \arctan(-v_{0x}/\omega x_0)$, which is positive if the quantity $-v_{0x}/\omega x_0$ (the argument of the arctangent function) is positive and negative if $-v_{0x}/\omega x_0$ is negative. In part (a) x_0 and v_{0x} are both positive, so $-v_{0x}/\omega x_0 < 0$ and $\phi < 0$. In part (b) x_0 is positive and v_{0x} is negative, so $-v_{0x}/\omega x_0 > 0$ and $\phi > 0$.

13.3 Answers: (a) (iii), (b) (v) To increase the total energy $E = \frac{1}{2}kA^2$ by a factor of 2, the amplitude A must increase by a factor of $\sqrt{2}$. Because the motion is SHM, changing the amplitude has no effect on the frequency.

13.4 Answer: (i) The oscillation period of a body of mass m attached to a hanging spring of force constant k is given by $T = 2\pi\sqrt{m/k}$, the same expression as for a body attached to a

horizontal spring. Neither m nor k changes when the apparatus is taken to Mars, so the period is unchanged. The only difference is that in equilibrium, the spring will stretch a shorter distance on Mars than on earth due to the weaker gravity.
13.5 Answer: no Just as for an object oscillating on a spring, at the equilibrium position the *speed* of the pendulum bob is instantaneously not changing (this is where the speed is maximum, so its derivative at this time is zero). But the *direction* of motion is changing because the pendulum bob follows a circular path. Hence the bob must have a component of acceleration perpendicular to the path and toward the center of the circle (see Section 3.4). To cause this acceleration at the equilibrium position when the string is vertical, the upward tension force at this position must be greater than the weight of the bob. This causes a net upward force on the bob and an upward acceleration toward the center of the circular path.
13.6 Answer: (i) The period of a physical pendulum is given by Eq. (13.39), $T = 2\pi\sqrt{I/mgd}$. The distance $d = L$ from the pivot to the center of gravity is the same for both the rod and the simple pendulum, as is the mass m. This means that for any displacement angle θ the same restoring torque acts on both the rod and the simple pendulum. However, the rod has a greater moment of inertia: $I_{rod} = \frac{1}{3}m(2L)^2 = \frac{4}{3}mL^2$ and $I_{simple} = mL^2$ (all the mass of the pendulum is a distance L from the pivot). Hence the rod has a longer period.
13.7 Answer: (ii) The oscillations are underdamped with a decreasing amplitude on each cycle of oscillation, like those graphed in Fig. 13.26. If the oscillations were undamped, they would continue indefinitely with the same amplitude. If they were critically damped or overdamped, the nose would not bob up and down but would return smoothly to the original equilibrium attitude without overshooting.
13.8 Answer: (i) Figure 13.28 shows that the curve of amplitude versus driving frequency moves upward at *all* frequencies as the value of the damping constant b is decreased. Hence for fixed values of k and m, the oscillator with the least damping (smallest value of b) will have the greatest response at any driving frequency.

PROBLEMS

For instructor-assigned homework, go to www.masteringphysics.com

Discussion Questions

Q13.1. An object is moving with SHM of amplitude A on the end of a spring. If the amplitude is doubled, what happens to the total distance the object travels in one period? What happens to the period? What happens to the maximum speed of the object? Discuss how these answers are related.
Q13.2. Think of several examples in everyday life of motions that are, at least approximately, simple harmonic. In what respects does each differ from SHM?
Q13.3. Does a tuning fork or similar tuning instrument undergo SHM? Why is this a crucial question for musicians?
Q13.4. A box containing a pebble is attached to an ideal horizontal spring and is oscillating on a friction-free air table. When the box has reached its maximum distance from the equilibrium point, the pebble is suddenly lifted out vertically without disturbing the box. Will the following characteristics of the motion increase, decrease, or remain the same in the subsequent motion of the box? Justify each answer. (a) frequency; (b) period; (c) amplitude; (d) the maximum kinetic energy of the box; (e) the maximum speed of the box.
Q13.5. If a uniform spring is cut in half, what is the force constant of each half? Justify your answer. How would the frequency of SHM using a half-spring differ from the frequency using the same mass and the entire spring?
Q13.6. The analysis of SHM in this chapter ignored the mass of the spring. How does the spring's mass change the characteristics of the motion?
Q13.7. Two identical gliders on an air track are connected by an ideal spring. Could such a system undergo SHM? Explain. How would the period compare with that of a single glider attached to a spring whose other end is rigidly attached to a stationary object? Explain.
Q13.8. You are captured by Martians, taken into their ship, and put to sleep. You awake some time later and find yourself locked in a small room with no windows. All the martians have left you with is your digital watch, your school ring, and your long silver-chain necklace. Explain how you can determine whether you are still on earth or have been transported to Mars.
Q13.9. The system shown in Fig. 13.17 is mounted in an elevator. What happens to the period of the motion (does it increase, decrease, or remain the same) if the elevator (a) accelerates upward at 5.0 m/s^2; (b) moves upward at a steady 5.0 m/s; (c) accelerates downward at 5.0 m/s^2? Justify your answers.
Q13.10. If a pendulum has a period of 2.5 s on earth, what would be its period in a space station orbiting the earth? If a mass hung from a vertical spring has a period of 5.0 s on earth, what would its period be in the space station? Justify each of your answers.
Q13.11. A simple pendulum is mounted in an elevator. What happens to the period of the pendulum (does it increase, decrease, or remain the same) if the elevator (a) accelerates upward at 5.0 m/s^2, (b) moves upward at a steady 5.0 m/s, (c) accelerates downward at 5.0 m/s^2, (d) accelerates downward at 9.8 m/s^2? Justify your answers.
Q13.12. What should you do to the length of the string of a simple pendulum to (a) double its frequency; (b) double its period; (c) double its angular frequency?
Q13.13. If a pendulum clock is taken to a mountaintop, does it gain or lose time, assuming it is correct at a lower elevation? Explain your answer.
Q13.14. When the amplitude of a simple pendulum increases, should its period increase or decrease? Give a qualitative argument; do not rely on Eq. (13.35). Is your argument also valid for a physical pendulum?
Q13.15. Why do short dogs (like Chihuahuas) walk with quicker strides than do tall dogs (like Great Danes)?
Q13.16. At what point in the motion of a simple pendulum is the string tension greatest? Least? In each case give the reasoning behind your answer.
Q13.17. Could a standard of time be based on the period of a certain standard pendulum? What advantages and disadvantages would such a standard have compared to the actual present-day standard discussed in Section 1.3?
Q13.18. For a simple pendulum, clearly distinguish between ω (the angular velocity) and ω (the angular frequency). Which is constant and which is variable?

Q13.19. A glider is attached to a fixed ideal spring and oscillates on a horizontal, friction-free air track. A coin is atop the glider and oscillating with it. At what points in the motion is the friction force on the coin greatest? At what points is it least? Justify your answers.

Q13.20. In designing structures in an earthquake-prone region, how should the natural frequencies of oscillation of a structure relate to typical earthquake frequencies? Why? Should the structure have a large or small amount of damping?

Exercises

Section 13.1 Describing Oscillation

13.1. A piano string sounds a middle A by vibrating primarily at 220 Hz. (a) Calculate the string's period and angular frequency. (b) Calculate the period and angular frequency for a soprano singing an A one octave higher, which is twice the frequency of the piano string.

13.2. If an object on a horizontal, frictionless surface is attached to a spring, displaced, and then released, it will oscillate. If it is displaced 0.120 m from its equilibrium position and released with zero initial speed, then after 0.800 s its displacement is found to be 0.120 m on the opposite side, and it has passed the equilibrium position once during this interval. Find (a) the amplitude; (b) the period; (c) the frequency.

13.3. The tip of a tuning fork goes through 440 complete vibrations in 0.500 s. Find the angular frequency and the period of the motion.

13.4. The displacement of an oscillating object as a function of time is shown in Fig. 13.30. What are (a) the frequency; (b) the amplitude; (c) the period; (d) the angular frequency of this motion?

Figure **13.30** Exercise 13.4.

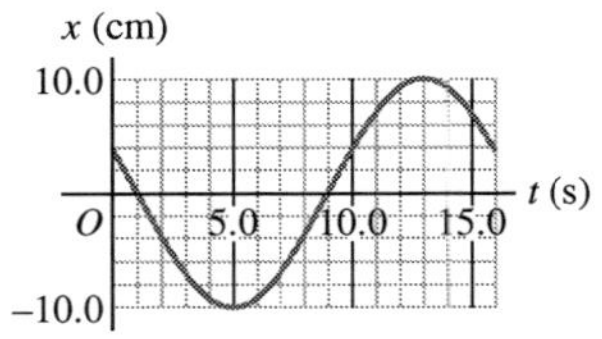

Section 13.2 Simple Harmonic Motion

13.5. A machine part is undergoing SHM with a frequency of 5.00 Hz and amplitude 1.80 cm. How long does it take the part to go from $x = 0$ to $x = -1.80$ cm?

13.6. In a physics lab, you attach a 0.200-kg air-track glider to the end of an ideal spring of negligible mass and start it oscillating. The elapsed time from when the glider first moves through the equilibrium point to the second time it moves through that point is 2.60 s. Find the spring's force constant.

13.7. When a body of unknown mass is attached to an ideal spring with force constant 120 N/m, it is found to vibrate with a frequency of 6.00 Hz. Find (a) the period of the motion; (b) the angular frequency; (c) the mass of the body.

13.8. When a 0.750-kg mass oscillates on an ideal spring, the frequency is 1.33 Hz. What will the frequency be if 0.220 kg are added to the original mass, and (b) subtracted from the original mass? Try to solve this problem *without* finding the force constant of the spring.

13.9. A harmonic oscillator consists of a 0.500-kg mass attached to an ideal spring with force constant 140 N/m. Find (a) the period; (b) the frequency; (c) the angular frequency of the oscillations.

13.10. Jerk. A guitar string vibrates at a frequency of 440 Hz. A point at its center moves in SHM with an amplitude of 3.0 mm and a phase angle of zero. (a) Write an equation for the position of the center of the string as a function of time. (b) What are the maximum values of the magnitudes of the velocity and acceleration of the center of the string? (c) The derivative of the acceleration with respect to time is a quantity called the *jerk.* Write an equation for the jerk of the center of the string as a function of time, and find the maximum value of the magnitude of the jerk.

13.11. A 2.00-kg, frictionless block is attached to an ideal spring with force constant 300 N/m. At $t = 0$ the spring is neither stretched nor compressed and the block is moving in the negative direction at 12.0 m/s. Find (a) the amplitude and (b) the phase angle. (c) Write an equation for the position as a function of time.

13.12. Repeat Exercise 13.11, but assume that at $t = 0$ the block has velocity −4.00 m/s and displacement +0.200 m.

13.13. The point of the needle of a sewing machine moves in SHM along the x-axis with a frequency of 2.5 Hz. At $t = 0$ its position and velocity components are +1.1 cm and −15 cm/s, respectively (a) Find the acceleration component of the needle at $t = 0$. (b) Write equations giving the position, velocity, and acceleration components of the point as a function of time.

13.14. An object is undergoing SHM with period 1.200 s and amplitude 0.600 m. At $t = 0$ the object is at $x = 0$. How far is the object from the equilibrium position when $t = 0.480$ s?

13.15. Weighing Astronauts. This procedure has actually been used to "weigh" astronauts in space. A 42.5-kg chair is attached to a spring and allowed to oscillate. When it is empty, the chair takes 1.30 s to make one complete vibration. But with an astronaut sitting in it, with her feet off the floor, the chair takes 2.54 s for one cycle. What is the mass of the astronaut?

13.16. A 0.400-kg object undergoing SHM has $a_x = -2.70\ \text{m/s}^2$ when $x = 0.300$ m. What is the time for one oscillation?

13.17. On a frictionless, horizontal air track, a glider oscillates at the end of an ideal spring of force constant 2.50 N/cm. The graph in Fig. 13.31 shows the acceleration of the glider as a function of time. Find (a) the mass of the glider; (b) the maximum displacement of the glider from the equilibrium point; (c) the maximum force the spring exerts on the glider.

Figure **13.31** Exercise 13.17.

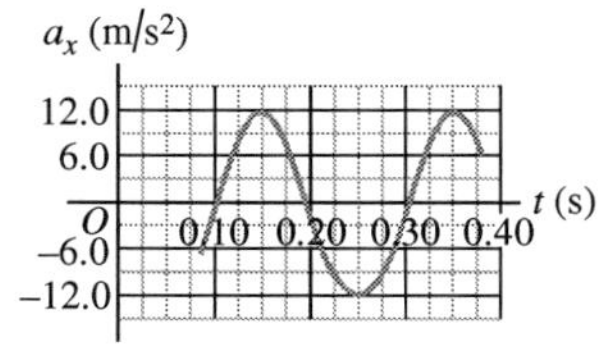

13.18. A 0.500-kg mass on a spring has velocity as a function of time given by $v_x(t) = (3.60\ \text{cm/s})\sin[(4.71\ \text{s}^{-1})t - \pi/2]$. What are (a) the period; (b) the amplitude; (c) the maximum acceleration of the mass; (d) the force constant of the spring?

13.19. A 1.50-kg mass on a spring has displacement as a function of time given by the equation

$$x(t) = (7.40\ \text{cm})\cos[(4.16\ \text{s}^{-1})t - 2.42].$$

Find (a) the time for one complete vibration; (b) the force constant of the spring; (c) the maximum speed of the mass; (d) the maximum force on the mass; (e) the position, speed, and acceleration of the mass at $t = 1.00$ s; (f) the force on the mass at that time.

13.20. An object is undergoing SHM with period 0.300 s and amplitude 6.00 cm. At $t = 0$ the object is instantaneously at rest at $x = 6.00$ cm. Calculate the time it takes the object to go from $x = 6.00$ cm to $x = -1.50$ cm.

Section 13.3 Energy in Simple Harmonic Motion

13.21. A tuning fork labeled 392 Hz has the tip of each of its two prongs vibrating with an amplitude of 0.600 mm. (a) What is the maximum speed of the tip of a prong? (b) A housefly (*Musca domestica*) with mass 0.0270 g is holding on to the tip of one of the prongs. As the prong vibrates, what is the fly's maximum kinetic energy? Assume that the fly's mass has a negligible effect on the frequency of oscillation.

13.22. A harmonic oscillator has angular frequency ω and amplitude A. (a) What are the magnitudes of the displacement and velocity when the elastic potential energy is equal to the kinetic energy? (Assume that $U = 0$ at equilibrium.) (b) How often does this occur in each cycle? What is the time between occurrences? (c) At an instant when the displacement is equal to $A/2$, what fraction of the total energy of the system is kinetic and what fraction is potential?

13.23. A 0.500-kg glider, attached to the end of an ideal spring with force constant $k = 450$ N/m, undergoes SHM with an amplitude of 0.040 m. Compute (a) the maximum speed of the glider; (b) the speed of the glider when it is at $x = -0.015$ m; (c) the magnitude of the maximum acceleration of the glider; (d) the acceleration of the glider at $x = -0.015$ m; (e) the total mechanical energy of the glider at any point in its motion.

13.24. A cheerleader waves her pom-pom in SHM with an amplitude of 18.0 cm and a frequency of 0.850 Hz. Find (a) the maximum magnitude of the acceleration and of the velocity; (b) the acceleration and speed when the pom-pom's coordinate is $x = +9.0$ cm; (c) the time required to move from the equilibrium position directly to a point 12.0 cm away. (d) Which of the quantities asked for in parts (a), (b), and (c) can be found using the energy approach used in Section 13.3, and which cannot? Explain.

13.25. For the situation described in part (a) of Example 13.5, what should be the value of the putty mass m so that the amplitude after the collision is one-half the original amplitude? For this value of m, what fraction of the original mechanical energy is converted into heat?

13.26. A 0.150-kg toy is undergoing SHM on the end of a horizontal spring with force constant $k = 300$ N/m. When the object is 0.0120 m from its equilibrium position, it is observed to have a speed of 0.300 m/s. What are (a) the total energy of the object at any point of its motion; (b) the amplitude of the motion; (c) the maximum speed attained by the object during its motion?

13.27. You are watching an object that is moving in SHM. When the object is displaced 0.600 m to the right of its equilibrium position, it has a velocity of 2.20 m/s to the right and an acceleration of 8.40 m/s^2 to the left. How much farther from this point will the object move before it stops momentarily and then starts to move back to the left?

13.28. On a horizontal, frictionless table, an open-topped 5.20-kg box is attached to an ideal horizontal spring having force constant 375 N/m. Inside the box is a 3.44-kg stone. The system is oscillating with an amplitude of 7.50 cm. When the box has reached its maximum speed, the stone is suddenly plucked vertically out of the box without touching the box. Find (a) the period and (b) the amplitude of the resulting motion of the box. (c) Without doing any calculations, is the new period greater or smaller than the original period? How do you know?

13.29. Inside a NASA test vehicle, a 3.50-kg ball is pulled along by a horizontal ideal spring fixed to a friction-free table. The force constant of the spring is 225 N/m. The vehicle has a steady acceleration of 5.00 m/s^2, and the ball is not oscillating. Suddenly, when the vehicle's speed has reached 45.0 m/s, its engines turn off, thus eliminating its acceleration but not its velocity. Find (a) the amplitude and (b) the frequency of the resulting oscillations of the ball. (c) What will be the ball's maximum speed relative to the vehicle?

Section 13.4 Applications of Simple Harmonic Motion

13.30. A proud deep-sea fisherman hangs a 65.0-kg fish from an ideal spring having negligible mass. The fish stretches the spring 0.120 m. (a) Find the force constant of the spring. The fish is now pulled down 5.00 cm and released. (b) What is the period of oscillation of the fish? (c) What is the maximum speed it will reach?

13.31. A 175-g glider on a horizontal, frictionless air track is attached to a fixed ideal spring with force constant 155 N/m. At the instant you make measurements on the glider, it is moving at 0.815 m/s and is 3.00 cm from its equilibrium point. Use *energy conservation* to find (a) the amplitude of the motion and (b) the maximum speed of the glider. (c) What is the angular frequency of the oscillations?

13.32. A thrill-seeking cat with mass 4.00 kg is attached by a harness to an ideal spring of negligible mass and oscillates vertically in SHM. The amplitude is 0.050 m, and at the highest point of the motion the spring has its natural unstretched length. Calculate the elastic potential energy of the spring (take it to be zero for the unstretched spring), the kinetic energy of the cat, the gravitational potential energy of the system relative to the lowest point of the motion, and the sum of these three energies when the cat is (a) at its highest point; (b) at its lowest point; (c) at its equilibrium position.

13.33. A 1.50-kg ball and a 2.00-kg ball are glued together with the lighter one below the heavier one. The upper ball is attached to a vertical ideal spring of force constant 165 N/m, and the system is vibrating vertically with amplitude 15.0 cm. The glue connecting the balls is old and weak, and it suddenly comes loose when the balls are at the lowest position in their motion. (a) Why is the glue more likely to fail at the *lowest* point than at any other point in the motion? (b) Find the amplitude and frequency of the vibrations after the lower ball has come loose.

13.34. A uniform, solid metal disk of mass 6.50 kg and diameter 24.0 cm hangs in a horizontal plane, supported at its center by a vertical metal wire. You find that it requires a horizontal force of 4.23 N tangent to the rim of the disk to turn it by 3.34°, thus twisting the wire. You now remove this force and release the disk from rest. (a) What is the torsion constant for the metal wire? (b) What are the frequency and period of the torsional oscillations of the disk? (c) Write the equation of motion for $\theta(t)$ for the disk.

13.35. A certain alarm clock ticks four times each second, with each tick representing half a period. The balance wheel consists of a thin rim with radius 0.55 cm, connected to the balance staff by thin spokes of negligible mass. The total mass of the balance wheel is 0.90 g. (a) What is the moment of inertia of the balance wheel about its shaft? (b) What is the torsion constant of the hairspring?

13.36. A thin metal disk with mass 2.00×10^{-3} kg and radius 2.20 cm is attached at its center to a long fiber (Fig. 13.32). The disk, when twisted and released, oscillates with a period of 1.00 s. Find the torsion constant of the fiber.

Figure **13.32** Exercise 13.36.

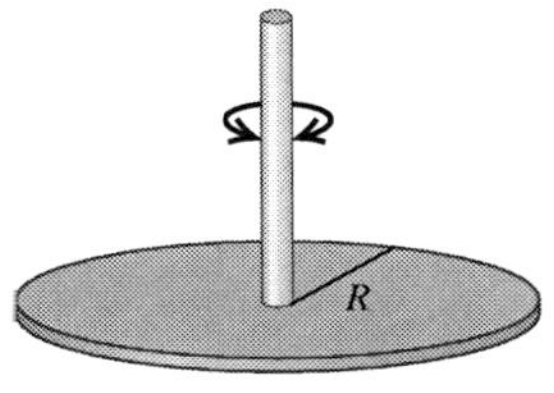

13.37. You want to find the moment of inertia of a complicated machine part about an axis through its center of mass. You suspend it from a wire along this axis. The wire has a torsion constant of $0.450 \text{ N} \cdot \text{m/rad}$ You twist the part a small amount about this axis and let it go, timing 125 oscillations in 265 s. What is the moment of inertia you want to find?

13.38. The balance wheel of a watch vibrates with an angular amplitude Θ, angular frequency ω, and phase angle $\phi = 0$. (a) Find expressions for the angular velocity $d\theta/dt$ and angular acceleration $d^2\theta/dt^2$ as functions of time. (b) Find the balance wheel's angular velocity and angular acceleration when its angular displacement is Θ, and when its angular displacement is $\Theta/2$ and θ is decreasing. (*Hint:* Sketch a graph of θ versus t.)

***13.39.** For the van der Waals interaction with potential-energy function given by Eq. (13.25), show that when the magnitude of the displacement x from equilibrium $(r = R_0)$ is small, the potential energy can be written approximately as $U \approx \frac{1}{2}kx^2 - U_0$. [*Hint:* In Eq. (13.25), let $r = R_0 + x$ and $u = x/R_0$. Then approximate $(1 + u)^n$ by the first *three* terms in Eq. (13.28).] How does k in this equation compare with the force constant in Eq. (13.29) for the force?

***13.40.** When displaced from equilibrium, the two hydrogen atoms in an H_2 molecule are acted on by a restoring force $F_x = -kx$ with $k = 580 \text{ N/m}$. Calculate the oscillation frequency of the H_2 molecule. (*Hint:* The mass of a hydrogen atom is 1.008 atomic mass units, or 1 u; see Appendix E. As in Example 13.7 (Section 13.4), use $m/2$ instead of m in the expression for f.)

Section 13.5 The Simple Pendulum

13.41. You pull a simple pendulum 0.240 m long to the side through an angle of 3.50° and release it. (a) How much time does it take the pendulum bob to reach its highest speed? (b) How much time does it take if the pendulum is released at an angle of 1.75° instead of 3.50°?

13.42. An 85.0-kg mountain climber plans to swing down, starting from rest, from a ledge using a light rope 6.50 m long. He holds one end of the rope, and the other end is tied higher up on a rock face. Since the ledge is not very far from the rock face, the rope makes a small angle with the vertical. At the lowest point of his swing, he plans to let go and drop a short distance to the ground. (a) How long after he begins his swing will the climber first reach his lowest point? (b) If he missed the first chance to drop off, how long after first beginning his swing will the climber reach his lowest point for the second time?

13.43. A building in San Francisco has light fixtures consisting of small 2.35-kg bulbs with shades hanging from the ceiling at the end of light thin cords 1.50 m long. If a minor earthquake occurs, how many swings per second will these fixtures make?

13.44. A Pendulum on Mars. A certain simple pendulum has a period on the earth of 1.60 s. What is its period on the surface of Mars, where $g = 3.71 \text{ m/s}^2$?

13.45. An apple weighs 1.00 N. When you hang it from the end of a long spring of force constant 1.50 N/m and negligible mass, it bounces up and down in SHM. If you stop the bouncing and let the apple swing from side to side through a small angle, the frequency of this simple pendulum is half the bounce frequency. (Because the angle is small, the back-and-forth swings do not cause any appreciable change in the length of the spring.) What is the unstretched length of the spring (with the apple removed)?

13.46. A small sphere with mass m is attached to a massless rod of length L that is pivoted at the top, forming a simple pendulum. The pendulum is pulled to one side so that the rod is at an angle Θ from the vertical, and released from rest. (a) In a diagram, show the pendulum just after it is released. Draw vectors representing the *forces* acting on the small sphere and the *acceleration* of the sphere. Accuracy counts! At this point, what is the linear acceleration of the sphere? (b) Repeat part (a) for the instant when the pendulum rod is at an angle $\Theta/2$ from the vertical. (c) Repeat part (a) for the instant when the pendulum rod is vertical. At this point, what is the linear speed of the sphere?

13.47. After landing on an unfamiliar planet, a space explorer constructs a simple pendulum of length 50.0 cm. She finds that the pendulum makes 100 complete swings in 136 s. What is the value of g on this planet?

13.48. A simple pendulum 2.00 m long swings through a maximum angle of 30.0° with the vertical. Calculate its period (a) assuming a small amplitude, and (b) using the first three terms of Eq. (13.35). (c) Which of the answers in parts (a) and (b) is more accurate? For the one that is less accurate, by what percent is it in error from the more accurate answer?

Section 13.6 The Physical Pendulum

13.49. A 1.80-kg connecting rod from a car engine is pivoted about a horizontal knife edge as shown in Fig. 13.33. The center of gravity of the rod was located by balancing and is 0.200 m from the pivot. When the rod is set into small-amplitude oscillation, it makes 100 complete swings in 120 s. Calculate the moment of inertia of the rod about the rotation axis through the pivot.

Figure **13.33** Exercise 13.49.

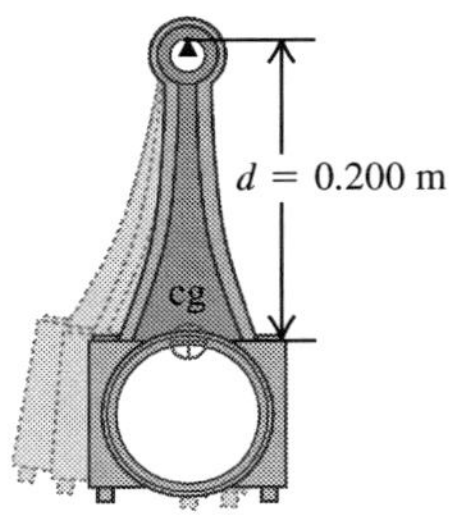

13.50. We want to hang a thin hoop on a horizontal nail and have the hoop make one complete small-angle oscillation each 2.0 s. What must the hoop's radius be?

13.51. Show that the expression for the period of a physical pendulum reduces to that of a simple pendulum if the physical pendulum consists of a particle with mass m on the end of a massless string of length L.

13.52. A 1.80-kg monkey wrench is pivoted 0.250 m from its center of mass and allowed to swing as a physical pendulum. The period for small-angle oscillations is 0.940 s. (a) What is the moment of inertia of the wrench about an axis through the pivot? (b) If the wrench is initially displaced 0.400 rad from its equilibrium position, what is the angular speed of the wrench as it passes through the equilibrium position?

13.53. Two pendulums have the same dimensions (length L) and total mass (m). Pendulum A is a very small ball swinging at the end of a uniform massless bar. In pendulum B, half the mass is in the ball and half is in the uniform bar. Find the period of each pendulum for small oscillations. Which one takes longer for a swing?

13.54. A holiday ornament in the shape of a hollow sphere with mass $M = 0.015$ kg and radius $R = 0.050$ m is hung from a tree limb by a small loop of wire attached to the surface of the sphere. If the ornament is displaced a small distance and released, it swings back and forth as a physical pendulum with negligible friction. Calculate its period. (*Hint:* Use the parallel-axis theorem to find the moment of inertia of the sphere about the pivot at the tree limb.)

13.55. The two pendulums shown in Fig. 13.34 each consist of a uniform solid ball of mass M supported by a massless string, but the ball for pendulum A is very tiny while the ball for pendulum B is much larger. Find the period of each pendulum for small displacements. Which ball takes longer to complete a swing?

Figure **13.34** Exercise 13.55.

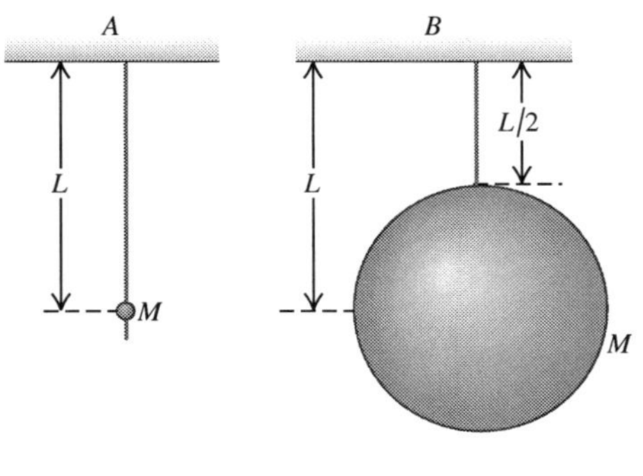

Section 13.7 Damped Oscillations

13.56. A 2.20-kg mass oscillates on a spring of force constant 250.0 N/m with a period of 0.615 s. (a) Is this system damped or not? How do you know? If it is damped, find the damping constant b. (b) Is the system undamped, underdamped, critically damped, or overdamped? How do you know?

13.57. An unhappy 0.300-kg rodent, moving on the end of a spring with force constant $k = 2.50$ N/m, is acted on by a damping force $F_x = -bv_x$. (a) If the constant b has the value 0.900 kg/s, what is the frequency of oscillation of the rodent? (b) For what value of the constant b will the motion be critically damped?

13.58. A 50.0-g hard-boiled egg moves on the end of a spring with force constant $k = 25.0$ N/m. Its initial displacement is 0.300 m. A damping force $F_x = -bv_x$ acts on the egg, and the amplitude of the motion decreases to 0.100 m in 5.00 s. Calculate the magnitude of the damping constant b.

13.59. The motion of an underdamped oscillator is described by Eq. (13.42). Let the phase angle ϕ be zero. (a) According to this equation, what is the value of x at $t = 0$? (b) What are the magnitude and direction of the velocity at $t = 0$? What does the result tell you about the slope of the graph of x versus t near $t = 0$? (c) Obtain an expression for the acceleration a_x at $t = 0$. For what value or range of values of the damping constant b (in terms of k and m) is the acceleration at $t = 0$ negative, zero, and positive? Discuss each case in terms of the shape of the graph of x versus t near $t = 0$.

Section 13.8 Forced Oscillations and Resonance

13.60. A sinusoidally varying driving force is applied to a damped harmonic oscillator of force constant k and mass m. If the damping constant has a value b_1, the amplitude is A_1 when the driving angular frequency equals $\sqrt{k/m}$. In terms of A_1, what is the amplitude for the same driving frequency and the same driving force amplitude F_{max}, if the damping constant is (a) $3b_1$ and (b) $b_1/2$?

13.61. A sinusoidally varying driving force is applied to a damped harmonic oscillator. (a) What are the units of the damping constant b? (b) Show that the quantity $\sqrt{km}$ has the same units as b. (c) In terms of F_{max} and k, what is the amplitude for $\omega_d = \sqrt{k/m}$ when (i) $b = 0.2\sqrt{km}$ and (ii) $b = 0.4\sqrt{km}$? Compare your results to Fig. 13.28.

13.62. An experimental package and its support structure, which are to be placed on board the International Space Station, act as an underdamped spring–mass system with a force constant of 2.1×10^6 N/m and mass 108 kg. A NASA requirement is that resonance for forced oscillations not occur for any frequency below 35 Hz. Does this package meet the requirement?

Problems

13.63. SHM in a Car Engine. The motion of the piston of an automobile engine (Fig. 13.35) is approximately simple harmonic. (a) If the stroke of an engine (twice the amplitude) is 0.100 m and the engine runs at 3500 rev/min, compute the acceleration of the piston at the endpoint of its stroke. (b) If the piston has mass 0.450 kg, what net force must be exerted on it at this point? (c) What are the speed and kinetic energy of the piston at the midpoint of its stroke? (d) What average power is required to accelerate the piston from rest to the speed found in part (c)? (e) If the engine runs at 7000 rev/min, what are the answers to parts (b), (c), and (d)?

Figure **13.35** Problem 13.63.

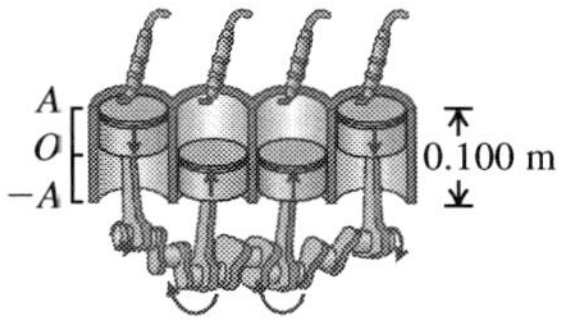

13.64. Four passengers with combined mass 250 kg compress the springs of a car with worn-out shock absorbers by 4.00 cm when they get in. Model the car and passengers as a single body on a single ideal spring. If the loaded car has a period of vibration of 1.08 s, what is the period of vibration of the empty car?

13.65. A glider is oscillating in SHM on an air track with an amplitude A_1. You slow it so that its amplitude is halved. What happens to its (a) period, frequency, and angular frequency; (b) total mechanical energy; (c) maximum speed; (d) speed at $x = \pm A_1/4$; (e) potential and kinetic energies at $x = \pm A_1/4$?

13.66. A child with poor table manners is sliding his 250-g dinner plate back and forth in SHM with an amplitude of 0.100 m on a horizontal surface. At a point 0.060 m away from equilibrium, the speed of the plate is 0.300 m/s. (a) What is the period? (b) What is the displacement when the speed is 0.160 m/s? (c) In the center of the dinner plate is a 10.0-g carrot slice. If the carrot slice is just on the verge of slipping at the endpoint of the path, what is the coefficient of static friction between the carrot slice and the plate?

13.67. A 1.50-kg, horizontal, uniform tray is attached to a vertical ideal spring of force constant 185 N/m and a 275-g metal ball is in the tray. The spring is below the tray, so it can oscillate up-and-down. The tray is then pushed down 15.0 cm below its equilibrium point (call this point A) and released from rest. (a) How high above point A will the tray be when the metal ball leaves the tray? (*Hint:* This does *not* occur when the ball and tray reach their maximum speeds.) (b) How much time elapses between releasing the system at point A and the ball leaving the tray? (c) How fast is the ball moving just as it leaves the tray?

13.68. A block with mass M rests on a frictionless surface and is connected to a horizontal spring of force constant k. The other end of the spring is attached to a wall (Fig. 13.36). A second block with mass m rests on top of the first block. The coefficient of static

Figure **13.36** Problem 13.68.

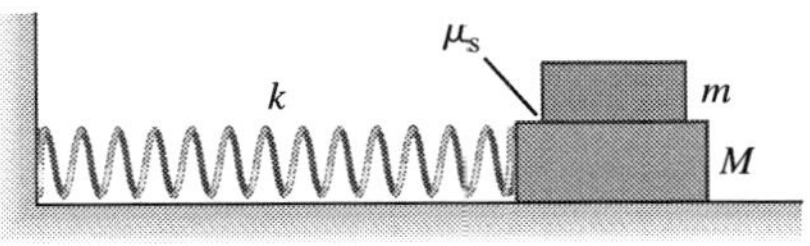

friction between the blocks is μ_s. Find the *maximum* amplitude of oscillation such that the top block will not slip on the bottom block.

13.69. A 10.0-kg mass is traveling to the right with a speed of 2.00 m/s on a smooth horizontal surface when it collides with and sticks to a second 10.0-kg mass that is initially at rest but is attached to a light spring with force constant 80.0 N/m. (a) Find the frequency, amplitude, and period of the subsequent oscillations. (b) How long does it take the system to return the first time to the position it had immediately after the collision?

13.70. A rocket is accelerating upward at 4.00 m/s^2 from the launchpad on the earth. Inside a small 1.50-kg ball hangs from the ceiling by a light 1.10-m wire. If the ball is displaced 8.50° from the vertical and released, find the amplitude and period of the resulting swings of this pendulum.

13.71. A square object of mass m is constructed of four identical uniform thin sticks, each of length L, attached together. This object is hung on a hook at its upper corner (Fig. 13.37). If it is rotated slightly to the left and then released, at what frequency will it swing back and forth?

Figure 13.37 Problem 13.71.

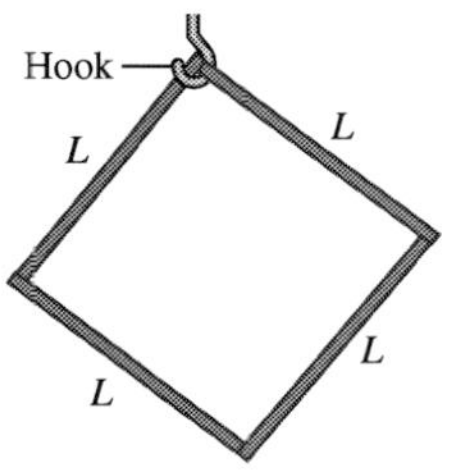

13.72. An object with mass 0.200 kg is acted on by an elastic restoring force with force constant 10.0 N/m. (a) Graph elastic potential energy U as a function of displacement x over a range of x from -0.300 m to $+0.300$ m. On your graph, let 1 cm = 0.05 J vertically and 1 cm = 0.05 m horizontally. The object is set into oscillation with an initial potential energy of 0.140 J and an initial kinetic energy of 0.060 J. Answer the following questions by referring to the graph. (b) What is the amplitude of oscillation? (c) What is the potential energy when the displacement is one-half the amplitude? (d) At what displacement are the kinetic and potential energies equal? (e) What is the value of the phase angle ϕ if the initial velocity is positive and the initial displacement is negative?

13.73. A 2.00-kg bucket containing 10.0 kg of water is hanging from a vertical ideal spring of force constant 125 N/m and oscillating up and down with an amplitude of 3.00 cm. Suddenly the bucket springs a leak in the bottom such that water drops out at a steady rate of 2.00 g/s. When the bucket is half full, find (a) the period of oscillation and (b) the rate at which the period is changing with respect to time. Is the period getting longer or shorter? (c) What is the shortest period this system can have?

13.74. A hanging wire is 1.80 m long. When a 60.0-kg steel ball is suspended from the wire, the wire stretches by 2.00 mm. If the ball is pulled down a small additional distance and released, at what frequency will it vibrate? Assume that the stress on the wire is less than the proportional limit (see Section 11.5).

13.75. A 5.00-kg partridge is suspended from a pear tree by an ideal spring of negligible mass. When the partridge is pulled down 0.100 m below its equilibrium position and released, it vibrates with a period of 4.20 s. (a) What is its speed as it passes through the equilibrium position? (b) What is its acceleration when it is 0.050 m above the equilibrium position? (c) When it is moving upward, how much time is required for it to move from a point 0.050 m below its equilibrium position to a point 0.050 m above it? (d) The motion of the partridge is stopped, and then it is removed from the spring. How much does the spring shorten?

13.76. A 0.0200-kg bolt moves with SHM that has an amplitude of 0.240 m and a period of 1.500 s. The displacement of the bolt is +0.240 m when $t = 0$. Compute (a) the displacement of the bolt when $t = 0.500$ s; (b) the magnitude and direction of the force acting on the bolt when $t = 0.500$ s; (c) the minimum time required for the bolt to move from its initial position to the point where $x = -0.180$ m; (d) the speed of the bolt when $x = -0.180$ m.

13.77. SHM of a Butcher's Scale. A spring of negligible mass and force constant $k = 400$ N/m is hung vertically, and a 0.200-kg pan is suspended from its lower end. A butcher drops a 2.2-kg steak onto the pan from a height of 0.40 m. The steak makes a totally inelastic collision with the pan and sets the system into vertical SHM. What are (a) the speed of the pan and steak immediately after the collision; (b) the amplitude of the subsequent motion; (c) the period of that motion?

13.78. A uniform beam is suspended horizontally by two identical vertical springs that are attached between the ceiling and each end of the beam. The beam has mass 225 kg, and a 175-kg sack of gravel sits on the middle of it. The beam is oscillating in SHM, with an amplitude of 40.0 cm and a frequency of 0.600 cycles/s. (a) The sack of gravel falls off the beam when the beam has its maximum upward displacement. What are the frequency and amplitude of the subsequent SHM of the beam? (b) If the gravel instead falls off when the beam has its maximum speed, what are the frequency and amplitude of the subsequent SHM of the beam?

13.79. On the planet Newtonia, a simple pendulum having a bob with mass 1.25 kg and a length of 185.0 cm takes 1.42 s, when released from rest, to swing through an angle of 12.5°, where it again has zero speed. The circumference of Newtonia is measured to be 51,400 km. What is the mass of the planet Newtonia?

13.80. A 40.0-N force stretches a vertical spring 0.250 m. (a) What mass must be suspended from the spring so that the system will oscillate with a period of 1.00 s? (b) If the amplitude of the motion is 0.050 m and the period is that specified in part (a), where is the object and in what direction is it moving 0.35 s after it has passed the equilibrium position, moving downward? (c) What force (magnitude and direction) does the spring exert on the object when it is 0.030 m below the equilibrium position, moving upward?

13.81. Don't Miss the Boat. While on a visit to Minnesota ("Land of 10,000 Lakes"), you sign up to take an excursion around one of the larger lakes. When you go to the dock where the 1500-kg boat is tied, you find that the boat is bobbing up and down in the waves, executing simple harmonic motion with amplitude 20 cm. The boat takes 3.5 s to make one complete up-and-down cycle. When the boat is at its highest point, its deck is at the same height as the stationary dock. As you watch the boat bob up and down, you (mass 60 kg) begin to feel a bit woozy, due in part to the previous night's dinner of lutefisk. As a result, you refuse to board the boat unless the level of the boat's deck is within 10 cm of the dock level. How much time do you have to board the boat comfortably during each cycle of up-and-down motion?

13.82. An interesting, though highly impractical example of oscillation is the motion of an object dropped down a hole that extends from one side of the earth, through its center, to the other side. With the assumption (not realistic) that the earth is a sphere of uniform density, prove that the motion is simple harmonic and find the period. [*Note:* The gravitational force on the object as a function of the object's distance r from the center of the earth was derived in Example 12.10 (Section 12.6). The motion is simple harmonic if the acceleration a_x and the displacement from equilibrium x are related by Eq. (13.8), and the period is then $T = 2\pi/\omega$.]

13.83. Two point masses m are held in place a distance d apart. Another point mass M is midway between them. M is then displaced a small distance x perpendicular to the line connecting the two fixed masses and released. (a) Show that the magnitude of the net gravitational force on M due to the fixed masses is given approximately by $F_{\text{net}} = \dfrac{16\,GmMx}{d^3}$ if $x \ll d$. What is the direction of this force? Is it a restoring force? (b) Show that the mass M will oscillate with an angular frequency of $(4/d)\sqrt{Gm/d}$ and period $\pi d/2\sqrt{d/Gm}$. (c) What would the period be if $m = 100$ kg and $d = 25.0$ cm? Does it seem that you could easily measure this period? What things prevent this experiment from easily being performed in an ordinary physics lab? (d) Will M oscillate if it is displaced from the center a small distance x toward either of the fixed masses? Why?

13.84. For a certain oscillator the net force on the body with mass m is given by $F_x = -cx^3$. (a) What is the potential energy function for this oscillator if we take $U = 0$ at $x = 0$? (b) One-quarter of a period is the time for the body to move from $x = 0$ to $x = A$. Calculate this time and hence the period. [*Hint:* Begin with Eq. (13.20), modified to include the potential-energy function you found in part (a), and solve for the velocity v_x as a function of x. Then replace v_x with dx/dt. Separate the variable by writing all fators containing x on one side and all factors containing t on the other side so that each side can be integrated. In the x-integral make the change of variable $u = x/A$. The resulting integral can be evaluated by numerical methods on a computer and has the value $\int_0^1 du/\sqrt{1 - u^4} = 1.31$.] (c) According to the result you obtained in part (b), does the period depend on the amplitude A of the motion? Are the oscillations simple harmonic?

13.85. Consider the circle of reference shown in Fig. 13.6. The x-component of the velocity of Q is the velocity of P. Compute this component, and show that the velocity of P is as given by Eq. (13.15).

***13.86. Diatomic Molecule.** Two identical atoms in a diatomic molecule vibrate as harmonic oscillators. However, their center of mass, midway between them, remains at rest. (a) Show that at any instant, the momenta of the atoms relative to the center of mass are $\vec{p}$ and $-\vec{p}$. (b) Show that the total kinetic energy K of the two atoms at any instant is the same as that of a single object with mass $m/2$ with a momentum of magnitude p. (Use $K = p^2/2m$.) This result shows why $m/2$ should be used in the expression for f in Example 13.7 (Section 13.4). (c) If the atoms are not identical but have masses m_1 and m_2, show that the result of part (a) still holds and the single object's mass in part (b) is $m_1 m_2/(m_1 + m_2)$. The quantity $m_1 m_2/(m_1 + m_2)$ is called the *reduced mass* of the system.

***13.87.** An approximation for the potential energy of a KCl molecule is $U = A[(R_0^7/8r^8) - 1/r]$, where $R_0 = 2.67 \times 10^{-10}$ m and $A = 2.31 \times 10^{-28}$ J $\cdot$ m. Using this approximation: (a) Show that the radial component of the force on each atom is $F_r = A[(R_0^7/r^9) - 1/r^2]$. (b) Show that R_0 is the equilibrium separation. (c) Find the minimum potential energy. (d) Use $r = R_0 + x$ and the first two terms of the binomial theorem (Eq. 13.28) to show that $F_r \approx -(7A/R_0^3)x$, so that the molecule's force constant is $k = 7A/R_0^3$. (e) With both the K and Cl atoms vibrating in opposite directions on opposite sides of the molecule's center of mass, $m_1 m_2/(m_1 + m_2) = 3.06 \times 10^{-26}$ kg is the mass to use in calculating the frequency (see Problem 13.86). Calculate the frequency of small-amplitude vibrations.

13.88. Two solid cylinders connected along their common axis by a short, light rod have radius R and total mass M and rest on a horizontal tabletop. A spring with force constant k has one end attached to a clamp and the other end attached to a frictionless ring at the center of mass of the cylinders (Fig. 13.38). The cylinders are pulled to the left a distance x, which stretches the spring, and released. There is sufficient friction between the tabletop and the cylinders for the cylinders to roll without slipping as they move back and forth on the end of the spring. Show that the motion of the center of mass of the cylinders is simple harmonic, and calculate its period in terms of M and k. [*Hint:* The motion is simple harmonic if a_x and x are related by Eq. (13.8), and the period then is $T = 2\pi/\omega$. Apply $\sum\tau_z = I_{\text{cm}}\alpha_z$ and $\sum F_x = Ma_{\text{cm-}x}$ to the cylinders in order to relate $a_{\text{cm-}x}$ and the displacement x of the cylinders from their equilibrium position.]

Figure **13.38** Problem 13.88.

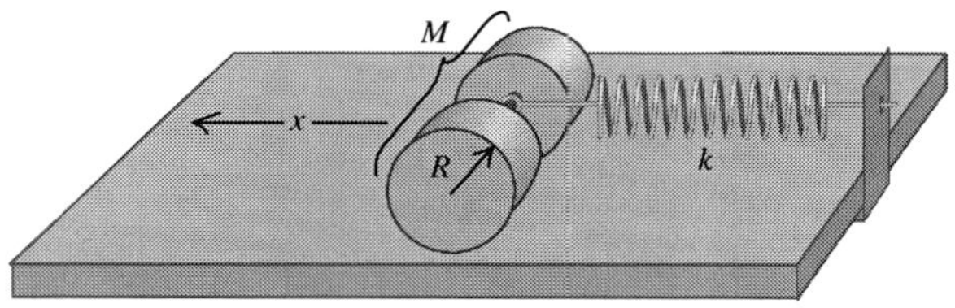

13.89. In Fig. 13.39 the upper ball is released from rest, collides with the stationary lower ball, and sticks to it. The strings are both 50.0 cm long. The upper ball has mass 2.00 kg, and it is initially 10.0 cm higher than the lower ball, which has mass 3.00 kg. Find the frequency and maximum angular displacement of the motion after the collision.

Figure **13.39** Problem 13.89.

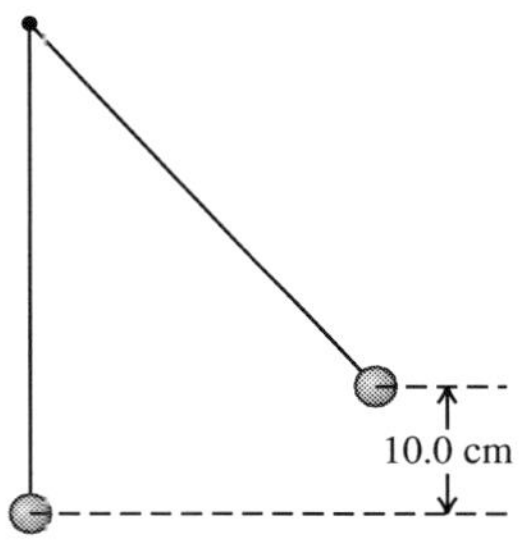

13.90. T. rex. Model the leg of the *T. rex* in Example 13.10 (Section 13.6) as two uniform rods, each 1.55 m long, joined rigidly end to end. Let the lower rod have mass M and the upper rod mass $2M$. The composite object is pivoted about the top of the upper rod. Compute the oscillation period of this object for small-amplitude oscillations. Compare your result to that of Example 13.10.

13.91. A slender, uniform, metal rod with mass M is pivoted without friction about an axis through its midpoint and perpendicular to the rod. A horizontal spring with force constant k is attached to the lower end of the rod, with the other end of the spring attached to a rigid support. If the rod is displaced by a small angle Θ from the vertical (Fig. 13.40) and released, show that it moves in angular SHM and calculate the period. (*Hint:* Assume that the angle Θ is small enough for the approximations $\sin\Theta \approx \Theta$ and $\cos\Theta \approx 1$ to be valid. The motion is simple harmonic if $d^2\theta/dt^2 = -\omega^2\theta$, and the period is then $T = 2\pi/\omega$.)

Figure **13.40** Problem 13.91.

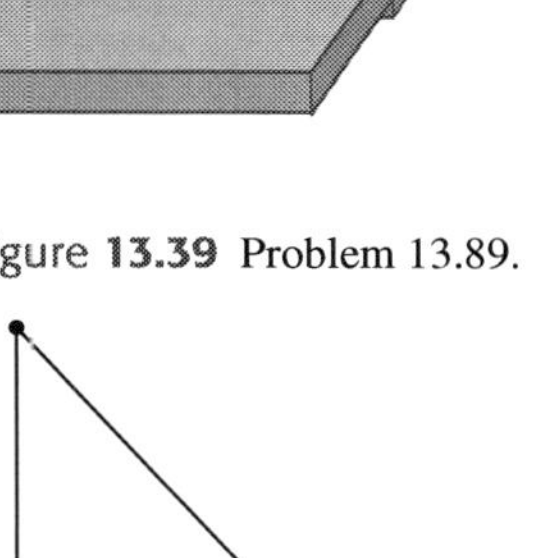

13.92. The Silently Ringing Bell Problem. A large bell is hung from a wooden beam so it can swing back and forth with negligible friction. The center of mass of the bell is 0.60 m below the pivot, the bell has mass 34.0 kg, and the moment of inertia of the bell about an axis at the pivot is 18.0 kg $\cdot$ m^2. The clapper is a small,

1.8-kg mass attached to one end of a slender rod that has length L and negligible mass. The other end of the rod is attached to the inside of the bell so it can swing freely about the same axis as the bell. What should be the length L of the clapper rod for the bell to ring silently—that is, for the period of oscillation for the bell to equal that for the clapper?

13.93. Two identical thin rods, each with mass m and length L, are joined at right angles to form an L-shaped object. This object is balanced on top of a sharp edge (Fig. 13.41). If the L-shaped object is deflected slightly, it oscillates. Find the frequency of oscillation.

Figure **13.41** Problem 13.93.

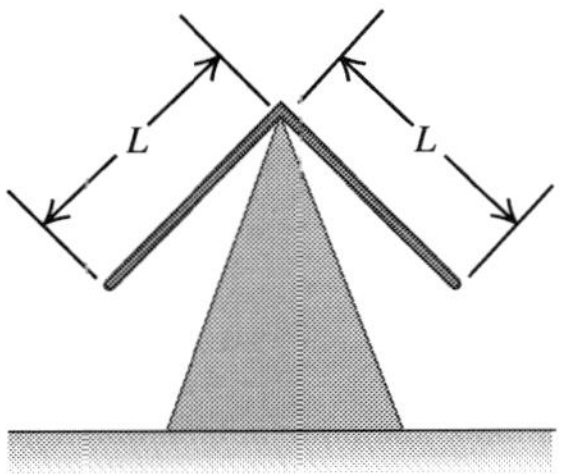

13.94. You want to construct a pendulum with a period of 4.00 s at a location where $g = 9.80\ \text{m/s}^2$. (a) What is the length of a *simple* pendulum having this period? (b) Suppose the pendulum must be mounted in a case that is not more than 0.50 m high. Can you devise a pendulum having a period of 4.00 s that will satisfy this requirement?

13.95. A uniform rod of length L oscillates through small angles about a point a distance x from its center. (a) Prove that its angular frequency is $\sqrt{gx/[(L^2/12) + x^2]}$. (b) Show that its maximum angular frequency occurs when $x = L/\sqrt{12}$. (c) What is the length of the rod if the maximum angular frequency is 2π rad/s?

Challenge Problems

13.96. Two springs, each with unstretched length 0.200 m, but with different force constants k_1 and k_2, are attached to opposite ends of a block with mass m on a level, frictionless surface. The outer ends of the springs are now attached to two pins P_1 and P_2, 0.100 m from the original positions of the ends of the springs (Fig. 13.42). Let $k_1 = 2.00\ \text{N/m}$, $k_2 = 6.00\ \text{N/m}$, and $m = 0.100\ \text{kg}$. (a) Find the length of each spring when the block is in its new equilibrium position after the springs have been attached to the pins. (b) Find the period of vibration of the block if it is slightly displaced from its new equilibrium position and released.

Figure **13.42** Challenge Problem 13.96.

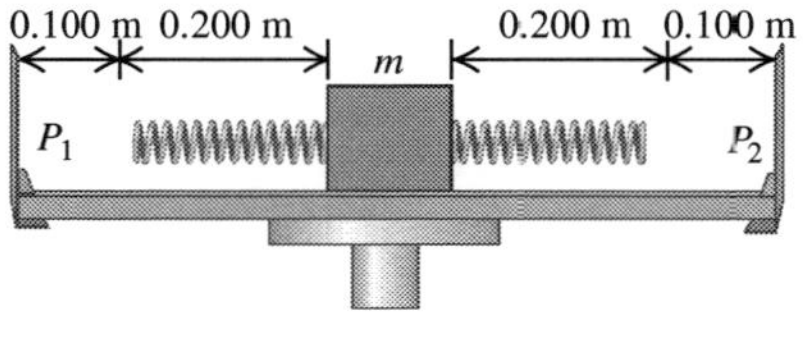

13.97. The Effective Force Constant of Two Springs. Two springs with the same unstretched length, but different force constants k_1 and k_2 are attached to a block with mass m on a level, frictionless surface. Calculate the effective force constant k_{eff} in each of the three cases (a), (b), and (c) depicted in Fig. 13.43. (The effective force constant is defined by $\sum F_x = -k_{\text{eff}}x$.) (d) An object with mass m, suspended from a uniform spring with a force constant k, vibrates with a frequency f_1. When the spring is cut in half and the same object is suspended from one of the halves, the frequency is f_2. What is the ratio f_2/f_1?

Figure **13.43** Challenge Problem 13.97.

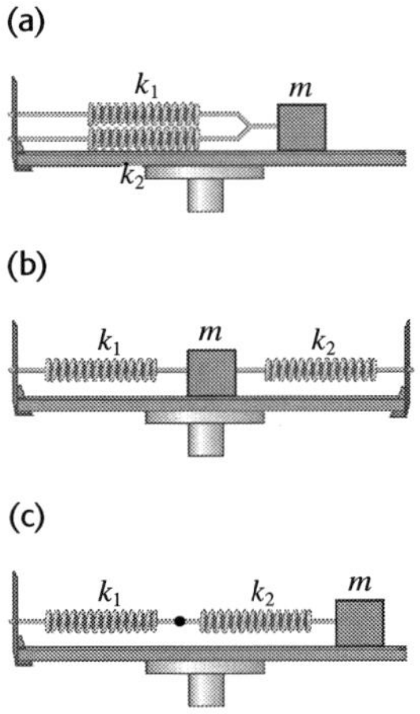

13.98. (a) What is the change ΔT in the period of a simple pendulum when the acceleration of gravity g changes by Δg? (*Hint:* The new period $T + \Delta T$ is obtained by substituting $g + \Delta g$ for g:

$$T + \Delta T = 2\pi\sqrt{\frac{L}{g + \Delta g}}$$

To obtain an approximate expression, expand the factor $(g + \Delta g)^{-1/2}$ using the binomial theorem (Appendix B) and keep only the first two terms:

$$(g + \Delta g)^{-1/2} = g^{-1/2} - \frac{1}{2}g^{-3/2}\Delta g + \cdots$$

The other terms contain higher powers of Δg and are very small if Δg is small.) Express your result as the *fractional* change in period $\Delta T/T$ in terms of the fractional change $\Delta g/g$. (b) A pendulum clock keeps correct time at a point where $g = 9.8000\ \text{m/s}^2$, but is found to lose 4.0 s each day at a higher elevation. Use the result of part (a) to find the approximate value of g at this new location.

13.99. A Spring with Mass. The preceding problems in this chapter have assumed that the springs had negligible mass. But of course no spring is completely massless. To find the effect of the spring's mass, consider a spring with mass M, equilibrium length L_0, and spring constant k. When stretched or compressed to a length L, the potential energy is $\frac{1}{2}kx^2$, where $x = L - L_0$. (a) Consider a spring, as described above, that has one end fixed and the other end moving with speed v. Assume that the speed of points along the length of the spring varies linearly with distance l from the fixed end. Assume also that the mass M of the spring is distributed uniformly along the length of the spring. Calculate the kinetic energy of the spring in terms of M and v. (*Hint:* Divide the spring into pieces of length dl; find the speed of each piece in terms of l, v, and L; find the mass of each piece in terms of dl, M, and L; and integrate from 0 to L. The result is *not* $\frac{1}{2}Mv^2$, since not all of the spring moves with the same speed.) (b) Take the time derivative of the conservation of energy equation, Eq. (13.21), for a mass m moving on the end of a *massless* spring. By comparing your results to Eq. (13.8), which defines ω, show that the angular frequency of oscillation is $\omega = \sqrt{k/m}$. (c) Apply the procedure of part (b) to obtain the angular frequency of oscillation ω of the spring considered in part (a). If the *effective mass* M' of the spring is defined by $\omega = \sqrt{k/M'}$, what is M' in terms of M?

13.100. A uniform, 1.00-m stick hangs from a horizontal axis at one end and oscillates as a physical pendulum. An object of small dimensions and with mass equal to that of the meter stick can be clamped to the stick at a distance y below the axis. Let T represent the period of the system with the body attached and T_0 the period of the meter stick alone. (a) Find the ratio T/T_0. Evaluate your expression for y ranging from 0 to 1.0 m in steps of 0.1 m, and graph T/T_0 versus y. (b) Is there any value of y, other than $y = 0$, for which $T = T_0$? If so, find it and explain why the period is unchanged when y has this value.

13.101. You measure the period of a physical pendulum about one pivot point to be T. Then you find another pivot point on the opposite side of the center of mass that gives the same period. The two points are separated by a distance L. Use the parallel-axis theorem to show that $g = L(2\pi/T)^2$. (This result shows a way that you can measure g without knowing the mass or any moments of inertia of the physical pendulum.)

13.102. Resonance in a Mechanical System. A mass m is attached to one end of a massless spring with a force constant k and an unstretched length l_0. The other end of the spring is free to turn about a nail driven into a frictionless, horizontal surface (Fig. 13.44). The mass is made to revolve in a circle with an angular frequency of revolution ω'. (a) Calculate the length l of the spring as a function of ω'. (b) What happens to the result in part (a) when ω' approaches the natural frequency $\omega = \sqrt{k/m}$ of the mass-spring system? (If your result bothers you, remember that massless springs and frictionless surfaces don't exist as such, but are only approximate descriptions of real springs and surfaces. Also, Hooke's law is only an approximation of the way real springs behave; the greater the elongation of the spring, the greater the deviation from Hooke's law.)

Figure **13.44** Challenge Problem 13.102.

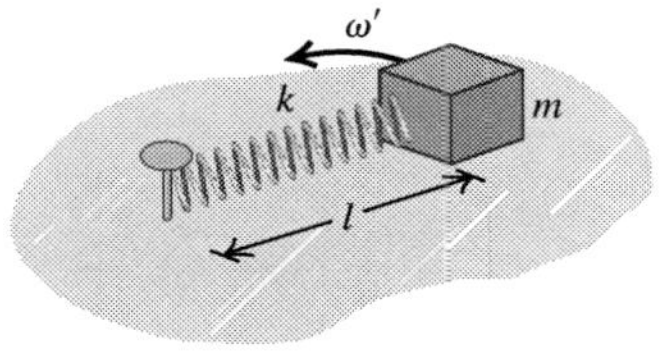

***13.103. Vibration of a Covalently Bonded Molecule.** Many diatomic (two-atom) molecules are bound together by *covalent bonds* that are much stronger than the van der Waals interaction. Examples include H_2, O_2, and N_2. Experiment shows that for many such molecules, the interaction can be described by a force of the form

$$F_r = A\left[e^{-2b(r-R_0)} - e^{-b(r-R_0)}\right]$$

where A and b are positive constants, r is the center-to-center separation of the atoms, and R_0 is the equilibrium separation. For the hydrogen molecule (H_2), $A = 2.97 \times 10^{-8}$ N, $b = 1.95 \times 10^{10}\ \text{m}^{-1}$, and $R_0 = 7.4 \times 10^{-11}$ m. Find the force constant for small oscillations around equilibrium. (*Hint:* Use the expansion for ϵ^x given in Appendix B.) Compare your result to the value given in Exercise 13.40.

MECHANICAL WAVES

15

? When an earthquake strikes, the news of the event travels through the body of the earth in the form of seismic waves. Which aspects of a seismic wave determine how much power is carried by the wave?

LEARNING GOALS

By studying this chapter, you will learn:

- What is meant by a mechanical wave, and the different varieties of mechanical waves.
- How to use the relationship among speed, frequency, and wavelength for a periodic wave.
- How to interpret and use the mathematical expression for a sinusoidal periodic wave.
- How to calculate the speed of waves on a rope or string.
- How to calculate the rate at which a mechanical wave transports energy.
- What happens when mechanical waves overlap and interfere.
- The properties of standing waves on a string, and how to analyze these waves.
- How stringed instruments produce sounds of specific frequencies.

Ripples on a pond, musical sounds, seismic tremors triggered by an earthquake—all these are *wave* phenomena. Waves can occur whenever a system is disturbed from equilibrium and when the disturbance can travel, or *propagate,* from one region of the system to another. As a wave propagates, it carries energy. The energy in light waves from the sun warms the surface of our planet; the energy in seismic waves can crack our planet's crust.

This chapter and the next are about mechanical waves—waves that travel within some material called a *medium.* (Chapter 16 is concerned with sound, an important type of mechanical wave.) We'll begin this chapter by deriving the basic equations for describing waves, including the important special case of *sinusoidal* waves in which the wave pattern is a repeating sine or cosine function. To help us understand waves in general, we'll look at the simple case of waves that travel on a stretched string or rope.

Waves on a string play an important role in music. When a musician strums a guitar or bows a violin, she makes waves that travel in opposite directions along the instrument's strings. What happens when these oppositely directed waves overlap is called *interference.* We'll discover that sinusoidal waves can occur on a guitar or violin string only for certain special frequencies, called *normal-mode frequencies,* determined by the properties of the string. The normal-mode frequencies of a stringed instrument determine the pitch of the musical sounds that the instrument produces. (In the next chapter we'll find that interference also helps explain the pitches of *wind* instruments such as flutes and pipe organs.)

Not all waves are mechanical in nature. *Electromagnetic* waves—including light, radio waves, infrared and ultraviolet radiation, and x rays—can propagate even in empty space, where there is *no* medium. We'll explore these and other nonmechanical waves in later chapters.

ActivPhysics

10.1 Properties of Mechanical Waves

15.1 Types of Mechanical Waves

A **mechanical wave** is a disturbance that travels through some material or substance called the **medium** for the wave. As the wave travels through the medium, the particles that make up the medium undergo displacements of various kinds, depending on the nature of the wave.

Figure 15.1 shows three varieties of mechanical waves. In Fig. 15.1a the medium is a string or rope under tension. If we give the left end a small upward shake or wiggle, the wiggle travels along the length of the string. Successive sections of string go through the same motion that we gave to the end, but at successively later times. Because the displacements of the medium are perpendicular or *transverse* to the direction of travel of the wave along the medium, this is called a **transverse wave.**

In Fig. 15.1b the medium is a liquid or gas in a tube with a rigid wall at the right end and a movable piston at the left end. If we give the piston a single back-and-forth motion, displacement and pressure fluctuations travel down the length of the medium. This time the motions of the particles of the medium are back and forth along the *same* direction that the wave travels. We call this a **longitudinal wave.**

In Fig. 15.1c the medium is a liquid in a channel, such as water in an irrigation ditch or canal. When we move the flat board at the left end forward and back once, a wave disturbance travels down the length of the channel. In this case the displacements of the water have *both* longitudinal and transverse components.

Each of these systems has an equilibrium state. For the stretched string it is the state in which the system is at rest, stretched out along a straight line. For the fluid in a tube it is a state in which the fluid is at rest with uniform pressure. And for the liquid in a trough it is a smooth, level water surface. In each case the wave motion is a disturbance from the equilibrium state that travels from one region of the medium to another. And in each case there are forces that tend to restore the system to its equilibrium position when it is displaced, just as the force of gravity tends to pull a pendulum toward its straight-down equilibrium position when it is displaced.

15.1 Three ways to make a wave that moves to the right. (a) The hand moves the string up and then returns, producing a transverse wave. (b) The piston moves to the right, compressing the gas or liquid, and then returns, producing a longitudinal wave. (c) The board moves to the right and then returns, producing a combination of longitudinal and transverse waves.

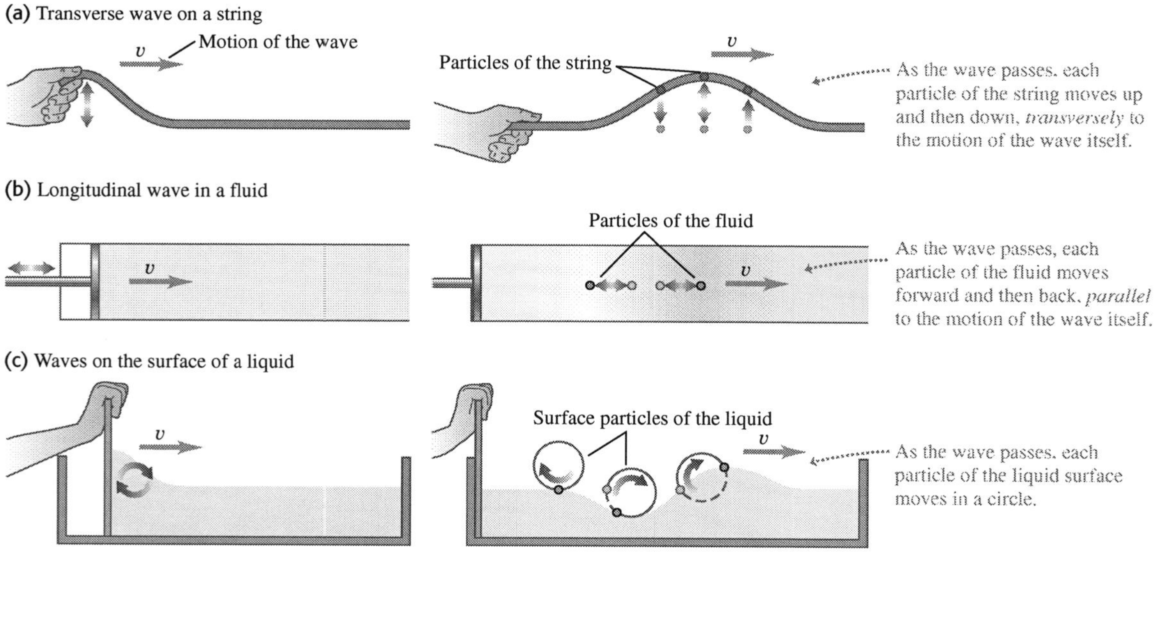

These examples have three things in common. First, in each case the disturbance travels or *propagates* with a definite speed through the medium. This speed is called the speed of propagation, or simply the **wave speed.** Its value is determined in each case by the mechanical properties of the medium. We will use the symbol v for wave speed. (The wave speed is *not* the same as the speed with which particles move when they are disturbed by the wave. We'll return to this point in Section 15.3.) Second, the medium itself does not travel through space; its individual particles undergo back-and-forth or up-and-down motions around their equilibrium positions. The overall pattern of the wave disturbance is what travels. Third, to set any of these systems into motion, we have to put in energy by doing mechanical work on the system. The wave motion transports this energy from one region of the medium to another. *Waves transport energy, but not matter, from one region to another* (Fig. 15.2).

15.2 "Doing the wave" at a sports stadium is an example of a mechanical wave: The disturbance propagates through the crowd, but there is no transport of matter (none of the spectators moves from one seat to another).

Test Your Understanding of Section 15.1 What type of wave is "the wave" shown in Fig. 15.2? (i) transverse; (ii) longitudinal; (iii) a combination of transverse and longitudinal.

15.2 Periodic Waves

The transverse wave on a stretched string in Fig. 15.1a is an example of a *wave pulse.* The hand shakes the string up and down just once, exerting a transverse force on it as it does so. The result is a single "wiggle," or pulse, that travels along the length of the string. The tension in the string restores its straight-line shape once the pulse has passed.

A more interesting situation develops when we give the free end of the string a repetitive, or *periodic,* motion. (You may want to review the discussion of periodic motion in Chapter 13 before going ahead.) Then each particle in the string also undergoes periodic motion as the wave propagates, and we have a **periodic wave.**

Periodic Transverse Waves

In particular, suppose we move the string up and down with *simple harmonic motion* (SHM) with amplitude A, frequency f, angular frequency $\omega = 2\pi f$, and period $T = 1/f = 2\pi/\omega$. Figure 15.3 shows one way to do this. The wave that results is a symmetrical sequence of *crests* and *troughs.* As we will see, periodic waves with simple harmonic motion are particularly easy to analyze; we call them **sinusoidal waves.** It also turns out that *any* periodic wave can be represented as a combination of sinusoidal waves. So this particular kind of wave motion is worth special attention.

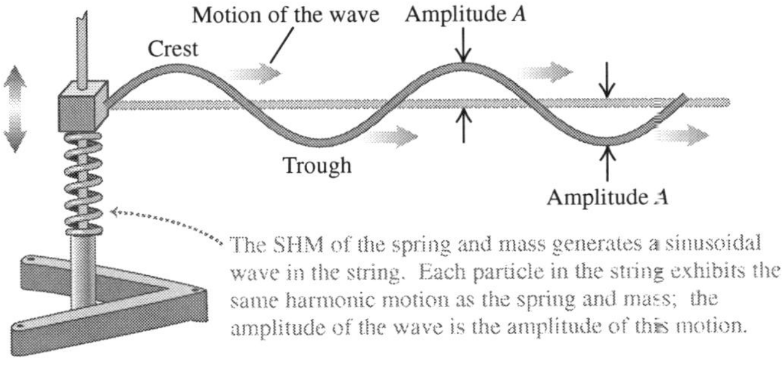

15.3 A block of mass m attached to a spring undergoes simple harmonic motion, producing a sinusoidal wave that travels to the right on the string. (In a real-life system a driving force would have to be applied to the block to replace the energy carried away by the wave.)

15.4 A sinusoidal transverse wave traveling to the right along a string. The vertical scale is exaggerated.

The string is shown at time intervals of $\frac{1}{8}$ period for a total of one period T. The highlighting shows the motion of one wavelength of the wave.

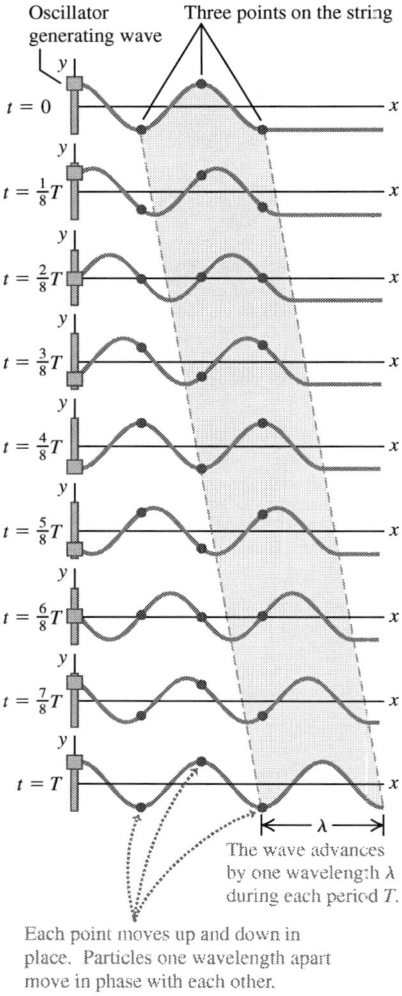

15.5 A series of drops falling into water produces a periodic wave that spreads radially outward. The wave crests and troughs are concentric circles. The wavelength λ is the radial distance between adjacent crests or adjacent troughs.

In Fig. 15.3 the wave that advances along the string is a *continuous succession* of transverse sinusoidal disturbances. Figure 15.4 shows the shape of a part of the string near the left end at time intervals of $\frac{1}{8}$ of a period, for a total time of one period. The wave shape advances steadily toward the right, as indicated by the highlighted area. As the wave moves, any point on the string (any of the red dots, for example) oscillates up and down about its equilibrium position with simple harmonic motion. *When a sinusoidal wave passes through a medium, every particle in the medium undergoes simple harmonic motion with the same frequency.*

CAUTION **Wave motion vs. particle motion** Be very careful to distinguish between the motion of the *transverse wave* along the string and the motion of a *particle* of the string. The wave moves with constant speed v *along* the length of the string, while the motion of the particle is simple harmonic and *transverse* (perpendicular) to the length of the string.

For a periodic wave, the shape of the string at any instant is a repeating pattern. The length of one complete wave pattern is the distance from one crest to the next, or from one trough to the next, or from any point to the corresponding point on the next repetition of the wave shape. We call this distance the **wavelength** of the wave, denoted by λ (the Greek letter lambda). The wave pattern travels with constant speed v and advances a distance of one wavelength λ in a time interval of one period T. So the wave speed v is given by $v = \lambda/T$ or, because $f = 1/T$,

$$v = \lambda f \qquad \text{(periodic wave)} \qquad (15.1)$$

The speed of propagation equals the product of wavelength and frequency. The frequency is a property of the *entire* periodic wave because all points on the string oscillate with the same frequency f.

Waves on a string propagate in just one dimension (in Fig. 15.4, along the x-axis). But the ideas of frequency, wavelength, and amplitude apply equally well to waves that propagate in two or three dimensions. Figure 15.5 shows a wave propagating in two dimensions on the surface of a tank of water. As with waves on a string, the wavelength is the distance from one crest to the next, and the amplitude is the height of a crest above the equilibrium level.

In many important situations including waves on a string, the wave speed v is determined entirely by the mechanical properties of the medium. In this case, increasing f causes λ to decrease so that the product $v = \lambda f$ remains the same, and waves of *all* frequencies propagate with the same wave speed. In this chapter we will consider *only* waves of this kind. (In later chapters we will study the propagation of light waves in matter for which the wave speed depends on frequency; this turns out to be the reason prisms break white light into a spectrum and raindrops create a rainbow.)

Periodic Longitudinal Waves

To understand the mechanics of a periodic *longitudinal* wave, we consider a long tube filled with a fluid, with a piston at the left end as in Fig. 15.1b. If we push the piston in, we compress the fluid near the piston, increasing the pressure in this region. This region then pushes against the neighboring region of fluid, and so on, and a wave pulse moves along the tube.

Now suppose we move the piston back and forth with simple harmonic motion, along a line parallel to the axis of the tube (Fig. 15.6). This motion forms regions in the fluid where the pressure and density are greater or less than the equilibrium values. We call a region of increased density a *compression;* a region of reduced density is a *rarefaction.* Figure 15.6 shows compressions as darkly shaded areas and rarefactions as lightly shaded areas. The wavelength is the distance from one compression to the next or from one rarefaction to the next.

15.6 Using an oscillating piston to make a sinusoidal longitudinal wave in a fluid.

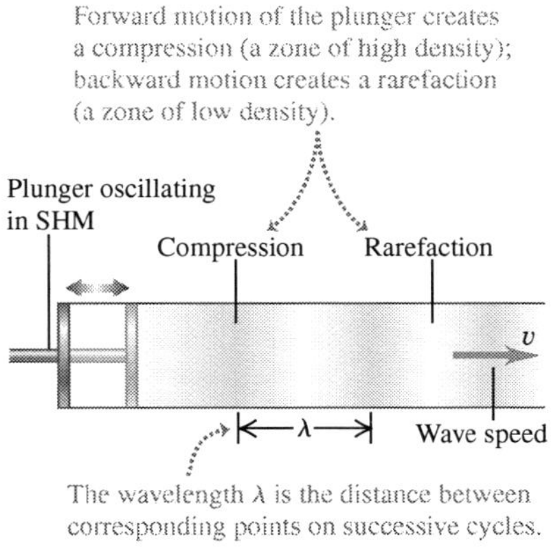

15.7 A sinusoidal longitudinal wave traveling to the right in a fluid. The wave has the same amplitude A and period T as the oscillation of the piston.

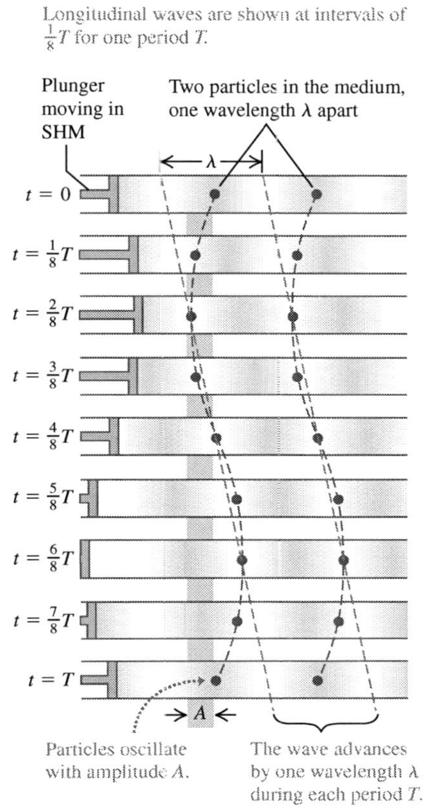

Figure 15.7 shows the wave propagating in the fluid-filled tube at time intervals of $\frac{1}{8}$ of a period, for a total time of one period. The pattern of compressions and rarefactions moves steadily to the right, just like the pattern of crests and troughs in a sinusoidal transverse wave (compare Fig. 15.4). Each particle in the fluid oscillates in SHM parallel to the direction of wave propagation (that is, left and right) with the same amplitude A and period T as the piston. The particles shown by the two red dots in Fig. 15.7 are one wavelength apart, and so oscillate in phase with each other.

Just like the sinusoidal transverse wave shown in Fig. 15.4, in one period T the longitudinal wave in Fig. 15.7 travels one wavelength λ to the right. Hence the fundamental equation $v = \lambda f$ holds for longitudinal waves as well as for transverse waves, and indeed for *all* types of periodic waves. Just as for transverse waves, in this chapter and the next we will consider only situations in which the speed of longitudinal waves does not depend on the frequency.

Example 15.1 Wavelength of a musical sound

Sound waves are longitudinal waves in air. The speed of sound depends on temperature; at 20°C it is 344 m/s (1130 ft/s). What is the wavelength of a sound wave in air at 20°C if the frequency is 262 Hz (the approximate frequency of middle C on a piano)?

SOLUTION

IDENTIFY: This problem involves the relationship among wave speed, wavelength, and frequency for a periodic wave. The target variable is the wavelength λ.

SET UP: The wave speed $v = 344$ m/s and the frequency $f = 262$ Hz are given, so we can use the relationship in Eq. (15.1) among v, λ, and f.

EXECUTE: We solve Eq. (15.1) for the target variable λ:

$$\lambda = \frac{v}{f} = \frac{344\ \text{m/s}}{262\ \text{Hz}} = \frac{344\ \text{m/s}}{262\ \text{s}^{-1}} = 1.31\ \text{m}$$

Note that the units of frequency are either hertz (Hz) or inverse seconds (s^{-1}).

EVALUATE: What happens to the wavelength if the frequency changes? The speed of sound waves is unaffected by changes in frequency, so the relationship $\lambda = v/f$ tells us that wavelength will change in inverse proportion to the frequency. As an example, the high C sung by coloratura sopranos is two octaves above middle C. Each octave corresponds to a factor of 2 in frequency, so the frequency of high C is four times that of middle C: $f = 4(262\ \text{Hz}) = 1048$ Hz. Hence the *wavelength* of high C is *one-fourth* as large: $\lambda = (1.31\ \text{m})/4 = 0.328$ m.

Test Your Understanding of Section 15.2 If you double the wavelength of a wave on a particular string, what happens to the wave speed v and the frequency f? (i) v doubles and f is unchanged; (ii) v is unchanged and f doubles; (iii) v becomes one-half as great and f is unchanged; (iv) v is unchanged and f becomes one-half as great; (v) none of these.

15.3 Mathematical Description of a Wave

Many characteristics of periodic waves can be described by using the concepts of wave speed, amplitude, period, frequency, and wavelength. Often, though, we need a more detailed description of the positions and motions of individual particles of the medium at particular times during wave propagation. For this description we need the concept of a *wave function,* a function that describes the position of any particle in the medium at any time. We will concentrate on *sinusoidal* waves, in which each particle undergoes simple harmonic motion about its equilibrium position.

As a specific example, let's look at waves on a stretched string. If we ignore the sag of the string due to gravity, the equilibrium position of the string is along a straight line. We take this to be the x-axis of a coordinate system. Waves on a string are *transverse;* during wave motion a particle with equilibrium position x is displaced some distance y in the direction perpendicular to the x-axis. The value of y depends on which particle we are talking about (that is, y depends on x) and also on the time t when we look at it. Thus y is a *function* of both x and t; $y = y(x, t)$. We call $y(x, t)$ the **wave function** that describes the wave. If we know this function for a particular wave motion, we can use it to find the displacement (from equilibrium) of any particle at any time. From this we can find the velocity and acceleration of any particle, the shape of the string, and anything else we want to know about the behavior of the string at any time.

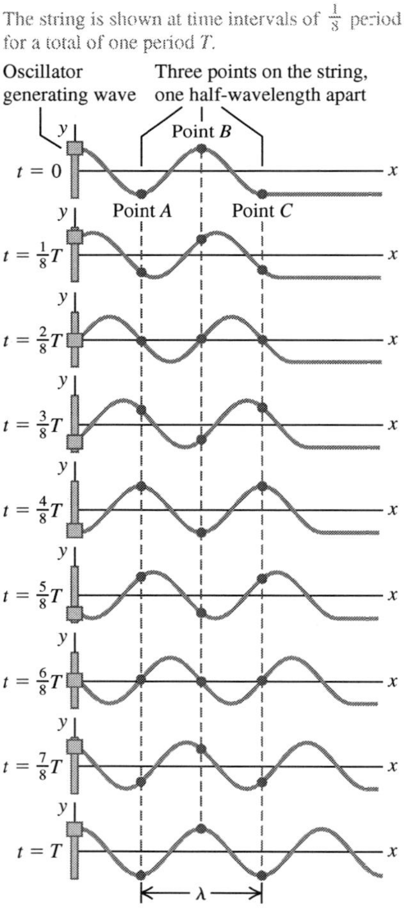

15.8 Tracking the oscillations of three points on a string as a sinusoidal wave propagates along it.

Wave Function for a Sinusoidal Wave

Let's see how to determine the form of the wave function for a sinusoidal wave. Suppose a sinusoidal wave travels from left to right (the direction of increasing x) along the string, as in Fig. 15.8. Every particle of the string oscillates with simple harmonic motion with the same amplitude and frequency. But the oscillations of particles at different points on the string are *not* all in step with each other. The particle at point B in Fig. 15.8 is at its maximum positive value of y at $t = 0$ and returns to $y = 0$ at $t = \frac{2}{8}T$; these same events occur for a particle at point A or point C at $t = \frac{4}{8}T$ and $t = \frac{6}{8}T$, exactly one half-period later. For any two particles of the string, the motion of the particle on the right (in terms of the wave, the "downstream" particle) lags behind the motion of the particle on the left by an amount proportional to the distance between the particles.

Hence the cyclic motions of various points on the string are out of step with each other by various fractions of a cycle. We call these differences *phase differences,* and we say that the *phase* of the motion is different for different points. For example, if one point has its maximum positive displacement at the same time that another has its maximum negative displacement, the two are a half-cycle out of phase. (This is the case for points A and B, or points B and C.)

Suppose that the displacement of a particle at the left end of the string $(x = 0)$, where the wave originates, is given by

$$y(x = 0, t) = A\cos\omega t = A\cos 2\pi f t \qquad (15.2)$$

That is, the particle oscillates in simple harmonic motion with amplitude A, frequency f, and angular frequency $\omega = 2\pi f$. The notation $y(x = 0, t)$ reminds us that the motion of this particle is a special case of the wave function $y(x, t)$ that describes the entire wave. At $t = 0$ the particle at $x = 0$ is at its maximum positive displacement $(y = A)$ and is instantaneously at rest (because the value of y is a maximum).

The wave disturbance travels from $x = 0$ to some point x to the right of the origin in an amount of time given by x/v, where v is the wave speed. So the motion of point x at time t is the same as the motion of point $x = 0$ at the earlier time $t - x/v$. Hence we can find the displacement of point x at time t by simply replacing t in Eq. (15.2) by $(t - x/v)$. When we do that, we find the following expression for the wave function:

$$y(x, t) = A\cos\left[\omega\left(t - \frac{x}{v}\right)\right]$$

Because $\cos(-\theta) = \cos\theta$, we can rewrite the wave function as

$$y(x, t) = A\cos\left[\omega\left(\frac{x}{v} - t\right)\right] = A\cos 2\pi f\left(\frac{x}{v} - t\right) \quad \text{(sinusoidal wave moving in } +x\text{-direction)} \quad (15.3)$$

The displacement $y(x, t)$ is a function of both the location x of the point and the time t. We could make Eq. (15.3) more general by allowing for different values of the phase angle, as we did for simple harmonic motion in Section 13.2, but for now we omit this.

We can rewrite the wave function given by Eq. (15.3) in several different but useful forms. We can express it in terms of the period $T = 1/f$ and the wavelength $\lambda = v/f$:

$$y(x, t) = A\cos 2\pi\left(\frac{x}{\lambda} - \frac{t}{T}\right) \quad \text{(sinusoidal wave moving in } +x\text{-direction)} \quad (15.4)$$

We get another convenient form of the wave function if we define a quantity k, called the **wave number:**

$$k = \frac{2\pi}{\lambda} \quad \text{(wave number)} \quad (15.5)$$

Substituting $\lambda = 2\pi/k$ and $f = \omega/2\pi$ into the wavelength-frequency relationship $v = \lambda f$ gives

$$\omega = vk \quad \text{(periodic wave)} \quad (15.6)$$

We can then rewrite Eq. (15.4) as

$$y(x, t) = A\cos(kx - \omega t) \quad \text{(sinusoidal wave moving in } +x\text{-direction)} \quad (15.7)$$

Which of these various forms for the wave function $y(x, t)$ we use in any specific problem is a matter of convenience. Note that ω has units rad/s, so for unit consistency in Eqs. (15.6) and (15.7) the wave number k must have the units rad/m. (Some physicists define the wave number as $1/\lambda$ rather than $2\pi/\lambda$. When reading other texts, be sure to determine how this term is defined.)

Graphing the Wave Function

The wave function $y(x, t)$ is graphed as a function of x for a specific time t in Fig. 15.9a. This graph gives the displacement y of a particle from its equilibrium position as a function of the coordinate x of the particle. If the wave is a transverse wave on a string, the graph in Fig. 15.9a represents the shape of the string at that instant, like a flash photograph of the string. In particular, at time $t = 0$,

$$y(x, t = 0) = A\cos kx = A\cos 2\pi\frac{x}{\lambda}$$

Figure 15.9b is a graph of the wave function versus time t for a specific coordinate x. This graph gives the displacement y of the particle at that coordinate as a

15.9 Two graphs of the wave function $y(x, t)$ in Eq. (15.7). **(a)** Graph of displacement y versus coordinate x at time $t = 0$. **(b)** Graph of displacement y versus time t at coordinate $x = 0$. The vertical scale is exaggerated in both (a) and (b).

(a) If we use Eq. (15.7) to plot y as a function of x for time $t = 0$, the curve shows the *shape* of the string at $t = 0$.

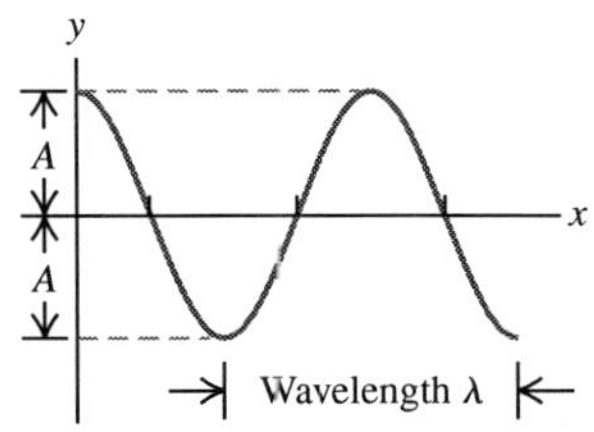

(b) If we use Eq. (15.7) to plot y as a function of t for position $x = 0$, the curve shows the *displacement* y of the particle at $x = 0$ as a function of time.

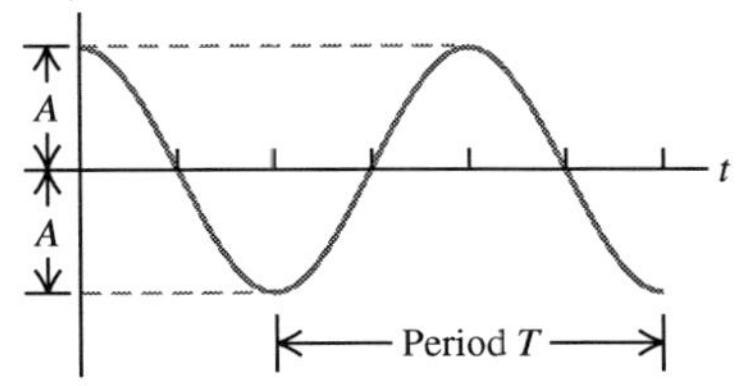

function of time; that is, it describes the motion of that particle. In particular, at the position $x = 0$,

$$y(x = 0, t) = A\cos(-\omega t) = A\cos\omega t = A\cos 2\pi\frac{t}{T}$$

This is consistent with our original statement about the motion at $x = 0$, Eq. (15.2).

CAUTION **Wave graphs** Although they may look the same at first glance, Figs. 15.9a and 15.9b are *not* identical. Figure 15.9a is a picture of the shape of the string at $t = 0$, while Fig. 15.9b is a graph of the displacement y of a particle at $x = 0$ as a function of time.

More on the Wave Function

We can modify Eqs. (15.3) through (15.7) to represent a wave traveling in the *negative* x-direction. In this case the displacement of point x at time t is the same as the motion of point $x = 0$ at the *later* time $(t + x/v)$, so in Eq. (15.2) we replace t by $(t + x/v)$. For a wave traveling in the negative x-direction,

$$y(x, t) = A\cos 2\pi f\left(\frac{x}{v} + t\right) = A\cos 2\pi\left(\frac{x}{\lambda} + \frac{t}{T}\right) = A\cos(kx + \omega t) \quad (15.8)$$

(sinusoidal wave moving in $-x$-direction)

In the expression $y(x, t) = A\cos(kx \pm \omega t)$ for a wave traveling in the $-x$ or $+x$-direction, the quantity $(kx \pm \omega t)$ is called the **phase.** It plays the role of an angular quantity (always measured in radians) in Eq. (15.7) or (15.8), and its value for any values of x and t determines what part of the sinusoidal cycle is occurring at a particular point and time. For a crest (where $y = A$ and the cosine function has the value 1), the phase could be 0, 2π, 4π, and so on; for a trough (where $y = -A$ and the cosine has the value -1), it could be π, 3π, 5π, and so on.

The wave speed is the speed with which we have to move along with the wave to keep alongside a point of a given phase, such as a particular crest of a wave on a string. For a wave traveling in the $+x$-direction, that means $kx - \omega t =$ constant. Taking the derivative with respect to t, we find $k\,dx/dt = \omega$, or

$$\frac{dx}{dt} = \frac{\omega}{k}$$

Comparing this with Eq. (15.6), we see that dx/dt is equal to the speed v of the wave. Because of this relationship, v is sometimes called the *phase velocity* of the wave. (*Phase speed* would be a better term.)

Problem-Solving Strategy 15.1 Mechanical Waves

MP

IDENTIFY *the relevant concepts:* Wave problems fall into two broad categories. *Kinematics* problems are concerned with describing wave motion; they involve wave speed v, wavelength λ (or wave number k), frequency f (or angular frequency ω), and amplitude A. They may also involve the position, velocity, and acceleration of individual particles in the medium. *Dynamics* problems also use concepts from Newton's laws such as force and mass. As an example, later in this chapter we'll encounter problems that involve the relationship of wave speed to the mechanical properties of the wave medium.

As always, make sure that you identify the target variable(s) for the problem. In some cases you'll be asked to find an expression for the wave function.

SET UP *the problem* using the following steps:

1. Make a list of the quantities whose values are given. To help you visualize the situation, you'll find it useful to sketch graphs of y versus x (like Fig. 15.9a) and of y versus t (like Fig. 15.9b). Label your graphs with the values of the known quantities.

2. Decide which equations you'll need to use. If any two of v, f, and λ are given, use Eq. (15.1) ($v = \lambda f$) to find the third quantity (see Example 15.1). If the problem involves the angular frequency ω and/or the wave number k, use the definitions of those quantities and Eq. (15.6) ($\omega = vk$). You may also need the various forms of the wave function given in Eqs. (15.3), (15.4), and (15.7).
3. If the wave speed isn't given and you don't have enough information to determine it using $v = \lambda f$, you may be able to use the relationship between v and the mechanical properties of the system. (In the next section we'll develop this relationship for waves on a string.)

EXECUTE *the solution* as follows: Solve for the unknown quantities using the equations you've selected. In some problems all you need to do is find the value of one of the wave variables.

If you're asked to determine the wave function, you need to know A and any two of v, λ, and f (or v, k, and ω). Once you have this information, you can use it in Eq. (15.3), (15.4), or (15.7) to get the specific wave function for the problem at hand. Once you have that, you can find the value of y at any point (value of x) and at any time by substituting into the wave function.

EVALUATE *your answer:* Look at your results with a critical eye. Check to see whether the values of v, f, and λ (or v, ω, and k) agree with the relationships given in Eq. (15.1) or (15.6). If you've calculated the wave function, check one or more special cases for which you can guess what the results ought to be

Example 15.2 Wave on a clothesline

Your cousin Throckmorton is playing with the clothesline. He unties one end, holds it taut, and wiggles the end up and down sinusoidally with frequency 2.00 Hz and amplitude 0.075 m. The wave speed is $v = 12.0\ \text{m/s}$. At time $t = 0$ the end has maximum positive displacement and is instantaneously at rest. Assume no wave bounces back from the far end to muddle up the pattern. (a) Find the amplitude, angular frequency, period, wavelength, and wave number of the wave. (b) Write a wave function describing the wave. (c) Write equations for the displacement as a function of time of Throckmorton's end of the clothesline and of a point 3.00 m from his end.

SOLUTION

IDENTIFY: This is a kinematics problem about the motion of the clothesline. Since Throcky moves his hand in a sinusoidal way, he produces a sinusoidal wave that propagates down the clothesline. Hence we can use all of the expressions we've developed in this section. Our target variables in part (a) are amplitude A, angular frequency ω, period T, wavelength λ, and wave number k, so we need to use the equations that relate these quantities. In parts (b) and (c) our target "variables" are actually expressions for displacement; to find these, we use the general equations for the wave function of a sinusoidal wave.

SET UP: A photograph of the clothesline at time $t = 0$ would look just like Fig. 15.9a, with the maximum displacement at $x = 0$ (the end that Throcky has in his hand). We take the positive x-direction to be the direction in which the wave propagates, so we can use Eqs. (15.4) and (15.7) to describe the displacement of the clothesline as a function of position x and time t. We also use the relationships $f = 1/T$, $\omega = 2\pi f$, $k = 2\pi/\lambda$, $v = \lambda f$, and $\omega = vk$.

EXECUTE: (a) The amplitude A of the wave is just the amplitude of the motion of the end of the clothesline, $A = 0.075\ \text{m}$. Similarly, the wave frequency is $f = 2.00\ \text{Hz}$, the same as the frequency of the end of the clothesline. The angular frequency is

$$\omega = 2\pi f = (2\pi\ \text{rad/cycle})(2.00\ \text{cycles/s}) = 4.00\pi\ \text{rad/s} = 12.6\ \text{rad/s}$$

The period is $T = 1/f = 0.500\ \text{s}$. We get the wavelength from Eq. (15.1):

$$\lambda = \frac{v}{f} = \frac{12.0\ \text{m/s}}{2.00\ \text{s}^{-1}} = 6.00\ \text{m}$$

We find the wave number from Eq. (15.5) or (15.6):

$$k = \frac{2\pi}{\lambda} = \frac{2\pi\ \text{rad}}{6.00\ \text{m}} = 1.05\ \text{rad/m} \qquad \text{or}$$

$$k = \frac{\omega}{v} = \frac{4.00\pi\ \text{rad/s}}{12.0\ \text{m/s}} = 1.05\ \text{rad/m}$$

(b) Since we found the values of A, T, and λ in part (a), we can write the wave function using Eq. (15.4):

$$\begin{aligned} y(x,t) &= A\cos 2\pi\left(\frac{x}{\lambda} - \frac{t}{T}\right) \\ &= (0.075\ \text{m})\cos 2\pi\left(\frac{x}{6.00\ \text{m}} - \frac{t}{0.500\ \text{s}}\right) \\ &= (0.075\ \text{m})\cos[(1.05\ \text{rad/m})x - (12.6\ \text{rad/s})t] \end{aligned}$$

We can also get this same equation from Eq. (15.7) by using the values of ω and k we obtained in part (a).

(c) With our choice of the positive x-direction, the two points in question are at $x = 0$ and $x = +3.00\ \text{m}$. For each point, we can find the displacement as a function of time by substituting these values of x into the wave function we found in part (b):

$$\begin{aligned} y(x = 0, t) &= (0.075\ \text{m})\cos 2\pi\left(\frac{0}{6.00\ \text{m}} - \frac{t}{0.500\ \text{s}}\right) \\ &= (0.075\ \text{m})\cos(12.6\ \text{rad/s})t \\ y(x = +3.00\ \text{m}, t) &= (0.075\ \text{m})\cos 2\pi\left(\frac{3.00\ \text{m}}{6.00\ \text{m}} - \frac{t}{0.500\ \text{s}}\right) \\ &= (0.075\ \text{m})\cos[\pi - (12.6\ \text{rad/s})t] \\ &= -(0.075\ \text{m})\cos(12.6\ \text{rad/s})t \end{aligned}$$

Continued

EVALUATE: In part (b), the quantity $(1.05\ \text{rad/m})x - (12.6\ \text{rad/s})t$ is the *phase* of a point x on the string at time t. The phases of the two points in part (c) differ by π because these points are separated by one half-wavelength $(\lambda/2 = (6.00\ \text{m})/2 = 3.00\ \text{m})$. Both points oscillate in SHM with the same frequency and amplitude, but their oscillations are one half-cycle out of phase. Thus, while a graph of y versus t for the point at $x = 0$ is a cosine curve (like Fig. 15.9b), a graph of y versus t for the point $x = 3.00$ m is a *negative* cosine (the same as a cosine curve shifted by one half-cycle).

Using the expression for $y(x = 0, t)$ in part (c), can you show that the end of the string at $x = 0$ is instantaneously at rest at $t = 0$, just as we stated at the beginning of this example? (*Hint:* Calculate the y-velocity at this point by taking the derivative of y with respect to t.)

Particle Velocity and Acceleration in a Sinusoidal Wave

From the wave function we can get an expression for the transverse velocity of any *particle* in a transverse wave. We call this v_y to distinguish it from the wave propagation speed v. To find the transverse velocity v_y at a particular point x, we take the derivative of the wave function $y(x, t)$ with respect to t, keeping x constant. If the wave function is

$$y(x, t) = A\cos(kx - \omega t)$$

then

$$v_y(x, t) = \frac{\partial y(x, t)}{\partial t} = \omega A \sin(kx - \omega t) \tag{15.9}$$

The ∂ in this expression is a modified d, used to remind us that $y(x, t)$ is a function of *two* variables and that we are allowing only one (t) to vary. The other (x) is constant because we are looking at a particular point on the string. This derivative is called a *partial derivative.* If you haven't reached this point yet in your study of calculus, don't fret; it's a simple idea.

Equation (15.9) shows that the transverse velocity of a particle varies with time, as we expect for simple harmonic motion. The maximum particle speed is ωA; this can be greater than, less than, or equal to the wave speed v, depending on the amplitude and frequency of the wave.

The *acceleration* of any particle is the *second* partial derivative of $y(x, t)$ with respect to t:

$$a_y(x, t) = \frac{\partial^2 y(x, t)}{\partial t^2} = -\omega^2 A\cos(kx - \omega t) = -\omega^2 y(x, t) \tag{15.10}$$

The acceleration of a particle equals $-\omega^2$ times its displacement, which is the result we obtained in Section 13.2 for simple harmonic motion.

We can also compute partial derivatives of $y(x, t)$ with respect to x, holding t constant. This corresponds to studying the shape of the string at one instant of time, like a flash photo. The first derivative $\partial y(x, t)/\partial x$ is the *slope* of the string at any point. The second partial derivative with respect to x is the *curvature* of the string:

$$\frac{\partial^2 y(x, t)}{\partial x^2} = -k^2 A\cos(kx - \omega t) = -k^2 y(x, t) \tag{15.11}$$

From Eqs. (15.10) and (15.11) and the relationship $\omega = vk$ we see that

$$\frac{\partial^2 y(x, t)/\partial t^2}{\partial^2 y(x, t)/\partial x^2} = \frac{\omega^2}{k^2} = v^2 \quad \text{and}$$
$$\frac{\partial^2 y(x, t)}{\partial x^2} = \frac{1}{v^2}\frac{\partial^2 y(x, t)}{\partial t^2} \quad \text{(wave equation)} \tag{15.12}$$

We've derived Eq. (15.12) for a wave traveling in the positive x-direction. You can use the same steps to show that the wave function for a sinusoidal wave propagating in the *negative* x-direction, $y(x, t) = A\cos(kx + \omega t)$, also satisfies this equation.

Equation (15.12), called the **wave equation,** is one of the most important equations in all of physics. Whenever it occurs, we know that a disturbance can propagate as a wave along the x-axis with wave speed v. The disturbance need not be a sinusoidal wave; we'll see in the next section that *any* wave on a string obeys Eq. (15.12), whether the wave is periodic or not (see also Problem 15.61). In Chapter 32 we will find that electric and magnetic fields satisfy the wave equation; the wave speed will turn out to be the speed of light, which will lead us to the conclusion that light is an electromagnetic wave.

Figure 15.10a shows the transverse velocity v_y and transverse acceleration a_y, given by Eqs. (15.9) and (15.10), for several points on a string as a sinusoidal wave passes along it. Note that at points where the string has an upward curvature $(\partial^2 y/\partial x^2 > 0)$, the acceleration of that point is positive $(a_y = \partial^2 y/\partial t^2 > 0)$; this follows from the wave equation, Eq. (15.12). For the same reason the acceleration is negative $(a_y = \partial^2 y/\partial t^2 < 0)$ at points where the string has a downward curvature $(\partial^2 y/\partial x^2 < 0)$, and the acceleration is zero $(a_y = \partial^2 y/\partial t^2 = 0)$ at points of inflection where the curvature is zero $(\partial^2 y/\partial x^2 = 0)$. We emphasize again that v_y and a_y are the *transverse* velocity and acceleration of points on the string; these points move along the y-direction, not along the propagation direction of the wave. Figure 15.10b shows the transverse motions of several points on the string.

The concept of wave function is equally useful with *longitudinal* waves, and everything we have said about wave functions can be adapted to this case. The quantity y still measures the displacement of a particle of the medium from its equilibrium position; the difference is that for a longitudinal wave, this displacement is *parallel* to the x-axis instead of perpendicular to it. We'll discuss longitudinal waves in detail in Chapter 16.

Test Your Understanding of Section 15.3 Figure 15.8 shows a sinusoidal wave of period T on a string at times $0, \frac{1}{8}T, \frac{2}{8}T, \frac{3}{8}T, \frac{4}{8}T, \frac{5}{8}T, \frac{6}{8}T, \frac{7}{8}T$, and T. (a) At which time is point A on the string moving upward with maximum speed? (b) At which time does point B on the string have the greatest upward acceleration? (c) At which time does point C on the string have a downward acceleration but an upward velocity?

15.10 (a) Another view of the wave at $t = 0$ in Fig. 15.9a. The vectors show the transverse velocity v_y and transverse acceleration a_y at several points on the string. (b) From $t = 0$ to $t = 0.05\,T$, a particle at point 1 is displaced to point 1′, a particle at point 2 is displaced to point 2′, and so on.

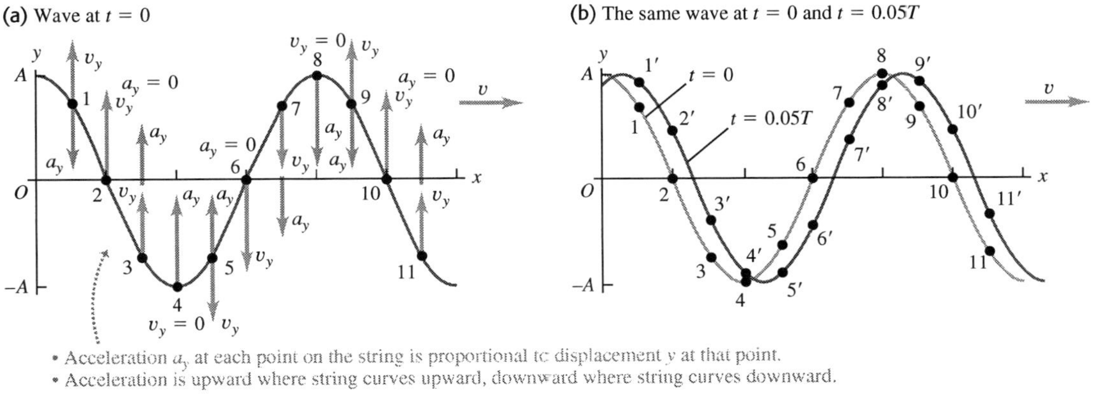

ActivPhysics ONLINE
10.2 Speed of Waves on a String

15.4 Speed of a Transverse Wave

One of the key properties of any wave is the wave *speed*. Light waves in air have a much greater speed of propagation than do sound waves in air (3.00×10^8 m/s versus 344 m/s); that's why you see the flash from a bolt of lightning before you hear the clap of thunder. In this section we'll see what determines the speed of propagation of one particular kind of wave: transverse waves on a string. The speed of these waves is important to understand in its own right because it is an essential part of analyzing stringed musical instruments, as we'll discuss later in this chapter. Furthermore, the speeds of many kinds of mechanical waves turn out to have the same basic mathematical expression as does the speed of waves on a string.

The physical quantities that determine the speed of transverse waves on a string are the *tension* in the string and its *mass per unit length* (also called *linear mass density*). We might guess that increasing the tension should increase the restoring forces that tend to straighten the string when it is disturbed, thus increasing the wave speed. We might also guess that increasing the mass should make the motion more sluggish and decrease the speed. Both these guesses turn out to be right. We'll develop the exact relationship among wave speed, tension, and mass per unit length by two different methods. The first is simple in concept and considers a specific wave shape; the second is more general but also more formal. Choose whichever you like better.

Wave Speed on a String: First Method

We consider a perfectly flexible string (Fig. 15.11). In the equilibrium position the tension is F, and the linear mass density (mass per unit length) is μ. (When portions of the string are displaced from equilibrium, the mass per unit length decreases a little, and the tension increases a little.) We ignore the weight of the string so that when the string is at rest in the equilibrium position, the string forms a perfectly straight line as in Fig. 15.11a.

Starting at time $t = 0$, we apply a constant upward force F_y at the left end of the string. We might expect that the end would move with constant acceleration; that would happen if the force were applied to a *point* mass. But here the effect of the force F_y is to set successively more and more mass in motion. The wave travels with constant speed v, so the division point P between moving and nonmoving portions moves with the same constant speed v (Fig. 15.11b).

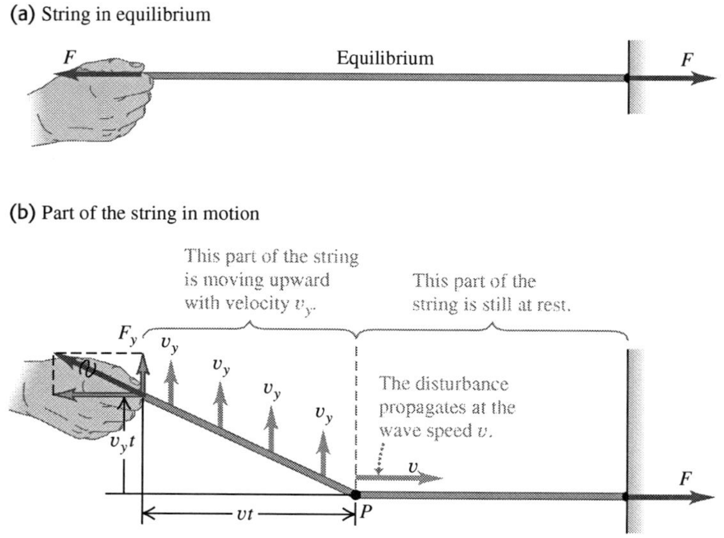

15.11 Propagation of a transverse wave on a string.

Figure 15.11b shows that all particles in the moving portion of the string move upward with constant *velocity* v_y, not constant acceleration. To see why this is so, we note that the *impulse* of the force F_y up to time t is $F_y t$. According to the impulse–momentum theorem (see Section 8.1), the impulse is equal to the change in the total transverse component of momentum $(mv_y - 0)$ of the moving part of the string. Because the system started with *no* transverse momentum, this is equal to the total momentum at time t:

$$F_y t = mv_y$$

The total momentum thus must increase proportionately with time. But since the division point P moves with constant speed, the length of string that is in motion and hence the total mass m in motion are also proportional to the time t that the force has been acting. So the *change* of momentum must be associated entirely with the increasing amount of mass in motion, not with an increasing velocity of an individual mass element. That is, mv_y changes because m, not v_y, changes.

At time t, the left end of the string has moved up a distance $v_y t$, and the boundary point P has advanced a distance vt. The total force at the left end of the string has components F and F_y. Why F? There is no motion in the direction along the length of the string, so there is no unbalanced horizontal force. Therefore F, the magnitude of the horizontal component, does not change when the string is displaced. In the displaced position the tension is $(F^2 + F_y^2)^{1/2}$ (greater than F), and the string stretches somewhat.

To derive an expression for the wave speed v, we again apply the impulse–momentum theorem to the portion of the string in motion at time t—that is, the portion to the left of P in Fig. 15.11b. The transverse *impulse* (transverse force times time) is equal to the change of transverse *momentum* of the moving portion (mass times transverse component of velocity). The impulse of the transverse force F_y in time t is $F_y t$. In Fig. 15.11b the right triangle whose vertex is at P, with sides $v_y t$ and vt, is similar to the right triangle whose vertex is at the position of the hand, with sides F_y and F. Hence

$$\frac{F_y}{F} = \frac{v_y t}{vt} \qquad F_y = F\frac{v_y}{v}$$

and

$$\text{Transverse impulse} = F_y t = F\frac{v_y}{v}t$$

The mass of the moving portion of the string is the product of the mass per unit length μ and the length vt, or μvt. The transverse momentum is the product of this mass and the transverse velocity v_y:

$$\text{Transverse momentum} = (\mu vt)v_y$$

We note again that the momentum increases with time *not* because mass is moving faster, as was usually the case in Chapter 8, but because *more mass* is brought into motion. But the impulse of the force F_y is still equal to the total change in momentum of the system. Applying this relationship, we obtain

$$F\frac{v_y}{v}t = \mu vtv_y$$

Solving this for v, we find

$$v = \sqrt{\frac{F}{\mu}} \quad \text{(speed of a transverse wave on a string)} \qquad (15.13)$$

15.12 These cables have a relatively large amount of mass per unit length (μ) and a low tension (F). If the cables are disturbed—say, by a bird landing on them—transverse waves will travel along them at a slow speed $v = \sqrt{F/\mu}$.

15.13 Free-body diagram for a segment of string. The force at each end of the string is tangent to the string at the point of application.

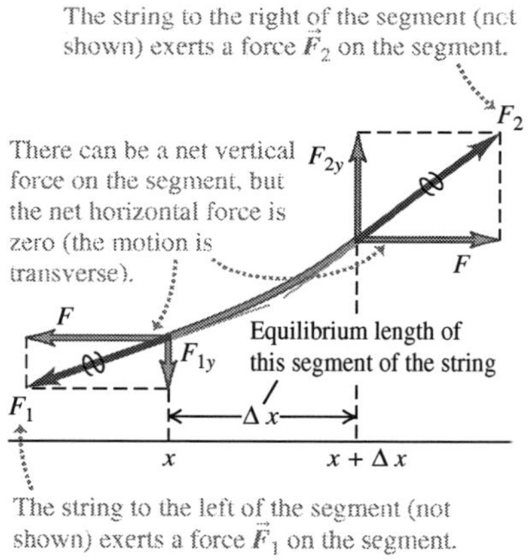

Equation (15.13) confirms our prediction that the wave speed v should increase when the tension F increases but decrease when the mass per unit length μ increases (Fig. 15.12).

Note that v_y does not appear in Eq. (15.13); thus the wave speed doesn't depend on v_y. Our calculation considered only a very special kind of pulse, but we can consider *any* shape of wave disturbance as a series of pulses with different values of v_y. So even though we derived Eq. (15.13) for a special case, it is valid for *any* transverse wave motion on a string, including the sinusoidal and other periodic waves we discussed in Section 15.3. Note also that the wave speed doesn't depend on the amplitude or frequency of the wave, in accordance with our assumptions in Section 15.3.

Wave Speed on a String: Second Method

Here is an alternative derivation of Eq. (15.13). If you aren't comfortable with partial derivatives, it can be omitted. We apply Newton's second law, $\sum\vec{F} = m\vec{a}$, to a small segment of string whose length in the equilibrium position is Δx (Fig. 15.13). The mass of the segment is $m = \mu\,\Delta x$; the forces at the ends are represented in terms of their x- and y-components. The x-components have equal magnitude F and add to zero because the motion is transverse and there is no component of acceleration in the x-direction. To obtain F_{1y} and F_{2y}, we note that the ratio F_{1y}/F is equal in magnitude to the *slope* of the string at point x and that F_{2y}/F is equal to the slope at point $x + \Delta x$. Taking proper account of signs, we find

$$\frac{F_{1y}}{F} = -\left(\frac{\partial y}{\partial x}\right)_x \qquad \frac{F_{2y}}{F} = \left(\frac{\partial y}{\partial x}\right)_{x+\Delta x} \tag{15.14}$$

The notation reminds us that the derivatives are evaluated at points x and $x + \Delta x$, respectively. From Eq. (15.14) we find that the net y-component of force is

$$F_y = F_{1y} + F_{2y} = F\left[\left(\frac{\partial y}{\partial x}\right)_{x+\Delta x} - \left(\frac{\partial y}{\partial x}\right)_x\right] \tag{15.15}$$

We now equate F_y from Eq. (15.15) to the mass $\mu\,\Delta x$ times the y-component of acceleration $\partial^2 y/\partial t^2$. We obtain

$$F\left[\left(\frac{\partial y}{\partial x}\right)_{x+\Delta x} - \left(\frac{\partial y}{\partial x}\right)_x\right] = \mu\,\Delta x\frac{\partial^2 y}{\partial t^2} \tag{15.16}$$

or, dividing by $F\,\Delta x$,

$$\frac{\left(\frac{\partial y}{\partial x}\right)_{x+\Delta x} - \left(\frac{\partial y}{\partial x}\right)_x}{\Delta x} = \frac{\mu}{F}\frac{\partial^2 y}{\partial t^2} \tag{15.17}$$

We now take the limit as $\Delta x \to 0$. In this limit, the left side of Eq. (15.17) becomes the derivative of $\partial y/\partial x$ with respect to x (at constant t)—that is, the *second* (partial) derivative of y with respect to x:

$$\frac{\partial^2 y}{\partial x^2} = \frac{\mu}{F}\frac{\partial^2 y}{\partial t^2} \tag{15.18}$$

Now, finally, comes the punch line of our story. Equation (15.18) has exactly the same form as the *wave equation,* Eq. (15.12), that we derived at the end of Section 15.3. That equation and Eq. (15.18) describe the very same wave motion, so they must be identical. Comparing the two equations, we see that for this to be so, we must have

$$v = \sqrt{\frac{F}{\mu}} \tag{15.19}$$

which is the same expression as Eq. (15.13).

In going through this derivation, we didn't make any special assumptions about the shape of the wave. Since our derivation led us to rediscover Eq. (15.12), the wave equation, we conclude that the wave equation is valid for waves on a string that have *any* shape.

The Speed of Mechanical Waves

Equation (15.13) or (15.19) gives the wave speed for only the special case of mechanical waves on a stretched string or rope. Remarkably, it turns out that for many types of mechanical waves, including waves on a string, the expression for wave speed has the same general form:

$$v = \sqrt{\frac{\text{Restoring force returning the system to equilibrium}}{\text{Inertia resisting the return to equilibrium}}}$$

To interpret this expression, let's look at the now-familiar case of waves on a string. The tension F in the string plays the role of the restoring force; it tends to bring the string back to its undisturbed, equilibrium configuration. The mass of the string—or, more properly, the linear mass density μ—provides the inertia that prevents the string from returning instantaneously to equilibrium. Hence we have $v = \sqrt{F/\mu}$ for the speed of waves on a string.

In Chapter 16 we'll see a similar expression for the speed of sound waves in a gas. Roughly speaking, the gas pressure provides the force that tends to return the gas to its undisturbed state when a sound wave passes through. The inertia is provided by the density, or mass per unit volume, of the gas.

Example 15.3 Calculating wave speed

One end of a nylon rope is tied to a stationary support at the top of a vertical mine shaft 80.0 m deep (Fig. 15.14). The rope is stretched taut by a box of mineral samples with mass 20.0 kg attached at the lower end. The mass of the rope is 2.00 kg. The geologist at the bottom of the mine signals to his colleague at the top by jerking the rope sideways. (a) What is the speed of a transverse wave on the rope? (b) If a point on the rope is given a transverse simple harmonic motion with a frequency of 2.00 Hz, how many cycles of the wave are there in the rope's length?

15.14 Sending signals along a vertical rope using transverse waves.

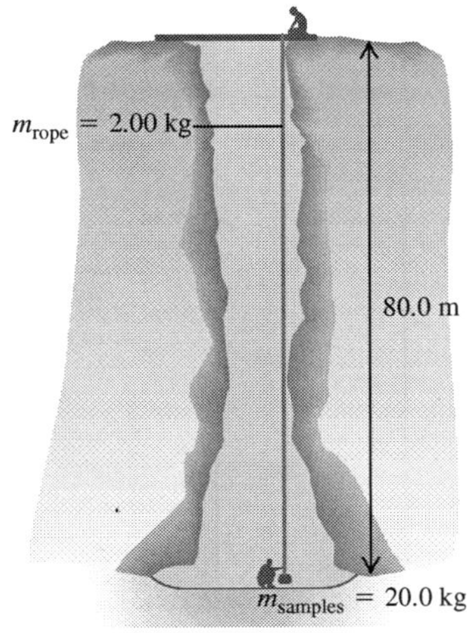

SOLUTION

IDENTIFY: In part (a) the target variable is the wave speed. This part involves *dynamics*—that is, the relationship between the wave speed and the properties of the rope (tension and liner mass density). Part (b) involves *kinematics,* since we need to know how wave speed, frequency, and wavelength are related. (The target variable is actually the number of wavelengths that fit into the length of the rope.)

We'll assume that the tension in the rope is provided by the weight of the box of samples. In fact, the weight of the rope itself contributes to the tension, which means that the tension is different at the top and bottom of the rope. We'll ignore this effect here, since the weight of the rope is small compared to the weight of the samples.

SET UP: We use the relationship $v = \sqrt{F/\mu}$ in part (a). If we neglect the weight of the rope itself, the tension F is just equal to the weight of the box. In part (b) we use the equation $v = f\lambda$ to find the wavelength, which we then compare to the 80.0-m length of the rope.

EXECUTE: (a) The tension in the rope (due to the sample box) is

$$F = m_{samples}g = (20.0\text{ kg})(9.80\text{ m/s}^2) = 196\text{ N}$$

Continued

and the mass per unit length of the rope is

$$\mu = \frac{m_{\text{rope}}}{L} = \frac{2.00 \text{ kg}}{80.0 \text{ m}} = 0.0250 \text{ kg/m}$$

Hence, from Eq. (15.13), the wave speed is

$$v = \sqrt{\frac{F}{\mu}} = \sqrt{\frac{196 \text{ N}}{0.0250 \text{ kg/m}}} = 88.5 \text{ m/s}$$

(b) From Eq. (15.1),

$$\lambda = \frac{v}{f} = \frac{88.5 \text{ m/s}}{2.00 \text{ s}^{-1}} = 44.3 \text{ m}$$

The length of the rope is 80.0 m, so the number of wave cycles in the rope is

$$\frac{80.0 \text{ m/s}}{44.3 \text{ m/cycle}} = 1.81 \text{ cycles}$$

EVALUATE: If we do account for the weight of the rope, the tension is greater at the top of the rope than at the bottom. Hence the wave speed increases and the wavelength increases as the wave travels up the rope. Can you verify that the wave speed at the top of the rope is 92.9 m/s?

Test Your Understanding of Section 15.4 The six strings of a guitar are the same length and under nearly the same tension, but they have different thicknesses. On which string do waves travel the fastest? (i) the thickest string; (ii) the thinnest string; (iii) the wave speed is the same on all strings.

15.5 Energy in Wave Motion

Every wave motion has *energy* associated with it. The energy we receive from sunlight and the destructive effects of ocean surf and earthquakes bear this out. To produce any of the wave motions we have discussed in this chapter, we have to apply a force to a portion of the wave medium; the point where the force is applied moves, so we do *work* on the system. As the wave propagates, each portion of the medium exerts a force and does work on the adjoining portion. In this way a wave can transport energy from one region of space to another.

As an example of energy considerations in wave motion, let's look again at transverse waves on a string. How is energy transferred from one portion of string to another? Picture a wave traveling from left to right (the positive x-direction) on the string, and consider a particular point a on the string (Fig. 15.15a). The string to the left of point a exerts a force on the string to the right of it, and vice versa. In Fig. 15.15b the string to the left of a has been removed, and the force it exerts at a is represented by the components F and F_y, as we did in Figs. 15.11 and 15.13. We note again that F_y/F is equal to the negative of the *slope* of the string at a, which is also given by $\partial y/\partial x$. Putting these together, we have

15.15 (a) Point a on a string carrying a wave from left to right. (b) The components of the force exerted on the part of the string to the right of point a by the part of the string to the left of point a.

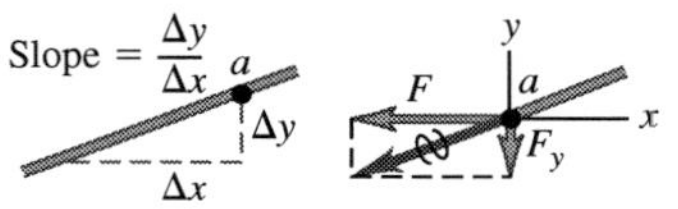

$$F_y(x, t) = -F\frac{\partial y(x, t)}{\partial x} \tag{15.20}$$

We need the negative sign because F_y is negative when the slope is positive. We write the vertical force as $F_y(x, t)$ as a reminder that its value may be different at different points along the string and at different times.

When point a moves in the y-direction, the force F_y does *work* on this point and therefore transfers energy into the part of the string to the right of a. The corresponding power P (rate of doing work) at the point a is the transverse force $F_y(x, t)$ at a times the transverse velocity $v_y(x, t) = \partial y(x, t)/\partial t$ of that point:

$$P(x, t) = F_y(x, t)v_y(x, t) = -F\frac{\partial y(x, t)}{\partial x}\frac{\partial y(x, t)}{\partial t} \tag{15.21}$$

This power is the *instantaneous* rate at which energy is transferred along the string. Its value depends on the position x on the string and on the time t. Note that energy is being transferred only at points where the string has a nonzero slope ($\partial y/\partial x$ is nonzero), so that there is a transverse component of the tension force, and where the string has a nonzero transverse velocity ($\partial y/\partial t$ is nonzero) so that the transverse force can do work.

Equation (15.21) is valid for *any* wave on a string, sinusoidal or not. For a sinusoidal wave with wave function given by Eq. (15.7), we have

$$\begin{aligned} y(x,t) &= A\cos(kx - \omega t) \\ \frac{\partial y(x,t)}{\partial x} &= -kA\sin(kx - \omega t) \\ \frac{\partial y(x,t)}{\partial t} &= \omega A\sin(kx - \omega t) \\ P(x,t) &= Fk\omega A^2 \sin^2(kx - \omega t) \end{aligned} \tag{15.22}$$

By using the relationships $\omega = vk$ and $v^2 = F/\mu$, we can also express Eq. (15.22) in the alternative form

$$P(x,t) = \sqrt{\mu F}\,\omega^2 A^2 \sin^2(kx - \omega t) \tag{15.23}$$

The $\sin^2$ function is never negative, so the instantaneous power in a sinusoidal wave is either positive (so that energy flows in the positive x-direction) or zero (at points where there is no energy transfer). Energy is never transferred in the direction opposite to the direction of wave propagation (Fig. 15.16).

The maximum value of the instantaneous power $P(x,t)$ occurs when the $\sin^2$ function has the value unity:

$$P_{max} = \sqrt{\mu F}\,\omega^2 A^2 \tag{15.24}$$

To obtain the *average* power from Eq. (15.23), we note that the *average* value of the $\sin^2$ function, averaged over any whole number of cycles, is $\frac{1}{2}$. Hence the average power is

$$P_{av} = \frac{1}{2}\sqrt{\mu F}\,\omega^2 A^2 \quad \text{(average power, sinusoidal wave on a string)} \tag{15.25}$$

15.16 The instantaneous power $P(x,t)$ in a sinusoidal wave as given by Eq. (15.23), shown as a function of time at coordinate $x = 0$. The power is never negative, which means that energy never flows opposite to the direction of wave propagation.

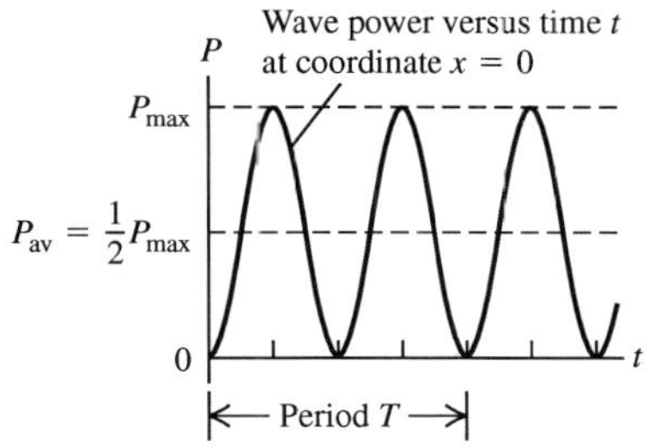

The average power is just one-half of the maximum instantaneous power (see Fig. 15.16).

The average rate of energy transfer is proportional to the square of the amplitude and to the square of the frequency. This proportionality is a general result for mechanical waves of all types, including seismic waves (see the photo that opens this chapter). For a mechanical wave, the rate of energy transfer quadruples if the frequency is doubled (for the same amplitude) or if the amplitude is doubled (for the same frequency).

Electromagnetic waves turn out to be a bit different. While the average rate of energy transfer in an electromagnetic wave is proportional to the square of the amplitude, just as for mechanical waves, it is independent of the value of ω.

Example 15.4 Power in a wave

(a) In Example 15.2, at what maximum rate does Throcky put energy into the clothesline? That is, what is his maximum instantaneous power? Assume that the linear mass density of the clothesline is $\mu = 0.250$ kg/m and that Throcky applies tension $F = 36.0$ N. (b) What is his average power? (c) As Throcky tires, the amplitude decreases. What is the average power when the amplitude has dropped to 7.50 mm?

SOLUTION

IDENTIFY: Our target variable in part (a) is the *maximum instantaneous* power, while the target variable in parts (b) and (c) is the *average* power. As we've seen, these two quantities have different values for a sinusoidal wave. We'll be able to calculate the values of both quantities because we know all the other properties of the wave from Example 15.2.

SET UP: For part (a) we use Eq. (15.24), and for parts (b) and (c) we use Eq. (15.25).

EXECUTE: (a) The maximum instantaneous power is

$$\begin{aligned} P_{max} &= \sqrt{\mu F}\,\omega^2 A^2 \\ &= \sqrt{(0.250\ \text{kg/m})(36.0\ \text{N})}\,(4.00\pi\ \text{rad/s})^2(0.075\ \text{m})^2 \\ &= 2.66\ \text{W} \end{aligned}$$

Continued

(b) From Eqs. (15.24) and (15.25), the average power is one-half of the maximum instantaneous power, so

$$P_{av} = \frac{1}{2}(2.66\ \text{W}) = 1.33\ \text{W}$$

(c) The new amplitude is $\frac{1}{10}$ of the value we used in parts (a) and (b). The average power is proportional to the *square* of the amplitude, so now the average power is

$$P_{av} = \left(\frac{1}{10}\right)^2 (1.33\ \text{W}) = 0.0133\ \text{W} = 13.3\ \text{mW}$$

EVALUATE: The *maximum* instantaneous power in part (a) occurs when the quantity $\sin^2(kx - \omega t)$ in Eq. (15.23) is equal to 1. At any given value of x, this happens twice per period of the wave—once when the sine function is equal to $+1$, and once when it's equal to -1. The *minimum* instantaneous power is zero; this occurs when $\sin(kx - \omega t) = 0$, which also happens twice per period.

Can you confirm that the given values of μ and F give the wave speed mentioned in Example 15.2?

Wave Intensity

Waves on a string carry energy in just one dimension of space (along the direction of the string). But other types of waves, including sound waves in air and seismic waves in the body of the earth, carry energy across all three dimensions of space. For waves that travel in three dimensions, we define the **intensity** (denoted by I) to be *the time average rate at which energy is transported by the wave, per unit area,* across a surface perpendicular to the direction of propagation. That is, intensity I is average power per unit area. It is usually measured in watts per square meter (W/m^2).

If waves spread out equally in all directions from a source, the intensity at a distance r from the source is inversely proportional to r^2 (Fig. 15.17). This follows directly from energy conservation. If the power output of the source is P, then the average intensity I_1 through a sphere with radius r_1 and surface area $4\pi r_1^2$ is

$$I_1 = \frac{P}{4\pi r_1^2}$$

The average intensity I_2 through a sphere with a different radius r_2 is given by a similar expression. If no energy is absorbed between the two spheres, the power P must be the same for both, and

$$4\pi r_1^2 I_1 = 4\pi r_2^2 I_2$$

$$\frac{I_1}{I_2} = \frac{r_2^2}{r_1^2} \quad \text{(inverse-square law for intensity)} \tag{15.26}$$

The intensity I at any distance r is therefore inversely proportional to r^2. This relationship is called the *inverse-square law* for intensity.

15.17 The greater the distance from a wave source, the greater the area over which the wave power is distributed and the smaller the wave intensity.

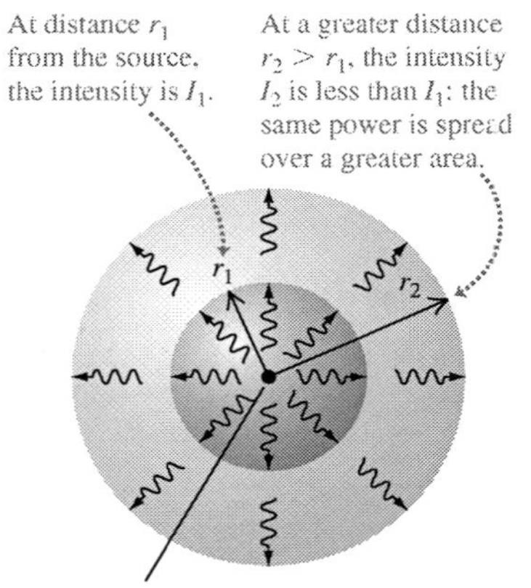

Example 15.5 The inverse-square law

A tornado warning siren on top of a tall pole radiates sound waves uniformly in all directions. At a distance of 15.0 m the intensity of the sound is $0.250\ \text{W/m}^2$. At what distance from the siren is the intensity $0.010\ \text{W/m}^2$?

SOLUTION

IDENTIFY: Because waves spread out equally in all directions, we can use the inverse-square law. Our target variable is a distance from the wave source.

SET UP: The relationship to use is Eq. (15.26). We are given the distance $r_1 = 15.0$ m at which the intensity is $I_1 = 0.250\ \text{W/m}^2$, and we want to find the distance r_2 at which the intensity is $I_2 = 0.010\ \text{W/m}^2$.

EXECUTE: We solve Eq. (15.26) for r_2:

$$r_2 = r_1\sqrt{\frac{I_1}{I_2}} = (15.0\ \text{m})\sqrt{\frac{0.250\ \text{W/m}^2}{0.010\ \text{W/m}^2}} = 75.0\ \text{m}$$

EVALUATE: As a check on our answer, note that r_2 is five times greater than r_1. By the inverse-square law, the intensity I_2 should be $1/5^2 = 1/25$ as great as I_1, and indeed it is.

By using the inverse-square law we've assumed that the sound waves travel in straight lines away from the siren. A more realistic solution of this problem would account for the reflection of sound waves from the ground. Such a solution is beyond our scope, however.

Test Your Understanding of Section 15.5 Four identical strings each carry a sinusoidal wave of frequency 10 Hz. The string tension and wave amplitude are different for different strings. Rank the following strings in order from highest to lowest value of the average wave power: (i) tension 10 N, amplitude 1.0 mm; (ii) tension 40 N, amplitude 1.0 mm; (iii) tension 10 N, amplitude 4.0 mm; (iv) tension 20 N, amplitude 2.0 mm.

15.6 Wave Interference, Boundary Conditions, and Superposition

Up to this point we've been discussing waves that propagate continuously in the same direction. But when a wave strikes the boundaries of its medium, all or part of the wave is *reflected.* When you yell at a building wall or a cliff face some distance away, the sound wave is reflected from the rigid surface and you hear an echo. When you flip the end of a rope whose far end is tied to a rigid support, a pulse travels the length of the rope and is reflected back to you. In both cases, the initial and reflected waves overlap in the same region of the medium. This overlapping of waves is called **interference.** (In general, the term "interference" refers to what happens when two or more waves pass through the same region at the same time.)

As a simple example of wave reflections and the role of the boundary of a wave medium, let's look again at transverse waves on a stretched string. What happens when a wave pulse or a sinusoidal wave arrives at the *end* of the string?

If the end is fastened to a rigid support, it is a *fixed* end that cannot move. The arriving wave exerts a force on the support; the reaction to this force, exerted *by* the support *on* the string, "kicks back" on the string and sets up a *reflected* pulse or wave traveling in the reverse direction. Figure 15.18 is a series of photographs showing the reflection of a pulse at the fixed end of a long coiled spring. The reflected pulse moves in the opposite direction from the initial, or *incident,* pulse, and its displacement is also opposite. Figure 15.19a illustrates this situation for a wave pulse on a string.

15.18 A series of images of a wave pulse, equally spaced in time from top to bottom. The pulse starts at the left in the top image, travels to the right, and is reflected from the fixed end at the right.

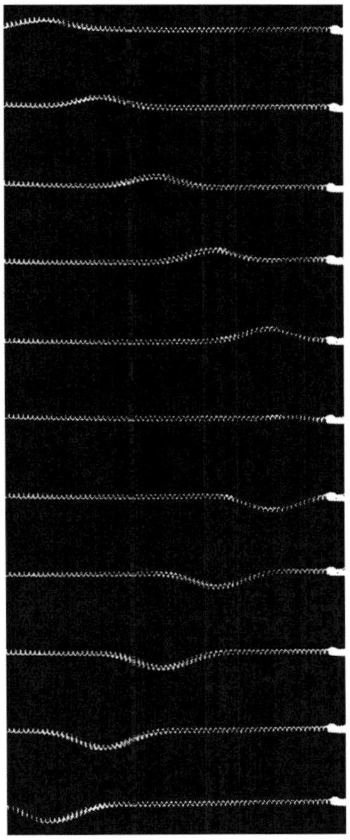

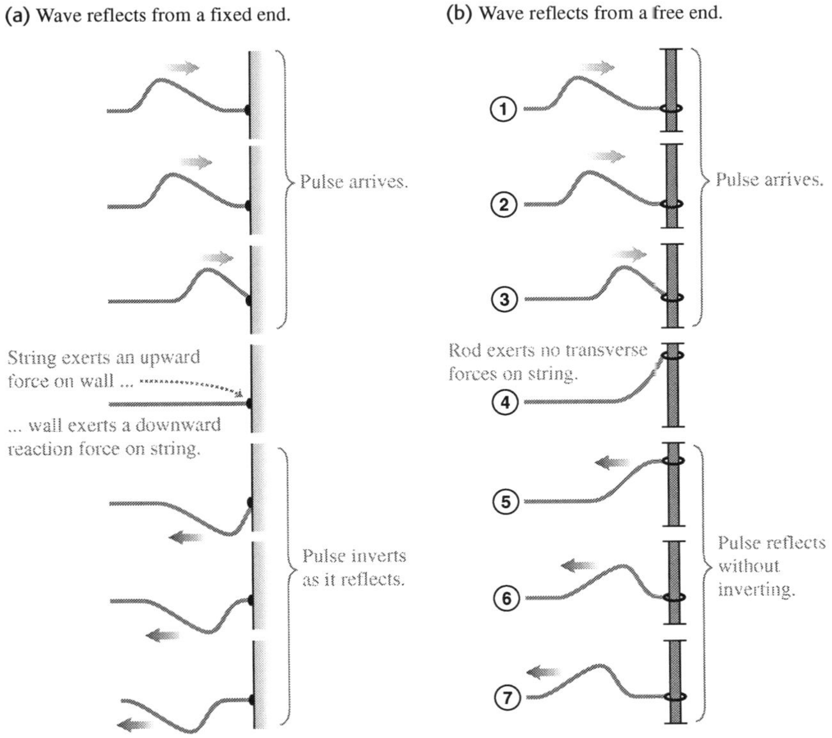

15.19 Reflection of a wave pulse (a) at a fixed end of a string and (b) at a free end. Time increases from top to bottom in each figure.

15.20 Overlap of two wave pulses—one right side up, one inverted—traveling in opposite directions. Time increases from top to bottom.

As the pulses overlap, the displacement of the string at any point is the algebraic sum of the displacements due to the individual pulses.

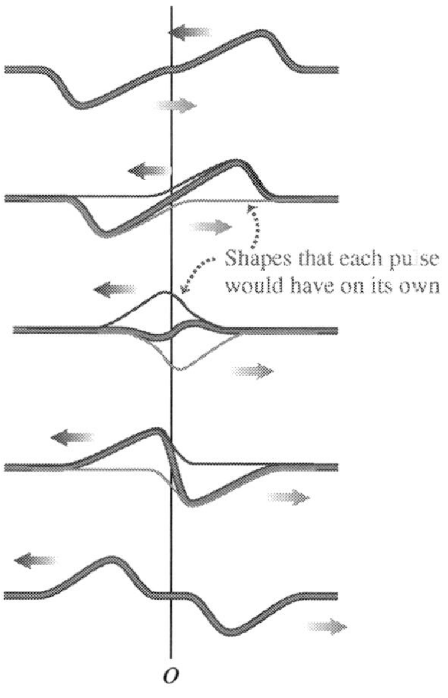

15.21 Overlap of two wave pulses—both right side up—traveling in opposite directions. Time increases from top to bottom. Compare to Fig. 15.20.

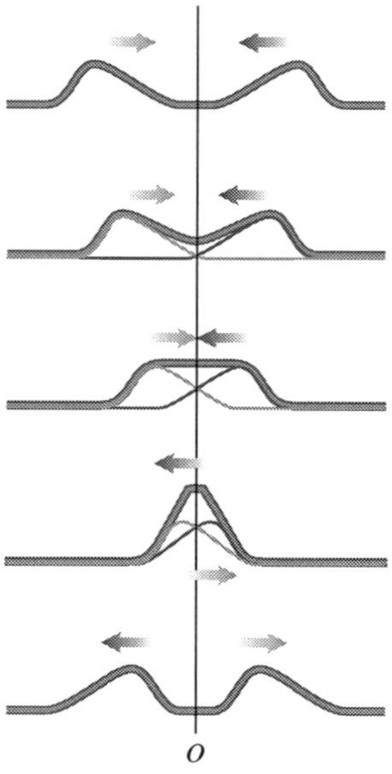

The opposite situation from an end that is held stationary is a *free* end, one that is perfectly free to move in the direction perpendicular to the length of the string. For example, the string might be tied to a light ring that slides on a frictionless rod perpendicular to the string, as in Fig. 15.19b. The ring and rod maintain the tension but exert no transverse force. When a wave arrives at this free end, the ring slides along the rod. The ring reaches a maximum displacement, and both it and the string come momentarily to rest, as in drawing 4 in Fig. 15.19b. But the string is now stretched, giving increased tension, so the free end of the string is pulled back down, and again a reflected pulse is produced (drawing 7). As for a fixed end, the reflected pulse moves in the opposite direction from the initial pulse, but now the direction of the displacement is the same as for the initial pulse. The conditions at the end of the string, such as a rigid support or the complete absence of transverse force, are called **boundary conditions.**

The formation of the reflected pulse is similar to the overlap of two pulses traveling in opposite directions. Figure 15.20 shows two pulses with the same shape, one inverted with respect to the other, traveling in opposite directions. As the pulses overlap and pass each other, the total displacement of the string is the *algebraic sum* of the displacements at that point in the individual pulses. Because these two pulses have the same shape, the total displacement at point O in the middle of the figure is zero at all times. Thus the motion of the left half of the string would be the same if we cut the string at point O, threw away the right side, and held the end at O fixed. The two pulses on the left side then correspond to the incident and reflected pulses, combining so that the total displacement at O is *always* zero. For this to occur, the reflected pulse must be inverted relative to the incident pulse.

Figure 15.21 shows two pulses with the same shape, traveling in opposite directions but *not* inverted relative to each other. The displacement at point O in the middle of the figure is not zero, but the slope of the string at this point is always zero. According to Eq. (15.20), this corresponds to the absence of any transverse force at this point. In this case the motion of the left half of the string would be the same as if we cut the string at point O and attached the end to a frictionless sliding ring (Fig. 15.19b) that maintains tension without exerting any transverse force. In other words, this situation corresponds to reflection of a pulse at a free end of a string at point O. In this case the reflected pulse is *not* inverted.

The Principle of Superposition

Combining the displacements of the separate pulses at each point to obtain the actual displacement is an example of the **principle of superposition:** When two waves overlap, the actual displacement of any point on the string at any time is obtained by adding the displacement the point would have if only the first wave were present and the displacement it would have if only the second wave were present. In other words, the wave function $y(x, t)$ that describes the resulting motion in this situation is obtained by *adding* the two wave functions for the two separate waves:

$$y(x, t) = y_1(x, t) + y_2(x, t) \quad \text{(principle of superposition)} \quad (15.27)$$

Mathematically, this additive property of wave functions follows from the form of the wave equation, Eq. (15.12) or (15.18), which every physically possible wave function must satisfy. Specifically, the wave equation is *linear;* that is, it contains the function $y(x, t)$ only to the first power (there are no terms involving $y(x, t)^2$, $y(x, t)^{1/2}$, etc.). As a result, if any two functions $y_1(x, t)$ and $y_2(x, t)$ satisfy the wave equation separately, their sum $y_1(x, t) + y_2(x, t)$ also satisfies it and is therefore a physically possible motion. Because this principle depends on the linearity of the wave equation and the corresponding linear-combination property of its solutions, it is also called the *principle of linear superposition.* For some physical systems, such as a medium that does not obey Hooke's law, the wave equation is *not* linear; this principle does not hold for such systems.

The principle of superposition is of central importance in all types of waves. When a friend talks to you while you are listening to music, you can distinguish the sound of speech and the sound of music from each other. This is precisely because the total sound wave reaching your ears is the algebraic sum of the wave produced by your friend's voice and the wave produced by the speakers of your stereo. If two sound waves did *not* combine in this simple linear way, the sound you would hear in this situation would be a hopeless jumble. Superposition also applies to electromagnetic waves (such as light) and many other types of waves.

Test Your Understanding of Section 15.6 Figure 15.22 shows two wave pulses with different shapes traveling in different directions along a string. Make a series of sketches like Fig. 15.21 showing the shape of the string as the two pulses approach, overlap, and then pass each other.

15.22 Two wave pulses with different shapes.

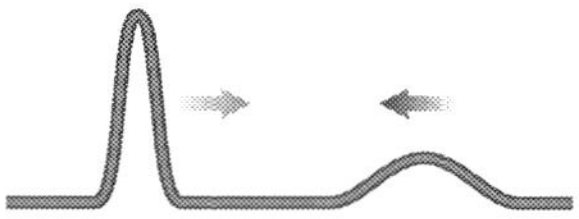

15.7 Standing Waves on a String

We have talked about the reflection of a wave *pulse* on a string when it arrives at a boundary point (either a fixed end or a free end). Now let's look at what happens when a *sinusoidal* wave is reflected by a fixed end of a string. We'll again approach the problem by considering the superposition of two waves propagating through the string, one representing the original or incident wave and the other representing the wave reflected at the fixed end.

Figure 15.23 shows a string that is fixed at its left end. Its right end is moved up and down in simple harmonic motion to produce a wave that travels to the left; the wave reflected from the fixed end travels to the right. The resulting motion when the two waves combine no longer looks like two waves traveling in opposite directions. The string appears to be subdivided into a number of segments, as in

15.23 (a)–(d) Time exposures of standing waves in a stretched string. From (a) to (d), the frequency of oscillation of the right-hand end increases and the wavelength of the standing wave decreases. (e) The extremes of the motion of the standing wave in part (b), with nodes at the center and at the ends. The right-hand end of the string moves very little compared to the antinodes and so is essentially a node.

(a) String is one-half wavelength long.

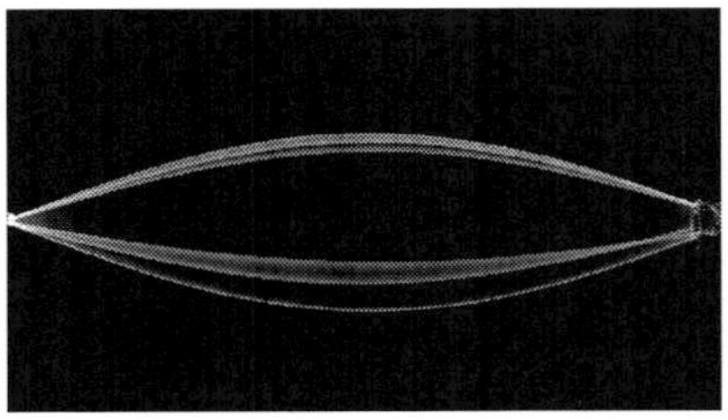

(b) String is one wavelength long.

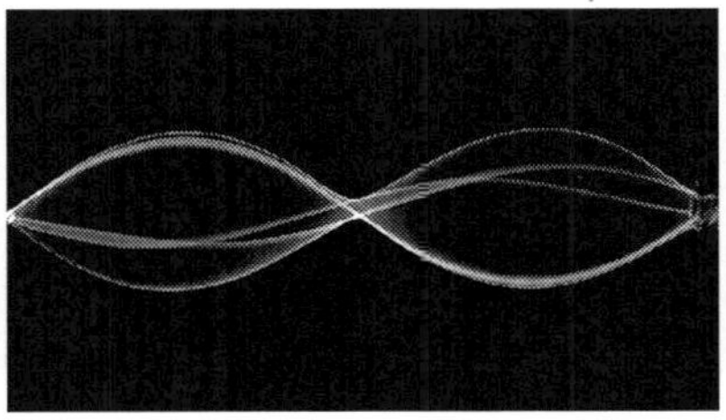

(c) String is one and a half wavelengths long.

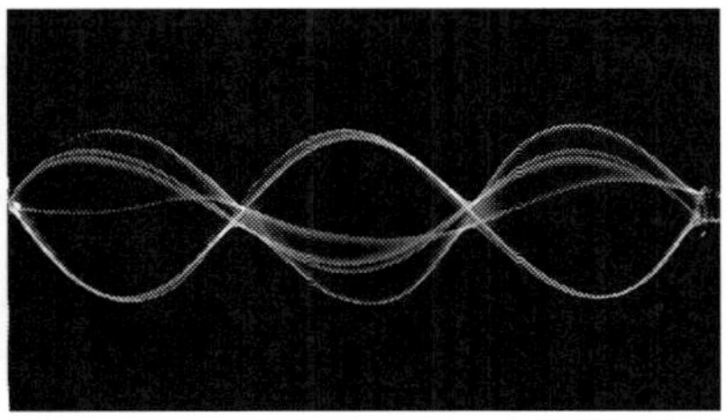

(d) String is two wavelengths long.

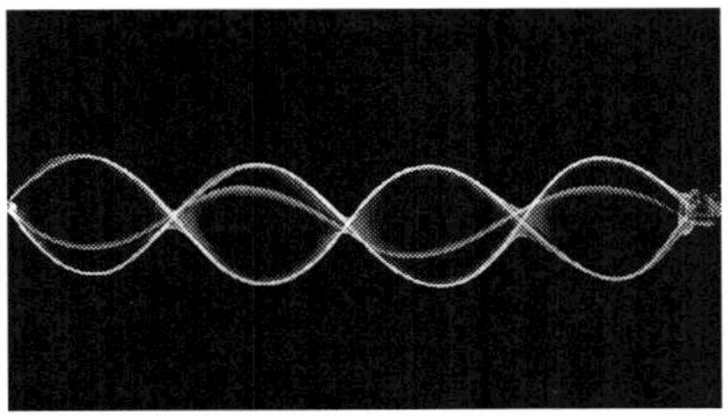

(e) The shape of the string in (b) at two different instants

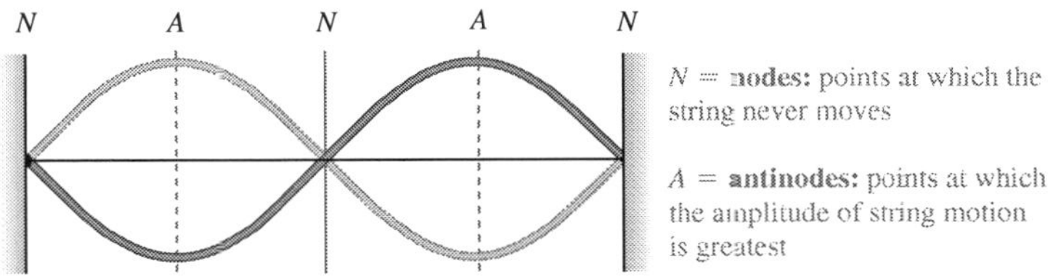

the time-exposure photographs of Figs. 15.23a, 15.23b, 15.23c, and 15.23d. Figure 15.23e shows two instantaneous shapes of the string in Fig. 15.23b. Let's compare this behavior with the waves we studied in Sections 15.1 through 15.5. In a wave that travels along the string, the amplitude is constant and the wave pattern moves with a speed equal to the wave speed. Here, instead, the wave pattern remains in the same position along the string and its amplitude fluctuates. There are particular points called **nodes** (labeled *N* in Fig. 15.23e) that never move at all. Midway between the nodes are points called **antinodes** (labeled *A* in Fig. 15.23e) where the amplitude of motion is greatest. Because the wave pattern doesn't appear to be moving in either direction along the string, it is called a **standing wave.** (To emphasize the difference, a wave that *does* move along the string is called a **traveling wave.**)

The principle of superposition explains how the incident and reflected waves combine to form a standing wave. In Fig. 15.24 the red curves show a wave traveling to the left. The blue curves show a wave traveling to the right with the same propagation speed, wavelength, and amplitude. The waves are shown at nine instants, $\frac{1}{16}$ of a period apart. At each point along the string, we add the displacements (the values of y) for the two separate waves; the result is the total wave on the string, shown in brown.

15.24 Formation of a standing wave. A wave traveling to the left (red curves) combines with a wave traveling to the right (blue curves) to form a standing wave (brown curves).

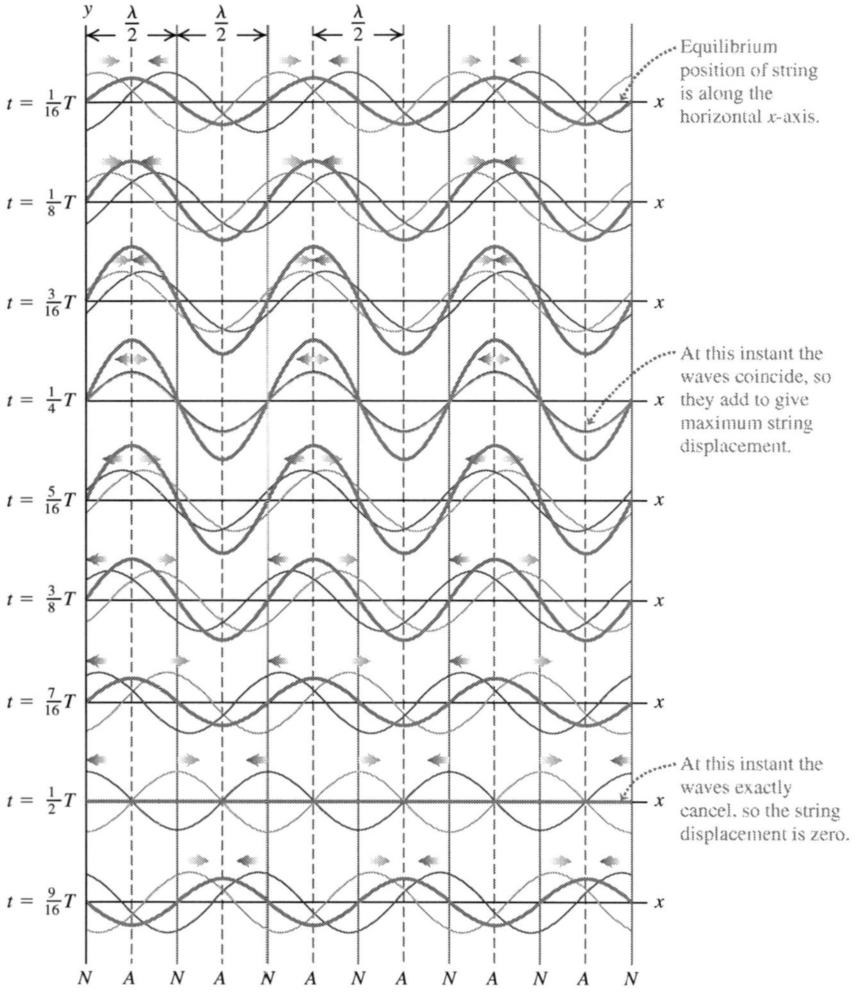

At certain instants, such as $t = \frac{1}{4}T$, the two wave patterns are exactly in phase with each other, and the shape of the string is a sine curve with twice the amplitude of either individual wave. At other instants, such as $t = \frac{1}{2}T$, the two waves are exactly out of phase with each other, and the total wave at that instant is zero. The resultant displacement is *always* zero at those places marked N at the bottom of Fig. 15.24. These are the *nodes.* At a node the displacements of the two waves in red and blue are always equal and opposite and cancel each other out. This cancellation is called **destructive interference.** Midway between the nodes are the points of *greatest* amplitude, or the *antinodes,* marked A. At the antinodes the displacements of the two waves in red and blue are always identical, giving a large resultant displacement; this phenomenon is called **constructive interference.** We can see from the figure that the distance between successive nodes or between successive antinodes is one half-wavelength, or $\lambda/2$.

We can derive a wave function for the standing wave of Fig. 15.24 by adding the wave functions $y_1(x, t)$ and $y_2(x, t)$ for two waves with equal amplitude, period, and wavelength traveling in opposite directions. Here $y_1(x, t)$ (the red curves in Fig. 15.24) represents an incoming, or *incident,* wave traveling to the left along the $+x$-axis, arriving at the point $x = 0$ and being reflected; $y_2(x, t)$ (the blue curves in Fig. 15.24) represents the *reflected* wave traveling to the right from $x = 0$. We noted in Section 15.6 that the wave reflected from a fixed end of a string is inverted, so we give a negative sign to one of the waves:

$$y_1(x, t) = -A\cos(kx + \omega t) \qquad \text{(incident wave traveling to the left)}$$
$$y_2(x, t) = A\cos(kx - \omega t) \qquad \text{(reflected wave traveling to the right)}$$

Note also that the change in sign corresponds to a shift in *phase* of 180° or π radians. At $x = 0$ the motion from the reflected wave is $A\cos\omega t$ and the motion from the incident wave is $-A\cos\omega t$, which we can also write as $A\cos(\omega t + \pi)$. From Eq. (15.27), the wave function for the standing wave is the sum of the individual wave functions:

$$y(x, t) = y_1(x, t) + y_2(x, t) = A[-\cos(kx + \omega t) + \cos(kx - \omega t)]$$

We can rewrite each of the cosine terms by using the identities for the cosine of the sum and difference of two angles: $\cos(a \pm b) = \cos a\cos b \mp \sin a\sin b$. Applying these and combining terms, we obtain the wave function for the standing wave:

$$y(x, t) = y_1(x, t) + y_2(x, t) = (2A\sin kx)\sin\omega t \qquad \text{or}$$

$$y(x, t) = (A_{SW}\sin kx)\sin\omega t \qquad \text{(standing wave on a string, fixed end at } x = 0) \qquad (15.28)$$

The standing wave amplitude A_{SW} is twice the amplitude A of either of the original traveling waves:

$$A_{SW} = 2A$$

Equation (15.28) has two factors: a function of x and a function of t. The factor $A_{SW}\sin kx$ shows that at each instant the shape of the string is a sine curve. But unlike a wave traveling along a string, the wave shape stays in the same position, oscillating up and down as described by the $\sin\omega t$ factor. This behavior is shown graphically by the brown curves in Fig. 15.24. Each point in the string still undergoes simple harmonic motion, but all the points between any successive pair of nodes oscillate *in phase.* This is in contrast to the phase differences between oscillations of adjacent points that we see with a wave traveling in one direction.

We can use Eq. (15.28) to find the positions of the nodes; these are the points for which $\sin kx = 0$, so the displacement is *always* zero. This occurs when $kx = 0, \pi, 2\pi, 3\pi, \ldots$, or, using $k = 2\pi/\lambda$,

$$x = 0, \frac{\pi}{k}, \frac{2\pi}{k}, \frac{3\pi}{k}, \ldots = 0, \frac{\lambda}{2}, \frac{2\lambda}{2}, \frac{3\lambda}{2}, \ldots \quad \text{(nodes of a standing wave on a string, fixed end at } x = 0) \tag{15.29}$$

In particular, there is a node at $x = 0$, as there should be, since this point is a fixed end of the string.

A standing wave, unlike a traveling wave, *does not* transfer energy from one end to the other. The two waves that form it would individually carry equal amounts of power in opposite directions. There is a local flow of energy from each node to the adjacent antinodes and back, but the *average* rate of energy transfer is zero at every point. If you evaluate the wave power given by Eq. (15.21) using the wave function of Eq. (15.28), you will find that the average power is zero (see Challenge Problem 15.84).

Problem-Solving Strategy 15.2 Standing Waves

MP

IDENTIFY *the relevant concepts:* As with traveling waves, it's useful to distinguish between the purely kinematic quantities, such as wave speed v, wavelength λ, and frequency f, and the dynamic quantities involving the properties of the medium, such as F and μ for transverse waves on a string. Once you decide what the target variable is, try to determine whether the problem is only kinematic in nature or whether the properties of the medium are also involved.

SET UP *the problem* using the following steps:

1. To visualize nodes and antinodes in standing waves, it is always helpful to draw diagrams. For a string you can draw the shape at one instant and label the nodes N and antinodes A. The distance between two adjacent nodes or two adjacent antinodes is always $\lambda/2$, and the distance between a node and the adjacent antinode is always $\lambda/4$.
2. Decide which equations you'll need to use. The wave function for the standing wave, like Eq. (15.28), is almost always useful.
3. You can compute the wave speed if you know either λ and f (or, equivalently, $k = 2\pi/\lambda$ and $\omega = 2\pi f$) or the properties of the medium (for a string, F and μ).

EXECUTE *the solution* as follows: Solve for the unknown quantities using the equations you've selected. Once you have the wave function, you can find the value of the displacement y at any point in the wave medium (value of x) and at any time. You can find the velocity of a particle in the wave medium by taking the partial derivative of y with respect to time. To find the acceleration of such a particle, take the second partial derivative of y with respect to time.

EVALUATE *your answer:* Compare your numerical answers with your diagram. Check that the wave function is compatible with the boundary conditions (for example, the displacement should be zero at a fixed end).

Example 15.6 Standing waves on a guitar string

One of the strings of a guitar lies along the x-axis when in equilibrium. The end of the string at $x = 0$ (the bridge of the guitar) is tied down. An incident sinusoidal wave, corresponding to the red curves in Fig. 15.24, travels along the string in the $-x$-direction at 143 m/s with an amplitude of 0.750 mm and a frequency of 440 Hz. This wave is reflected from the fixed end at $x = 0$, and the superposition of the incident traveling wave and the reflected traveling wave forms a standing wave. (a) Find the equation giving the displacement of a point on the string as a function of position and time. (b) Locate the points on the string that don't move at all. (c) Find the amplitude, maximum transverse velocity, and maximum transverse acceleration at the points of maximum oscillation.

SOLUTION

IDENTIFY: This is a *kinematics* problem in which we are asked to describe the motion of the string (see Problem-Solving Strategy 15.1 in Section 15.3). The target variables are the wave function of the standing wave in part (a), the locations of the points that don't move, or the *nodes* in part (b), and the maximum values of displacement y, transverse velocity v_y, and transverse acceleration a_y. in part (c). (Waves on a string are transverse waves, so *transverse* means "in the direction of the displacement"—that is, in the y-direction.) To find these quantities we use the expression that we derived in this section for a standing wave on a string with a fixed end, as well as other relationships from Sections 15.2 and 15.3.

SET UP: Since there is a fixed end at $x = 0$, we may use Eqs. (15.28) and (15.29) to describe this standing wave. We also use the relationships among ω, k, f, λ, and the wave speed v.

EXECUTE: (a) To use Eq. (15.28) we need the values of A_{SW}, ω, and k. The amplitude of the incident wave is $A = 0.750\text{ mm} =$

7.50×10^{-4} m; the reflected wave has the same amplitude, and the standing wave amplitude is $A_{SW} = 2A = 1.50 \times 10^{-3}$ m. The angular frequency ω and wave number k are

$$\omega = 2\pi f = (2\pi \text{ rad})(440 \text{ s}^{-1}) = 2760 \text{ rad/s}$$

$$k = \frac{\omega}{v} = \frac{2760 \text{ rad/s}}{143 \text{ m/s}} = 19.3 \text{ rad/m}$$

Then Eq. (15.28) gives

$$y(x, t) = (A_{SW} \sin kx) \sin \omega t$$
$$= [(1.50 \times 10^{-3} \text{ m}) \sin(19.3 \text{ rad/m})x] \sin(2760 \text{ rad/s})t$$

(b) The positions of the nodes are given by Eq. (15.29): $x = 0$, $\lambda/2$, λ, $3\lambda/2$, ... The wavelength is

$$\lambda = \frac{v}{f} = \frac{143 \text{ m/s}}{440 \text{ Hz}} = 0.325 \text{ m}$$

so the nodes are at the following distances from the fixed end:

$$x = 0, 0.163 \text{ m}, 0.325 \text{ m}, 0.488 \text{ m}, \ldots$$

(c) From the expression in part (a) for $y(x, t)$, we see that the maximum displacement from equilibrium is 1.50×10^{-3} m = 1.50 mm, which is just twice the amplitude of the incident wave. This maximum occurs at the *antinodes,* which are midway between adjacent nodes (that is, at $x = 0.081$ m, 0.244 m, 0.406 m, ...).

For a particle on the string at any point x, the transverse (y-) velocity is

$$v_y(x, t) = \frac{\partial y(x, t)}{\partial t}$$
$$= [(1.50 \times 10^{-3} \text{ m}) \sin(19.3 \text{ rad/m})x] \times [(2760 \text{ rad/s}) \cos(2760 \text{ rad/s})t]$$
$$= [(4.15 \text{ m/s}) \sin(19.3 \text{ rad/m})x] \cos(2760 \text{ rad/s})t$$

At an antinode, $\sin(19.3 \text{ rad/m})x = \pm 1$ and the transverse velocity varies in value between 4.15 m/s and −4.15 m/s. As is always the case in simple harmonic motion, the maximum velocity occurs when the particle is passing through the equilibrium position ($y = 0$).

The transverse acceleration $a_y(x, t)$ is the first partial derivative of $v_y(x, t)$ with respect to time (that is, the *second* partial derivative of $y(x, t)$ with respect to time). We leave the calculation to you; the result is

$$a_y(x, t) = \frac{\partial v_y(x, t)}{\partial t} = \frac{\partial^2 y(x, t)}{\partial t^2}$$
$$= [(-1.15 \times 10^4 \text{ m/s}^2) \sin(19.3 \text{ rad/m})x] \times \sin(2760 \text{ rad/s})t$$

At the antinodes, the transverse acceleration varies in value between $+1.15 \times 10^4 \text{ m/s}^2$ and $-1.15 \times 10^4 \text{ m/s}^2$.

EVALUATE: The maximum transverse velocity at an antinode is quite respectable (about 15 km/h, or 9.3 mi/h). But the maximum transverse acceleration is tremendous, 1170 times the acceleration due to gravity! Guitar strings are made of sturdy stuff to be able to withstand such acceleration.

Guitar strings are actually tied down at *both* ends. We'll see the consequences of this in the next section.

Test Your Understanding of Section 15.7 Suppose the frequency of the standing wave in Example 15.6 were doubled from 440 Hz to 880 Hz. Would all of the nodes for $f = 440$ Hz also be nodes for $f = 880$ Hz? If so, would there be additional nodes for $f = 880$ Hz? If not, which nodes are absent for $f = 880$ Hz?

15.8 Normal Modes of a String

When we described standing waves on a string rigidly held at one end, as in Fig. 15.23, we made no assumptions about the length of the string or about what was happening at the other end. Let's now consider a string of a definite length L, rigidly held at *both* ends. Such strings are found in many musical instruments, including pianos, violins, and guitars. When a guitar string is plucked, a wave is produced in the string; this wave is reflected and re-reflected from the ends of the string, making a standing wave. This standing wave on the string in turn produces a sound wave in the air, with a frequency determined by the properties of the string. This is what makes stringed instruments so useful in making music.

To understand these properties of standing waves on a string fixed at both ends, let's first examine what happens when we set up a sinusoidal wave on such a string. The standing wave that results must have a node at *both* ends of the string. We saw in the preceding section that adjacent nodes are one half-wavelength ($\lambda/2$) apart, so the length of the string must be $\lambda/2$, or $2(\lambda/2)$, or $3(\lambda/2)$, or in general some integer number of half-wavelengths:

$$L = n\frac{\lambda}{2} \quad (n = 1, 2, 3, \ldots) \quad \text{(string fixed at both ends)} \quad (15.30)$$

ActivPhysics ONLINE

10.4 Standing Waves on Strings
10.5 Tuning a Stringed Instrument: Standing Waves
10.6 String Mass and Standing Waves

15.25 Each string of a violin naturally oscillates at one or more of its harmonic frequencies, producing sound waves in the air with the same frequencies.

That is, if a string with length L is fixed at both ends, a standing wave can exist only if its wavelength satisfies Eq. (15.30).

Solving this equation for λ and labeling the possible values of λ as λ_n, we find

$$\lambda_n = \frac{2L}{n} \quad (n = 1, 2, 3, \dots) \qquad \text{(string fixed at both ends)} \qquad (15.31)$$

Waves can exist on the string if the wavelength is *not* equal to one of these values, but there cannot be a steady wave pattern with nodes and antinodes, and the total wave cannot be a standing wave. Equation (15.31) is illustrated by the standing waves shown in Figs. 15.23a, 15.23b, 15.23c, and 15.23d; these represent $n = 1, 2, 3$, and 4, respectively.

Corresponding to the series of possible standing-wave wavelengths λ_n is a series of possible standing-wave frequencies f_n, each related to its corresponding wavelength by $f_n = v/\lambda_n$. The smallest frequency f_1 corresponds to the largest wavelength (the $n = 1$ case), $\lambda_1 = 2L$:

$$f_1 = \frac{v}{2L} \qquad \text{(string fixed at both ends)} \qquad (15.32)$$

This is called the **fundamental frequency.** The other standing-wave frequencies are $f_2 = 2v/2L$, $f_3 = 3v/2L$, and so on. These are all integer multiples of the fundamental frequency f_1, such as $2f_1$, $3f_1$, $4f_1$, and so on, and we can express *all* the frequencies as

$$f_n = n\frac{v}{2L} = nf_1 \quad (n = 1, 2, 3, \dots) \qquad \text{(string fixed at both ends)} \qquad (15.33)$$

These frequencies are called **harmonics,** and the series is called a **harmonic series.** Musicians sometimes call f_2, f_3, and so on **overtones;** f_2 is the second harmonic or the first overtone, f_3 is the third harmonic or the second overtone, and so on. The first harmonic is the same as the fundamental frequency (Fig. 15.25).

For a string with fixed ends at $x = 0$ and $x = L$, the wave function $y(x, t)$ of the nth standing wave is given by Eq. (15.28) (which satisfies the condition that there is a node at $x = 0$), with $\omega = \omega_n = 2\pi f_n$ and $k = k_n = 2\pi/\lambda_n$:

$$y_n(x, t) = A_{SW} \sin k_n x \sin \omega_n t \qquad (15.34)$$

You can easily show that this wave function has nodes at both $x = 0$ and $x = L$, as it must.

A **normal mode** of an oscillating system is a motion in which all particles of the system move sinusoidally with the same frequency. For a system made up of a string of length L fixed at both ends, each of the wavelengths given by Eq. (15.31) corresponds to a possible normal-mode pattern and frequency. There are infinitely many normal modes, each with its characteristic frequency and vibration pattern. Figure 15.26 shows the first four normal-mode patterns and their associ-

15.26 The first four normal modes of a string fixed at both ends. (Compare these to the photographs in Fig. 15.23.)

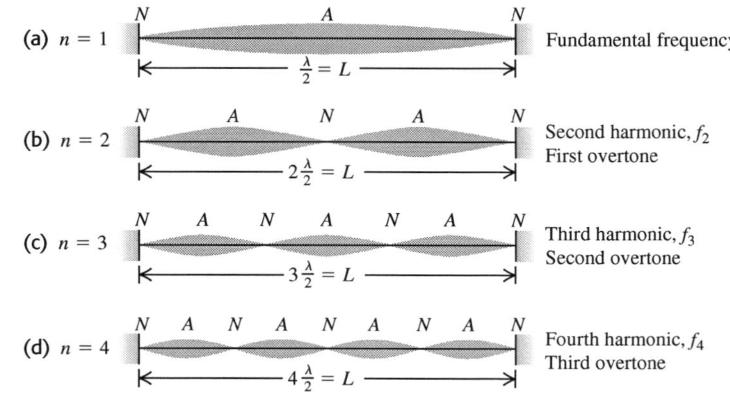

ated frequencies and wavelengths; these correspond to Eq. (15.34) with $n = 1, 2, 3$, and 4. By contrast, a harmonic oscillator, which has only one oscillating particle, has only one normal mode and one characteristic frequency. The string fixed at both ends has infinitely many normal modes because it is made up of a very large (effectively infinite) number of particles. More complicated oscillating systems also have infinite numbers of normal modes, though with more complex normal-mode patterns than a string (Fig. 15.27).

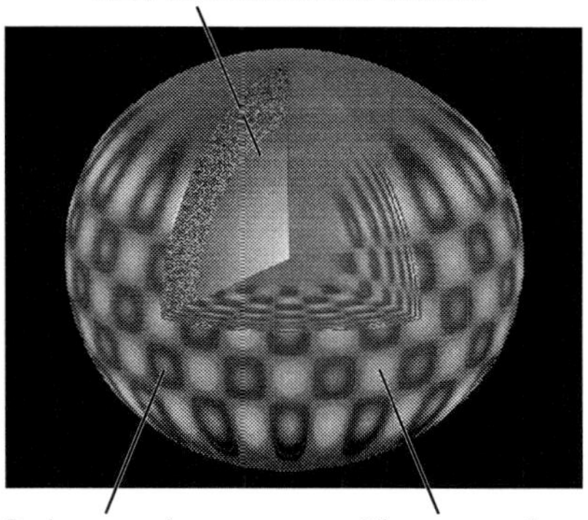

15.27 Astronomers have discovered that the sun oscillates in several different normal modes. This computer simulation shows one such mode.

Complex Standing Waves

If we could displace a string so that its shape is the same as one of the normal-mode patterns and then release it, it would vibrate with the frequency of that mode. Such a vibrating string would displace the surrounding air with the same frequency, producing a traveling sinusoidal sound wave that your ears would perceive as a pure tone. But when a string is struck (as in a piano) or plucked (as is done to guitar strings), the shape of the displaced string is *not* as simple as one of the patterns in Fig. 15.26. The fundamental as well as many overtones are present in the resulting vibration. This motion is therefore a combination or *superposition* of many normal modes. Several simple-harmonic motions of different frequencies are present simultaneously, and the displacement of any point on the string is the sum (or superposition) of displacements associated with the individual modes. The sound produced by the vibrating string is likewise a superposition of traveling sinusoidal sound waves, which you perceive as a rich, complex tone with the fundamental frequency f_1. The standing wave on the string and the traveling sound wave in the air have similar **harmonic content** (the extent to which frequencies higher than the fundamental are present). The harmonic content depends on how the string is initially set into motion. If you pluck the strings of an acoustic guitar in the normal location over the sound hole, the sound that you hear has a different harmonic content than if you pluck the strings next to the fixed end on the guitar body.

ActivPhysics Online
10.10 Complex Waves: Fourier Analysis

It is possible to represent every possible motion of the string as some superposition of normal-mode motions. Finding this representation for a given vibration pattern is called *harmonic analysis.* The sum of sinusoidal functions that represents a complex wave is called a *Fourier series.* Figure 15.28 shows how a standing wave that is produced by plucking a guitar string of length L at a point $L/4$ from one end can be represented as a combination of sinusoidal functions.

15.28 When a guitar string is plucked (pulled into a triangular shape) and released, a standing wave results. The standing wave is well represented (except at the sharp maximum point) by the sum of just three sinusoidal functions. Including additional sinusoidal functions further improves the representation.

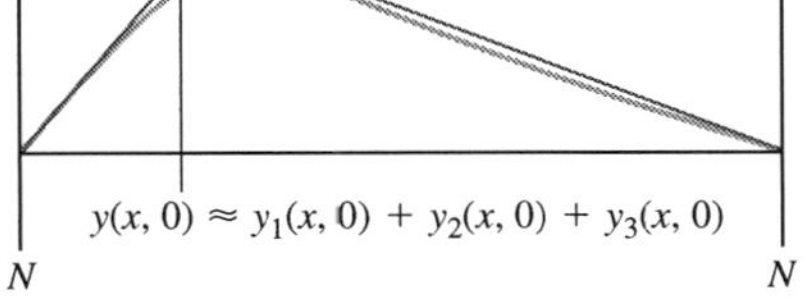

Standing Waves and String Instruments

As we have seen, the fundamental frequency of a vibrating string is $f_1 = v/2L$. The speed v of waves on the string is determined by Eq. (15.13), $v = \sqrt{F/\mu}$. Combining these equations, we find

$$f_1 = \frac{1}{2L}\sqrt{\frac{F}{\mu}} \quad \text{(string fixed at both ends)} \qquad (15.35)$$

This is also the fundamental frequency of the sound wave created in the surrounding air by the vibrating string. Familiar musical instruments show how f_1 depends on the properties of the string. The inverse dependence of frequency on length L is illustrated by the long strings of the bass (low-frequency) section of the piano or the bass viol compared with the shorter strings on the treble section of the piano or the violin (Fig. 15.29). The pitch of a violin or guitar is usually varied by pressing a string against the fingerboard with the fingers to change the length L of the vibrating portion of the string. Increasing the tension F increases the wave speed v and thus increases the frequency (and the pitch). All string instruments are "tuned" to the correct frequencies by varying the tension; you tighten the string to raise the pitch. Finally, increasing the mass per unit length μ decreases the wave

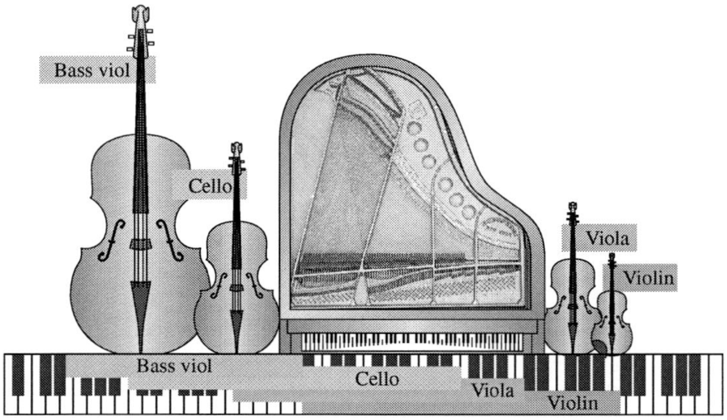

15.29 Comparing the range of a concert grand piano to the ranges of a bass viol, a cello, a viola, and a violin. In all cases, longer strings produce bass notes and shorter strings produce treble notes.

speed and thus the frequency. The lower notes on a steel guitar are produced by thicker strings, and one reason for winding the bass strings of a piano with wire is to obtain the desired low frequency from a relatively short string.

Wind instruments such as saxophones and trombones also have normal modes. As for stringed instruments, the frequencies of these normal modes determine the pitch of the musical tones that these instruments produce. We'll discuss these instruments and many other aspects of sound in Chapter 16.

Example 15.7 A giant bass viol

In an effort to get your name in the *Guinness Book of World Records,* you set out to build a bass viol with strings that have a length of 5.00 m between fixed points. One string has a linear mass density of 40.0 g/m and a fundamental frequency of 20.0 Hz (the lowest frequency that the human ear can hear). Calculate (a) the tension of this string, (b) the frequency and wavelength on the string of the second harmonic, and (c) the frequency and wavelength on the string of the second overtone.

SOLUTION

IDENTIFY: The target variable in part (a) is the string tension; we find this from the expression for the fundamental frequency of the string, which involves the tension. In parts (b) and (c) the target variables are the frequency and wavelength of different harmonics. We determine these from the given length of the string and the fundamental frequency.

SET UP: For part (a), the equation to use is Eq. (15.35); it involves the known values of f_1, L, and μ as well as the target variable F. We solve parts (b) and (c) using Eqs. (15.31) and (15.33).

EXECUTE: (a) We solve Eq. (15.35) for the string tension F:

$$F = 4\mu L^2 f_1^2 = 4(40.0 \times 10^{-3}\ \text{kg/m})(5.00\ \text{m})^2(20.0\ \text{s}^{-1})^2$$
$$= 1600\ \text{N} = 360\ \text{lb}$$

(b) The second harmonic is denoted by $n = 2$. From Eq. (15.33), the second harmonic frequency is

$$f_2 = 2f_1 = 2(20.0\ \text{Hz}) = 40.0\ \text{Hz}$$

From Eq. (15.31), the wavelength on the string of the second harmonic is

$$\lambda_2 = \frac{2L}{2} = 5.00\ \text{m}$$

(c) The second overtone is the "second tone over" (above) the fundamental—that is, $n = 3$. Its frequency and wavelength are

$$f_3 = 3f_1 = 3(20.0\ \text{Hz}) = 60.0\ \text{Hz}$$
$$\lambda_3 = \frac{2L}{3} = 3.33\ \text{m}$$

EVALUATE: The tension in part (a) is a bit larger than in a real bass viol, for which the string tension is typically a few hundred newtons. The wavelengths in parts (b) and (c) are equal to the length of the string and two-thirds the length of the string, respectively; these results agree with the drawings of standing waves in Fig. 15.26.

Example 15.8 From waves on a string to sound waves in air

What are the frequency and wavelength of the sound waves produced in the air when the string in Example 15.7 is vibrating at its fundamental frequency? The speed of sound in air at 20°C is 344 m/s.

SOLUTION

IDENTIFY: Our target variables are f and λ for the sound wave produced by the bass viol, *not* for the standing wave on the string.

However, when the string vibrates at a particular frequency, the surrounding air is forced to vibrate at the same frequency. So the frequency of the sound wave is the same as that of the standing wave on the string. The relationship $\lambda = v/f$ shows that the *wavelength* of the sound wave is typically different from the wavelength of the standing wave on the string, because the two waves have different speeds.

SET UP: The only equation we need is $v = \lambda f$. We apply this to both the standing wave on the string (speed v_{string}) and the traveling sound wave (speed v_{sound}).

EXECUTE: The sound wave frequency is the same as the standing-wave fundamental frequency: $f = f_1 = 20.0$ Hz. The wavelength of the sound wave is

$$\lambda_{1(\text{sound})} = \frac{v_{\text{sound}}}{f_1} = \frac{344\ \text{m/s}}{20.0\ \text{Hz}} = 17.2\ \text{m}$$

EVALUATE: Note that $\lambda_{1(\text{sound})}$ is greater than the wavelength of the standing wave on the string, $\lambda_{1(\text{string})} = 2L = 2(5.00\ \text{m}) = 10.0$ m. This is because the speed of sound is greater than the speed of waves on the string, $v_{\text{string}} = \lambda_{1(\text{string})} f_1 = (10.0\ \text{m})(20.0\ \text{Hz}) = 200\ \text{m/s}$. Hence, for *any* normal mode on this string, the sound wave that is produced has the same frequency as the wave on the string but a wavelength that is greater by a factor of $v_{\text{sound}}/v_{\text{string}} = (344\ \text{m/s})/(200\ \text{m/s}) = 1.72$.

Test Your Understanding of Section 15.8 While a guitar string is vibrating, you gently touch the midpoint of the string to ensure that the string does not vibrate at that point. Which normal modes *cannot* be present on the string while you are touching it in this way?

CHAPTER 15 SUMMARY

Waves and their properties: A wave is any disturbance from an equilibrium condition that propagates from one region to another. A mechanical wave always travels within some material called the medium. The wave disturbance propagates at the wave speed v, which depends on the type of wave and the properties of the medium.

In a periodic wave, the motion of each point of the medium is periodic. A sinusoidal wave is a special periodic wave in which each point moves in simple harmonic motion. For any periodic wave, the frequency f is the number of cycles per unit time, the period T is the time for one cycle, the wavelength λ is the distance over which the wave pattern repeats, and the amplitude A is the maximum displacement of a particle in the medium. The product of λ and f equals the wave speed. (See Example 15.1.)

$$v = \lambda f \quad (15.1)$$

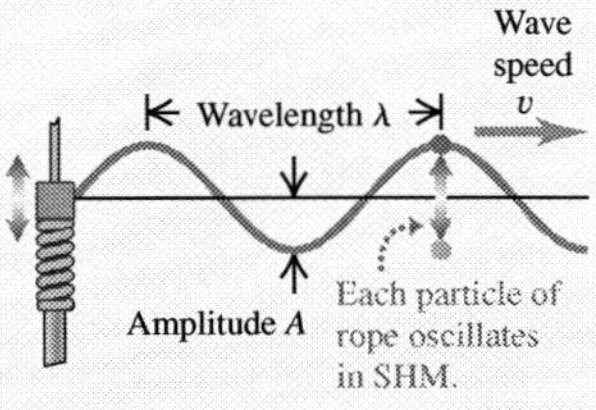

Wave functions and wave dynamics: The wave function $y(x, t)$ describes the displacements of individual particles in the medium. Equations (15.3), (15.4), and (15.7) give the wave equation for a sinusoidal wave traveling in the $+x$-direction. If the wave is moving in the $-x$-direction, the minus signs in the cosine functions are replaced by plus signs. (See Example 15.2.)

The wave function obeys a partial differential equation called the wave equation, Eq. (15.12).

The speed of transverse waves on a string depends on the tension F and mass per unit length μ. (See Example 15.3.)

$$y(x,t) = A\cos\left[\omega\left(\frac{x}{v} - t\right)\right] = A\cos 2\pi f\left(\frac{x}{v} - t\right) \quad (15.3)$$

$$y(x,t) = A\cos 2\pi\left(\frac{x}{\lambda} - \frac{t}{T}\right) \quad (15.4)$$

$$y(x,t) = A\cos(kx - \omega t) \quad (15.7)$$

where $k = 2\pi/\lambda$ and $\omega = 2\pi f = vk$

$$\frac{\partial^2 y(x,t)}{\partial x^2} = \frac{1}{v^2}\frac{\partial^2 y(x,t)}{\partial t^2} \quad (15.12)$$

$$v = \sqrt{\frac{F}{\mu}} \quad \text{(waves on a string)} \quad (15.13)$$

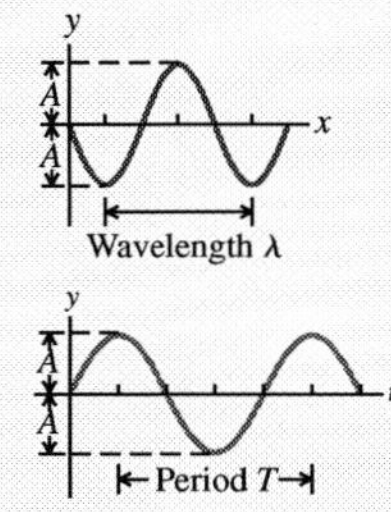

Wave power: Wave motion conveys energy from one region to another. For a sinusoidal mechanical wave, the average power P_{av} is proportional to the square of the wave amplitude and the square of the frequency. For waves that spread out in three dimensions, the wave intensity I is inversely proportional to the distance from the source. (See Examples 15.4 and 15.5.)

$$P_{av} = \frac{1}{2}\sqrt{\mu F}\,\omega^2 A^2 \quad (15.25)$$

(average power, sinusoidal wave)

$$\frac{I_1}{I_2} = \frac{r_2^2}{r_1^2} \quad (15.26)$$

(inverse-square law for intensity)

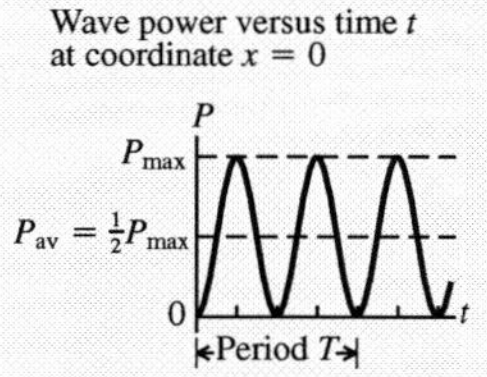

Wave superposition: A wave that reaches a boundary of the medium in which it propagates is reflected. The principle of superposition states that the total wave displacement at any point where two or more waves overlap is the sum of the displacements of the individual waves.

$$y(x,t) = y_1(x,t) + y_2(x,t) \quad (15.27)$$

(principle of superposition)

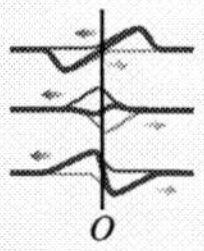

Standing waves on a string: When a sinusoidal wave is reflected from a fixed or free end of a stretched string, the incident and reflected waves combine to form a standing sinusoidal wave with nodes and antinodes. Adjacent nodes are spaced a distance $\lambda/2$ apart, as are adjacent antinodes. (See Example 15.6.)

When both ends of a string with length L are held fixed, standing waves can occur only when L is an integer multiple of $\lambda/2$. Each frequency with its associated vibration pattern is called a normal mode. The lowest frequency f_1 is called the fundamental frequency. (See Examples 15.7 and 15.8.)

$$y(x, t) = (A_{SW} \sin kx) \sin \omega t \quad (15.28)$$
(standing on a wave on a string, fixed end at $x = 0$)

$$f_n = n\frac{v}{2L} = nf_1 \ (n = 1, 2, 3, \dots) \quad (15.33)$$

$$f_1 = \frac{1}{2L}\sqrt{\frac{F}{\mu}} \quad (15.35)$$
(string fixed at both ends)

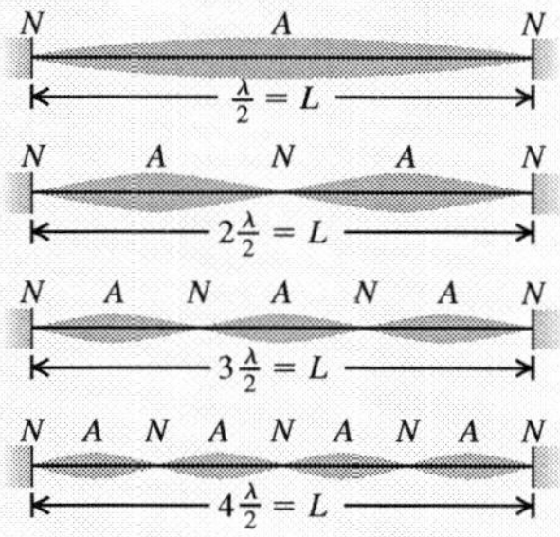

Key Terms

mechanical wave, *488*
medium, *488*
transverse wave, *488*
longitudinal wave, *488*
wave speed, *489*
periodic wave, *489*
sinusoidal wave, *489*
wavelength, *490*
wave function, *492*
wave number, *493*
phase, *494*
wave equation, *497*
intensity, *504*
interference, *505*
boundary condition, *506*
principle of superposition *506*
node, *508*
antinode, *508*
standing wave, *508*
traveling wave, *508*
destructive interference, *509*
constructive interference, *509*
fundamental frequency, *512*
harmonics, *512*
harmonic series, *512*
overtone, *512*
normal mode, *512*
harmonic content, *513*

Answer to Chapter Opening Question ?

The power of a mechanical wave depends on its frequency and amplitude [see Eq. (15.25)].

Answers to Test Your Understanding Questions

15.1 Answer: (i) The "wave" travels horizontally from one spectator to the next along each row of the stadium, but the displacement of each spectator is vertically upward. Since the displacement is perpendicular to the direction in which the wave travels, the wave is transverse.

15.2 Answer: (iv) The speed of waves on a string, v, does not depend on the wavelength. We can rewrite the relationship $v = \lambda f$ as $f = v/\lambda$, which tells us that if the wavelength λ doubles, the frequency f becomes one-half as great.

15.3 Answers: (a) $\frac{2}{8}T$, (b) $\frac{4}{8}T$, (c) $\frac{5}{8}T$ Since the wave is sinusoidal, each point on the string oscillates in simple harmonic motion (SHM). Hence we can apply all of the ideas from Chapter 13 about SHM to the wave depicted in Fig. 15.8. (a) A particle in SHM has its maximum speed when it is passing through the equilibrium position ($y = 0$ in Fig. 15.8). The particle at point A is moving upward through this position at $t = \frac{2}{8}T$. (b) In vertical SHM the greatest *upward* acceleration occurs when a particle is at its maximum *downward* displacement. This occurs for the particle at point B at $t = \frac{4}{8}T$. (c) A particle in vertical SHM has a *downward* acceleration when its displacement is *upward*. The particle at C has an upward displacement and is moving downward at $t = \frac{5}{8}T$.

15.4 Answer: (ii) The relationship $v = \sqrt{F/\mu}$ [Eq. (15.13)] says that the wave speed is greatest on the string with the smallest linear mass density. This is the thinnest string, which has the smallest amount of mass m and hence the smallest linear mass density $\mu = m/L$ (all strings are the same length).

15.5 Answer: (iii), (iv), (ii), (i) Equation (15.25) says that the average power in a sinusoidal wave on a string is $P_{av} = \frac{1}{2}\sqrt{\mu F}\omega^2 A^2$. All four strings are identical, so all have the same mass, the same length, and the same linear mass density μ. The frequency f is the same for each wave, as is the angular frequency $\omega = 2\pi f$. Hence the average wave power for each string is proportional to the square root of the string tension F and the square of the amplitude A. Compared to string (i), the average power in each string is (ii) $\sqrt{4} = 2$ times greater; (ii) $4^2 = 16$ times greater; and (iv) $\sqrt{2}(2)^2 = 4\sqrt{2}$ times greater.

15.6 Answer:

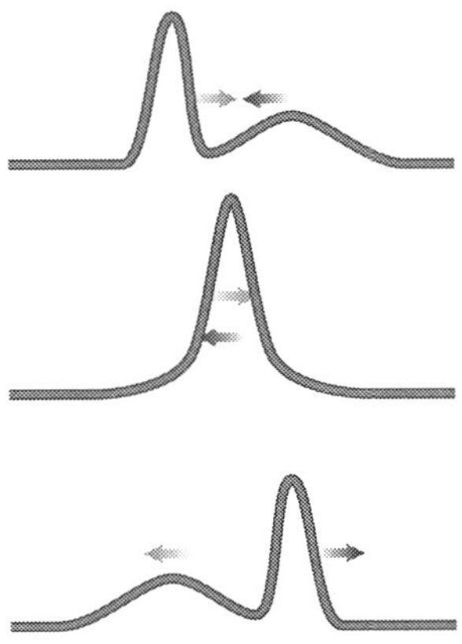

15.7 Answers: yes, yes Doubling the frequency makes the wavelength half as large. Hence the spacing between nodes (equal to $\lambda/2$) is also half as large. There are nodes at all of the previous positions, but there is also a new node between every pair of old nodes.

15.8 Answers: $n = 1, 3, 5, \ldots$ When you touch the string at its center, you are demanding that there be a node at the center. Hence only standing waves with a node at $x = L/2$ are allowed. From Figure 15.26 you can see that the normal modes $n = 1, 3, 5, \ldots$ cannot be present.

PROBLEMS

For instructor-assigned homework, go to www.masteringphysics.com

Discussion Questions

Q15.1. Two waves travel on the same string. Is it possible for them to have (a) different frequencies; (b) different wavelengths; (c) different speeds; (d) different amplitudes; (e) the same frequency but different wavelengths? Explain your reasoning.

Q15.2. Under a tension F, it takes 2.00 s for a pulse to travel the length of a taut wire. What tension is required (in terms of F) for the pulse to take 6.00 s instead?

Q15.3. What kinds of energy are associated with waves on a stretched string? How could you detect such energy experimentally?

Q15.4. The amplitude of a wave decreases gradually as the wave travels down a long, stretched string. What happens to the energy of the wave when this happens?

Q15.5. For the wave motions discussed in this chapter, does the speed of propagation depend on the amplitude? What makes you say this?

Q15.6. The speed of ocean waves depends on the depth of the water; the deeper the water, the faster the wave travels. Use this to explain why ocean waves crest and "break" as they near the shore.

Q15.7. Is it possible to have a longitudinal wave on a stretched string? Why or why not? Is it possible to have a transverse wave on a steel rod? Again, why or why not? If your answer is yes in either case, explain how you would create such a wave.

Q15.8. An echo is sound reflected from a distant object, such as a wall or a cliff. Explain how you can determine how far away the object is by timing the echo.

Q15.9. Why do you see lightning before you hear the thunder? A familiar rule of thumb is to start counting slowly, once per second, when you see the lightning; when you hear the thunder, divide the number you have reached by 3 to obtain your distance from the lightning in kilometers (or divide by 5 to obtain your distance in miles). Why does this work, or does it?

Q15.10. For transverse waves on a string, is the wave speed the same as the speed of any part of the string? Explain the difference between these two speeds. Which one is constant?

Q15.11. Children make toy telephones by sticking each end of a long string through a hole in the bottom of a paper cup and knotting it so it will not pull out. When the spring is pulled taut, sound can be transmitted from one cup to the other. How does this work? Why is the transmitted sound louder than the sound traveling through air for the same distance?

Q15.12. The four strings on a violin have different thicknesses, but are all under approximately the same tension. Do waves travel faster on the thick strings or the thin strings? Why? How does the fundamental vibration frequency compare for the thick versus the thin strings?

Q15.13. A sinusoidal wave can be described by a cosine function, which is negative just as often as positive. So why isn't the average power delivered by this wave zero?

Q15.14. Two strings of different mass per unit length μ_1 and μ_2 are tied together and stretched with a tension F. A wave travels along the string and passes the discontinuity in μ. Which of the following wave properties will be the same on both sides of the discontinuity, and which ones will change? speed of the wave; frequency; wavelength. Explain the physical reasoning behind each of your answers.

Q15.15. A long rope with mass m is suspended from the ceiling and hangs vertically. A wave pulse is produced at the lower end of the rope, and the pulse travels up the rope. Does the speed of the wave pulse change as it moves up the rope, and if so, does it increase or decrease?

Q15.16. In a transverse wave on a string, the motion of the string is perpendicular to the length of the string. How, then, is it possible for energy to move along the length of the string?

Q15.17. Both wave intensity and gravitation obey inverse-square laws. Do they do so for the same reason? Discuss the reason for each of these inverse-square laws as well as you can.

Q15.18. Energy can be transferred along a string by wave motion. However, in a standing wave on a string, no energy can ever be transferred past a node. Why not?

Q15.19. Can a standing wave be produced on a string by superposing two waves traveling in opposite directions with the same frequency but different amplitudes? Why or why not? Can a standing wave be produced by superposing two waves traveling in opposite directions with different frequencies but the same amplitude? Why or why not?

Q15.20. If you stretch a rubber band and pluck it, you hear a (somewhat) musical tone. How does the frequency of this tone change as you stretch the rubber band further? (Try it!) Does this agree with Eq. (15.35) for a string fixed at both ends? Explain.

Q15.21. A musical interval of an *octave* corresponds to a factor of 2 in frequency. By what factor must the tension in a guitar or violin string be increased to raise its pitch one octave? To raise it two octaves? Explain your reasoning. Is there any danger in attempting these changes in pitch?

Q15.22. By touching a string lightly at its center while bowing, a violinist can produce a note exactly one octave above the note to which the string is tuned—that is, a note with exactly twice the frequency. Why is this possible?

Q15.23. As we discussed in Section 15.1, water waves are a combination of longitudinal and transverse waves. Defend the following statement: "When water waves hit a vertical wall, the wall is a node of the longitudinal displacement but an antinode of the transverse displacement."

Q15.24. Violins are short instruments, while cellos and basses are long. In terms of the frequency of the waves they produce, explain why this is so.

Q15.25. What is the purpose of the frets on a guitar? In terms of the frequency of the vibration of the strings, explain their use.

Exercises

Section 15.2 Periodic Waves

15.1. The speed of sound in air at 20°C is 344 m/s. (a) What is the wavelength of a sound wave with a frequency of 784 Hz, corresponding to the note G_5 on a piano, and how many milliseconds does each vibration take? (b) What is the wavelength of a sound wave one octave higher than the note in part (a)?

15.2. Audible Sound. Provided the amplitude is sufficiently great, the human ear can respond to longitudinal waves over a range of frequencies from about 20.0 Hz to about 20.0 kHz. (a) If you were to mark the beginning of each complete wave pattern with a red dot for the long-wavelength sound and a blue dot for the short-wavelength sound, how far apart would the red dots be, and how far apart would the blue dots be? (b) In reality would adjacent dots in each set be far enough apart for you to easily measure their separation with a meterstick? (c) Suppose you repeated part (a) in water, where sound travels at 1480 m/s. How far apart would the dots be in each set? Could you readily measure their separation with a meterstick?

15.3. Tsunami! On December 26, 2004, a great earthquake occurred off the coast of Sumatra and triggered immense waves (tsunami) that killed some 200,000 people. Satellites observing these waves from space measured 800 km from one wave crest to the next and a period between waves of 1.0 hour. What was the speed of these waves in m/s and km/h? Does your answer help you understand why the waves caused such devastation?

15.4. Ultrasound Imaging. Sound having frequencies above the range of human hearing (about 20,000 Hz) is called *ultrasound.* Waves above this frequency can be used to penetrate the body and to produce images by reflecting from surfaces. In a typical ultrasound scan, the waves travel through body tissue with a speed of 1500 m/s. For a good, detailed image, the wavelength should be no more than 1.0 mm. What frequency sound is required for a good scan?

15.5. Visible Light. Light is a wave, but not a mechanical wave. The quantities that oscillate are electric and magnetic fields. Light visible to humans has wavelengths between 400 nm (violet) and 700 nm (red), and all light travels through vacuum at speed $c = 3.00 \times 10^8$ m/s. (a) What are the limits of the frequency and period of visible light? (b) Could you time a single light vibration with a stopwatch?

Section 15.3 Mathematical Description of a Wave

15.6. A certain transverse wave is described by

$$y(x,t) = (6.50\ \text{mm})\cos 2\pi\left(\frac{x}{28.0\ \text{cm}} - \frac{t}{0.0360\ \text{s}}\right)$$

Determine the wave's (a) amplitude; (b) wavelength; (c) frequency; (d) speed of propagation; (e) direction of propagation.

15.7. Transverse waves on a string have wave speed 8.00 m/s, amplitude 0.0700 m, and wavelength 0.320 m. The waves travel in the $-x$-direction, and at $t = 0$ the $x = 0$ end of the string has its maximum upward displacement. (a) Find the frequency, period, and wave number of these waves. (b) Write a wave function describing the wave. (c) Find the transverse displacement of a particle at $x = 0.360$ m at time $t = 0.150$ s. (d) How much time must elapse from the instant in part (c) until the particle at $x = 0.360$ m next has maximum upward displacement?

15.8. A water wave traveling in a straight line on a lake is described by the equation

$$y(x,t) = (3.75\ \text{cm})\cos(0.450\ \text{cm}^{-1}x + 5.40\ \text{s}^{-1}t)$$

where y is the displacement perpendicular to the undisturbed surface of the lake. (a) How much time does it take for one complete wave pattern to go past a fisherman in a boat at anchor, and what horizontal distance does the wave crest travel in that time? (b) What are the wave number and the number of waves per second that pass the fisherman? (c) How fast does a wave crest travel past the fisherman, and what is the maximum speed of his cork floater as the wave causes it to bob up and down?

15.9. Which of the following wave functions satisfies the wave equation, Eq. (15.12)? (a) $y(x,t) = A\cos(kx + \omega t)$; (b) $y(x,t) = A\sin(kx + \omega t)$; (c) $y(x,t) = A(\cos kx + \cos\omega t)$. (d) For the wave of part (b), write the equations for the transverse velocity and transverse acceleration of a particle at point x.

15.10. A wave on a string is described by $y(x,t) = A\cos(kx - \omega t)$. (a) Graph y, v_y, and a_y as functions of x for time $t = 0$. (b) Consider the following points on the string: (i) $x = 0$; (ii) $x = \pi/4k$; (iii) $x = \pi/2k$; (iv) $x = 3\pi/4k$; (v) $x = \pi/k$; (vi) $x = 5\pi/4k$; (vii) $x = 3\pi/2k$; (viii) $x = 7\pi/4k$. For a particle at each of these points at $t = 0$, describe in words whether the particle is moving and in what direction, and whether the particle is speeding up, slowing down, or instantaneously not accelerating.

15.11. A sinusoidal wave is propagating along a stretched string that lies along the x-axis. The displacement of the string as a function of time is graphed in Fig.15.30 for particles at $x = 0$ and at $x = 0.0900$ m. (a) What is the amplitude of the wave? (b) What is the period of the wave? (c) You are told that the two points $x = 0$ and $x = 0.0900$ m are within one wavelength of each other. If the wave is moving in the $+x$-direction, determine the wavelength and the wave speed. (d) If instead the wave is moving in the $-x$-direction, determine the wavelength and the wave speed. (e) Would it be possible to determine definitively the wavelength in parts (c) and (d) if you were not told that the two points were within one wavelength of each other? Why or why not?

Figure 15.30 Exercise 15.11.

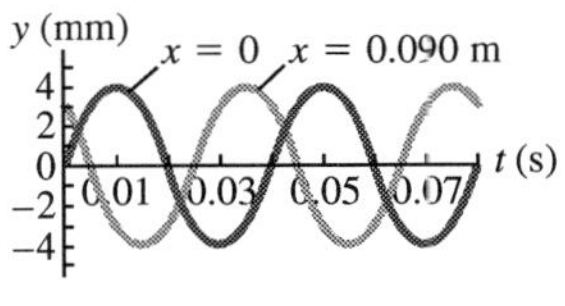

15.12. Speed of Propagation vs. Particle Speed. (a) Show that Eq. (15.3) may be written as

$$y(x,t) = A\cos\left[\frac{2\pi}{\lambda}(x - vt)\right]$$

(b) Use $y(x,t)$ to find an expression for the transverse velocity v_y of a particle in the string on which the wave travels. (c) Find the maximum speed of a particle of the string. Under what circumstances is this equal to the propagation speed v? Less than v? Greater than v?

15.13. A transverse wave on a string has amplitude 0.300 cm, wavelength 12.0 cm, and speed 6.00 cm/s. It is represented by $y(x,t)$ as given in Exercise 15.12. (a) At time $t = 0$, compute y at

1.5-cm intervals of x (that is, at $x = 0, x = 1.5$ cm, $x = 3.0$ cm, and so on) from $x = 0$ to $x = 12.0$ cm. Graph the results. This is the shape of the string at time $t = 0$. (b) Repeat the calculations for the same values of x at times $t = 0.400$ s and $t = 0.800$ s. Graph the shape of the string at these instants. In what direction is the wave traveling?

Section 15.4 Speed of a Transverse Wave

15.14. With what tension must a rope with length 2.50 m and mass 0.120 kg be stretched for transverse waves of frequency 40.0 Hz to have a wavelength of 0.750 m?

15.15. One end of a horizontal rope is attached to a prong of an electrically driven tuning fork that vibrates the rope transversely at 120 Hz. The other end passes over a pulley and supports a 1.50-kg mass. The linear mass density of the rope is 0.0550 kg/m. (a) What is the speed of a transverse wave on the rope? (b) What is the wavelength? (c) How would your answers to parts (a) and (b) change if the mass were increased to 3.00 kg?

15.16. A 1.50-m string of weight 1.25 N is tied to the ceiling at its upper end, and the lower end supports a weight W. When you pluck the string slightly, the waves traveling up the string obey the equation

$$y(x, t) = (8.50\text{ mm})\cos(172\text{ m}^{-1}x - 2730\text{ s}^{-1}t)$$

(a) How much time does it take a pulse to travel the full length of the string? (b) What is the weight W? (c) How many wavelengths are on the string at any instant of time? (d) What is the equation for waves traveling *down* the string?

15.17. A thin, 75.0-cm wire has a mass of 16.5 g. One end is tied to a nail, and the other end is attached to a screw that can be adjusted to vary the tension in the wire. (a) To what tension (in newtons) must you adjust the screw so that a transverse wave of wavelength 3.33 cm makes 875 vibrations per second? (b) How fast would this wave travel?

15.18. Weighty Rope. If in Example 15.3 (Section 15.4) we do *not* neglect the weight of the rope, what is the wave speed (a) at the bottom of the rope; (b) at the middle of the rope; (c) at the top of the rope?

15.19. A simple harmonic oscillator at the point $x = 0$ generates a wave on a rope. The oscillator operates at a frequency of 40.0 Hz and with an amplitude of 3.00 cm. The rope has a linear mass density of 50.0 g/m and is stretched with a tension of 5.00 N. (a) Determine the speed of the wave. (b) Find the wavelength. (c) Write the wave function $y(x, t)$ for the wave. Assume that the oscillator has its maximum upward displacement at time $t = 0$. (d) Find the maximum transverse acceleration of points on the rope. (e) In the discussion of transverse waves in this chapter, the force of gravity was ignored. Is that a reasonable approximation for this wave? Explain.

Section 15.5 Energy in Wave Motion

15.20. A piano wire with mass 3.00 g and length 80.0 cm is stretched with a tension of 25.0 N. A wave with frequency 120.0 Hz and amplitude 1.6 mm travels along the wire. (a) Calculate the average power carried by the wave. (b) What happens to the average power if the wave amplitude is halved?

15.21. A jet plane at take-off can produce sound of intensity 10.0 W/m^2 at 30.0 m away. But you prefer the tranquil sound of normal conversation, which is 1.0 μW/m^2. Assume that the plane behaves like a point source of sound. (a) What is the closest distance you should live from the airport runway to preserve your peace of mind? (b) What intensity from the jet does your friend experience if she lives twice as far from the runway as you do? (c) What power of sound does the jet produce at take-off?

15.22. Threshold of Pain. You are investigating the report of a UFO landing in an isolated portion of New Mexico, and you encounter a strange object that is radiating sound waves uniformly in all directions. Assume that the sound comes from a point source and that you can ignore reflections. You are slowly walking toward the source. When you are 7.5 m from it, you measure its intensity to be 0.11 W/m^2. An intensity of 1.0 W/m^2 is often used as the "threshold of pain." How much closer to the source can you move before the sound intensity reaches this threshold?

15.23. Energy Output. By measurement you determine that sound waves are spreading out equally in all directions from a point source and that the intensity is 0.026 W/m^2 at a distance of 4.3 m from the source. (a) What is the intensity at a distance of 3.1 m from the source? (b) How much sound energy does the source emit in one hour if its power output remains constant?

15.24. A fellow student with a mathematical bent tells you that the wave function of a traveling wave on a thin rope is $y(x, t) = 2.30\text{ mm}\cos[(6.98\text{ rad/m})x + (742\text{ rad/s})t]$. Being more practical, you measure the rope to have a length of 1.35 m and a mass of 0.00338 kg. You are then asked to determine the following: (a) amplitude; (b) frequency; (c) wavelength; (d) wave speed; (e) direction the wave is traveling; (f) tension in the rope; (g) average power transmitted by the wave.

15.25. What is the total power output of the siren in Example 15.5?

Section 15.6 Wave Interference, Boundary Conditions, and Superposition

15.26. Reflection. A wave pulse on a string has the dimensions shown in Fig.15.31 at $t = 0$. The wave speed is 40 cm/s. (a) If point O is a fixed end, draw the total wave on the string at $t = 15$ ms, 20 ms, 25 ms, 30 ms, 35 ms, 40 ms, and 45 ms. (b) Repeat part (a) for the case in which point O is a free end.

Figure **15.31** Exercise 15.26.

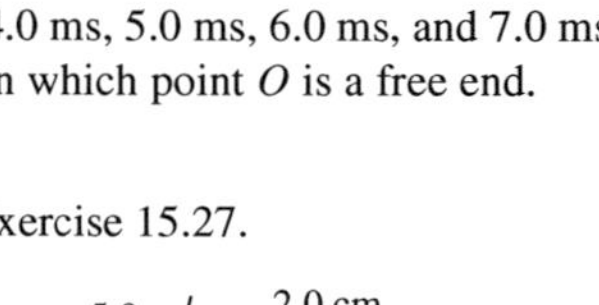

15.27. Reflection. A wave pulse on a string has the dimensions shown in Fig.15.32 at $t = 0$. The wave speed is 5.0 m/s. (a) If point O is a fixed end, draw the total wave on the string at $t = 1.0$ ms, 2.0 ms, 3.0 ms, 4.0 ms, 5.0 ms, 6.0 ms, and 7.0 ms. (b) Repeat part (a) for the case in which point O is a free end.

Figure **15.32** Exercise 15.27.

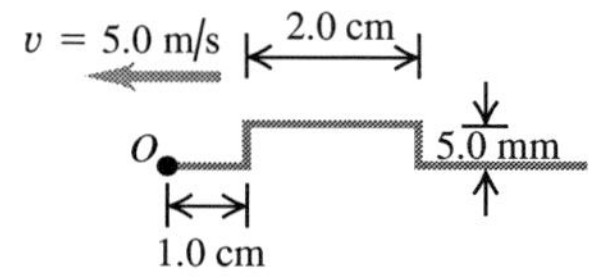

15.28. Interference of Triangular Pulses. Two triangular wave pulses are traveling toward each other on a stretched string as shown in Fig.15.33. Each pulse is identical to the other and travels at 2.00 cm/s. The leading edges of the pulses are 1.00 cm apart at $t = 0$. Sketch the shape of the string at $t = 0.250$ s, $t = 0.500$ s, $t = 0.750$ s, $t = 1.000$ s, and $t = 1.250$ s.

Figure **15.33** Exercise 15.28.

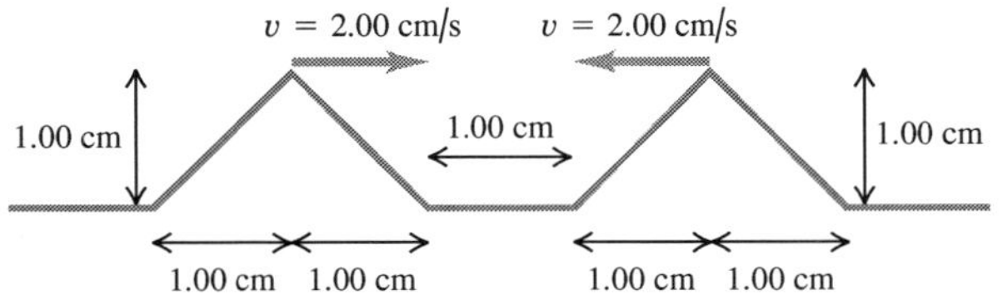

15.29. Suppose that the left-traveling pulse in Exercise 15.28 is *below* the level of the unstretched string instead of above it. Make the same sketches that you did in that exercise.

15.30. Two pulses are moving in opposite directions at 1.0 cm/s on a taut string, as shown in Fig. 15.34. Each square is 1.0 cm. Sketch the shape of the string at the end of (a) 6.0 s; (b) 7.0 s; (c) 8.0 s.

Figure **15.34** Exercise 15.30.

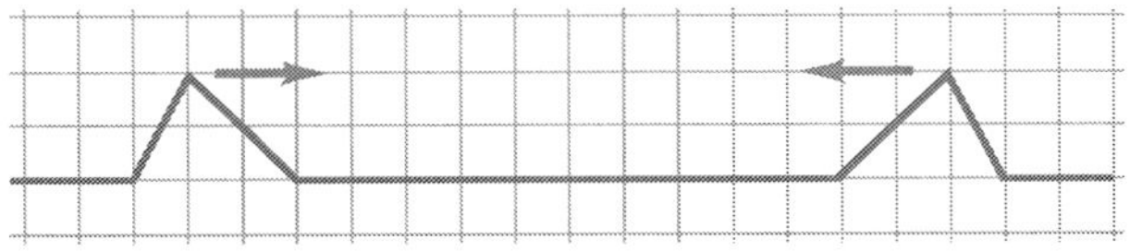

15.31. Interference of Rectangular Pulses. Figure 15.35 shows two rectangular wave pulses on a stretched string traveling toward each other. Each pulse is traveling with a speed of 1.00 mm/s and has the height and width shown in the figure. If the leading edges of the pulses are 8.00 mm apart at $t = 0$, sketch the shape of the string at $t = 4.00$ s, $t = 6.00$ s, and $t = 10.0$ s.

Figure **15.35** Exercise 15.31.

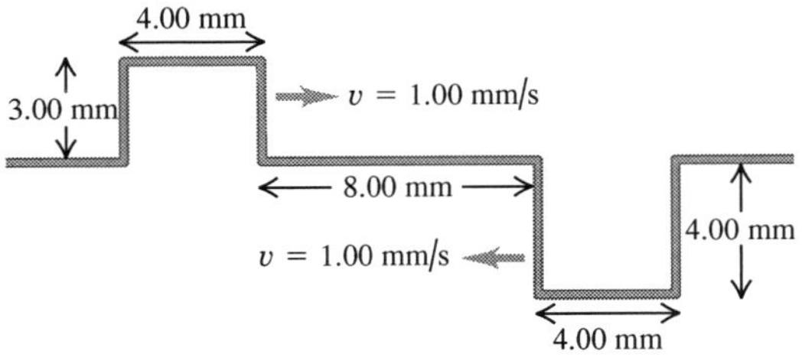

15.32. Two traveling waves moving on a string are identical except for opposite velocities. They obey the equation $y(x, t) = A\sin(kx \pm \omega t)$, where the plus-or-minus sign in the argument depends on the direction the wave is traveling. (a) Show that the vibrating string is described by the equation $y_{net}(x, t) = 2A\sin kx \cos \omega t$. (*Hint:* Use the trigonometric formulas for $\sin(a \pm b)$.) (b) Show that the string never moves at the places along it for which $x = n\lambda/2$, where n is a nonnegative integer.

Section 15.7 Standing Waves on a String

15.33. Standing waves on a wire are described by Eq. (15.28), with $A_{SW} = 2.50$ mm, $\omega = 942$ rad/s, and $k = 0.750\pi$ rad/m. The left end of the wire is at $x = 0$. At what distances from the left end are (a) the nodes of the standing wave and (b) the antinodes of the standing wave?

15.34. Adjacent antinodes of a standing wave on a string are 15.0 cm apart. A particle at an antinode oscillates in simple harmonic motion with amplitude 0.850 cm and period 0.0750 s. The string lies along the $+x$-axis and is fixed at $x = 0$. (a) How far apart are the adjacent nodes? (b) What are the wavelength, amplitude, and speed of the two traveling waves that form this pattern? (c) Find the maximum and minimum transverse speeds of a point at an antinode. (d) What is the shortest distance along the string between a node and an antinode?

15.35. Wave Equation and Standing Waves. (a) Prove by direct substitution that $y(x, t) = (A_{SW}\sin kx)\sin \omega t$ is a solution of the wave equation, Eq. (15.12), for $v = \omega/k$. (b) Explain why the relationship $v = \omega/k$ for *traveling* waves also applies to *standing* waves.

15.36. Give the details of the derivation of Eq. (15.28) from $y_1(x, t) + y_2(x, t) = A[-\cos(kx + \omega t) + \cos(kx - \omega t)]$.

15.37. Let $y_1(x, t) = A\cos(k_1x - \omega_1t)$ and $y_2(x, t) = A\cos(k_2x - \omega_2t)$ be two solutions to the wave equation, Eq. (15.12), for the same v. Show that $y(x, t) = y_1(x, t) + y_2(x, t)$ is also a solution to the wave equation.

Section 15.8 Normal Modes of a String

15.38. A 1.50-m-long rope is stretched between two supports with a tension that makes the speed of transverse waves 48.0 m/s. What are the wavelength and frequency of (a) the fundamental; (b) the second overtone; (c) the fourth harmonic?

15.39. A wire with mass 40.0 g is stretched so that its ends are tied down at points 80.0 cm apart. The wire vibrates in its fundamental mode with frequency 60.0 Hz and with an amplitude at the antinodes of 0.300 cm. (a) What is the speed of propagation of transverse waves in the wire? (b) Compute the tension in the wire. (c) Find the maximum transverse velocity and acceleration of particles in the wire.

15.40. A piano tuner stretches a steel piano wire with a tension of 800 N. The steel wire is 0.400 m long and has a mass of 3.00 g. (a) What is the frequency of its fundamental mode of vibration? (b) What is the number of the highest harmonic that could be heard by a person who is capable of hearing frequencies up to 10,000 Hz?

15.41. A thin, taut string tied at both ends and oscillating in its third harmonic has its shape described by the equation $y(x, t) = (5.60\text{ cm})\sin[(0.0340\text{ rad/cm})x]\sin[(50.0\text{ rad/s})t]$, where the origin is at the left end of the string, the x-axis is along the string and the y-axis is perpendicular to the string. (a) Draw a sketch that shows the standing wave pattern. (b) Find the amplitude of the two traveling waves that make up this standing wave. (c) What is the length of the string? (d) Find the wavelength, frequency, period, and speed of the traveling waves. (e) Find the maximum transverse speed of a point on the string. (f) What would be the equation $y(x, t)$ for this string if it were vibrating in its eighth harmonic?

15.42. The wave function of a standing wave is $y(x, t) = 4.44\text{ mm}\sin[(32.5\text{ rad/m})x]\sin[(754\text{ rad/s})t]$. For the two traveling waves that make up this standing wave, find the (a) amplitude; (b) wavelength; (c) frequency; (d) wave speed; (e) wave functions.

(f) From the information given, can you determine which harmonic this is? Explain.

15.43. Consider again the rope and traveling wave of Exercise 15.24. Assume that the ends of the rope are held fixed and that this traveling wave and the reflected wave are traveling in the opposite direction. (a) What is the wave function $y(x, t)$ for the standing wave that is produced? (b) In which harmonic is the standing wave oscillating? (c) What is the frequency of the fundamental oscillation?

15.44. One string of a certain musical instrument is 75.0 cm long and has a mass of 8.75 g. It is being played in a room where the speed of sound is 344 m/s. (a) To what tension must you adjust the string so that, when vibrating in its second overtone, it produces sound of wavelength 3.35 cm? (b) What frequency sound does this string produce in its fundamental mode of vibration?

15.45. The portion of the string of a certain musical instrument between the bridge and upper end of the finger board (that part of the string that is free to vibrate) is 60.0 cm long, and this length of the string has mass 2.00 g. The string sounds an A_4 note (440 Hz) when played. (a) Where must the player put a finger (what distance x from bridge) to play a D_5 note (587 Hz)? (See Fig. 15.36.) For both the A_4 and D_5 notes, the string vibrates in its fundamental mode. (b) Without retuning, is it possible to play a G_4 note (392 Hz) on this string? Why or why not?

Figure 15.36 Exercise 15.45.

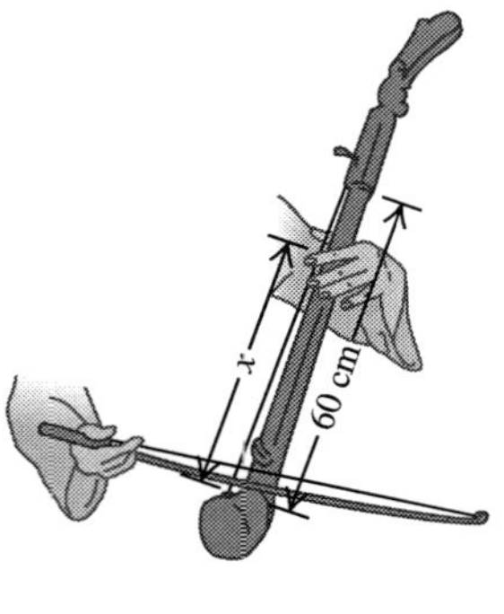

15.46. (a) A horizontal string tied at both ends is vibrating in its fundamental mode. The traveling waves have speed v, frequency f, amplitude A, and wavelength λ. Calculate the maximum transverse velocity and maximum transverse acceleration of points located at (i) $x = \lambda/2$, (ii) $x = \lambda/4$, and (iii) $x = \lambda/8$ from the left-hand end of the string. (b) At each of the points in part (a), what is the amplitude of the motion? (c) At each of the points in part (a), how much time does it take the string to go from its largest upward displacement to its largest downward displacement?

15.47. Guitar String. One of the 63.5-cm-long strings of an ordinary guitar is tuned to produce the note B_3 (frequency 245 Hz) when vibrating in its fundamental mode. (a) Find the speed of transverse waves on this string. (b) If the tension in this string is increased by 1.0%, what will be the new fundamental frequency of the string? (c) If the speed of sound in the surrounding air is 344 m/s, find the frequency and wavelength of the sound wave produced in the air by the vibration of the B_3 string. How do these compare to the frequency and wavelength of the standing wave on the string?

15.48. Waves on a Stick. A flexible stick 2.0 m long is not fixed in any way and is free to vibrate. Make clear drawings of this stick vibrating in its first three harmonics, and then use your drawings to find the wavelengths of each of these harmonics. (*Hint:* Should the ends be nodes or antinodes?)

Problems

15.49. A transverse sine wave with an amplitude of 2.50 mm and a wavelength of 1.80 m travels from left to right along a long, horizontal, stretched string with a speed of 36.0 m/s. Take the origin at the left end of the undisturbed string. At time $t = 0$ the left end of the string has its maximum upward displacement. (a) What are the frequency, angular frequency, and wave number of the wave? (b) What is the function $y(x, t)$ that describes the wave? (c) What is $y(t)$ for a particle at the left end of the string? (d) What is $y(t)$ for a particle 1.35 m to the right of the origin? (e) What is the maximum magnitude of transverse velocity of any particle of the string? (f) Find the transverse displacement and the transverse velocity of a particle 1.35 m to the right of the origin at time $t = 0.0625$ s.

15.50. A transverse wave on a rope is given by

$$y(x, t) = (0.750\ \text{cm})\cos\pi[(0.400\ \text{cm}^{-1})x + (250\ \text{s}^{-1})t]$$

(a) Find the amplitude, period, frequency, wavelength, and speed of propagation. (b) Sketch the shape of the rope at these values of t: 0, 0.0005 s, 0.0010 s. (c) Is the wave traveling in the $+x$- or $-x$-direction? (d) The mass per unit length of the rope is 0.0500 kg/m. Find the tension. (e) Find the average power of this wave.

15.51. Three pieces of string, each of length L, are joined together end to end, to make a combined string of length $3L$. The first piece of string has mass per unit length μ_1, the second piece has mass per unit length $\mu_2 = 4\mu_1$, and the third piece has mass per unit length $\mu_3 = \mu_1/4$. (a) If the combined string is under tension F, how much time does it take a transverse wave to travel the entire length $3L$? Give your answer in terms of L, F, and μ_1. (b) Does your answer to part (a) depend on the order in which the three pieces are joined together? Explain.

15.52. A 1750-N irregular beam is hanging horizontally by its ends from the ceiling by two vertical wires (A and B), each 1.25 m long and weighing 2.50 N. The center of gravity of this beam is one-third of the way along the beam from the end where wire A is attached. If you pluck both strings at the same time at the beam, what is the time delay between the arrival of the two pulses at the ceiling? Which pulse arrives first?

15.53. Ant Joy Ride. You place your pet ant Klyde (mass m) on top of a horizontal, stretched rope, where he holds on tightly. The rope has mass M and length L and is under tension F. You start a sinusoidal transverse wave of wavelength λ and amplitude A propagating along the rope. The motion of the rope is in a vertical plane. Klyde's mass is so small that his presence has no effect on the propagation of the wave. (a) What is Klyde's top speed as he oscillates up and down? (b) Klyde enjoys the ride and begs for more. You decide to double his top speed by changing the tension while keeping the wavelength and amplitude the same. Should the tension be increased or decreased, and by what factor?

15.54. Weightless Ant. An ant with mass m is standing peacefully on top of a horizontal, stretched rope. The rope has mass per unit length μ and is under tension F. Without warning, Cousin Throckmorton starts a sinusoidal transverse wave of wavelength λ propagating along the rope. The motion of the rope is in a vertical plane. What minimum wave amplitude will make the ant become momentarily weightless? Assume that m is so small that the presence of the ant has no effect on the propagation of the wave.

15.55. When a transverse sinusoidal wave is present on a string, the particles of the string undergo SHM. This is the same motion as that of a mass m attached to an ideal spring of force constant k', for which the angular frequency of oscillation was found in Chapter 13 to be $\omega = \sqrt{k'/m}$. Consider a string with tension F and mass per unit length μ, along which is propagating a sinusoidal wave with amplitude A and wavelength λ. (a) Find the "force constant" k' of the restoring force that acts on a short segment of the string of length Δx (where $\Delta x \ll \lambda$). (b) How does the "force constant" calculated in part (b) depend on F, μ, A, and λ? Explain the physical reasons this should be so.

15.56. A 5.00-m, 0.732-kg wire is used to support two uniform 235-N posts of equal length (Fig. 15.37). Assume that the wire is essentially horizontal and that the speed of sound is 344 m/s. A strong wind is blowing, causing the wire to vibrate in its 7th overtone. What are the frequency and wavelength of the sound this wire produces?

Figure **15.37** Problem 15.56.

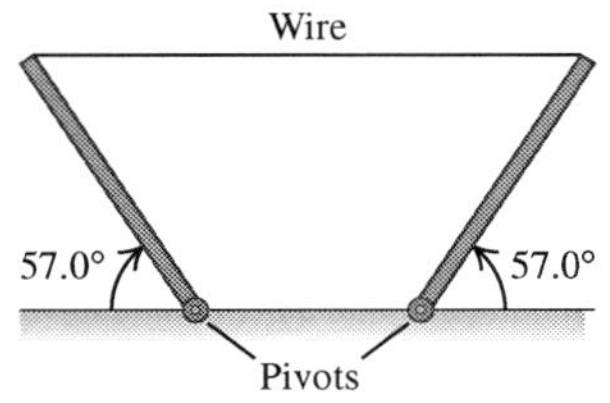

15.57. A Nonsinusoidal Wave. The shape of a wave on a string at a specific instant is shown in Fig. 15.38. The wave is propagating to the right, in the $+x$-direction. (a) Determine the direction of the transverse *velocity* of each of the six labeled points on the string. If the velocity is zero, state it as such. Explain your reasoning. (b) Determine the direction of the transverse *acceleration* of each of the six labeled points on the string. Explain your reasoning. (c) How would your answers be affected if the wave were propagating to the left, in the $-x$-direction?

Figure **15.38** Problem 15.57.

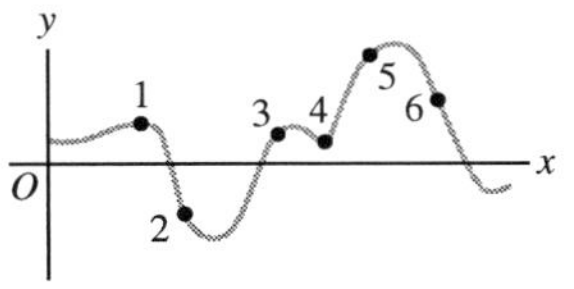

15.58. A continuous succession of sinusoidal wave pulses are produced at one end of a very long string and travel along the length of the string. The wave has frequency 40.0 Hz, amplitude 5.00 mm, and wavelength 0.600 m. (a) How long does it take the wave to travel a distance of 8.00 m along the length of the string? (b) How long does it take a point on the string to travel a distance of 8.00 m, once the wave train has reached the point and set it into motion? (c) In parts (a) and (b), how does the time change if the amplitude is doubled?

15.59. Two-Dimensional Waves. A stretched string lies along the x-axis. The string is displaced along both the y- and z-directions, so that the transverse displacement of the string is given by

$$y(x, t) = A\cos(kx - \omega t) \qquad z(x, t) = A\sin(kx - \omega t)$$

(a) Draw a graph of z versus y for a particle on the string at $x = 0$. This shows the trajectory of the particle as seen by an observer on the $+x$-axis looking back toward $x = 0$. Indicate the position of the particle at $t = 0$, $t = \pi/2\omega$, $t = \pi/\omega$, and $t = 3\pi/2\omega$. (b) Find the velocity vector of a particle at an *arbitrary* position x on the string. Show that this represents the tangential velocity of a particle moving in a circle of radius A with angular velocity ω, and show that the speed of the particle is constant (i.e., the particle is in uniform circular motion). (See Problem 3.75.) (c) Find the acceleration vector of the particle in part (b). Show that the acceleration is always directed toward the center of the circle and that its magnitude is $a = \omega^2 A$. Explain these results in terms of uniform circular motion. Suppose that the displacement of the string was instead given by

$$y(x, t) = A\cos(kx - \omega t) \qquad z(x, t) = -A\sin(kx - \omega t)$$

Describe how the motion of a particle at x would be different from the motion described in part (a).

15.60. A vertical, 1.20-m length of 18-gauge (diameter of 1.024 mm) copper wire has a 100.0-N ball hanging from it. (a) What is the wavelength of the third harmonic for this wire? (b) A 500.0-N ball now *replaces* the original ball. What is the change in the wavelength of the third harmonic caused by replacing the light ball with the heavy one? (*Hint:* See Table 11.1 for Young's modulus.)

15.61. Waves of Arbitrary Shape. (a) Explain why *any* wave described by a function of the form $y(x, t) = f(x - vt)$ moves in the $+x$-direction with speed v. (b) Show that $y(x, t) = f(x - vt)$ satisfies the wave equation, no matter what the functional form of f. To do this, write $y(x, t) = f(u)$, where $u = x - vt$. Then, to take partial derivatives of $y(x, t)$, use the chain rule:

$$\frac{\partial y(x,t)}{\partial t} = \frac{df(u)}{du}\frac{\partial u}{\partial t} = \frac{df(u)}{du}(-v)$$
$$\frac{\partial y(x,t)}{\partial x} = \frac{df(u)}{du}\frac{\partial u}{\partial x} = \frac{df(u)}{du}$$

(c) A wave pulse is described by the function $y(x, t) = De^{-(Bx - Ct)^2}$, where B, C, and D are all positive constants. What is the speed of this wave?

15.62. Equation (15.7) for a sinusoidal wave can be made more general by including a phase angle ϕ, where $0 \le \phi \le 2\pi$ (in radians). Then the wave function $y(x, t)$ becomes

$$y(x, t) = A\cos(kx - \omega t + \phi)$$

(a) Sketch the wave as a function of x at $t = 0$ for $\phi = 0$, $\phi = \pi/4$, $\phi = \pi/2$, $\phi = 3\pi/4$, and $\phi = 3\pi/2$. (b) Calculate the transverse velocity $v_y = \partial y/\partial t$. (c) At $t = 0$, a particle on the string at $x = 0$ has displacement $y = A/\sqrt{2}$. Is this enough information to determine the value of ϕ? In addition, if you are told that a particle at $x = 0$ is moving toward $y = 0$ at $t = 0$, what is the value of ϕ? (d) Explain in general what you must know about the wave's behavior at a given instant to determine the value of ϕ.

15.63. (a) Show that Eq. (15.25) can also be written as $P_{av} = \frac{1}{2}Fk\omega A^2$, where k is the wave number of the wave. (b) If the tension F in the string is quadrupled while the amplitude A is kept the same, how must k and ω each change to keep the average power constant? [*Hint:* Recall Eq. (15.6).]

15.64. Energy in a Triangular Pulse. A triangular wave pulse on a taut string travels in the positive x-direction with speed v. The tension in the string is F, and the linear mass density of the string is μ. At $t = 0$, the shape of the pulse is given by

$$y(x, 0) = \begin{cases} 0 & \text{if } x < -L \\ h(L + x)/L & \text{for } -L < x < 0 \\ h(L - x)/L & \text{for } 0 < x < L \\ 0 & \text{for } x > L \end{cases}$$

(a) Draw the pulse at $t = 0$. (b) Determine the wave function $y(x, t)$ at all times t. (c) Find the instantaneous power in the wave. Show that the power is zero except for $-L < (x - vt) < L$ and that in this interval the power is constant. Find the value of this constant power.

15.65. A sinusoidal transverse wave travels on a string. The string has length 8.00 m and mass 6.00 g. The wave speed is 30.0 m/s, and the wavelength is 0.200 m. (a) If the wave is to have an average power of 50.0 W, what must be the amplitude of the wave? (b) For this same string, if the amplitude and wavelength are the same as in part (a), what is the average power for the wave if the tension is increased such that the wave speed is doubled?

15.66. Instantaneous Power in a Wave. (a) Graph $y(x, t)$ as given by Eq. (15.7) as a function of x for a given time t (say, $t = 0$). On the same axes, make a graph of the instantaneous power $P(x, t)$ as given by Eq. (15.23). (b) Explain the connection between the slope of the graph of $y(x, t)$ versus x and the value of $P(x, t)$. In particular, explain what is happening at points where $P = 0$, where there is no instantaneous energy transfer. (c) The quantity $P(x, t)$ always has the same sign. What does this imply about the direction of energy flow? (d) Consider a wave moving in the $-x$-direction, for which $y(x, t) = A\cos(kx + \omega t)$. Calculate $P(x, t)$ for this wave, and make a graph of $y(x, t)$ and $P(x, t)$ as functions of x for a given time t (say, $t = 0$). What differences arise from reversing the direction of the wave?

15.67. A metal wire, with density ρ and Young's modulus Y, is stretched between rigid supports. At temperature T, the speed of a transverse wave is found to be v_1. When the temperature is increased to $T + \Delta T$, the speed decreases to $v_2 < v_1$. Determine the coefficient of linear expansion of the wire.

15.68. A vibrating string 50.0 cm long is under a tension of 1.00 N. The results from five successive stroboscopic pictures are shown in Fig. 15.39. The strobe rate is set at 5000 flashes per minute, and observations reveal that the maximum displacement occurred at flashes 1 and 5 with no other maxima in between. (a) Find the period, frequency, and wavelength for the traveling waves on this string. (b) In what normal mode (harmonic) is the string vibrating? (c) What is the speed of the traveling waves on the string? (d) How fast is point P moving when the string is in (i) position 1 and (ii) position 3? (e) What is the mass of this string? (Section 15.3).

Figure 15.39 Problem 15.68.

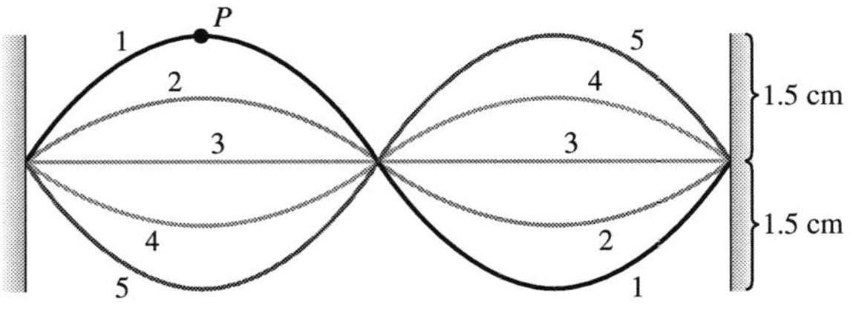

15.69. Clothesline Nodes. Cousin Throckmorton is once again playing with the clothesline in Example 15.2 (Section 15.3). One end of the clothesline is attached to a vertical post. Throcky holds the other end loosely in his hand, so that the speed of waves on the clothesline is a relatively slow 0.720 m/s. He finds several frequencies at which he can oscillate his end of the clothesline so that a light clothespin 45.0 cm from the post doesn't move. What are these frequencies?

15.70. A guitar string is vibrating in its fundamental mode, with nodes at each end. The length of the segment of the string that is free to vibrate is 0.386 m. The maximum transverse acceleration of a point at the middle of the segment is 8.40×10^3 m/s^2 and the maximum transverse velocity is 3.80 m/s. (a) What is the amplitude of this standing wave? (b) What is the wave speed for the transverse traveling waves on this string?

15.71. As shown in Exercise 15.35, a standing wave given by Eq. (15.28) satisfies the wave equation Eq. (15.12). (a) Show that a standing wave given by Eq. (15.28) also satisfies the equation

$$\frac{\partial^2 y(x, t)}{\partial t^2} = -\omega^2 y(x, t)$$

Interpret this equation in terms of what you know about simple harmonic motion. (b) Does a traveling wave given by $y(x, t) = A\cos(kx - \omega t)$ also satisfy the equation in part (a)? Interpret this result.

15.72. (a) The red and blue waves in Fig. 15.20 combine so that the displacement of the string at O is always zero. To show this mathematically for a wave of arbitrary shape, consider a wave moving to the right along the string in Fig. 15.20 (shown in blue) that, at time T, is given by $y_1(x, T) = f(x)$, where f is some function of x. (The form of $f(x)$ determines the shape of the wave.) If the point O corresponds to $x = 0$, explain why, at time T, the wave moving to the left in Fig. 15.20 (shown in red) is given by the function $y_2(x, T) = -f(-x)$. (b) Show that the total wave function $y(x, T) = y_1(x, T) + y_2(x, T)$ is zero at O, independent of the form of the function $f(x)$. (c) The red and blue waves in Fig. 15.21 combine so that the slope of the string at O is always zero. To show this mathematically for a wave of arbitrary shape, again let the wave moving to the right in Fig. 15.21 (shown in blue) be given by $y_1(x, T) = f(x)$ at time T. Explain why the wave moving to the left (shown in red) at this same time T is given by $y_2(x, T) = f(-x)$. (d) Show that the total wave function $y(x, T) = y_1(x, T) + y_2(x, T)$ has zero slope at O, independent of the form of the function $f(x)$, as long as $f(x)$ has a finite first derivative.

15.73. A string that lies along the $+x$-axis has a free end at $x = 0$. (a) By using steps similar to those used to derive Eq. (15.28), show that an incident traveling wave $y_1(x, t) = A\cos(kx + \omega t)$ gives rise to a standing wave $y(x, t) = 2A\cos\omega t\cos kx$. (b) Show that the standing wave has an antinode at its free end $(x = 0)$. (c) Find the maximum displacement, maximum speed, and maximum acceleration of the free end of the string.

15.74. A string with both ends held fixed is vibrating in its third harmonic. The waves have a speed of 192 m/s and a frequency of 240 Hz. The amplitude of the standing wave at an antinode is 0.400 cm. (a) Calculate the amplitude at points on the string a distance of (i) 40.0 cm; (ii) 20.0 cm; and (iii) 10.0 cm from the left end of the string. (b) At each point in part (a), how much time does it take the string to go from its largest upward displacement to its largest downward displacement? (c) Calculate the maximum transverse velocity and the maximum transverse acceleration of the string at each of the points in part (a).

15.75. A uniform cylindrical steel wire, 55.0 cm long and 1.14 mm in diameter, is fixed at both ends. To what tension must it be adjusted so that, when vibrating in its first overtone, it produces the note D# of frequency 311 Hz? Assume that it stretches an insignificant amount. (*Hint:* See Table 14.1.)

15.76. Holding Up Under Stress. A string or rope will break apart if it is placed under too much tensile stress [Eq. (11.8)]. Thicker ropes can withstand more tension without breaking because the thicker the rope, the greater the cross-sectional area and the smaller the stress. One type of steel has density 7800 kg/m^3 and will break if the tensile stress exceeds 7.0×10^8 N/m^2. You want to make a guitar string from 4.0 g of this type of steel. In use, the guitar string must be able to withstand a tension of 900 N without breaking. Your job is the following: (a) Determine the maximum length and minimum radius the string can have. (b) Determine the highest possible fundamental frequency of standing waves on this string, if the entire length of the string is free to vibrate.

15.77. Combining Standing Waves. A guitar string of length L is plucked in such a way that the total wave produced is the sum of the fundamental and the second harmonic. That is, the standing wave is given by

$$y(x, t) = y_1(x, t) + y_2(x, t)$$

where

$$y_1(x, t) = C\sin\omega_1 t\sin k_1 x$$
$$y_2(x, t) = C\sin\omega_2 t\sin k_2 x$$

with $\omega_1 = vk_1$ and $\omega_2 = vk_2$. (a) At what values of x are the nodes of y_1? (b) At what values of x are the nodes of y_2? (c) Graph the total wave at $t = 0$, $t = \frac{1}{8}f_1$, $t = \frac{1}{4}f_1$, $t = \frac{3}{8}f_1$, and $t = \frac{1}{2}f_1$. (d) Does the sum of the two standing waves y_1 and y_2 produce a standing wave? Explain.

15.78. When a massive aluminum sculpture is hung from a steel wire, the fundamental frequency for transverse standing waves on the wire is 250.0 Hz. The sculpture (but not the wire) is then completely submerged in water. (a) What is the new fundamental frequency? (*Hint:* See Table 14.1.) (b) Why is it a good approximation to treat the wire as being fixed at both ends?

15.79. Tuning an Instrument. A musician tunes the C-string of her instrument to a fundamental frequency of 65.4 Hz. The vibrating portion of the string is 0.600 m long and has a mass of 14.4 g. (a) With what tension must the musician stretch it? (b) What percent increase in tension is needed to increase the frequency from 65.4 Hz to 73.4 Hz, corresponding to a rise in pitch from C to D?

Challenge Problems

15.80. Longitudinal Waves on a Spring. A long spring such as a Slinky™ is often used to demonstrate longitudinal waves. (a) Show that if a spring that obeys Hooke's law has mass m, length L, and force constant k', the speed of longitudinal waves on the spring is $v = L\sqrt{k'/m}$. (b) Evaluate v for a spring with $m = 0.250$ kg, $L = 2.00$ m, and $k' = 1.50$ N/m.

15.81. (a) Show that for a wave on a string, the kinetic energy *per unit length of string* is

$$u_k(x, t) = \frac{1}{2}\mu v_y^2(x, t) = \frac{1}{2}\mu\left(\frac{\partial y(x, t)}{\partial t}\right)^2$$

where μ is the mass per unit length. (b) Calculate $u_k(x, t)$ for a sinusoidal wave given by Eq. (15.7). (c) There is also elastic potential energy in the string, associated with the work required to deform and stretch the string. Consider a short segment of string at position x that has unstretched length Δx, as in Fig. 15.13. Ignoring the (small) curvature of the segment, its slope is $\partial y(x, t)/\partial x$. Assume that the displacement of the string from equilibrium is small, so that $\partial y/\partial x$ has a magnitude much less than unity. Show that the stretched length of the segment is approximately

$$\Delta x\left[1 + \frac{1}{2}\left(\frac{\partial y(x, t)}{\partial x}\right)^2\right]$$

(*Hint:* Use the relationship $\sqrt{1 + u} \approx 1 + \frac{1}{2}u$, valid for $|u| \ll 1$.) (d) The potential energy stored in the segment equals the work done by the string tension F (which acts along the string) to stretch the segment from its unstretched length Δx to the length calculated in part (c). Calculate this work and show that the potential energy *per unit length of string* is

$$u_p(x, t) = \frac{1}{2}F\left(\frac{\partial y(x, t)}{\partial x}\right)^2$$

(e) Calculate $u_p(x, t)$ for a sinusoidal wave given by Eq. (15.7). (f) Show that $u_k(x, t) = u_p(x, t)$ for all x and t. (g) Show $y(x, t)$, $u_k(x, t)$, and $u_p(x, t)$ as functions of x for $t = 0$ in one graph with all three functions on the same axes. Explain why u_k and u_p are maximum where y is zero, and vice versa. (h) Show that the instantaneous power in the wave, given by Eq. (15.22), is equal to the total energy per unit length multiplied by the wave speed v. Explain why this result is reasonable.

15.82. A deep-sea diver is suspended beneath the surface of Loch Ness by a 100-m-long cable that is attached to a boat on the surface (Fig. 15.40). The diver and his suit have a total mass of 120 kg and a volume of 0.0800 m^3. The cable has a diameter of 2.00 cm and a linear mass density of $\mu = 1.10$ kg/m. The diver thinks he sees something moving in the murky depths and jerks the end of the cable back and forth to send transverse waves up the cable as a signal to his companions in the boat. (a) What is the tension in the cable at its lower end, where it is attached to the diver? Do not forget to include the buoyant force that the water (density 1000 kg/m^3) exerts on him. (b) Calculate the tension in the cable a distance x above the diver. The buoyant force on the cable must be included in your calculation. (c) The speed of transverse waves on the cable is given by $v = \sqrt{F/\mu}$ (Eq. 15.13). The speed therefore varies along the cable, since the tension is not constant. (This expression neglects the damping force that the water exerts on the moving cable.) Integrate to find the time required for the first signal to reach the surface.

Figure 15.40 Challenge Problem 15.82.

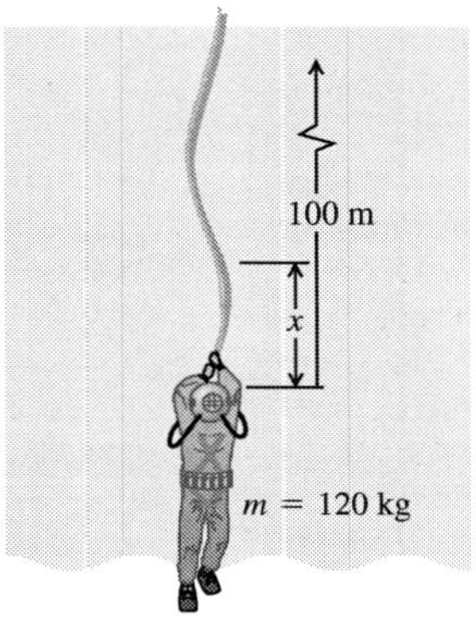

15.83. A uniform rope with length L and mass m is held at one end and whirled in a horizontal circle with angular velocity ω. You can ignore the force of gravity on the rope. Find the time required for a transverse wave to travel from one end of the rope to the other.

15.84. Instantaneous Power in a Standing Wave. From Eq. (15.21), the instantaneous rate at which a wave transmits energy along a string (instantaneous power) is

$$P(x, t) = -F\frac{\partial y(x, t)}{\partial x}\frac{\partial y(x, t)}{\partial t}$$

where F is the tension. (a) Evaluate $P(x, t)$ for a standing wave of the form given by Eq. (15.28). (b) Show that for all values of x, the *average* power P_{av} carried by the standing wave is zero. (Equation (15.25) does not apply here. Can you see why?) (c) For a standing wave given by Eq. (15.28), graph $P(x, t)$ and the displacement $y(x, t)$ as functions of x for $t = 0$, $t = \pi/4\omega$, $t = \pi/2\omega$, and $t = 3\pi/4\omega$. (Positive $P(x, t)$ means energy is flowing in the $+x$-direction; negative $P(x, t)$ means the flow is in the

$-x$-direction.) (d) The *kinetic* energy per unit length of string is greatest where the string has the greatest transverse speed, and the *potential* energy per unit length of string is greatest where the string has the steepest slope (because there the string is stretched the most). (See Challenge Problem 15.81.) Using these ideas, discuss the flow of energy along the string.

15.85. Out of Tune. The B-string of a guitar is made of steel (density 7800 kg/m^3), is 63.5 cm long, and has diameter 0.406 mm. The fundamental frequency is $f = 247.0$ Hz. (a) Find the string tension. (b) If the tension F is changed by a small amount ΔF, the frequency f changes by a small amount Δf. Show that

$$\frac{\Delta f}{f} = \frac{1}{2}\frac{\Delta f}{f}$$

(c) The string is tuned as in part (a) when its temperature is 18.5°C. Strenuous playing can make the temperature of the string rise, changing its vibration frequency. Find Δf if the temperature of the string rises to 29.5°C. The steel string has a Young's modulus of 2.00×10^{11} Pa and a coefficient of linear expansion of $1.20 \times 10^{-5}\ (\text{C}°)^{-1}$. Assume that the temperature of the body of the guitar remains constant. Will the vibration frequency rise or fall?

SOUND AND HEARING

16

?Most people like to listen to music, but hardly anyone likes to listen to noise. What is the physical difference between musical sound and noise?

LEARNING GOALS

By studying this chapter, you will learn:

- How to describe a sound wave in terms of either particle displacements or pressure fluctuations.
- How to calculate the speed of sound waves in different materials.
- How to calculate the intensity of a sound wave.
- What determines the particular frequencies of sound produced by an organ or a flute.
- How resonance occurs in musical instruments.
- What happens when sound waves from different sources overlap.
- How to describe what happens when two sound waves of slightly different frequencies are combined.
- Why the pitch of a siren changes as it moves past you.

Of all the mechanical waves that occur in nature, the most important in our everyday lives are longitudinal waves in a medium—usually air—called *sound* waves. The reason is that the human ear is tremendously sensitive and can detect sound waves even of very low intensity. Besides their use in spoken communication, our ears allow us to pick up a myriad of cues about our environment, from the welcome sound of a meal being prepared to the warning sound of an approaching car. The ability to hear an unseen nocturnal predator was essential to the survival of our ancestors, so it is no exaggeration to say that we humans owe our existence to our highly evolved sense of hearing.

Up to this point we have described mechanical waves primarily in terms of displacement; however, a description of sound waves in terms of *pressure* fluctuations is often more appropriate, largely because the ear is primarily sensitive to changes in pressure. We'll study the relationships among displacement, pressure fluctuation, and intensity and the connections between these quantities and human sound perception.

When a source of sound or a listener moves through the air, the listener may hear a different frequency from the one emitted by the source. This is the Doppler effect, which has important applications in medicine and technology.

16.1 Sound Waves

The most general definition of **sound** is a longitudinal wave in a medium. Our main concern in this chapter is with sound waves in air, but sound can travel through any gas, liquid, or solid. You may be all too familiar with the propagation of sound through a solid if your neighbor's stereo speakers are right next to your wall.

The simplest sound waves are sinusoidal waves, which have definite frequency, amplitude, and wavelength. The human ear is sensitive to waves in the frequency range from about 20 to 20,000 Hz, called the **audible range,** but we also use the

16.1 A sinusoidal longitudinal wave traveling to the right in a fluid. (Compare to Fig. 15.7.)

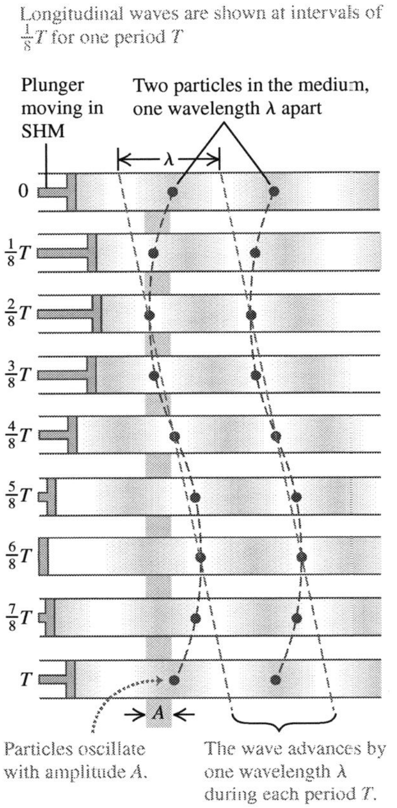

16.2 As a sound wave propagates along the x-axis, the left and right ends undergo different displacements y_1 and y_2.

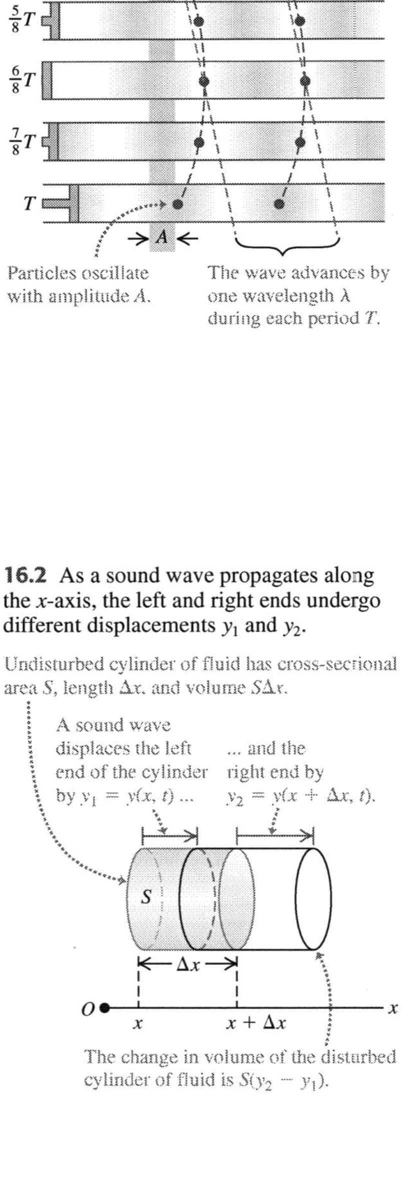

term "sound" for similar waves with frequencies above **(ultrasonic)** and below **(infrasonic)** the range of human hearing.

Sound waves usually travel out in all directions from the source of sound, with an amplitude that depends on the direction and distance from the source. We'll return to this point in the next section. For now, we concentrate on the idealized case of a sound wave that propagates in the positive x-direction only. As we discussed in Section 15.3, such a wave is described by a wave function $y(x, t)$, which gives the instantaneous displacement y of a particle in the medium at position x at time t. If the wave is sinusoidal, we can express it using Eq. (15.7):

$$y(x, t) = A\cos(kx - \omega t) \qquad \text{(sound wave propagating in the } +x\text{-direction)} \tag{16.1}$$

Remember that in a longitudinal wave the displacements are *parallel* to the direction of travel of the wave, so distances x and y are measured parallel to each other, not perpendicular as in a transverse wave. The amplitude A is the maximum displacement of a particle in the medium from its equilibrium position (Fig. 16.1). Hence A is also called the **displacement amplitude.**

Sound Waves As Pressure Fluctuations

Sound waves may also be described in terms of variations of *pressure* at various points. In a sinusoidal sound wave in air, the pressure fluctuates above and below atmospheric pressure p_a in a sinusoidal variation with the same frequency as the motions of the air particles. The human ear operates by sensing such pressure variations. A sound wave entering the ear canal exerts a fluctuating pressure on one side of the eardrum; the air on the other side of the eardrum, vented to the outside by the Eustachian tube, is at atmospheric pressure. The pressure difference on the two sides of the eardrum sets it into motion. Microphones and similar devices also usually sense pressure differences, not displacements, so it is very useful to develop a relationship between these two descriptions.

Let $p(x, t)$ be the instantaneous pressure fluctuation in a sound wave at any point x at time t. That is, $p(x, t)$ is the amount by which the pressure *differs* from normal atmospheric pressure p_a. Think of $p(x, t)$ as the *gauge pressure* defined in Section 14.2; it can be either positive or negative. The *absolute* pressure at a point is then $p_a + p(x, t)$.

To see the connection between the pressure fluctuation $p(x, t)$ and the displacement $y(x, t)$ in a sound wave propagating in the $+x$-direction, consider an imaginary cylinder of a wave medium (gas, liquid, or solid) with cross-sectional area S and axis along the direction of propagation (Fig. 16.2). When no sound wave is present, the cylinder has length Δx and volume $V = S\Delta x$, as shown by the shaded volume in Fig. 16.2. When a wave is present, at time t the end of the cylinder that is initially at x is displaced by $y_1 = y(x, t)$, and the end that is initially at $x + \Delta x$ is displaced by $y_2 = y(x + \Delta x, t)$; this is shown by the red lines. If $y_2 > y_1$, as shown in Fig. 16.2, the cylinder's volume has increased, which causes a decrease in pressure. If $y_2 < y_1$, the cylinder's volume has decreased and the pressure has increased. If $y_2 = y_1$, the cylinder is simply shifted to the left or right; there is no volume change and no pressure fluctuation. The pressure fluctuation depends on the *difference* between the displacement at neighboring points in the medium.

Quantitatively, the change in volume ΔV of the cylinder is

$$\Delta V = S(y_2 - y_1) = S[y(x + \Delta x, t) - y(x, t)]$$

In the limit as $\Delta x \rightarrow 0$, the fractional change in volume dV/V (volume change divided by original volume) is

$$\frac{dV}{V} = \lim_{\Delta x \rightarrow 0} \frac{S[y(x + \Delta x, t) - y(x, t)]}{S\Delta x} = \frac{\partial y(x, t)}{\partial x} \tag{16.2}$$

The fractional volume change is related to the pressure fluctuation by the bulk modulus B, which by definition [Eq. (11.13)] is $B = -p(x,t)/(dV/V)$ (see Section 11.4). Solving for $p(x,t)$, we have

$$p(x,t) = -B\frac{\partial y(x,t)}{\partial x} \tag{16.3}$$

The negative sign arises because when $\partial y(x,t)/\partial x$ is positive, the displacement is greater at $x + \Delta x$ than at x, corresponding to an increase in volume and a *decrease* in pressure.

When we evaluate $\partial y(x,t)/\partial x$ for the sinusoidal wave of Eq. (16.1), we find

$$p(x,t) = BkA\sin(kx - \omega t) \tag{16.4}$$

Figure 16.3 shows $y(x,t)$ and $p(x,t)$ for a sinusoidal sound wave at $t = 0$. It also shows how individual particles of the wave are displaced at this time. While $y(x,t)$ and $p(x,t)$ describe the same wave, these two functions are one-quarter cycle out of phase: At any time, the displacement is greatest where the pressure fluctuation is zero, and vice versa. In particular, note that the compressions (points of greatest pressure and density) and rarefactions (points of lowest pressure and density) are points of *zero* displacement.

CAUTION **Graphs of a sound wave** Keep in mind that the graphs in Fig. 16.3 show the wave at only *one* instant of time. Because the wave is propagating in the $+x$-direction, as time goes by the wave patterns in the functions $y(x,t)$ and $p(x,t)$ move to the right at the wave speed $v = \omega/k$. Hence the positions of the compressions and rarefactions also move to the right at this same speed. The particles, by contrast, simply oscillate back and forth in simple harmonic motion as shown in Fig. 16.1.

Equation (16.4) shows that the quantity BkA represents the maximum pressure fluctuation. We call this the **pressure amplitude,** denoted by p_{max}:

$$p_{max} = BkA \quad \text{(sinusoidal sound wave)} \tag{16.5}$$

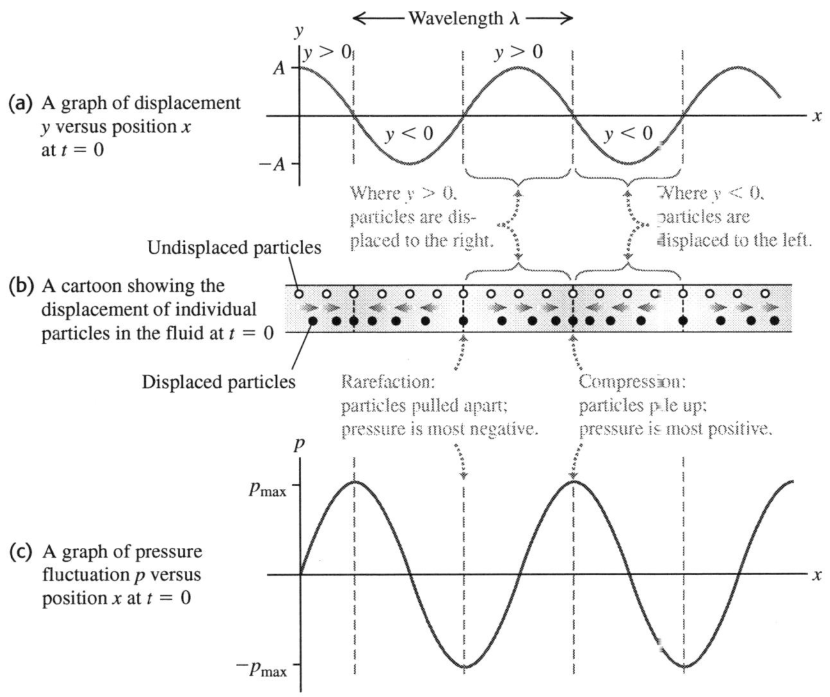

16.3 Three ways to describe a sound wave.

The pressure amplitude is directly proportional to the displacement amplitude A, as we might expect, and it also depends on wavelength. Waves of shorter wavelength λ (larger wave number $k = 2\pi/\lambda$) have greater pressure variations for a given amplitude because the maxima and minima are squeezed closer together. A medium with a large value of bulk modulus B requires a relatively large pressure amplitude for a given displacement amplitude because large B means a less compressible medium; that is, greater pressure change is required for a given volume change.

Example 16.1 Amplitude of a sound wave in air

In a sinusoidal sound wave of moderate loudness the maximum pressure variations are of the order of 3.0×10^{-2} Pa above and below atmospheric pressure p_a (nominally 1.013×10^5 Pa at sea level). Find the corresponding maximum displacement if the frequency is 1000 Hz. In air at normal atmospheric pressure and density, the speed of sound is 344 m/s and the bulk modulus is 1.42×10^5 Pa.

SOLUTION

IDENTIFY: This problem involves the relationship between two different ways of describing a sound wave: in terms of displacement and in terms of pressure. The target variable is the displacement amplitude A.

SET UP: We are given the pressure amplitude p_{max}, wave speed v, frequency f, and bulk modulus B. Our target variable A is related to p_{max} by Eq. (16.5). We also use the relationship $\omega = vk$ [Eq. (15.6)] to determine the wave number k from v and the angular frequency $\omega = 2\pi f$.

EXECUTE: From Eq. (16.5), the maximum displacement is $A = p_{max}/Bk$. From Eq. (15.6), the wave number is

$$k = \frac{\omega}{v} = \frac{2\pi f}{v} = \frac{(2\pi \text{ rad})(1000 \text{ Hz})}{344 \text{ m/s}} = 18.3 \text{ rad/m}$$

Then

$$A = \frac{p_{max}}{Bk} = \frac{3.0 \times 10^{-2} \text{ Pa}}{(1.42 \times 10^5 \text{ Pa})(18.3 \text{ rad/m})}$$
$$= 1.2 \times 10^{-8} \text{ m}$$

EVALUATE: This displacement amplitude is only about $\frac{1}{100}$ the size of a human cell. Remember that the ear actually senses pressure fluctuations; it detects these minuscule displacements only indirectly.

Example 16.2 Amplitude of a sound wave in the inner ear

When a sound wave enters the ear, it sets the eardrum into oscillation, which in turn causes oscillation of the three tiny bones in the middle ear called the *ossicles* (Fig. 16.4). This oscillation is finally transmitted to the fluid-filled inner ear; the motion of the fluid disturbs hair cells within the inner ear, which transmit nerve impulses to the brain with the information that a sound is present. The moving part of the eardrum has an area of about 43 mm^2, and the area of the stapes (the smallest of the ossicles) where it connects to the inner ear is about 3.2 mm^2. For the sound in Example 16.1, determine (a) the pressure amplitude and (b) the displacement amplitude of the wave in the fluid of the inner ear. The speed of sound in this fluid is about 1500 m/s.

SOLUTION

IDENTIFY: Although the sound wave is now traveling in liquid (mostly water) rather than gas, the same principles and relationships among the properties of the wave apply.

SET UP: We can safely neglect the mass of the ossicles (about 58 mg $= 5.8 \times 10^{-5}$ kg), so the force exerted by the ossicles on the fluid in the inner ear is the same as the force exerted on the eardrum and ossicles by the sound wave in air. (We used this same idea in Chapters 4 and 5 when we said that the tension is the same at either end of a massless rope.) Hence the pressure amplitude p_{max} is greater in the inner ear than in the outside air because the same force is exerted on a smaller area (the area of the stapes versus the area of the eardrum).

Given the pressure amplitude in the inner ear, we find the displacement amplitude using Eq. (16.5). The values of B and k are different than in the air. To determine k, note that the wave in the inner ear has the same angular frequency ω as the wave in the air

16.4 The anatomy of the human ear. The middle ear is the size of a small marble; the ossicles (incus, malleus, and stapes) are the smallest bones in the human body.

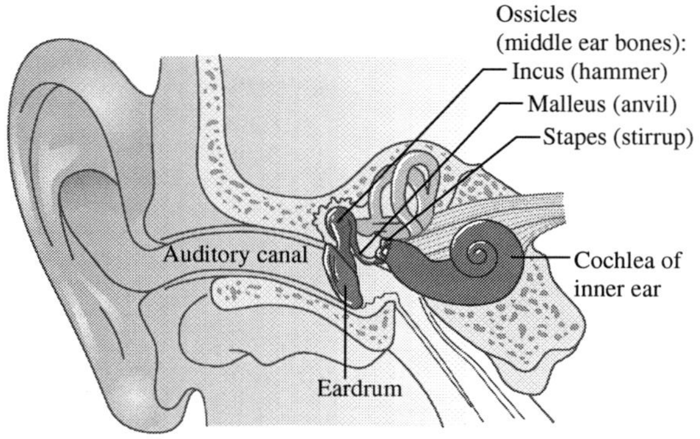

because the air, eardrum, ossicles, and inner-ear fluid all oscillate together. But because the wave speed v is greater in the inner ear than in the air $(1500 \text{ m/s}$ versus $344 \text{ m/s})$, the wave number $k = \omega/v$ is smaller.

EXECUTE: (a) Using the area of the eardrum and pressure amplitude found in Example 16.1, the maximum force exerted by the sound wave in air on the eardrum is $F_{max} = p_{max(air)}S_{eardrum}$. Hence the pressure amplitude in the inner ear fluid is

$$p_{max(inner\ ear)} = \frac{F_{max}}{S_{stapes}} = \frac{p_{max(air)}S_{eardrum}}{S_{stapes}}$$
$$= \frac{(3.0 \times 10^{-2}\text{ Pa})(43\text{ mm}^2)}{3.2\text{ mm}^2} = 0.40\text{ Pa}$$

(b) To find the maximum displacement, we again use the relationship $A = p_{max}/Bk$ as in Example 16.1. The fluid in the inner ear is mostly water, which has a much greater bulk modulus than air because water is much more difficult to compress. From Table 11.2 the compressibility of water (unfortunately also called k) equals $45.8 \times 10^{-11}\text{ Pa}^{-1}$, so $B_{fluid} = 1/(45.8 \times 10^{-11}\text{ Pa}^{-1}) = 2.18 \times 10^9\text{ Pa}$.

We determine the value of the wave number k using the value of ω from Example 16.1 and $v = 1500\text{ m/s}$ for the inner-ear fluid. Hence

$$k_{inner\ ear} = \frac{\omega}{v_{inner\ ear}} = \frac{(2\pi\text{ rad})(1000\text{ Hz})}{1500\text{ m/s}} = 4.2\text{ rad/m}$$

When we put everything together, the maximum displacement of the fluid in the inner ear is

$$A_{inner\ ear} = \frac{p_{max(inner\ ear)}}{B_{fluid}k_{inner\ ear}} = \frac{0.40\text{ Pa}}{(2.18 \times 10^9\text{ Pa})(4.2\text{ rad/m})}$$
$$= 4.4 \times 10^{-11}\text{ m}$$

EVALUATE: The result in part (a) shows that the effect of the ossicles is to increase the pressure amplitude in the inner ear by a factor of $(43\text{ mm}^2)/(3.2\text{ mm}^2) = 13$. This amplification factor helps give the human ear its great sensitivity.

The displacement amplitude in the inner ear is even smaller than in the air. What really matters in the inner ear, however, is the *pressure* amplitude, since the pressure variations within the fluid cause the forces that set the hair cells into motion.

Perception of Sound Waves

The physical characteristics of a sound wave are directly related to the perception of that sound by a listener. For a given frequency, the greater the pressure amplitude of a sinusoidal sound wave, the greater the perceived **loudness.** The relationship between pressure amplitude and loudness is not a simple one, and it varies from one person to another. One important factor is that the ear is not equally sensitive to all frequencies in the audible range. A sound at one frequency may seem louder than one of equal pressure amplitude at a different frequency. At 1000 Hz the minimum pressure amplitude that can be perceived with normal hearing is about 3×10^{-5} Pa; to produce the same loudness at 200 Hz or 15,000 Hz requires about 3×10^{-4} Pa. Perceived loudness also depends on the health of the ear. A loss of sensitivity at the high-frequency end usually happens naturally with age but can be further aggravated by excessive noise levels. Studies have shown that many young rock musicians have suffered permanent ear damage and have hearing that is typical of persons 65 years of age. Headphones for personal music players used at high volume pose similar threats to hearing. Be careful!

The frequency of a sound wave is the primary factor in determining the **pitch** of a sound, the quality that lets us classify the sound as "high" or "low." The higher the frequency of a sound (within the audible range), the higher the pitch that a listener will perceive. Pressure amplitude also plays a role in determining pitch. When a listener compares two sinusoidal sound waves with the same frequency but different pressure amplitudes, the one with the greater pressure amplitude is usually perceived as louder but also as slightly lower in pitch.

Musical sounds have wave functions that are more complicated than a simple sine function. The pressure fluctuation in the sound wave produced by a clarinet is shown in Fig. 16.5a. The pattern is so complex because the column of air in a wind instrument like a clarinet vibrates at a fundamental frequency and at many harmonics at the same time. (In Section 15.8, we described this same behavior for a string that has been plucked, bowed, or struck. We'll examine the physics of wind instruments in Section 16.5.) The sound wave produced in the surrounding air has a similar amount of each harmonic—that is, a similar *harmonic content.* Figure 16.5b shows the harmonic content of the sound of a clarinet. The mathematical process of translating a pressure–time graph like Fig. 16.5a into a graph of harmonic content like Fig. 16.5b is called *Fourier analysis.*

ActivPhysics Online
10.10 Complex Waves: Fourier Analysis

16.5 Different representations of the sound of (a), (b) a clarinet and (c), (d) an alto recorder. (Graphs adapted from R.E. Berg and D.G. Stork, *The Physics of Sound,* Prentice-Hall, 1982.)

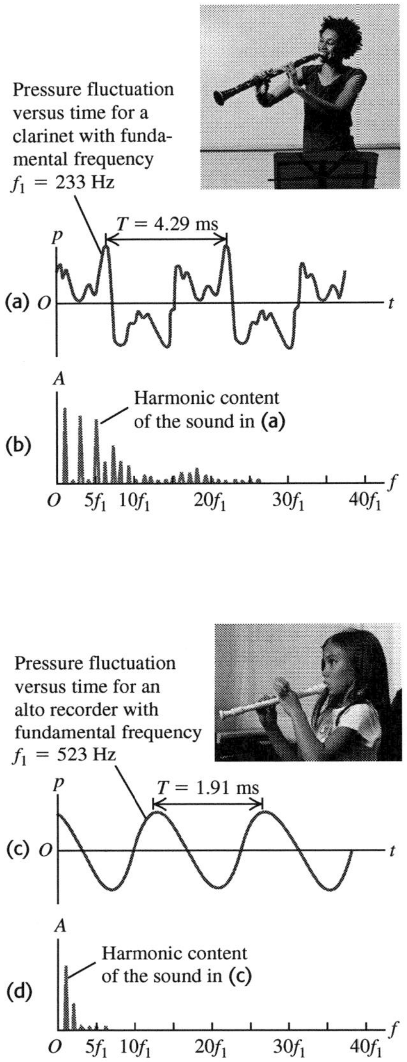

Two tones produced by different instruments might have the same fundamental frequency (and thus the same pitch) but sound different because of different harmonic content. The difference in sound is called *tone color, quality,* or **timbre** and is often described in subjective terms such as reedy, golden, round, mellow, and tinny. A tone that is rich in harmonics, like the clarinet tone in Figs. 16.5a and 16.5b, usually sounds thin and "stringy" or "reedy," while a tone containing mostly a fundamental, like the alto recorder tone in Figs. 16.5c and 16.5d, is more mellow and flutelike. The same principle applies to the human voice, which is another example of a wind instrument; the vowels "a" and "e" sound different because of differences in harmonic content.

Another factor in determining tone quality is the behavior at the beginning (*attack*) and end (*decay*) of a tone. A piano tone begins with a thump and then dies away gradually. A harpsichord tone, in addition to having different harmonic content, begins much more quickly with a click, and the higher harmonics begin before the lower ones. When the key is released, the sound also dies away much more rapidly with a harpsichord than with a piano. Similar effects are present in other musical instruments. With wind and string instruments the player has considerable control over the attack and decay of the tone, and these characteristics help to define the unique characteristics of each instrument.

Unlike the tones made by musical instruments or the vowels in human speech, **noise** is a combination of *all* frequencies, not just frequencies that are integer multiples of a fundamental frequency. (An extreme case is "white noise," which contains equal amounts of all frequencies across the audible range.) Examples include the sound of the wind and the hissing sound you make in saying the consonant "s." ?

Test Your Understanding of Section 16.1 You use an electronic signal generator to produce a sinusoidal sound wave in air. You then increase the frequency of the wave from 100 Hz to 400 Hz while keeping the pressure amplitude constant. What effect does this have on the displacement amplitude of the sound wave? (i) It becomes four times greater; (ii) it becomes twice as great; (iii) it is unchanged; (iv) it becomes $\frac{1}{2}$ as great; (v) it becomes $\frac{1}{4}$ as great.

16.2 Speed of Sound Waves

We found in Section 15.4 that the speed v of a transverse wave on a string depends on the string tension F and the linear mass density μ: $v = \sqrt{F/\mu}$. What, we may ask, is the corresponding expression for the speed of sound waves in a gas or liquid? On what properties of the medium does the speed depend?

We can make an educated guess about these questions by remembering a claim that we made in Section 15.4: For mechanical waves in general, the expression for the wave speed is of the form

$$v = \sqrt{\frac{\text{Restoring force returning the system to equilibrium}}{\text{Inertia resisting the return to equilibrium}}}$$

A sound wave in a bulk fluid causes compressions and rarefactions of the fluid, so the restoring-force term in the above expression must be related to how easy or difficult it is to compress the fluid. This is precisely what the bulk modulus B of the medium tells us. According to Newton's second law, inertia is related to mass. The "massiveness" of a bulk fluid is described by its density, or mass per unit volume, ρ. (The corresponding quantity for a string is the mass per unit length, μ.) Hence we expect that the speed of sound waves should be of the form $v = \sqrt{B/\rho}$.

To check our guess, we'll derive the speed of sound waves in a fluid in a pipe. This is a situation of some importance, since all musical wind instruments are fundamentally pipes in which a longitudinal wave (sound) propagates in a fluid (air)

(Fig. 16.6). Human speech works on the same principle; sound waves propagate in your vocal tract, which is basically an air-filled pipe connected to the lungs at one end (your larynx) and to the outside air at the other end (your mouth). The steps in our derivation are completely parallel to those we used in Section 15.4 to find the speed of transverse waves, so you'll find it useful to review that section.

16.6 When a wind instrument like this French horn is played, sound waves propagate through the air within the instrument's pipes. The properties of the sound that emerges from the large bell depend on the speed of these waves.

Speed of Sound in a Fluid

Figure 16.7 shows a fluid (either liquid or gas) with density ρ in a pipe with cross-sectional area A. In the equilibrium state, the fluid is under a uniform pressure p. In Fig. 16.7a the fluid is at rest. We take the x-axis along the length of the pipe. This is also the direction in which we make a longitudinal wave propagate, so the displacement y is also measured along the pipe, just as in Section 16.1 (see Fig. 16.2).

At time $t = 0$ we start the piston at the left end moving toward the right with constant speed v_y. This initiates a wave motion that travels to the right along the length of the pipe, in which successive sections of fluid begin to move and become compressed at successively later times.

Figure 16.7b shows the fluid at time t. All portions of fluid to the left of point P are moving to the right with speed v_y, and all portions to the right of P are still at rest. The boundary between the moving and stationary portions travels to the right with a speed equal to the speed of propagation or wave speed v. At time t the piston has moved a distance $v_y t$, and the boundary has advanced a distance vt. As with a transverse disturbance in a string, we can compute the speed of propagation from the impulse–momentum theorem.

16.7 A sound wave propagating in a fluid confined to a tube. **(a)** Fluid in equilibrium. **(b)** A time t after the piston begins moving to the right at speed v_y, the fluid between the piston and point P is in motion. The speed of sound waves is v.

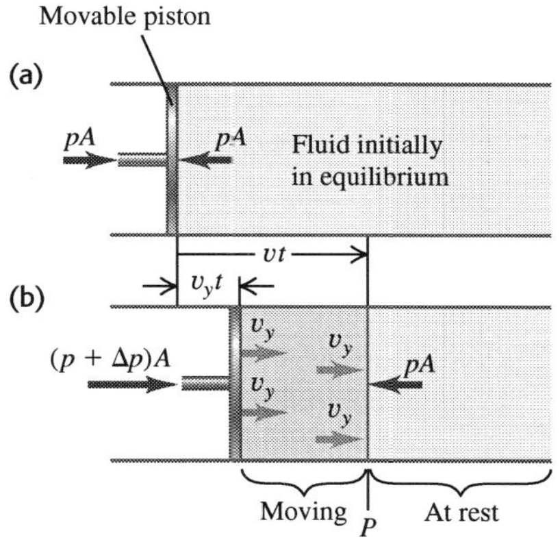

The quantity of fluid set in motion in time t is the amount that originally occupied a section of the cylinder with length vt, cross-sectional area A, and volume vtA. The mass of this fluid is ρvtA, and its longitudinal momentum (that is, momentum along the length of the pipe) is

$$\text{Longitudinal momentum} = (\rho vtA)v_y$$

Next we compute the increase of pressure, Δp, in the moving fluid. The original volume of the moving fluid, Avt, has decreased by an amount $Av_y t$. From the definition of the bulk modulus B, Eq. (11.13) in Section 11.5,

$$B = \frac{-\text{Pressure change}}{\text{Fractional volume change}} = \frac{-\Delta p}{-Av_y t/Avt}$$

$$\Delta p = B\frac{v_y}{v}$$

The pressure in the moving fluid is $p + \Delta p$ and the force exerted on it by the piston is $(p + \Delta p)A$. The net force on the moving fluid (see Fig. 16.7b) is ΔpA, and the longitudinal impulse is

$$\text{Longitudinal impulse} = \Delta pAt = B\frac{v_y}{v}At$$

Because the fluid was at rest at time $t = 0$, the change in momentum up to time t is equal to the momentum at that time. Applying the impulse–momentum theorem (Section 8.1), we find

$$B\frac{v_y}{v}At = \rho vtAv_y \quad (16.6)$$

When we solve this expression for v, we get

$$v = \sqrt{\frac{B}{\rho}} \quad \text{(speed of a longitudinal wave in a fluid)} \quad (16.7)$$

which agrees with our educated guess. Thus the speed of propagation of a longitudinal pulse in a fluid depends only on the bulk modulus B and the density ρ of the medium.

While we derived Eq. (16.7) for waves in a pipe, it also applies to longitudinal waves in a bulk fluid. Thus the speed of sound waves traveling in air or water is determined by this equation.

Speed of Sound in a Solid

When a longitudinal wave propagates in a *solid* rod or bar, the situation is somewhat different. The rod expands sideways slightly when it is compressed longitudinally, while a fluid in a pipe with constant cross section cannot move sideways. Using the same kind of reasoning that led us to Eq. (16.7), we can show that the speed of a longitudinal pulse in the rod is given by

$$v = \sqrt{\frac{Y}{\rho}} \quad \text{(speed of a longitudinal wave in a solid rod)} \qquad (16.8)$$

where Y is Young's modulus, defined in Section 11.4.

CAUTION **Solid rods vs. bulk solids** Equation (16.8) applies only to a rod or bar whose sides are free to bulge and shrink a little as the wave travels. It does not apply to longitudinal waves in a *bulk* solid, since in these materials, sideways motion in any element of material is prevented by the surrounding material. The speed of longitudinal waves in a bulk solid depends on the density, the bulk modulus, and the *shear* modulus; a full discussion is beyond the scope of this book.

As with the derivation for a transverse wave on a string, Eqs. (16.7) and (16.8) are valid for sinusoidal and other periodic waves, not just for the special case discussed here.

Table 16.1 lists the speed of sound in several bulk materials. Sound waves travel more slowly in lead than in aluminum or steel because lead has a lower bulk modulus and shear modulus and a higher density.

Table 16.1 Speed of Sound in Various Bulk Materials

Material	Speed of Sound (m/s)
Gases	
Air (20°C)	344
Helium (20°C)	999
Hydrogen (20°C)	1330
Liquids	
Liquid helium (4 K)	211
Mercury (20°C)	1451
Water (0°C)	1402
Water (20°C)	1482
Water (100°C)	1543
Solids	
Aluminum	6420
Lead	1960
Steel	5941

Example 16.3 Wavelength of sonar waves

A ship uses a sonar system to detect underwater objects (Fig. 16.8). The system emits underwater sound waves and measures the time interval for the reflected wave (echo) to return to the detector. Determine the speed of sound waves in water using Eq. (16.7) and find the wavelength of a 262-Hz wave.

SOLUTION

IDENTIFY: Our target variables are the speed and wavelength of a sound wave in a fluid.

16.8 A sonar system uses underwater sound waves to detect and locate submerged objects.

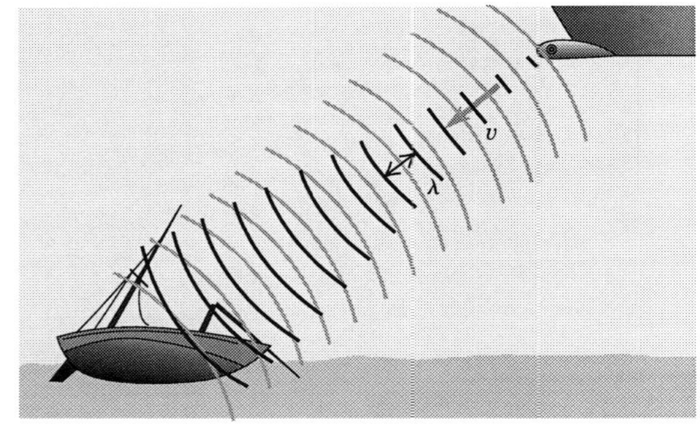

SET UP: To use Eq. (16.7) for the wave speed, we find the bulk modulus of water from the compressibility (Table 11.2) and the density ($\rho = 1.00 \times 10^3\ \text{kg/m}^3$). Given the speed and the frequency $f = 262$ Hz, we find the wavelength from the relationship $v = f\lambda$.

EXECUTE: From Table 11.2 we find that the compressibility of water, which is the reciprocal of the bulk modulus, is $k = 45.8 \times 10^{-11}\ \text{Pa}^{-1}$. Thus $B = (1/45.8) \times 10^{11}$ Pa. We obtain

$$v = \sqrt{\frac{B}{\rho}} = \sqrt{\frac{(1/45.8) \times 10^{11}\ \text{Pa}}{1.00 \times 10^3\ \text{kg/m}^3}} = 1480\ \text{m/s}$$

The wavelength is given by

$$\lambda = \frac{v}{f} = \frac{1480\ \text{m/s}}{262\ \text{s}^{-1}} = 5.65\ \text{m}$$

EVALUATE: The calculated value of v agrees well with the experimental value in Table 16.1. Though water is far denser than air (ρ is larger), it is also far more incompressible (B is larger) and the speed $v = \sqrt{B/\rho}$ turns out to be more than four times the speed of sound in air at ordinary temperatures.

We found in Example 15.1 (Section 15.2) that a wave of this frequency in air has a wavelength of 1.31 m. The speed of sound in water is greater than in air, so the wavelength $\lambda = v/f$ must be greater as well; this is just as we calculated.

Dolphins emit high-frequency sound waves (typically 100,000 Hz) and use the echoes for guidance and for hunting. The corresponding wavelength in water is 1.48 cm. With this high-frequency "sonar" system they can sense objects that are roughly as small as the wavelength (but not much smaller). *Ultrasonic imaging* is a medical technique that uses exactly the same physical principle; sound waves of very high frequency and very short wavelength, called *ultrasound,* are scanned over the human body, and the "echoes" from interior organs are used to create an image. With ultrasound of frequency 5 MHz $= 5 \times 10^6$ Hz, the wavelength in water (the primary constituent of the body) is 0.3 mm, and features as small as this can be discerned in the image. Ultrasound is used for the study of heart-valve action, detection of tumors, and prenatal examinations (Fig. 16.9). Ultrasound is more sensitive than x rays in distinguishing various kinds of tissues and does not have the radiation hazards associated with x rays.

Example 16.4 Speed of a longitudinal wave

What is the speed of longitudinal waves in a lead rod?

SOLUTION

IDENTIFY: Waves of this kind are made by clamping the rod in place and striking one end face-on with a hammer. The amplitude of the resulting waves is so small as to be nearly invisible to the naked eye. But the question is about the *speed* of the waves, which does not depend on the amplitude. Note that we cannot simply use the value for lead from Table 16.1, as that value refers to the speed of sound in a *bulk* material, not a rod.

SET UP: Equation (16.8) applies to this situation. We find the values of Young's modulus Y and the density ρ from Tables 11.1 and 14.1, respectively.

EXECUTE: We find $Y = 1.6 \times 10^{10}$ Pa and $\rho = 11.3 \times 10^3\ \mathrm{kg/m^3}$ (that is, 11.3 times the density of water), so

$$v = \sqrt{\frac{Y}{\rho}} = \sqrt{\frac{1.6 \times 10^{10}\ \mathrm{Pa}}{11.3 \times 10^3\ \mathrm{kg/m^3}}} = 1.2 \times 10^3\ \mathrm{m/s}$$

EVALUATE: This is more than three times the speed of sound in air, but slower than the speed of sound in *bulk* lead (see Table 16.1). The reason is that for lead the bulk modulus is greater than Young's modulus.

Speed of Sound in a Gas

Most of the sound waves that we encounter on a daily basis propagate in air. To use Eq. (16.7) to find the speed of sound waves in air, we must keep in mind that the bulk modulus of a gas depends on the pressure of the gas: The greater the pressure applied to a gas to compress it, the more it resists further compression and hence the greater the bulk modulus. (That's why specific values of the bulk modulus for gases are not given in Table 11.1.) The expression for the bulk modulus of a gas for use in Eq. (16.7) is

$$B = \gamma p_0 \tag{16.9}$$

where p_0 is the equilibrium pressure of the gas. The quantity γ (the Greek letter gamma) is called the *ratio of heat capacities.* It is a dimensionless number that characterizes the thermal properties of the gas. (We'll learn more about this quantity in Chapter 19.) As an example, the ratio of heat capacities for air is $\gamma = 1.40$. At normal atmospheric pressure $p_0 = 1.01 \times 10^5$ Pa, so $B = (1.40)(1.013 \times 10^5\ \mathrm{Pa}) = 1.42 \times 10^5$ Pa. This value is minuscule compared to the bulk modulus of a typical solid (see Table 11.1), which is approximately 10^{10} to 10^{11} Pa. This shouldn't be surprising: It's simply a statement that air is far easier to compress than steel.

The density ρ of a gas also depends on the pressure, which in turn depends on the temperature. It turns out that the ratio B/ρ for a given type of ideal gas does *not* depend on the pressure at all, only the temperature. From Eq. (16.7), this means that the speed of sound in a gas is fundamentally a function of temperature T:

$$v = \sqrt{\frac{\gamma RT}{M}} \quad \text{(speed of sound in an ideal gas)} \tag{16.10}$$

16.9 This three-dimensional image of a fetus in the womb was made using a sequence of ultrasound scans. Each individual scan reveals a two-dimensional "slice" through the fetus; a large number of these slices were then combined digitally to yield this remarkable view.

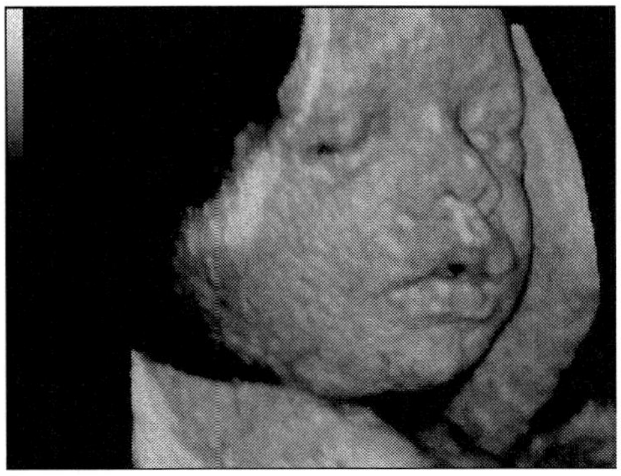

ActivPhysics ONLINE
10.3 Speed of Sound in a Gas

This expression incorporates several quantities that you may recognize from your study of ideal gases in chemistry and which we will study in Chapters 17, 18, and 19. The temperature T is the *absolute* temperature in kelvins (K), equal to the Celsius temperature plus 273.15; thus 20.00°C corresponds to $T = 293.15$ K. The quantity M is the *molar mass,* or mass per mole of the substance of which the gas is composed. The *gas constant* R has the same value for all gases. The current best numerical value of R is

$$R = 8.314472(15)\ \text{J/mol} \cdot \text{K}$$

which for practical calculations we can write as 8.314 J/mol · K.

For any particular gas, γ, R, and M are constants, and the wave speed is proportional to the square root of the absolute temperature. We will see in Chapter 18 that Eq. (16.10) is almost identical to the expression for the average speed of molecules in an ideal gas. This shows that sound speeds and molecular speeds are closely related; exploring that relationship in detail would be beyond our scope.

Example 16.5 Speed of sound in air

Compute the speed of sound waves in air at room temperature $(T = 20°\text{C})$ and find the range of wavelengths in air to which the human ear (which can hear frequencies in the range of 20–20,000 Hz) is sensitive. The mean molar mass for air (a mixture of principally nitrogen and oxygen) is 28.8×10^{-3} kg/mol and the ratio of heat capacities is $\gamma = 1.40$.

SOLUTION

IDENTIFY: This problem uses the relationship between the properties of a gas and the speed of sound in that gas.

SET UP: We use Eq. (16.10) to find the sound speed and the relationship $v = f\lambda$ to determine the wavelength that corresponds to each frequency.

EXECUTE: At $T = 20°\text{C} = 293$ K we find

$$v = \sqrt{\frac{\gamma RT}{M}}$$

$$= \sqrt{\frac{(1.40)(8.314\ \text{J/mol} \cdot \text{K})(293\ \text{K})}{28.8 \times 10^{-3}\ \text{kg/mol}}} = 344\ \text{m/s}$$

Using this value of v and the expression $\lambda = v/f$, we find that at 20°C a 20-Hz note corresponds to a wavelength of 17 m and a 20,000-Hz note corresponds to a wavelength of 1.7 cm.

EVALUATE: Our calculated value of v agrees with the measured speed of sound at this temperature to within 0.3%.

It's interesting to note that bats can hear much higher frequencies. Like dolphins, bats use high-frequency sound waves for navigation. A typical frequency is 100 kHz; the corresponding wavelength in air at 20°C is about 3.4 mm. It turns out that sound waves with such short wavelengths are more easily reflected from small objects than are long-wavelength waves. This is important for certain species of bats that track the reflected sound from the small flying insects that they eat.

In this discussion we have ignored the *molecular* nature of a gas and have treated it as a continuous medium. A gas is actually composed of molecules in random motion, separated by distances that are large in comparison with their diameters. The vibrations that constitute a wave in a gas are superposed on the random thermal motion. At atmospheric pressure, a molecule travels an average distance of about 10^{-7} m between collisions, while the displacement amplitude of a faint sound may be only 10^{-9} m. We can think of a gas with a sound wave passing through as being comparable to a swarm of bees; the swarm as a whole oscillates slightly while individual insects move about through the swarm, apparently at random.

Test Your Understanding of Section 16.2 Mercury is 13.6 times denser than water. Based on Table 16.1, at 20°C which of these liquids has the greater bulk modulus? (i) mercury; (ii) water; (iii) both are about the same; (iv) not enough information is given to decide.

16.3 Sound Intensity

Traveling sound waves, like all other traveling waves, transfer energy from one region of space to another. We saw in Section 15.5 that a useful way to describe the energy carried by a sound wave is through the *wave intensity* I, equal to the time average rate at which energy is transported per unit area across a surface perpendicular to the direction of propagation. Let's see how to express the intensity of a sound wave in terms of the displacement amplitude A or pressure amplitude p_{max}.

Intensity and Displacement Amplitude

For simplicity, let us consider a sound wave propagating in the $+x$-direction so that we can use our expressions from Section 16.1 for the displacement $y(x, t)$ and pressure fluctuation $p(x, t)$—Eqs. (16.1) and (16.4), respectively. In Section 6.4 we saw that power equals the product of force and velocity [see Eq. (6.18)]. So the power per unit area in this sound wave equals the product of $p(x, t)$ (force per unit area) and the *particle* velocity $v_y(x, t)$. The particle velocity $v_y(x, t)$ is the velocity at time t of that portion of the wave medium at coordinate x. Using Eqs. (16.1) and (16.4), we find

$$\begin{aligned} v_y(x, t) &= \frac{\partial y(x, t)}{\partial t} = \omega A \sin(kx - \omega t) \\ p(x, t)v_y(x, t) &= [BkA \sin(kx - \omega t)][\omega A \sin(kx - \omega t)] \\ &= B\omega kA^2 \sin^2(kx - \omega t) \end{aligned}$$

CAUTION **Wave velocity vs. particle velocity** Remember that the velocity of the wave as a whole is *not* the same as the particle velocity. While the wave continues to move in the direction of propagation, individual particles in the wave medium merely slosh back and forth, as shown in Fig. (16.1). Furthermore, the maximum speed of a particle of the medium can be very different from the wave speed.

The intensity is, by definition, the time average value of $p(x, t)v_y(x, t)$. For any value of x the average value of the function $\sin^2(kx - \omega t)$ over one period $T = 2\pi/\omega$ is $\frac{1}{2}$, so

$$I = \frac{1}{2}B\omega kA^2 \qquad (16.11)$$

By using the relationships $\omega = vk$ and $v^2 = B/\rho$, we can transform Eq. (16.11) into the form

$$I = \frac{1}{2}\sqrt{\rho B}\,\omega^2 A^2 \quad \text{(intensity of a sinusoidal sound wave)} \qquad (16.12)$$

This equation shows why in a stereo system, a low-frequency woofer has to vibrate with much larger amplitude than a high-frequency tweeter to produce the same sound intensity.

Intensity and Pressure Amplitude

It is usually more useful to express I in terms of the pressure amplitude p_{max}. Using Eq. (16.5) and the relationship $\omega = vk$, we find

$$I = \frac{\omega p_{max}^2}{2Bk} = \frac{v p_{max}^2}{2B} \qquad (16.13)$$

By using the wave speed relationship $v^2 = B/\rho$, we can also write Eq. (16.13) in the alternative forms

$$I = \frac{p_{max}^2}{2\rho v} = \frac{p_{max}^2}{2\sqrt{\rho B}} \quad \text{(intensity of a sinusoidal sound wave)} \qquad (16.14)$$

16.10 By cupping your hands like this, you direct the sound waves emerging from your mouth so that they don't propagate to the sides. Hence the intensity decreases with distance more slowly than the inverse-square law would predict, and you can be heard at greater distances.

You should verify these expressions (see Exercise 16.16). Comparison of Eqs. (16.12) and (16.14) shows that sinusoidal sound waves of the same intensity but different frequency have different displacement amplitudes A but the *same* pressure amplitude p_{max}. This is another reason it is usually more convenient to describe a sound wave in terms of pressure fluctuations, not displacement.

The *total* average power carried across a surface by a sound wave equals the product of the intensity at the surface and the surface area if the intensity over the surface is uniform. The average total sound power emitted by a person speaking in an ordinary conversational tone is about 10^{-5} W, while a loud shout corresponds to about 3×10^{-2} W. If all the residents of New York City were to talk at the same time, the total sound power would be about 100 W, equivalent to the electric power requirement of a medium-sized light bulb. On the other hand, the power required to fill a large auditorium or stadium with loud sound is considerable (see Example 16.8.)

If the sound source emits waves in all directions equally, the intensity decreases with increasing distance r from the source according to the inverse-square law: The intensity is proportional to $1/r^2$. We discussed this law and its consequences in Section 15.5. If the sound goes predominantly in one direction, the inverse-square law does not apply and the intensity decreases with distance more slowly than $1/r^2$ (Fig. 16.10).

The inverse-square relationship also does not apply indoors because sound energy can reach a listener by reflection from the walls and ceiling. Indeed, part of the architect's job in designing an auditorium is to tailor these reflections so that the intensity is as nearly uniform as possible over the entire auditorium.

Problem-Solving Strategy 16.1 Sound Intensity

IDENTIFY *the relevant concepts:* The relationships between intensity and amplitude of a sound wave are rather straightforward. Quite a few other quantities are involved in these relationships, however, so it's particularly important to decide which is your target variable.

SET UP *the problem* using the following steps:

1. Sort the various physical quantities into categories. The amplitude is described by A or p_{max}, and the frequency f can be determined from ω, k, or λ. These quantities are related through the wave speed v, which in turn is determined by the properties of the medium: B and ρ for a liquid; γ, T, and M for a gas.
2. Determine which quantities are given and which are the unknown target variables. Then start looking for relationships that take you where you want to go.

EXECUTE *the solution* as follows: Use the equations you've selected to solve for the target variables. Be certain that all of the quantities are expressed in the correct units. In particular, if temperature is used to calculate the speed of sound in a gas, make sure that it is expressed in kelvins (Celsius temperature plus 273.15).

EVALUATE *your answer:* There are multiple relationships among the quantities that describe a wave. Try using an alternative one to check your results.

Example 16.6 Intensity of a sound wave in air

Find the intensity of the sound wave in Example 16.1, with $p_{max} = 3.0 \times 10^{-2}$ Pa. Assume the temperature is 20°C so that the density of air is $\rho = 1.20\ \text{kg/m}^3$ and the speed of sound is $v = 344$ m/s.

SOLUTION

IDENTIFY: Our target variable is the intensity I of the sound wave. We are given the pressure amplitude p_{max} of the wave as well as the density ρ and wave speed v for the medium.

SET UP: We can determine I from p_{max}, ρ, and v using Eq. (16.14).

EXECUTE: From Eq. (16.14),

$$I = \frac{p_{max}{}^2}{2\rho v} = \frac{(3.0 \times 10^{-2}\ \text{Pa})^2}{2(1.20\ \text{kg/m}^3)(344\ \text{m/s})}$$
$$= 1.1 \times 10^{-6}\ \text{J/(s}\cdot\text{m}^2) = 1.1 \times 10^{-6}\ \text{W/m}^2$$

EVALUATE: This seems like a very low intensity, but in fact it is well within the range of sound intensities encountered on a daily basis. A very loud sound wave at the threshold of pain has a pressure amplitude of about 30 Pa and an intensity of about 1W/m^2. The pressure amplitude of the faintest sound wave that can be heard is about 3×10^{-5} Pa, and the corresponding intensity is about $10^{-12}\ \text{W/m}^2$. (Try these values of p_{max} in Eq. (16.14) to check that the corresponding intensities are as we have stated.)

Example 16.7 Same intensity, different frequencies

A 20-Hz sound wave has the same intensity as the 1000-Hz sound wave in Examples 16.1 and 16.6. What are the displacement amplitude and pressure amplitude of the 20-Hz sound wave?

SOLUTION

IDENTIFY: We are given the intensity and frequency of the sound wave. Our target variables are the displacement amplitude A and pressure amplitude p_{max}.

SET UP: Given the intensity, we can use Eq. (16.12) to determine the displacement amplitude A. Note that we are not given the value of B. However, ρ and B depend on only the properties of the medium, not the amplitude or frequency, so their values will cancel out if we equate the intensities at 20 Hz and at 1000 Hz. We can also use Eq. (16.14) to find the pressure amplitude at 20 Hz.

EXECUTE: Inspection of Eq. (16.12) shows that if a wave in a given medium (same ρ and B) has the same intensity I at two different frequencies, then the product ωA must have the same value for both frequencies. From Example 16.1, $A = 1.2 \times 10^{-8}$ m at 1000 Hz, so

$$(20\ \text{Hz})A_{20} = (1000\ \text{Hz})(1.2 \times 10^{-8}\ \text{m})$$
$$A_{20} = 6.0 \times 10^{-7}\ \text{m} = 0.60\ \mu\text{m}$$

Do you understand why we didn't have to convert the frequencies to angular frequencies?

Since the intensity is the same for both frequencies, Eq. (16.14) shows that the *pressure* amplitude p_{max} must also be the same for both. Hence $p_{max} = 3.0 \times 10^{-2}$ Pa for $f = 20$ Hz.

EVALUATE: Our result reinforces the idea that pressure amplitude offers a more convenient description of a sound wave than displacement amplitude. Note also that using Eq. (16.5) and $k = \omega/v$, we get $p_{max} = BkA = (B/v)\omega A$; the bulk modulus B and wave speed v depend on only the medium, so we again conclude that the product ωA must have the same value for both frequencies.

Example 16.8 "Play it loud!"

For an outdoor concert we want the sound intensity at a distance of 20 m from the speaker array to be 1 W/m^2. Assuming that the sound waves have the same intensity in all directions, what acoustic power output is needed from the speaker array?

SOLUTION

IDENTIFY: This example uses the definition of intensity as power per unit area. Here the total power is the target variable, and the area in question is a hemisphere centered on the speaker array.

SET UP: We make the assumptions that the speakers are near ground level and that the acoustic power is spread uniformly over a hemisphere 20 m in radius (that is, we assume that none of the acoustic power is directed into the ground). The surface area of this hemisphere is equal to $(\frac{1}{2})(4\pi)(20\ \text{m})^2$, or about 2500 m^2. The required power is the product of this area and the intensity.

EXECUTE: The speaker array power is

$$(1\ \text{W/m}^2)(2500\ \text{m}^2) = 2500\ \text{W} = 2.5\ \text{kW}$$

EVALUATE: The electrical power input to the speaker would need to be considerably larger because the efficiency of such devices is not very high (typically a few percent for ordinary speakers, and up to 25% for horn-type speakers).

The Decibel Scale

Because the ear is sensitive over a broad range of intensities, a *logarithmic* intensity scale is usually used. The **sound intensity level** β of a sound wave is defined by the equation

$$\beta = (10\ \text{dB}) \log \frac{I}{I_0} \quad \text{(definition of sound intensity level)} \quad (16.15)$$

In this equation, I_0 is a reference intensity, chosen to be 10^{-12} W/m^2, approximately the threshold of human hearing at 1000 Hz. Recall that "log" means the logarithm to base 10. Sound intensity levels are expressed in **decibels,** abbreviated dB. A decibel is $\frac{1}{10}$ of a *bel,* a unit named for Alexander Graham Bell (the inventor of the telephone). The bel is inconveniently large for most purposes, and the decibel is the usual unit of sound intensity level.

If the intensity of a sound wave equals I_0 or 10^{-12} W/m^2, its sound intensity level is 0 dB. An intensity of 1 W/m^2 corresponds to 120 dB. Table 16.2 gives the sound intensity levels in decibels of several familiar sounds. You can use Eq. (16.15) to check the value of sound intensity level β given for each intensity in the table.

Table 16.2 Sound Intensity Levels from Various Sources (Representative Values)

Source or Description of Sound	Sound Intensity Level, β (dB)	Intensity, I (W/m^2)
Military jet aircraft 30 m away	140	10^2
Threshold of pain	120	1
Riveter	95	3.2×10^{-3}
Elevated train	90	10^{-3}
Busy street traffic	70	10^{-5}
Ordinary conversation	65	3.2×10^{-6}
Quiet automobile	50	10^{-7}
Quiet radio in home	40	10^{-8}
Average whisper	20	10^{-10}
Rustle of leaves	10	10^{-11}
Threshold of hearing at 1000 Hz	0	10^{-12}

Because the ear is not equally sensitive to all frequencies in the audible range, some sound-level meters weight the various frequencies unequally. One such scheme leads to the so-called dBA scale; this scale deemphasizes the low and very high frequencies, where the ear is less sensitive than at midrange frequencies.

Example 16.9 Temporary deafness

A 10-minute exposure to 120-dB sound will typically shift your threshold of hearing at 1000 Hz from 0 dB up to 28 dB for a while. Ten years of exposure to 92-dB sound will cause a *permanent* shift up to 28 dB. What intensities correspond to 28 dB and 92 dB?

SOLUTION

IDENTIFY: We are given two different sound intensity levels in dB, and we want to find the corresponding intensities in W/m^2.

SET UP: We can solve Eq. (16.15) to find the intensity I that corresponds to each value of the sound intensity level β.

EXECUTE: We rearrange Eq. (16.15) by dividing both sides by 10 dB and then using the relationship $10^{\log x} = x$:

$$I = I_0 10^{\beta/(10\text{ dB})}$$

When $\beta = 28$ dB,

$$I = (10^{-12}\text{ W/m}^2)10^{(28\text{ dB}/10\text{ dB})}$$
$$= (10^{-12}\text{ W/m}^2)10^{2.8} = 6.3 \times 10^{-10}\text{ W/m}^2$$

Similarly, for $\beta = 92$ dB,

$$I = (10^{-12}\text{ W/m}^2)10^{(92\text{ dB}/10\text{ dB})} = 1.6 \times 10^{-3}\text{ W/m}^2$$

EVALUATE: If your answers are a factor of 10 too large, you may have entered 10×10^{-12} in your calculator instead of 1×10^{-12}. Be careful!

Example 16.10 A bird sings in a meadow

Consider an idealized model with a bird (treated as a point source) emitting constant sound power, with intensity inversely proportional to the square of the distance from the bird. By how many decibels does the sound intensity level drop when you move twice as far away from the bird?

SOLUTION

IDENTIFY: Because the decibel scale is logarithmic, the *difference* between two sound intensity levels (the target variable) corresponds to the *ratio* of the corresponding intensities. The ratio of the intensities comes from the inverse-square law.

SET UP: We label the two points P_1 and P_2 (Fig. 16.11). We use Eq. (16.15), the definition of sound intensity level, twice (once at each point). We use Eq. (15.26), the statement of the inverse-square law, to relate the intensities at the two points.

16.11 When you double your distance from a point source of sound, by how much does the sound intensity level decrease?

EXECUTE: The difference in sound intensity level, $\beta_2 - \beta_1$, is given by

$$\beta_2 - \beta_1 = (10\text{ dB})\left(\log\frac{I_2}{I_0} - \log\frac{I_1}{I_0}\right)$$
$$= (10\text{ dB})[(\log I_2 - \log I_0) - (\log I_1 - \log I_0)]$$
$$= (10\text{ dB})\log\frac{I_2}{I_1}$$

Now we use the reciprocal of Eq. (15.26); $I_2/I_1 = r_1^2/r_2^2$, so

$$\beta_2 - \beta_1 = (10\text{ dB})\log\frac{r_1^2}{r_2^2} = (10\text{ dB})\log\frac{r_1^2}{(2r_1)^2}$$
$$= (10\text{ dB})\log\frac{1}{4} = -6.0\text{ dB}$$

A decrease in intensity of a factor of 4 corresponds to a 6-dB decrease in sound intensity level.

EVALUATE: Our result is negative, which tells us (correctly) that the sound intensity level is less at P_2 than at P_1. The 6-dB difference doesn't depend on the value of the sound intensity level at P_1. If P_1 is relatively close to the bird so that $\beta_1 = 56$ dB, then at a point P_2 twice as far away, $\beta_2 = 50$ dB; if point P_1 is more distant from the bird so that $\beta_1 = 28$ dB, then at a point P_2 twice as far away, $\beta_2 = 22$ dB.

It's interesting to note that the perceived *loudness* of a sound is not directly proportional to its intensity. As an example, most people usually interpret an increase of 8 to 10 dB in sound intensity level (corresponding to an intensity that increases by a factor of 6 to 10) as a doubling of loudness.

Test Your Understanding of Section 16.3 You double the intensity of a sound wave in air while leaving the frequency unchanged. (The pressure, density, and temperature of the air remain unchanged as well.) What effect does this have on the displacement amplitude, pressure amplitude, bulk modulus, sound speed, and sound intensity level?

16.4 Standing Sound Waves and Normal Modes

When longitudinal (sound) waves propagate in a fluid in a pipe with finite length, the waves are reflected from the ends in the same way that transverse waves on a string are reflected at its ends. The superposition of the waves traveling in opposite directions again forms a standing wave. Just as for transverse standing waves on a string (see Section 15.7), standing sound waves (normal modes) in a pipe can be used to create sound waves in the surrounding air. This is the operating principle of the human voice as well as many musical instruments, including woodwinds, brasses, and pipe organs.

Transverse waves on a string, including standing waves, are usually described only in terms of the displacement of the string. But, as we have seen, sound waves in a fluid may be described either in terms of the displacement of the fluid or in terms of the pressure variation in the fluid. To avoid confusion, we'll use the terms **displacement node** and **displacement antinode** to refer to points where particles of the fluid have zero displacement and maximum displacement, respectively.

We can demonstrate standing sound waves in a column of gas using an apparatus called Kundt's tube (Fig. 16.12). A horizontal glass tube a meter or so long

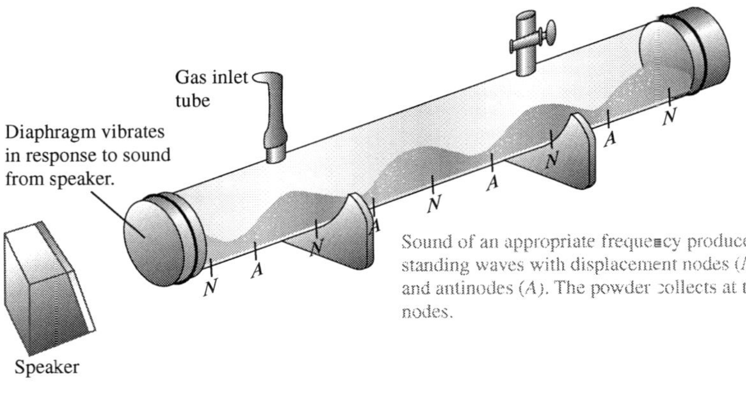

16.12 Demonstrating standing sound waves using a Kundt's tube. The blue shading represents the density of the gas at an instant when the gas pressure at the displacement nodes is a maximum or a minimum.

16.13 In a standing sound wave, a displacement node N is a pressure antinode (a point where the pressure fluctuates the most) and a displacement antinode A is a pressure node (a point where the pressure does not fluctuate at all).

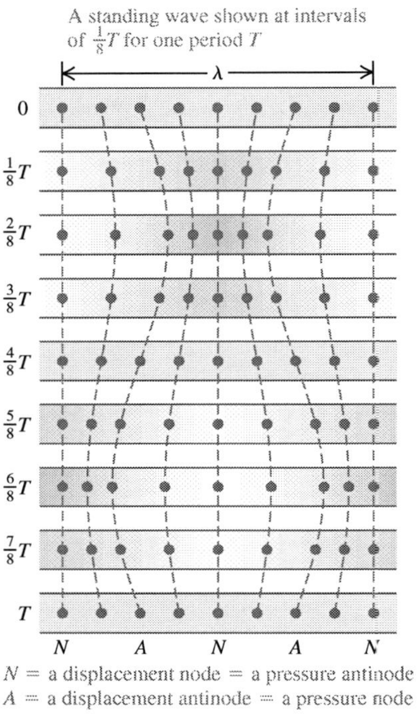

is closed at one end and has a flexible diaphragm at the other end that can transmit vibrations. A nearby loudspeaker is driven by an audio oscillator and amplifier; this produces sound waves that force the diaphragm to vibrate sinusoidally with a frequency that we can vary. The sound waves within the tube are reflected at the other, closed end of the tube. We spread a small amount of light powder uniformly along the bottom of the tube. As we vary the frequency of the sound, we pass through frequencies at which the amplitude of the standing waves becomes large enough for the powder to be swept along the tube at those points where the gas is in motion. The powder therefore collects at the displacement nodes (where the gas is not moving). Adjacent nodes are separated by a distance equal to $\lambda/2$, and we can measure this distance. Given the wavelength, we can use this experiment to determine the wave speed: We read the frequency f from the oscillator dial, and we can then calculate the speed v of the waves from the relationship $v = \lambda f$.

Figure 16.13 shows the motions of nine different particles within a gas-filled tube in which there is a standing sound wave. A particle at a displacement node (N) does not move, while a particle at a displacement antinode (A) oscillates with maximum amplitude. Note that particles on opposite sides of a displacement node vibrate in opposite phase. When these particles approach each other, the gas between them is compressed and the pressure rises; when they recede from each other, there is an expansion and the pressure drops. Hence at a displacement *node* the gas undergoes the maximum amount of compression and expansion, and the variations in pressure and density above and below the average have their maximum value. By contrast, particles on opposite sides of a displacement *antinode* vibrate *in phase;* the distance between the particles is nearly constant, and there is *no* variation in pressure or density at a displacement antinode.

We use the term **pressure node** to describe a point in a standing sound wave at which the pressure and density do not vary and the term **pressure antinode** to describe a point at which the variations in pressure and density are greatest. Using these terms, we can summarize our observations about standing sound waves as follows:

A pressure node is always a displacement antinode, and a pressure antinode is always a displacement node.

Figure 16.12 depicts a standing sound wave at an instant at which the pressure variations are greatest; the blue shading shows that the density and pressure of the gas have their maximum and minimum values at the displacement nodes (labeled N).

When reflection takes place at a *closed* end of a pipe (an end with a rigid barrier or plug), the displacement of the particles at this end must always be zero, analogous to a fixed end of a string. Thus a closed end of a pipe is a displacement node and a pressure antinode; the particles do not move, but the pressure variations are maximum. An *open* end of a pipe is a pressure node because it is open to the atmosphere, where the pressure is constant. Because of this, an open end is always a displacement *antinode,* in analogy to a free end of a string; the particles oscillate with maximum amplitude, but the pressure does not vary. (Strictly speaking, the pressure node actually occurs somewhat beyond an open end of a pipe. But if the diameter of the pipe is small in comparison to the wavelength, which is true for most musical instruments, this effect can safely be neglected.) Thus longitudinal waves in a column of fluid are reflected at the closed and open ends of a pipe in the same way that transverse waves in a string are reflected at fixed and free ends, respectively.

Conceptual Example 16.11 The sound of silence

A directional loudspeaker aims a sound wave of wavelength λ at a wall (Fig. 16.14). At what distances from the wall could you stand and hear no sound at all?

SOLUTION

Your ear detects pressure variations in the air; increases or decreases in the pressure outside your eardrum cause it to move slightly in or out, a motion that generates an electrical signal that is sent to the brain. (If you've ever had trouble getting your ears to "pop" on a drive up into the mountains or on an airline flight, you're familiar with just how sensitive your ears are to pressure changes.) Hence you will hear no sound if your ear is at a pressure node, which is a displacement antinode. The wall is a displacement node; the distance from a node to an adjacent antinode is $\lambda/4$, and the distance from one antinode to the next is $\lambda/2$ (Fig. 16.14). Hence the distances d from the wall at which no sound will be heard are

$d = \lambda/4$
(first displacement antinode and pressure node)

$d = \lambda/4 + \lambda/2 = 3\lambda/4$
(second displacement antinode and pressure node)

$d = 3\lambda/4 + \lambda/2 = 5\lambda/4$
(third displacement antinode and pressure node)

and so on. If the loudspeaker is not highly directional, this effect is hard to notice because of multiple reflections of sound waves from the floor, ceiling, and other points on the walls.

16.14 When a sound wave is directed at a wall, it interferes with the reflected wave to create a standing wave. The *N*'s and *A*'s are *displacement* nodes and antinodes.

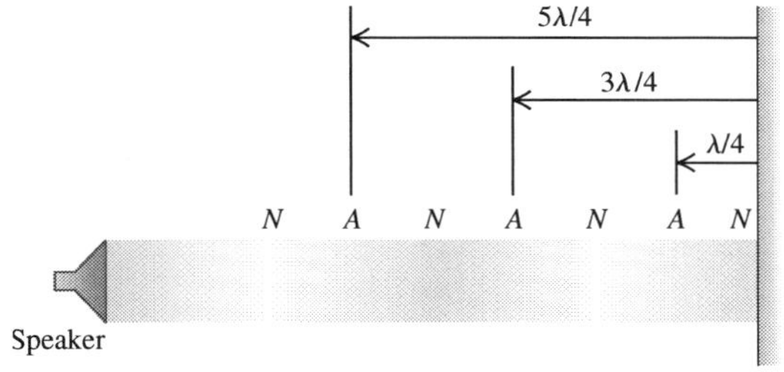

Organ Pipes and Wind Instruments

The most important application of standing sound waves is the production of musical tones by wind instruments. Organ pipes are one of the simplest examples (Fig. 16.15). Air is supplied by a blower, at a gauge pressure typically of the order of 10^3 Pa $(10^{-2}$ atm$)$, to the bottom end of the pipe (Fig. 16.16). A stream of air emerges from the narrow opening at the edge of the horizontal surface and is directed against the top edge of the opening, which is called the *mouth* of the pipe. The column of air in the pipe is set into vibration, and there is a series of possible normal modes, just as with the stretched string. The mouth always acts as an open end; thus it is a pressure node and a displacement antinode. The other end of the pipe (at the top in Fig. 16.16) may be either open or closed.

16.15 Organ pipes of different sizes produce tones with different frequencies.

In Fig. 16.17, both ends of the pipe are open, so both ends are pressure nodes and displacement antinodes. An organ pipe that is open at both ends is called an *open pipe*. The fundamental frequency f_1 corresponds to a standing-wave pattern with a displacement antinode at each end and a displacement node in the middle (Fig. 16.17a). The distance between adjacent antinodes is always equal to one half-wavelength, and in this case that is equal to the length L of the pipe; $\lambda/2 = L$. The corresponding frequency, obtained from the relationship $f = v/\lambda$, is

$$f_1 = \frac{v}{2L} \quad \text{(open pipe)} \tag{16.16}$$

Figures 16.17b and 16.17c show the second and third harmonics (first and second overtones); their vibration patterns have two and three displacement nodes, respectively. For these, a half-wavelength is equal to $L/2$ and $L/3$, respectively, and the frequencies are twice and three times the fundamental, respectively. That is, $f_2 = 2f_1$ and $f_3 = 3f_1$. For *every* normal mode of an open pipe the length L must be an integer number of half-wavelengths, and the possible wavelengths λ_n are given by

$$L = n\frac{\lambda_n}{2} \quad \text{or} \quad \lambda_n = \frac{2L}{n} \quad (n = 1, 2, 3, \ldots) \quad \text{(open pipe)} \tag{16.17}$$

16.16 Cross sections of an organ pipe at two instants one half-period apart. The N's and A's are *displacement* nodes and antinodes; as the blue shading shows, these are points of maximum pressure variation and zero pressure variation, respectively.

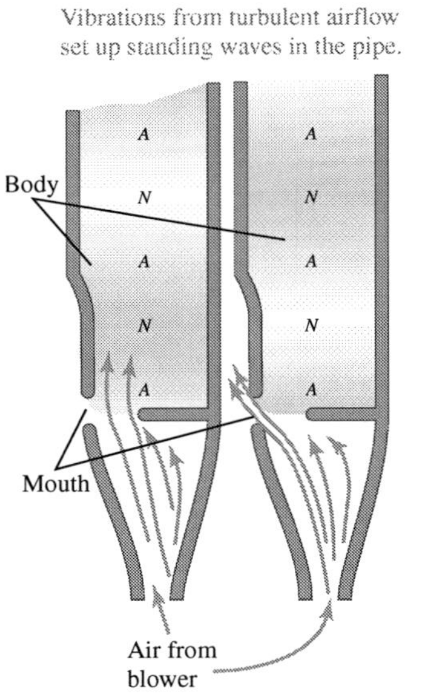

The corresponding frequencies f_n are given by $f_n = v/\lambda_n$, so all the normal-mode frequencies for a pipe that is open at both ends are given by

$$f_n = \frac{nv}{2L} \quad (n = 1, 2, 3, \dots) \quad \text{(open pipe)} \tag{16.18}$$

The value $n = 1$ corresponds to the fundamental frequency, $n = 2$ to the second harmonic (or first overtone), and so on. Alternatively, we can say

$$f_n = nf_1 \quad (n = 1, 2, 3, \dots) \quad \text{(open pipe)} \tag{16.19}$$

with f_1 given by Eq. (16.16).

Figure 16.18 shows a pipe that is open at the left end but closed at the right end. This is called a *stopped pipe.* The left (open) end is a displacement antinode

16.17 A cross section of an open pipe showing the first three normal modes. The shading indicates the pressure variations. The red curves are graphs of the displacement along the pipe axis at two instants separated in time by one half-period. The N's and A's are the *displacement* nodes and antinodes; interchange these to show the *pressure* nodes and antinodes.

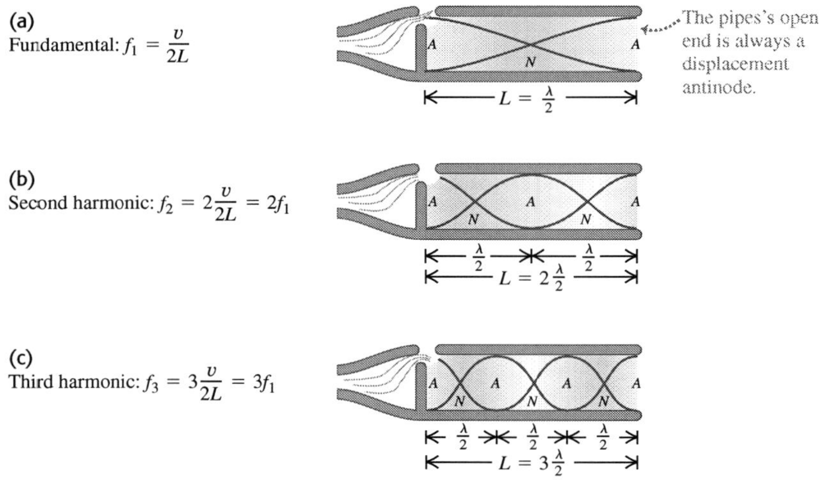

16.18 A cross section of a stopped pipe showing the first three normal modes as well as the *displacement* nodes and antinodes. Only odd harmonics are possible.

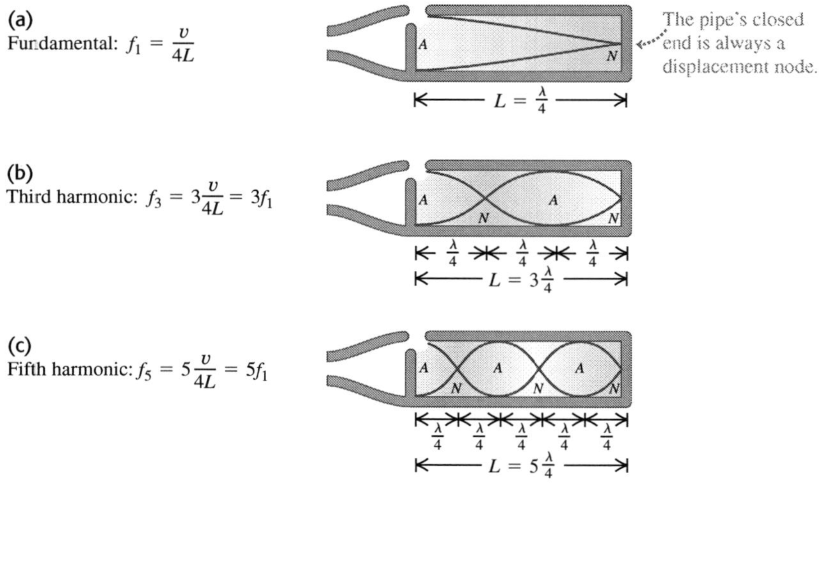

(pressure node), but the right (closed) end is a displacement node (pressure antinode). The distance between a node and the adjacent antinode is always one quarter-wavelength. Figure 16.18a shows the lowest-frequency mode; the length of the pipe is a quarter-wavelength $(L = \lambda_1/4)$. The fundamental frequency is $f_1 = v/\lambda_1$, or

$$f_1 = \frac{v}{4L} \quad \text{(stopped pipe)} \tag{16.20}$$

This is one-half the fundamental frequency for an *open* pipe of the same length. In musical language, the *pitch* of a closed pipe is one octave lower (a factor of 2 in frequency) than that of an open pipe of the same length. Figure 16.18b shows the next mode, for which the length of the pipe is *three-quarters* of a wavelength, corresponding to a frequency $3f_1$. For Fig. 16.18c, $L = 5\lambda/4$ and the frequency is $5f_1$. The possible wavelengths are given by

$$L = n\frac{\lambda_n}{4} \quad \text{or} \quad \lambda_n = \frac{4L}{n} \quad (n = 1, 3, 5, \ldots) \quad \text{(stopped pipe)} \tag{16.21}$$

The normal-mode frequencies are given by $f_n = v/\lambda_n$, or

$$f_n = \frac{nv}{4L} \quad (n = 1, 3, 5, \ldots) \quad \text{(stopped pipe)} \tag{16.22}$$

or

$$f_n = nf_1 \quad (n = 1, 3, 5, \ldots) \quad \text{(stopped pipe)} \tag{16.23}$$

with f_1 given by Eq. (16.20). We see that the second, fourth, and all *even* harmonics are missing. In a pipe that is closed at one end, the fundamental frequency is $f_1 = v/4L$, and only the odd harmonics in the series $(3f_1, 5f_1, \ldots)$ are possible.

A final possibility is a pipe that is closed at *both* ends, with displacement nodes and pressure antinodes at both ends. This wouldn't be of much use as a musical instrument because there would be no way for the vibrations to get out of the pipe.

Example 16.12 A tale of two pipes

On a day when the speed of sound is 345 m/s, the fundamental frequency of a stopped organ pipe is 220 Hz. (a) How long is this stopped pipe? (b) The second *overtone* of this pipe has the same wavelength as the third *harmonic* of an open pipe. How long is the open pipe?

SOLUTION

IDENTIFY: This problem uses the relationship between the length and normal-mode frequencies of an open pipe (Fig. 16.17) and a stopped pipe (Fig. 16.18).

SET UP: Since this is a stopped pipe (open at one end and closed at the other), the normal-mode frequencies are given by Eq. (16.22). We use this to determine the length L from the frequency in part (a). In part (b), we must make a comparison to an open pipe, for which the frequencies are given by Eq. (16.18).

EXECUTE: (a) For a stopped pipe, $f_1 = v/4L$, so the length of the stopped pipe is

$$L_{\text{stopped}} = \frac{v}{4f_1} = \frac{345\ \text{m/s}}{4(220\ \text{s}^{-1})} = 0.392\ \text{m}$$

(b) The frequency of the first overtone of a stopped pipe is $f_3 = 3f_1$, and the frequency of the second overtone is $f_5 = 5f_1$:

$$f_5 = 5f_1 = 5(220\ \text{Hz}) = 1100\ \text{Hz}$$

If the wavelengths are the same, the frequencies are the same, so the frequency of the third harmonic of the open pipe is also 1100 Hz. The third harmonic of an open pipe is at $3f_1 = 3(v/2L)$. If this equals 1100 Hz, then

$$1100\ \text{Hz} = 3\left(\frac{345\ \text{m/s}}{2L_{\text{open}}}\right) \quad \text{and} \quad L_{\text{open}} = 0.470\ \text{m}$$

EVALUATE: The stopped pipe is 0.392 m long and has a fundamental frequency of 220 Hz; the open pipe is longer, 0.470 m, but has a higher fundamental frequency of $(1100\ \text{Hz})/3 = 367$ Hz. If this seems like a contradiction, you should again compare Figs. 16.17a and 16.18a.

In an organ pipe in actual use, several modes are always present at once; the motion of the air is a superposition of these modes. This situation is analogous to a string that is struck or plucked, as in Fig. 15.28. Just as for a vibrating string, a complex standing wave in the pipe produces a traveling sound wave in the surrounding air with a harmonic content similar to that of the standing wave. A very narrow pipe produces a sound wave rich in higher harmonics, which we hear as a thin and "stringy" tone; a fatter pipe produces mostly the fundamental mode, heard as a softer, more flutelike tone. The harmonic content also depends on the shape of the pipe's mouth.

We have talked about organ pipes, but this discussion is also applicable to other wind instruments. The flute and the recorder are directly analogous. The most significant difference is that those instruments have holes along the pipe. Opening and closing the holes with the fingers changes the effective length L of the air column and thus changes the pitch. Any individual organ pipe, by comparison, can play only a single note. The flute and recorder behave as *open* pipes, while the clarinet acts as a *stopped* pipe (closed at the reed end, open at the bell).

Equations (16.18) and (16.22) show that the frequencies of any wind instrument are proportional to the speed of sound v in the air column inside the instrument. As Eq. (16.10) shows, v depends on temperature; it increases when temperature increases. Thus the pitch of all wind instruments rises with increasing temperature. An organ that has some of its pipes at one temperature and others at a different temperature is bound to sound out of tune.

Test Your Understanding of Section 16.4 If you connect a hose to one end of a metal pipe and blow compressed air into it, the pipe produces a musical tone. If instead you blow compressed helium into the pipe at the same pressure and temperature, will the pipe produce (i) the same tone, (ii) a higher-pitch tone, or (iii) a lower-pitch tone?

16.19 (a) The air in an open pipe is forced to oscillate at the same frequency as the sinusoidal sound waves coming from the loudspeaker. (b) The resonance curve of the open pipe graphs the amplitude of the standing sound wave in the pipe as a function of the driving frequency.

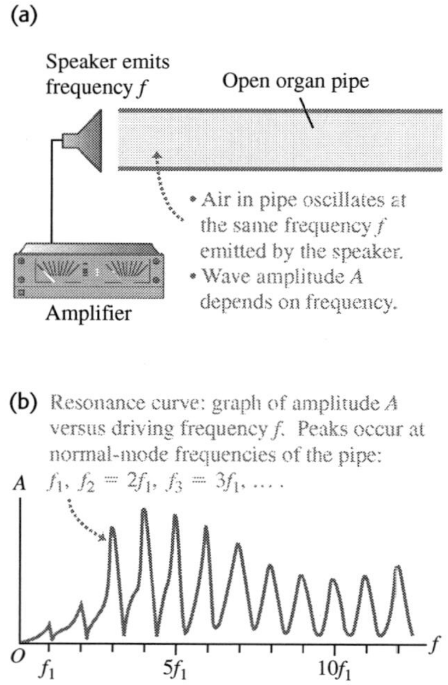

16.5 Resonance and Sound

Many mechanical systems have normal modes of oscillation. As we have seen, these include columns of air (as in an organ pipe) and stretched strings (as in a guitar; see Section 15.8). In each mode, every particle of the system oscillates with simple harmonic motion at the same frequency as the mode. Air columns and stretched strings have an infinite series of normal modes, but the basic concept is closely related to the simple harmonic oscillator, discussed in Chapter 13, which has only a single normal mode (that is, only one frequency at which it oscillates after being disturbed).

Suppose we apply a periodically varying force to a system that can oscillate. The system is then forced to oscillate with a frequency equal to the frequency of the applied force (called the *driving frequency*). This motion is called a *forced oscillation.* We talked about forced oscillations of the harmonic oscillator in Section 13.8, and we suggest that you review that discussion. In particular, we described the phenomenon of mechanical **resonance.** A simple example of resonance is pushing Cousin Throckmorton on a swing. The swing is a pendulum; it has only a single normal mode, with a frequency determined by its length. If we push the swing periodically with this frequency, we can build up the amplitude of the motion. But if we push with a very different frequency, the swing hardly moves at all.

Resonance also occurs when a periodically varying force is applied to a system with many normal modes. An example is shown in Fig. 16.19a. An open organ pipe is placed next to a loudspeaker that is driven by an amplifier and emits pure sinusoidal sound waves of frequency f, which can be varied by adjusting the amplifier. The air in the pipe is forced to vibrate with the same frequency f as the *driving force* provided by the loudspeaker. In general the amplitude of this motion is relatively small, and the air inside the pipe will not move in any of the normal-mode patterns shown in Fig. 16.17. But if the frequency f of the force is close to one of the normal-mode frequencies, the air in the pipe moves in the normal-mode pattern for that frequency, and the amplitude can become quite large. Figure 16.19b shows the amplitude of oscillation of the air in the pipe as a function of the driving frequency f. The shape of this graph is called the **resonance curve** of the pipe; it has peaks where f equals the normal-mode frequencies of the pipe. The detailed shape of the resonance curve depends on the geometry of the pipe.

If the frequency of the force is precisely *equal* to a normal-mode frequency, the system is in resonance, and the amplitude of the forced oscillation is maximum. If there were no friction or other energy-dissipating mechanism, a driving force at a normal-mode frequency would continue to add energy to the system, and the amplitude would increase indefinitely. In such an idealized case the peaks in the resonance curve of Fig. 16.19b would be infinitely high. But in any real system there is always some dissipation of energy, or damping, as we discussed in Section 13.8; the amplitude of oscillation in resonance may be large, but it cannot be infinite.

The "sound of the ocean" you hear when you put your ear next to a large seashell is due to resonance. The noise of the outside air moving past the seashell is a mixture of sound waves of almost all audible frequencies, which forces the air inside the seashell to oscillate. The seashell behaves like an organ pipe, with a set of normal-mode frequencies; hence the inside air oscillates most strongly at those frequencies, producing the seashell's characteristic sound. To hear a similar phenomenon, uncap a full bottle of your favorite beverage and blow across the open top. The noise is provided by your breath blowing across the top, and the "organ pipe" is the column of air inside the bottle above the surface of the liquid. If you take a drink and repeat the experiment, you will hear a lower tone because the "pipe" is longer and the normal-mode frequencies are lower.

Resonance also occurs when a stretched string is forced to oscillate (see Section 15.8). Suppose that one end of a stretched string is held fixed while the other is given a transverse sinusoidal motion with small amplitude, setting up standing waves. If the frequency of the driving mechanism is *not* equal to one of the normal-mode frequencies of the string, the amplitude at the antinodes is fairly small. However, if the frequency is equal to any one of the normal-mode frequencies, the string is in resonance, and the amplitude at the antinodes is very much larger than that at the driven end. The driven end is not precisely a node, but it lies much closer to a node than to an antinode when the string is in resonance. The photographs in Fig. 15.23 were made this way, with the left end of the string fixed and the right end oscillating vertically with small amplitude; large-amplitude standing waves resulted when the frequency of oscillation of the right end was equal to the fundamental frequency or to one of the first three overtones.

It is easy to demonstrate resonance with a piano. Push down the damper pedal (the right-hand pedal) so that the dampers are lifted and the strings are free to vibrate, and then sing a steady tone into the piano. When you stop singing, the piano seems to continue to sing the same note. The sound waves from your voice excite vibrations in the strings that have natural frequencies close to the frequencies (fundamental and harmonics) present in the note you sang.

A more spectacular example is a singer breaking a wine glass with her amplified voice. A good-quality wine glass has normal-mode frequencies that you can hear by tapping it. If the singer emits a loud note with a frequency corresponding exactly to one of these normal-mode frequencies, large-amplitude oscillations can build up and break the glass (Fig. 16.20).

Resonance is a very important concept, not only in mechanical systems but in all areas of physics. In Chapter 31 we will see examples of resonance in electric circuits.

16.20 The frequency of the sound from this trumpet exactly matches one of the normal-mode frequencies of the goblet. The resonant vibrations of the goblet have such large amplitude that the goblet tears itself apart.

Example 16.13 An organ–guitar duet

A stopped organ pipe is sounded near a guitar, causing one of the strings to vibrate with large amplitude. We vary the tension of the string until we find the maximum amplitude. The string is 80% as long as the stopped pipe. If both the pipe and the string vibrate at their fundamental frequency, calculate the ratio of the wave speed on the string to the speed of sound in air.

SOLUTION

IDENTIFY: The large response of the string is an example of resonance. It occurs because the organ pipe and the guitar string have the same fundamental frequency.

SET UP: Letting the subscripts a and s stand for the air in the pipe and the string, respectively, the condition for resonance is $f_{1a} = f_{1s}$. Equation (16.20) gives the fundamental frequency for a stopped pipe, while the fundamental frequency for a guitar string held at both ends is given by Eq. (15.32). These expressions involve the wave speed in air (v_a) and on the string (v_s) and the lengths of the pipe and string; we are given that $L_s = 0.80L_a$, and our target variable is the ratio v_s/v_a.

EXECUTE: From Eqs. (16.20) and (15.32), $f_{1a} = v_a/4L_a$ and $f_{1s} = v_s/2L_s$. Setting these equal to each other, we find

$$\frac{v_a}{4L_a} = \frac{v_s}{2L_s}$$

Substituting $L_s = 0.80L_a$ and rearranging, we get

$$\frac{v_s}{v_a} = 0.40$$

EVALUATE: As an example, if the speed of sound in air is 345 m/s, the wave speed on the string is $(0.40)(345\text{ m/s}) = 138\text{ m/s}$. Note that while the standing waves in the pipe and on the string have the same frequency, they have different *wavelengths* $\lambda = v/f$ because the two media have different wave speeds v. Which standing wave has the greater wavelength?

Test Your Understanding of Section 16.5 A stopped organ pipe of length L has a fundamental frequency of 220 Hz. For which of the following organ pipes will there be a resonance if a tuning fork of frequency 660 Hz is sounded next to the pipe? (There may be more than one correct answer.) (i) a stopped organ pipe of length L; (ii) a stopped organ pipe of length $2L$; (iii) an open organ pipe of length L; (iii) an open organ pipe of length $2L$.

16.6 Interference of Waves

Wave phenomena that occur when two or more waves overlap in the same region of space are grouped under the heading *interference*. As we have seen, standing waves are a simple example of an interference effect: Two waves traveling in opposite directions in a medium combine to produce a standing wave pattern with nodes and antinodes that do not move.

16.21 Two speakers driven by the same amplifier. Constructive interference occurs at point P, and destructive interference occurs at point Q.

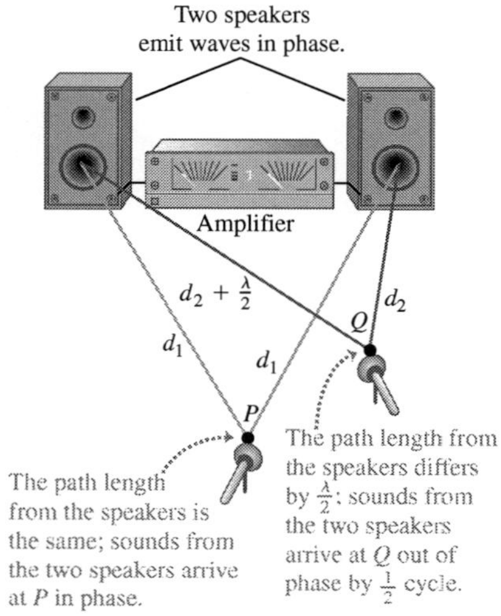

Figure 16.21 shows an example of another type of interference that involves waves that spread out in space. Two speakers, driven in phase by the same amplifier, emit identical sinusoidal sound waves with the same constant frequency. We place a microphone at point P in the figure, equidistant from the speakers. Wave crests emitted from the two speakers at the same time travel equal distances and arrive at point P at the same time; hence the waves arrive in phase, and there is constructive interference. The total wave amplitude at P is twice the amplitude from each individual wave, and we can measure this combined amplitude with the microphone.

Now let's move the microphone to point Q, where the distances from the two speakers to the microphone differ by a half-wavelength. Then the two waves arrive a half-cycle out of step, or *out of phase;* a positive crest from one speaker arrives at the same time as a negative crest from the other. Destructive interference takes place, and the amplitude measured by the microphone is much *smaller* than when only one speaker is present. If the amplitudes from the two speakers are equal, the two waves cancel each other out completely at point Q, and the total amplitude there is zero.

CAUTION Interference and traveling waves Although this situation bears some resemblance to standing waves in a pipe, the total wave in Fig. 16.21 is a *traveling* wave, not a standing wave. To see why, recall that in a standing wave there is no net flow of energy in any direction. By contrast, in Fig. 16.21 there is an overall flow of energy from the speakers into the surrounding air; this is characteristic of a traveling wave. The interference between the waves from the two speakers simply causes the energy flow to be *channeled* into certain directions (for example, toward P) and away from other directions (for example, away from Q). You can see another difference between Fig. 16.21 and a standing wave by considering a point, such as Q, where destructive interference occurs. Such a point is *both* a displacement node *and* a pressure node because there is no wave at all at this point. Compare this to a standing wave, in which a pressure node is a displacement antinode and vice versa.

Constructive interference occurs wherever the distances traveled by the two waves differ by a whole number of wavelengths, $0, \lambda, 2\lambda, 3\lambda, \ldots$; in all these cases the waves arrive at the microphone in phase (Fig. 16.22a). If the distances from the two speakers to the microphone differ by any half-integer number of wavelengths, $\lambda/2, 3\lambda/2, 5\lambda/2, \ldots$, the waves arrive at the microphone out of phase and there will be destructive interference (Fig. 16.22b). In this case, little or no sound energy flows toward the microphone directly in front of the speakers. The energy is instead directed to the sides, where constructive interference occurs.

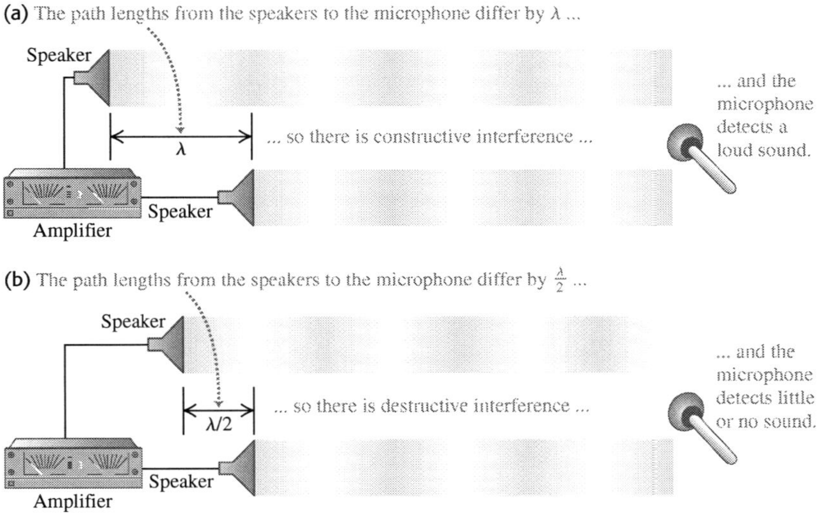

16.22 Two speakers driven by the same amplifier, emitting waves in phase. Only the waves directed toward the microphone are shown, and they are separated for clarity. **(a)** Constructive interference occurs when the path difference is $0, \lambda, 2\lambda, 3\lambda, \ldots$. **(b)** Destructive interference occurs when the path difference is $\lambda/2, 3\lambda/2, 5\lambda/2, \ldots$.

Example 16.14 Loudspeaker interference

Two small loudspeakers, A and B (Fig. 16.23), are driven by the same amplifier and emit pure sinusoidal waves in phase. If the speed of sound is 350 m/s, (a) for what frequencies does constructive interference occur at point P? (b) For what frequencies does destructive interference occur at point P?

SOLUTION

IDENTIFY: The nature of the interference at P depends on the difference in path lengths from points A and B to P and how this difference compares to the wavelength.

SET UP: We calculate the path lengths from A to P and from B to P using the Pythagorean theorem. Constructive interference occurs

16.23 What sort of interference occurs at P?

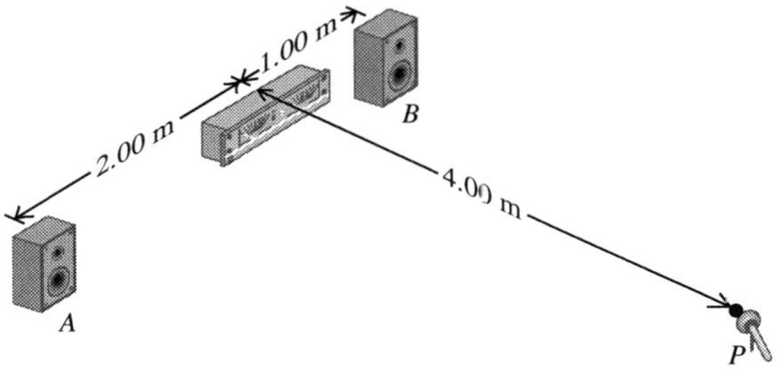

Continued

when the difference in path lengths equals a whole number of wavelengths, while destructive interference occurs when the path length difference is a half-integer number of wavelengths. To find the corresponding frequencies, we use the relationship $v = f\lambda$.

EXECUTE: The distance from speaker A to point P is $[(2.00\text{ m})^2 + (4.00\text{ m})^2]^{1/2} = 4.47\text{ m}$, and the distance from speaker B to point P is $[(1.00\text{ m})^2 + (4.00\text{ m})^2]^{1/2} = 4.12\text{ m}$. The path difference is $d = 4.47\text{ m} - 4.12\text{ m} = 0.35\text{ m}$.

(a) Constructive interference occurs when the path difference is $d = 0, \lambda, 2\lambda, \ldots$ or $d = 0, v/f, 2v/f, \ldots = nv/f$. So the possible frequencies are

$$f_n = \frac{nv}{d} = n\frac{350\text{ m/s}}{0.35\text{ m}} \qquad (n = 1, 2, 3, \ldots)$$
$$= 1000\text{ Hz}, 2000\text{ Hz}, 3000\text{ Hz}, \ldots$$

(b) Destructive interference occurs when the path difference is $d = \lambda/2, 3\lambda/2, 5\lambda/2, \ldots$ or $d = v/2f, 3v/2f, 5v/2f, \ldots$. The possible frequencies are

$$f_n = \frac{nv}{2d} = n\frac{350\text{ m/s}}{2(0.35\text{ m})} \qquad (n = 1, 2, 3, \ldots)$$
$$= 500\text{ Hz}, 1500\text{ Hz}, 2500\text{ Hz}, \ldots$$

EVALUATE: As we increase the frequency, the sound at point P alternates between large and small amplitudes; the maxima and minima occur at the frequencies we have found. It can be hard to notice this effect in an ordinary room because of multiple reflections from the walls, floor, and ceiling. Such an experiment is best done either outdoors or in an anechoic chamber, which has walls that absorb almost all sound and thereby eliminate reflections.

Experiments closely analogous to the one in Example 16.14, but using light, have provided both strong evidence for the wave nature of light and a means of measuring its wavelengths. We will discuss these experiments in detail in Chapter 35.

Interference effects are used to control noise from very loud sound sources such as gas-turbine power plants or jet engine test cells. The idea is to use additional sound sources that in some regions of space interfere destructively with the unwanted sound and cancel it out. Microphones in the controlled area feed signals back to the sound sources, which are continuously adjusted for optimum cancellation of noise in the controlled area.

Test Your Understanding of Section 16.6 Suppose that speaker A in Fig. 16.23 emits a sinusoidal sound wave of frequency 500 Hz and speaker B emits a sinusoidal sound wave of frequency 1000 Hz. What sort of interference will there be between these two waves? (i) constructive interference at various points, including point P, and destructive interference at various other points; (ii) destructive interference at various points, including point P, and constructive interference at various points; (iii) neither (i) nor (ii).

16.7 Beats

ActivPhysics Online
10.7 Beats and Beat Frequency

In Section 16.6 we talked about *interference* effects that occur when two different waves with the same frequency overlap in the same region of space. Now let's look at what happens when we have two waves with equal amplitude but slightly different frequencies. This occurs, for example, when two tuning forks with slightly different frequencies are sounded together, or when two organ pipes that are supposed to have exactly the same frequency are slightly "out of tune."

Consider a particular point in space where the two waves overlap. The displacements of the individual waves at this point are plotted as functions of time in Fig. 16.24a. The total length of the time axis represents 1 second, and the frequencies are 16 Hz (blue graph) and 18 Hz (red graph). Applying the principle of superposition, we add the two displacements at each instant of time to find the total displacement at that time. The result is the graph of Fig. 16.24b. At certain times the two waves are in phase; their maxima coincide and their amplitudes add. But because of their slightly different frequencies, the two waves cannot be in phase at all times. Indeed, at certain times (like $t = 0.50$ s in Fig. 16.24) the two waves are exactly *out* of phase. The two waves then cancel each other, and the total amplitude is zero.

The resultant wave in Fig. 16.24b looks like a single sinusoidal wave with a varying amplitude that goes from a maximum to zero and back. In this example

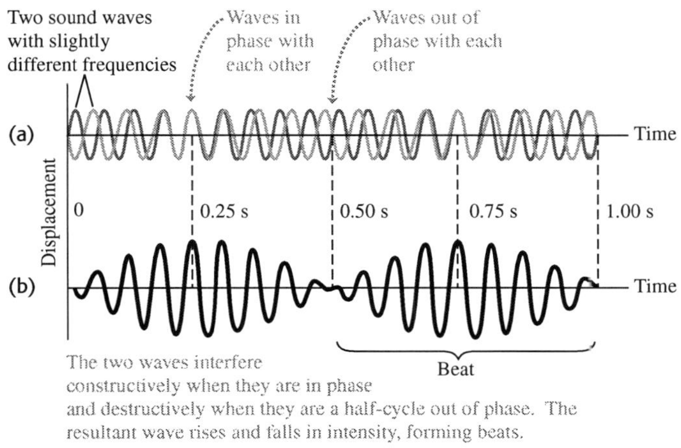

16.24 Beats are fluctuations in amplitude produced by two sound waves of slightly different frequency, here 16 Hz and 18 Hz. (a) Individual waves. (b) Resultant wave formed by superposition of the two waves. The beat frequency is 18 Hz − 16 Hz = 2 Hz.

the amplitude goes through two maxima and two minima in 1 second, so the frequency of this amplitude variation is 2 Hz. The amplitude variation causes variations of loudness called **beats,** and the frequency with which the loudness varies is called the **beat frequency.** In this example the beat frequency is the *difference* of the two frequencies. If the beat frequency is a few hertz, we hear it as a waver or pulsation in the tone.

We can prove that the beat frequency is *always* the difference of the two frequencies f_a and f_b. Suppose f_a is larger than f_b; the corresponding periods are T_a and T_b, with $T_a < T_b$. If the two waves start out in phase at time $t = 0$, they are again in phase when the first wave has gone through exactly one more cycle than the second. This happens at a value of t equal to T_{beat}, the *period* of the beat. Let n be the number of cycles of the first wave in time T_{beat}; then the number of cycles of the second wave in the same time is $(n - 1)$, and we have the relationships

$$T_{beat} = nT_a \qquad \text{and} \qquad T_{beat} = (n - 1)T_b$$

Eliminating n between these two equations, we find

$$T_{beat} = \frac{T_a T_b}{T_b - T_a}$$

The reciprocal of the beat period is the beat *frequency,* $f_{beat} = 1/T_{beat}$, so

$$f_{beat} = \frac{T_b - T_a}{T_a T_b} = \frac{1}{T_a} - \frac{1}{T_b}$$

and finally

$$f_{beat} = f_a - f_b \qquad \text{(beat frequency)} \qquad (16.24)$$

As claimed, the beat frequency is the difference of the two frequencies. In using Eq. (16.24), remember that f_a is the higher frequency.

An alternative way to derive Eq. (16.24) is to write functions to describe the curves in Fig. 16.24a and then add them. Suppose that at a certain position the two waves are given by $y_a(t) = A\sin 2\pi f_a t$ and $y_b(t) = -A\sin 2\pi f_b t$. We use the trigonometric identity

$$\sin a - \sin b = 2\sin\frac{1}{2}(a - b)\cos\frac{1}{2}(a + b)$$

We can then express the total wave $y(t) = y_a(t) + y_b(t)$ as

$$y_a(t) + y_b(t) = \left[2A\sin\frac{1}{2}(2\pi)(f_a - f_b)t\right]\cos\frac{1}{2}(2\pi)(f_a + f_b)t$$

The amplitude factor (the quantity in brackets) varies slowly with frequency $\frac{1}{2}(f_a - f_b)$. The cosine factor varies with a frequency equal to the *average* frequency $\frac{1}{2}(f_a + f_b)$. The *square* of the amplitude factor, which is proportional to the intensity that the ear hears, goes through two maxima and two minima per cycle. So the beat frequency f_{beat} that is heard is twice the quantity $\frac{1}{2}(f_a - f_b)$, or just $f_a - f_b$, in agreement with Eq. (16.24).

Beats between two tones can be heard up to a beat frequency of about 6 or 7 Hz. Two piano strings or two organ pipes differing in frequency by 2 or 3 Hz sound wavery and "out of tune," although some organ stops contain two sets of pipes deliberately tuned to beat frequencies of about 1 to 2 Hz for a gently undulating effect. Listening for beats is an important technique in tuning all musical instruments.

At frequency differences greater than about 6 or 7 Hz, we no longer hear individual beats, and the sensation merges into one of *consonance* or *dissonance,* depending on the frequency ratio of the two tones. In some cases the ear perceives a tone called a *difference tone,* with a pitch equal to the beat frequency of the two tones. For example, if you listen to a whistle that produces sounds at 1800 Hz and 1900 Hz when blown, you will hear not only these tones but also a much lower 100-Hz tone.

The engines on multiengine propeller aircraft have to be synchronized so that the propeller sounds don't cause annoying beats, which are heard as loud throbbing sounds (Fig. 16.25). On some planes this is done electronically; on others the pilot does it by ear, just like tuning a piano.

16.25 If the two propellers on this airplane are not precisely synchronized, the pilots, passengers, and listeners on the ground will hear beats.

Test Your Understanding of Section 16.7 One tuning fork vibrates at 440 Hz, while a second tuning fork vibrates at an unknown frequency. When both tuning forks are sounded simultaneously, you hear a tone that rises and falls in intensity three times per second. What is the frequency of the second tuning fork? (i) 434 Hz; (ii) 437 Hz; (iii) 443 Hz; (iv) 446 Hz; (v) either 434 Hz or 446 Hz; (vi) either 437 Hz or 443 Hz.

16.8 The Doppler Effect

ActivPhysics ONLINE

10.8 Doppler Effect: Conceptual Introduction
10.9 Doppler Effect: Problems

You've probably noticed that when a car approaches you with its horn sounding, the pitch seems to drop as the car passes. This phenomenon, first described by the 19th-century Austrian scientist Christian Doppler, is called the **Doppler effect.** When a source of sound and a listener are in motion relative to each other, the frequency of the sound heard by the listener is not the same as the source frequency. A similar effect occurs for light and radio waves; we'll return to this later in this section.

To analyze the Doppler effect for sound, we'll work out a relationship between the frequency shift and the velocities of source and listener relative to the medium (usually air) through which the sound waves propagate. To keep things simple, we consider only the special case in which the velocities of both source and listener lie along the line joining them. Let v_S and v_L be the velocity components along this line for the source and the listener, respectively, relative to the medium. We choose the positive direction for both v_S and v_L to be the direction from the listener L to the source S. The speed of sound relative to the medium, v, is always considered positive.

Moving Listener

Let's think first about a listener L moving with velocity v_L toward a stationary source S (Fig. 16.26). The source emits a sound wave with frequency f_S and wavelength $\lambda = v/f_S$. The figure shows four wave crests, separated by equal dis-

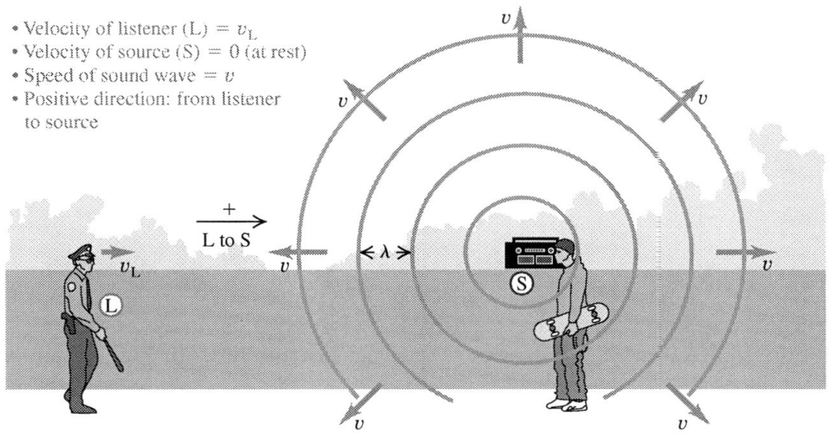

16.26 A listener moving toward a stationary source hears a frequency that is higher than the source frequency. This is because the relative speed of listener and wave is greater than the wave speed v.

tances λ. The wave crests approaching the moving listener have a speed of propagation *relative to the listener* of $(v + v_L)$. So the frequency f_L with which the crests arrive at the listener's position (that is, the frequency the listener hears) is

$$f_L = \frac{v + v_L}{\lambda} = \frac{v + v_L}{v/f_S} \tag{16.25}$$

or

$$f_L = \left(\frac{v + v_L}{v}\right) f_S = \left(1 + \frac{v_L}{v}\right) f_S \quad \text{(moving listener, stationary source)} \tag{16.26}$$

So a listener moving toward a source $(v_L > 0)$, as in Fig. 16.26, hears a higher frequency (higher pitch) than does a stationary listener. A listener moving away from the source $(v_L < 0)$ hears a lower frequency (lower pitch)

Moving Source and Moving Listener

Now suppose the source is also moving, with velocity v_S (Fig. 16.27). The wave speed relative to the wave medium (air) is still v; it is determined by the properties of the medium and is not changed by the motion of the source. But the wavelength is no longer equal to v/f_S. Here's why. The time for emission of one cycle of the wave is the period $T = 1/f_S$. During this time, the wave travels a distance $vT = v/f_S$ and the source moves a distance $v_S T = v_S/f_S$. The wavelength is the distance between successive wave crests, and this is determined by the *relative* displacement of source and wave. As Fig. 16.27 shows, this is different in front of

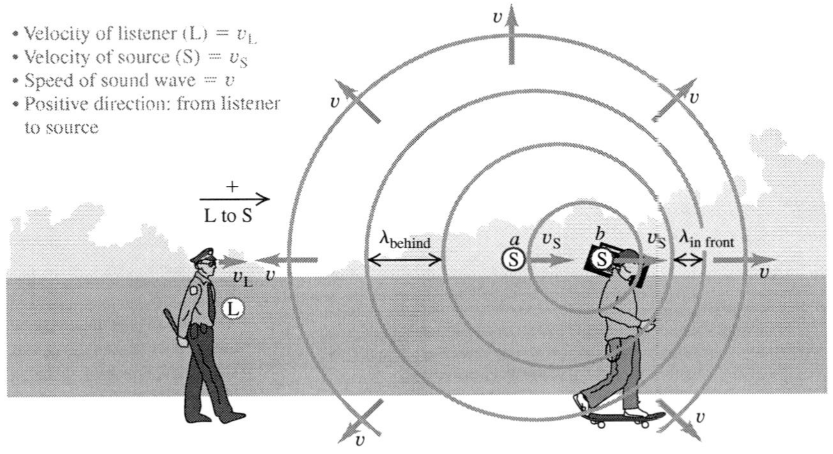

16.27 Wave crests emitted by a moving source are crowded together in front of the source (to the right of this source) and stretched out behind it (to the left of this source).

and behind the source. In the region to the right of the source in Fig. 16.27 (that is, in front of the source), the wavelength is

$$\lambda_{\text{in front}} = \frac{v}{f_S} - \frac{v_S}{f_S} = \frac{v - v_S}{f_S} \qquad \text{(wavelength in front of a moving source)} \tag{16.27}$$

In the region to the left of the source (that is, behind the source), it is

$$\lambda_{\text{behind}} = \frac{v + v_S}{f_S} \qquad \text{(wavelength behind a moving source)} \tag{16.28}$$

The waves in front of and behind the source are compressed and stretched out, respectively, by the motion of the source.

To find the frequency heard by the listener behind the source, we substitute Eq.(16.28) into the first form of Eq. (16.25):

$$f_L = \frac{v + v_L}{\lambda_{\text{behind}}} = \frac{v + v_L}{(v + v_S)/f_S}$$

$$f_L = \frac{v + v_L}{v + v_S} f_S \qquad \text{(Doppler effect, moving source and moving listener)} \tag{16.29}$$

This expresses the frequency f_L heard by the listener in terms of the frequency f_S of the source.

Although we derived it for the particular situation shown in Fig. 16.27, Eq. (16.29) includes *all* possibilities for motion of source and listener (relative to the medium) along the line joining them. If the listener happens to be at rest in the medium, v_L is zero. When both source and listener are at rest or have the same velocity relative to the medium, $v_L = v_S$ and $f_L = f_S$. Whenever the direction of the source or listener velocity is opposite to the direction from the listener toward the source (which we have defined as positive), the corresponding velocity to be used in Eq. (16.29) is negative.

As an example, the frequency heard by a listener at rest $(v_L = 0)$ is $f_L = [v/(v + v_S)]f_S$. If the source is moving toward the listener (in the negative direction), then $v_S < 0, f_L > f_S$, and the listener hears a higher frequency than that emitted by the source. If instead the source is moving away from the listener (in the positive direction), then $v_S > 0, f_L < f_S$, and the listener hears a lower frequency. This explains the change in pitch that you hear from the siren of an ambulance as it passes you (Fig. 16.28).

16.28 The Doppler effect explains why the siren on a fire engine or ambulance has a high pitch $(f_L > f_S)$ when it is approaching you $(v_S < 0)$ and a low pitch $(f_L < f_S)$ when it is moving away $(v_S > 0)$.

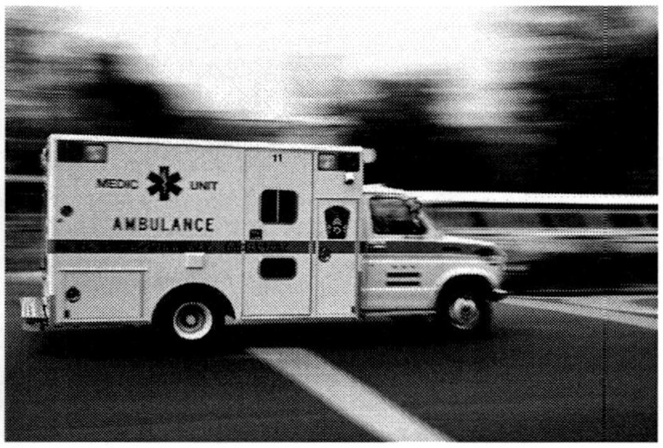

Problem-Solving Strategy 16.2 Doppler Effect

IDENTIFY *the relevant concepts:* The Doppler effect is relevant whenever the source of waves, the wave detector (listener), or both are in motion.

SET UP *the problem* using the following steps:

1. Establish a coordinate system. Define the positive direction to be the direction from the listener to the source, and make sure you know the signs of all relevant velocities. A velocity in the direction from the listener toward the source is positive; a velocity in the opposite direction is negative. Also, the velocities must all be measured relative to the air in which the sound is traveling.
2. Use consistent notation to identify the various quantities: subscript S for source, L for listener.
3. Determine which unknown quantities are your target variables.

EXECUTE *the solution* as follows:

1. Use Eq. (16.29) to relate the frequencies at the source and the listener, the sound speed, and the velocities of the source and the listener. If the source is moving, you can find the wavelength measured by the listener using Eq.(16.27) or (16.28).
2. When a wave is reflected from a surface, either stationary or moving, the analysis can be carried out in two steps. In the first, the surface plays the role of listener; the frequency with which the wave crests arrive at the surface is f_L. Then think of the surface as a new source, emitting waves with this same frequency f_L. Finally, determine what frequency is heard by a listener detecting this new wave.

EVALUATE *your answer:* Ask whether your final result makes sense. If the source and the listener are moving toward each other, $f_L > f_S$; if they are moving apart, $f_L < f_S$. If the source and the listener have no relative motion, $f_L = f_S$.

Example 16.15 Doppler effect I: Wavelengths

A police siren emits a sinusoidal wave with frequency $f_S = 300$ Hz. The speed of sound is 340 m/s. (a) Find the wavelength of the waves if the siren is at rest in the air. (b) If the siren is moving at 30 m/s (108 km/h, or 67 mi/h), find the wavelengths of the waves in front of and behind the source.

SOLUTION

IDENTIFY: The Doppler effect is not involved in part (a), since neither the source nor the listener is moving. In part (b), the source is in motion and we must invoke the Doppler effect.

SET UP: Figure 16.29 shows the situation. We use the relationship $v = \lambda f$ to determine the wavelength when the police siren is at rest. When it is in motion, we find the wavelength on either side of the siren using Eqs. (16.27) and (16.28).

EXECUTE: (a) When the source is at rest,

$$\lambda = \frac{v}{f_S} = \frac{340\ \text{m/s}}{300\ \text{Hz}} = 1.13\ \text{m}$$

16.29 Our sketch for this problem.

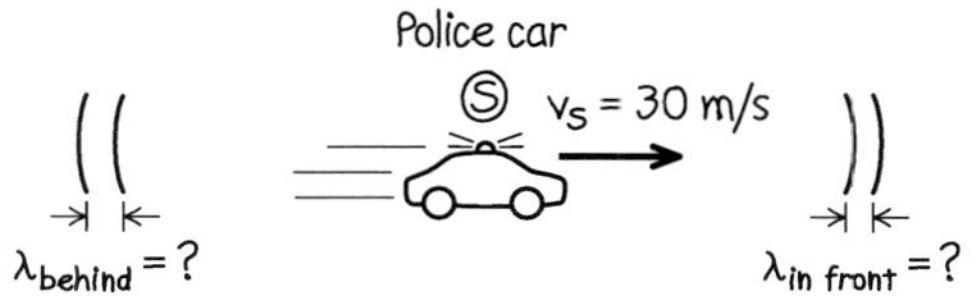

(b) The situation is shown in Fig. 16.29. From Eq. (16.27), in front of the siren,

$$\lambda_{\text{in front}} = \frac{v - v_S}{f_S} = \frac{340\ \text{m/s} - 30\ \text{m/s}}{300\ \text{Hz}} = 1.03\ \text{m}$$

From Eq. (16.28), behind the siren,

$$\lambda_{\text{behind}} = \frac{v + v_S}{f_S} = \frac{340\ \text{m/s} + 30\ \text{m/s}}{300\ \text{Hz}} = 1.23\ \text{m}$$

EVALUATE: The wavelength is less in front of the siren and greater behind the siren, as it should be.

Example 16.16 Doppler effect II: Frequencies

If a listener L is at rest and the siren in Example 16.15 is moving away from L at 30 m/s, what frequency does the listener hear?

SOLUTION

IDENTIFY: Our target variable is the frequency f_L heard by the listener, who is behind the moving source.

SET UP: Figure 16.30 shows the situation. We know $f_S = 300$ Hz from Example 16.15, and we have $v_L = 0$ and $v_S = 30$ m/s. (The source velocity v_S is positive because the siren is moving in the same direction as the direction from listener to source.)

16.30 Our sketch for this problem.

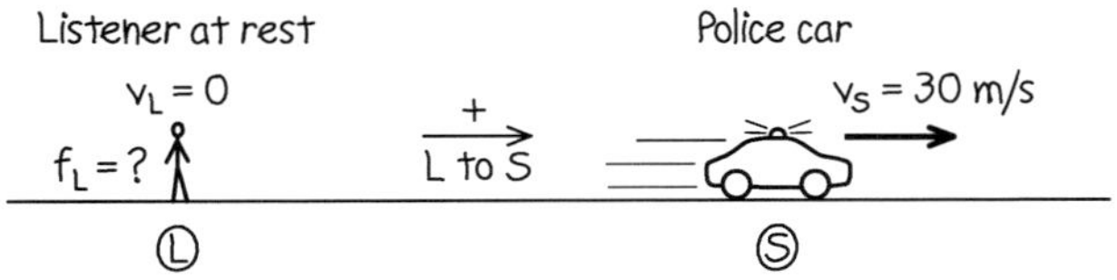

EXECUTE: From Eq. (16.29),

$$f_L = \frac{v}{v + v_S} f_S = \frac{340\ \text{m/s}}{340\ \text{m/s} + 30\ \text{m/s}}(300\ \text{Hz}) = 276\ \text{Hz}$$

EVALUATE: The source and listener are moving apart, so the frequency f_L heard by the listener is less than the frequency f_S emitted by the source.

Here's an alternative approach we can use to check our result. From Example 16.15, the wavelength behind the source (which is where the listener in Fig. 16.30 is located) is 1.23 m, so

$$f_L = \frac{v}{\lambda} = \frac{340\ \text{m/s}}{1.23\ \text{m}} = 276\ \text{Hz}$$

Even though the source is moving, the wave speed v relative to the stationary listener is unchanged.

Example 16.17 Doppler effect III: A moving listener

If the siren is at rest and the listener is moving away from the siren at 30 m/s, what frequency does the listener hear?

SOLUTION

IDENTIFY: Again our target variable is the frequency f_L heard by the listener, but now the listener is in motion and the source is at rest.

SET UP: Figure 16.31 shows the situation. The positive direction (from listener to source) is still from left to right, so $v_L = -30$ m/s.

EXECUTE: From Eq. (16.29),

$$f_L = \frac{v + v_L}{v} f_S = \frac{340\ \text{m/s} + (-30\ \text{m/s})}{340\ \text{m/s}}(300\ \text{Hz}) = 274\ \text{Hz}$$

16.31 Our sketch for this problem.

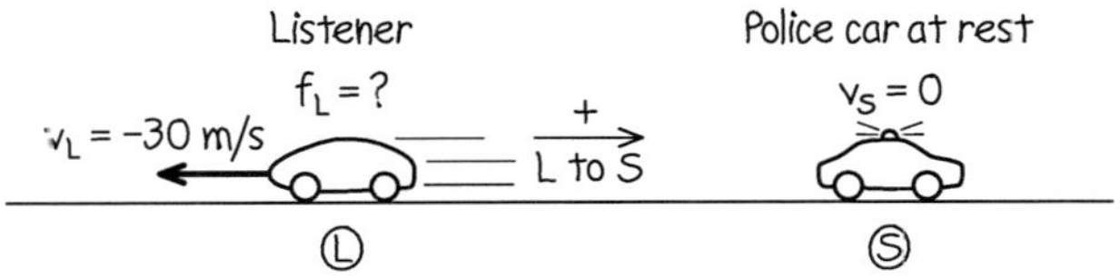

EVALUATE: Again the frequency heard by the listener is less than the source frequency. Note that the *relative velocity* of source and listener is the same as in Example 16.16, but the Doppler shift is different because the velocities relative to the *air* are different.

Example 16.18 Doppler effect IV: Moving source, moving listener

If the siren is moving away from the listener with a speed of 45 m/s relative to the air and the listener is moving toward the siren with a speed of 15 m/s relative to the air, what frequency does the listener hear?

SOLUTION

IDENTIFY: Now *both* the listener and the source are in motion. Once again our target variable is the frequency f_L heard by the listener.

SET UP: Figure 16.32 shows the situation. Both the source velocity $v_S = 45\ \text{m/s}$ and the listener's velocity $v_L = 15\ \text{m/s}$ are positive because both velocity vectors point in the direction from listener to source.

16.32 Our sketch for this problem.

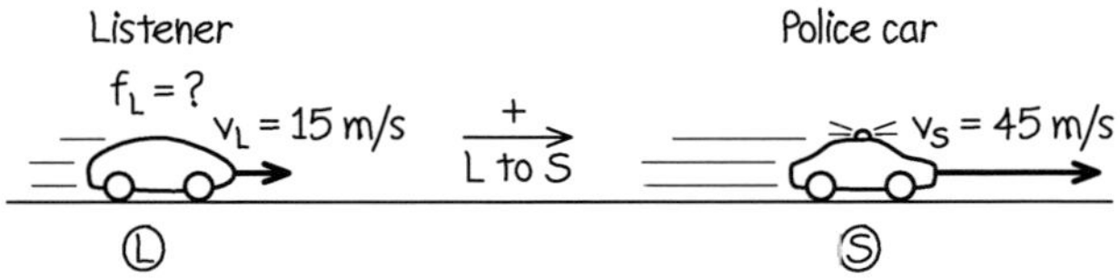

EXECUTE: Once again using Eq. (16.29), we find

$$f_L = \frac{v + v_L}{v + v_S} f_S = \frac{340\ \text{m/s} + 15\ \text{m/s}}{340\ \text{m/s} + 45\ \text{m/s}}(300\ \text{Hz})$$
$$= 277\ \text{Hz}$$

EVALUATE: The frequency heard by the listener is again less than the source frequency, but the value is different than in the preceding two examples, even though the source and listener move away from each other at 30 m/s in all three cases. The *sign* of the Doppler shift of frequency (that is, whether f_L is less than or greater than f_S) depends on how the source and the listener are moving relative to each other; to determine the *value* of the Doppler shift of frequency, you must know the velocities of source and listener relative to the air.

Example 16.19 Doppler effect V: A double Doppler shift

The police car with its 300-Hz siren is moving toward a warehouse at 30 m/s, intending to crash through the door. What frequency does the driver of the police car hear reflected from the warehouse?

SOLUTION

IDENTIFY: In this situation there are *two* Doppler shifts, as shown in Fig. 16.33. In the first shift, the warehouse is the stationary "listener." The frequency of sound reaching the warehouse, which we call f_W, is greater than 300 Hz because the source is approaching. In the second shift, the warehouse acts as a source of sound with frequency f_W, and the listener is the driver of the police car; she hears a frequency greater than f_W because she is approaching the source.

16.33 Two stages of the sound wave's motion from the police car to the warehouse and back to the police car.

(a) Sound travels from police car's siren (source S) to warehouse ("listener" L).

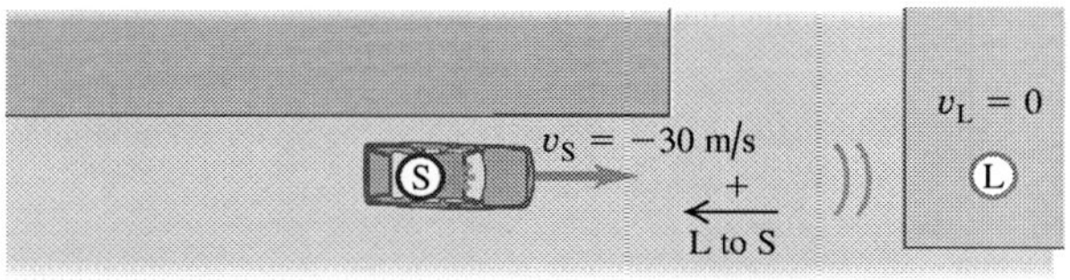

(b) Reflected sound travels from warehouse (source S) to police car (listener L).

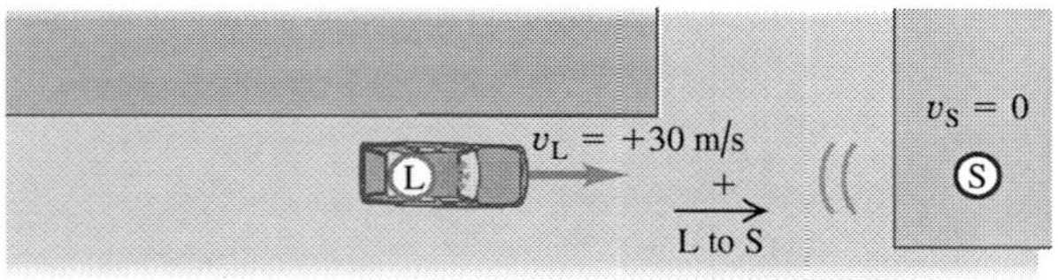

SET UP: To determine f_W, we use Eq. (16.29) with f_L replaced by f_W. For this part of the problem, $v_L = v_W = 0$ (the warehouse is at rest) and $v_S = -30\ \text{m/s}$ (the siren is moving in the negative direction from source to listener).

To determine the frequency heard by the driver, which is our target variable, we again use Eq. (16.29) but now with f_S replaced by f_W. For this second part of the problem, $v_S = 0$ because the stationary warehouse is the source and the velocity of the listener (the driver) is $v_L = +30\ \text{m/s}$. (The listener's velocity is positive because it is in the direction from listener to source.)

EXECUTE: The frequency reaching the warehouse is

$$f_W = \frac{v}{v + v_S} f_S = \frac{340\ \text{m/s}}{340\ \text{m/s} + (-30\ \text{m/s})}(300\ \text{Hz}) = 329\ \text{Hz}$$

Then the frequency heard by the driver is

$$f_L = \frac{v + v_L}{v} f_W = \frac{340\ \text{m/s} + 30\ \text{m/s}}{340\ \text{m/s}}(329\ \text{Hz}) = 358\ \text{Hz}$$

EVALUATE: Because there are two Doppler shifts, the reflected sound heard by the driver has an even higher frequency than the sound heard by a stationary listener in the warehouse.

Doppler Effect for Electromagnetic Waves

In the Doppler effect for sound, the velocities v_L and v_S are always measured relative to the *air* or whatever medium we are considering. There is also a Doppler effect for *electromagnetic* waves in empty space, such as light waves or radio waves. In this case there is no medium that we can use as a reference to measure velocities, and all that matters is the *relative* velocity of source and receiver. (By contrast, the Doppler effect for sound does not depend simply on this relative velocity, as discussed in Example 16.18.)

To derive the expression for the Doppler frequency shift for light, we have to use the special theory of relativity. We will discuss this in Chapter 37, but for now we quote the result without derivation. The wave speed is the speed of light, usually denoted by c, and it is the same for both source and receiver. In the frame of reference in which the receiver is at rest, the source is moving away from the receiver with velocity v. (If the source is *approaching* the receiver, v is negative.) The source frequency is again f_S. The frequency f_R measured by the receiver R (the frequency of arrival of the waves at the receiver) is then given by

$$f_R = \sqrt{\frac{c - v}{c + v}}\, f_S \quad \text{(Doppler effect for light)} \qquad (16.30)$$

When v is positive, the source is moving directly *away* from the receiver and f_R is always *less* than f_S; when v is negative, the source is moving directly *toward* the receiver and f_R is *greater* than f_S. The qualitative effect is the same as for sound, but the quantitative relationship is different.

A familiar application of the Doppler effect for radio waves is the radar device mounted on the side window of a police car to check other cars' speeds. The electromagnetic wave emitted by the device is reflected from a moving car, which acts as a moving source, and the wave reflected back to the device is Doppler-shifted in frequency. The transmitted and reflected signals are combined to produce beats, and the speed can be computed from the frequency of the beats. Similar techniques ("Doppler radar") are used to measure wind velocities in the atmosphere.

The Doppler effect is also used to track satellites and other space vehicles. In Fig. 16.34 a satellite emits a radio signal with constant frequency f_S. As the satellite orbits past, it first approaches and then moves away from the receiver; the frequency f_R of the signal received on earth changes from a value greater than f_S to a value less than f_S as the satellite passes overhead.

16.34 Change of velocity component along the line of sight of a satellite passing a tracking station. The frequency received at the tracking station changes from high to low as the satellite passes overhead.

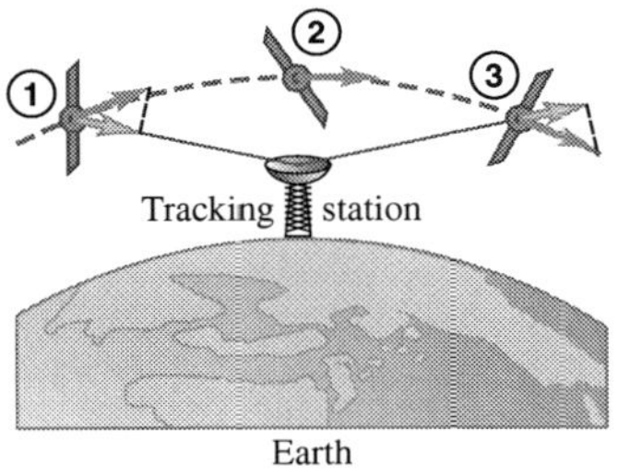

The Doppler effect for electromagnetic waves, including visible light, is important in astronomy. Astronomers compare wavelengths of light from distant stars to those emitted by the same elements on earth. For example, in a binary star system, in which two stars orbit about their common center of mass, one star's light is Doppler-shifted to higher frequencies when that star is moving toward an observer on earth and to lower frequencies when it is moving away. Measurements of the frequency shifts reveal information about the orbits and masses of the stars that make up the binary system.

Light from most galaxies is shifted toward the longer-wavelength or red end of the visible spectrum, an effect called the *red shift*. This is often described as a Doppler shift resulting from motion of these galaxies away from us. However, from the point of view of the general theory of relativity, it is something much more fundamental: It is associated with the expansion of space itself. Distant galaxies have large red shifts because their light has been in transit for a long time and the wavelength of that light has shared in the expansion of all the space through which it moved. Extrapolating this expansion backward to 13.7 billion years ago leads to the "Big Bang" picture. The Big Bang was not an explosion in space but the initial rapid expansion of space itself.

Test Your Understanding of Section 16.8 You are at an outdoor concert with a wind blowing at 10 m/s from the performers toward you. Is the sound you hear Doppler-shifted? If so, is it shifted to lower or higher frequencies?

*16.9 Shock Waves

You may have experienced "sonic booms" caused by an airplane flying overhead faster than the speed of sound. We can see qualitatively why this happens from Fig. 16.35. Let v_S denote the *speed* of the airplane relative to the air, so that it is always positive. The motion of the airplane through the air produces sound; if v_S is less than the speed of sound v, the waves in front of the airplane are crowded together with a wavelength given by Eq. (16.27):

$$\lambda_{\text{in front}} = \frac{v - v_S}{f_S}$$

As the speed v_S of the airplane approaches the speed of sound v, the wavelength approaches zero and the wave crests pile up on each other (Fig. 16.35a). The airplane must exert a large force to compress the air in front of it; by Newton's third law, the air exerts an equally large force back on the airplane. Hence there is a large increase in aerodynamic drag (air resistance) as the airplane approaches the speed of sound, a phenomenon known as the "sound barrier."

When v_S is greater in magnitude than v, the source of sound is **supersonic,** and Eqs. (16.27) and (16.29) for the Doppler effect no longer describe the sound wave in front of the source. Figure 16.35b shows a cross section of what happens. As the airplane moves, it displaces the surrounding air and produces sound. A series of wave crests is emitted from the nose of the airplane; each spreads out in a circle centered at the position of the airplane when it emitted the crest. After a time t the crest emitted from point S_1 has spread to a circle with radius vt, and the airplane has moved a greater distance $v_S t$ to position S_2. You can see that the circular crests interfere constructively at points along the blue line that makes an angle α with the direction of the airplane velocity, leading to a very-large-amplitude wave crest along this line. This large-amplitude crest is called a **shock wave** (Fig. 16.35c).

From the right triangle in Fig. 16.35b we can see that the angle α is given by

$$\sin\alpha = \frac{vt}{v_S t} = \frac{v}{v_S} \quad \text{(shock wave)} \qquad (16.31)$$

In this relationship, v_S is the *speed* of the source (the magnitude of its velocity) relative to the air and is always positive. The ratio v_S/v is called the **Mach number.** It is greater than unity for all supersonic speeds, and $\sin\alpha$ in Eq. (16.31) is the reciprocal of the Mach number. The first person to break the sound barrier

16.35 Wave crests around a sound source S moving (a) slightly slower than the speed of sound v and (b) faster than the sound speed v. (c) This photograph shows a T-38 jet airplane moving at 1.1 times the speed of sound. Separate shock waves are produced by the nose, wings, and tail. The angles of these waves vary because the air speeds up and slows down as it moves around the airplane, so the relative speed v_S of the airplane and air is different for shock waves produced at different points.

(a) Sound source S (airplane) moving at nearly the speed of sound

(b) Sound source moving faster than the speed of sound

(c) Shock waves around a supersonic airplane

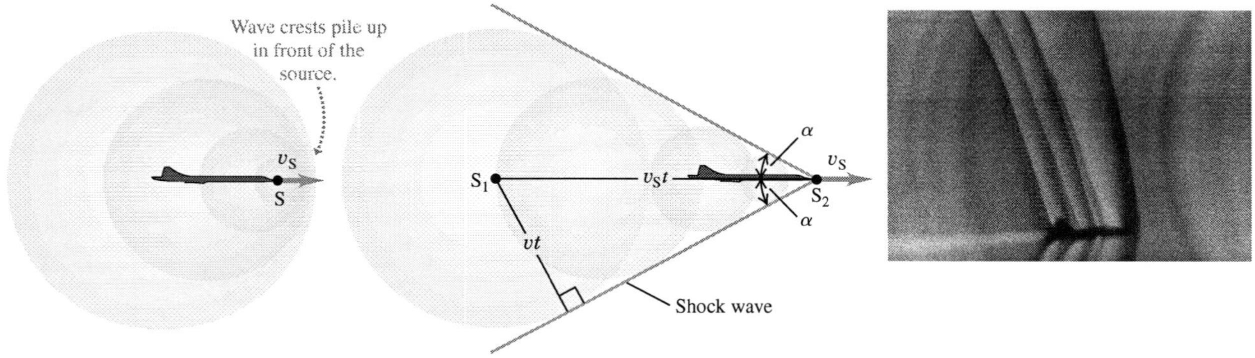

16.36 The first supersonic airplane, the Bell X-1, was shaped much like a 50-caliber bullet—which was known to be able to travel faster than sound.

was Capt. Chuck Yeager of the U.S. Air Force, flying the Bell X-1 at Mach 1.06 on October 14, 1947 (Fig. 16.36).

Shock waves are actually three-dimensional; a shock wave forms a *cone* around the direction of motion of the source. If the source (possibly a supersonic jet airplane or a rifle bullet) moves with constant velocity, the angle α is constant, and the shock-wave cone moves along with the source. It's the arrival of this shock wave that causes the sonic boom you hear after a supersonic airplane has passed by. The larger the airplane, the stronger the sonic boom; the shock wave produced at ground level by the (now retired) Concorde supersonic airliner flying at 12,000 m (40,000 ft) caused a sudden jump in air pressure of about 20 Pa. In front of the shock-wave cone, there is no sound. Inside the cone a stationary listener hears the Doppler-shifted sound of the airplane moving away.

CAUTION **Shock waves** We emphasize that a shock wave is produced *continuously* by any object that moves through the air at supersonic speed, not only at the instant that it "breaks the sound barrier." The sound waves that combine to form the shock wave, as in Fig. 16.35b, are created by the motion of the object itself, not by any sound source that the object may carry. The cracking noises of a bullet and of the tip of a circus whip are due to their supersonic motion. A supersonic jet airplane may have very loud engines, but these do not cause the shock wave. Indeed, a space shuttle makes a very loud sonic boom when coming in for a landing; its engines are out of fuel at this point, so it is a supersonic glider.

Shock waves have applications outside of aviation. They are used to break up kidney stones and gallstones without invasive surgery, using a technique with the impressive name *extracorporeal shock-wave lithotripsy.* A shock wave produced outside the body is focused by a reflector or acoustic lens so that as much of it as possible converges on the stone. When the resulting stresses in the stone exceed its tensile strength, it breaks into small pieces and can be eliminated. This technique requires accurate determination of the location of the stone, which may be done using ultrasonic imaging techniques (see the discussion of ultrasonic imaging on page 535).

Example 16.20 Sonic boom from a supersonic airplane

An airplane is flying at Mach 1.75 at an altitude of 8000 m, where the speed of sound is 320 m/s. How long after the plane passes directly overhead will you hear the sonic boom?

SOLUTION

IDENTIFY: The shock wave forms a cone trailing backward from the airplane, so the problem is really asking for how much time elapses from when the airplane flies overhead to when the shock wave reaches you.

SET UP: Figure 16.37 shows the situation just as the shock wave reaches you at point L. A time t (our target variable) has elapsed since the airplane passed overhead, during which the airliner flying at speed v_S has traveled a distance $v_S t$. We use trigonometry to solve for t.

EXECUTE: From Eq. (16.31) the angle α of the shock cone is

$$\alpha = \arcsin \frac{1}{1.75} = 34.8°$$

The speed of the plane is the speed of sound multiplied by the Mach number:

$$v_S = (1.75)(320 \text{ m/s}) = 560 \text{ m/s}$$

Continued

16.37 You hear a sonic boom when the shock wave reaches you at L (*not* just when the plane breaks the sound barrier). A listener to the right of L has not yet heard the sonic boom but will shortly; a listener to the left of L has already heard the sonic boom.

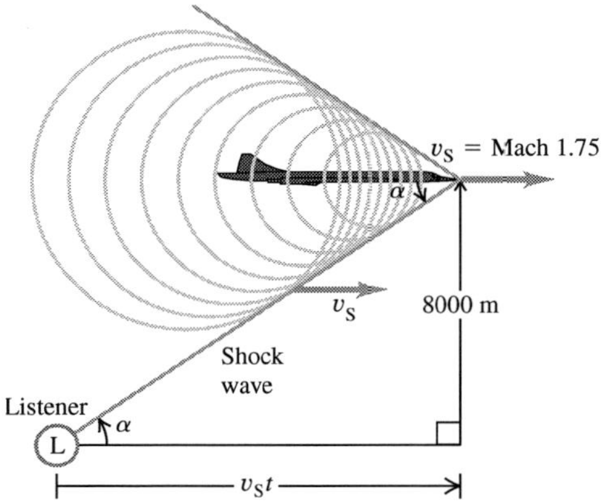

From Fig. 16.37 we have

$$\tan \alpha = \frac{8000\text{ m}}{v_S t}$$

$$t = \frac{8000\text{ m}}{(560\text{ m/s})(\tan 34.8°)} = 20.5\text{ s}$$

EVALUATE: You hear the boom 20.5 s after the airplane passes overhead, and at that time it has traveled $(560\text{ m/s})(20.5\text{ s}) =$ 11.5 km past the straight-overhead point.

In this calculation we assumed that the speed of sound is the same at all altitudes, so $\alpha = \arcsin v/v_S$ is a constant and the shock wave forms a perfect cone. In fact, the speed of sound decreases with increasing altitude. How would this affect the result?

Test Your Understanding of Section 16.9 What would you hear if you were directly behind (to the left of) the supersonic airplane in Fig. 16.37? (i) a sonic boom; (ii) the sound of the airplane, Doppler-shifted to higher frequencies; (iii) the sound of the airplane, Doppler-shifted to lower frequencies; (iv) nothing.

CHAPTER 16 SUMMARY

Sound waves: Sound consists of longitudinal waves in a medium. A sinusoidal sound wave is characterized by its frequency f and wavelength λ (or angular frequency ω and wave number k) and by its displacement amplitude A. The pressure amplitude p_{max} is directly proportional to the displacement amplitude, the wave number, and the bulk modulus B of the wave medium. (See Examples 16.1 and 16.2.)

The speed of a sound wave in a fluid depends on the bulk modulus B and density ρ. If the fluid is an ideal gas, the speed can be expressed in terms of the temperature T, molar mass M, and ratio of heat capacities γ of the gas. The speed of longitudinal waves in a solid rod depends on the density and Young's modulus Y. (See Examples 16.3–16.5.)

$$p_{max} = BkA \qquad (16.5)$$
(sinusoidal sound wave)

$$v = \sqrt{\frac{B}{\rho}} \qquad (16.7)$$
(longitudinal wave in a fluid)

$$v = \sqrt{\frac{\gamma RT}{M}} \qquad (16.10)$$
(sound wave in a ideal gas)

$$v = \sqrt{\frac{Y}{\rho}} \qquad (16.8)$$
(longitudinal wave in a solid rod)

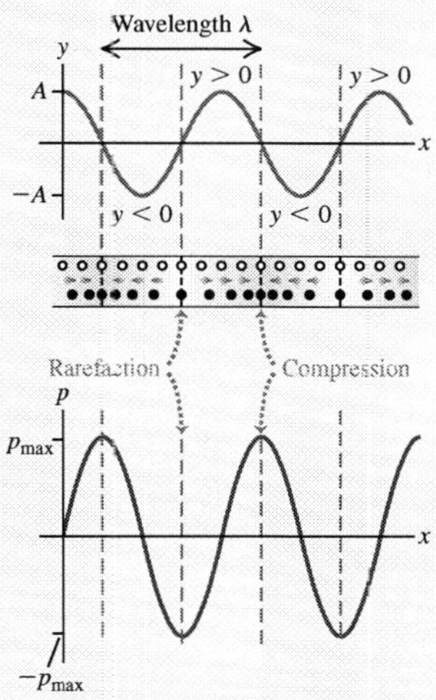

Intensity and sound intensity level: The intensity I of a sound wave is the time average rate at which energy is transported by the wave, per unit area. For a sinusoidal wave, the intensity can be expressed in terms of the displacement amplitude A or the pressure amplitude p_{max}. (See Examples 16.6–16.8.)

The sound intensity level β of a sound wave is a logarithmic measure of its intensity. It is measured relative to I_0, an arbitrary intensity defined to be $10^{-12}\ \text{W/m}^2$. Sound intensity levels are expressed in decibels (dB). (See Examples 16.9 and 16.10.)

$$I = \frac{1}{2}\sqrt{\rho B}\,\omega^2 A^2 = \frac{p_{max}^{\ 2}}{2\rho v} = \frac{p_{max}^{\ 2}}{2\sqrt{\rho B}} \qquad (16.12), (16.14)$$
(intensity of a sinusoidal sound wave)

$$\beta = (10\ \text{dB}) \log \frac{I}{I_0} \qquad (16.15)$$
(definition of sound intensity level)

Standing sound waves: Standing sound waves can be set up in a pipe or tube. A closed end is a displacement node and a pressure antinode; an open end is a displacement antinode and a pressure node. For a pipe of length L open at both ends, the normal-mode frequencies are integer multiples of the sound speed divided by $2L$. For a stopped pipe (one that is open at only one end), the normal-mode frequencies are the odd multiples of the sound speed divided by $4L$. (See Examples 16.11 and 16.12.)

A pipe or other system with normal-mode frequencies can be driven to oscillate at any frequency. A maximum response, or resonance, occurs if the driving frequency is close to one of the normal-mode frequencies of the system. (See Example 16.13.)

$$f_n = \frac{nv}{2L} \qquad (n = 1, 2, 3, \ldots) \qquad (16.18)$$
(open pipe)

$$f_n = \frac{nv}{4L} \qquad (n = 1, 3, 5, \ldots) \qquad (16.22)$$
(stopped pipe)

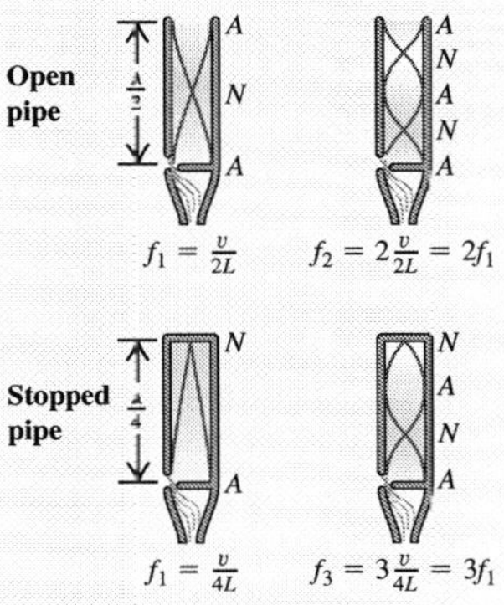

Interference: When two or more waves overlap in the same region of space, the resulting effects are called interference. The resulting amplitude can be either larger or smaller than the amplitude of each individual wave, depending on whether the waves are in phase (constructive interference) or out of phase (destructive interference). (See Example 16.14.)

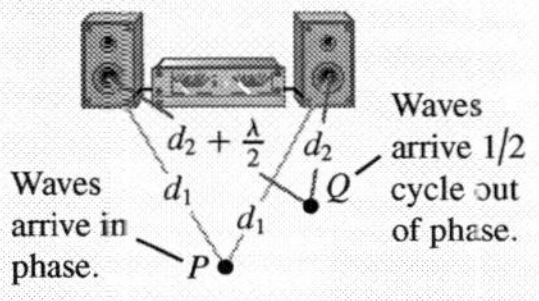

Beats: Beats are heard when two tones with slightly different frequencies f_a and f_b are sounded together. The beat frequency f_{beat} is the difference between f_a and f_b.

$$f_{beat} = f_a - f_b \quad (16.24)$$
(beat frequency)

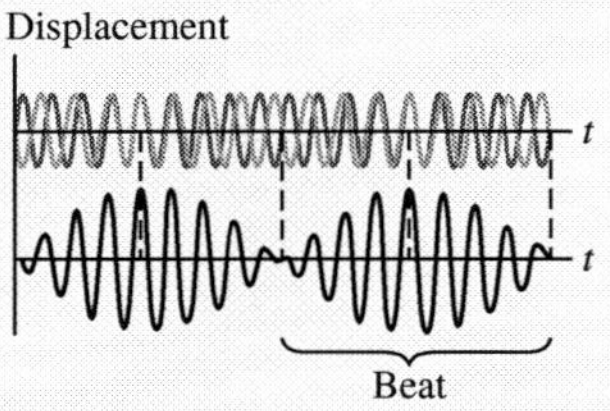

Doppler effect: The Doppler effect for sound is the frequency shift that occurs when there is motion of a source of sound, a listener, or both, relative to the medium. The source and listener frequencies f_S and f_L are related by the source and listener velocities v_S and v_L relative to the medium and to the speed of sound v. (See Examples 16.15–16.19)

$$f_L = \frac{v + v_L}{v + v_S} f_S \quad (16.29)$$
(Doppler effect, moving source and moving listener)

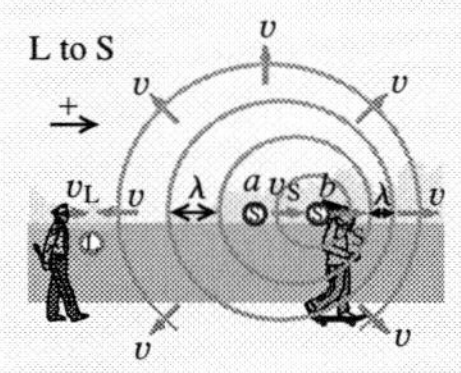

***Shock waves:** A sound source moving with a speed v_S greater than the speed of sound v creates a shock wave. The wave front is a cone with angle α. (See Example 16.20.)

$$\sin\alpha = \frac{v}{v_S} \quad (16.31)$$

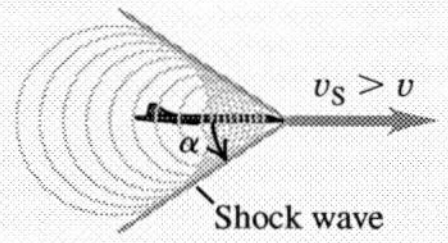

Key Terms

sound, *527*
audible range, *527*
ultrasonic, *528*
infrasonic, *528*
displacement amplitude, *528*
pressure amplitude, *529*
loudness, *531*
pitch, *531*
timbre, *532*
noise, *532*
sound intensity level, *539*
decibel, *539*
displacement node, *541*
displacement antinode, *541*
pressure node, *542*
pressure antinode, *542*
resonance, *546*
resonance curve, *547*
beats, *551*
beat frequency, *551*
Doppler effect, *552*
*supersonic, *558*
*shock wave, *558*
*Mach number, *558*

Answer to Chapter Opening Question ?

Both musical sounds and noise are made up of a combination of sinusoidal sound waves. The difference is that the frequencies of the sine waves in musical sound are all integer multiples of a fundamental frequency, while *all* frequencies are present in noise.

Answers to Test Your Understanding Questions

16.1 Answer: (v) From Eq. (16.5), the displacement amplitude is $A = p_{max}/Bk$. The pressure amplitude p_{max} and bulk modulus B remain the same, but the frequency f increases by a factor of 4. Hence the wave number $k = \omega/v = 2\pi f/v$ also increases by a factor of 4. Since A is inversely proportional to k, the displacement amplitude becomes $\frac{1}{4}$ as great. In other words, at higher frequency a smaller maximum displacement is required to produce the same maximum pressure fluctuation.

16.2 Answer: (i) From Eq. (16.7), the speed of longitudinal waves (sound) in a fluid is $v = \sqrt{B/\rho}$. We can rewrite this to give an expression for the bulk modulus B in terms of the fluid density ρ and the sound speed v: $B = \rho v^2$. At 20°C the speed of sound in mercury is slightly less than in water (1451 m/s versus 1482 m/s), but the density of mercury is greater than that of water by a large factor (13.6). Hence the bulk modulus of mercury is greater than that of water by a factor of $(13.6)(1451/1482)^2 = 13.0$.

16.3 Answer: A and p_{max} increase by a factor of $\sqrt{2}$, B and v are unchanged, β increases by 3.0 dB Equations (16.9) and (16.10) show that the bulk modulus B and sound speed v remain the same because the physical properties of the air are unchanged. From Eqs. (16.12) and (16.14), the intensity is proportional to the square of the displacement amplitude or the square of the pressure amplitude. Hence doubling the intensity means that A and p_{max} both increase by a factor of $\sqrt{2}$. Example 16.10 shows that *multiplying* the intensity by a factor of $2(I_2/I_1 = 2)$ corresponds to *adding* to the sound intensity level by $(10\text{ dB})\log(I_2/I_1) = (10\text{ dB})\log 2 = 3.0\text{ dB}$.

16.4 Answer: (ii) Helium is less dense and has a lower molar mass than air, so sound travels faster in helium than in air. The normal-mode frequencies for a pipe are proportional to the sound speed v, so the frequency and hence the pitch increase when the air in the pipe is replaced with helium.

16.5 Answer: (i) and (iv) There will be a resonance if 660 Hz is one of the pipe's normal-mode frequencies. A stopped organ pipe has normal-mode frequencies that are odd multiples of its fundamental frequency [see Eq. (16.22) and Fig. 16.18]. Hence pipe (i), which has fundamental frequency 220 Hz, also has a normal-mode frequency of $3(220\text{ Hz}) = 660$ Hz. Pipe (ii) has twice the length of pipe (i); from Eq. (16.20), the fundamental frequency of a stopped pipe is inversely proportional to the length, so pipe (ii) has a fundamental frequency of $(\frac{1}{2})(220\text{ Hz}) = 110$ Hz. Its other normal-mode frequencies are 330 Hz, 550 Hz, 770 Hz, . . . , so a 660-Hz tuning fork will not cause resonance. Pipe (iii) is an open pipe of the same length as pipe (i), so its fundamental frequency is twice as great as for pipe (i) [compare Eqs. (16.16) and (16.20)], or $2(220\text{ Hz}) = 440$ Hz. Its other normal-mode frequencies are integer multiples of the fundamental frequency [see Eq. (16.19)], or 880 Hz, 1320 Hz, . . . , none of which match the 660-Hz frequency of the tuning fork. Pipe (iv) is also an open pipe but with twice the length of pipe (iii) [see Eq. (16.18)], so its normal-mode frequencies are one-half those of pipe (iii): 220 Hz, 440 Hz, 660 Hz, . . . , so the third harmonic will resonate with the tuning fork.

16.6 Answer: (iii) Constructive and destructive interference between two waves can occur only if the two waves have the same frequency. In this case the frequencies are different, so there are no points where the two waves always reinforce each other (constructive interference) or always cancel each other (destructive interference).

16.7 Answer: (vi) The beat frequency is 3 Hz, so the difference between the two tuning fork frequencies is also 3 Hz. Hence the second tuning fork vibrates at a frequency of either 443 Hz or 437 Hz. You can distinguish between the two possibilities by comparing the pitches of the two tuning forks sounded one at a time: the frequency is 437 Hz if the second tuning fork has a lower pitch and 443 Hz if it has a higher pitch.

16.8 Answer: no The air (the medium for sound waves) is moving from the source toward the listener. Hence, relative to the air, both the source and the listener are moving in the direction from listener to source. So both velocities are positive and $v_S = v_L = +10\text{ m/s}$. The equality of these two velocities means that the numerator and the denominator in Eq. (16.29) are the same, so $f_L = f_S$ and there is *no* Doppler shift.

16.9 Answer: (iii) Figure 16.37 shows that there are sound waves inside the cone of the shock wave. Behind the airplane the wave crests are spread apart, just as they are behind the moving source in Fig. 16.27. Hence the waves that reach you have an increased wavelength and a lower frequency.

PROBLEMS

For instructor-assigned homework, go to **www.masteringphysics.com**

Discussion Questions

Q16.1. When sound travels from air into water, does the frequency of the wave change? The speed? The wavelength? Explain your reasoning.

Q16.2. The hero of a western movie listens for an oncoming train by putting his ear to the track. Why does this method give an earlier warning of the approach of a train than just listening in the usual way?

Q16.3. Would you expect the pitch (or frequency) of an organ pipe to increase or decrease with increasing temperature? Explain.

Q16.4. In most modern wind instruments the pitch is changed by using keys or valves to change the length of the vibrating air column. The bugle, however, has no valves or keys, yet it can play many notes. How might this be possible? Are there restrictions on what notes a bugle can play?

Q16.5. Symphonic musicians always "warm up" their wind instruments by blowing into them before a performance. What purpose does this serve?

Q16.6. In a popular and amusing science demonstration, a person inhales helium and then his voice becomes high and squeaky. Why does this happen? (*Warning:* Inhaling too much helium can cause unconsciousness or death.)

Q16.7. Lane dividers on highways sometimes have regularly spaced ridges or ripples. When the tires of a moving car roll along such a divider, a musical note is produced. Why? Explain how this phenomenon could be used to measure the car's speed.

Q16.8. The tone quality of an acoustic guitar is different when the strings are plucked near the bridge (the lower end of the strings) than when they are plucked near the sound hole (close to the center of the strings). Why?

Q16.9. Which has a more direct influence on the loudness of a sound wave: the *displacement* amplitude or the *pressure* amplitude? Explain your reasoning.

Q16.10. If the pressure amplitude of a sound wave is halved, by what factor does the intensity of the wave decrease? By what factor must the pressure amplitude of a sound wave be increased in order to increase the intensity by a factor of 16? Explain.

Q16.11. Does the sound intensity level β obey the inverse-square law? Why?

Q16.12. A small fraction of the energy in a sound wave is absorbed by the air through which the sound passes. How does this modify the inverse-square relationship between intensity and distance from the source? Explain your reasoning.

Q16.13. A wire under tension and vibrating in its first overtone produces sound of wavelength λ. What is the new wavelength of the sound (in terms of λ) if the tension is doubled?

Q16.14. A small metal band is slipped onto one of the tines of a tuning fork. As this band is moved closer and closer to the end of the tine, what effect does this have on the wavelength and frequency of the sound the tine produces? Why?

Q16.15. An organist in a cathedral plays a loud chord and then releases the keys. The sound persists for a few seconds and gradually dies away. Why does it persist? What happens to the sound energy when the sound dies away?

Q16.16. Two vibrating tuning forks have identical frequencies, but one is stationary and the other is mounted at the rim of a rotating platform. What does a listener hear? Explain.

Q16.17. A large church has part of the organ in the front of the church and part in the back. A person walking rapidly down the aisle while both segments are playing at once reports that the two segments sound out of tune. Why?

Q16.18. A sound source and a listener are both at rest on the earth, but a strong wind is blowing from the source toward the listener. Is there a Doppler effect? Why or why not?

Q16.19. Can you think of circumstances in which a Doppler effect would be observed for surface waves in water? For elastic waves propagating in a body of water deep below the surface? If so, describe the circumstances and explain your reasoning. If not, explain why not.

Q16.20. Stars other than our sun normally appear featureless when viewed through telescopes. Yet astronomers can readily use the light from these stars to determine that they are rotating and even measure the speed of their surface. How do you think they can do this?

Q16.21. If you wait at a railroad crossing as a train approaches and passes, you hear a Doppler shift in its sound. But if you listen closely, you hear that the change in frequency is continuous; it does not suddenly go from one high frequency to another low frequency. Instead the frequency *smoothly* (but rather quickly) changes from high to low as the train passes. Why does this smooth change occur?

Q16.22. In case 1, a source of sound approaches a stationary observer at speed v. In case 2, the observer moves toward the stationary source at the same speed v. If the source is always producing the same frequency sound, will the observer hear the same frequency in both cases, since the relative speed is the same each time? Why not?

***Q16.23.** Does an aircraft make a sonic boom only at the instant its speed exceeds Mach 1? Explain your reasoning.

***Q16.24.** If you are riding in a supersonic aircraft, what do you hear? Explain your reasoning. In particular, do you hear a continuous sonic boom? Why or why not?

***Q16.25.** A jet airplane is flying at a constant altitude at a steady speed v_S greater than the speed of sound. Describe what observers at points A, B, and C hear at the instant shown in Fig. 16.38, when the shock wave has just reached point B. Explain your reasoning.

Figure 16.38 Question Q16.25.

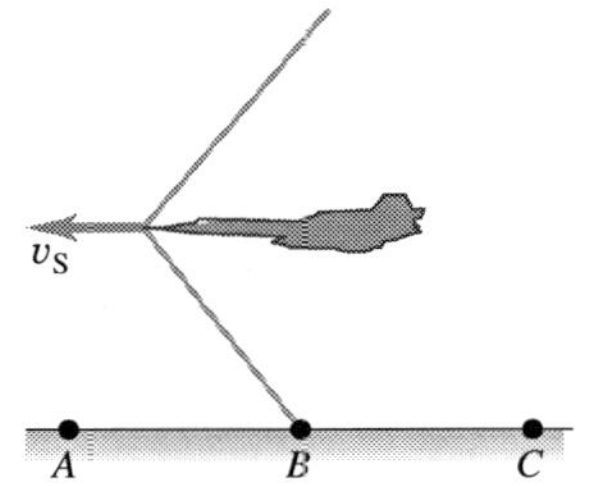

Exercises

Unless indicated otherwise, assume the speed of sound in air to be $v = 344 \text{m/s}$.

Section 16.1 Sound Waves

16.1. Example 16.1 (Section 16.1) showed that for sound waves in air with frequency 1000 Hz, a displacement amplitude of 1.2×10^{-8} m produces a pressure amplitude of 3.0×10^{-2} Pa. (a) What is the wavelength of these waves? (b) For 1000-Hz waves in air, what displacement amplitude would be needed for the pressure amplitude to be at the pain threshold, which is 30 Pa? (c) For what wavelength and frequency will waves with a displacement amplitude of 1.2×10^{-8} m produce a pressure amplitude of 1.5×10^{-3} Pa?

16.2. Example 16.1 (Section 16.1) showed that for sound waves in air with frequency 1000 Hz, a displacement amplitude of 1.2×10^{-8} m produces a pressure amplitude of 3.0×10^{-2} Pa. Water at 20°C has a bulk modulus of 2.2×10^9 Pa, and the speed of sound in water at this temperature is 1480 m/s. For 1000-Hz sound waves in 20°C water, what displacement amplitude is produced if the pressure amplitude is 3.0×10^{-2} Pa? Explain why your answer is much less than 1.2×10^{-8} Pa.

16.3. Consider a sound wave in air that has displacement amplitude 0.0200 mm. Calculate the pressure amplitude for frequencies of (a) 150 Hz; (b) 1500 Hz; (c) 15,000 Hz. In each case compare the result to the pain threshold, which is 30 Pa.

16.4. A loud factory machine produces sound having a displacement amplitude of 1.00 μm, but the frequency of this sound can be adjusted. In order to prevent ear damage to the workers, the maximum pressure amplitude of the sound waves is limited to 10.0 Pa. Under the conditions of this factory, the bulk modulus of air is 1.42×10^5 Pa. What is the highest-frequency sound to which this machine can be adjusted without exceeding the prescribed limit? Is this frequency audible to the workers?

Section 16.2 Speed of Sound Waves

16.5. (a) In a liquid with density 1300 kg/m^3, longitudinal waves with frequency 400 Hz are found to have wavelength 8.00 m. Calculate the bulk modulus of the liquid. (b) A metal bar with a length of 1.50 m has density 6400 kg/m^3. Longitudinal sound waves take 3.90×10^{-4} s to travel from one end of the bar to the other. What is Young's modulus for this metal?

16.6. A major earthquake centered on Loma Prieta, California, near San Francisco, occurred at 5:04 P.M. local time on October 17, 1989 (in UTC, Coordinated Universal Time, 0h 4m 15s on October 18, 1989). The primary seismic waves (P waves) from such an earthquake are longitudinal waves that travel through the earth's crust. P waves were detected at Caracas, Venezuela, at 0h 13m 54s UTC; at Kevo, Finland, at 0h 15m 35s UTC; and at Vienna, Austria, at 0h 17m 02s UTC. The distances the P waves traveled from Loma Prieta were 6280 km to Caracas, 8690 km to Kevo, and 9650 km to Vienna. a) Use the arrival times to calculate the average speed of the P waves that traveled to these three cities. How can you account for any differences between the average speeds? (b) The average density of the earth's crust is about 3.3 g/cm^3. Use this value to calculate the bulk modulus of the earth's crust along the path traveled by the P waves to each of the three cities. How do your answers compare to the bulk moduli in Table 11.1?

16.7. A submerged scuba diver hears the sound of a boat horn directly above her on the surface of the lake. At the same time, a friend on dry land 22.0 m from the boat also hears the horn (Fig. 16.39). The horn is 1.2 m above the surface of the water. What is the distance (labeled by "?" in Fig. 16.39) from the horn to the diver? Both air and water are at 20°C.

Figure 16.39 Exercise 16.7.

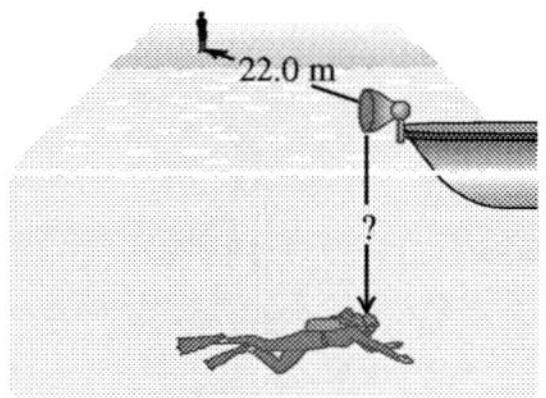

16.8. At a temperature of 27.0°C, what is the speed of longitudinal waves in (a) hydrogen (molar mass 2.02 g/mol)? (b) helium (molar mass 4.00 g/mol); (c) argon (molar mass 39.9 g/mol)? See Table 19.1 for values of γ. (d) Compare your answers for parts (a), (b), and (c) with the speed in air at the same temperature.

16.9. An oscillator vibrating at 1250 Hz produces a sound wave that travels through an ideal gas at 325 m/s when the gas temperature is 22.0°C. For a certain experiment, you need to have the same oscillator produce sound of wavelength 28.5 cm in this gas. What should the gas temperature be to achieve this wavelength?

16.10. (a) Show that the fractional change in the speed of sound (dv/v) due to a very small temperature change dT is given by $dv/v = \frac{1}{2} dT/T$. (*Hint:* Start with Eq. 16.10.) (b) The speed of sound in air at 20°C is found to be 344 m/s. Use the result in part (a) to find the change in the speed of sound for a 1.0°C change in air temperature.

16.11. An 80.0-m-long brass rod is struck at one end. A person at the other end hears two sounds as a result of two longitudinal waves, one traveling in the metal rod and the other traveling in the air. What is the time interval between the two sounds? The speed of sound in air is 344 m/s; relevant information about brass can be found in Table 11.1 and Table 14.1.

16.12. What is the difference between the speed of longitudinal waves in air at 27.0°C and their speed at −13.0°C?

16.13. What must be the stress (F/A) in a stretched wire of a material whose Young's modulus is Y for the speed of longitudinal waves to equal 30 times the speed of transverse waves?

Section 16.3 Sound Intensity

16.14. Use information from Table 16.2 to answer the following questions about sound in air. At 20°C the bulk modulus for air is 1.42×10^5 Pa and its density is 1.20 kg/m^3. At this temperature, what are the pressure amplitude (in Pa and atm) and the displacement amplitude (in m and nm) (a) for the softest sound a person can normally hear at 1000 Hz and (b) for the sound from a riveter at the same frequency? (c) How much energy per second does each wave deliver to a square 5.00 mm on a side?

16.15. Longitudinal Waves in Different Fluids. (a) A longitudinal wave propagating in a water-filled pipe has intensity $3.00 \times 10^{-6} \text{ W/m}^2$ and frequency 3400 Hz. Find the amplitude A and wavelength λ of the wave. Water has density 1000 kg/m^3 and bulk modulus 2.18×10^9 Pa. (b) If the pipe is filled with air at pressure 1.00×10^5 Pa and density 1.20 kg/m^3, what will be the amplitude A and wavelength λ of a longitudinal wave with the same intensity and frequency as in part (a)? (c) In which fluid is the amplitude larger, water or air? What is the ratio of the two amplitudes? Why is this ratio so different from 1.00?

16.16. Derive Eq. (16.14) from the equations that precede it.

16.17. A sound wave in air at 20°C has a frequency of 150 Hz and a displacement amplitude of 5.00×10^{-3} mm. For this sound wave calculate the (a) pressure amplitude (in Pa); (b) intensity (in W/m^2); (c) sound intensity level (in decibels).

16.18. (a) What is the sound intensity level in a car when the sound intensity is $0.500 \ \mu\text{W/m}^2$? (b) What is the sound intensity level in the air near a jackhammer when the pressure amplitude of the sound is 0.150 Pa and the temperature is 20.0°C?

16.19. For a person with normal hearing, the faintest sound that can be heard at a frequency of 400 Hz has a pressure amplitude of about 6.0×10^{-5} Pa. Calculate the (a) intensity; (b) sound intensity level; (c) displacement amplitude of this sound wave at 20°C.

16.20. The intensity due to a number of independent sound sources is the sum of the individual intensities. (a) When four quadruplets cry simultaneously, how many decibels greater is the sound intensity level than when a single one cries? (b) To increase the sound intensity level again by the same number of decibels as in part (a), how many more crying babies are required?

16.21. A baby's mouth is 30 cm from her father's ear and 1.50 m from her mother's ear. What is the difference between the sound intensity levels heard by the father and by the mother?

16.22. The Sacramento City Council recently adopted a law to reduce the allowed sound intensity level of the much despised leaf blowers from their current level of about 95 dB to 70 dB. With the new law, what is the ratio of the new allowed intensity to the previously allowed intensity?

16.23. (a) By what factor must the sound intensity be increased to raise the sound intensity level by 13.0 dB? (b) Explain why you don't need to know the original sound intensity.

Section 16.4 Standing Sound Waves and Normal Modes

16.24. The fundamental frequency of a pipe that is open at both ends is 594 Hz. (a) How long is this pipe? If one end is now closed, find (b) the wavelength and (c) the frequency of the new fundamental.

16.25. Standing sound waves are produced in a pipe that is 1.20 m long. For the fundamental and first two overtones, determine the locations along the pipe (measured from the left end) of the displacement nodes and the pressure nodes if (a) the pipe is open at both ends and (b) the pipe is closed at the left end and open at the right end.

16.26. Find the fundamental frequency and the frequency of the first three overtones of a pipe 45.0 cm long (a) if the pipe is open at both ends and (b) if the pipe is closed at one end. Use $v = 344 \text{ m/s}$. (c) For each of these cases, what is the number of the highest harmonic that may be heard by a person who can hear frequencies from 20 Hz to 20,000 Hz?

16.27. The Human Voice. The human vocal tract is a pipe that extends about 17 cm from the lips to the vocal folds (also called "vocal cords") near the middle of your throat. The vocal folds behave rather like the reed of a clarinet, and the vocal tract acts like a stopped pipe. Estimate the first three standing-wave frequencies of the vocal tract. Use $v = 344 \text{ m/s}$. (The answers are only an estimate, since the position of lips and tongue affects the motion of air in the vocal tract.)

16.28. Human Ear. The auditory canal of the ear (see Fig. 16.4) is filled with air. One end is open, and the other end is closed by the eardrum. A particular person's auditory canal is 2.40 cm long and can be modeled as a pipe. (a) What are the fundamental frequency and wavelength of this person's auditory canal? Is this sound audible? (b) Find the frequency of the highest *audible* harmonic of this person's canal. Which harmonic is this?

16.29. A certain pipe produces a fundamental frequency of 262 Hz in air. (a) If the pipe is filled with helium at the same temperature, what fundamental frequency does it produce? (The molar mass of air is 28.8 g/mol, and the molar mass of helium is 4.00 g/mol.) (b) Does your answer to part (a) depend on whether the pipe is open or stopped? Why or why not?

16.30. Singing in the Shower. A pipe closed at both ends can have standing waves inside of it, but you normally don't hear them because little of the sound can get out. But you *can* hear them if you are *inside* the pipe, such as someone singing in the shower. (a) Show that the wavelengths of standing waves in a pipe of length L that is closed at both ends are $\lambda_n = 2L/n$ and the frequencies are given by $f_n = nv/2L = nf_1$, where $n = 1, 2, 3, \ldots$. (b) Modeling it as a pipe, find the frequency of the fundamental and the first two overtones for a shower 2.50 m tall. Are these frequencies audible?

Section 16.5 Resonance and Sound

16.31. You blow across the open mouth of an empty test tube and produce the fundamental standing wave of the air column inside the test tube. The speed of sound in air is 344 m/s and the test tube acts as a stopped pipe. (a) If the length of the air column in the test tube is 14.0 cm, what is the frequency of this standing wave? (b) What is the frequency of the fundamental standing wave in the air column if the test tube is half filled with water?

16.32. You have a stopped pipe of adjustable length close to a taut 85.0-cm, 7.25-g wire under a tension of 4110 N. You want to adjust the length of the pipe so that, when it produces sound at its fundamental frequency, this sound causes the wire to vibrate in its second *overtone* with very large amplitude. How long should the pipe be?

Section 16.6 Interference of Waves

16.33. Two loudspeakers, A and B (Fig. 16.40), are driven by the same amplifier and emit sinusoidal waves in phase. Speaker B is 2.00 m to the right of speaker A. Consider point Q along the extension of the line connecting the speakers, 1.00 m to the right of speaker B. Both speakers emit sound waves that travel directly from the speaker to point Q. (a) What is the lowest frequency for which *constructive* interference occurs at point Q? (b) What is the lowest frequency for which *destructive* interference occurs at point Q?

Figure **16.40** Exercises 16.33 and 16.34.

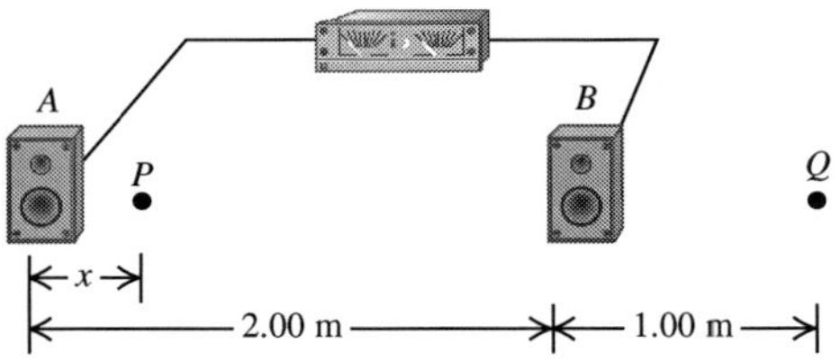

16.34. Two loudspeakers, A and B (Fig. 16.40), are driven by the same amplifier and emit sinusoidal waves in phase. Speaker B is 2.00 m to the right of speaker A. The frequency of the sound waves produced by the loudspeakers is 206 Hz. Consider point P between the speakers and along the line connecting them, a distance x to the right of speaker A. Both speakers emit sound waves that travel directly from the speaker to point P. (a) For what values of x will *destructive* interference occur at point P? (b) For what values of x will *constructive* interference occur at point P? (c) Interference effects like those in parts (a) and (b) are almost never a factor in listening to home stereo equipment. Why not?

16.35. Two loudspeakers, A and B, are driven by the same amplifier and emit sinusoidal waves in phase. Speaker B is 12.0 m to the right of speaker A. The frequency of the waves emitted by each speaker is 688 Hz. You are standing between the speakers, along the line connecting them, and are at a point of constructive interference. How far must you walk toward speaker B to move to a point of destructive interference?

16.36. Two loudspeakers, A and B, are driven by the same amplifier and emit sinusoidal waves in phase. The frequency of the waves emitted by each speaker is 172 Hz. You are 8.00 m from A. What is the closest you can be to B and be at a point of destructive interference?

16.37. Two loudspeakers, A and B, are driven by the same amplifier and emit sinusoidal waves in phase. The frequency of the waves emitted by each speaker is 860 Hz. Point P is 12.0 m from A and 13.4 m from B. Is the interference at P constructive or destructive? Give the reasoning behind your answer.

Section 16.7 Beats

16.38. Two guitarists attempt to play the same note of wavelength 6.50 cm at the same time, but one of the instruments is slightly out of tune and plays a note of wavelength 6.52 cm instead. What is the frequency of the beat these musicians hear when they play together?

16.39. Two organ pipes, open at one end but closed at the other, are each 1.14 m long. One is now lengthened by 2.00 cm. Find the frequency of the beat they produce when playing together in their fundamental.

16.40. Two identical taut strings under the same tension F produce a note of the same fundamental frequency f_0. The tension in one of them is now increased by a very small amount ΔF. (a) If they are played together in their fundamental, show that the frequency of the beat produced is $f_{beat} = f_0 (\Delta F / 2F)$. (b) Two identical violin strings, when in tune and stretched with the same tension, have a fundamental frequency of 440.0 Hz. One of the strings is retuned by increasing its tension. When this is done, 1.5 beats per second are heard when both strings are plucked simultaneously at their centers. By what percentage was the string tension changed?

Section 16.8 The Doppler Effect

16.41. On the planet Arrakis a male ornithoid is flying toward his mate at 25.0 m/s while singing at a frequency of 1200 Hz. If the stationary female hears a tone of 1240 Hz, what is the speed of sound in the atmosphere of Arrakis?

16.42. In Example 16.19 (Section 16.8), suppose the police car is moving away from the warehouse at 20 m/s. What frequency does the driver of the police car hear reflected from the warehouse?

16.43. Two train whistles, A and B, each have a frequency of 392 Hz. A is stationary and B is moving toward the right (away from A) at a speed of 35.0 m/s. A listener is between the two whistles and is moving toward the right with a speed of 15.0 m/s (Fig. 16.41). No wind is blowing. (a) What is the frequency from A as heard by the listener? (b) What is the frequency from B as heard by the listener? (c) What is the beat frequency detected by the listener?

Figure **16.41** Exercise 16.43.

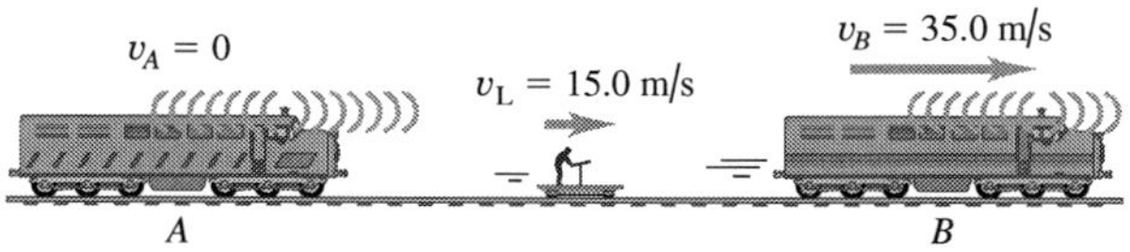

16.44. A railroad train is traveling at 25.0 m/s in still air. The frequency of the note emitted by the locomotive whistle is 400 Hz. What is the wavelength of the sound waves a) in front of the locomotive and (b) behind the locomotive? What is the frequency of the sound heard by a stationary listener c) in front of the locomotive and (d) behind the locomotive?

16.45. A swimming duck paddles the water with its feet once every 1.6 s, producing surface waves with this period. The duck is moving at constant speed in a pond where the speed of surface waves is 0.32 m/s, and the crests of the waves ahead of the duck are spaced 0.12 m apart. (a) What is the duck's speed? (b) How far apart are the crests behind the duck?

16.46. Moving Source vs. Moving Listener. (a) A sound source producing 1.00-kHz waves moves toward a stationary listener at one-half the speed of sound. What frequency will the listener hear? (b) Suppose instead that the source is stationary and the listener moves toward the source at one-half the speed of sound. What frequency does the listener hear? How does your answer compare to that in part (a)? Explain on physical grounds why the two answers differ.

16.47. A car alarm is emitting sound waves of frequency 520 Hz. You are on a motorcycle, traveling directly away from the car. How fast must you be traveling if you detect a frequency of 490 Hz?

16.48. A railroad train is traveling at 30.0 m/s in still air. The frequency of the note emitted by the train whistle is 262 Hz. What frequency is heard by a passenger on a train moving in the opposite direction to the first at 18.0 m/s and (a) approaching the first; and (b) receding from the first?

16.49. Doppler Radar. A giant thunderstorm is moving toward a weather station at 45.0 mi/h (20.1 m/s). If the station sends a radar beam of frequency 200.0 MHz toward the storm, what is the difference in frequency between the emitted beam and the beam reflected back from the storm? Be careful to carry plenty of significant figures! (*Hint:* The storm reflects the same frequency that it receives.)

16.50. Extrasolar Planets. In the not-too-distant future, it should be possible to detect the presence of planets moving around other stars by measuring the Doppler shift in the infrared light they emit. If a planet is going around its star at 50.00 km/s while emitting infrared light of frequency 3.330×10^{14} Hz, what frequency light will be received from this planet when it is moving directly away from us? (*Note:* Infrared light is light having wavelengths longer than those of visible light.)

16.51. How fast (as a percentage of light speed) would a star have to be moving so that the frequency of the light we receive from it is 10.0% higher than the frequency of the light it is emitting? Would it be moving away from us or toward us? (Assume it is moving either directly away from us or directly toward us.)

*Section 16.9 Shock Waves

***16.52.** The shock-wave cone created by the space shuttle at one instant during its reentry into the atmosphere makes an angle of 58.0° with its direction of motion. The speed of sound at this altitude is 331 m/s. (a) What is the Mach number of the shuttle at this instant, and (b) how fast (in m/s and mi/h) is it traveling relative to the atmosphere? (c) What would be its Mach number and the angle of its shock-wave cone if it flew at the same speed but at low altitude where the speed of sound is 344 m/s?

***16.53.** A jet plane flies overhead at Mach 1.70 and at a constant altitude of 950 m. (a) What is the angle α of the shock-wave cone? (b) How much time after the plane passes directly overhead do you hear the sonic boom? Neglect the variation of the speed of sound with altitude.

Problems

16.54. (a) Defend the following statement: "In a sinusoidal sound wave, the pressure variation given by Eq. (16.4) is greatest where the displacement given by Eq. (16.1) is zero." (b) For a sinusoidal sound wave given by Eq. (16.1) with amplitude $A = 10.0 \mu\text{m}$ and wavelength $\lambda = 0.250$ m, graph the displacement y and pressure fluctuation p as functions of x at time $t = 0$. Show at least two wavelengths of the wave on your graphs. (c) The displacement y in a *non*sinusoidal sound wave is shown in Fig. 16.42 as a function of x for $t = 0$. Draw a graph showing the pressure fluctuation p in this wave as a function of x at $t = 0$. This sound wave has the same 10.0-μm amplitude as the wave in part (b). Does it have the same pressure amplitude? Why or why not? (d) Is the statement in part (a) necessarily true if the sound wave is *not* sinusoidal? Explain your reasoning.

Figure **16.42** Problem 16.54.

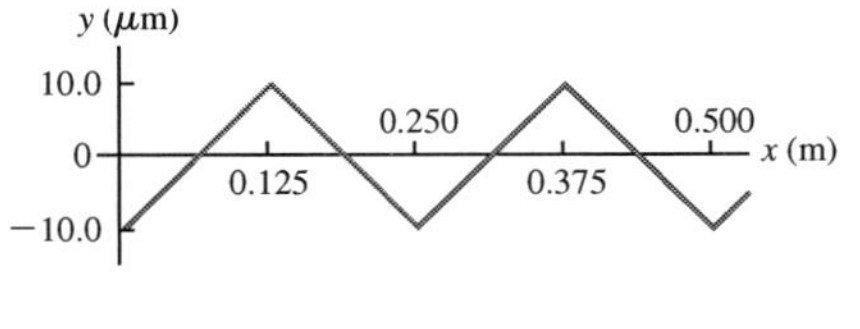

16.55. A soprano and a bass are singing a duet. While the soprano sings an $A^{\#}$ at 932 Hz, the bass sings an $A^{\#}$ but three octaves lower. In this concert hall, the density of air is 1.20 kg/m^3 and its bulk modulus is 1.42×10^5 Pa. In order for their notes to have the same sound intensity level, what must be (a) the ratio of the pressure amplitude of the bass to that of the soprano, and (b) the ratio of the displacement amplitude of the bass to that of the soprano? (c) What displacement amplitude (in m and nm) does the soprano produce to sing her $A^{\#}$ at 72.0 dB?

16.56. The sound from a trumpet radiates uniformly in all directions in 20°C air. At a distance of 5.00 m from the trumpet the sound intensity level is 52.0 dB. The frequency is 587 Hz. (a) What is the pressure amplitude at this distance? (b) What is the displacement amplitude? (c) At what distance is the sound intensity level 30.0 dB?

16.57. A Thermometer. Suppose you have a tube of length L containing a gas whose temperature you want to take, but you cannot get inside the tube. One end is closed, and the other end is open but a small speaker producing sound of variable frequency is at that end. You gradually increase the frequency of the speaker until the sound from the tube first becomes very loud. With further increase of the frequency, the loudness decreases but then gets very loud again at still higher frequencies. Call f_0 the lowest frequency at which the sound is very loud. (a) Show that the absolute temperature of this gas is given by $T = 16ML^2f_0^2/\gamma R$, where M is the molar mass of the gas, γ is the ratio of its heat capacities, and R is the ideal gas constant. (b) At what frequency above f_0 will the sound from the tube next reach a maximum in loudness? (c) How could you determine the speed of sound in this tube at temperature T?

16.58. A uniform 165-N bar is supported horizontally by two identical wires A and B (Fig. 16.43). A small 185-N cube of lead is placed $\frac{3}{4}$ of the way from A to B. The wires are each 75.0 cm long and have a mass of 5.50 g. If both of them are simultaneously plucked at the center, what is the frequency of the beats that they will produce when vibrating in their fundamental?

Figure **16.43** Problem 16.58.

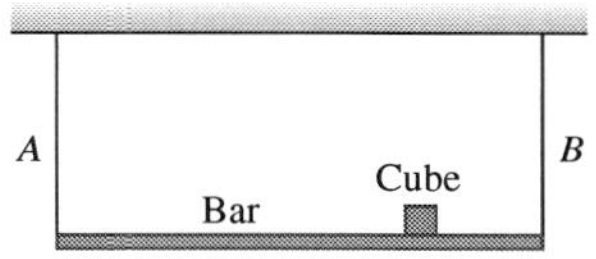

16.59. A person is playing a small flute 10.75 cm long, open at one end and closed at the other, near a taut string having a fundamental frequency of 600.0 Hz. If the speed of sound is 344.0 m/s, for which harmonics of the flute will the string resonate? In each case, which harmonic of the string is in resonance?

16.60. A New Musical Instrument. You have designed a new musical instrument of very simple construction. Your design consists of a metal tube with length L and diameter $L/10$. You have stretched a string of mass per unit length μ across the open end of the tube. The other end of the tube is closed. To produce the musical effect you're looking for, you want the frequency of the third-harmonic standing wave on the string to be the same as the fundamental frequency for sound waves in the air column in the tube. The speed of sound waves in this air column is v_s. (a) What must be the tension of the string to produce the desired effect? (b) What happens to the sound produced by the instrument if the tension is changed to twice the value calculated in part (a)? (c) For the tension calculated in part (a), what other harmonics of the string, if any, are in resonance with standing waves in the air column?

16.61. An organ pipe has two successive harmonics with frequencies 1372 and 1764 Hz. (a) Is this an open or a stopped pipe? Explain. (b) What two harmonics are these? (c) What is the length of the pipe?

16.62. Longitudinal Standing Waves in a Solid. Longitudinal standing waves can be produced in a solid rod by holding it at some point between the fingers of one hand and stroking it with the other hand. The rod oscillates with antinodes at both ends. (a) Why are the ends antinodes and not nodes? (b) The fundamental frequency can be obtained by stroking the rod while it is held at its center. Explain why this is the *only* place to hold the rod to obtain the fundamental. (c) Calculate the fundamental frequency of a steel rod of length 1.50 m (see Table 16.1). (d) What is the next possible standing-wave frequency of this rod? Where should the rod be held to excite a standing wave of this frequency?

16.63. (a) Determine the first three normal-mode frequencies for a pipe of length L that is closed at *both* ends. Explain your reasoning. (b) Use the results of part (a) to estimate the normal-mode frequencies of a shower stall. Explain the connection between these frequencies and the observation that your singing voice probably sounds better in the shower, especially when you sing at certain frequencies.

16.64. One type of steel has a density of $7.8 \times 10^3\ \text{kg/m}^3$ and a breaking stress of $7.0 \times 10^8\ \text{N/m}^2$. A cylindrical guitar string is to be made of 4.00 g of this steel. (a) What are the length and radius of the longest and thinnest string that can be placed under a tension of 900 N without breaking? (b) What is the highest fundamental frequency that this string could have?

16.65. A long tube contains air at a pressure of 1.00 atm and a temperature of 77.0°C. The tube is open at one end and closed at the other by a movable piston. A tuning fork near the open end is vibrating with a frequency of 500 Hz. Resonance is produced when the piston is at distances 18.0, 55.5, and 93.0 cm from the open end. (a) From these measurements, what is the speed of sound in air at 77.0°C? (b) From the result of part (a), what is the value of γ? (c) These data show that a displacement antinode is slightly outside of the open end of the tube. How far outside is it?

16.66. Human Hearing. The auditory canal of the human ear (see Fig. 16.4) extends about 2.5 cm from the outside ear to the eardrum. (a) Explain why the human ear is especially sensitive to sounds at frequencies around 3500 Hz. Use $v = 344\ \text{m/s}$. (b) Would you expect the ear to be especially sensitive to frequencies around 7000 Hz? Around 10,500 Hz? Why or why not?

16.67. A platinum wire (density $21.4\ \text{g/cm}^3$) is 225 μm in diameter and 0.450 m long. One end of the wire is attached to the ceiling, while a 420-g mass is attached to the other end so that the wire hangs vertically under tension. If a vibrating tuning fork of just the right frequency is held next to the wire, the wire begins to vibrate as well. (a) What tuning-fork frequencies will cause this to happen? You may assume that the bottom end of the wire (to which the mass is attached) is essentially stationary, and that the tension in the wire is essentially constant along its length. (b) Justify the assumptions made in part (a).

16.68. The frequency of the note F_4 is 349 Hz. (a) If an organ pipe is open at one end and closed at the other, what length must it have for its fundamental mode to produce this note at 20.0°C? (b) At what air temperature will the frequency be 370 Hz, corresponding to a rise in pitch from F to F#? (Ignore the change in length of the pipe due to the temperature change.)

16.69. A standing wave with a frequency of 1100 Hz in a column of methane (CH_4) at 20.0°C produces nodes that are 0.200 m apart. What is the value of γ for methane? (The molar mass of methane is 16.0 g/mol.)

16.70. Two identical loudspeakers are located at points A and B, 2.00 m apart. The loudspeakers are driven by the same amplifier and produce sound waves with a frequency of 784 Hz. Take the speed of sound in air to be 344 m/s. A small microphone is moved out from point B along a line perpendicular to the line connecting A and B (line BC in Fig. 16.44). (a) At what distances from B will there be *destructive* interference? (b) At what distances from B will there be *constructive* interference? (c) If the frequency is made low enough, there will be no positions along the line BC at which destructive interference occurs. How low must the frequency be for this to be the case?

Figure **16.44** Problem 16.70.

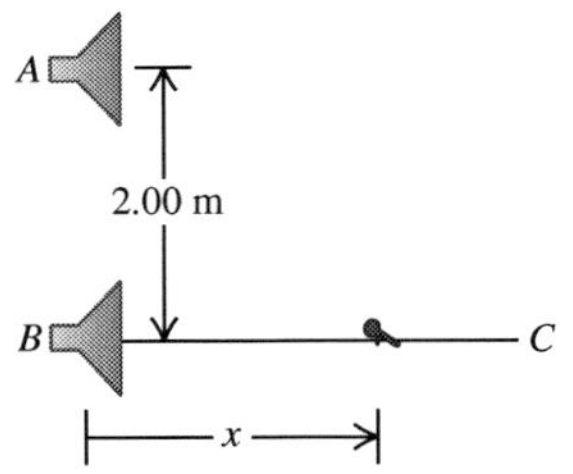

16.71. Wagnerian Opera. A man marries a great Wagnerian soprano but, alas, he discovers he cannot stand Wagnerian opera. In order to save his eardrums, the unhappy man decides he must silence his larklike wife for good. His plan is to tie her to the front of his car and send car and soprano speeding toward a brick wall. This soprano is quite shrewd, however, having studied physics in her student days at the music conservatory. She realizes that this wall has a resonant frequency of 600 Hz, which means that if a continuous sound wave of this frequency hits the wall, it will fall down, and she will be saved to sing more Isoldes. The car is heading toward the wall at a high speed of 30 m/s. (a) At what frequency must the soprano sing so that the wall will crumble? (b) What frequency will the soprano hear reflected from the wall just before it crumbles?

16.72. A bat flies toward a wall, emitting a steady sound of frequency 2.00 kHz. This bat hears its own sound plus the sound reflected by the wall. How fast should the bat fly in order to hear a beat frequency of 10.0 Hz?

16.73. A small sphere of radius R is arranged to pulsate so that its radius varies in simple harmonic motion between a minimum of $R - \Delta R$ and a maximum of $R + \Delta R$ with frequency f. This produces sound waves in the surrounding air of density ρ and bulk modulus B. (a) Find the intensity of sound waves at the surface of the sphere. (The amplitude of oscillation of the sphere is the same as that of the air at the surface of the sphere.) (b) Find the total acoustic power radiated by the sphere. (c) At a distance $d \gg R$ from the center of the sphere, find the amplitude, pressure amplitude, and intensity of the sound wave.

16.74. Ultrasound in Medicine. A 2.00-MHz sound wave travels through a pregnant woman's abdomen and is reflected from the fetal heart wall of her unborn baby. The heart wall is moving toward the sound receiver as the heart beats. The reflected sound is then mixed with the transmitted sound, and 85 beats per second are detected. The speed of sound in body tissue is 1500 m/s. Calculate the speed of the fetal heart wall at the instant this measurement is made.

16.75. The sound source of a ship's sonar system operates at a frequency of 22.0 kHz. The speed of sound in water (assumed to be at a uniform 20°C) is 1482 m/s. (a) What is the wavelength of the waves emitted by the source? (b) What is the difference in frequency between the directly radiated waves and the waves reflected from a whale traveling directly toward the ship at 4.95 m/s? The ship is at rest in the water.

16.76. A police siren of frequency f_{siren} is attached to a vibrating platform. The platform and siren oscillate up and down in simple harmonic motion with amplitude A_p and frequency f_p. (a) Find the maximum and minimum sound frequencies that you would hear at a position directly above the siren. (b) At what point in the motion of the platform is the maximum frequency heard? The minimum frequency? Explain.

16.77. Horseshoe bats (genus *Rhinolophus*) emit sounds from their nostrils and then listen to the frequency of the sound reflected from their prey to determine the prey's speed. (The "horseshoe" that gives the bat its name is a depression around the nostrils that acts like a focusing mirror, so that the bat emits sound in a narrow beam like a flashlight.) A *Rhinolophus* flying at speed v_{bat} emits sound of frequency f_{bat}; the sound it hears reflected from an insect flying toward it has a higher frequency f_{refl}. (a) Show that the speed of the insect is

$$v_{insect} = v\left[\frac{f_{refl}(v - v_{bat}) - f_{bat}(v + v_{bat})}{f_{refl}(v - v_{bat}) + f_{bat}(v + v_{bat})}\right]$$

where v is the speed of sound. (b) If $f_{bat} = 80.7\text{kHz}$, $f_{refl} = 83.5\text{kHz}$, and $v_{bat} = 3.9$ m/s, calculate the speed of the insect.

16.78. (a) Show that Eq. (16.30) can be written as

$$f_R = f_S\left(1 - \frac{v}{c}\right)^{1/2}\left(1 + \frac{v}{c}\right)^{-1/2}$$

(b) Use the binomial theorem to show that if $v \ll c$, this is approximately equal to

$$f_R = f_S\left(1 - \frac{v}{c}\right)$$

(c) A pilotless reconnaissance aircraft emits a radio signal with a frequency of 243 MHz. It is flying directly toward a test engineer on the ground. The engineer detects beats between the received signal and a local signal also of frequency 243 MHz. The beat frequency is 46.0 Hz. What is the speed of the aircraft? (Radio waves travel at the speed of light, $c = 3.00 \times 10^8$ m/s.)

16.79. Supernova! The gas cloud known as the Crab Nebula can be seen with even a small telescope. It is the remnant of a *supernova*, a cataclysmic explosion of a star. The explosion was seen on the earth on July 4, 1054 C.E. The streamers glow with the characteristic red color of heated hydrogen gas. In a laboratory on the earth, heated hydrogen produces red light with frequency 4.568×10^{14} Hz; the red light received from streamers in the Crab Nebula pointed toward the earth has frequency 4.586×10^{14} Hz. (a) Estimate the speed with which the outer edges of the Crab Nebula are expanding. Assume that the speed of the center of the nebula relative to the earth is negligible. (You may use the formulas derived in Problem 16.78. The speed of light is 3.00×10^8 m/s.) (b) Assuming that the expansion speed has been constant since the supernova explosion, estimate the diameter of the Crab Nebula. Give your answer in meters and in light-years. (c) The angular diameter of the Crab Nebula as seen from earth is about 5 arc minutes (1 arc minute $= \frac{1}{60}$ degree). Estimate the distance (in light-years) to the Crab Nebula, and estimate the year in which the supernova explosion actually took place.

16.80. Weather Forecasting. A radar installation used for monitoring thunderstorms emits radio waves at a frequency of 2800 MHz. A line of thunderstorms is approaching the installation at 42.0 km/h. (a) In the frame of reference of the thunderstorms, is the frequency of the radio waves greater than or less than 2800 MHz? Why? By what amount does the frequency differ from 2800 MHz? (You may use the formulas derived in Problem 16.78. Radio waves travel at the speed of light, $c = 3.00 \times 10^8$ m/s.) (b) Radio waves are reflected from the water drops in the thunderstorms, and the reflected waves are detected back at the installation. As measured by the receiver at the installation, is the frequency of these reflected waves greater or less than 2800 MHz? Why? By what amount does the frequency of the reflected waves differ from 2800 MHz?

16.81. A woman stands at rest in front of a large, smooth wall. She holds a vibrating tuning fork of frequency f_0 directly in front of her (between her and the wall). (a) The woman now runs toward the wall with speed v_W. She detects beats due to the interference between the sound waves reaching her directly from the fork and those reaching her after being reflected from the wall. How many beats per second will she detect? (*Note:* If the beat frequency is too large, the woman may have to use some instrumentation other than her ears to detect and count the beats.) (b) If the woman instead runs away from the wall, holding the tuning fork at her back so it is between her and the wall, how many beats per second will she detect?

***16.82.** On a clear day you see a jet plane flying overhead. From the apparent size of the plane, you determine that it is flying at a constant altitude h. You hear the sonic boom at time T after the plane passes directly overhead. Show that if the speed of sound v is the same at all altitudes, the speed of the plane is

$$v_S = \frac{hv}{\sqrt{h^2 - v^2T^2}}$$

(*Hint:* Trigonometric identities will be useful.)

Challenge Problems

16.83. Figure 16.45 shows the pressure fluctuation p of a nonsinusoidal sound wave as a function of x for $t = 0$. The wave is traveling in the $+x$-direction. (a) Graph the pressure fluctuation p as a function of t for $x = 0$. Show at least two cycles of oscillation. (b) Graph the displacement y in this sound wave as a function of x at $t = 0$. At $x = 0$, the displacement at $t = 0$ is zero. Show at least two wavelengths of the wave. (c) Graph the displacement y as a function of t for $x = 0$. Show at least two cycles of oscillation. (d) Calculate the maximum velocity and the maximum acceleration of an element of the air through which this sound wave is traveling. (e) Describe how the cone of a loudspeaker must move as a function of time to produce the sound wave in this problem.

Figure 16.45 Challenge Problem 16.83.

16.84. Two loudspeakers, A and B, radiate sound uniformly in all directions in air at 20°C. The acoustic power output from A is 8.00×10^{-4} W, and from B it is 6.00×10^{-5} W. Both loudspeakers are vibrating in phase at a frequency of 172 Hz. (a) Determine the difference in phase of the two signals at a point C along the line joining A and B, 3.00 m from B and 4.00 m from A (Fig. 16.46). (b) Determine the intensity and sound intensity level at C from speaker A if speaker B is turned off and the intensity and sound intensity level at point C from speaker B if speaker A is turned off. (c) With both speakers on, what are the intensity and sound intensity level at C?

Figure **16.46** Challenge Problem 16.84.

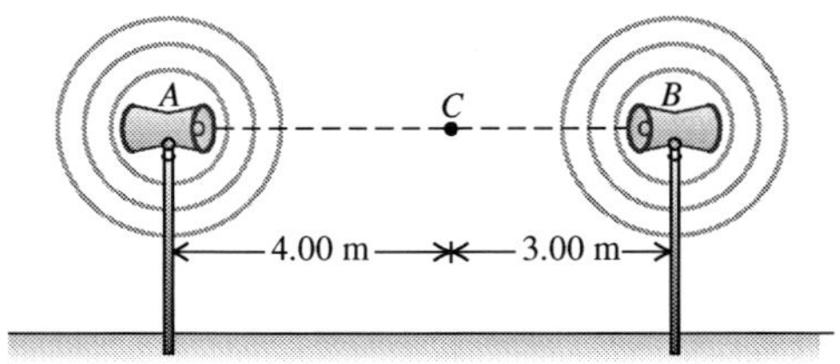

33 THE NATURE AND PROPAGATION OF LIGHT

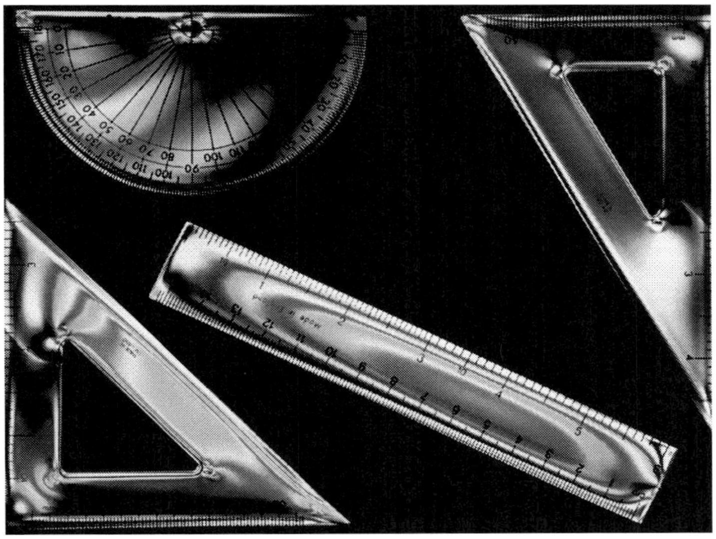

? These drafting tools are made of clear plastic, but a rainbow of colors appears when they are placed between two special filters called polarizers. How does this cause the colors?

LEARNING GOALS

By studying this chapter, you will learn:

- What light rays are, and how they are related to wave fronts.
- The laws that govern the reflection and refraction of light.
- The circumstances under which light is totally reflected at an interface.
- How to make polarized light out of ordinary light.
- How Huygens's principle helps us analyze reflection and refraction.

Blue lakes, ochre deserts, green forests, and multicolored rainbows can be enjoyed by anyone who has eyes with which to see them. But by studying the branch of physics called **optics,** which deals with the behavior of light and other electromagnetic waves, we can reach a deeper appreciation of the visible world. A knowledge of the properties of light allows us to understand the blue color of the sky and the design of optical devices such as telescopes, microscopes, cameras, eyeglasses, and the human eye. The same basic principles of optics also lie at the heart of modern developments such as the laser, optical fibers, holograms, optical computers, and new techniques in medical imaging.

The importance of optics to physics, and to science and engineering in general, is so great that we will devote the next four chapters to its study. In this chapter we begin with a study of the laws of reflection and refraction and the concepts of dispersion, polarization, and scattering of light. Along the way we compare the various possible descriptions of light in terms of particles, rays, or waves, and we introduce Huygens's principle, an important link that connects the ray and wave viewpoints. In Chapter 34 we'll use the ray description of light to understand how mirrors and lenses work, and we'll see how mirrors and lenses are used in optical instruments such as cameras, microscopes, and telescopes. We'll explore the wave characteristics of light further in Chapters 35 and 36.

33.1 The Nature of Light

Until the time of Isaac Newton (1642–1727), most scientists thought that light consisted of streams of particles (called *corpuscles*) emitted by light sources. Galileo and others tried (unsuccessfully) to measure the speed of light. Around

1665, evidence of *wave* properties of light began to be discovered. By the early 19th century, evidence that light is a wave had grown very persuasive.

In 1873, James Clerk Maxwell predicted the existence of electromagnetic waves and calculated their speed of propagation, as we learned in Chapter 32. This development, along with the experimental work of Heinrich Hertz starting in 1887, showed conclusively that light is indeed an electromagnetic wave.

The Two Personalities of Light

The wave picture of light is not the whole story, however. Several effects associated with emission and absorption of light reveal a particle aspect, in that the energy carried by light waves is packaged in discrete bundles called *photons* or *quanta.* These apparently contradictory wave and particle properties have been reconciled since 1930 with the development of quantum electrodynamics, a comprehensive theory that includes *both* wave and particle properties. The *propagation* of light is best described by a wave model, but understanding emission and absorption requires a particle approach.

33.1 An electric heating element emits primarily infrared radiation. But if its temperature is high enough, it also emits a discernible amount of visible light.

The fundamental sources of all electromagnetic radiation are electric charges in accelerated motion. All bodies emit electromagnetic radiation as a result of thermal motion of their molecules; this radiation, called *thermal radiation,* is a mixture of different wavelengths. At sufficiently high temperatures, all matter emits enough visible light to be self-luminous; a very hot body appears "red-hot" (Fig. 33.1) or "white-hot." Thus hot matter in any form is a light source. Familiar examples are a candle flame, hot coals in a campfire, the coils in an electric room heater, and an incandescent lamp filament (which usually operates at a temperature of about 3000°C).

Light is also produced during electrical discharges through ionized gases. The bluish light of mercury-arc lamps, the orange-yellow of sodium-vapor lamps, and the various colors of "neon" signs are familiar. A variation of the mercury-arc lamp is the *fluorescent* lamp (see Fig. 30.7). This light source uses a material called a *phosphor* to convert the ultraviolet radiation from a mercury arc into visible light. This conversion makes fluorescent lamps more efficient than incandescent lamps in transforming electrical energy into light.

33.2 Ophthalmic surgeons use lasers for repairing detached retinas and for cauterizing blood vessels in retinopathy. Pulses of blue-green light from an argon laser are ideal for this purpose, since they pass harmlessly through the transparent part of the eye but are absorbed by red pigments in the retina.

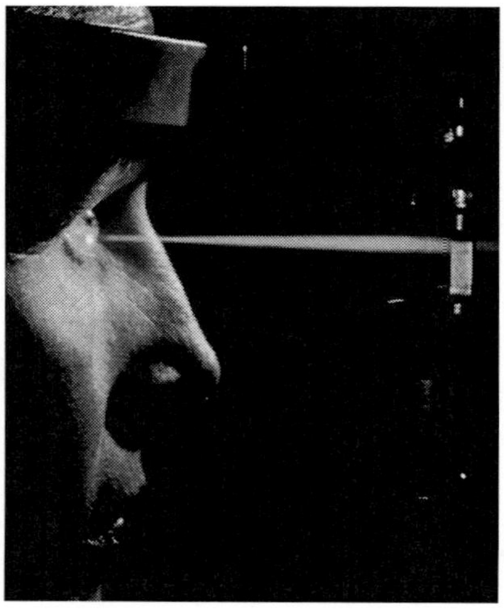

A light source that has attained prominence in the last forty years is the *laser.* In most light sources, light is emitted independently by different atoms within the source; in a laser, by contrast, atoms are induced to emit light in a cooperative, coherent fashion. The result is a very narrow beam of radiation that can be enormously intense and that is much more nearly *monochromatic,* or single-frequency, than light from any other source. Lasers are used by physicians for microsurgery, in CD players and computers to scan the information encoded on a compact disc or CD-ROM, in industry to cut through steel and to fuse high-melting-point materials, and in many other applications (Fig. 33.2).

No matter what its source, electromagnetic radiation travels in vacuum at the same speed. As we saw in Sections 1.3 and 32.1, the speed of light in vacuum is defined to be

$$c = 2.99792458 \times 10^8 \text{ m/s}$$

or 3.00×10^8 m/s to three significant figures. The duration of one second is defined by the cesium clock (see Section 1.3), so one meter is defined to be the distance that light travels in 1/299,792,458 s.

Waves, Wave Fronts, and Rays

We often use the concept of a **wave front** to describe wave propagation. We introduced this concept in Section 32.2 to describe the leading edge of a wave. More generally, we define a wave front as *the locus of all adjacent points at which the phase of vibration of a physical quantity associated with the wave is the same.* That is, at any instant, all points on a wave front are at the same part of the cycle of their variation.

When we drop a pebble into a calm pool, the expanding circles formed by the wave crests, as well as the circles formed by the wave troughs between them, are wave fronts. Similarly, when sound waves spread out in still air from a pointlike source, or when electromagnetic radiation spreads out from a pointlike emitter, any spherical surface that is concentric with the source is a wave front, as shown in Fig. 33.3. In diagrams of wave motion we usually draw only parts of a few wave fronts, often choosing consecutive wave fronts that have the same phase and thus are one wavelength apart, such as crests of water waves. Similarly, a diagram for sound waves might show only the "pressure crests," the surfaces over which the pressure is maximum, and a diagram for electromagnetic waves might show only the "crests" on which the electric or magnetic field is maximum.

We will often use diagrams that show the shapes of the wave fronts or their cross sections in some reference plane. For example, when electromagnetic waves are radiated by a small light source, we can represent the wave fronts as spherical surfaces concentric with the source or, as in Fig. 33.4a, by the circular intersections of these surfaces with the plane of the diagram. Far away from the source, where the radii of the spheres have become very large, a section of a spherical surface can be considered as a plane, and we have a *plane* wave like those discussed in Sections 32.2 and 32.3 (Fig, 33.4b).

To describe the directions in which light propagates, it's often convenient to represent a light wave by **rays** rather than by wave fronts. Rays were used to describe light long before its wave nature was firmly established. In a particle theory of light, rays are the paths of the particles. From the wave viewpoint *a ray is an imaginary line along the direction of travel of the wave.* In Fig. 33.4a the rays are the radii of the spherical wave fronts, and in Fig. 33.4b they are straight lines perpendicular to the wave fronts. When waves travel in a homogeneous isotropic material (a material with the same properties in all regions and in all directions), the rays are always straight lines normal to the wave fronts. At a boundary surface between two materials, such as the surface of a glass plate in air, the wave speed and the direction of a ray may change, but the ray segments in the air and in the glass are straight lines.

The next several chapters will give you many opportunities to see the interplay of the ray, wave, and particle descriptions of light. The branch of optics for which the ray description is adequate is called **geometric optics;** the branch dealing specifically with wave behavior is called **physical optics.** This chapter and the following one are concerned mostly with geometric optics. In Chapters 35 and 36 we will study wave phenomena and physical optics.

33.3 Spherical wave fronts of sound spread out uniformly in all directions from a point source in a motionless medium, such as still air, that has the same properties in all regions and in all directions. Electromagnetic waves in vacuum also spread out as shown here.

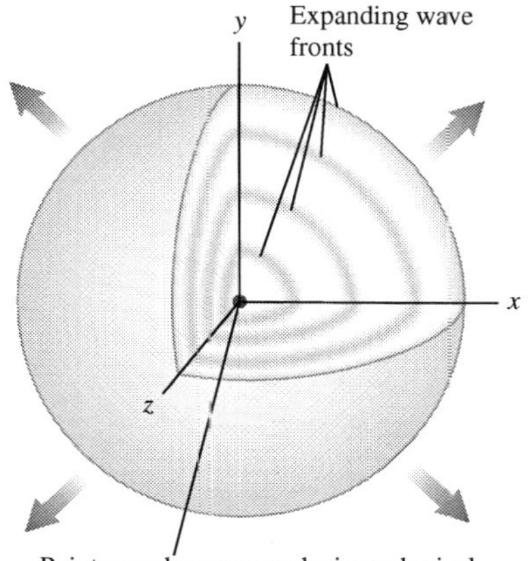

33.4 Wave fronts (blue) and rays (purple).

(a)

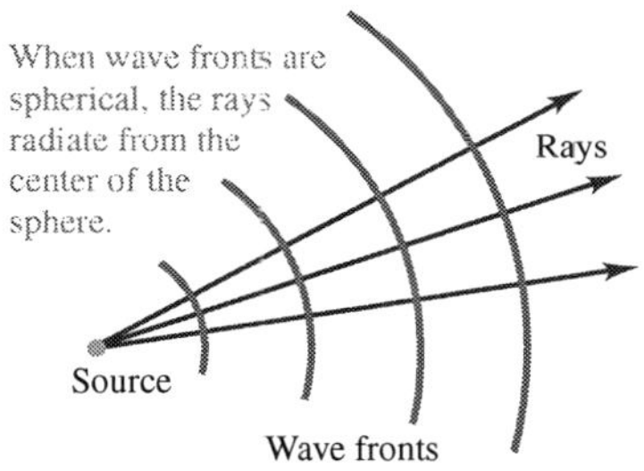

(b)

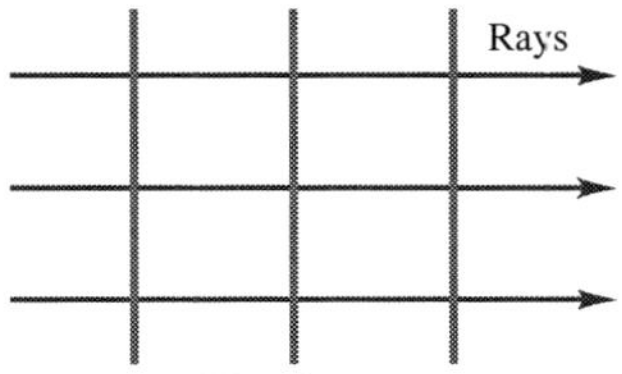

Test Your Understanding of Section 33.1 Some crystals are *not* isotropic: Light travels through the crystal at a higher speed in some directions than in others. In a crystal in which light travels at the same speed in the x- and z-directions but at a faster speed in the y-direction, what would be the shape of the wave fronts produced by a light source at the origin? (i) spherical, like those shown in Fig. 33.3; (ii) ellipsoidal, flattened along the y-axis; (iii) ellipsoidal, stretched out along the y-axis. ▮

33.2 Reflection and Refraction

In this section we'll use the *ray* model of light to explore two of the most important aspects of light propagation: **reflection** and **refraction.** When a light wave strikes a smooth interface separating two transparent materials (such as air and glass or water and glass), the wave is in general partly *reflected* and partly *refracted* (transmitted) into the second material, as shown in Fig. 33.5a. For example, when you look into a restaurant window from the street, you see a reflection of the street scene, but a person inside the restaurant can look out through the window at the same scene as light reaches him by refraction.

ActivPhysics ONLINE

15.1 Reflection and Refraction
15.3 Refraction Applications

33.5 (a) A plane wave is in part reflected and in part refracted at the boundary between two media (in this case, air and glass). The light that reaches the inside of the coffee shop is refracted twice, once entering the glass and once exiting the glass. (b), (c) How light behaves at the interface between the air outside the coffee shop (material a) and the glass (material b). For the case shown here, material b has a larger index of refraction that material a ($n_b > n_a$) and the angle θ_b is smaller than θ_a.

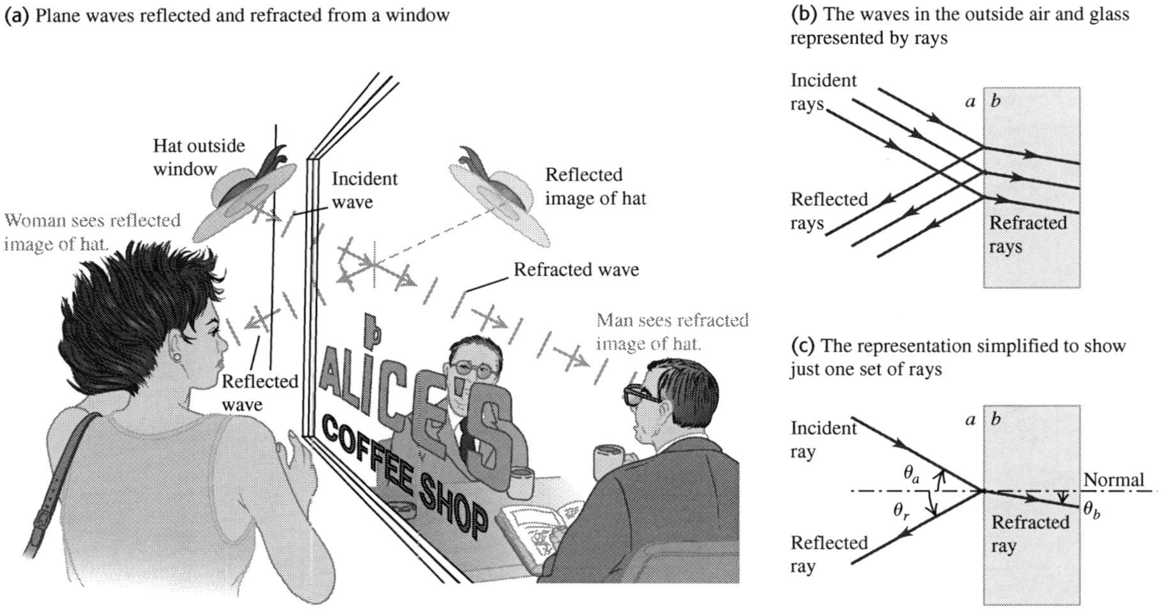

The segments of plane waves shown in Fig. 33.5a can be represented by bundles of rays forming *beams* of light (Fig. 33.5b). For simplicity we often draw only one ray in each beam (Fig. 33.5c). Representing these waves in terms of rays is the basis of geometric optics. We begin our study with the behavior of an individual ray.

33.6 Two types of reflection.

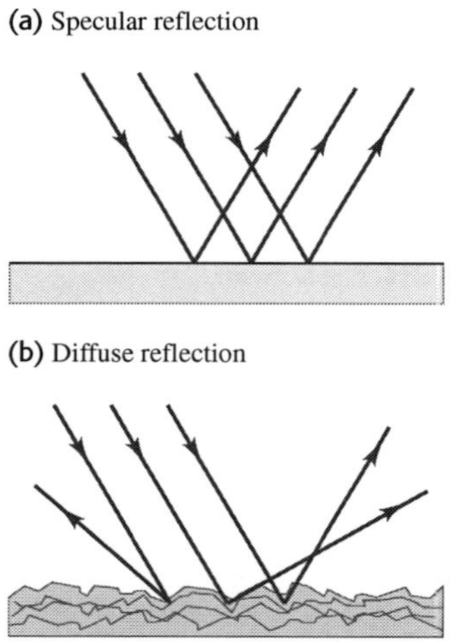

We describe the directions of the incident, reflected, and refracted (transmitted) rays at a smooth interface between two optical materials in terms of the angles they make with the *normal* (perpendicular) to the surface at the point of incidence, as shown in Fig. 33.5c. If the interface is rough, both the transmitted light and the reflected light are scattered in various directions, and there is no single angle of transmission or reflection. Reflection at a definite angle from a very smooth surface is called **specular reflection** (from the Latin word for "mirror"); scattered reflection from a rough surface is called **diffuse reflection.** This distinction is shown in Fig. 33.6. Both kinds of reflection can occur with either transparent materials or *opaque* materials that do not transmit light. The vast majority of objects in your environment (including clothing, plants, other people, and this book) are visible to you because they reflect light in a diffuse manner from their surfaces. Our primary concern, however, will be with specular reflection from a very smooth surface such as highly polished glass, plastic, or metal. Unless stated otherwise, when referring to "reflection" we will always mean *specular* reflection.

The **index of refraction** of an optical material (also called the **refractive index**), denoted by n, plays a central role in geometric optics. It is the ratio of the speed of light c in vacuum to the speed v in the material:

$$n = \frac{c}{v} \quad \text{(index of refraction)} \qquad (33.1)$$

Light always travels *more slowly* in a material than in vacuum, so the value of n in anything other than vacuum is always greater than unity. For vacuum, $n = 1$.

Since n is a ratio of two speeds, it is a pure number without units. (The relationship of the value of n to the electric and magnetic properties of a material is described in Section 32.3.)

CAUTION **Wave speed and index of refraction** Keep in mind that the wave speed v is *inversely* proportional to the index of refraction n. The greater the index of refraction in a material, the *slower* the wave speed in that material. Failure to remember this point can lead to serious confusion!

The Laws of Reflection and Refraction

Experimental studies of the directions of the incident, reflected, and refracted rays at a smooth interface between two optical materials lead to the following conclusions (Fig. 33.7):

1. **The incident, reflected, and refracted rays and the normal to the surface all lie in the same plane.** The plane of the three rays is perpendicular to the plane of the boundary surface between the two materials. We always draw ray diagrams so that the incident, reflected, and refracted rays are in the plane of the diagram.
2. **The angle of reflection θ_r is equal to the angle of incidence θ_a for all wavelengths and for any pair of materials.** That is, in Fig. 33.5c,

$$\theta_r = \theta_a \quad \text{(law of reflection)} \tag{33.2}$$

This relationship, together with the observation that the incident and reflected rays and the normal all lie in the same plane, is called the **law of reflection.**

3. For monochromatic light and for a given pair of materials, a and b, on opposite sides of the interface, **the ratio of the sines of the angles θ_a and θ_b, where both angles are measured from the normal to the surface, is equal to the inverse ratio of the two indexes of refraction:**

$$\frac{\sin\theta_a}{\sin\theta_b} = \frac{n_b}{n_a} \tag{33.3}$$

or

$$n_a \sin\theta_a = n_b \sin\theta_b \quad \text{(law of refraction)} \tag{33.4}$$

This experimental result, together with the observation that the incident and refracted rays and the normal all lie in the same plane, is called the **law of refraction** or **Snell's law,** after the Dutch scientist Willebrord Snell (1591–1626). There is some doubt that Snell actually discovered it. The discovery that $n = c/v$ came much later.

While these results were first observed experimentally, they can be derived theoretically from a wave description of light. We do this in Section 33.7.

Equations (33.3) and (33.4) show that when a ray passes from one material (a) into another material (b) having a larger index of refraction ($n_b > n_a$) and hence a slower wave speed, the angle θ_b with the normal is *smaller* in the second material than the angle θ_a in the first; hence the ray is bent *toward* the normal (Fig. 33.8a). When the second material has a *smaller* index of refraction than the first material ($n_b < n_a$) and hence a faster wave speed, the ray is bent *away from* the normal (Fig. 33.8b).

No matter what the materials on either side of the interface, in the case of *normal* incidence the transmitted ray is not bent at all (Fig. 33.8c). In this case $\theta_a = 0$ and $\sin\theta_a = 0$, so from Eq. (33.4) θ_b is also equal to zero, so the transmitted ray is

33.7 The laws of reflection and refraction.

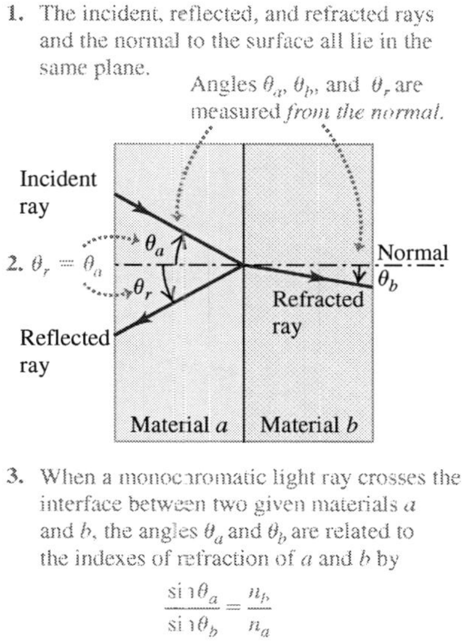

33.8 Refraction and reflection in three cases. (a) Material b has a larger index of refraction than material a. (b) Material b has a smaller index of refraction than material a. (c) The incident light ray is normal to the interface between the materials.

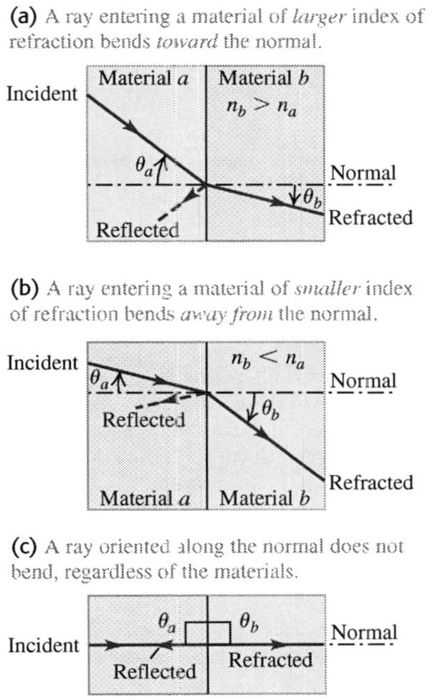

33.9 (a) This ruler is actually straight, but it appears to bend at the surface of the water. (b) Light rays from any submerged object bend away from the normal when they emerge into the air. As seen by an observer above the surface of the water, the object appears to be much closer to the surface than it actually is.

(a) A straight ruler half-immersed in water

(b) Why the ruler appears bent

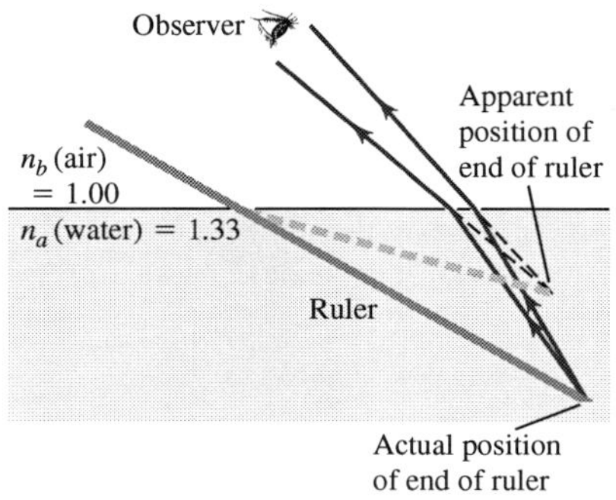

also normal to the interface. Equation (33.2) shows that θ_r, too, is equal to zero, so the reflected ray travels back along the same path as the incident ray.

The law of refraction explains why a partially submerged ruler or drinking straw appears bent; light rays coming from below the surface change in direction at the air–water interface, so the rays appear to be coming from a position above their actual point of origin (Fig. 33.9). A similar effect explains the appearance of the setting sun (Fig. 33.10).

An important special case is refraction that occurs at an interface between vacuum, for which the index of refraction is unity by definition, and a material. When a ray passes from vacuum into a material (b), so that $n_a = 1$ and $n_b > 1$, the ray is always bent *toward* the normal. When a ray passes from a material into vacuum, so that $n_a > 1$ and $n_b = 1$, the ray is always bent *away from* the normal.

The laws of reflection and refraction apply regardless of which side of the interface the incident ray comes from. If a ray of light approaches the interface in

33.10 (a) The index of refraction of air is slightly greater than 1, so light rays from the setting sun bend downward when they enter our atmosphere. (The effect is exaggerated in this figure.) (b) Stronger refraction occurs for light coming from the lower limb of the sun (the part that appears closest to the horizon), which passes through denser air in the lower atmosphere. As a result, the setting sun appears flattened vertically. (See Problem 33.55.)

(a)

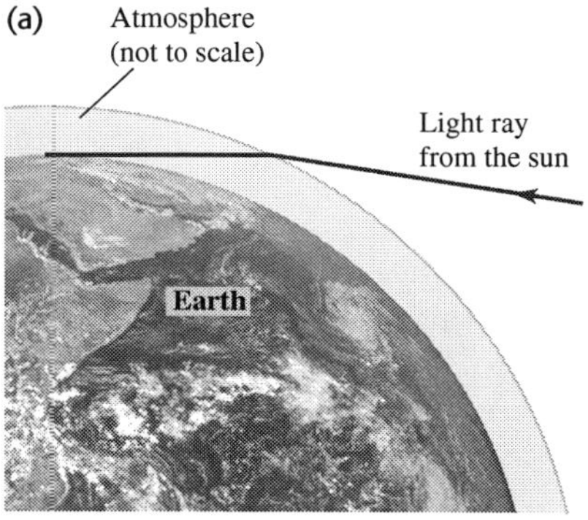

(b)

Fig. 33.8a or 33.8b from the right rather than from the left, there are again reflected and refracted rays; these two rays, the incident ray, and the normal to the surface again lie in the same plane. Furthermore, the path of a refracted ray is *reversible;* it follows the same path when going from b to a as when going from a to b. [You can verify this using Eq. (33.4).] Since reflected and incident rays make the same angle with the normal, the path of a reflected ray is also reversible. That's why when you see someone's eyes in a mirror, they can also see you.

The *intensities* of the reflected and refracted rays depend on the angle of incidence, the two indexes of refraction, and the polarization (that is, the direction of the electric-field vector) of the incident ray. The fraction reflected is smallest at normal incidence ($\theta_a = 0°$), where it is about 4% for an air–glass interface. This fraction increases with increasing angle of incidence to 100% at grazing incidence, when $\theta_a = 90°$.

It's possible to use Maxwell's equations to predict the amplitude, intensity, phase, and polarization states of the reflected and refracted waves. Such an analysis is beyond our scope, however.

The index of refraction depends not only on the substance but also on the wavelength of the light. The dependence on wavelength is called *dispersion;* we will consider it in Section 33.4. Indexes of refraction for several solids and liquids are given in Table 33.1 for a particular wavelength of yellow light.

The index of refraction of air at standard temperature and pressure is about 1.0003, and we will usually take it to be exactly unity. The index of refraction of a gas increases as its density increases. Most glasses used in optical instruments have indexes of refraction between about 1.5 and 2.0. A few substances have larger indexes; one example is diamond, with 2.417.

Table 33.1 Index of Refraction for Yellow Sodium Light $\lambda_0 = 589$ nm

Substance	Index of Refraction, n
Solids	
Ice (H_2O)	1.309
Fluorite (CaF_2)	1.434
Polystyrene	1.49
Rock salt (NaCl)	1.544
Quartz (SiO_2)	1.544
Zircon ($ZrO_2 \cdot SiO_2$)	1.923
Diamond (C)	2.417
Fabulite ($SrTiO_3$)	2.409
Rutile (TiO_2)	2.62
Glasses (typical values)	
Crown	1.52
Light flint	1.58
Medium flint	1.62
Dense flint	1.66
Lanthanum flint	1.80
Liquids at 20°C	
Methanol (CH_3OH)	1.329
Water (H_2O)	1.333
Ethanol (C_2H_5OH)	1.36
Carbon tetrachloride (CCl_4)	1.460
Turpentine	1.472
Glycerine	1.473
Benzene	1.501
Carbon disulfide (CS_2)	1.628

Index of Refraction and the Wave Aspects of Light

We have discussed how the direction of a light ray changes when it passes from one material to another material with a different index of refraction. It's also important to see what happens to the *wave* characteristics of the light when this happens.

First, the frequency f of the wave does not change when passing from one material to another. That is, the number of wave cycles arriving per unit time must equal the number leaving per unit time; this is a statement that the boundary surface cannot create or destroy waves.

Second, the wavelength λ of the wave *is* different in general in different materials. This is because in any material, $v = \lambda f$; since f is the same in any material as in vacuum and v is always less than the wave speed c in vacuum, λ is also correspondingly reduced. Thus the wavelength λ of light in a material is *less than* the wavelength λ_0 of the same light in vacuum. From the above discussion, $f = c/\lambda_0 = v/\lambda$. Combining this with Eq. (33.1), $n = c/v$, we find

$$\lambda = \frac{\lambda_0}{n} \quad \text{(wavelength of light in a material)} \qquad (33.5)$$

When a wave passes from one material into a second material with larger index of refraction, so that $n_b > n_a$, the wave speed decreases. The wavelength $\lambda_b = \lambda_0/n_b$ in the second material is then shorter than the wavelength $\lambda_a = \lambda_0/n_a$ in the first material. If instead the second material has a smaller index of refraction than the first material, so that $n_b < n_a$, then the wave speed increases. Then the wavelength λ_b in the second material is longer than the wavelength λ_a in the first material. This makes intuitive sense; the waves get "squeezed" (the wavelength gets shorter) if the wave speed decreases and get "stretched" (the wavelength gets longer) if the wave speed increases.

Problem-Solving Strategy 33.1 Reflection and Refraction

IDENTIFY *the relevant concepts:* You need to use the ideas of this section, called *geometric optics,* whenever light encounters a boundary between two different materials. In general, part of the light is reflected back into the first material and part is refracted into the second material. These ideas apply to electromagnetic radiation of all frequencies and wavelengths, not just visible light.

SET UP *the problem* using the following steps:

1. In geometric optics problems involving rays and angles, *always* start by drawing a large, neat diagram. Label all known angles and indexes of refraction.
2. Determine the target variables.

EXECUTE *the solution* as follows:

1. Apply the laws of reflection, Eq. (33.2), and refraction, Eq. (33.4). Remember to always measure the angles of incidence, reflection, and refraction from the *normal* to the surface where the reflection and refraction occur, *never* from the surface itself.
2. You will often have to use some simple geometry or trigonometry in working out angular relationships. The sum of the interior angles in a triangle is 180°, an angle and its complement differ by 180°, and so on. Ask yourself, "What information am I given?", "What do I need to know in order to find this angle?", or "What other angles or other quantities can I compute using the information given in the problem?"
3. Remember that the frequency of the light does not change when it moves from one material to another, but the wavelength changes in accordance with Eq. (33.5).

EVALUATE *your answer:* In problems that involve refraction, check that the direction of refraction makes sense. If the second material has a higher index of refraction than the first material, the refracted ray bends toward the normal and the refracted angle is smaller than the incident angle. If the first material has the higher index of refraction, the refracted ray bends away from the normal and the refracted angle is larger than the incident angle. Do your results agree with these rules?

Example 33.1 Reflection and refraction

In Fig. 33.11, material *a* is water and material *b* is a glass with index of refraction 1.52. If the incident ray makes an angle of 60° with the normal, find the directions of the reflected and refracted rays.

SOLUTION

IDENTIFY: This is a problem in geometric optics. We are given the incident angle and the index of refraction of each material, and we need to find the reflected and refracted angles.

33.11 Reflection and refraction of light passing from water to glass.

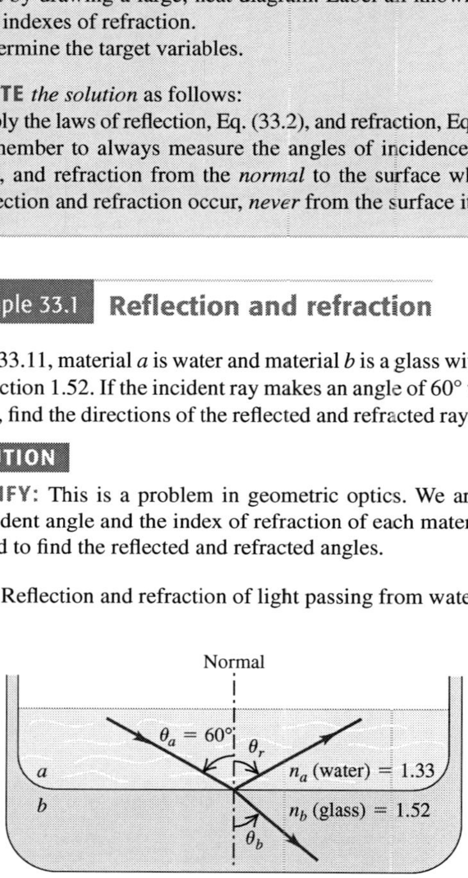

SET UP: Figure 33.11 shows the rays and angles for this situation. The target variables are the reflected angle θ_r and the refracted angle θ_b. Since n_b is greater than n_a, the refracted angle must be smaller than the incident angle θ_a; this is shown in the figure.

EXECUTE: According to Eq. (33.2), the angle the reflected ray makes with the normal is the same as that of the incident ray, so $\theta_r = \theta_a = 60.0°$.

To find the direction of the refracted ray, we use Snell's law, Eq.(33.4), with $n_a = 1.33$, $n_b = 1.52$, and $\theta_a = 60.0°$. We find

$$n_a \sin\theta_a = n_b \sin\theta_b$$

$$\sin\theta_b = \frac{n_a}{n_b}\sin\theta_a = \frac{1.33}{1.52}\sin 60.0° = 0.758$$

$$\theta_b = 49.3°$$

EVALUATE: The second material has a larger refractive index than the first, just like the situation shown in Fig. 33.8a. Hence, the refracted ray is bent toward the normal as the wave slows down upon entering the second material, and $\theta_b < \theta_a$.

Example 33.2 Index of refraction in the eye

The wavelength of the red light from a helium-neon laser is 633 nm in air but 474 nm in the aqueous humor inside your eyeball. Calculate the index of refraction of the aqueous humor and the speed and frequency of the light in this substance.

SOLUTION

IDENTIFY: The key ideas here are the relationship between index of refraction n and wave speed v and the relationship between index of refraction and wavelength λ.

SET UP: We use the definition of index of refraction given by Eq. (33.1), $n = c/v$, as well as Eq. (33.5), $\lambda = \lambda_0/n$. It will also be helpful to use the relationship $v = \lambda f$ among wave speed, wavelength, and frequency.

EXECUTE: The index of refraction of air is very close to unity, so we assume that the wavelengths in air and vacuum are the same. Then the wavelength λ in the material is given by Eq. (33.5) with $\lambda_0 = 633$ nm:

$$\lambda = \frac{\lambda_0}{n} \qquad n = \frac{\lambda_0}{\lambda} = \frac{633\text{ nm}}{474\text{ nm}} = 1.34$$

This is about the same index of refraction as for water. Then $n = c/v$ gives

$$v = \frac{c}{n} = \frac{3.00 \times 10^8 \text{ m/s}}{1.34} = 2.25 \times 10^8 \text{ m/s}$$

Finally, from $v = \lambda f$,

$$f = \frac{v}{\lambda} = \frac{2.25 \times 10^8 \text{ m/s}}{474 \times 10^{-9} \text{ m}} = 4.74 \times 10^{14} \text{ Hz}$$

EVALUATE: Note that while the speed and wavelength have different values in air and in the aqueous humor, the *frequency* in air, f_0, is the same as the frequency f in the aqueous humor:

$$f_0 = \frac{c}{\lambda_0} = \frac{3.00 \times 10^8 \text{ m/s}}{633 \times 10^{-9} \text{ m}} = 4.74 \times 10^{14} \text{ Hz}$$

This illustrates the general rule that when a light wave passes from one material into another, the wave frequency is unchanged.

Example 33.3 A twice-reflected ray

Two mirrors are perpendicular to each other. A ray traveling in a plane perpendicular to both mirrors is reflected from one mirror, then the other, as shown in Fig. 33.12. What is the ray's final direction relative to its original direction?

SOLUTION

IDENTIFY: This problem involves only the law of reflection.

SET UP: There are two reflections in this situation, so we must apply the law of reflection twice.

EXECUTE: For mirror 1 the angle of incidence is θ_1, and this equals the angle of reflection. The sum of interior angles in the triangle shown in the figure is 180°, so we see that the angles of incidence and reflection for mirror 2 are both $90° - \theta_1$. The total change in direction of the ray after both reflections is therefore $2(90° - \theta_1) + 2\theta_1 = 180°$. That is, the ray's final direction is opposite to its original direction.

EVALUATE: An alternative viewpoint is that specular reflection reverses the sign of the component of light velocity perpendicular to the surface but leaves the other components unchanged. We invite you to verify this in detail. You should also be able to use this result to show that when a ray of light is successively reflected by three mirrors forming a corner of a cube (a "corner reflector"), its final direction is again opposite to its original direction. This principle is widely used in tail-light lenses and bicycle reflectors to improve their night-time visibility. Apollo astronauts placed arrays of corner reflectors on the moon. By use of laser beams reflected from these arrays, the earth–moon distance has been measured to within 0.15 m.

33.12 A ray moving in the xy-plane. The first reflection changes the sign of the y-component of its velocity, and the second reflection changes the sign of the x-component. For a different ray with a z-component of velocity, a third mirror (perpendicular to the two shown) could be used to change the sign of that component.

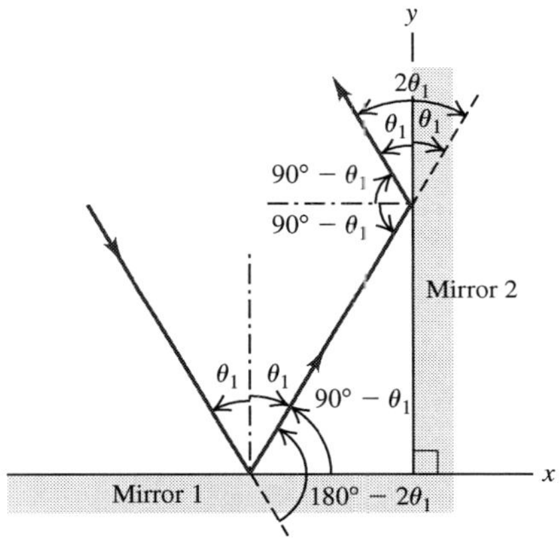

Test Your Understanding of Section 33.2 You are standing on the shore of a lake. You spot a tasty fish swimming some distance below the lake surface. (a) If you want to spear the fish, should you aim the spear (i) above, (ii) below, or (iii) directly at the apparent position of the fish? (b) If instead you use a high-power laser to simultaneously kill and cook the fish, should you aim the laser (i) above, (ii) below, or (iii) directly at the apparent position of the fish?

33.3 Total Internal Reflection

ActivPhysics Online
15.2 Total Internal Reflection

We have described how light is partially reflected and partially transmitted at an interface between two materials with different indexes of refraction. Under certain circumstances, however, *all* of the light can be reflected back from the interface, with none of it being transmitted, even though the second material is transparent. Figure 33.13a shows how this can occur. Several rays are shown radiating from a point source in material a with index of refraction n_a. The rays strike the surface of a second material b with index n_b, where $n_a > n_b$. (For

33.13 (a) Total internal reflection. The angle of incidence for which the angle of refraction is 90° is called the critical angle: this is the case for ray 3. The reflected portions of rays 1, 2, and 3 are omitted for clarity. (b) Rays of laser light enter the water in the fishbowl from above; they are reflected at the bottom by mirrors tilted at slightly different angles. One ray undergoes total internal reflection at the air–water interface.

(a) Total internal reflection

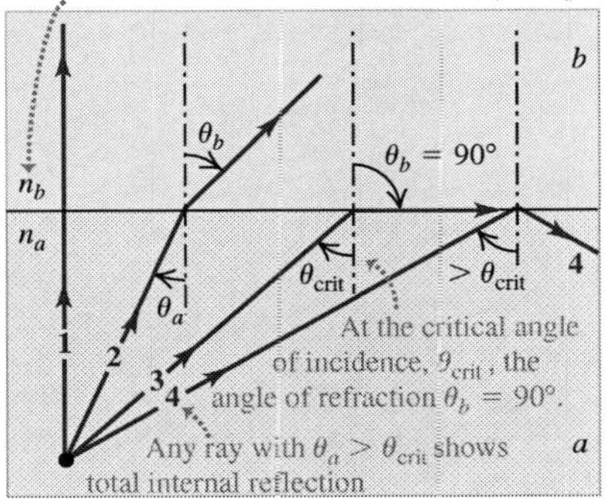

(b) Total internal reflection demonstrated with a laser, mirrors, and water in a fishbowl

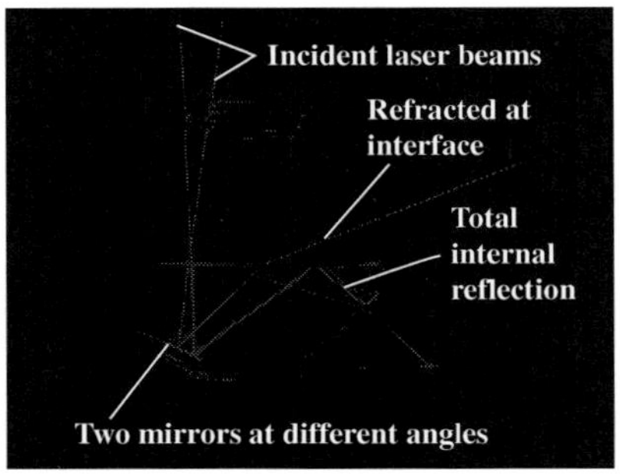

instance, materials a and b could be water and air, respectively.) From Snell's law of refraction,

$$\sin\theta_b = \frac{n_a}{n_b}\sin\theta_a$$

Because n_a/n_b is greater than unity, sin θ_b is larger than $\sin\theta_a$; the ray is bent *away from* the normal. Thus there must be some value of θ_a *less than* 90° for which Snell's law gives $\sin\theta_b = 1$ and $\theta_b = 90°$. This is shown by ray 3 in the diagram, which emerges just grazing the surface at an angle of refraction of 90°. Compare the diagram in Fig. 33.13a to the photograph of light rays in Fig. 33.13b.

The angle of incidence for which the refracted ray emerges tangent to the surface is called the **critical angle,** denoted by θ_{crit}. (A more detailed analysis using Maxwell's equations shows that as the incident angle approaches the critical angle, the transmitted intensity approaches zero.) If the angle of incidence is *larger* than the critical angle, the sine of the angle of refraction, as computed by Snell's law, would have to be greater than unity, which is impossible. Beyond the critical angle, the ray *cannot* pass into the upper material; it is trapped in the lower material and is completely reflected at the boundary surface. This situation, called **total internal reflection,** occurs only when a ray is incident on the interface with a second material whose index of refraction is *smaller* than that of the material in which the ray is traveling.

We can find the critical angle for two given materials by setting $\theta_b = 90°(\sin\theta_b = 1)$ in Snell's law. We then have

$$\sin\theta_{crit} = \frac{n_b}{n_a} \quad \text{(critical angle for total internal reflection)} \qquad (33.6)$$

Total internal reflection will occur if the angle of incidence θ_a is larger than or equal to θ_{crit}.

Applications of Total Internal Reflection

Total internal reflection finds numerous uses in optical technology. As an example, consider glass with index of refraction $n = 1.52$. If light propagating within this glass encounters a glass–air interface, the critical angle is

$$\sin\theta_{crit} = \frac{1}{1.52} = 0.658 \qquad \theta_{crit} = 41.1°$$

(a) Total internal reflection in a Porro prism

(b) Binoculars use Porro prisms to reflect the light to each eyepiece.

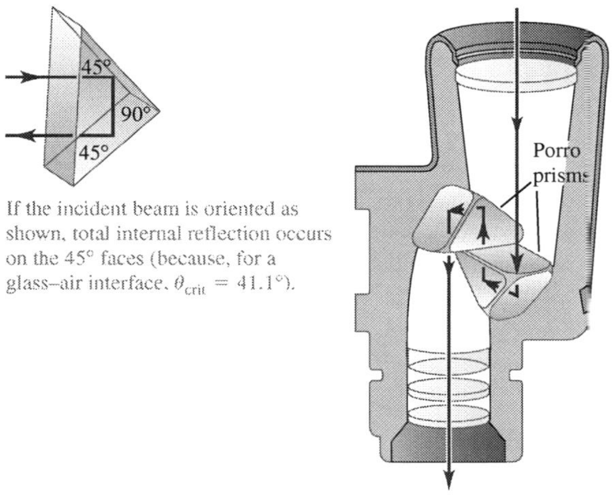

33.14 (a) Total internal reflection in a Porro prism. (b) A combination of two Porro prisms in binoculars.

The light will be *totally reflected* if it strikes the glass–air surface at an angle of 41.1° or larger. Because the critical angle is slightly smaller than 45°, it is possible to use a prism with angles of 45°–45°–90° as a totally reflecting surface. As reflectors, totally reflecting prisms have some advantages over metallic surfaces such as ordinary coated-glass mirrors. While no metallic surface reflects 100% of the light incident on it, light can be *totally* reflected by a prism. The reflecting properties of a prism have the additional advantages of being permanent and unaffected by tarnishing.

A 45°–45°–90° prism, used as in Fig. 33.14a, is called a *Porro* prism. Light enters and leaves at right angles to the hypotenuse and is totally reflected at each of the shorter faces. The total change of direction of the rays is 180°. Binoculars often use combinations of two Porro prisms, as in Fig. 33.14b.

When a beam of light enters at one end of a transparent rod (Fig. 33.15), the light can be totally reflected internally if the index of refraction of the rod is greater than that of the surrounding material. The light is "trapped" within the rod even if the rod is curved, provided that the curvature is not too great. Such a rod is sometimes called a *light pipe.* A bundle of fine glass or plastic fibers behaves in the same way and has the advantage of being flexible. A bundle may consist of thousands of individual fibers, each of the order of 0.002 to 0.01 mm in diameter. If the fibers are assembled in the bundle so that the relative positions of the ends are the same (or mirror images) at both ends, the bundle can transmit an image, as shown in Fig. 33.16.

Fiber-optic devices have found a wide range of medical applications in instruments called *endoscopes,* which can be inserted directly into the bronchial tubes, the bladder, the colon, and so on, for direct visual examination. A bundle of fibers can be enclosed in a hypodermic needle for study of tissues and blood vessels far beneath the skin.

Fiber optics also have applications in communication systems, in which they are used to transmit a modulated laser beam. The rate at which information can be transmitted by a wave (light, radio, or whatever) is proportional to the frequency. To see qualitatively why this is so, consider modulating (modifying) the wave by chopping off some of the wave crests. Suppose each crest represents a binary digit, with a chopped-off crest representing a zero and an unmodified crest representing a one. The number of binary digits we can transmit per unit time is thus proportional to the frequency of the wave. Infrared and visible-light waves have much higher frequency than do radio waves, so a modulated laser beam can transmit an enormous amount of information through a single fiber-optic cable.

33.15 A transparent rod with refractive index greater than that of the surrounding material.

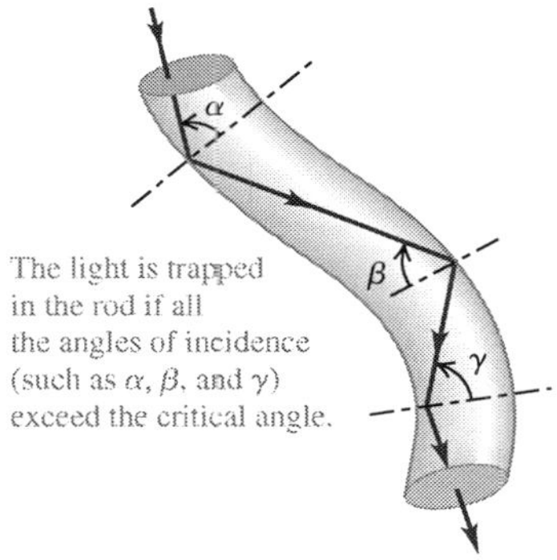

33.16 Image transmission by a bundle of optical fibers.

33.17 To maximize their brilliance, diamonds are cut so that there is total internal reflection on their back surfaces.

Another advantage of optical fibers is that they can be made thinner than conventional copper wire, so more fibers can be bundled together in a cable of a given diameter. Hence more distinct signals (for instance, different phone lines) can be sent over the same cable. Because fiber-optic cables are electrical insulators, they are immune to electrical interference from lightning and other sources, and they don't allow unwanted currents between source and receiver. For these and other reasons, fiber-optic cables are playing an increasingly important role in long-distance telephone, television, and Internet communication.

Total internal reflection also plays an important role in the design of jewelry. The brilliance of diamond is due in large measure to its very high index of refraction $(n = 2.417)$ and correspondingly small critical angle. Light entering a cut diamond is totally internally reflected from facets on its back surface, and then emerges from its front surface (Fig. 33.17). "Imitation diamond" gems, such as cubic zirconia, are made from less expensive crystalline materials with comparable indexes of refraction.

Conceptual Example 33.4 A leaky periscope

A periscope for a submarine uses two totally reflecting 45°–45°–90° prisms with total internal reflection on the sides adjacent to the 45° angles. It springs a leak, and the bottom prism is covered with water. Explain why the periscope no longer works.

SOLUTION

The critical angle for water $(n_b = 1.33)$ on glass $(n_a = 1.52)$ is

$$\theta_{\text{crit}} = \arcsin\frac{1.33}{1.52} = 61.0^\circ$$

The 45° angle of incidence for a totally reflecting prism is *smaller than* the 61° critical angle, so total internal reflection does not occur at the glass–water boundary. Most of the light is transmitted into the water, and very little is reflected back into the prism.

Test Your Understanding of Section 33.3 In which of the following situations is there total internal reflection? (i) Light propagating in water $(n = 1.33)$ strikes a water–air interface at an incident angle of 70°; (ii) light propagating in glass $(n = 1.52)$ strikes a glass–water interface at an incident angle of 70°; (iii) light propagating in water strikes a water–glass interface at an incident angle of 70°.

*33.4 Dispersion

Ordinary white light is a superposition of waves with wavelengths extending throughout the visible spectrum. The speed of light *in vacuum* is the same for all wavelengths, but the speed in a material substance is different for different wavelengths. Therefore the index of refraction of a material depends on wavelength. The dependence of wave speed and index of refraction on wavelength is called **dispersion.**

33.18 Variation of index of refraction n with wavelength for different transparent materials. The horizontal axis shows the wavelength λ_0 of the light *in vacuum;* the wavelength in the material is equal to $\lambda = \lambda_0/n$.

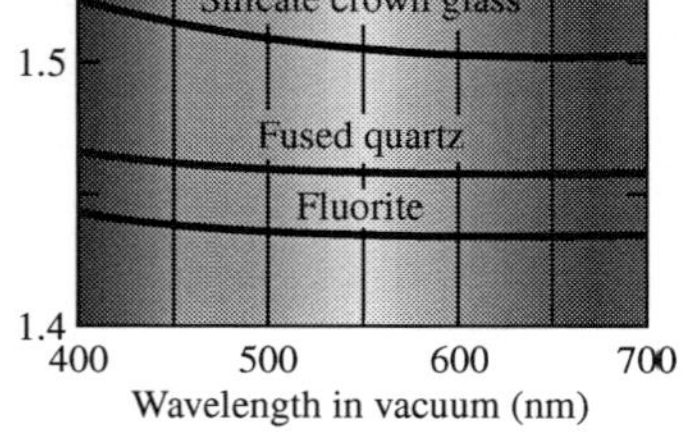

Figure 33.18 shows the variation of index of refraction n with wavelength for a few common optical materials. Note that the horizontal axis of this figure is the wavelength of the light *in vacuum,* λ_0; the wavelength in the material is given by Eq. (33.5), $\lambda = \lambda_0/n$. In most materials the value of *n decreases* with increasing wavelength and decreasing frequency, and thus *n increases* with decreasing wavelength and increasing frequency. In such a material, light of longer wavelength has greater speed than light of shorter wavelength.

Figure 33.19 shows a ray of white light incident on a prism. The deviation (change of direction) produced by the prism increases with increasing index of refraction and frequency and decreasing wavelength. Violet light is deviated most, and red is deviated least; other colors are in intermediate positions. When it comes out of the prism, the light is spread out into a fan-shaped beam, as shown.

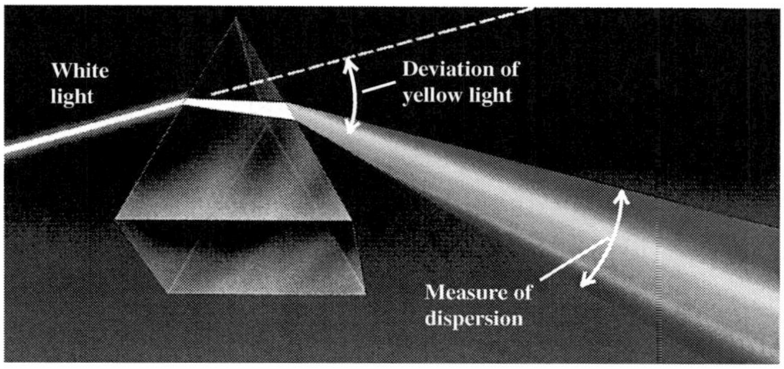

33.19 Dispersion of light by a prism. The band of colors is called a spectrum.

The light is said to be *dispersed* into a spectrum. The amount of dispersion depends on the *difference* between the refractive indexes for violet light and for red light. From Fig. 33.18 we can see that for a substance such as fluorite, the difference between the indexes for red and violet is small, and the dispersion will also be small. A better choice of material for a prism whose purpose is to produce a spectrum would be silicate flint glass, for which there is a larger difference in the value of n between red and violet.

As we mentioned in Section 33.3, the brilliance of diamond is due in part to its unusually large refractive index; another important factor is its large dispersion, which causes white light entering a diamond to emerge as a multicolored spectrum. Crystals of rutile and of strontium titanate, which can be produced synthetically, have about eight times the dispersion of diamond.

Rainbows

When you experience the beauty of a rainbow, as in Fig. 33.20a, you are seeing the combined effects of dispersion, refraction, and reflection. Sunlight comes from behind you, enters a water droplet, is (partially) reflected from the back surface of the droplet, and is refracted again upon exiting the droplet (Fig. 33.20b). A light ray that enters the middle of the raindrop is reflected straight back. All other rays exit the raindrop within an angle Δ of that middle ray, with many rays "piling up" at the angle Δ. What you see is a disk of light of angular radius Δ centered on the down-sun point (the point in the sky opposite the sun); due to the "piling up" of light rays, the disk is brightest around its rim, which we see as a rainbow (Fig. 33.20c). Because no light reaches your eye from angles larger than Δ, the sky looks dark outside the rainbow (see Fig. 33.20a). The value of the angle Δ depends on the index of refraction of the water that makes up the raindrops, which in turn depends on the wavelength (Fig. 33.20d). The bright disk of red light is slightly larger than that for orange light, which in turn is slightly larger than that for yellow light, and so on. As a result, you see the rainbow as a band of colors.

In many cases you can see a second, larger rainbow. It is the result of dispersion, refraction, and *two* reflections from the back surface of the droplet (Fig. 33.20e). Each time a light ray hits the back surface, part of the light is refracted out of the drop (not shown in Fig. 33.20); after two such hits, relatively little light is left inside the drop, which is why the secondary rainbow is noticeably fainter than the primary rainbow. Just as a mirror held up to a book reverses the printed letters, so the second reflection reverses the sequence of colors in the secondary rainbow. You can see this effect in Fig 33.20a.

33.5 Polarization

Polarization is a characteristic of all transverse waves. This chapter is about light, but to introduce some basic polarization concepts, let's go back to the transverse waves on a string that we studied in Chapter 15. For a string that in

ActivPhysics ONLINE
16.9 Physical Optics: Polarization

33.20 How rainbows form.

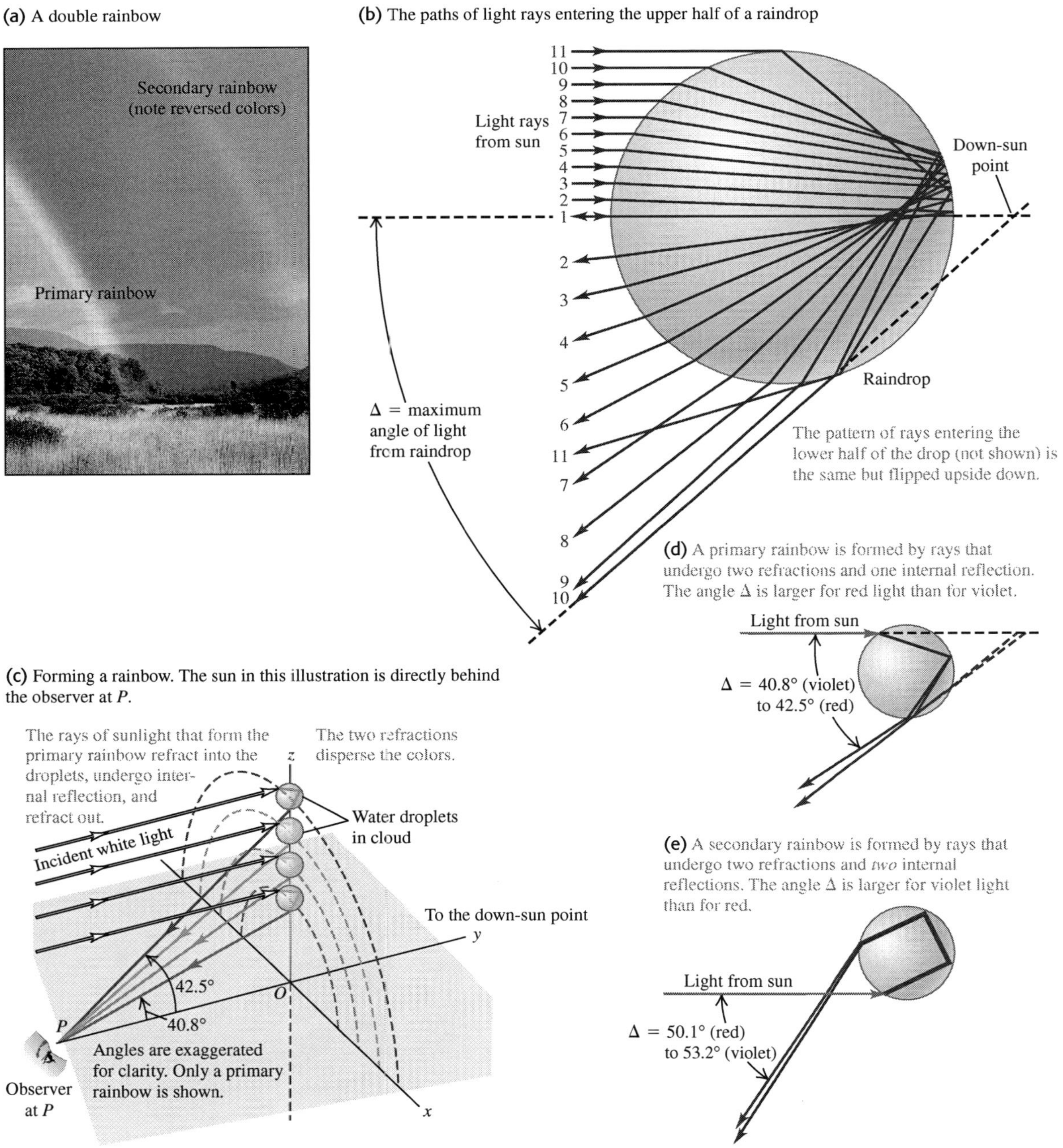

equilibrium lies along the x-axis, the displacements may be along the y-direction, as in Fig. 33.21a. In this case the string always lies in the xy-plane. But the displacements might instead be along the z-axis, as in Fig. 33.21b; then the string always lies in the xz-plane.

When a wave has only y-displacements, we say that it is **linearly polarized** in the y-direction; a wave with only z-displacements is linearly polarized in the z-direction. For mechanical waves we can build a **polarizing filter,** or **polarizer,** that permits only waves with a certain polarization direction to pass. In Fig. 33.21c the string can slide vertically in the slot without friction, but no hori-

33.21 (a), (b) Polarized waves on a string. (c) Making a polarized wave on a string from an unpolarized one using a polarizing filter.

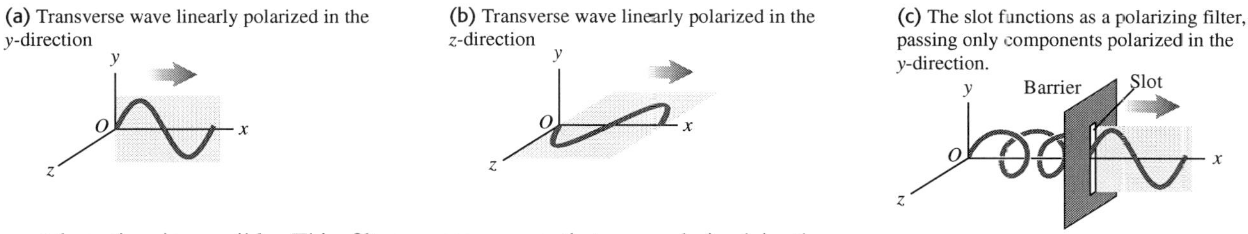

zontal motion is possible. This filter passes waves that are polarized in the y-direction but blocks those that are polarized in the z-direction.

This same language can be applied to electromagnetic waves, which also have polarization. As we learned in Chapter 32, an electromagnetic wave is a *transverse* wave; the fluctuating electric and magnetic fields are perpendicular to each other and to the direction of propagation. We always define the direction of polarization of an electromagnetic wave to be the direction of the *electric*-field vector $\vec{E}$, not the magnetic field, because many common electromagnetic-wave detectors respond to the electric forces on electrons in materials, not the magnetic forces. Thus the electromagnetic wave described by Eq. (32.17),

$$\vec{E}(x, t) = \hat{j}E_{\max}\cos(kx - \omega t)$$
$$\vec{B}(x, t) = \hat{k}B_{\max}\cos(kx - \omega t)$$

is said to be polarized in the y-direction because the electric field has only a y-component.

CAUTION **The meaning of "polarization"** It's unfortunate that the same word "polarization" that is used to describe the direction of $\vec{E}$ in an electromagnetic wave is also used to describe the shifting of electric charge within a body, such as in response to a nearby charged body; we described this latter kind of polarization in Section 21.2 (see Fig. 21.7). You should remember that while these two concepts have the same name, they do not describe the same phenomenon.

Polarizing Filters

Waves emitted by a radio transmitter are usually linearly polarized. The vertical antennas that are used for radio broadcasting emit waves that, in a horizontal plane around the antenna, are polarized in the vertical direction (parallel to the antenna) (Fig. 33.22a). Rooftop TV antennas have horizontal elements in the United States and vertical elements in Great Britain because the transmitted waves have different polarizations.

(a)

(b)

33.22 (a) Electrons in the red and white broadcast antenna oscillate vertically, producing vertically polarized electromagnetic waves that propagate away from the antenna in the horizontal direction. (The small gray antennas are for relaying cellular phone signals.) (b) No matter how this light bulb is oriented, the random motion of electrons in the filament produces unpolarized light waves.

The situation is different for visible light. Light from ordinary sources, such as incandescent light bulbs and fluorescent light fixtures, is *not* polarized (Fig. 33.22b). The "antennas" that radiate light waves are the molecules that make up the sources. The waves emitted by any one molecule may be linearly polarized, like those from a radio antenna. But any actual light source contains a tremendous number of molecules with random orientations, so the emitted light is a random mixture of waves linearly polarized in all possible transverse directions. Such light is called **unpolarized light** or **natural light.** To create polarized light from unpolarized natural light requires a filter that is analogous to the slot for mechanical waves in Fig. 33.21c.

Polarizing filters for electromagnetic waves have different details of construction, depending on the wavelength. For microwaves with a wavelength of a few centimeters, a good polarizer is an array of closely spaced, parallel conducting wires that are insulated from each other. (Think of a barbecue grill with the outer metal ring replaced by an insulating one.) Electrons are free to move along the length of the conducting wires and will do so in response to a wave whose $\vec{E}$ field is parallel to the wires. The resulting currents in the wires dissipate energy by I^2R heating; the dissipated energy comes from the wave, so whatever wave passes through the grid is greatly reduced in amplitude. Waves with $\vec{E}$ oriented perpendicular to the wires pass through almost unaffected, since electrons cannot move through the air between the wires. Hence a wave that passes through such a filter will be predominantly polarized in the direction perpendicular to the wires.

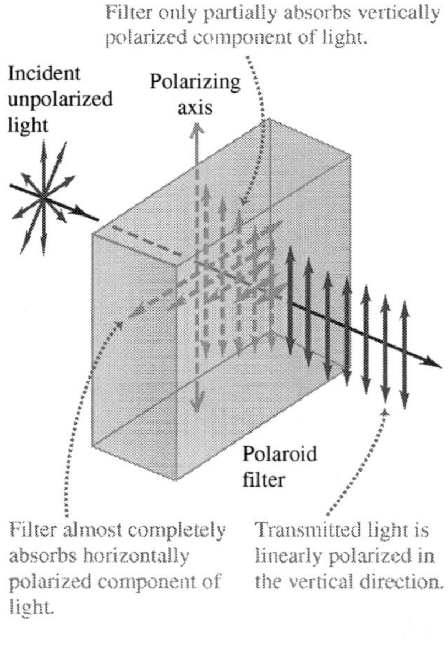

33.23 A Polaroid filter is illuminated by unpolarized natural light (shown by $\vec{E}$ vectors that point in all directions perpendicular to the direction of propagation). The transmitted light is linearly polarized along the polarizing axis (shown by $\vec{E}$ vectors along the polarization direction only).

The most common polarizing filter for visible light is a material known by the trade name Polaroid, widely used for sunglasses and polarizing filters for camera lenses. Developed originally by the American scientist Edwin H. Land, this material incorporates substances that have **dichroism,** a selective absorption in which one of the polarized components is absorbed much more strongly than the other (Fig. 33.23). A Polaroid filter transmits 80% or more of the intensity of a wave that is polarized parallel to a certain axis in the material, called the **polarizing axis,** but only 1% or less for waves that are polarized perpendicular to this axis. In one type of Polaroid filter, long-chain molecules within the filter are oriented with their axis perpendicular to the polarizing axis; these molecules preferentially absorb light that is polarized along their length, much like the conducting wires in a polarizing filter for microwaves.

Using Polarizing Filters

An *ideal* polarizing filter (polarizer) passes 100% of the incident light that is polarized in the direction of the filter's polarizing axis but completely blocks all light that is polarized perpendicular to this axis. Such a device is an unattainable idealization, but the concept is useful in clarifying the basic ideas. In the following discussion we will assume that all polarizing filters are ideal. In Fig. 33.24 unpolarized light is incident on a flat polarizing filter. The polarizing axis is represented by the blue line. The $\vec{E}$ vector of the incident wave can be represented in terms of components parallel and perpendicular to the polarizer axis; only the component of $\vec{E}$ parallel to the polarizing axis is transmitted. Hence the light emerging from the polarizer is linearly polarized parallel to the polarizing axis.

When unpolarized light is incident on an ideal polarizer as in Fig. 33.24, the intensity of the transmitted light is *exactly half* that of the incident unpolarized light, no matter how the polarizing axis is oriented. Here's why: We can resolve the $\vec{E}$ field of the incident wave into a component parallel to the polarizing axis and a component perpendicular to it. Because the incident light is a random mixture of all states of polarization, these two components are, on average, equal. The ideal polarizer transmits only the component that is parallel to the polarizing axis, so half the incident intensity is transmitted.

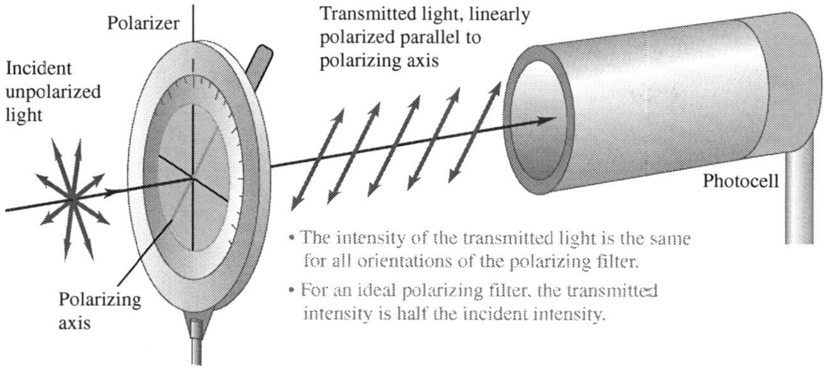

33.24 Unpolarized natural light is incident on the polarizing filter. The photocell measures the intensity of the transmitted linearly polarized light.

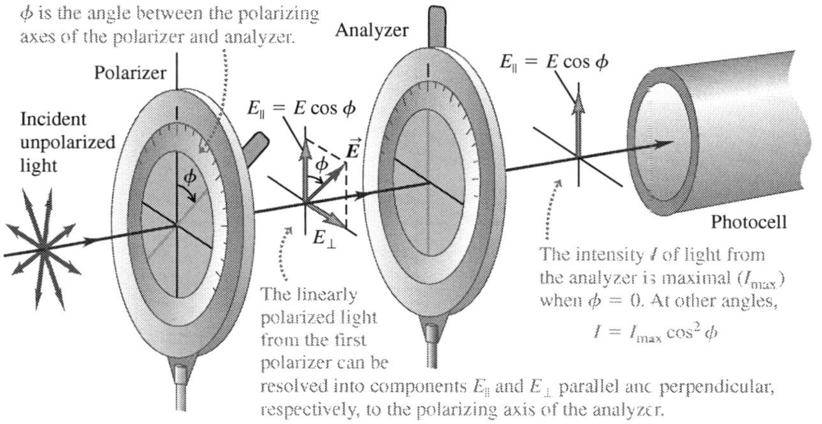

33.25 An ideal analyzer transmits only the electric field component parallel to its transmission direction (that is, its polarizing axis).

What happens when the linearly polarized light emerging from a polarizer passes through a second polarizer, as in Fig. 33.25? Consider the general case in which the polarizing axis of the second polarizer, or *analyzer,* makes an angle ϕ with the polarizing axis of the first polarizer. We can resolve the linearly polarized light that is transmitted by the first polarizer into two components, as shown in Fig. 33.25, one parallel and the other perpendicular to the axis of the analyzer. Only the parallel component, with amplitude $E\cos\phi$, is transmitted by the analyzer. The transmitted intensity is greatest when $\phi = 0$, and it is zero when polarizer and analyzer are *crossed* so that $\phi = 90°$ (Fig. 33.26). To determine the direction of polarization of the light transmitted by the first polarizer, rotate the analyzer until the photocell in Fig. 33.25 measures zero intensity; the polarization axis of the first polarizer is then perpendicular to that of the analyzer.

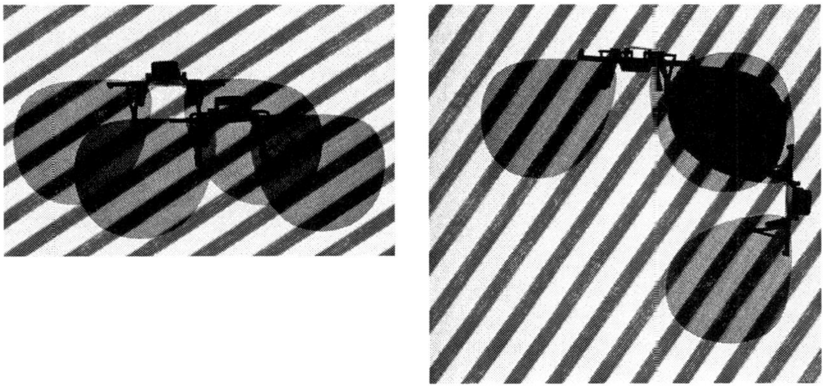

33.26 These photos show the view through Polaroid sunglasses whose polarizing axes are (left) aligned ($\phi = 0$) and (right) perpendicular ($\phi = 90°$). The transmitted intensity is greatest when the axes are aligned; it is zero when the axes are perpendicular.

To find the transmitted intensity at intermediate values of the angle ϕ, we recall from our energy discussion in Section 32.4 that the intensity of an electromagnetic wave is proportional to the *square* of the amplitude of the wave [see Eq.(32.29)]. The ratio of transmitted to incident *amplitude* is $\cos\phi$, so the ratio of transmitted to incident *intensity* is $\cos^2\phi$. Thus the intensity of the light transmitted through the analyzer is

$$I = I_{max}\cos^2\phi \quad \text{(Malus's law, polarized light passing through an analyzer)} \qquad (33.7)$$

where I_{max} is the maximum intensity of light transmitted (at $\phi = 0$) and I is the amount transmitted at angle ϕ. This relationship, discovered experimentally by Etienne Louis Malus in 1809, is called **Malus's law.** Malus's law applies *only* if the incident light passing through the analyzer is already linearly polarized.

Problem-Solving Strategy 33.2 Linear Polarization

MP

IDENTIFY *the relevant concepts:* Remember that in all electromagnetic waves, including light waves, the direction of the $\vec{E}$ field is the direction of polarization and is perpendicular to the propagation direction. When working with polarizers, you are really dealing with components of $\vec{E}$ parallel and perpendicular to the polarizing axis. Everything you know about components of vectors is applicable here.

SET UP *the problem* using the following steps:

1. Just as for problems in geometric optics, you should *always* start by drawing a large, neat diagram. Label all known angles, including the angles of any and all polarizing axes.
2. Determine the target variables.

EXECUTE *the solution* as follows:

1. Remember that a polarizer lets pass only electric-field components parallel to its polarizing axis.
2. If the incident light is linearly polarized and has amplitude E and intensity I_{max}, the light that passes through an ideal polarizer has amplitude $E\cos\phi$ and intensity $I_{max}\cos^2\phi$, where ϕ is the angle between the incident polarization direction and the filter's polarizing axis.
3. Unpolarized light is a random mixture of all possible polarization states, so on the average it has equal components in any two perpendicular directions. When passed through an ideal polarizer, unpolarized light becomes linearly polarized light with half the incident intensity. Partially linearly polarized light is a superposition of linearly polarized and unpolarized light.
4. The intensity (average power per unit area) of a wave is proportional to the *square* of its amplitude. If you find that two waves differ in amplitude by a certain factor, their intensities differ by the square of that factor.

EVALUATE *your answer:* Check your answer for any obvious errors. If your results say that light emerging from a polarizer has greater intensity than the incident light, something's wrong: a polarizer can't add energy to a light wave.

Example 33.5 Two polarizers in combination

In Fig. 33.25 the incident unpolarized light has intensity I_0. Find the intensities transmitted by the first and second polarizers if the angle between the axes of the two filters is 30°.

SOLUTION

IDENTIFY: This problem involves a polarizer (a polarizing filter on which unpolarized light shines, producing polarized light) and an analyzer (a second polarizing filter on which the polarized light shines).

SET UP: The diagram has already been drawn for us in Fig. 33.25. We are given the intensity I_0 of the incident natural light and the angle $\phi = 30°$ between the polarizing axes. Our target variables are the intensities of the light emerging from the first polarizer and of the light emerging from the second polarizer.

EXECUTE: As we explained above, the intensity of the linearly polarized light transmitted by the first filter is $I_0/2$. According to Eq. (33.7) with $\phi = 30°$, the second filter reduces the intensity by a factor of $\cos^2 30° = \frac{3}{4}$. Thus the intensity transmitted by the second polarizer is

$$\left(\frac{I_0}{2}\right)\left(\frac{3}{4}\right) = \frac{3}{8}I_0$$

EVALUATE: Note that the intensity decreases after each passage through a polarizer. The only situation in which the transmitted intensity does *not* decrease is if the polarizer is ideal (so it absorbs none of the light that passes through it) and if the incident light is linearly polarized along the polarizing axis, so $\phi = 0$.

33.27 When light is incident on a reflecting surface at the polarizing angle, the reflected light is linearly polarized.

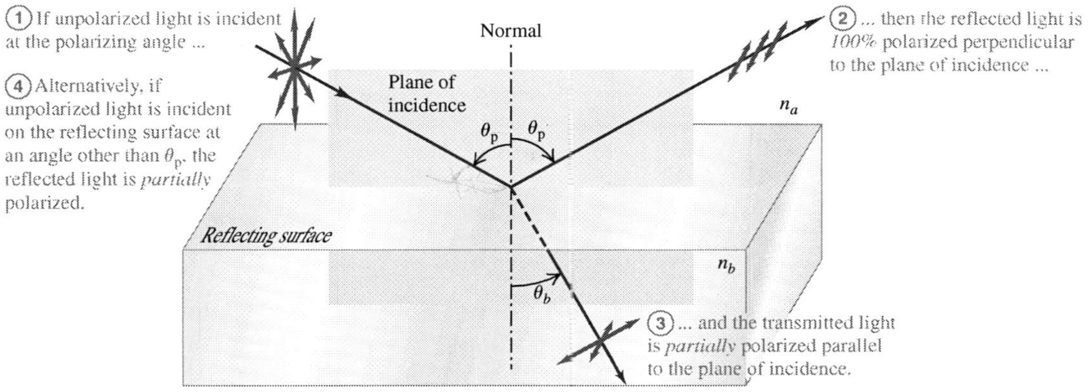

Polarization by Reflection

Unpolarized light can be polarized, either partially or totally, by *reflection.* In Fig. 33.27, unpolarized natural light is incident on a reflecting surface between two transparent optical materials; the plane containing the incident and reflected rays and the normal to the surface is called the **plane of incidence.** For most angles of incidence, waves for which the electric-field vector $\vec{E}$ is perpendicular to the plane of incidence (that is, parallel to the reflecting surface) are reflected more strongly than those for which $\vec{E}$ lies in this plane. In this case the reflected light is *partially polarized* in the direction perpendicular to the plane of incidence.

But at one particular angle of incidence, called the **polarizing angle** θ_p, the light for which $\vec{E}$ lies in the plane of incidence is *not reflected at all* but is completely refracted. At this same angle of incidence the light for which $\vec{E}$ is perpendicular to the plane of incidence is partially reflected and partially refracted. The *reflected* light is therefore *completely* polarized perpendicular to the plane of incidence, as shown in Fig. 33.27. The *refracted* (transmitted) light is *partially* polarized parallel to this plane; the refracted light is a mixture of the component parallel to the plane of incidence, all of which is refracted, and the remainder of the perpendicular component.

In 1812 the British scientist Sir David Brewster discovered that when the angle of incidence is equal to the polarizing angle θ_p, the reflected ray and the refracted ray are perpendicular to each other (Fig. 33.28). In this case the angle of refraction θ_b becomes the complement of θ_p, so $\theta_b = 90° - \theta_p$. From the law of refraction,

$$n_a \sin\theta_p = n_b \sin\theta_b$$

so we find

$$n_a \sin\theta_p = n_b \sin(90° - \theta_p) = n_b \cos\theta_p$$

$$\tan\theta_p = \frac{n_b}{n_a} \quad \text{(Brewster's law for the polarizing angle)} \qquad (33.8)$$

This relationship is known as **Brewster's law.** Although discovered experimentally, it can also be *derived* from a wave model using Maxwell's equations.

Polarization by reflection is the reason polarizing filters are widely used in sunglasses (Fig. 33.26). When sunlight is reflected from a horizontal surface, the plane of incidence is vertical, and the reflected light contains a preponderance of

33.28 The significance of the polarizing angle. The open circles represent a component of $\vec{E}$ that is perpendicular to the plane of the figure (the plane of incidence) and parallel to the surface between the two materials.

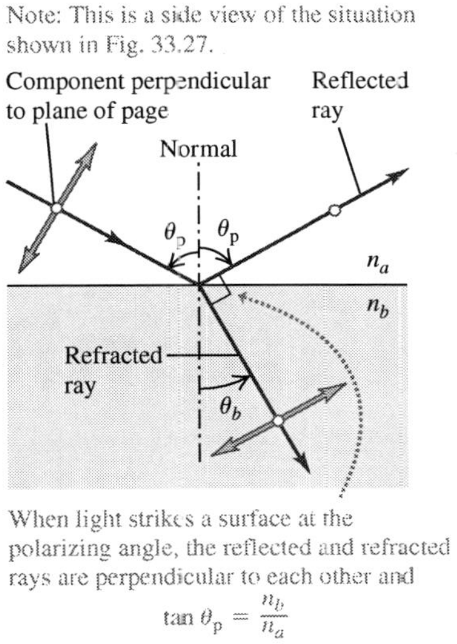

light that is polarized in the horizontal direction. When the reflection occurs at a smooth asphalt road surface or the surface of a lake, it causes unwanted glare. Vision can be improved by eliminating this glare. The manufacturer makes the polarizing axis of the lens material vertical, so very little of the horizontally polarized light reflected from the road is transmitted to the eyes. The glasses also reduce the overall intensity of the transmitted light to somewhat less than 50% of the intensity of the unpolarized incident light.

Example 33.6 Reflection from a swimming pool's surface

Sunlight reflects off the smooth surface of an unoccupied swimming pool. (a) At what angle of reflection is the light completely polarized? (b) What is the corresponding angle of refraction for the light that is transmitted (refracted) into the water? (c) At night an underwater floodlight is turned on in the pool. Repeat parts (a) and (b) for rays from the floodlight that strike the smooth surface from below.

SOLUTION

IDENTIFY: This problem involves polarization by reflection at an air–water interface in parts (a) and (b) and at a water–air interface in part (c).

SET UP: Figure 33.29 shows our sketches of the light rays for the situation during the day [parts (a) and (b)] and at night [part (c)]. In part (a) we're looking for the polarizing angle for light that is first in air, then in water; we find this with Brewster's law, Eq. (33.8). In part (b) we want the angle of the refracted light for this situation. In part (c) we again want the polarizing angle, but for light that is first in water, then in air. Again we use Eq. (33.8) to determine this angle.

33.29 Our sketches for this problem.

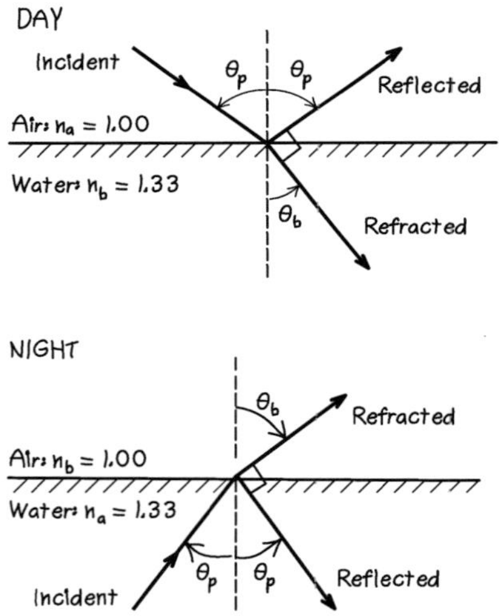

EXECUTE: (a) The top part of Fig. 33.29 shows the situation during the day. Since the light moves from air toward water, we have $n_a = 1.00$ (air) and $n_b = 1.33$ (water). From Eq. (33.8),

$$\theta_p = \arctan\frac{n_b}{n_a} = \arctan\frac{1.33}{1.00} = 53.1^\circ$$

(b) The incident light is at the polarizing angle, so the reflected and refracted rays are perpendicular; hence

$$\theta_p + \theta_b = 90^\circ$$
$$\theta_b = 90^\circ - 53.1^\circ = 36.9^\circ$$

(c) The situation at night is shown in the bottom part of Fig. 33.29. Now the light is *first* in the water, then in the air, so $n_a = 1.33$ and $n_b = 1.00$. Again using Eq. (33.8), we have

$$\theta_p = \arctan\frac{1.00}{1.33} = 36.9^\circ$$
$$\theta_b = 90^\circ - 36.9^\circ = 53.1^\circ$$

EVALUATE: We can check our answer in part (b) using Snell's law, $n_a \sin\theta_a = n_b \sin\theta_b$, or

$$\sin\theta_b = \frac{n_a \sin\theta_p}{n_b} = \frac{1.00 \sin 53.1^\circ}{1.33} = 0.600$$
$$\theta_b = 36.9^\circ$$

Note that the two polarizing angles found in parts (a) and (c) add to 90°. This is *not* an accident; can you see why?

Circular and Elliptical Polarization

Light and other electromagnetic radiation can also have *circular* or *elliptical* polarization. To introduce these concepts, let's return once more to mechanical waves on a stretched string. In Fig. 33.21, suppose the two linearly polarized waves in parts (a) and (b) are in phase and have equal amplitude. When they are superposed, each point in the string has simultaneous y- and z-displacements of equal magnitude. A little thought shows that the resultant wave lies in a plane oriented at 45° to the y- and z-axes (i.e., in a plane making a 45° angle with the

xy- and xz-planes). The amplitude of the resultant wave is larger by a factor of $\sqrt{2}$ than that of either component wave, and the resultant wave is linearly polarized.

But now suppose the two equal-amplitude waves differ in phase by a quarter-cycle. Then the resultant motion of each point corresponds to a superposition of two simple harmonic motions at right angles, with a quarter-cycle phase difference. The y-displacement at a point is greatest at times when the z-displacement is zero, and vice versa. The motion of the string as a whole then no longer takes place in a single plane. It can be shown that each point on the rope moves in a *circle* in a plane parallel to the yz-plane. Successive points on the rope have successive phase differences, and the overall motion of the string has the appearance of a rotating helix. This is shown to the left of the polarizing filter in Fig. 33.21c. This particular superposition of two linearly polarized waves is called **circular polarization.** By convention, the wave is said to be *right circularly polarized* when the sense of motion of a particle of the string, to an observer looking *backward* along the direction of propagation, is *clockwise;* the wave is said to be *left circularly polarized* if the sense of motion is the reverse.

Figure 33.30 shows the analogous situation for an electromagnetic wave. Two sinusoidal waves of equal amplitude, polarized in the y- and z-directions and with a quarter-cycle phase difference, are superposed. The result is a wave in which the $\vec{E}$ vector at each point has a constant magnitude but *rotates* around the direction of propagation. The wave in Fig. 33.30 is propagating toward you and the $\vec{E}$ vector appears to be rotating clockwise, so it is called a *right circularly polarized* electromagnetic wave. If instead the $\vec{E}$ vector of a wave coming toward you appears to be rotating counterclockwise, it is called a *left circularly polarized* electromagnetic wave.

If the phase difference between the two component waves is something other than a quarter-cycle, or if the two component waves have different amplitudes, then each point on the string traces out not a circle but an *ellipse.* The resulting wave is said to be **elliptically polarized.**

For electromagnetic waves with radio frequencies, circular or elliptical polarization can be produced by using two antennas at right angles, fed from the same transmitter but with a phase-shifting network that introduces the appropriate phase difference. For light, the phase shift can be introduced by use of a material that exhibits *birefringence*—that is, has different indexes of refraction for different directions of polarization. A common example is calcite $(CaCO_3)$. When a

33.30 Circular polarization of an electromagnetic wave moving toward you parallel to the x-axis. The y-component of $\vec{E}$ lags the z-component by a quarter-cycle. This phase difference results in right circular polarization.

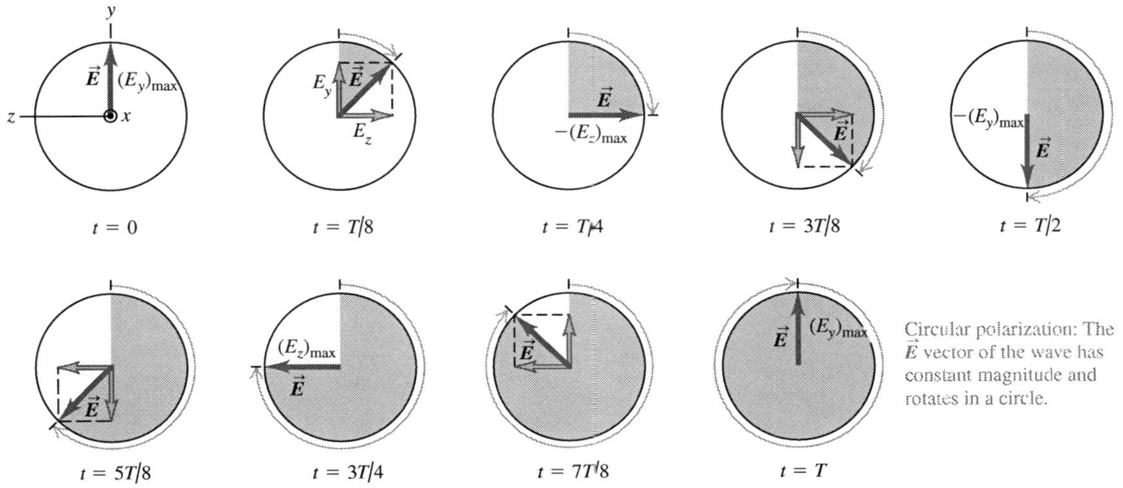

calcite crystal is oriented appropriately in a beam of unpolarized light, its refractive index (for a wavelength in vacuum of 589 nm) is 1.658 for one direction of polarization and 1.486 for the perpendicular direction. When two waves with equal amplitude and with perpendicular directions of polarization enter such a material, they travel with different speeds. If they are in phase when they enter the material, then in general they are no longer in phase when they emerge. If the crystal is just thick enough to introduce a quarter-cycle phase difference, then the crystal converts linearly polarized light to circularly polarized light. Such a crystal is called a *quarter-wave plate.* Such a plate also converts circularly polarized light to linearly polarized light. Can you prove this? (See Problem 33.43.)

33.31 Photoelastic stress analysis of a model of a cross section of a Gothic cathedral. The masonry construction that was used for this kind of building had great strength in compression but very little in tension (see Section 11.4). Inadequate buttressing and high winds sometimes caused tensile stresses in normally compressed structural elements, leading to some spectacular collapses.

Photoelasticity

Some optical materials that are not normally birefringent become so when they are subjected to mechanical stress. This is the basis of the science of *photoelasticity.* Stresses in girders, boiler plates, gear teeth, and cathedral pillars can be analyzed by constructing a transparent model of the object, usually of a plastic material, subjecting it to stress, and examining it between a polarizer and an analyzer in the crossed position. Very complicated stress distributions can be studied by these optical methods. ?

Figure 33.31 is a photograph of a photoelastic model under stress. The polarized light that enters the model can be thought of as having a component along each of the two directions of the birefringent plastic. Since these two components travel through the plastic at different speeds, the light that emerges from the other side of the model can have a different overall direction of polarization. Hence some of this transmitted light will be able to pass through the analyzer even though its polarization axis is at a 90° angle to the polarizer's axis, and the stressed areas in the plastic will appear as bright spots. The amount of birefringence is different for different wavelengths and hence different colors of light; the color that appears at each location in Fig. 33.31 is that for which the transmitted light is most nearly polarized along the analyzer's polarization axis.

Test Your Understanding of Section 33.5 You are taking a photograph of a sunlit high-rise office building. In order to minimize the reflections from the building's windows, you place a polarizing filter on the camera lens. How should you orient the filter? (i) with the polarizing axis vertical; (ii) with the polarizing axis horizontal; (iii) either orientation will minimize the reflections just as well; (iv) neither orientation will have any effect. MP

*33.6 Scattering of Light

The sky is blue. Sunsets are red. Skylight is partially polarized; that's why the sky looks darker from some angles than from others when it is viewed through Polaroid sunglasses. As we will see, a single phenomenon is responsible for all of these effects.

When you look at the daytime sky, the light that you see is sunlight that has been absorbed and then re-radiated in a variety of directions. This process is called **scattering.** (If the earth had no atmosphere, the sky would appear as black in the daytime as it does at night, just as it does to an astronaut in space or on the moon; you would see the sun's light only if you looked directly at it, and the stars would be visible in the daytime.) Figure 33.32 shows some of the details of the scattering process. Sunlight, which is unpolarized, comes from the left along the x-axis and passes over an observer looking vertically upward along the y-axis. (We are viewing the situation from the side.) Consider the molecules of the earth's atmosphere located at point O. The electric field in the beam of sunlight sets the electric charges in these molecules into vibration.

33.32 When the sunbathing observer on the left looks up, he sees blue, polarized sunlight that has been scattered by air molecules. The observer on the right sees reddened, unpolarized light when he looks at the sun.

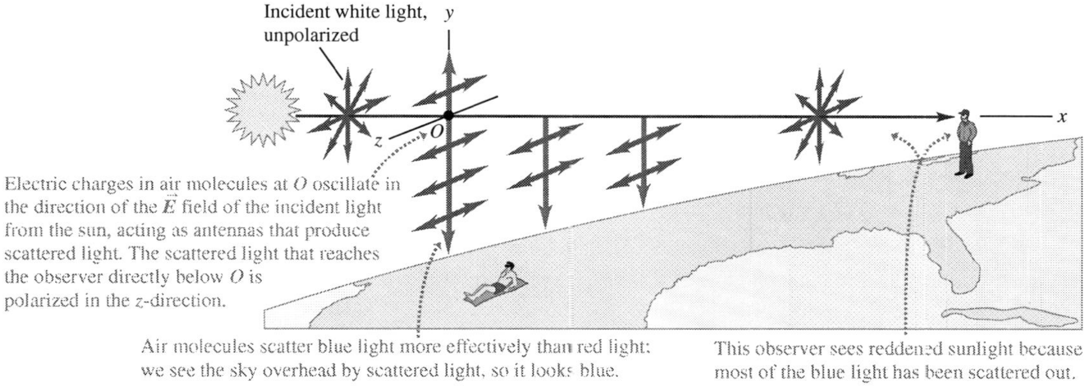

Since light is a transverse wave, the direction of the electric field in any component of the sunlight lies in the yz-plane, and the motion of the charges takes place in this plane. There is no field, and hence no motion of charges, in the direction of the x-axis.

An incident light wave sets the electric charges in the molecules at point O vibrating along the line of $\vec{E}$. We can resolve this vibration into two components, one along the y-axis and the other along the z-axis. Each component in the incident light produces the equivalent of two molecular "antennas," oscillating with the same frequency as the incident light and lying along the y- and z-axes.

We mentioned in Chapter 32 that an oscillating charge, like those in an antenna, does not radiate in the direction of its oscillation. (See Fig. 32.3 in Section 32.1.) Thus the "antenna" along the y-axis does not send any light to the observer directly below it, although it does emit light in other directions. Therefore the only light reaching this observer comes from the other molecular "antenna," corresponding to the oscillation of charge along the z-axis. This light is linearly polarized, with its electric field along the z-axis (parallel to the "antenna"). The red vectors on the y-axis below point O in Fig. 33.32 show the direction of polarization of the light reaching the observer.

As the original beam of sunlight passes though the atmosphere, its intensity decreases as its energy goes into the scattered light. Detailed analysis of the scattering process shows that the intensity of the light scattered from air molecules increases in proportion to the fourth power of the frequency (inversely to the fourth power of the wavelength). Thus the intensity ratio for the two ends of the visible spectrum is $(700\ \text{nm}/400\ \text{nm})^4 = 9.4$. Roughly speaking, scattered light contains nine times as much blue light as red, and that's why the sky is blue.

Clouds contain a high concentration of water droplets or ice crystals, which also scatter light. Because of this high concentration, light passing through the cloud has many more opportunities for scattering than does light passing through a clear sky. Thus light of *all* wavelengths is eventually scattered out of the cloud, so the cloud looks white (Fig. 33.33). Milk looks white for the same reason; the scattering is due to fat globules in the milk. If you dilute milk by mixing it with enough water, the concentration of fat globules will be so low that only blue light will be substantially scattered; the dilute solution will look blue, not white. (Nonfat milk, which also has a very low concentration of globules, looks somewhat bluish for this same reason.)

Near sunset, when sunlight has to travel a long distance through the earth's atmosphere, a substantial fraction of the blue light is removed by scattering. White light minus blue light looks yellow or red. This explains the yellow or red hue that we so often see from the setting sun (and that is seen by the observer at the far right of Fig. 33.32).

33.33 Clouds are white because they efficiently scatter sunlight of all wavelengths.

Because skylight is partially polarized, polarizers are useful in photography. The sky in a photograph can be darkened by orienting the polarizer axis to be perpendicular to the predominant direction of polarization of the scattered light. The most strongly polarized light comes from parts of the sky that are 90° away from the sun—for example, from directly overhead when the sun is on the horizon at sunrise or sunset.

33.7 Huygens's Principle

The laws of reflection and refraction of light rays that we introduced in Section 33.2 were discovered experimentally long before the wave nature of light was firmly established. However, we can *derive* these laws from wave considerations and show that they are consistent with the wave nature of light. The same kind of analysis that we use here will be of central importance in Chapters 35 and 36 in our discussion of physical optics.

We begin with a principle called **Huygens's principle.** This principle, stated originally by the Dutch scientist Christiaan Huygens in 1678, is a geometrical method for finding, from the known shape of a wave front at some instant, the shape of the wave front at some later time. Huygens assumed that **every point of a wave front may be considered the source of secondary wavelets that spread out in all directions with a speed equal to the speed of propagation of the wave.** The new wave front at a later time is then found by constructing a surface *tangent* to the secondary wavelets or, as it is called, the *envelope* of the wavelets. All the results that we obtain from Huygens's principle can also be obtained from Maxwell's equations. Thus it is not an independent principle, but it is often very convenient for calculations with wave phenomena.

33.34 Applying Huygens's principle to wave front AA' to construct a new wave front BB'.

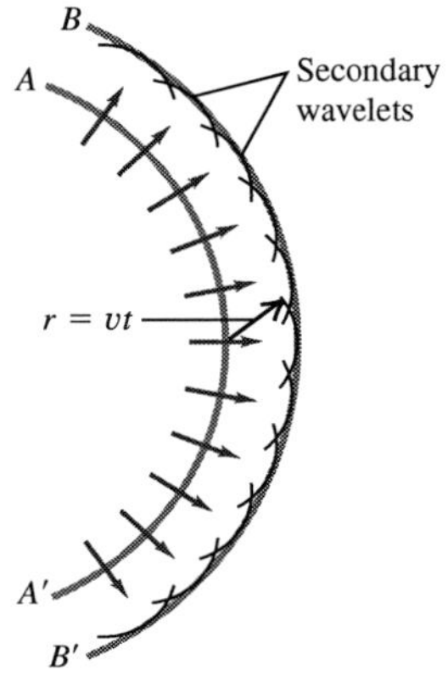

Huygens's principle is shown in Fig. 33.34. The original wave front AA' is traveling outward from a source, as indicated by the arrows. We want to find the shape of the wave front after a time interval t. Let v be the speed of propagation of the wave; then in time t it travels a distance vt. We construct several circles (traces of spherical wavelets) with radius $r = vt$, centered at points along AA'. The trace of the envelope of these wavelets, which is the new wave front, is the curve BB'. We are assuming that the speed v is the same at all points and in all directions.

33.35 Using Huygens's principle to derive the law of reflection.

(a) Successive positions of a plane wave AA' as it is reflected from a plane surface

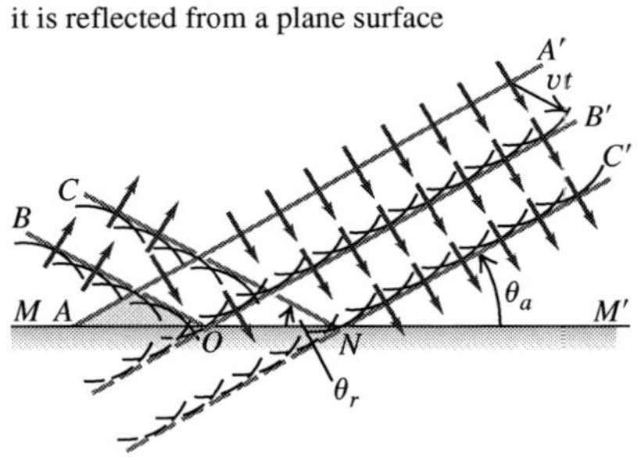

(b) Magnified portion of (a)

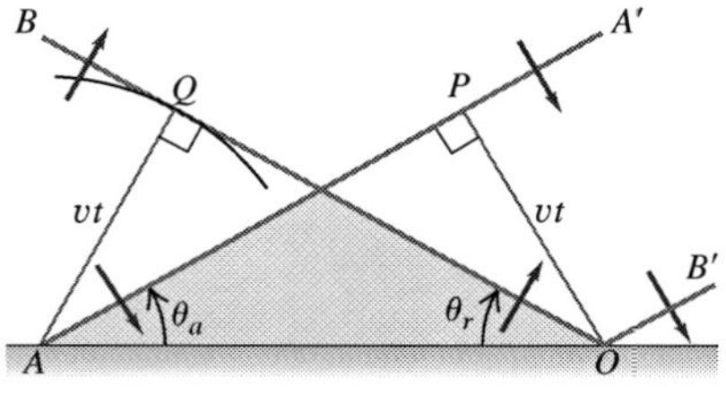

Reflection and Huygens's Principle

To derive the law of reflection from Huygens's principle, we consider a plane wave approaching a plane reflecting surface. In Fig. 33.35a the lines AA', OB', and NC' represent successive positions of a wave front approaching the surface MM'. Point A on the wave front AA' has just arrived at the reflecting surface. We can use Huygens's principle to find the position of the wave front after a time interval t. With points on AA' as centers, we draw several secondary wavelets with radius vt. The wavelets that originate near the upper end of AA' spread out unhindered, and their envelope gives the portion OB' of the new wave front. If the reflecting surface were not there, the wavelets originating near the lower end of AA' would similarly reach the positions shown by the broken circular arcs. Instead, these wavelets strike the reflecting surface.

The effect of the reflecting surface is to *change the direction* of travel of those wavelets that strike it, so the part of a wavelet that would have penetrated the surface actually lies to the left of it, as shown by the full lines. The first such wavelet is centered at point A; the envelope of all such reflected wavelets is the portion OB of the wave front. The trace of the entire wave front at this instant is the bent line BOB'. A similar construction gives the line CNC' for the wave front after another interval t.

From plane geometry the angle θ_a between the incident *wave front* and the *surface* is the same as that between the incident *ray* and the *normal* to the surface and is therefore the angle of incidence. Similarly, θ_r is the angle of reflection. To find the relationship between these angles, we consider Fig. 33.35b. From O we draw $OP = vt$, perpendicular to AA'. Now OB, by construction, is tangent to a circle of radius vt with center at A. If we draw AQ from A to the point of tangency, the triangles APO and OQA are congruent because they are right triangles with the side AO in common and with $AQ = OP = vt$. The angle θ_a therefore equals the angle θ_r, and we have the law of reflection.

Refraction and Huygens's Principle

We can derive the law of *refraction* by a similar procedure. In Fig. 33.36a we consider a wave front, represented by line AA', for which point A has just arrived at the boundary surface SS' between two transparent materials a and b, with indexes of refraction n_a and n_b and wave speeds v_a and v_b. (The *reflected* waves are not shown in the figure; they proceed as in Fig. 33.35.) We can apply Huygens's principle to find the position of the refracted wave fronts after a time t.

With points on AA' as centers, we draw several secondary wavelets. Those originating near the upper end of AA' travel with speed v_a and, after a time interval t, are spherical surfaces of radius $v_a t$. The wavelet originating at point A, however, is traveling in the second material b with speed v_b and at time t is a spherical surface of radius $v_b t$. The envelope of the wavelets from the original wave front is the plane whose trace is the bent line BOB'. A similar construction leads to the trace CPC' after a second interval t.

The angles θ_a and θ_b between the surface and the incident and refracted wave fronts are the angle of incidence and the angle of refraction, respectively. To find the relationship between these angles, refer to Fig. 33.36b. We draw $OQ = v_a t$, perpendicular to AQ, and we draw $AB = v_b t$, perpendicular to BO. From the right triangle AOQ,

$$\sin\theta_a = \frac{v_a t}{AO}$$

and from the right triangle AOB,

$$\sin\theta_b = \frac{v_b t}{AO}$$

Combining these, we find

$$\frac{\sin\theta_a}{\sin\theta_b} = \frac{v_a}{v_b} \tag{33.9}$$

We have defined the index of refraction n of a material as the ratio of the speed of light c in vacuum to its speed v in the material: $n_a = c/v_a$ and $n_b = c/v_b$. Thus

$$\frac{n_b}{n_a} = \frac{c/v_b}{c/v_a} = \frac{v_a}{v_b}$$

and we can rewrite Eq. (33.9) as

$$\frac{\sin\theta_a}{\sin\theta_b} = \frac{n_b}{n_a} \quad \text{or}$$

$$n_a \sin\theta_a = n_b \sin\theta_b$$

which we recognize as Snell's law, Eq. (33.4). So we have derived Snell's law from a wave theory. Alternatively, we may choose to regard Snell's law as an experimental result that defines the index of refraction of a material; in that case

33.36 Using Huygens's principle to derive the law of refraction. The case $v_b < v_a$ $(n_b > n_a)$ is shown.

(a) Successive positions of a plane wave AA' as it is refracted by a plane surface

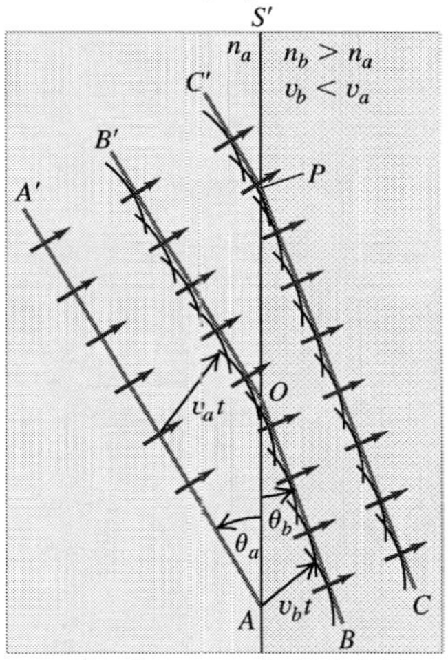

(b) Magnified portion of (a)

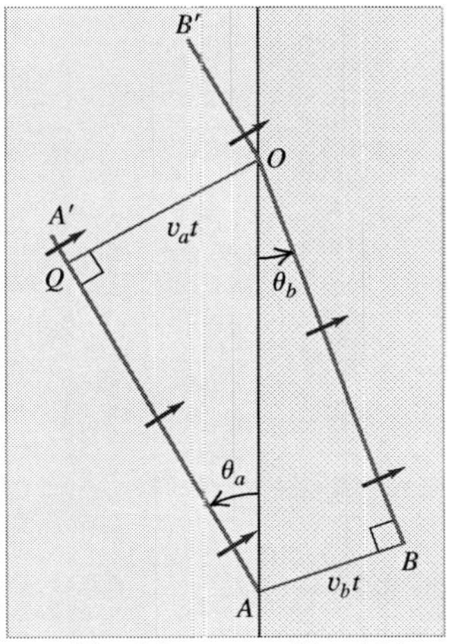

33.37 How mirages are formed.

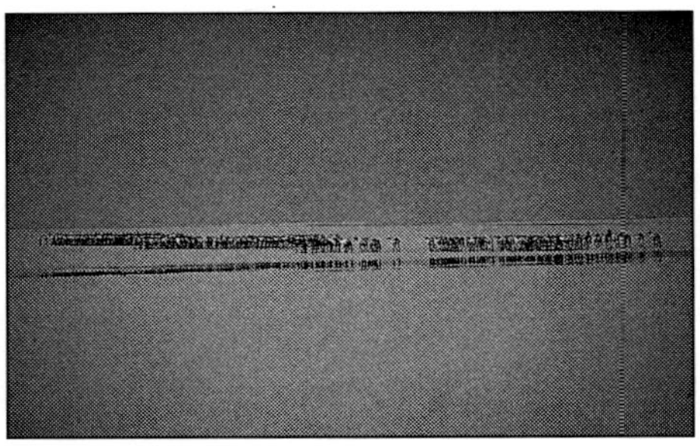

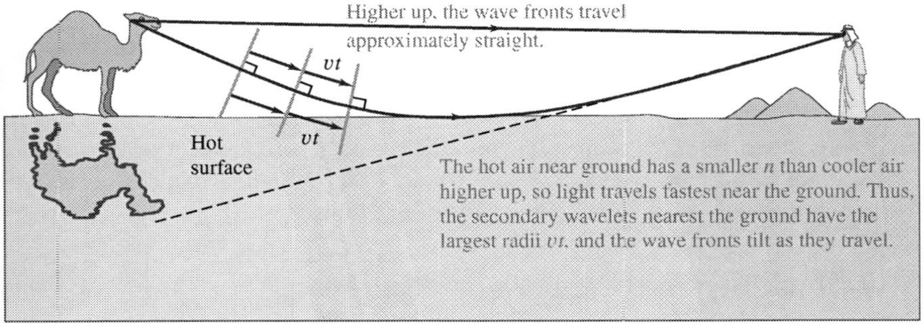

this analysis helps to confirm the relationship $v = c/n$ for the speed of light in a material.

Mirages offer an interesting example of Huygens's principle in action. When the surface of pavement or desert sand is heated intensely by the sun, a hot, less dense, smaller-n layer of air forms near the surface. The speed of light is slightly greater in the hotter air near the ground, the Huygens wavelets have slightly larger radii, the wave fronts tilt slightly, and rays that were headed toward the surface with a large angle of incidence (near 90°) can be bent up as shown in Fig. 33.37. Light farther from the ground is bent less and travels nearly in a straight line. The observer sees the object in its natural position, with an inverted image below it, as though seen in a horizontal reflecting surface. Even when the turbulence of the heated air prevents a clear inverted image from being formed, the mind of the thirsty traveler can interpret the apparent reflecting surface as a sheet of water.

It is important to keep in mind that Maxwell's equations are the fundamental relationships for electromagnetic wave propagation. But it is a remarkable fact that Huygens's principle anticipated Maxwell's analysis by two centuries. Maxwell provided the theoretical underpinning for Huygens's principle. Every point in an electromagnetic wave, with its time-varying electric and magnetic fields, acts as a source of the continuing wave, as predicted by Ampere's and Faraday's laws.

Test Your Understanding of Section 33.7 Sound travels faster in warm air than in cold air. Imagine a weather front that runs north-south, with warm air to the west of the front and cold air to the east. A sound wave traveling in a northeast direction in the warm air encounters this front. How will the direction of this sound wave change when it passes into the cold air? (i) The wave direction will deflect toward the north; (ii) the wave direction will deflect toward the east; (iii) the wave direction will be unchanged.

CHAPTER 33 SUMMARY

Light and its properties: Light is an electromagnetic wave. When emitted or absorbed, it also shows particle properties. It is emitted by accelerated electric charges. The speed of light is a fundamental physical constant.

A wave front is a surface of constant phase; wave fronts move with a speed equal to the propagation speed of the wave. A ray is a line along the direction of propagation, perpendicular to the wave fronts. Representation of light by rays is the basis of geometric optics.

When light is transmitted from one material to another, the frequency of the light is unchanged, but the wavelength and wave speed can change. The index of refraction n of a material is the ratio of the speed of light in vacuum c to the speed v in the material. If λ_0 is the wavelength in vacuum, the same wave has a shorter wavelength λ in a medium with index of refraction n. (See Example 33.2.)

The variation of index of refraction n with wavelength λ is called dispersion. Usually n decreases with increasing λ.

$$n = \frac{c}{v} \quad (33.1)$$

$$\lambda = \frac{\lambda_0}{n} \quad (33.5)$$

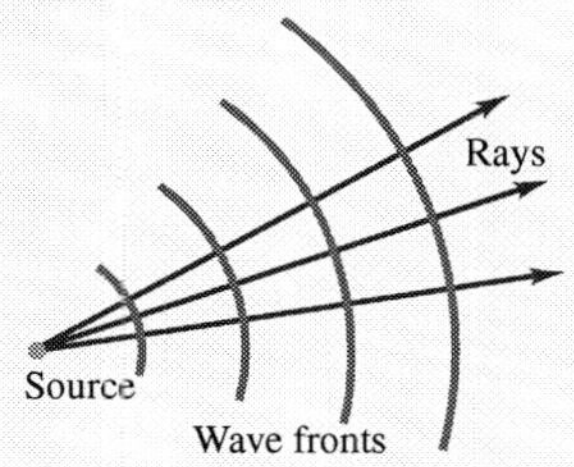

Reflection and refraction: At a smooth interface between two optical materials, the incident, reflected, and refracted rays and the normal to the interface all lie in a single plane called the plane of incidence. The law of reflection states that the angles of incidence and reflection are equal. The law of refraction relates the angles of incidence and refraction to the indexes of refraction of the materials. Angles of incidence, reflection, and refraction are always measured from the normal to the surface. (See Examples 33.1 and 33.3.)

$$\theta_r = \theta_a \quad (33.2)$$
(law of reflection)

$$n_a \sin\theta_a = n_b \sin\theta_b \quad (33.4)$$
(law of refraction)

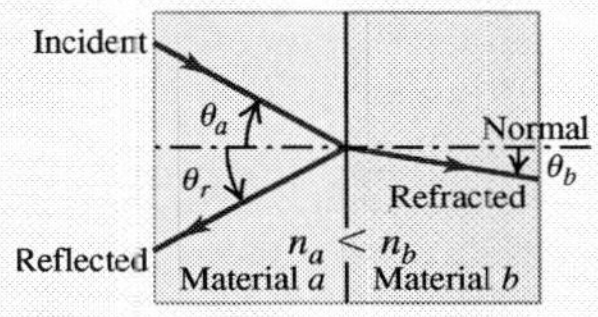

Total internal reflection: When a ray travels in a material of greater index of refraction n_a toward a material of smaller index n_b, total internal reflection occurs at the interface when the angle of incidence exceeds a critical angle θ_{crit}. (See Example 33.4.)

$$\sin\theta_{crit} = \frac{n_b}{n_a} \quad (33.6)$$

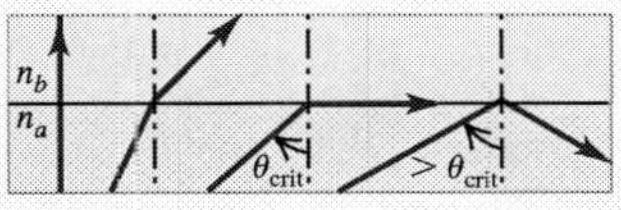

Polarization of light: The direction of polarization of a linearly polarized electromagnetic wave is the direction of the $\vec{E}$ field. A polarizing filter passes waves that are linearly polarized along its polarizing axis and blocks waves polarized perpendicularly to that axis. When polarized light of intensity I_{max} is incident on a polarizing filter used as an analyzer, the intensity I of the light transmitted through the analyzer depends on the angle ϕ between the polarization direction of the incident light and the polarizing axis of the analyzer. (See Example 33.5.)When two linearly polarized waves with a phase difference are superposed, the result is circularly or elliptically polarized light. In this case the $\vec{E}$ vector is not confined to a plane containing the direction of propagation, but rather describes circles or ellipses in planes perpendicular to the propagation direction.

Light is scattered by air molecules. The scattered light is partially polarized.

$$I = I_{max}\cos^2\phi \quad (33.7)$$
(Malus's law)

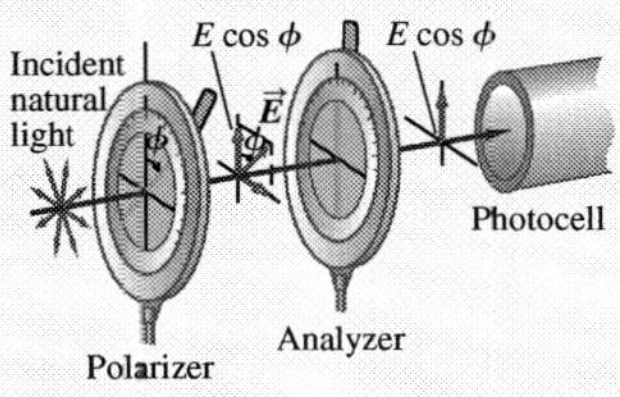

Polarization by reflection: When unpolarized light strikes an interface between two materials, Brewster's law states that the reflected light is completely polarized perpendicular to the plane of incidence (parallel to the interface) if the angle of incidence equals the polarizing angle θ_p. (See Example 33.6.)

$$\tan\theta_p = \frac{n_b}{n_a} \quad (33.8)$$
(Brewster's law)

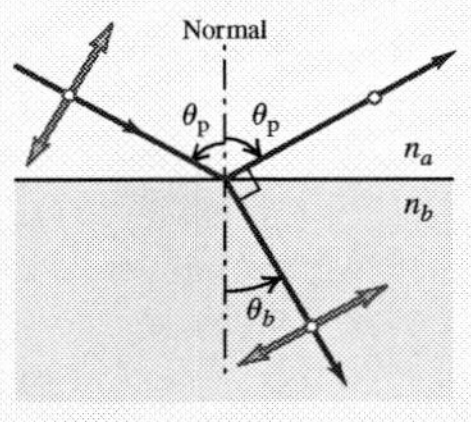

Huygens's principle: Huygens's principle states that if the position of a wave front at one instant is known, then the position of the front at a later time can be constructed by imagining the front as a source of secondary wavelets. Huygens's principle can be used to derive the laws of reflection and refraction.

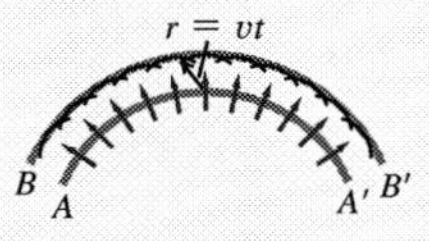

Key Terms

optics, *1121*
wave front, *1122*
ray, *1123*
geometric optics, *1123*
physical optics, *1123*
reflection, *1123*
refraction, *1123*
specular reflection, *1124*
diffuse reflection, *1124*
index of refraction (refractive index), *1124*
law of reflection, *1125*
law of refraction (Snell's law), *1125*
critical angle, *1130*
total internal reflection, *1130*
dispersion, *1132*
linear polarization, *1134*
polarizing filter (polarizer), *1134*
unpolarized light (natural light), *1135*
dichroism, *1136*
polarizing axis, *1136*
Malus's law, *1138*
plane of incidence, *1139*
polarizing angle, *1139*
Brewster's law, *1139*
circular polarization, *1141*
elliptical polarization, *1141*
scattering, *1142*
Huygens's principle, *1144*

Answer to Chapter Opening Question ?

This is the same effect as shown in Fig. 33.31. The drafting tools are placed between two polarizing filters whose polarizing axes are perpendicular. In places where the clear plastic is under stress, the plastic becomes birefringent; that is, light travels through it at a speed that depends on its polarization. The result is that the light that emerges from the plastic has a different polarization than the light that enters. A spot on the plastic appears bright if the emerging light has the same polarization as the second polarizing filter. The amount of birefringence depends on the wavelength of the light as well as the amount of stress on the plastic, so different colors are seen at different locations on the plastic.

Answers to Test Your Understanding Questions

33.1 Answer: (iii) The waves go farther in the y-direction in a given amount of time than in the other directions, so the wave fronts are elongated in the y-direction.

33.2 Answers: (a) (ii), (b) (iii) As shown in the figure, light rays coming from the fish bend away from the normal when they pass from the water $(n = 1.33)$ into the air $(n = 1.00)$. As a result, the fish appears to be higher in the water than it actually is. Hence you should aim a spear *below* the apparent position of the fish. If you use a laser beam, you should aim *at* the apparent position of the fish: The beam of laser light takes the same path from you to the fish as ordinary light takes from the fish to you (though in the opposite direction).

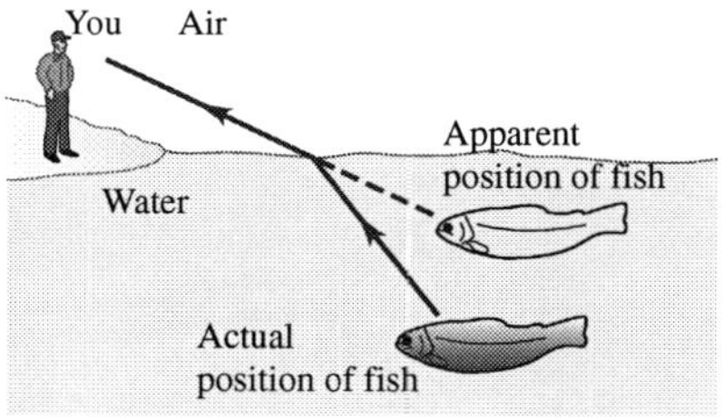

33.3 Answers: (i), (ii) Total internal reflection can occur only if two conditions are met: n_b must be less than n_a, and the critical angle θ_{crit} (where $\sin\theta_{crit} = n_b/n_a$) must be smaller than the angle of incidence θ_a. In the first two cases both conditions are met: The critical angles are (i) $\theta_{crit} = \sin^{-1}(1/1.33) = 48.8°$ and (ii) $\theta_{crit} = \sin^{-1}(1.33/1.52) = 61.0°$, both of which are smaller than $\theta_a = 70°$. In the third case $n_b = 1.52$ is greater than $n_a = 1.33$, so total internal reflection cannot occur for any incident angle.

33.5 Answer: (ii) The sunlight reflected from the windows of the high-rise building is partially polarized in the vertical direction, since each window lies in a vertical plane. The Polaroid filter in front of the lens is oriented with its polarizing axis perpendicular to the dominant direction of polarization of the reflected light.
33.7 Answer: (ii) Huygens's principle applies to waves of all kinds, including sound waves. Hence this situation is exactly like that shown in Fig. 33.36, with material a representing the warm air, material b representing the cold air in which the waves travel more slowly, and the interface between the materials representing the weather front. North is toward the top of the figure and east is toward the right, so Fig. 33.36 shows that the rays (which indicate the direction of propagation) deflect toward the east.

PROBLEMS

For instructor-assigned homework, go to **www.masteringphysics.com**

Discussion Questions

Q33.1. Light requires about 8 minutes to travel from the sun to the earth. Is it delayed appreciably by the earth's atmosphere? Explain.
Q33.2. Sunlight or starlight passing through the earth's atmosphere is always bent toward the vertical. Why? Does this mean that a star is not really where it appears to be? Explain.
Q33.3. A beam of light goes from one material into another. On *physical* grounds, explain *why* the wavelength changes but the frequency and period do not.
Q33.4. A student claimed that, because of atmospheric refraction (see Discussion Question Q33.2), the sun can be seen after it has set and that the day is therefore longer than it would be if the earth had no atmosphere. First, what does she mean by saying that the sun can be seen after it has set? Second, comment on the validity of her conclusion.
Q33.5. When hot air rises from a radiator or heating duct, objects behind it appear to shimmer or waver. What causes this?
Q33.6. Devise straightforward experiments to measure the speed of light in a given glass using (a) Snell's law; (b) total internal reflection; (c) Brewster's law.
Q33.7. Sometimes when looking at a window, you see two reflected images slightly displaced from each other. What causes this?
Q33.8. If you look up from underneath toward the surface of the water in your aquarium, you may see an upside-down reflection of your pet fish in the surface of the water. Explain how this can happen.
Q33.9. A ray of light in air strikes a glass surface. Is there a range of angles for which total reflection occurs? Explain.
Q33.10. When light is incident on an interface between two materials, the angle of the refracted ray depends on the wavelength, but the angle of the reflected ray does not. Why should this be?
Q33.11. A salesperson at a bargain counter claims that a certain pair of sunglasses has Polaroid filters; you suspect that the glasses are just tinted plastic. How could you find out for sure?
Q33.12. Does it make sense to talk about the polarization of a *longitudinal* wave, such as a sound wave? Why or why not?
Q33.13. How can you determine the direction of the polarizing axis of a single polarizer?
Q33.14. It has been proposed that automobile windshields and headlights should have polarizing filters to reduce the glare of oncoming lights during night driving. Would this work? How should the polarizing axes be arranged? What advantages would this scheme have? What disadvantages?
Q33.15. When a sheet of plastic food wrap is placed between two crossed polarizers, no light is transmitted. When the sheet is stretched in one direction, some light passes through the crossed polarizers. What is happening?
Q33.16. If you sit on the beach and look at the ocean through Polaroid sunglasses, the glasses help to reduce the glare from sunlight reflecting off the water. But if you lie on your side on the beach, there is little reduction in the glare. Explain why there is a difference.
Q33.17. When unpolarized light is incident on two crossed polarizers, no light is transmitted. A student asserted that if a third polarizer is inserted between the other two, some transmission will occur. Does this make sense? How can adding a third filter *increase* transmission?
Q33.18. For the old "rabbit-ear" style TV antennas, it's possible to alter the quality of reception considerably simply by changing the orientation of the antenna. Why?
Q33.19. In Fig. 33.32, since the light that is scattered out of the incident beam is polarized, why is the transmitted beam not also partially polarized?
Q33.20. You are sunbathing in the late afternoon when the sun is relatively low in the western sky. You are lying flat on your back, looking straight up through Polaroid sunglasses. To minimize the amount of sky light reaching your eyes, how should you lie: with your feet pointing north, east, south, west, or in some other direction? Explain your reasoning.
Q33.21. Light scattered from blue sky is strongly polarized because of the nature of the scattering process described in Section 33.6. But light scattered from white clouds is usually *not* polarized. Why not?
Q33.22. Atmospheric haze is due to water droplets or smoke particles ("smog"). Such haze reduces visibility by scattering light, so that the light from distant objects becomes randomized and images become indistinct. Explain why visibility through haze can be improved by wearing red-tinted sunglasses, which filter out blue light.
Q33.23. The explanation given in Section 33.6 for the color of the setting sun should apply equally well to the *rising* sun, since sunlight travels the same distance through the atmosphere to reach your eyes at either sunrise or sunset. Typically, however, sunsets are redder than sunrises. Why? (*Hint:* Particles of all kinds in the atmosphere contribute to scattering.)
Q33.24. Huygens's principle also applies to sound waves. During the day, the temperature of the atmosphere decreases with increasing altitude above the ground. But at night, when the ground cools, there is a layer of air just above the surface in which the temperature *increases* with altitude. Use this to explain why sound waves from distant sources can be heard more clearly at night than in the daytime. (*Hint:* The speed of sound increases with increasing temperature. Use the ideas displayed in Fig. 33.37 for light.)
Q33.25. Can water waves be reflected and refracted? Give examples. Does Huygens's principle apply to water waves? Explain.

Exercises

Section 33.2 Reflection and Refraction

33.1. Two plane mirrors intersect at right angles. A laser beam strikes the first of them at a point 11.5 cm from their point of intersection, as shown in Fig. 33.38 For what angle of incidence at the first mirror will this ray strike the midpoint of the second mirror (which is 28.0 cm long) after reflecting from the first mirror?

Figure **33.38** Exercise 33.1.

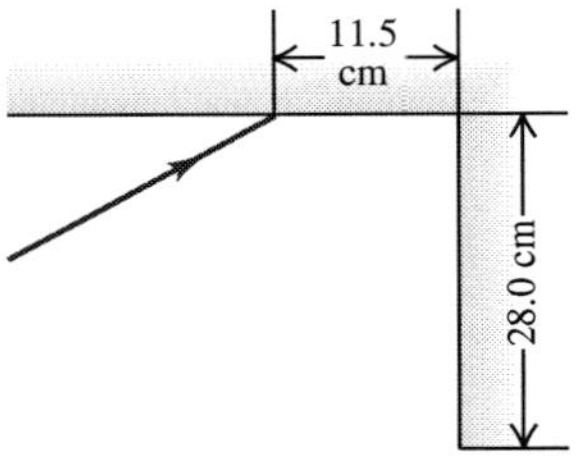

33.2. Three plane mirrors intersect at right angles. A beam of laser light strikes the first of them at an angle θ with respect to the normal (Fig. 33.39). (a) Show that when this ray is reflected off of the other two mirrors and crosses the original ray, the angle α between these two rays will be $\alpha = 180° - 2\theta$. (b) For what angle θ will the two rays be perpendicular when they cross?

Figure **33.39** Exercise 33.2

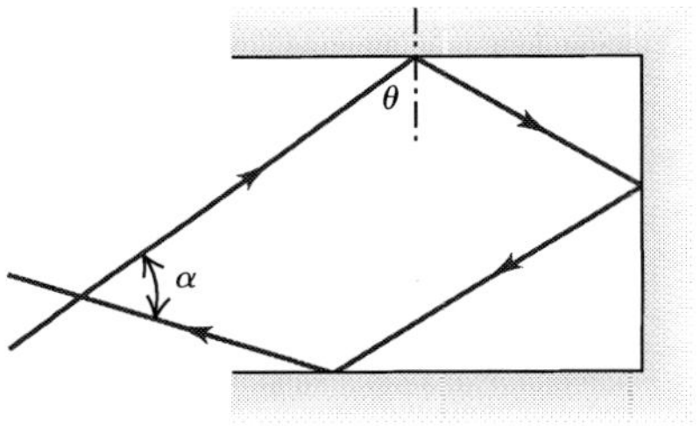

33.3. A beam of light has a wavelength of 650 nm in vacuum. (a) What is the speed of this light in a liquid whose index of refraction at this wavelength is 1.47? (b) What is the wavelength of these waves in the liquid?

33.4. Light with a frequency of 5.80×10^{14} Hz travels in a block of glass that has an index of refraction of 1.52. What is the wavelength of the light (a) in vacuum and (b) in the glass?

33.5. A light beam travels at 1.94×10^8 m/s in quartz. The wavelength of the light in quartz is 355 nm. (a) What is the index of refraction of quartz at this wavelength? (b) If this same light travels through air, what is its wavelength there?

33.6. Light of a certain frequency has a wavelength of 438 nm in water. What is the wavelength of this light in benzene?

33.7. A parallel beam of light in air makes an angle of 47.5° with the surface of a glass plate having a refractive index of 1.66. (a) What is the angle between the reflected part of the beam and the surface of the glass? (b) What is the angle between the refracted beam and the surface of the glass?

33.8. Using a fast-pulsed laser and electronic timing circuitry, you find that light travels 2.50 m within a plastic rod in 11.5 ns. What is the refractive index of the plastic?

33.9. Light traveling in air is incident on the surface of a block of plastic at an angle of 62.7° to the normal and is bent so that it makes a 48.1° angle with the normal in the plastic. Find the speed of light in the plastic.

33.10. (a) A tank containing methanol has walls 2.50 cm thick made of glass of refractive index 1.550. Light from the outside air strikes the glass at a 41.3° angle with the normal to the glass. Find the angle the light makes with the normal in the methanol. (b) The tank is emptied and refilled with an unknown liquid. If light incident at the same angle as in part (a) enters the liquid in the tank at an angle of 20.2° from the normal, what is the refractive index of the unknown liquid?

33.11. (a) Light passes through three parallel slabs of different thicknesses and refractive indexes. The light is incident in the first slab and finally refracts into the third slab. Show that the middle slab has no effect on the final direction of the light. That is, show that the direction of the light in the third slab is the same as if the light had passed directly from the first slab into the third slab. (b) Generalize this result to a stack of N slabs. What determines the final direction of the light in the last slab?

33.12. A horizontal, parallel-sided plate of glass having a refractive index of 1.52 is in contact with the surface of water in a tank. A ray coming from above in air makes an angle of incidence of 35.0° with the normal to the top surface of the glass. (a) What angle does the ray refracted into the water make with the normal to the surface? (b) What is the dependence of this angle on the refractive index of the glass?

33.13. In a material having an index of refraction n, a light ray has frequency f, wavelength λ, and speed v. What are the frequency, wavelength, and speed of this light (a) in vacuum and (b) in a material having refractive index n'? In each case, express your answers in terms of *only* f, λ, v, n, and n'.

33.14. Prove that a ray of light reflected from a plane mirror rotates through an angle of 2θ when the mirror rotates through an angle θ about an axis perpendicular to the plane of incidence.

33.15. A ray of light is incident on a plane surface separating two sheets of glass with refractive indexes 1.70 and 1.58. The angle of incidence is 62.0°, and the ray originates in the glass with $n = 1.70$. Compute the angle of refraction.

33.16. In Example 33.1 the water–glass interface is horizontal. If instead this interface were tilted 15.0° above the horizontal, with the right side higher than the left side, what would be the angle from the vertical of the ray in the glass? (The ray in the water still makes an angle of 60.0° with the vertical.)

Section 33.3 Total Internal Reflection

33.17. Light Pipe. Light enters a solid pipe made of plastic having an index of refraction of 1.60. The light travels parallel to the upper part of the pipe (Fig. 33.40). You want to cut the face AB so that all the light will reflect back into the pipe after it first strikes that face. (a) What is the largest that θ can be if the pipe is in air? (b) If the pipe is immersed in water of refractive index 1.33, what is the largest that θ can be?

Figure **33.40** Exercise 33.17

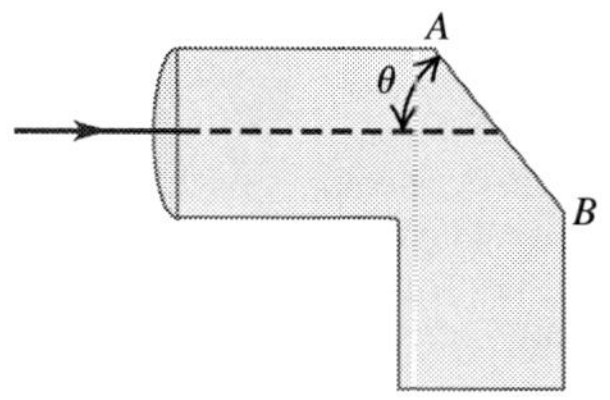

33.18. A beam of light is traveling inside a solid glass cube having index of refraction 1.53. It strikes the surface of the cube from the inside. (a) If the cube is in air, at what minimum angle with the normal inside the glass will this light *not* enter the air at this surface? (b) What would be the minimum angle in part (a) if the cube were immersed in water?

33.19. The critical angle for total internal reflection at a liquid–air interface is 42.5°. (a) If a ray of light traveling in the liquid has an angle of incidence at the interface of 35.0°, what angle does the refracted ray in the air make with the normal? (b) If a ray of light traveling in air has an angle of incidence at the interface of 35.0°, what angle does the refracted ray in the liquid make with the normal?

33.20. At the very end of Wagner's series of operas *Ring of the Nibelung,* Brünnhilde takes the golden ring from the finger of the dead Siegfried and throws it into the Rhine, where it sinks to the bottom of the river. Assuming that the ring is small enough compared to the depth of the river to be treated as a point and that the Rhine is 10.0 m deep where the ring goes in, what is the area of the largest circle at the surface of the water over which light from the ring could escape from the water?

33.21. A ray of light is traveling in a glass cube that is totally immersed in water. You find that if the ray is incident on the glass–water interface at an angle to the normal larger than 48.7°, no light is refracted into the water. What is the refractive index of the glass?

33.22. Light is incident along the normal on face *AB* of a glass prism of refractive index 1.52, as shown in Fig. 33.41. Find the largest value the angle α can have without any light refracted out of the prism at face *AC* if (a) the prism is immersed in air and (b) the prism is immersed in water.

Figure **33.41** Exercise 33.22.

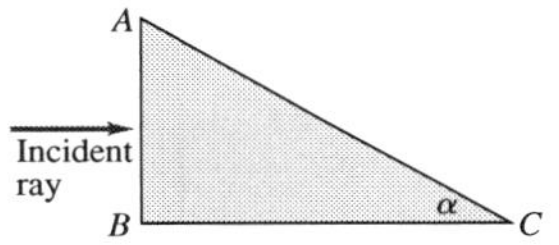

33.23. A ray of light in diamond (index of refraction 2.42) is incident on an interface with air. What is the *largest* angle the ray can make with the normal and not be totally reflected back into the diamond?

Section 33.4 Dispersion

33.24. A beam of light strikes a sheet of glass at an angle of 57.0° with the normal in air. You observe that red light makes an angle of 38.1° with the normal in the glass, while violet light makes a 36.7° angle. (a) What are the indexes of refraction of this glass for these colors of light? (b) What are the speeds of red and violet light in the glass?

Section 33.5 Polarization

33.25. A beam of unpolarized light of intensity I_0 passes through a series of ideal polarizing filters with their polarizing directions turned to various angles as shown in Fig. 33.42. (a) What is the light intensity (in terms of I_0) at points *A*, *B*, and *C*? (b) If we remove the middle filter, what will be the light intensity at point *C*?

Figure **33.42** Exercise 33.25.

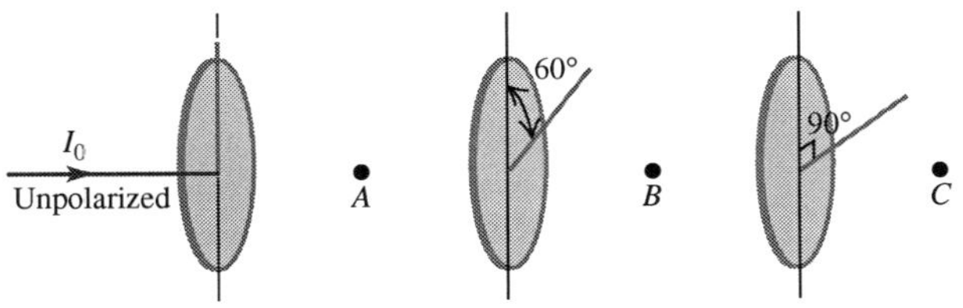

33.26. Light traveling in water strikes a glass plate at an angle of incidence of 53.0°; part of the beam is reflected and part is refracted. If the reflected and refracted portions make an angle of 90.0° with each other, what is the index of refraction of the glass?

33.27. A parallel beam of unpolarized light in air is incident at an angle of 54.5° (with respect to the normal) on a plane glass surface. The reflected beam is completely linearly polarized. (a) What is the refractive index of the glass? (b) What is the angle of refraction of the transmitted beam?

33.28. Light of original intensity I_0 passes through two ideal polarizing filters having their polarizing axes oriented as shown in Fig. 33.43. You want to adjust the angle ϕ so that the intensity at point *P* is equal to $I_0/10$. (a) If the original light is unpolarized, what should ϕ be? (b) If the original light is linearly polarized in the same direction as the polarizing axis of the first polarizer the light reaches, what should ϕ be?

Figure **33.43** Exercise 33.28.

33.29. A beam of polarized light passes through a polarizing filter. When the angle between the polarizing axis of the filter and the direction of polarization of the light is θ, the intensity of the emerging beam is I. If you now want the intensity to be $I/2$, what should be the angle (in terms of θ) between the polarizing angle of the filter and the original direction of polarization of the light?

33.30. The refractive index of a certain glass is 1.66. For what incident angle is light reflected from the surface of this glass completely polarized if the glass is immersed in (a) air and (b) water?

33.31. Unpolarized light of intensity 20.0 W/cm^2 is incident on two polarizing filters. The axis of the first filter is at an angle of 25.0° counterclockwise from the vertical (viewed in the direction the light is traveling), and the axis of the second filter is at 62.0° counterclockwise from the vertical. What is the intensity of the light after it has passed through the second polarizer?

33.32. A polarizer and an analyzer are oriented so that the maximum amount of light is transmitted. To what fraction of its maximum value is the intensity of the transmitted light reduced when the analyzer is rotated through (a) 22.5°; (b) 45.0°; (c) 67.5°?

33.33. Three Polarizing Filters. Three polarizing filters are stacked with the polarizing axes of the second and third at 45.0° and 90.0°, respectively, with that of the first. (a) If unpolarized light of intensity I_0 is incident on the stack, find the intensity and state of polarization of light emerging from each filter. (b) If the second filter is removed, what is the intensity of the light emerging from each remaining filter?

33.34. Three polarizing filters are stacked, with the polarizing axis of the second and third filters at 23.0° and 62.0°, respectively, to that of the first. If unpolarized light is incident on the stack, the light has intensity 75.0 W/cm^2 after it passes through the stack. If the incident intensity is kept constant, what is the intensity of the light after it has passed through the stack if the second polarizer is removed?

*Section 33.6 Scattering of Light

33.35. A beam of white light passes through a uniform thickness of air. If the intensity of the scattered light in the middle of the green part of the visible spectrum is I, find the intensity (in terms of I) of scattered light in the middle of (a) the red part of the spectrum and (b) the violet part of the spectrum. Consult Table 32.1.

Section 33.7 Huygens's Principle

33.36. Bending Around Corners. Traveling particles do not bend around corners, but waves do. To see why, suppose that a plane wave front strikes the edge of a sharp object traveling perpendicular to the surface (Fig. 33.44). Use Huygens's principle to show that this wave will bend around the upper edge of the object. (*Note:* This effect, called *diffraction,* can easily be seen for water waves, but it also occurs for light, as you will see in Chapters 35 and 36. However due to the very short wavelength of visible light, it is not so apparent in daily life.)

Figure 33.44 Exercise 33.36.

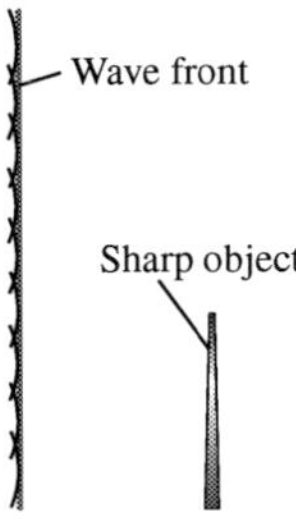

Problems

33.37. The Corner Reflector. An inside corner of a cube is lined with mirrors to make a corner reflector (see Example 33.3 in Section 33.2). A ray of light is reflected successively from each of three mutually perpendicular mirrors; show that its final direction is always exactly opposite to its initial direction.

33.38. A light beam is directed parallel to the axis of a hollow cylindrical tube. When the tube contains only air, it takes the light 8.72 ns to travel the length of the tube, but when the tube is filled with a transparent jelly, it takes the light 2.04 ns longer to travel its length. What is the refractive index of this jelly?

33.39. Light traveling in a material of refractive index n_1 is incident at angle θ_1 with respect to the normal at the interface with a slab of material that has parallel faces and refractive index n_2. After the light passes through this material, it is refracted into a material with refractive index n_3 and in this third material it makes an angle of θ_3 with the normal. (a) Find θ_3 in terms of θ_1 and the refractive indexes of the materials. (b) The ray in the third material is now reversed, so that it is incident on the n_3-to-n_2 interface with the angle θ_3 found in part (a). Show that when the light refracts into the material with refractive index n_1, the angle it makes with the normal is angle θ_1. This shows that the refracted ray is reversible. (c) Are reflected rays also reversible? Explain.

33.40. In a physics lab, light with wavelength 490 nm travels in air from a laser to a photocell in 17.0 ns. When a slab of glass 0.840 m thick is placed in the light beam, with the beam incident along the normal to the parallel faces of the slab, it takes the light 21.2 ns to travel from the laser to the photocell. What is the wavelength of the light in the glass?

33.41. A ray of light is incident in air on a block of a transparent solid whose index of refraction is n. If $n = 1.38$, what is the *largest* angle of incidence θ_a for which total internal reflection will occur at the vertical face (point A shown in Fig. 33.45)?

Figure 33.45 Problem 33.41.

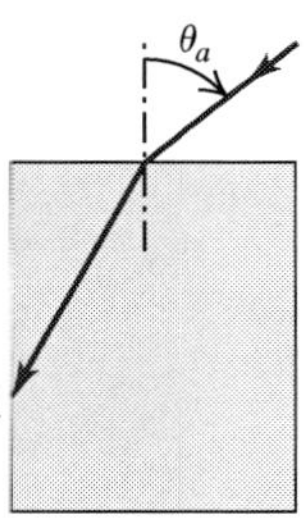

33.42. A light ray in air strikes the right-angle prism shown in Fig. 33.46. This ray consists of two different wavelengths. When it emerges at face AB, it has been split into two different rays that diverge from each other by 8.50°. Find the index of refraction of the prism for each of the two wavelengths.

Figure 33.46 Problem 33.42.

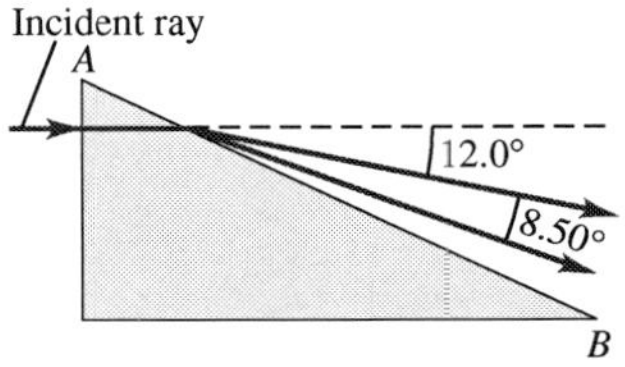

33.43. A quarter-wave plate converts linearly polarized light to circularly polarized light. Prove that a quarter-wave plate also converts circularly polarized light to linearly polarized light.

33.44. A glass plate 2.50 mm thick, with an index of refraction of 1.40, is placed between a point source of light with wavelength 540 nm (in vacuum) and a screen. The distance from source to screen is 1.80 cm. How many wavelengths are there between the source and the screen?

33.45. Old photographic plates were made of glass with a light-sensitive emulsion on the front surface. This emulsion was somewhat transparent. When a bright point source is focused on the front of the plate, the developed photograph will show a halo around the image of the spot. If the glass plate is 3.10 mm thick and the halos have an inner radius of 5.34 mm, what is the index of refraction of the glass? (*Hint:* Light from the spot on the front surface is scattered in all directions by the emulsion. Some of it is then totally reflected at the back surface of the plate and returns to the front surface.)

33.46. After a long day of driving you take a late-night swim in a motel swimming pool. When you go to your room, you realize that you have lost your room key in the pool. You borrow a powerful flashlight and walk around the pool, shining the light into it. The light shines on the key, which is lying on the bottom of the pool, when the flashlight is held 1.2 m above the water surface and is directed at the surface a horizontal distance of 1.5 m from the edge (Fig. 33.47). If the water here is 4.0 m deep, how far is the key from the edge of the pool?

Figure 33.47 Problem 33.46.

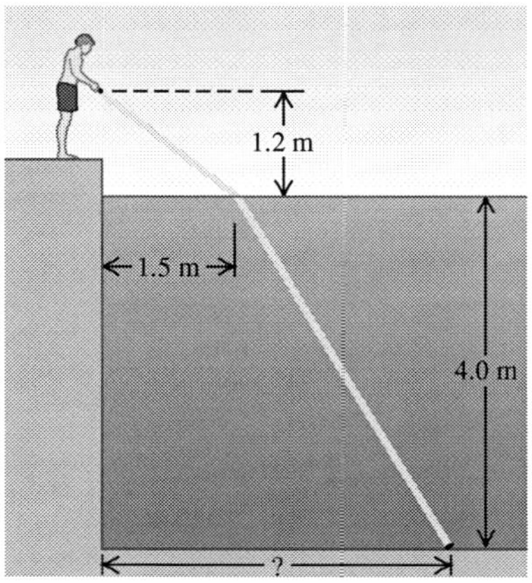

33.47. You sight along the rim of a glass with vertical sides so that the top rim is lined up with the opposite edge of the bottom (Fig. 33.48a). The glass is a thin-walled, hollow cylinder 16.0 cm high with a top and bottom of the glass diameter of 8.0 cm. While you keep your eye in the same position, a friend fills the glass with a transparent liquid, and you then see a dime that is lying at the center of the bottom of the glass (Fig. 33.48b). What is the index of refraction of the liquid?

Figure **33.48** Problem 33.47.

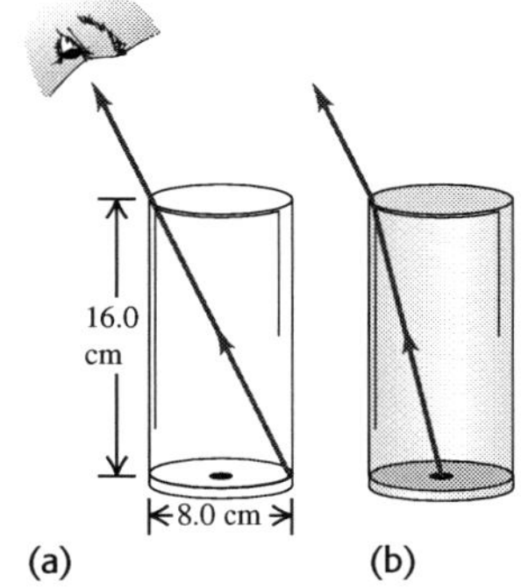

33.48. A beaker with a mirrored bottom is filled with a liquid whose index of refraction is 1.63. A light beam strikes the top surface of the liquid at an angle of 42.5° from the normal. At what angle from the normal will the beam exit from the liquid after traveling down through the liquid, reflecting from the mirrored bottom, and returning to the surface?

33.49. A thin layer of ice $(n = 1.309)$ floats on the surface of water $(n = 1.333)$ in a bucket. A ray of light from the bottom of the bucket travels upward through the water. (a) What is the largest angle with respect to the normal that the ray can make at the ice–water interface and still pass out into the air above the ice? (b) What is this angle after the ice melts?

33.50. A $45°-45°-90°$ prism is immersed in water. A ray of light is incident normally on one of its shorter faces. What is the minimum index of refraction that the prism must have if this ray is to be totally reflected within the glass at the long face of the prism?

33.51. The prism shown in Fig. 33.49 has a refractive index of 1.66, and the angles *A* are 25.0°. Two light rays *m* and *n* are parallel as they enter the prism. What is the angle between them after they emerge?

Figure **33.49** Problem 33.51.

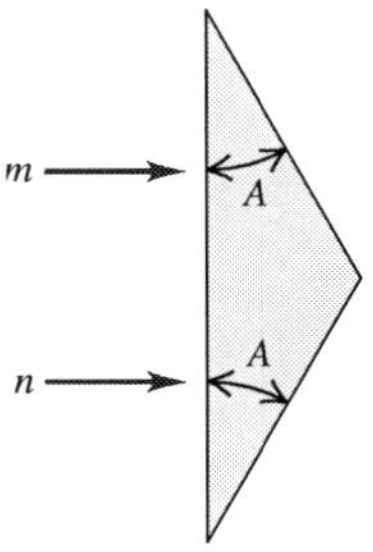

33.52. Light is incident normally on the short face of a $30°-60°-90°$ prism (Fig. 33.50). A drop of liquid is placed on the hypotenuse of the prism. If the index of the prism is 1.62, find the maximum index that the liquid may have if the light is to be totally reflected.

Figure **33.50** Problem 33.52.

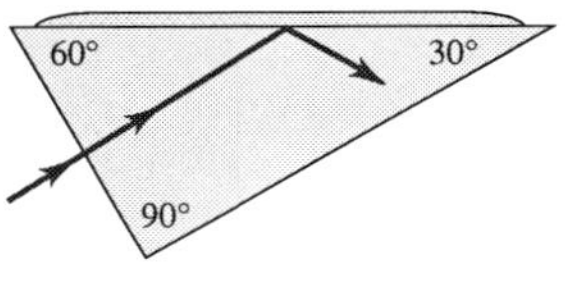

33.53. A horizontal cylindrical tank 2.20 m in diameter is half full of water. The space above the water is filled with a pressurized gas of unknown refractive index. A small laser can move along the curved bottom of the water and aims a light beam toward the center of the water surface (Fig. 33.51). You observe that when the laser has moved a distance $S = 1.09$ m or more (measured along the curved surface) from the lowest point in the water, no light enters the gas. (a) What is the index of refraction of the gas? (b) How long does it take the light beam to travel from the laser to the rim of the tank when (i) $S > 1.09$ m and (ii) $S < 1.09$ m?

Figure **33.51** Problem 33.53.

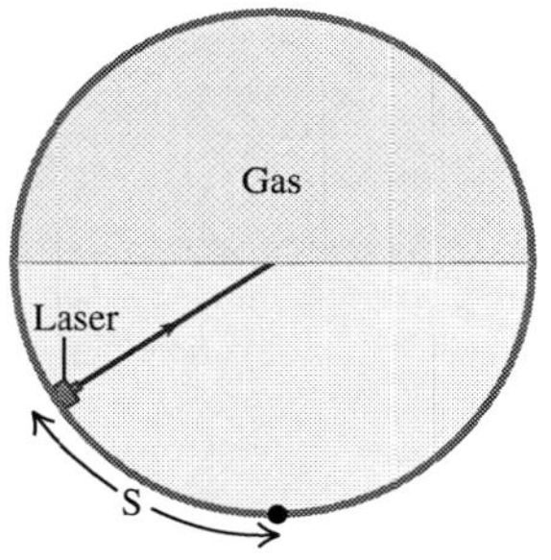

33.54. A large cube of glass has a metal reflector on one face and water on an adjoining face (Fig. 33.52). A light beam strikes the reflector, as shown. You observe that as you gradually increase the angle of the light beam, if $\theta \geq 59.2°$ no light enters the water. What is the speed of light in this glass?

Figure **33.52** Problem 33.54.

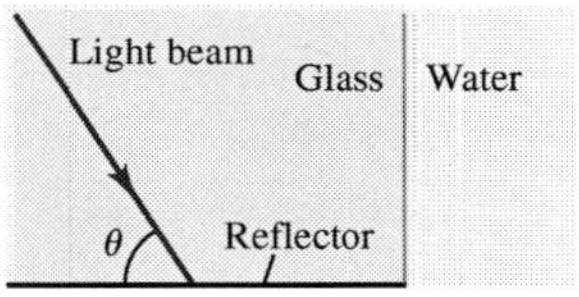

33.55. When the sun is either rising or setting and appears to be just on the horizon, it is in fact *below* the horizon. The explanation for this seeming paradox is that light from the sun bends slightly when entering the earth's atmosphere, as shown in Fig. 33.53. Since our perception is based on the idea that light travels in straight lines, we perceive the light to be coming from an apparent position that is an angle δ above the sun's true position. (a) Make the simplifying assumptions that the atmosphere has uniform density, and hence uniform index of refraction n, and extends to a height h above the earth's surface, at which point it abruptly stops. Show that the angle δ is given by

$$\delta = \arcsin\left(\frac{nR}{R + h}\right) - \arcsin\left(\frac{R}{R + h}\right)$$

where $R = 6378$ km is the radius of the earth. (b) Calculate δ using $n = 1.0003$ and $h = 20$ km. How does this compare to the

Figure **33.53** Problem 33.55.

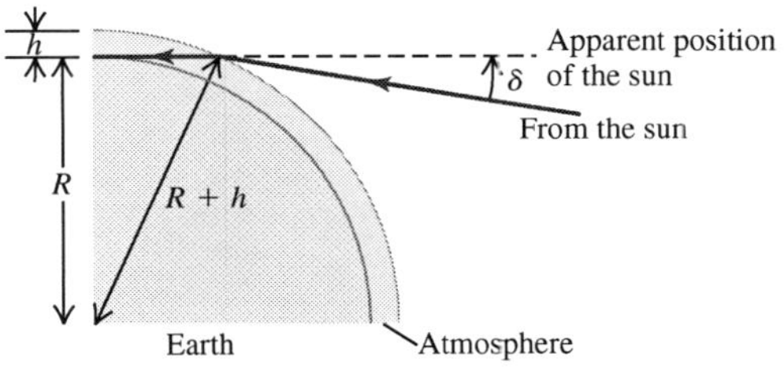

angular radius of the sun, which is about one quarter of a degree? (In actuality a light ray from the sun bends gradually, not abruptly, since the density and refractive index of the atmosphere change gradually with altitude.)

33.56. Fermat's Principle of Least Time. A ray of light traveling with speed c leaves point 1 shown in Fig. 33.54 and is reflected to point 2. The ray strikes the reflecting surface a horizontal distance x from point 1. (a) Show that the time t required for the light to travel from 1 to 2 is

$$t = \frac{\sqrt{y_1^2 + x^2} + \sqrt{y_2^2 + (l - x)^2}}{c}$$

(b) Take the derivative of t with respect to x. Set the derivative equal to zero to show that this time reaches its *minimum* value when $\theta_1 = \theta_2$, which is the law of reflection and corresponds to the actual path taken by the light. This is an example of Fermat's *principle of least time,* which states that among all possible paths between two points, the one actually taken by a ray of light is that for which the time of travel is a *minimum.* (In fact, there are some cases in which the time is a maximum rather than a minimum.)

Figure **33.54** Problem 33.56.

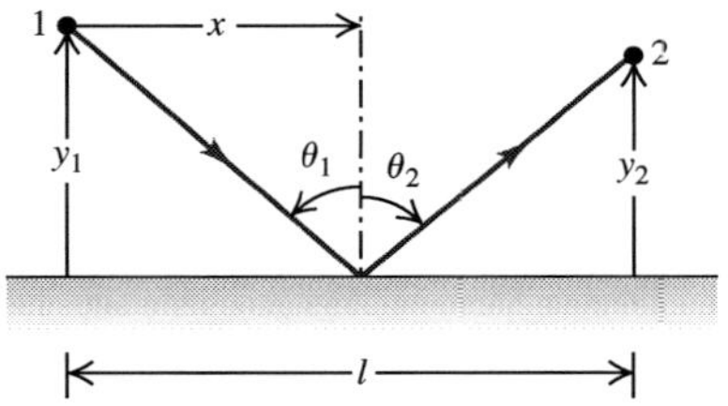

33.57. A ray of light goes from point A in a medium in which the speed of light is v_1 to point B in a medium in which the speed is v_2 (Fig. 33.55). The ray strikes the interface a horizontal distance x to the right of point A. (a) Show that the time required for the light to go from A to B is

$$t = \frac{\sqrt{h_1^2 + x^2}}{v_1} + \frac{\sqrt{h_2^2 + (l - x)^2}}{v_2}$$

(b) Take the derivative of t with respect to x. Set this derivative equal to zero to show that this time reaches its *minimum* value when $n_1 \sin\theta_1 = n_2 \sin\theta_2$. This is Snell's law, and corresponds to the actual path taken by the light. This is another example of Fermat's principle of least time (see Problem 33.56).

Figure **33.55** Problem 33.57.

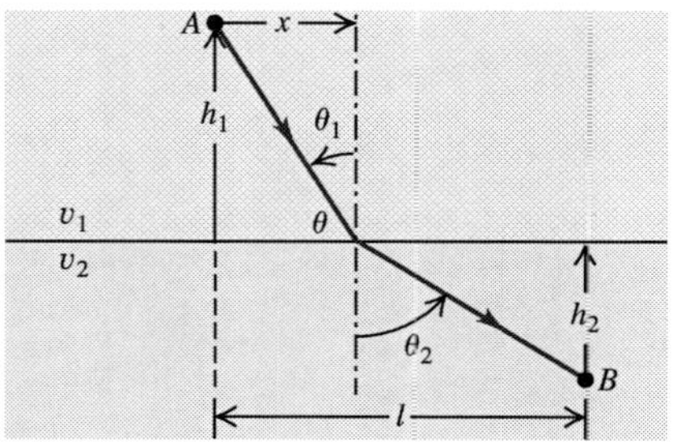

33.58. Light is incident in air at an angle θ_a (Fig. 33.56) on the upper surface of a transparent plate, the surfaces of the plate being plane and parallel to each other. (a) Prove that $\theta_a = \theta'_a$. (b) Show that this is true for any number of different parallel plates. (c) Prove that the lateral displacement d of the emergent beam is given by the relationship

$$d = t\frac{\sin(\theta_a - \theta'_b)}{\cos\theta'_b}$$

where t is the thickness of the plate. (d) A ray of light is incident at an angle of 66.0° on one surface of a glass plate 2.40 cm thick with an index of refraction 1.80. The medium on either side of the plate is air. Find the lateral displacement between the incident and emergent rays.

Figure **33.56** Problem 33.58.

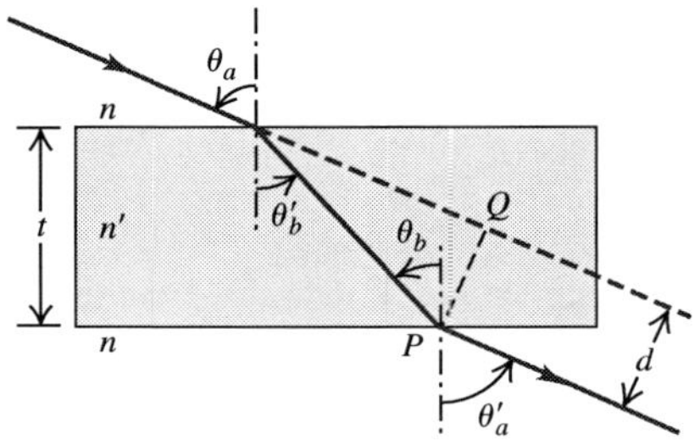

33.59. Light traveling downward is incident on a horizontal film of thickness t, as shown in Fig. 33.57. The incident ray splits into two rays, A and B. Ray A reflects from the top of the film. Ray B reflects from the bottom of the film and then refracts back into the material that is above the film. If the film has parallel faces, show that rays A and B end up parallel to each other.

Figure **33.57** Problem 33.59.

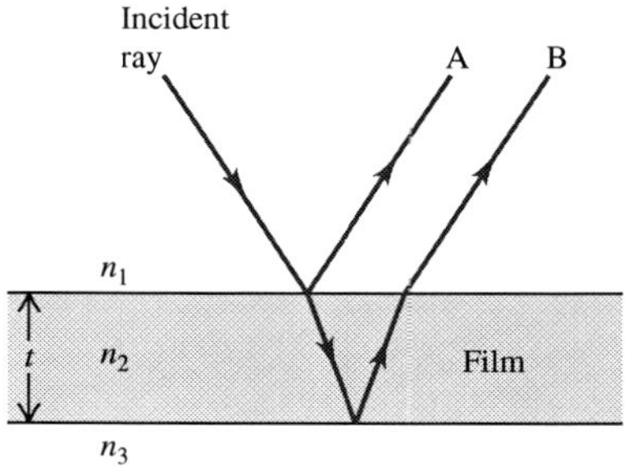

33.60. A thin beam of white light is directed at a flat sheet of silicate flint glass at an angle of 20.0° to the surface of the sheet. Due to dispersion in the glass, the beam is spread out as shown in a spectrum in Fig. 33.58. The refractive index of silicate flint glass versus wavelength is graphed in Fig. 33.18. (a) The rays a and b shown in Fig. 33.58 correspond to the extremes of the visible spectrum. Which corresponds to red and which to violet? Explain your reasoning. (b) For what thickness d of the glass sheet will the spectrum be 1.0 mm wide, as shown (see Problem 33.58)?

Figure **33.58** Problem 33.60.

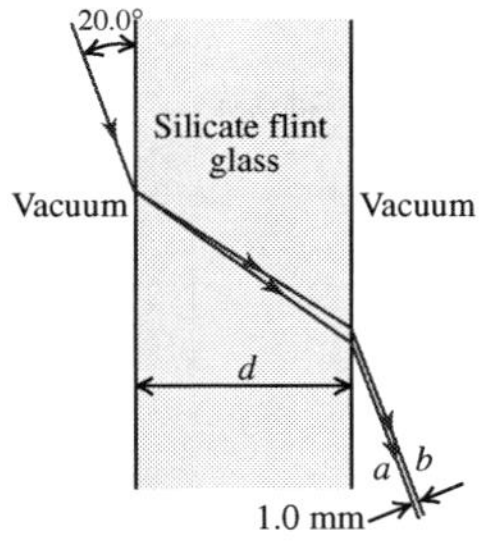

33.61. Angle of Deviation. The incident angle θ_a shown in Fig. 33.59 is chosen so that the light passes symmetrically through the prism, which has refractive index n and apex angle A. (a) Show that the angle of deviation δ (the angle between the initial and final directions of the ray) is given by

$$\sin\frac{A+\delta}{2} = n\sin\frac{A}{2}$$

(When the light passes through symmetrically, as shown, the angle of deviation is a minimum.) (b) Use the result of part (a) to find the angle of deviation for a ray of light passing symmetrically through a prism having three equal angles ($A = 60.0°$) and $n = 1.52$. (c) A certain glass has a refractive index of 1.61 for red light (700 nm) and 1.66 for violet light (400 nm). If both colors pass through symmetrically, as described in part (a), and if $A = 60.0°$, find the difference between the angles of deviation for the two colors.

Figure **33.59** Problem 33.61.

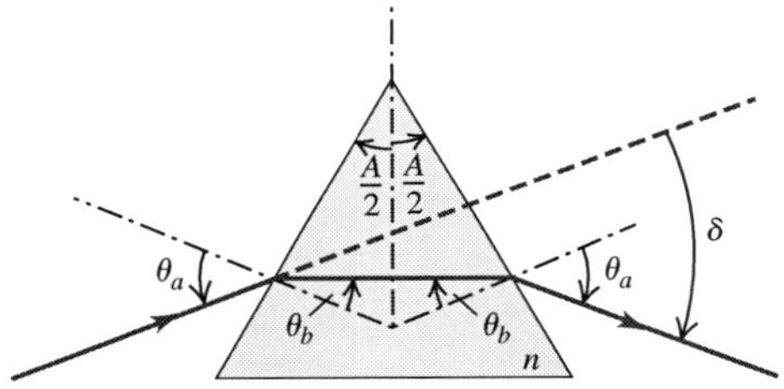

33.62. A beam of unpolarized sunlight strikes the vertical plastic wall of a water tank at an unknown angle. Some of the light reflects from the wall and enters the water (Fig. 33.60). The refractive index of the plastic wall is 1.61. If the light that has been reflected from the wall into the water is observed to be completely polarized, what angle does this beam make with the normal inside the water?

Figure **33.60** Problem 33.62.

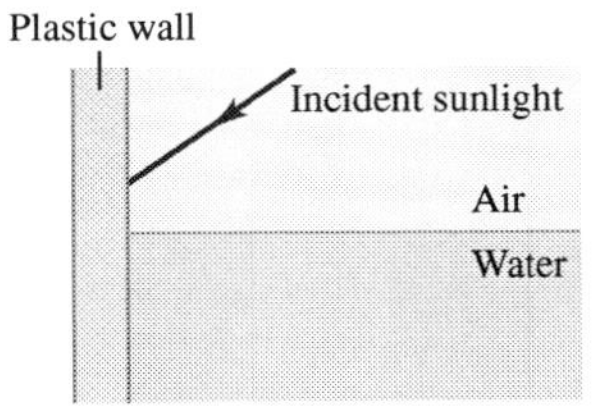

33.63. A beam of light traveling horizontally is made of an unpolarized component with intensity I_0 and a polarized component with intensity I_p. The plane of polarization of the polarized component is oriented at an angle of θ with respect to the vertical. The data in the table give the intensity measured through a polarizer with an orientation of ϕ with respect to the vertical. (a) What is the orientation of the polarized component? (That is, what is the angle θ?) (b) What are the values of I_0 and I_p?

φ (°)	I_{total} (W/m^2)	ϕ (°)	I_{total} (W/m^2)
0	18.4	100	8.6
10	21.4	110	6.3
20	23.7	120	5.2
30	24.8	130	5.2
40	24.8	140	6.3
50	23.7	150	8.6
60	21.4	160	11.6
70	18.4	170	15.0
80	15.0	180	18.4
90	11.6		

33.64. A certain birefringent material has indexes of refraction n_1 and n_2 for the two perpendicular components of linearly polarized light passing through it. The corresponding wavelengths are $\lambda_1 = \lambda_0/n_1$ and λ_0/n_2, where λ_0 is the wavelength in vacuum. (a) If the crystal is to function as a quarter-wave plate, the number of wavelengths of each component within the material must differ by $\frac{1}{4}$ Show that the minimum thickness for a quarter-wave plate is

$$d = \frac{\lambda_0}{4(n_1 - n_2)}$$

(b) Find the minimum thickness of a quarter-wave plate made of siderite ($FeO \cdot CO_2$) if the indexes of refraction are $n_1 = 1.875$ and $n_2 = 1.635$ and the wavelength in vacuum is $\lambda_0 = 589$ nm.

Challenge Problems

33.65. Consider two vibrations of equal amplitude and frequency but differing in phase, one along the x-axis,

$$x = a\sin(\omega t - \alpha)$$

and the other along the y-axis,

$$y = a\sin(\omega t - \beta)$$

These can be written as follows:

$$\frac{x}{a} = \sin\omega t\cos\alpha - \cos\omega t\sin\alpha \quad (1)$$

$$\frac{y}{a} = \sin\omega t\cos\beta - \cos\omega t\sin\beta \quad (2)$$

(a) Multiply Eq. (1) by $\sin\beta$ and Eq. (2) by $\sin\alpha$, and then subtract the resulting equations. (b) Multiply Eq. (1) by $\cos\beta$ and Eq. (2) by $\cos\alpha$, and then subtract the resulting equations. (c) Square and add the results of parts (a) and (b). (d) Derive the equation $x^2 + y^2 - 2xy\cos\delta = a^2\sin^2\delta$, where $\delta = \alpha - \beta$. (e) Use the above result to justify each of the diagrams in Fig. 33.61 (next page). In the figure, the angle given is the phase difference between two simple harmonic motions of the same frequency and amplitude, one horizontal (along the x-axis) and the other vertical (along the y-axis). The figure thus shows the resultant motion from the superposition of the two perpendicular harmonic motions.

Figure **33.61** Challenge Problem 33.65.

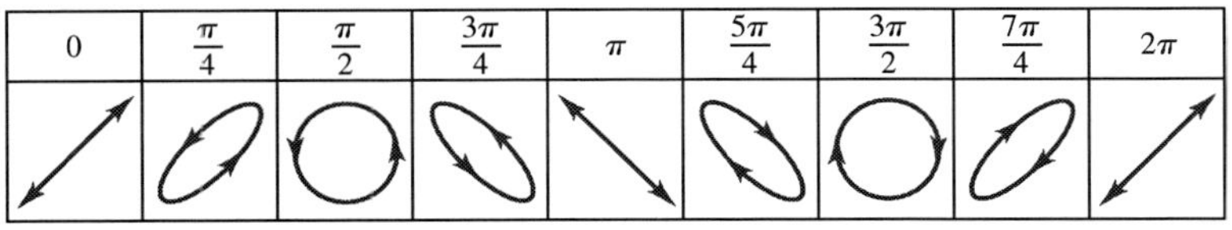

33.66. A rainbow is produced by the reflection of sunlight by spherical drops of water in the air. Figure 33.62 shows a ray that refracts into a drop at point A, is reflected from the back surface of the drop at point B, and refracts back into the air at point C. The angles of incidence and refraction, θ_a and θ_b, are shown at points A and C, and the angles of incidence and reflection, θ_a and θ_r, are shown at point B. (a) Show that $\theta_a^B = \theta_b^A$, $\theta_a^C = \theta_b^A$, and $\theta_b^C = \theta_a^A$. (b) Show that the angle in radians between the ray before it enters the drop at A and after it exits at C (the total angular deflection of the ray) is $\Delta = 2\theta_a^A - 4\theta_b^A + \pi$. (*Hint:* Find the angular deflections that occur at A, B, and C, and add them to get Δ.) (c) Use Snell's law to write Δ in terms of θ_a^A and n, the refractive index of the water in the drop. (d) A rainbow will form when the angular deflection Δ is *stationary* in the incident angle θ_a^A—that is, when $d\Delta/d\theta_a^A = 0$. If this condition is satisfied, all the rays with incident angles close to θ_a^A will be sent back in the same direction, producing a bright zone in the sky. Let θ_1 be the value of θ_a^A for which this occurs. Show that $\cos^2\theta_1 = \frac{1}{3}(n^2 - 1)$. (*Hint:* You may find the derivative formula $d(\arcsin u(x))/dx = (1 - u^2)^{-1/2}(du/dx)$ helpful.) (e) The index of refraction in water is 1.342 for violet light and 1.330 for red light. Use the results of parts (c) and (d) to find θ_1 and Δ for violet and red light. Do your results agree with the angles shown in Fig. 33.20d? When you view the rainbow, which color, red or violet, is higher above the horizon?

Figure **33.62** Challenge Problem 33.66.

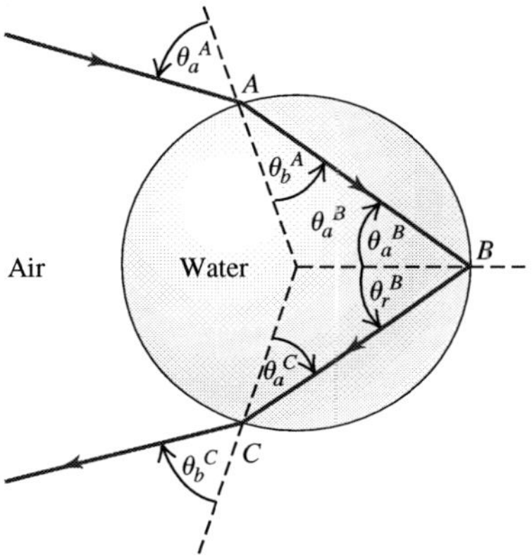

33.67. A *secondary rainbow* is formed when the incident light undergoes two internal reflections in a spherical drop of water as shown in Fig. 33.20e. (See Challenge Problem 33.66.) (a) In terms of the incident angle θ_a^A and the refractive index n of the drop, what is the angular deflection Δ of the ray? That is, what is the angle between the ray before it enters the drop and after it exits? (b) What is the incident angle θ_2 for which the derivative of Δ with respect to the incident angle θ_a^A is zero? (c) The indexes of refraction for red and violet light in water are given in part (e) of Challenge Problem 33.66. Use the results of parts (a) and (b) to find θ_2 and Δ for violet and red light. Do your results agree with the angles shown in Fig. 33.20e? When you view a secondary rainbow, is red or violet higher above the horizon? Explain.

GEOMETRIC OPTICS

34

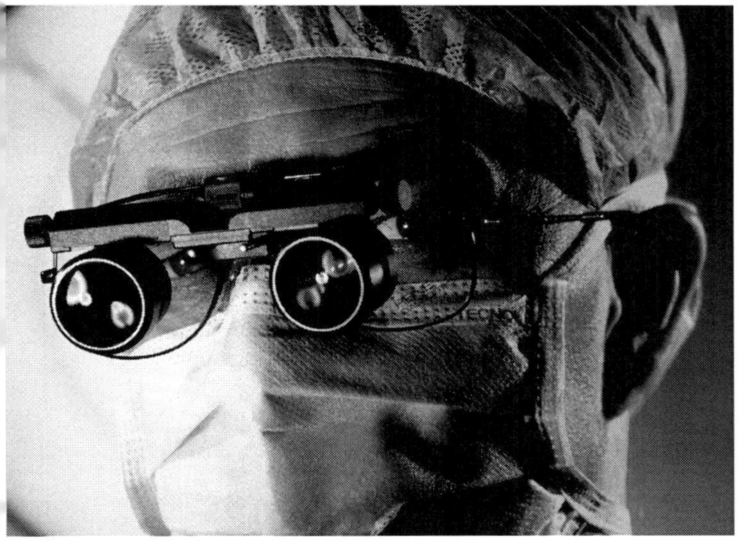

? How do magnifying lenses work? At what distance from the object being examined do they provide the sharpest view?

LEARNING GOALS

By studying this chapter, you will learn:

- How a plane mirror forms an image.
- Why concave and convex mirrors form different kinds of image.
- How images can be formed by a curved interface between two transparent materials.
- What aspects of a lens determine the type of image that it produces.
- What determines the field of view of a camera lens.
- What causes various defects in human vision, and how they can be corrected.
- The principle of the simple magnifier.
- How microscopes and telescopes work.

Your reflection in the bathroom mirror, the view of the moon through a telescope, the patterns seen in a kaleidoscope—all of these are examples of *images*. In each case the object that you're looking at appears to be in a different place than its actual position: Your reflection is on the other side of the mirror, the moon appears to be much closer when seen through a telescope, and objects seen in a kaleidoscope seem to be in many places at the same time. In each case, light rays that come from a point on an object are deflected by reflection or refraction (or a combination of the two), so they converge toward or appear to diverge from a point called an *image point.* Our goal in this chapter is to see how this is done and to explore the different kinds of images that can be made with simple optical devices.

To understand images and image formation, all we need are the ray model of light, the laws of reflection and refraction, and some simple geometry and trigonometry. The key role played by geometry in our analysis is the reason for the name *geometric optics* that is given to the study of how light rays form images. We'll begin our analysis with one of the simplest image-forming optical devices, a plane mirror. We'll go on to study how images are formed by curved mirrors, by refracting surfaces, and by thin lenses. Our results will lay the foundation for understanding many familiar optical instruments, including camera lenses, magnifiers, the human eye, microscopes, and telescopes.

34.1 Reflection and Refraction at a Plane Surface

Before discussing what is meant by an image, we first need the concept of **object** as it is used in optics. By an *object* we mean anything from which light rays radiate. This light could be emitted by the object itself if it is *self-luminous,* like the glowing filament of a light bulb. Alternatively, the light could be emitted by

ActivPhysics Online
15.4 Geometric Optics: Plane Mirrors

34.1 Light rays radiate from a point object P in all directions.

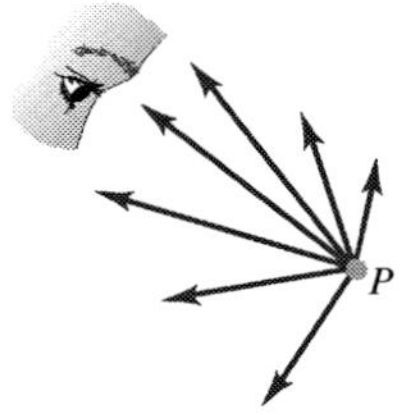

34.2 Light rays from the object at point P are reflected from a plane mirror. The reflected rays entering the eye look as though they had come from image point P'.

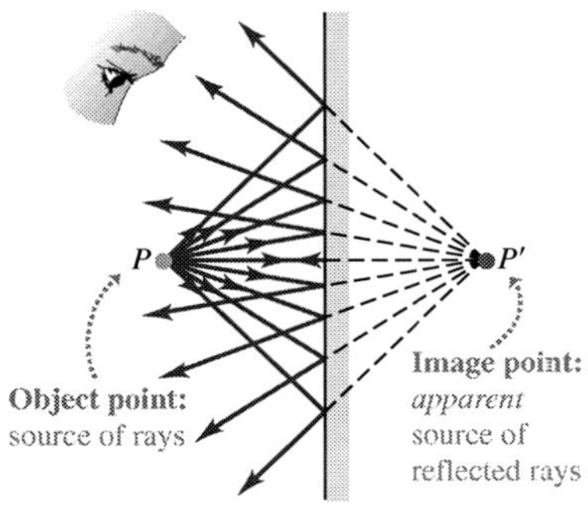

34.3 Light rays from the object at point P are refracted at the plane interface. The refracted rays entering the eye look as though they had come from image point P'.

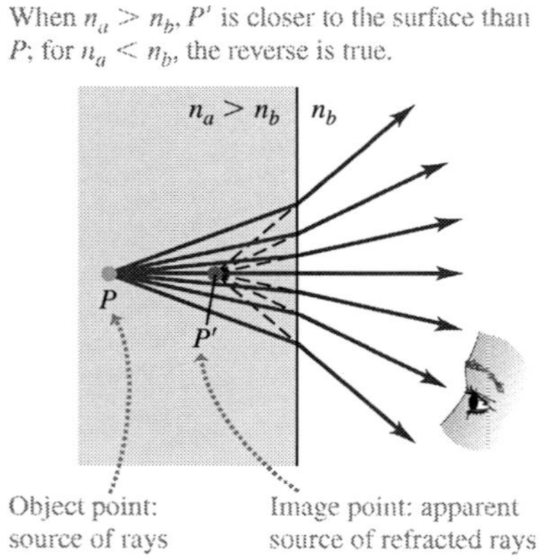

another source (such as a lamp or the sun) and then reflected from the object; an example is the light you see coming from the pages of this book. Figure 34.1 shows light rays radiating in all directions from an object at a point P. For an observer to see this object directly, there must be no obstruction between the object and the observer's eyes. Note that light rays from the object reach the observer's left and right eyes at different angles; these differences are processed by the observer's brain to infer the *distance* from the observer to the object.

The object in Fig. 34.1 is a **point object** that has no physical extent. Real objects with length, width, and height are called **extended objects.** To start with, we'll consider only an idealized point object, since we can always think of an extended object as being made up of a very large number of point objects.

Suppose some of the rays from the object strike a smooth, plane reflecting surface (Fig. 34.2). This could be the surface of a material with a different index of refraction, which reflects part of the incident light, or a polished metal surface that reflects almost 100% of the light that strikes it. We will always draw the reflecting surface as a black line with a shaded area behind it, as in Fig. 34.2. Bathroom mirrors have a thin sheet of glass that lies in front of and protects the reflecting surface; we'll ignore the effects of this thin sheet.

According to the law of reflection, all rays striking the surface are reflected at an angle from the normal equal to the angle of incidence. Since the surface is plane, the normal is in the same direction at all points on the surface, and we have *specular* reflection. After the rays are reflected, their directions are the same as though they had come from point P'. We call point P an *object point* and point P' the corresponding *image point,* and we say that the reflecting surface forms an **image** of point P. An observer who can see only the rays reflected from the surface, and who doesn't know that he's seeing a reflection, *thinks* that the rays originate from the image point P'. The image point is therefore a convenient way to describe the directions of the various reflected rays, just as the object point P describes the directions of the rays arriving at the surface *before* reflection.

If the surface in Fig. 34.2 were *not* smooth, the reflection would be *diffuse,* and rays reflected from different parts of the surface would go in uncorrelated directions (see Fig. 33.6b). In this case there would not be a definite image point P' from which all reflected rays seem to emanate. You can't see your reflection in the surface of a tarnished piece of metal because its surface is rough; polishing the metal smoothes the surface so that specular reflection occurs and a reflected image becomes visible.

An image is also formed by a plane *refracting* surface, as shown in Fig. 34.3. Rays coming from point P are refracted at the interface between two optical materials. When the angles of incidence are small, the final directions of the rays after refraction are the same as though they had come from point P', as shown, and again we call P' an *image point.* In Section 33.2 we described how this effect makes underwater objects appear closer to the surface than they really are (see Fig. 33.9).

In both Figs. 34.2 and 34.3 the rays do not actually pass through the image point P'. Indeed, if the mirror in Fig. 34.2 is opaque, there is no light at all on its right side. If the outgoing rays don't actually pass through the image point, we call the image a **virtual image.** Later we will see cases in which the outgoing rays really *do* pass through an image point, and we will call the resulting image a **real image.** The images that are formed on a projection screen, on the photographic film in a camera, and on the retina of your eye are real images.

Image Formation by a Plane Mirror

Let's concentrate for now on images produced by *reflection;* we'll return to refraction later in the chapter. To find the precise location of the virtual image P' that a plane mirror forms of an object at P, we use the construction shown in Fig. 34.4. The figure shows two rays diverging from an object point P at a distance s to the left of a plane mirror. We call s the **object distance.** The ray PV is incident normally on the mirror (that is, it is perpendicular to the mirror surface), and it returns along its original path.

The ray PB makes an angle θ with PV. It strikes the mirror at an angle of incidence θ and is reflected at an equal angle with the normal. When we extend the two reflected rays backward, they intersect at point P', at a distance s' behind the mirror. We call s' the **image distance.** The line between P and P' is perpendicular to the mirror. The two triangles PVB and $P'VB$ are congruent, so P and P' are at equal distances from the mirror, and s and s' have equal magnitudes. The image point P' is located exactly opposite the object point P as far *behind* the mirror as the object point is from the front of the mirror.

We can repeat the construction of Fig. 34.4 for each ray diverging from P. The directions of *all* the outgoing reflected rays are the same as though they had originated at point P', confirming that P' is the *image* of P. No matter where the observer is located, she will always see the image at the point P'.

34.4 Construction for determining the location of the image formed by a plane mirror. The image point P' is as far behind the mirror as the object point P is in front of it.

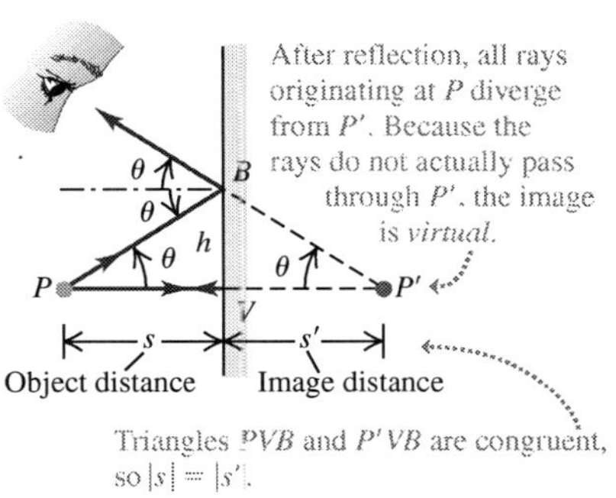

Sign Rules

Before we go further, let's introduce some general sign rules. These may seem unnecessarily complicated for the simple case of an image formed by a plane mirror, but we want to state the rules in a form that will be applicable to *all* the situations we will encounter later. These will include image formation by a plane or spherical reflecting or refracting surface, or by a pair of refracting surfaces forming a lens. Here are the rules:

1. **Sign rule for the object distance:** When the object is on the same side of the reflecting or refracting surface as the incoming light, the object distance s is positive; otherwise, it is negative.
2. **Sign rule for the image distance:** When the image is on the same side of the reflecting or refracting surface as the outgoing light, the image distance s' is positive; otherwise, it is negative.
3. **Sign rule for the radius of curvature of a spherical surface:** When the center of curvature C is on the same side as the outgoing light, the radius of curvature is positive; otherwise, it is negative.

34.5 For both of these situations, the object distance s is positive (rule 1) and the image distance s' is negative (rule 2).

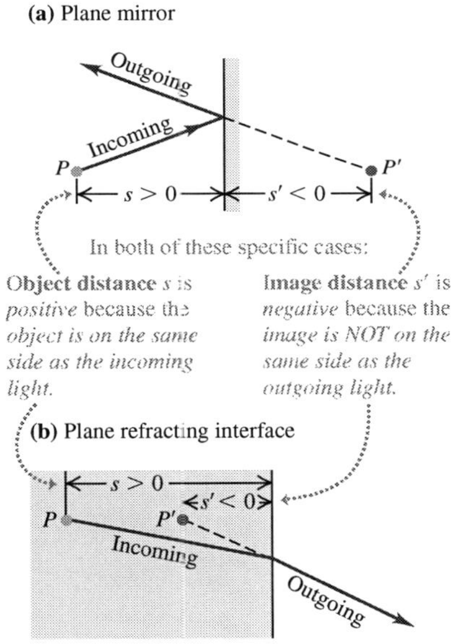

Figure 34.5 illustrates rules 1 and 2 for two different situations. For a mirror the incoming and outgoing sides are always the same; for example, in Figs. 34.2, 34.4, and 34.5a they are both the left side. For the refracting surfaces in Figs. 34.3 and 34.5b the incoming and outgoing sides are on the left and right sides, respectively, of the interface between the two materials. (Note that other textbooks may use different rules.)

In Figs. 34.4 and 34.5a the object distance s is *positive* because the object point P is on the incoming side (the left side) of the reflecting surface. The image distance s' is *negative* because the image point P' is *not* on the outgoing side (the left side) of the surface. The object and image distances s and s' are related simply by

$$s = -s' \qquad \text{(plane mirror)} \tag{34.1}$$

For a plane reflecting or refracting surface, the radius of curvature is infinite and not a particularly interesting or useful quantity; in these cases we really don't need sign rule 3. But this rule will be of great importance when we study image formation by *curved* reflecting and refracting surfaces later in the chapter.

34.6 Construction for determining the height of an image formed by reflection at a plane reflecting surface.

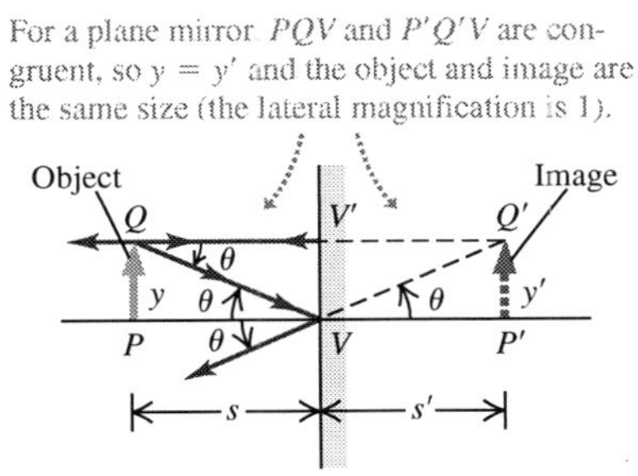

Image of an Extended Object: Plane Mirror

Next we consider an *extended* object with finite size. For simplicity we often consider an object that has only one dimension, like a slender arrow, oriented parallel to the reflecting surface; an example is the arrow PQ in Fig. 34.6. The distance from the head to the tail of an arrow oriented in this way is called its *height;* in Fig. 34.6 the height is y. The image formed by such an extended object is an extended image; to each point on the object, there corresponds a point on the image. Two of the rays from Q are shown; *all* the rays from Q appear to diverge from its image point Q' after reflection. The image of the arrow is the line $P'Q'$, with height y'. Other points of the object PQ have image points between P' and Q'. The triangles PQV and $P'Q'V$ are congruent, so the object PQ and image $P'Q'$ have the same size and orientation, and $y = y'$.

The ratio of image height to object height, y'/y, in *any* image-forming situation is called the **lateral magnification** m; that is,

$$m = \frac{y'}{y} \quad \text{(lateral magnification)} \tag{34.2}$$

Thus for a plane mirror the lateral magnification m is unity. When you look at yourself in a plane mirror, your image is the same size as the real you.

In Fig. 34.6 the image arrow points in the *same* direction as the object arrow; we say that the image is **erect.** In this case, y and y' have the same sign, and the lateral magnification m is positive. The image formed by a plane mirror is always erect, so y and y' have both the same magnitude and the same sign; from Eq. (34.2) the lateral magnification of a plane mirror is always $m = +1$. Later we will encounter situations in which the image is **inverted;** that is, the image arrow points in the direction *opposite* to that of the object arrow. For an inverted image, y and y' have *opposite* signs, and the lateral magnification m is *negative.*

The object in Fig. 34.6 has only one dimension. Figure 34.7 shows a *three-dimensional* object and its three-dimensional virtual image formed by a plane mirror. The object and image are related in the same way as a left hand and a right hand.

34.7 The image formed by a plane mirror is virtual, erect, and reversed. It is the same size as the object.

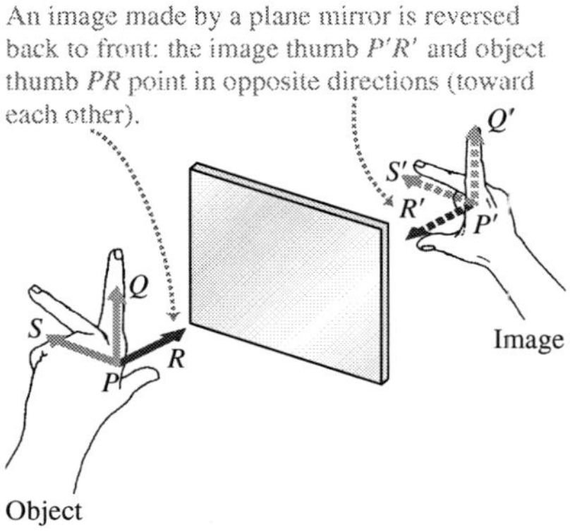

34.8 The image formed by a plane mirror is reversed; the image of a right hand is a left hand, and so on. (The hand is resting on a horizontal mirror.) Are images of the letters H and A reversed?

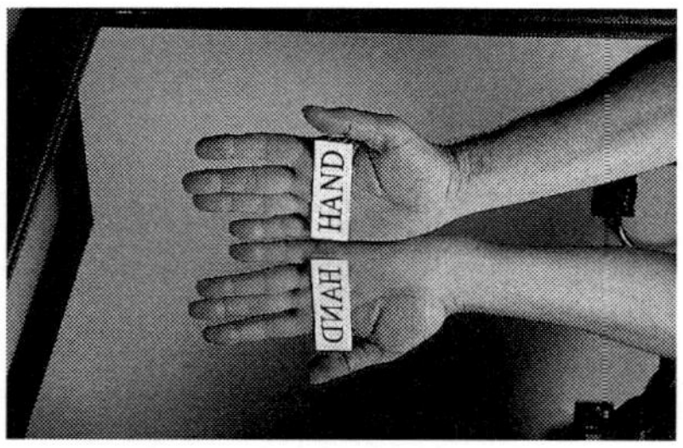

CAUTION **Reflections in a plane mirror** At this point, you may be asking, "Why does a plane mirror reverse images left and right but not top and bottom?" This question is quite misleading! As Fig. 34.7 shows, the up-down image $P'Q'$ and the left-right image $P'S'$ are parallel to their objects and are not reversed at all. Only the front-back image $P'R'$ is reversed relative to PR. Hence it's most correct to say that a plane mirror reverses *back to front.* To verify this object-image relationship, point your thumbs along PR and $P'R'$, your forefingers along PQ and $P'Q'$, and your middle fingers along PS and $P'S'$. When an object and its image are related in this way, the image is said to be **reversed;** this means that *only* the front-back dimension is reversed.

The reversed image of a three-dimensional object formed by a plane mirror is the same *size* as the object in all its dimensions. When the transverse dimensions of object and image are in the same direction, the image is erect. Thus a plane mirror always forms an erect but reversed image. Figure 34.8 illustrates this point.

An important property of all images formed by reflecting or refracting surfaces is that an *image* formed by one surface or optical device can serve as the *object* for a second surface or device. Figure 34.9 shows a simple example. Mirror 1 forms an image P_1' of the object point P, and mirror 2 forms another image P_2', each in the way we have just discussed. But in addition, the image P_1' formed by mirror 1 serves an object for mirror 2, which then forms an image of this object at point P_3' as shown. Similarly, mirror 1 uses the image P_2' formed by mirror 2 as an object and forms an image of it. We leave it to you to show that this image point is also at P_3'. The idea that an image formed by one device can act as the object for a second device is of great importance in geometric optics. We will use it later in this chapter to locate the image formed by two successive curved-surface refractions in a lens. This idea will help us to understand image formation by combinations of lenses, as in a microscope or a refracting telescope.

34.9 Images P_1' and P_2' are formed by a single reflection of each ray from the object at P. Image P_3', located by treating either of the other images as an object, is formed by a double reflection of each ray.

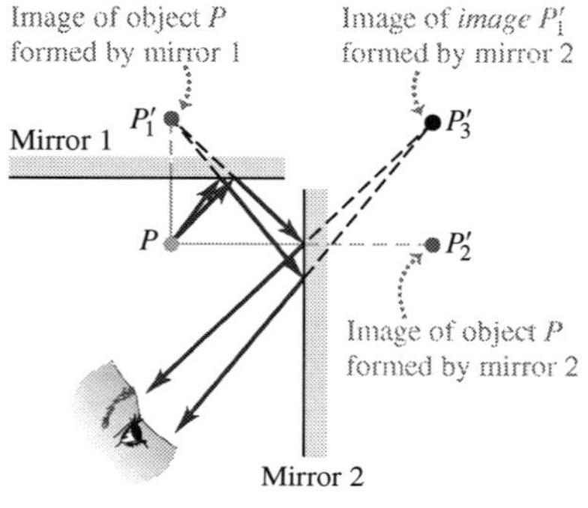

Test Your Understanding of Section 34.1 If you walk directly toward a plane mirror at a speed v, at what speed does your image approach you? (i) slower than v; (ii) v; (iii) faster than v but slower than $2v$; (iv) $2v$; (v) faster than $2v$.

34.2 Reflection at a Spherical Surface

A plane mirror produces an image that is the same size as the object. But there are many applications for mirrors in which the image and object must be of different sizes. A magnifying mirror used when applying makeup gives an image that is *larger* than the object, and surveillance mirrors (used in stores to help spot shoplifters) give an image that is *smaller* than the object. There are also applications of mirrors in which a *real* image is desired, so light rays do indeed pass through the image point P'. A plane mirror by itself cannot perform any of these tasks. Instead, *curved* mirrors are used.

34.10 (a) A concave spherical mirror forms a real image of a point object P on the mirror's optic axis. (b) The eye sees some of the outgoing rays and perceives them as having come from P'.

(a) Construction for finding the position P' of an image formed by a concave spherical mirror

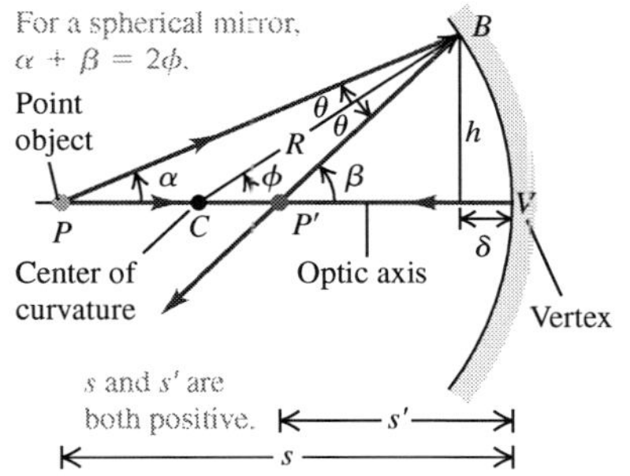

(b) The paraxial approximation, which holds for rays with small α

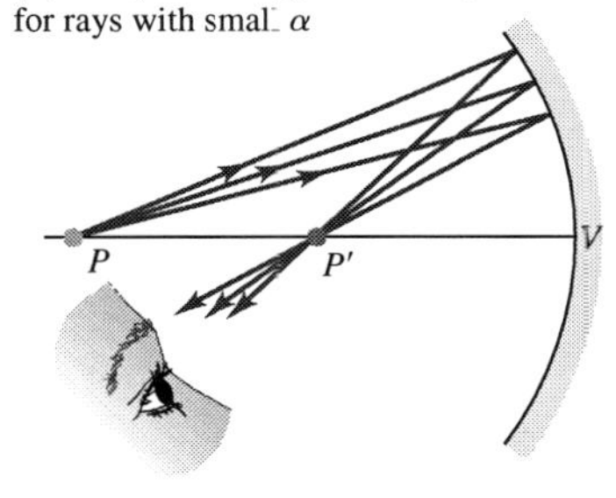

All rays from P that have a small angle α pass through P', forming a real image.

34.11 The sign rule for the radius of a spherical mirror.

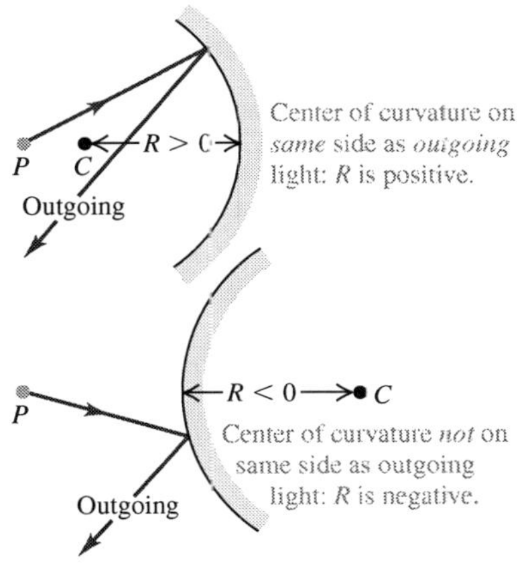

Image of a Point Object: Spherical Mirror

We'll consider the special (and easily analyzed) case of image formation by a *spherical* mirror. Figure 34.10a shows a spherical mirror with radius of curvature R, with its concave side facing the incident light. The **center of curvature** of the surface (the center of the sphere of which the surface is a part) is at C, and the **vertex** of the mirror (the center of the mirror surface) is at V. The line CV is called the **optic axis.** Point P is an object point that lies on the optic axis; for the moment, we assume that the distance from P to V is greater than R.

Ray PV, passing through C, strikes the mirror normally and is reflected back on itself. Ray PB, at an angle α with the axis, strikes the mirror at B, where the angles of incidence and reflection are θ. The reflected ray intersects the axis at point P'. We will show shortly that *all* rays from P intersect the axis at the *same* point P', as in Fig. 34.10b, provided that the angle α is small. Point P' is therefore the *image* of object point P. Unlike the reflected rays in Fig. 34.1, the reflected rays in Fig. 34.10b actually do intersect at point P', then diverge from P' *as if* they had originated at this point. Thus P' is a *real* image.

To see the usefulness of having a real image, suppose that the mirror is in a darkened room in which the only source of light is a self-luminous object at P. If you place a small piece of photographic film at P', all the rays of light coming from point P that reflect off the mirror will strike the same point P' on the film; when developed, the film will show a single bright spot, representing a sharply focused image of the object at point P. This principle is at the heart of most astronomical telescopes, which use large concave mirrors to make photographs of celestial objects. With a *plane* mirror like that in Fig. 34.2, placing a piece of film at the image point P' would be a waste of time; the light rays never actually pass through the image point, and the image can't be recorded on film. Real images are *essential* for photography.

Let's now find the location of the real image point P' in Fig. 34.10a and prove the assertion that all rays from P intersect at P' (provided that their angle with the optic axis is small). The object distance, measured from the vertex V, is s; the image distance, also measured from V, is s'. The signs of s, s', and the radius of curvature R are determined by the sign rules given in Section 34.1. The object point P is on the same side as the incident light, so according to sign rule 1, s is positive. The image point P' is on the same side as the reflected light, so according to sign rule 2, the image distance s' is also positive. The center of curvature C is on the same side as the reflected light, so according to sign rule 3, R, too, is positive; R is always positive when reflection occurs at the *concave* side of a surface Fig. 34.11).

We now use the following theorem from plane geometry: An exterior angle of a triangle equals the sum of the two opposite interior angles. Applying this theorem to triangles PBC and $P'BC$ in Fig. 34.10a, we have

$$\phi = \alpha + \theta \qquad \beta = \phi + \theta$$

Eliminating θ between these equations gives

$$\alpha + \beta = 2\phi \tag{34.3}$$

We may now compute the image distance s'. Let h represent the height of point B above the optic axis, and let δ represent the short distance from V to the foot of this vertical line. We now write expressions for the tangents of α, β, and ϕ, remembering that s, s', and R are all positive quantities:

$$\tan\alpha = \frac{h}{s - \delta} \qquad \tan\beta = \frac{h}{s' - \delta} \qquad \tan\phi = \frac{h}{R - \delta}$$

These trigonometric equations cannot be solved as simply as the corresponding algebraic equations for a plane mirror. However, *if the angle α is small,* the angles β and ϕ are also small. The tangent of an angle that is much less than one radian is nearly equal to the angle itself (measured in radians), so we can replace $\tan\alpha$ by α, and so on, in the equations above. Also, if α is small, we can neglect the distance δ compared with s', s, and R. So for small angles we have the following approximate relationships:

$$\alpha = \frac{h}{s} \qquad \beta = \frac{h}{s'} \qquad \phi = \frac{h}{R}$$

Substituting these into Eq. (34.3) and dividing by h, we obtain a general relationship among s, s', and R:

$$\frac{1}{s} + \frac{1}{s'} = \frac{2}{R} \quad \text{(object–image relationship, spherical mirror)} \tag{34.4}$$

This equation does not contain the angle α. Hence *all* rays from P that make sufficiently small angles with the axis intersect at P' after they are reflected; this verifies our earlier assertion. Such rays, nearly parallel to the axis and close to it, are called **paraxial rays.** (The term **paraxial approximation** is often used for the approximations we have just described.) Since all such reflected light rays converge on the image point, a concave mirror is also called a *converging mirror.*

Be sure you understand that Eq. (34.4), as well as many similar relationships that we will derive later in this chapter and the next, is only *approximately* correct. It results from a calculation containing approximations, and it is valid only for paraxial rays. If we increase the angle α that a ray makes with the optic axis, the point P' where the ray intersects the optic axis moves somewhat closer to the vertex than for a paraxial ray. As a result, a spherical mirror, unlike a plane mirror, does not form a precise point image of a point object; the image is "smeared out." This property of a spherical mirror is called **spherical aberration.** When the primary mirror of the Hubble Space Telescope (Fig. 34.12a) was manufactured, tiny errors were made in its shape that led to an unacceptable amount of spherical aberration (Fig. 34.12b). The performance of the telescope improved dramatically after the installation of corrective optics (Fig. 34.12c).

If the radius of curvature becomes infinite $(R = \infty)$, the mirror becomes *plane,* and Eq. (34.4) reduces to Eq. (34.1) for a plane reflecting surface.

34.12 (a), (b) Soon after the Hubble Space Telescope (HST) was placed in orbit in 1990, it was discovered that the concave primary mirror (also called the *objective mirror*) was too shallow by about 1/50 the width of a human hair, leading to spherical aberration of the star's image. (c) After corrective optics were installed in 1993, the effects of spherical aberration were almost completely eliminated.

(a) The 2.4-m-diameter primary mirror of the Hubble Space Telescope

(b) A star seen with the original mirror

(c) The same star with corrective optics

Focal Point and Focal Length

When the object point P is very far from the spherical mirror $(s = \infty)$, the incoming rays are parallel. (The star shown in Fig. 34.12c is an example of such a distant object.) From Eq. (34.4) the image distance s' in this case is given by

$$\frac{1}{\infty} + \frac{1}{s'} = \frac{2}{R} \qquad s' = \frac{R}{2}$$

The situation is shown in Fig. 34.13a. The beam of incident parallel rays converges, after reflection from the mirror, to a point F at a distance $R/2$ from the vertex of the mirror. The point F at which the incident parallel rays converge is called the **focal point;** we say that these rays are brought to a focus. The distance from the vertex to the focal point, denoted by f, is called the **focal length.** We see that f is related to the radius of curvature R by

$$f = \frac{R}{2} \quad \text{(focal length of a spherical mirror)} \qquad (34.5)$$

The opposite situation is shown in Fig. 34.13b. Now the *object* is placed at the focal point F, so the object distance is $s = f = R/2$. The image distance s' is again given by Eq. (34.4):

$$\frac{2}{R} + \frac{1}{s'} = \frac{2}{R} \qquad \frac{1}{s'} = 0 \qquad s' = \infty$$

With the object at the focal point, the reflected rays in Fig. 34.13b are parallel to the optic axis; they meet only at a point infinitely far from the mirror, so the image is at infinity.

Thus the focal point F of a spherical mirror has the properties that (1) any incoming ray parallel to the optic axis is reflected through the focal point and (2) any incoming ray that passes through the focal point is reflected parallel to the optic axis. For spherical mirrors these statements are true only for paraxial rays. For parabolic mirrors these statements are *exactly* true; this is why parabolic mirrors are preferred for astronomical telescopes. Spherical or parabolic mirrors are used in flashlights and headlights to form the light from the bulb into a parallel beam. Some solar-power plants use an array of plane mirrors to simulate an approximately spherical concave mirror; light from the sun is collected by the mirrors and directed to the focal point, where a steam boiler is placed. (The concepts of focal point and focal length also apply to lenses, as we'll see in Section 34.4.)

We will usually express the relationship between object and image distances for a mirror, Eq. (34.4), in terms of the focal length f:

$$\frac{1}{s} + \frac{1}{s'} = \frac{1}{f} \quad \text{(object–image relationship, spherical mirror)} \qquad (34.6)$$

34.13 The focal point and focal length of a concave mirror.

(a) All parallel rays incident on a spherical mirror reflect through the focal point.

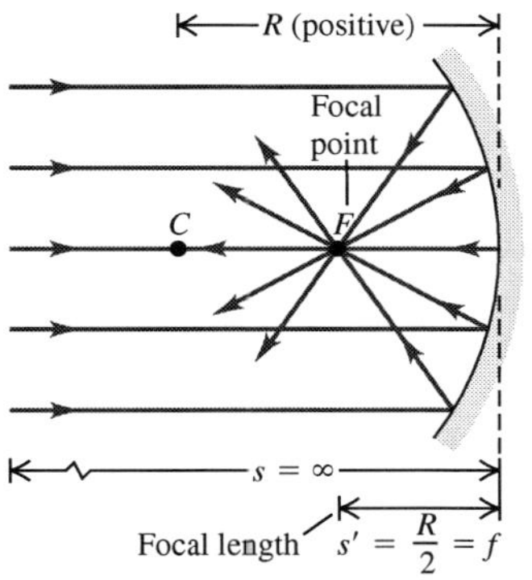

(b) Rays diverging from the focal point reflect to form parallel outgoing rays.

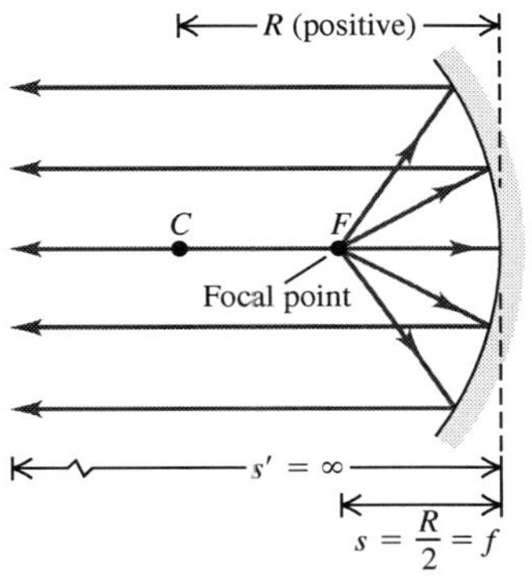

Image of an Extended Object: Spherical Mirror

Now suppose we have an object with *finite* size, represented by the arrow PQ in Fig. 34.14, perpendicular to the optic axis CV. The image of P formed by paraxial rays is at P'. The object distance for point Q is very nearly equal to that for point P, so the image $P'Q'$ is nearly straight and perpendicular to the axis. Note that the object and image arrows have different sizes, y and y', respectively, and that they have opposite orientation. In Eq. (34.2) we defined the *lateral magnification m* as the ratio of image size y' to object size y:

$$m = \frac{y'}{y}$$

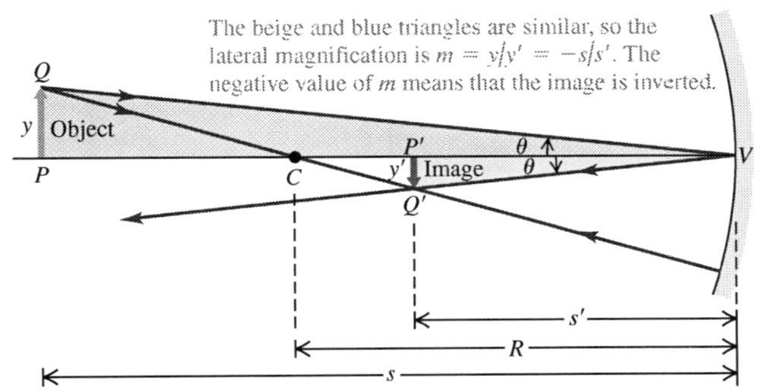

34.14 Construction for determining the position, orientation, and height of an image formed by a concave spherical mirror.

ActivPhysics ONLINE

Because triangles PVQ and $P'VQ'$ in Fig. 34.14 are *similar,* we also have the relationship $y/s = -y'/s'$. The negative sign is needed because object and image are on opposite sides of the optic axis; if y is positive, y' is negative. Therefore

$$m = \frac{y'}{y} = -\frac{s'}{s} \qquad \text{(lateral magnification, spherical mirror)} \tag{34.7}$$

If m is positive, the image is erect in comparison to the object; if m is negative, the image is *inverted* relative to the object, as in Fig. 34.14. For a *plane* mirror, $s = -s'$, so $y' = y$ and $m = +1$; since m is positive, the image is erect, and since $|m| = 1$, the image is the same size as the object.

CAUTION **Lateral magnification can be less than 1** Although the ratio of image size to object size is called the *lateral magnification,* the image formed by a mirror or lens may be larger than, smaller than, or the same size as the object. If it is smaller, then the lateral magnification is less than unity in absolute value: $|m| < 1$. The image formed by an astronomical telescope mirror or a camera lens is usually *much* smaller than the object. For example, the image of the bright star shown in Fig. 34.12c is just a few millimeters across, while the star itself is hundreds of thousands of kilometers in diameter.

In our discussion of concave mirrors we have so far considered only objects that lie *outside* or at the focal point, so that the object distance s is greater than or equal to the (positive) focal length f. In this case the image point is on the same side of the mirror as the outgoing rays, and the image is real and inverted. If an object is placed *inside* the focal point of a concave mirror, so that $s < f$, the resulting image is *virtual* (that is, the image point is on the opposite side of the mirror from the object), *erect,* and *larger* than the object. Mirrors used when applying makeup (referred to at the beginning of this section) are concave mirrors; in use, the distance from the face to the mirror is less than the focal length, and an enlarged, erect image is seen. You can prove these statements about concave mirrors by applying Eqs. (34.6) and (34.7) (see Exercise 34.11). We'll also be able to verify these results later in this section, after we've learned some graphical methods for relating the positions and sizes of the object and the image.

Example 34.1 Image formation by a concave mirror I

A concave mirror forms an image, on a wall 3.00 m from the mirror, of the filament of a headlight lamp 10.0 cm in front of the mirror. (a) What are the radius of curvature and focal length of the mirror? (b) What is the height of the image if the height of the object is 5.00 mm?

SOLUTION

IDENTIFY: This problem uses the ideas developed in this section. Our target variables are the radius of curvature R, focal length f, and image height y'.

SET UP: Figure 34.15 shows the situation. We are given the distances from the mirror to the object (s) and from the mirror to the image (s'). We use the object–image relationship given by Eq. (34.6) to determine the focal length f, and then find the radius of curvature R using Eq. (34.5). Equation (34.7) lets us calculate the image height y' from the distances s and s' and the object height y.

EXECUTE: (a) Both the object and the image are on the concave side of the mirror (the reflective side), so both object distance and

34.15 Our sketch for this problem.

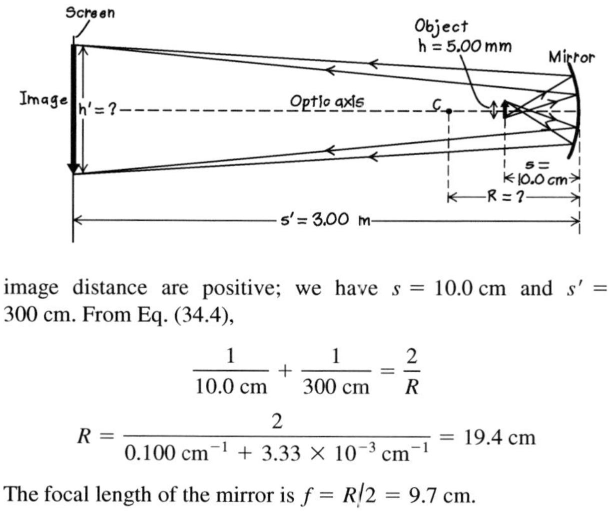

image distance are positive; we have $s = 10.0$ cm and $s' = 300$ cm. From Eq. (34.4),

$$\frac{1}{10.0\text{ cm}} + \frac{1}{300\text{ cm}} = \frac{2}{R}$$

$$R = \frac{2}{0.100\text{ cm}^{-1} + 3.33 \times 10^{-3}\text{ cm}^{-1}} = 19.4\text{ cm}$$

The focal length of the mirror is $f = R/2 = 9.7$ cm.

(b) From Eq. (34.7) the lateral magnification is

$$m = \frac{y'}{y} = -\frac{s'}{s} = -\frac{300\text{ cm}}{10.0\text{ cm}} = -30.0$$

Because m is negative, the image is inverted. The height of the image is 30.0 times the height of the object, or $(30.0)(5.00\text{ mm}) = 150\text{ mm}$.

EVALUATE: Note that the object is placed just outside the focal point ($s = 10.0$ cm compared to $f = 9.7$ cm). This is very similar to what is done in automobile headlights. With the filament close to the focal point, the concave mirror produces a beam of nearly parallel rays.

Conceptual Example 34.2 Image formation by a concave mirror II

In Example 34.1, suppose that the left half of the mirror's reflecting surface is covered with nonreflective soot. What effect will this have on the image of the filament?

SOLUTION

It would be natural to guess that the image would now show only half of the filament. But in fact the image will still show the *entire* filament. The explanation can be seen by examining Fig. 34.10b. Light rays coming from any object point P are reflected from *all* parts of the mirror and converge on the corresponding image point P'. If part of the mirror surface is made nonreflective or is removed altogether, the light rays from the remaining reflective surface still form an image of every part of the object.

The only effect of reducing the reflecting area is that the image becomes dimmer because less light energy reaches the image point. In our example the reflective area of the mirror is reduced by one-half, and the image will be one-half as bright. *Increasing* the reflective area makes the image brighter. To make reasonably bright images of distant stars, astronomical telescopes use mirrors that are up to several meters in diameter. Figure 34.12a shows an example.

Convex Mirrors

In Fig. 34.16a the *convex* side of a spherical mirror faces the incident light. The center of curvature is on the side opposite to the outgoing rays; according to sign rule 3 in Section 34.1, R is negative (see Fig. 34.11). Ray PB is reflected, with the angles of incidence and reflection both equal to θ. The reflected ray, projected backward, intersects the axis at P'. As with a concave mirror, *all* rays from P that are reflected by the mirror diverge from the same point P', provided that the angle α is small. Therefore P' is the image of P. The object distance s is positive, the image distance s' is negative, and the radius of curvature R is *negative* for a *convex* mirror.

Figure 34.16b shows two rays diverging from the head of the arrow PQ and the virtual image $P'Q'$ of this arrow. The same procedure that we used for a concave mirror can be used to show that for a convex mirror,

$$\frac{1}{s} + \frac{1}{s'} = \frac{2}{R}$$

and the lateral magnification is

$$m = \frac{y'}{y} = -\frac{s'}{s}$$

34.16 Image formation by a convex mirror.

(a) Construction for finding the position of an image formed by a convex mirror

(b) Construction for finding the magnification of an image formed by a convex mirror

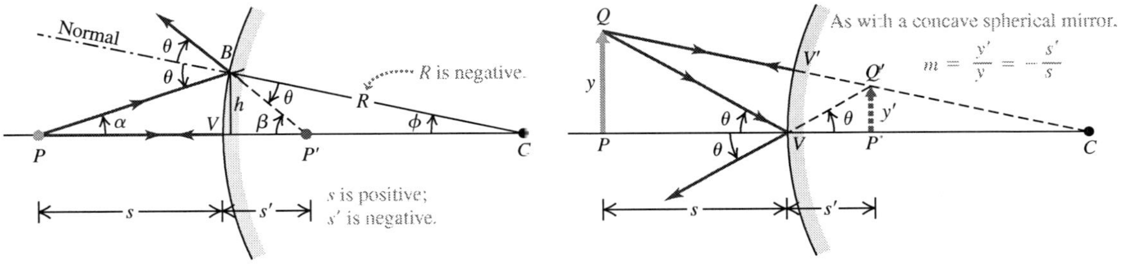

34.17 The focal point and focal length of a convex mirror.

(a) Paraxial rays incident on a convex spherical mirror diverge from a virtual focal point.

(b) Rays aimed at the virtual focal point are parallel to the axis after reflection.

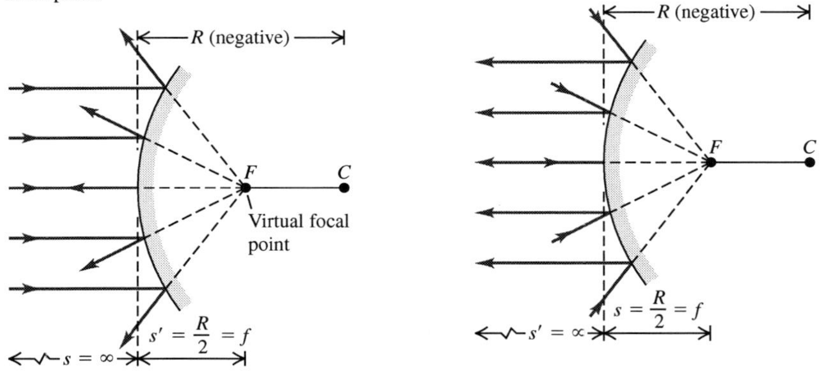

These expressions are exactly the same as Eqs. (34.4) and (34.7) for a concave mirror. Thus when we use our sign rules consistently, Eqs. (34.4) and (34.7) are valid for both concave and convex mirrors.

When R is negative (convex mirror), incoming rays that are parallel to the optic axis are not reflected through the focal point F. Instead, they diverge as though they had come from the point F at a distance f *behind* the mirror, as shown in Fig. 34.17a. In this case, f is the focal length, and F is called a *virtual focal point.* The corresponding image distance s' is negative, so both f and R are negative, and Eq. (34.5), $f = R/2$, holds for convex as well as concave mirrors. In Fig. 34.17b the incoming rays are converging as though they would meet at the virtual focal point F, and they are reflected parallel to the optic axis.

In summary, Eqs. (34.4) through (34.7), the basic relationships for image formation by a spherical mirror, are valid for both concave and convex mirrors, provided that we use the sign rules consistently.

Example 34.3 Santa's image problem

Santa checks himself for soot, using his reflection in a shiny silvered Christmas tree ornament 0.750 m away (Fig. 34.18a). The diameter of the ornament is 7.20 cm. Standard reference works state that he is a "right jolly old elf," so we estimate his height to be 1.6 m. Where and how tall is the image of Santa formed by the ornament? Is it erect or inverted?

SOLUTION

IDENTIFY: Santa is the object, and the surface of the ornament closest to Santa acts as a convex mirror. The relationships among object distance, image distance, focal length, and magnification are the same as for concave mirrors, provided we use the sign rules consistently.

34.18 (a) The ornament forms a virtual, reduced, erect image of Santa. (b) Our sketch of two of the rays forming the image.

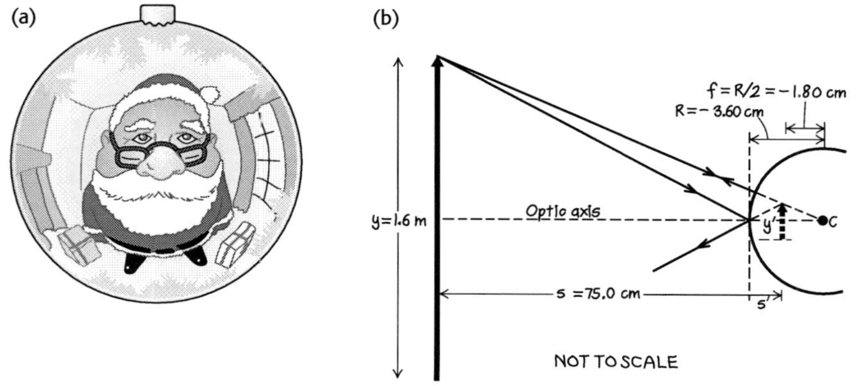

SET UP: Figure 34.18b shows the situation. Since the mirror is convex, its radius of curvature and focal length are negative. The object distance is $s = 0.750\text{ m} = 75.0\text{ cm}$ and Santa's height is $y = 1.6\text{ m}$. We use Eq. (34.6) to determine the image distance s', and then use Eq. (34.7) to find the lateral magnification m and hence the image height y'. The sign of m tells us whether the image is erect or inverted.

EXECUTE: The radius of the convex mirror (half the diameter) is $R = -(7.20\text{ cm})/2 = -3.60\text{ cm}$, and the focal length is $f = R/2 = -1.80\text{ cm}$. From Eq. (34.6),

$$\frac{1}{s'} = \frac{1}{f} - \frac{1}{s} = \frac{1}{-1.80\text{ cm}} - \frac{1}{75.0\text{ cm}}$$

$$s' = -1.76\text{ cm}$$

Because s' is negative, the image is behind the mirror—that is, on the side opposite to the outgoing light (Fig. 34.18b)—and it is virtual. The image is about halfway between the front surface of the ornament and its center.

The lateral magnification m is given by Eq. (34.7):

$$m = \frac{y'}{y} = -\frac{s'}{s} = -\frac{-1.76\text{ cm}}{75.0\text{ cm}} = 0.0234$$

Because m is positive, the image is erect. It is only about 0.0234 as tall as Santa himself:

$$y' = my = (0.0234)(1.6\text{ m}) = 3.8 \times 10^{-2}\text{ m} = 3.8\text{ cm}$$

EVALUATE: When the object distance s is positive, a convex mirror *always* forms an erect, virtual, reduced, reversed image. For this reason, convex mirrors are used for shoplifting surveillance in stores, at blind intersections, and as "wide-angle" rear-view mirrors for cars and trucks (including those that bear the legend "Objects in mirror are closer than they appear").

Graphical Methods for Mirrors

In Examples 34.1 and 34.3, we used Eqs. (34.6) and (34.7) to find the position and size of the image formed by a mirror. We can also determine the properties of the image by a simple *graphical* method. This method consists of finding the point of intersection of a few particular rays that diverge from a point of the object (such as point Q in Fig. 34.19) and are reflected by the mirror. Then (neglecting aberrations) *all* rays from this object point that strike the mirror will intersect at the same point. For this construction we always choose an object point that is *not* on the optic axis. Four rays that we can usually draw easily are shown in Fig. 34.19. These are called **principal rays.**

1. *A ray parallel to the axis,* after reflection, passes through the focal point F of a concave mirror or appears to come from the (virtual) focal point of a convex mirror.
2. *A ray through (or proceeding toward) the focal point F* is reflected parallel to the axis.
3. *A ray along the radius* through or away from the center of curvature C intersects the surface normally and is reflected back along its original path.
4. *A ray to the vertex V* is reflected forming equal angles with the optic axis.

34.19 The graphical method of locating an image formed by spherical mirror. The colors of the rays are for identification only; they do not refer to specific colors of light.

(a) Principal rays for concave mirror

(b) Principal rays for convex mirror

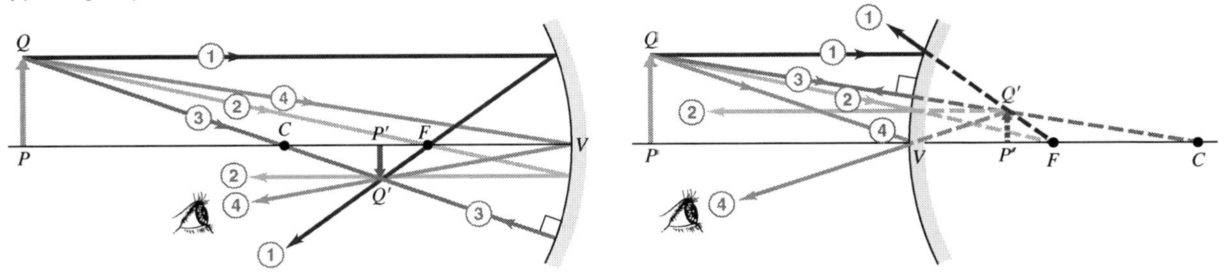

(1) Ray parallel to axis reflects through focal point.
(2) Ray through focal point reflects parallel to axis.
(3) Ray through center of curvature intersects the surface normally and reflects along its original path.
(4) Ray to vertex reflects symmetrically around optic axis.

(1) Reflected parallel ray appears to come from focal point.
(2) Ray toward focal point reflects parallel to axis.
(3) As with concave mirror: Ray radial to center of curvature intersects the surface normally and reflects along its original path.
(4) As with concave mirror: Ray to vertex reflects symmetrically around optic axis.

Once we have found the position of the image point by means of the intersection of any two of these principal rays (1, 2, 3, 4), we can draw the path of any other ray from the object point to the same image point.

CAUTION **Principal rays are not the only rays** Although we've emphasized the principal rays, in fact *any* ray from the object that strikes the mirror will pass through the image point (for a real image) or appear to originate from the image point (for a virtual image). Usually, you only need to draw the principal rays, because these are all you need to locate the image.

Problem-Solving Strategy 34.1 Image Formation by Mirrors

IDENTIFY *the relevant concepts:* There are two different and complementary ways to solve problems involving image formation by mirrors. One approach uses equations, while the other involves drawing a principal-ray diagram. A successful problem solution uses *both* approaches.

SET UP *the problem:* Determine the target variables. The three key quantities are the focal length, object distance, and image distance; typically you'll be given two of these and will have to determine the third.

EXECUTE *the solution* as follows:

1. The principal-ray diagram is to geometric optics what the free-body diagram is to mechanics. In any problem involving image formation by a mirror, *always* draw a principal-ray diagram first if you have enough information. (The same advice should be followed when dealing with lenses in the following sections.)
2. It is usually best to orient your diagrams consistently with the incoming rays traveling from left to right. Don't draw a lot of other rays at random; stick with the principal rays, the ones you know something about. Use a ruler and measure distances carefully! A freehand sketch will *not* give good results.
3. If your principal rays don't converge at a real image point, you may have to extend them straight backward to locate a virtual image point, as in Fig. 34.19b. We recommend drawing the extensions with broken lines. Another useful aid is to color-code the different principal rays, as is done in Fig. 34.19.
4. Check your results using Eq. (34.6), $1/s + 1/s' = 1/f$, and the lateral magnification equation, Eq. (34.7). The results you find using this equation must be consistent with your principal-ray diagram; if not, double check both your calculations and your diagram.
5. Pay careful attention to signs on object and image distances, radii of curvature, and object and image heights. A negative sign on any of these quantities *always* has significance. Use the equations and the sign rules carefully and consistently, and they will tell you the truth! Note that the *same* sign rules (given in Section 34.1) work for all four cases in this chapter: reflection and refraction from plane and spherical surfaces.

EVALUATE *your answer:* You've already checked your results by using both diagrams and equations. But it always helps to take a look back and ask yourself, "Do these results make sense?"

Example 34.4 Concave mirror, different object distances

A concave mirror has a radius of curvature with absolute value 20 cm. Find graphically the image of an object in the form of an arrow perpendicular to the axis of the mirror at each of the following object distances: (a) 30 cm, (b) 20 cm, (c) 10 cm, and (d) 5 cm. Check the construction by *computing* the size and lateral magnification of each image.

SOLUTION

IDENTIFY: This problem asks us to use *both* graphical methods and calculations to find the image made by a mirror. This is a good practice to follow in all problems that involve image formation.

SET UP: We are given the radius of curvature $R = 20$ cm (positive since the mirror is concave) and hence the focal length $f = R/2 = 10$ cm. In each case we are told the object distance s and are asked to find the image distance s' and the lateral magnification $m = -s'/s$.

EXECUTE: Figure 34.20 shows the principal-ray diagrams for the four cases. Study each of these diagrams carefully, comparing each numbered ray with the description above. Several points are worth noting. First, in (b) the object and image distances are equal. Ray 3 cannot be drawn in this case because a ray from Q through the center of curvature C does not strike the mirror. Ray 2 cannot be drawn in (c) because a ray from Q toward F also does not strike the mirror. In this case the outgoing rays are parallel, corresponding to an infinite image distance. In (d) the outgoing rays have no real intersection point; they must be extended backward to find the point from which they appear to diverge—that is, from the *virtual image point* Q'. The case shown in (d) illustrates the general observation that an object placed inside the focal point of a concave mirror produces a virtual image.

Measurements of the figures, with appropriate scaling, give the following approximate image distances: (a) 15 cm; (b) 20 cm; (c) ∞ or $-\infty$ (because the outgoing rays are parallel and do not converge at any finite distance); (d) -10 cm. To *compute* these distances, we use Eq. (34.6) with $f = 10$ cm:

$$\text{(a)}\quad \frac{1}{30\text{ cm}} + \frac{1}{s'} = \frac{1}{10\text{ cm}} \qquad s' = 15\text{ cm}$$

$$\text{(b)}\quad \frac{1}{20\text{ cm}} + \frac{1}{s'} = \frac{1}{10\text{ cm}} \qquad s' = 20\text{ cm}$$

(c) $\dfrac{1}{10\text{ cm}} + \dfrac{1}{s'} = \dfrac{1}{10\text{ cm}}$ $\quad s' = \infty\ (\text{or } -\infty)$

(d) $\dfrac{1}{5\text{ cm}} + \dfrac{1}{s'} = \dfrac{1}{10\text{ cm}}$ $\quad s' = -10\text{ cm}$

In (a) and (b) the image is real; in (d) it is virtual. In (c) the image is formed at infinity.

The lateral magnifications measured from the figures are approximately (a) $-\frac{1}{2}$; (b) -1; (c) ∞ or $-\infty$ (because the image distance is infinite); (d) $+2$. *Computing* the magnifications from Eq. (34.7), we find:

(a) $m = -\dfrac{15\text{ cm}}{30\text{ cm}} = -\dfrac{1}{2}$

(b) $m = -\dfrac{20\text{ cm}}{20\text{ cm}} = -1$

(c) $m = -\dfrac{\infty\text{ cm}}{10\text{ cm}} = -\infty\ (\text{or } +\infty)$

(d) $m = -\dfrac{-10\text{ cm}}{5\text{ cm}} = +2$

In (a) and (b) the image is inverted; in (d) it is erect.

EVALUATE: Notice the trend as the object is moved closer to the mirror. When the object is far from the mirror, as in Fig. 34.20a, the image is smaller than the object, inverted, and real. As the object distance decreases, the image moves farther from the mirror and increases in size (Fig. 34.20b). When the object is at the focal point, the image is at infinity (Fig. 34.20c). If the object is moved inside the focal point, the image becomes larger than the object, erect, and virtual (Fig. 34.20d). You can test these conclusions by looking at objects reflected in the concave bowl of a metal spoon.

34.20 Using principal-ray diagrams to locate the image $P'Q'$ made by a concave mirror.

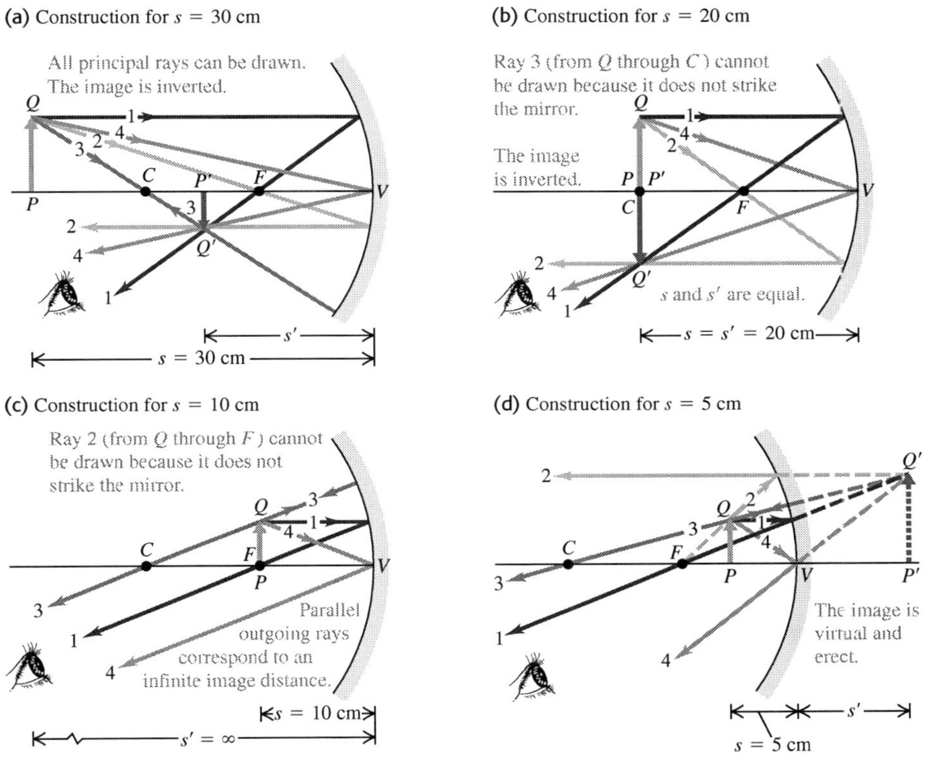

Test Your Understanding of Section 34.2 A cosmetics mirror is designed so that your reflection appears right-side up and enlarged. (a) Is the mirror concave or convex? (b) To see an enlarged image, what should be the distance from the mirror (of focal length f) to your face? (i) $|f|$; (ii) less than $|f|$; (iii) greater than $|f|$.

34.3 Refraction at a Spherical Surface

As we mentioned in Section 34.1, images can be formed by refraction as well as by reflection. To begin with, let's consider refraction at a spherical surface—that is, at a spherical interface between two optical materials with different indexes of refraction. This analysis is directly applicable to some real optical systems, such as the human eye. It also provides a stepping-stone for the analysis of lenses, which usually have *two* spherical (or nearly spherical) surfaces.

Image of a Point Object: Spherical Refracting Surface

In Fig. 34.21 a spherical surface with radius R forms an interface between two materials with different indexes of refraction n_a and n_b. The surface forms an image P' of an object point P; we want to find how the object and image distances (s and s') are related. We will use the same sign rules that we used for spherical mirrors. The center of curvature C is on the outgoing side of the surface, so R is positive. Ray PV strikes the vertex V and is perpendicular to the surface (that is, to the plane that is tangent to the surface at the point of incidence V). It passes into the second material without deviation. Ray PB, making an angle α with the axis, is incident at an angle θ_a with the normal and is refracted at an angle θ_b. These rays intersect at P', a distance s' to the right of the vertex. The figure is drawn for the case $n_a < n_b$. The object and image distances are both positive.

We are going to prove that if the angle α is small, *all* rays from P intersect at the same point P', so P' is the *real image* of P. We use much the same approach as we did for spherical mirrors in Section 34.2. We again use the theorem that an exterior angle of a triangle equals the sum of the two opposite interior angles; applying this to the triangles PBC and $P'BC$ gives

$$\theta_a = \alpha + \phi \qquad \phi = \beta + \theta_b \tag{34.8}$$

From the law of refraction,

$$n_a \sin\theta_a = n_b \sin\theta_b$$

Also, the tangents of α, β, and ϕ are

$$\tan\alpha = \frac{h}{s + \delta} \qquad \tan\beta = \frac{h}{s' - \delta} \qquad \tan\phi = \frac{h}{R - \delta} \tag{34.9}$$

For paraxial rays, θ_a and θ_b are both small in comparison to a radian, and we may approximate both the sine and tangent of either of these angles by the angle itself (measured in radians). The law of refraction then gives

$$n_a\theta_a = n_b\theta_b$$

Combining this with the first of Eqs. (34.8), we obtain

$$\theta_b = \frac{n_a}{n_b}(\alpha + \phi)$$

When we substitute this into the second of Eqs. (34.8), we get

$$n_a\alpha + n_b\beta = (n_b - n_a)\phi \tag{34.10}$$

Now we use the approximations $\tan\alpha = \alpha$, and so on, in Eqs. (34.9) and also neglect the small distance δ; those equations then become

$$\alpha = \frac{h}{s} \qquad \beta = \frac{h}{s'} \qquad \phi = \frac{h}{R}$$

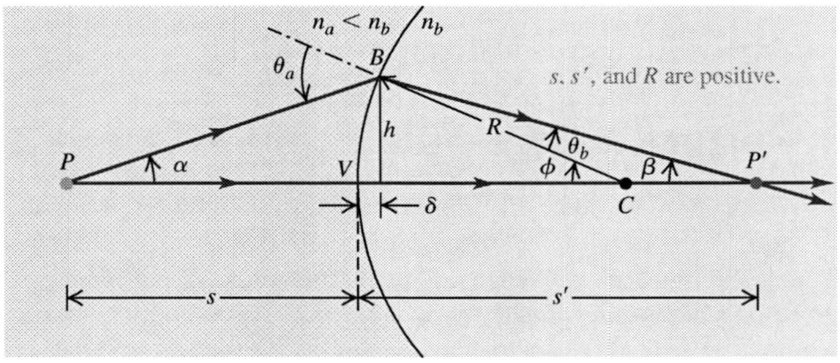

34.21 Construction for finding the position of the image point P' of a point object P formed by refraction at a spherical surface. The materials to the left and right of the interface have refractive indexes n_a and n_b, respectively. In the case shown here, $n_a < n_b$.

Finally, we substitute these into Eq. (34.10) and divide out the common factor h. We obtain

$$\frac{n_a}{s} + \frac{n_b}{s'} = \frac{n_b - n_a}{R} \quad \text{(object–image relationship, spherical refracting surface)} \tag{34.11}$$

This equation does not contain the angle α, so the image distance is the same for *all* paraxial rays emanating from P; this proves our assertion that P' is the image of P.

To obtain the lateral magnification m for this situation, we use the construction in Fig. 34.22. We draw two rays from point Q, one through the center of curvature C and the other incident at the vertex V. From the triangles PQV and $P'Q'V$,

$$\tan\theta_a = \frac{y}{s} \qquad \tan\theta_b = \frac{-y'}{s'}$$

and from the law of refraction,

$$n_a \sin\theta_a = n_b \sin\theta_b$$

For small angles,

$$\tan\theta_a = \sin\theta_a \qquad \tan\theta_b = \sin\theta_b$$

so finally

$$\frac{n_a y}{s} = -\frac{n_b y'}{s'} \qquad \text{or}$$

$$m = \frac{y'}{y} = -\frac{n_a s'}{n_b s} \quad \text{(lateral magnification, spherical refracting surface)} \tag{34.12}$$

Equations (34.11) and (34.12) can be applied to both convex and concave refracting surfaces, provided that you use the sign rules consistently. It doesn't matter whether n_b is greater or less than n_a. To verify these statements, you should construct diagrams like Figs. 34.21 and 34.22 for the following three cases: (i) $R > 0$ and $n_a > n_b$, (ii) $R < 0$ and $n_a < n_b$, and (iii) $R < 0$ and $n_a > n_b$. Then in each case, use your diagram to again derive Eqs. (34.11) and (34.12).

Here's a final note on the sign rule for the radius of curvature R of a surface. For the convex reflecting surface in Fig. 34.16, we considered R negative, but the convex *refracting* surface in Fig. 34.21 has a *positive* value of R. This may seem inconsistent, but it isn't. The rule is that R is positive if the center of curvature C is on the outgoing side of the surface and negative if C is on the other side. For the convex reflecting surface in Fig. 34.16, R is negative because point C is to the right of the surface but outgoing rays are to the left. For the convex refracting surface in Fig. 34.21, R is positive because both C and the outgoing rays are to the right of the surface.

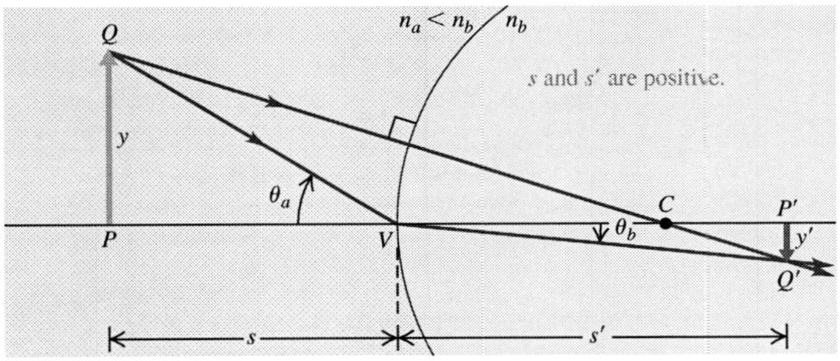

34.22 Construction for determining the height of an image formed by refraction at a spherical surface. In the case shown here, $n_a < n_b$.

34.23 Light rays refract as they pass through the curved surfaces of these water droplets.

Refraction at a curved surface is one reason gardeners avoid watering plants at midday. As sunlight enters a water drop resting on a leaf (Fig. 34.23), the light rays are refracted toward each other as in Figs. 34.21 and 34.22. The sunlight that strikes the leaf is therefore more concentrated and able to cause damage.

An important special case of a spherical refracting surface is a *plane* surface between two optical materials. This corresponds to setting $R = \infty$ in Eq. (34.11). In this case,

$$\frac{n_a}{s} + \frac{n_b}{s'} = 0 \qquad \text{(plane refracting surface)} \tag{34.13}$$

To find the lateral magnification m for this case, we combine this equation with the general relationship, Eq. (34.12), obtaining the simple result

$$m = 1$$

That is, the image formed by a *plane* refracting surface always has the same lateral size as the object, and it is always erect.

An example of image formation by a plane refracting surface is the appearance of a partly submerged drinking straw or canoe paddle. When viewed from some angles, the object appears to have a sharp bend at the water surface because the submerged part appears to be only about three-quarters of its actual distance below the surface. (We commented on the appearance of a submerged object in Section 33.2; see Fig. 33.9.)

Example 34.5 Image formation by refraction I

A cylindrical glass rod in air (Fig. 34.24) has index of refraction 1.52. One end is ground to a hemispherical surface with radius $R = 2.00$ cm. (a) Find the image distance of a small object on the axis of the rod, 8.00 cm to the left of the vertex. (b) Find the lateral magnification.

SOLUTION

IDENTIFY: This problem uses the ideas of refraction at a curved surface. Our target variables are the image distance s' and the lateral magnification m.

SET UP: Here material a is air $(n_a = 1.00)$ and material b is the glass of which the rod is made $(n_b = 1.52)$. We are given $s =$ 8.00 cm; the radius of the spherical surface is positive $(R = +2.00\text{ cm})$ because the center of curvature is on the outgoing side of the surface. We use Eq. (34.11) to determine the image distance and Eq. (34.12) to find the lateral magnification.

34.24 The glass rod in air forms a real image.

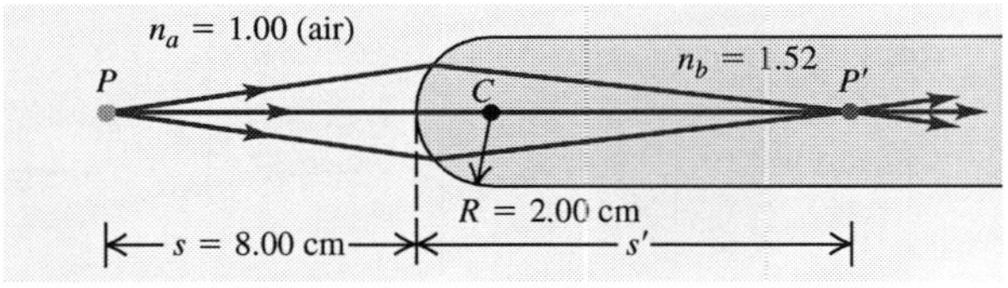

EXECUTE: (a) From Eq. (34.11),

$$\frac{1.00}{8.00\text{ cm}} + \frac{1.52}{s'} = \frac{1.52 - 1.00}{+2.00\text{ cm}}$$
$$s' = +11.3\text{ cm}$$

(b) From Eq. (34.12),

$$m = -\frac{n_a s'}{n_b s} = -\frac{(1.00)(11.3\text{ cm})}{(1.52)(8.00\text{ cm})} = -0.929$$

EVALUATE: Because the image distance s' is positive, the image is formed 11.3 cm to the *right* of the vertex (on the outgoing side), as shown in Fig. 34.24. The value of m tells us that the image is somewhat smaller than the object, and it is inverted. If the object is an arrow 1.000 mm high, pointing upward, the image is an arrow 0.929 mm high, pointing downward.

Example 34.6 Image formation by refraction II

The glass rod in Example 34.5 is immersed in water (index of refraction $n = 1.33$), as shown in Fig. 34.25. The other quantities have the same values as before. Find the image distance and lateral magnification.

SOLUTION

IDENTIFY: The situation is the same as in Example 34.5 except that now $n_a = 1.33$.

SET UP: As in Example 34.5, we use Eqs. (34.11) and (34.12) to determine s' and m, respectively.

EXECUTE: From Eq. (34.11),

$$\frac{1.33}{8.00\text{ cm}} + \frac{1.52}{s'} = \frac{1.52 - 1.33}{+2.00\text{ cm}}$$
$$s' = -21.3\text{ cm}$$

34.25 When immersed in water, the glass rod forms a virtual image.

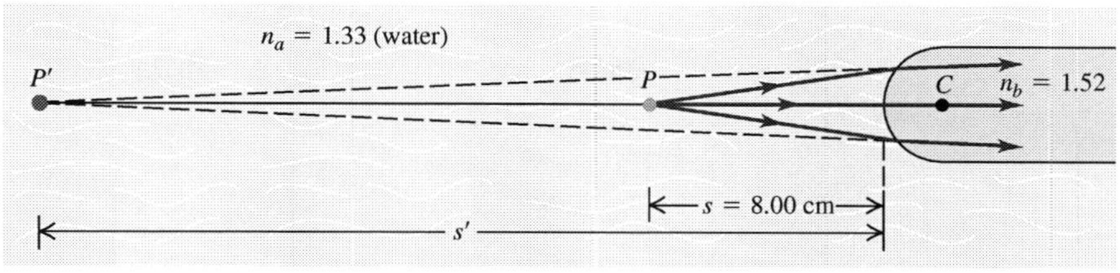

The lateral magnification in this case is

$$m = -\frac{(1.33)(-21.3\text{ cm})}{(1.52)(8.00\text{ cm})} = +2.33$$

EVALUATE: The negative value of s' means that after the rays are refracted by the surface, they are not converging but *appear* to diverge from a point 21.3 cm to the *left* of the vertex. We saw a similar case in the reflection of light from a convex mirror; we call the point a *virtual image.* In this example the surface forms a virtual image 21.3 cm to the left of the vertex. The vertical image is erect (because m is positive) and 2.33 times as large as the object.

Example 34.7 Apparent depth of a swimming pool

Swimming pool owners know that the pool always looks shallower than it really is and that it is important to identify the deep parts conspicuously so that people who can't swim won't jump into water that's over their heads. If a nonswimmer looks straight down into water that is actually 2.00 m (about 6 ft, 7 in.) deep, how deep does it appear to be?

SOLUTION

IDENTIFY: The surface of the water acts as a plane refracting surface.

SET UP: Figure 34.26 shows the situation. To determine the apparent depth of the pool, we imagine that there is an arrow PQ painted on the bottom of the pool. The refracting surface of the pool forms a virtual image $P'Q'$ of this arrow. We use Eq. (34.13) to find the depth of this arrow; it tells us the apparent depth of the pool.

EXECUTE: The object distance is the actual depth of the pool, $s = 2.00$ m. Material a is the water $(n_a = 1.33)$ and material b is air $(n_b = 1.00)$. The position of the image is given by Eq. (34.13):

$$\frac{n_a}{s} + \frac{n_b}{s'} = \frac{1.33}{2.00\text{ m}} + \frac{1.00}{s'} = 0$$
$$s' = -1.50\text{ m}$$

The image distance is negative. From the sign rules in Section 34.1, this means that the image is virtual and on the incoming side of the refracting surface—that is, on the same side as the object. The apparent depth is 1.50 m (about 4 ft, 11 in.), or only three-quarters of the actual depth. A 6-ft nonswimmer who didn't allow for this effect would be in trouble.

EVALUATE: Recall that the lateral magnification for a plane refracting surface is $m = 1$. Hence the image $P'Q'$ of the arrow is the same *horizontal length* as the actual arrow PQ. Only its depth is different. You can see this effect in Fig. 34.27.

34.26 Arrow $P'Q'$ is the virtual image of the underwater arrow PQ. The angles of the ray with the vertical are exaggerated for clarity.

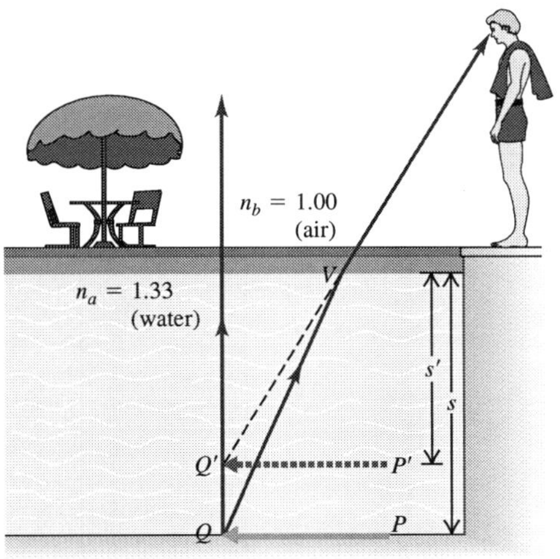

34.27 The submerged portion of this straw appears to be at a shallower depth (closer to the surface) than it actually is.

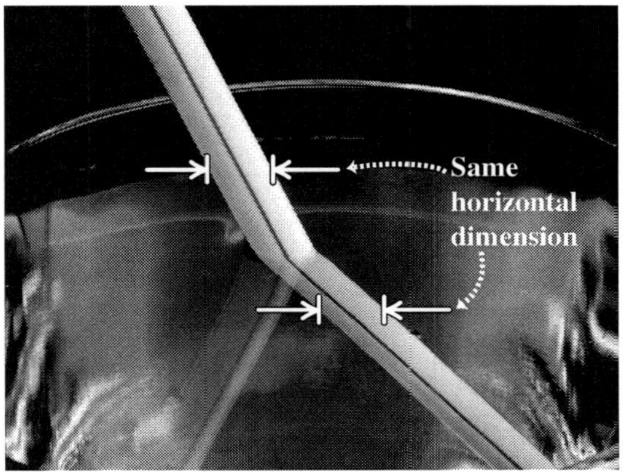

Test Your Understanding of Section 34.3 The water droplets in Fig. 34.23 have radius of curvature R and index of refraction $n = 1.33$. Can they form an image of the sun on the leaf?

34.4 Thin Lenses

ActivPhysics ONLINE
15.9 Thin Lens Ray Diagram
15.10 Converging Thin Lenses
15.11 Diverging Thin Lenses

The most familiar and widely used optical device (after the plane mirror) is the *lens.* A lens is an optical system with two refracting surfaces. The simplest lens has two *spherical* surfaces close enough together that we can neglect the distance between them (the thickness of the lens); we call this a **thin lens.** If you wear eyeglasses or contact lenses while reading, you are viewing these words through a pair of thin lenses. We can analyze thin lenses in detail using the results of Section 34.3 for refraction by a single spherical surface. However, we postpone this analysis until later in the section so that we can first discuss the properties of thin lenses.

Properties of a Lens

A lens of the shape shown in Fig. 34.28 has the property that when a beam of rays parallel to the axis passes through the lens, the rays converge to a point F_2 (Fig. 34.28a) and form a real image at that point. Such a lens is called a **converging lens.** Similarly, rays passing through point F_1 emerge from the lens as a beam of parallel rays (Fig. 34.28b). The points F_1 and F_2 are called the first and second *focal points,* and the distance f (measured from the center of the lens) is called the *focal length.* Note the similarities between the two focal points of a converging lens and the single focal point of a concave mirror (Fig. 34.13). As for a concave mirror, the focal length of a converging lens is defined to be a *positive* quantity, and such a lens is also called a *positive lens.*

The central horizontal line in Fig. 34.28 is called the *optic axis,* as with spherical mirrors. The centers of curvature of the two spherical surfaces lie on and define the optic axis. The two focal lengths in Fig. 34.28, both labeled f, *are always equal* for a thin lens, even when the two sides have different curvatures. We will derive this somewhat surprising result later in the section, when we derive the relationship of f to the index of refraction of the lens and the radii of curvature of its surfaces.

34.28 F_1 and F_2 are the first and second focal points of a converging thin lens. The numerical value of f is positive.

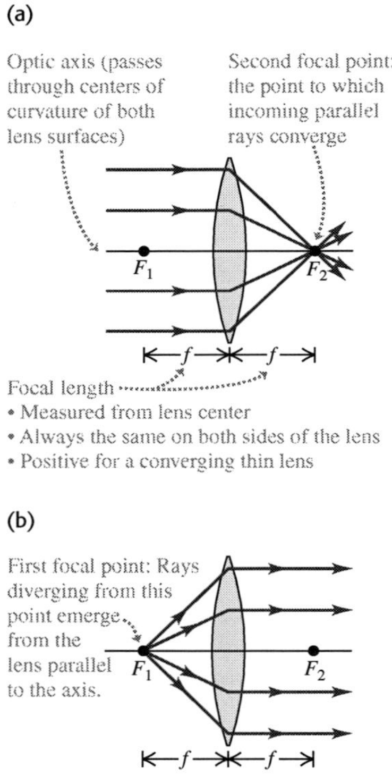

Image of an Extended Object: Converging Lens

Like a concave mirror, a converging lens can form an image of an extended object. Figure 34.29 shows how to find the position and lateral magnification of an image made by a thin converging lens. Using the same notation and sign rules as before, we let s and s' be the object and image distances, respectively, and let y and y' be the object and image heights. Ray QA, parallel to the optic axis before refraction, passes through the second focal point F_2 after refraction. Ray QOQ' passes undeflected straight through the center of the lens because at the center the two surfaces are parallel and (we have assumed) very close together.

34.29 Construction used to find image position for a thin lens. To emphasize that the lens is assumed to be very thin, the ray QAQ' is shown as bent at the midplane of the lens rather than at the two surfaces and ray QOQ' is shown as a straight line.

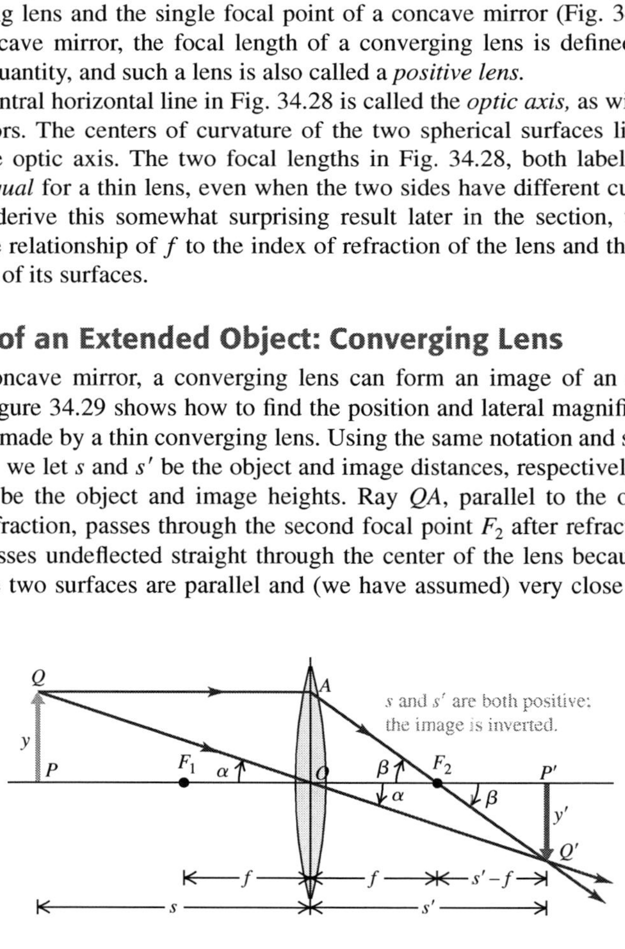

There is refraction where the ray enters and leaves the material but no net change in direction.

The two angles labeled α in Fig. 34.29 are equal. Therefore the two right triangles PQO and $P'Q'O$ are *similar,* and ratios of corresponding sides are equal. Thus

$$\frac{y}{s} = -\frac{y'}{s'} \quad \text{or} \quad \frac{y'}{y} = -\frac{s'}{s} \tag{34.14}$$

(The reason for the negative sign is that the image is below the optic axis and y' is negative.) Also, the two angles labeled β are equal, and the two right triangles OAF_2 and $P'Q'F_2$ are similar, so

$$\frac{y}{f} = -\frac{y'}{s'-f} \quad \text{or}$$

$$\frac{y'}{y} = -\frac{s'-f}{f} \tag{34.15}$$

We now equate Eqs. (34.14) and (34.15), divide by s', and rearrange to obtain

$$\frac{1}{s} + \frac{1}{s'} = \frac{1}{f} \quad \text{(object–image relationship, thin lens)} \tag{34.16}$$

This analysis also gives the lateral magnification $m = y'/y$ for the lens; from Eq. (34.14),

$$m = -\frac{s'}{s} \quad \text{(lateral magnification, thin lens)} \tag{34.17}$$

The negative sign tells us that when s and s' are both positive, as in Fig. 34.29, the image is *inverted,* and y and y' have opposite signs.

Equations (34.16) and (34.17) are the basic equations for thin lenses. They are *exactly* the same as the corresponding equations for spherical mirrors, Eqs. (34.6) and (34.7). As we will see, the same sign rules that we used for spherical mirrors are also applicable to lenses. In particular, consider a lens with a positive focal length (a converging lens). When an object is outside the first focal point F_1 of this lens (that is, when $s > f$), the image distance s' is positive (that is, the image is on the same side as the outgoing rays); this image is real and inverted, as in Fig. 34.29. An object placed inside the first focal point of a converging lens, so that $s < f$, produces an image with a negative value of s'; this image is located on the same side of the lens as the object and is virtual, erect, and larger than the object. You can verify these statements algebraically using Eqs. (34.16) and (34.17); we'll also verify them in the next section, using graphical methods analogous to those introduced for mirrors in Section 34.2.

Figure 34.30 shows how a lens forms a three-dimensional image of a three-dimensional object. Point R is nearer the lens than point P. From Eq. (34.16),

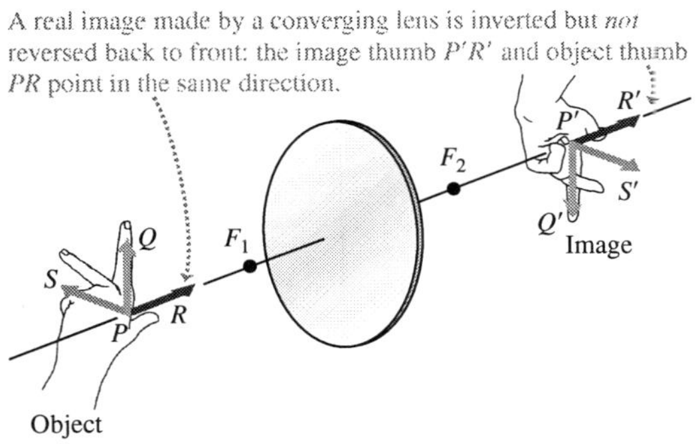

34.30 The image $S'P'Q'R'$ of a three-dimensional object $SPQR$ is not reversed by a lens.

34.31 F_2 and F_1 are the second and first focal points of a diverging thin lens, respectively. The numerical value of f is negative.

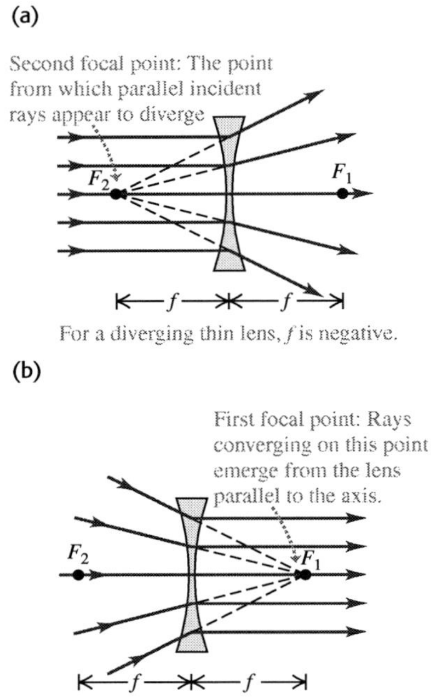

image point R' is farther from the lens than is image point P', and the image $P'R'$ points in the same direction as the object PR. Arrows $P'S'$ and $P'Q'$ are reversed relative to PS and PQ.

Let's compare Fig. 34.30 with Fig. 34.7, which shows the image formed by a plane *mirror.* We note that the image formed by the lens is inverted, but it is *not* reversed front to back along the optic axis. That is, if the object is a left hand, its image is also a left hand. You can verify this by pointing your left thumb along PR, your left forefinger along PQ, and your left middle finger along PS. Then rotate your hand 180°, using your thumb as an axis; this brings the fingers into coincidence with $P'Q'$ and $P'S'$. In other words, an *inverted* image is equivalent to an image that has been rotated by 180° about the lens axis.

Diverging Lenses

So far we have been discussing *converging* lenses. Figure 34.31 shows a **diverging lens;** the beam of parallel rays incident on this lens *diverges* after refraction. The focal length of a diverging lens is a negative quantity, and the lens is also called a *negative lens.* The focal points of a negative lens are reversed, relative to those of a positive lens. The second focal point, F_2, of a negative lens is the point from which rays that are originally parallel to the axis *appear to diverge* after refraction, as in Fig. 34.31a. Incident rays converging toward the first focal point F_1, as in Fig. 34.31b, emerge from the lens parallel to its axis. Comparing with Section 34.2, you can see that a diverging lens has the same relationship to a converging lens as a convex mirror has to a concave mirror.

Equations (34.16) and (34.17) apply to *both* positive and negative lenses. Figure 34.32 shows various types of lenses, both converging and diverging. Here's an important observation: *Any lens that is thicker at its center than at its edges is a converging lens with positive f; and any lens that is thicker at its edges than at its center is a diverging lens with negative f* (provided that the lens has a greater index of refraction than the surrounding material). We can prove this using the *lensmaker's equation,* which it is our next task to derive.

34.32 Various types of lenses.

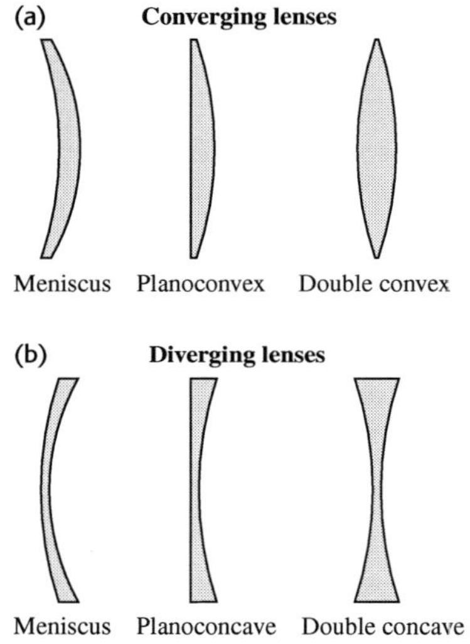

The Lensmaker's Equation

We'll now derive Eq. (34.16) in more detail and at the same time derive the *lensmaker's equation,* which is a relationship among the focal length f, the index of refraction n of the lens, and the radii of curvature R_1 and R_2 of the lens surfaces. We use the principle that an image formed by one reflecting or refracting surface can serve as the object for a second reflecting or refracting surface.

We begin with the somewhat more general problem of two spherical interfaces separating three materials with indexes of refraction n_a, n_b, and n_c, as shown in Fig. 34.33. The object and image distances for the first surface are s_1 and s_1', and those for the second surface are s_2 and s_2'. We assume that the lens is thin, so that the distance t between the two surfaces is small in comparison with the object and

34.33 The image formed by the first surface of a lens serves as the object for the second surface. The distances s_1' and s_2 are taken to be equal; this is a good approximation if the lens thickness t is small.

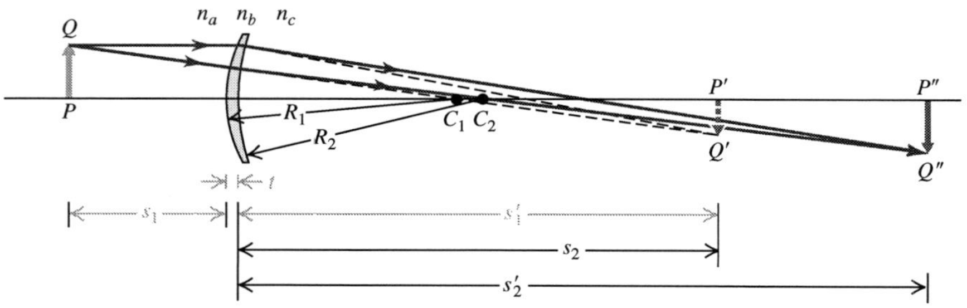

image distances and can therefore be neglected. This is usually the case with eyeglass lenses (Fig. 34.34). Then s_2 and s_1' have the same magnitude but opposite sign. For example, if the first image is on the outgoing side of the first surface, s_1' is positive. But when viewed as an object for the second surface, the first image is *not* on the incoming side of that surface. So we can say that $s_2 = -s_1'$.

We need to use the single-surface equation, Eq. (34.11), twice, once for each surface. The two resulting equations are

$$\frac{n_a}{s_1} + \frac{n_b}{s_1'} = \frac{n_b - n_a}{R_1}$$

$$\frac{n_b}{s_2} + \frac{n_c}{s_2'} = \frac{n_c - n_b}{R_2}$$

Ordinarily, the first and third materials are air or vacuum, so we set $n_a = n_c = 1$. The second index n_b is that of the lens, which we can call simply n. Substituting these values and the relationship $s_2 = -s_1'$, we get

$$\frac{1}{s_1} + \frac{n}{s_1'} = \frac{n-1}{R_1}$$

$$-\frac{n}{s_1'} + \frac{1}{s_2'} = \frac{1-n}{R_2}$$

To get a relationship between the initial object position s_1 and the final image position s_2', we add these two equations. This eliminates the term n/s_1', and we obtain

$$\frac{1}{s_1} + \frac{1}{s_2'} = (n-1)\left(\frac{1}{R_1} - \frac{1}{R_2}\right)$$

Finally, thinking of the lens as a single unit, we call the object distance simply s instead of s_1, and we call the final image distance s' instead of s_2'. Making these substitutions, we have

$$\frac{1}{s} + \frac{1}{s'} = (n-1)\left(\frac{1}{R_1} - \frac{1}{R_2}\right) \quad (34.18)$$

Now we compare this with the other thin-lens equation, Eq. (34.16). We see that the object and image distances s and s' appear in exactly the same places in both equations and that the focal length f is given by

$$\frac{1}{f} = (n-1)\left(\frac{1}{R_1} - \frac{1}{R_2}\right) \quad \text{(lensmaker's equation for a thin lens)} \quad (34.19)$$

This is the **lensmaker's equation.** In the process of rederiving the relationship between object distance, image distance, and focal length for a thin lens, we have also derived an expression for the focal length f of a lens in terms of its index of refraction n and the radii of curvature R_1 and R_2 of its surfaces. This can be used to show that all the lenses in Fig. 34.32a are converging lenses with positive focal lengths and that all the lenses in Fig. 34.32b are diverging lenses with negative focal lengths (see Exercise 34.30).

We use all our sign rules from Section 34.1 with Eqs. (34.18) and (34.19). For example, in Fig. 34.35, s, s', and R_1 are positive, but R_2 is negative.

It is not hard to generalize Eq. (34.19) to the situation in which the lens is immersed in a material with an index of refraction greater than unity. We invite you to work out the lensmaker's equation for this more general situation.

We stress that the paraxial approximation is indeed an approximation! Rays that are at sufficiently large angles to the optic axis of a spherical lens will not be brought to the same focus as paraxial rays; this is the same problem of spherical

34.34 These eyeglass lenses satisfy the thin-lens approximation: Their thickness is small compared to the object and image distances.

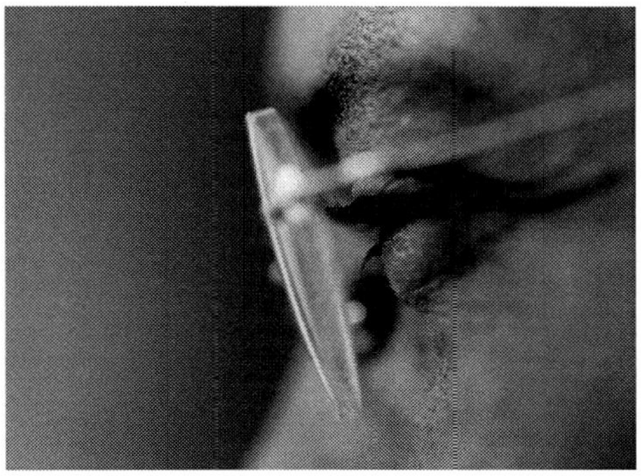

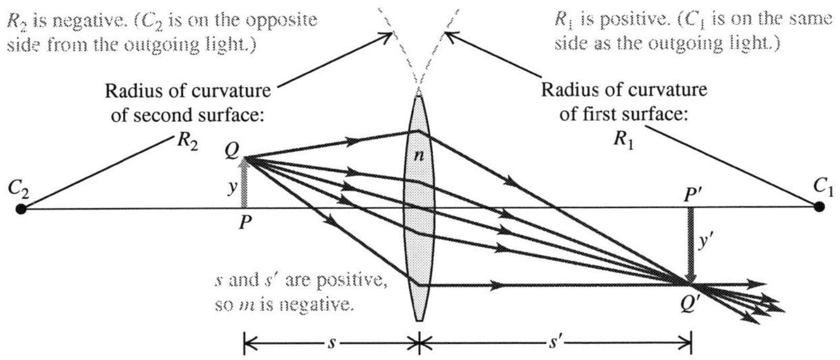

34.35 A converging thin lens with a positive focal length f.

aberration that plagues spherical *mirrors* (see Section 34.2). To avoid this and other limitations of thin spherical lenses, lenses of more complicated shape are used in precision optical instruments.

Example 34.8 Determining the focal length of a lens

(a) Suppose the absolute values of the radii of curvature of the lens surfaces in Fig. 34.35 are both equal to 10 cm and the index of refraction is $n = 1.52$. What is the focal length f of the lens? (b) Suppose the lens in Fig. 34.31 also has $n = 1.52$, and the absolute values of the radii of curvature of its lens surfaces are also both equal to 10 cm. What is the focal length of this lens?

SOLUTION

IDENTIFY: We are asked to find the focal length of (a) a lens that is convex on both sides (Fig. 34.35) and (b) a lens that is concave on both sides (Fig. 34.31).

SET UP: We use the lensmaker's equation, Eq. (34.19), to determine the focal length in each situation. We take account of whether the surfaces are convex or concave by paying careful attention to the signs of the radii of curvature R_1 and R_2.

EXECUTE: (a) Figure 34.35 shows that the center of curvature of the first surface (C_1) is on the outgoing side of the lens, while the center of curvature of the second surface (C_2) is on the *incoming* side. Hence R_1 is positive but R_2 is negative: $R_1 = +10$ cm, $R_2 = -10$ cm. From Eq. (34.19),

$$\frac{1}{f} = (1.52 - 1)\left(\frac{1}{+10\text{ cm}} - \frac{1}{-10\text{ cm}}\right)$$

$$f = 9.6\text{ cm}$$

(b) For a double-concave lens the center of curvature of the first surface is on the *incoming* side, while the center of curvature of the second surface is on the outgoing side. Hence R_1 is negative and R_2 is positive: $R_1 = -10$ cm, $R_2 = +10$ cm. Again using Eq. (34.19),

$$\frac{1}{f} = (1.52 - 1)\left(\frac{1}{-10\text{ cm}} - \frac{1}{+10\text{ cm}}\right)$$

$$f = -9.6\text{ cm}$$

EVALUATE: In part (a) the focal length is positive, so this is a converging lens; this makes sense, since the lens is thicker at its center than at its edges. In part (b) the focal length is *negative,* so this is a *diverging* lens; this also makes sense, since the lens is thicker at its edges than at its center.

Graphical Methods for Lenses

We can determine the position and size of an image formed by a thin lens by using a graphical method very similar to the one we used in Section 34.2 for spherical mirrors. Again we draw a few special rays called *principal rays* that diverge from a point of the object that is *not* on the optic axis. The intersection of these rays, after they pass through the lens, determines the position and size of the image. In using this graphical method, we will consider the entire deviation of a ray as occurring at the midplane of the lens, as shown in Fig. 34.36. This is consistent with the assumption that the distance between the lens surfaces is negligible.

The three principal rays whose paths are usually easy to trace for lenses are shown in Fig. 34.36:

1. *A ray parallel to the axis* emerges from the lens in a direction that passes through the second focal point F_2 of a converging lens, or appears to come from the second focal point of a diverging lens.
2. *A ray through the center of the lens* is not appreciably deviated; at the center of the lens the two surfaces are parallel, so this ray emerges at essentially the same angle at which it enters and along essentially the same line.

34.36 The graphical method of locating an image formed by a thin lens. The colors of the rays are for identification only; they do not refer to specific colors of light. (Compare Fig. 34.19 for spherical mirrors.)

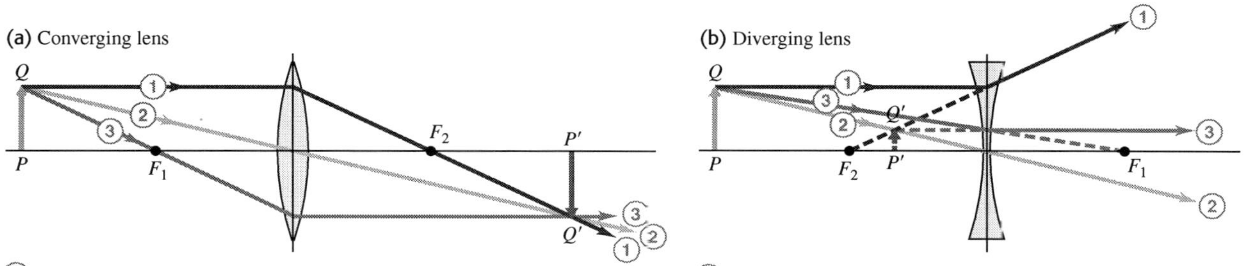

(1) Parallel incident ray refracts to pass through second focal point F_2.
(2) Ray through center of lens does not deviate appreciably.
(3) Ray through the first focal point F_1 emerges parallel to the axis.

(1) Parallel incident ray appears after refraction to have come from the second focal point F_2.
(2) Ray through center of lens does not deviate appreciably.
(3) Ray aimed at the first focal point F_1 emerges parallel to the axis.

3. *A ray through (or proceeding toward) the first focal point* F_1 emerges parallel to the axis.

When the image is real, the position of the image point is determined by the intersection of any two rays 1, 2, and 3 (Fig. 34.36a). When the image is virtual, we extend the diverging outgoing rays backward to their intersection point to find the image point (Fig. 34.36b).

CAUTION **Principal rays are not the only rays** Keep in mind that *any* ray from the object that strikes the lens will pass through the image point (for a real image) or appear to originate from the image point (for a virtual image). (We made a similar comment about image formation by mirrors in Section 34.2.) We've emphasized the principal rays because they're the only ones you need to draw to locate the image.

Figure 34.37 shows principal-ray diagrams for a converging lens for several object distances. We suggest you study each of these diagrams very carefully, comparing each numbered ray with the above description.

34.37 Formation of images by a thin converging lens for various object distances. The principal rays are numbered. (Compare Fig. 34.20 for a concave spherical mirror.)

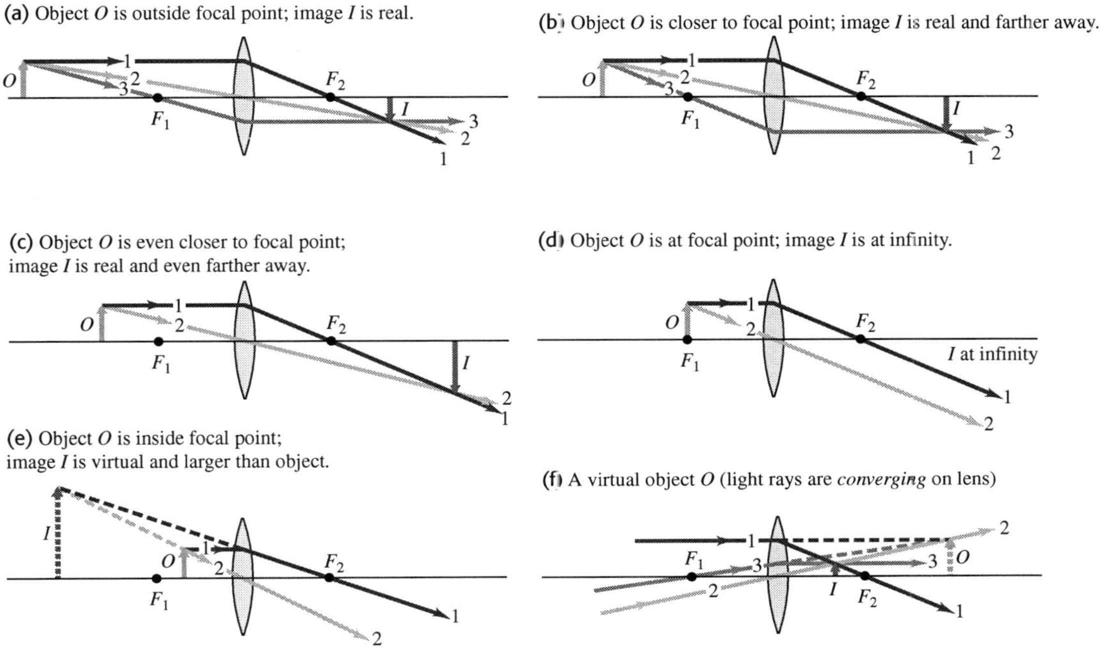

Parts (a), (b), and (c) of Fig. 34.37 help explain what happens in focusing a camera. For a photograph to be in sharp focus, the film must be at the position of the real image made by the camera's lens. The image distance increases as the object is brought closer, so the film is moved farther behind the lens (i.e., the lens is moved farther in front of the film). In Fig. 34.37d the object is at the focal point; ray 3 can't be drawn because it doesn't pass through the lens. In Fig. 34.37e the object distance is less than the focal length. The outgoing rays are divergent, and the image is *virtual;* its position is located by extending the outgoing rays backward, so the image distance s' is negative. Note also that the image is erect and larger than the object. (We'll see the usefulness of this in Section 34.6.) Figure 34.37f corresponds to a *virtual object.* The incoming rays do not diverge from a real object, but are *converging* as though they would meet at the tip of the virtual object O on the right side; the object distance s is negative in this case. The image is real and is located between the lens and the second focal point. This situation can arise if the rays that strike the lens in Fig. 34.37f emerge from another converging lens (not shown) to the left of the figure.

Problem-Solving Strategy 34.2 Image Formation by Thin Lenses

MP

IDENTIFY *the relevant concepts:* Problem-Solving Strategy 34.1 (Section 34.2) for mirrors is equally applicable here, and you should review it now. As for mirrors, you should solve problems involving image formation by lenses using *both* equations and a principal-ray diagram.

SET UP *the problem:* As always, determine the target variables.

EXECUTE *the solution* as follows:

1. Always begin with a principal-ray diagram if you have enough information. Orient your diagrams consistently so that light travels from left to right. Don't just sketch these diagrams; draw the rays with a ruler and measure the distances carefully.
2. Draw the principal rays so they bend at the midplane of the lens, as shown in Fig. 34.36. For a lens there are only three principal rays, compared to four for a mirror. Be sure to draw *all three* whenever possible. The intersection of any two determines the image, but if the third doesn't pass through the same intersection point, you know you have made a mistake. Redundancy can be useful in spotting errors.
3. If the outgoing principal rays don't converge at a real image point, the image is virtual. Then you have to extend the outgoing rays backward to find the virtual image point, which lies on the *incoming* side of the lens.
4. The same sign rules we have used for mirrors and single refracting surfaces (see Section 34.1) are also applicable for thin lenses. Be extremely careful to get your signs right and to interpret the signs of results correctly.
5. Use Eqs. (34.16) and (34.17) to confirm by calculation your graphical results for the image position and size. This gives an extremely useful consistency check.
6. The *image* from one lens or mirror may serve as the *object* for another. In that case, be careful in finding the object and image *distances* for this intermediate image; be sure you include the distance between the two elements (lenses and/or mirrors) correctly.

EVALUATE *your answer:* Cast a critical eye on your diagrams and calculations to make certain that your results are consistent.

Example 34.9 Image location and magnification with a converging lens

A converging lens has a focal length of 20 cm. Find graphically the image location for an object at each of the following distances from the lens: (a) 50 cm; (b) 20 cm; (c) 15 cm; (d) −40 cm. Determine the magnification in each case. Check your results by calculating the image position and lateral magnification from Eqs. (34.16) and (34.17), respectively.

SOLUTION

IDENTIFY: This problem illustrates the usefulness of both graphical and computational methods for problems with thin lenses, just as for problems with curved mirrors.

SET UP: In each case we are given the focal length $f = 20$ cm and the value of the object distance s. Our target variables are the image distance s' and the lateral magnification $m = -s'/s$.

EXECUTE: The appropriate principal-ray diagrams are shown in (a) Fig. 34.37a, (b) Fig. 34.37d, (c) Fig. 34.37e, and (d) Fig. 34.37f. The approximate image distances, from measurements of these diagrams, are 35 cm, $-\infty$, −40 cm, and 15 cm, and the approximate magnifications are $-\frac{2}{3}$, $+\infty$, and +3, and $+\frac{1}{3}$, respectively.

Calculating the image positions from Eq. (34.16), we find

$$\text{(a)}\ \frac{1}{50\text{ cm}} + \frac{1}{s'} = \frac{1}{20\text{ cm}} \qquad s' = 33.3\text{ cm}$$

$$\text{(b)}\ \frac{1}{20\text{ cm}} + \frac{1}{s'} = \frac{1}{20\text{ cm}} \qquad s' = \pm\infty$$

$$\text{(c)}\ \frac{1}{15\text{ cm}} + \frac{1}{s'} = \frac{1}{20\text{ cm}} \qquad s' = -60\text{ cm}$$

$$\text{(d)}\ \frac{1}{-40\text{ cm}} + \frac{1}{s'} = \frac{1}{20\text{ cm}} \qquad s' = 13.3\text{ cm}$$

The graphical results are fairly close to these except for part (c); the accuracy of the diagram in Fig. 34.37e is limited because the rays extended backward have nearly the same direction.

From Eq. (34.17) the lateral magnifications are

$$\text{(a)}\ m = -\frac{33.3\ \text{cm}}{50\ \text{cm}} = -\frac{2}{3}$$

$$\text{(b)}\ m = -\frac{\pm\infty\ \text{cm}}{20\ \text{cm}} = \pm\infty$$

$$\text{(c)}\ m = -\frac{-60\ \text{cm}}{15\ \text{cm}} = +4$$

$$\text{(d)}\ m = -\frac{13.3\ \text{cm}}{-40\ \text{cm}} = +\frac{1}{3}$$

EVALUATE: Note that s' is positive in parts (a) and (d) but negative in part (c). This makes sense: The image is real in parts (a) and (d) but virtual in part (c). The light rays that emerge from the lens in part (b) are parallel and never converge, so the image can be regarded as being at either $+\infty$ or $-\infty$.

The values of magnification tell us that the image is inverted in part (a) and erect in parts (c) and (d), in accordance with the principal-ray diagrams. The infinite value of magnification in part (b) is another way of saying that the image is formed infinitely far away.

Example 34.10 Image formation by a diverging lens

You are given a thin diverging lens. You find that a beam of parallel rays spreads out after passing through the lens, as though all the rays came from a point 20.0 cm from the center of the lens. You want to use this lens to form an erect virtual image that is $\frac{1}{3}$ the height of the object. (a) Where should the object be placed? (b) Draw a principal-ray diagram.

SOLUTION

IDENTIFY: The observation with parallel rays shows that the focal length is $f = -20$ cm. We want the lateral magnification to be $m = +\frac{1}{3}$ (positive because the image is to be erect).

SET UP: We use the given information to determine the ratio s'/s from Eq. (34.17) and then determine the object distance s with Eq. (34.16).

EXECUTE: (a) From Eq. (34.17), $m = +\frac{1}{3} = -s'/s$, so $s' = -s/3$. If we insert this result into Eq. (34.16), we find

$$\frac{1}{s} + \frac{1}{-s/3} = \frac{1}{-20.0\ \text{cm}}$$

$$s = 40.0\ \text{cm}$$

$$s' = -\frac{s}{3} = -\frac{40.0\ \text{cm}}{3} = -13.3\ \text{cm}$$

The image distance is negative, so the object and image are on the same side of the lens.

(b) Figure 34.38 is a principal-ray diagram for this problem, with the rays numbered in the same way as in Fig. 34.36b.

EVALUATE: A diverging lens is often mounted in the front door of a home. It provides the occupant of the home with an erect, reduced image of anyone standing outside the door. The occupant can see the outside person's entire height and decide whether to let him or her in.

34.38 Principal-ray diagram for an image formed by a thin diverging lens.

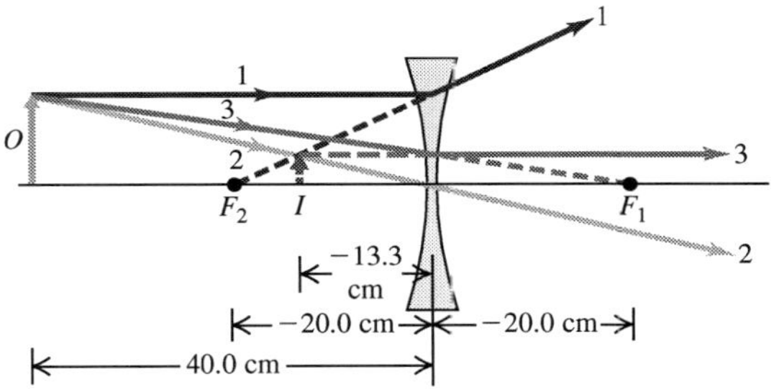

Example 34.11 An image of an image

An object 8.0 cm high is placed 12.0 cm to the left of a converging lens of focal length 8.0 cm. A second converging lens of focal length 6.0 cm is placed 36.0 cm to the right of the first lens. Both lenses have the same optic axis. Find the position, size, and orientation of the image produced by the two lenses in combination. (Combinations of converging lenses are used in telescopes and microscopes, to be discussed in Section 34.7.)

SOLUTION

IDENTIFY: The situation is shown in Fig. 34.39. The object O lies outside the first focal point F_1 of the first lens, so this lens produces a real image I. The light rays that strike the second lens diverge from this real image just as if I was a material object. Hence the *image* made by the *first* lens acts as an *object* for the *second* lens. Our goal is to determine the properties of the final image made by the second lens.

SET UP: We use both graphical and computational methods to determine the properties of the final image.

EXECUTE: In Fig. 34.39 (next page) we have drawn principal rays 1, 2, and 3 from the head of the object arrow O to find the position of the first image I and principal rays 1′, 2′, and 3′ from the head of the first image to find the position of the second image I' made by the second lens (even though rays 2′ and 3′ don't actually exist in this case). Note that the image is inverted *twice,* once by each lens, so the second image I' has the same orientation as the original object.

Continued

To *calculate* the position and size of the second image I', we must first find the position and size of the first image I. Applying Eq. (34.16), $1/s + 1/s' = 1/f$, to the first lens gives

$$\frac{1}{12.0\text{ cm}} + \frac{1}{s'_1} = \frac{1}{8.0\text{ cm}} \qquad s'_1 = +24.0\text{ cm}$$

The first image I is 24.0 cm to the right of the first lens. The lateral magnification is $m_1 = -(24.0\text{ cm})/(12.0\text{ cm}) = -2.00$, so the height of the first image is $(-2.0)(8.0\text{ cm}) = -16.0\text{ cm}$.

The first image is $36.0\text{ cm} - 24.0\text{ cm} = 12.0\text{ cm}$ to the left of the second lens, so the object distance for the second lens is $+12.0$ cm. Using Eq. (34.16) for the second lens gives the position of the second and final image:

$$\frac{1}{12.0\text{ cm}} + \frac{1}{s'_2} = \frac{1}{6.0\text{ cm}} \qquad s'_2 = +12.0\text{ cm}$$

The final image is 12.0 cm to the right of the second lens and 48.0 cm to the right of the first lens. The magnification produced by the second converging lens is $m_2 = -(12.0\text{ cm})/(12.0\text{ cm}) = -1.0$.

EVALUATE: The value of m_2 means that the final image is just as large as the first image but has the opposite orientation. This is also shown in the principal-ray diagram.

34.39 Principal-ray diagram for a combination of two converging lenses. The first lens makes a real image of the object. This real image acts as an object for the second lens.

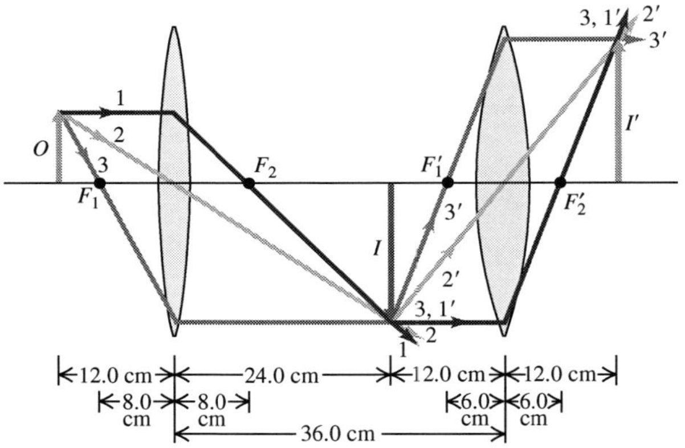

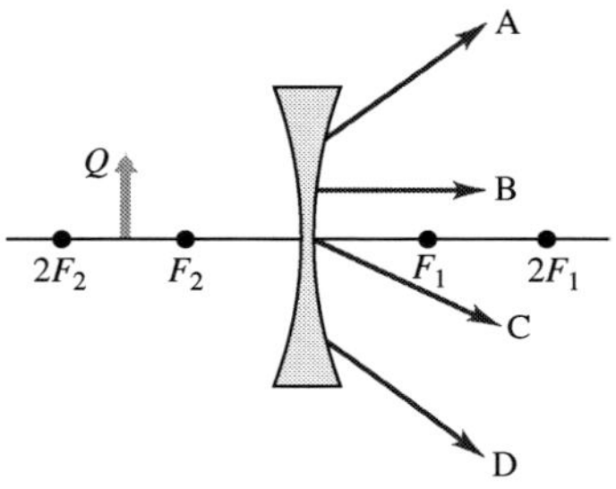

Test Your Understanding of Section 34.4 A diverging lens and an object are positioned as shown in the figure at left. Which of the rays A, B, C, and D could emanate from point Q at the top of the object?

34.5 Cameras

The concept of *image,* which is so central to understanding simple mirror and lens systems, plays an equally important role in the analysis of optical instruments (also called *optical devices*). Among the most common optical devices are cameras, which make an image of an object and record it either electronically or on film.

The basic elements of a **camera** are a light-tight box ("camera" is a Latin word meaning "a room or enclosure"), a converging lens, a shutter to open the lens for a prescribed length of time, and a light-sensitive recording medium (Fig. 34.40). In a digital camera this is an electronic detector called a charge-coupled device (CCD) array; in an older camera, this is photographic film. The lens forms an inverted real image on the recording medium of the object being photographed. High-quality camera lenses have several elements, permitting partial correction of various *aberrations,* including the dependence of

34.40 Key elements of a digital camera.

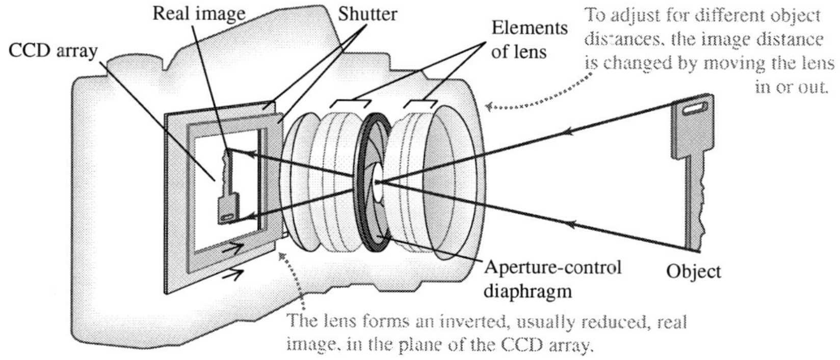

34.41 (a), (b), (c) Three photographs taken with the same camera from the same position in the Boston Public Garden using lenses with focal lengths $f = 28$ mm, 105 mm, and 300 mm. Increasing the focal length increases the image size proportionately. (d) The larger the value of f, the smaller the angle of view. The angles shown here are for a camera with image area 24 mm × 36 mm (corresponding to 35-mm film) and refer to the angle of view along the diagonal dimension of the film.

(a) $f = 28$ mm (b) $f = 105$ mm (c) $f = 300$ mm

(d) The angles of view for the photos in (a)–(c)

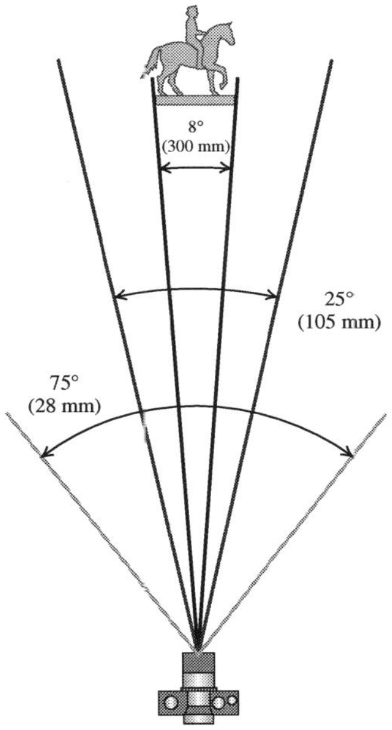

index of refraction on wavelength and the limitations imposed by the paraxial approximation.

When the camera is in proper *focus,* the position of the recording medium coincides with the position of the real image formed by the lens. The resulting photograph will then be as sharp as possible. With a converging lens, the image distance increases as the object distance decreases (see Figs. 34.41a, 34.41b, and 34.41c, and the discussion in Section 34.4). Hence in "focusing" the camera, we move the lens closer to the film for a distant object and farther from the film for a nearby object.

Camera Lenses: Focal Length

The choice of the focal length f for a camera lens depends on the film size and the desired angle of view. Figure 34.41 shows three photographs taken on 35-mm film with the same camera at the same position, but with lenses of different focal lengths. A lens of long focal length, called a *telephoto* lens, gives a small angle of view and a large image of a distant object (such as the statue in Fig. 34.41c); a lens of short focal length gives a small image and a wide angle of view (as in Fig. 34.41a) and is called a *wide-angle* lens. To understand this behavior, recall that the focal length is the distance from the lens to the image when the object is infinitely far away. In general, for *any* object distance, using a lens of longer focal length gives a greater image distance. This also increases the height of the image; as was discussed in Section 34.4, the ratio of the image height y' to the object height y (the *lateral magnification*) is equal in absolute value to the ratio of image distance s' to the object distance s [Eq. (34.17)]:

$$m = \frac{y'}{y} = -\frac{s'}{s}$$

With a lens of short focal length, the ratio s'/s is small, and a distant object gives only a small image. When a lens with a long focal length is used, the image of this same object may entirely cover the area of the film. Hence the longer the focal length, the narrower the angle of view (Fig. 34.41d).

Camera Lenses: f-Number

For the film to record the image properly, the total light energy per unit area reaching the film (the "exposure") must fall within certain limits. This is controlled by the *shutter* and the *lens aperture.* The shutter controls the time interval during which light enters the lens. This is usually adjustable in steps corresponding to factors of about 2, often from 1 s to $\frac{1}{1000}$ s.

The intensity of light reaching the film is proportional to the area viewed by the camera lens and to the effective area of the lens. The size of the area that the lens "sees" is proportional to the square of the angle of view of the lens, and so is roughly proportional to $1/f^2$. The effective area of the lens is controlled by means of an adjustable lens aperture, or *diaphragm,* a nearly circular hole with variable diameter D; hence the effective area is proportional to D^2. Putting these factors together, we see that the intensity of light reaching the film with a particular lens is proportional to D^2/f^2. The light-gathering capability of a lens is commonly expressed by photographers in terms of the ratio f/D, called the ***f*-number** of the lens:

$$f\text{-number} = \frac{\text{Focal length}}{\text{Aperture diameter}} = \frac{f}{D} \tag{34.20}$$

For example, a lens with a focal length $f = 50$ mm and an aperture diameter $D = 25$ mm is said to have an f-number of 2, or "an aperture of $f/2$." The light intensity reaching the film is *inversely* proportional to the square of the f-number.

For a lens with a variable-diameter aperture, increasing the diameter by a factor of $\sqrt{2}$ changes the f-number by $1/\sqrt{2}$ and increases the intensity at the film by a factor of 2. Adjustable apertures usually have scales labeled with successive numbers (often called *f-stops*) related by factors of $\sqrt{2}$, such as

$$f/2 \quad f/2.8 \quad f/4 \quad f/5.6 \quad f/8 \quad f/11 \quad f/16$$

and so on. The larger numbers represent smaller apertures and exposures, and each step corresponds to a factor of 2 in intensity (Fig. 34.42). The actual *exposure* (total amount of light reaching the film) is proportional to both the aperture area and the time of exposure. Thus $f/4$ and $\frac{1}{500}$ s, $f/5.6$ and $\frac{1}{250}$ s, and $f/8$ and $\frac{1}{125}$ s all correspond to the same exposure.

34.42 A camera lens with an adjustable diaphragm.

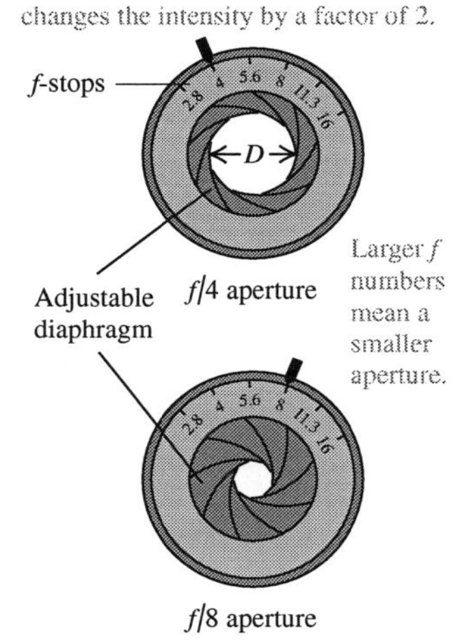

Zoom Lenses and Projectors

Many photographers use a *zoom lens,* which is not a single lens but a complex collection of several lens elements that give a continuously variable focal length, often over a range as great as 10 to 1. Figures 34.43a and 34.43b show a simple system with variable focal length, and Fig. 34.43c shows a typical zoom lens for a single-lens reflex camera. Zoom lenses give a range of image sizes of a given object. It is an enormously complex problem in optical design to keep the image in focus and maintain a constant f-number while the focal length changes. When you vary the focal length of a typical zoom lens, two groups of elements move within the lens and a diaphragm opens and closes.

A *projector* for viewing slides, digital images, or motion pictures operates very much like a camera in reverse. In a movie projector, a lamp shines on the

34.43 A simple zoom lens uses a converging lens and a diverging lens in tandem. **(a)** When the two lenses are close together, the combination behaves like a single lens of long focal length. **(b)** If the two lenses are moved farther apart, the combination behaves like a short-focal-length lens. **(c)** A typical zoom lens for a single-lens reflex camera, containing twelve elements arranged in four groups.

(a) Zoom lens set for long focal length

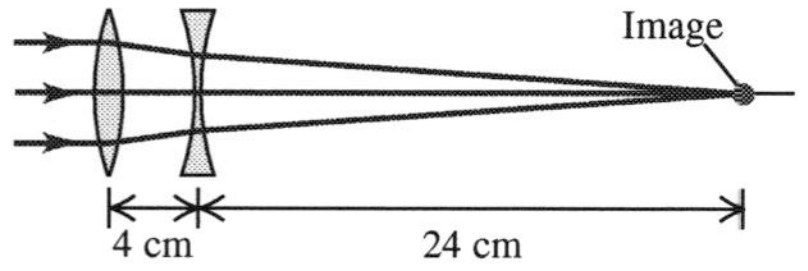

(b) Zoom lens set for short focal length

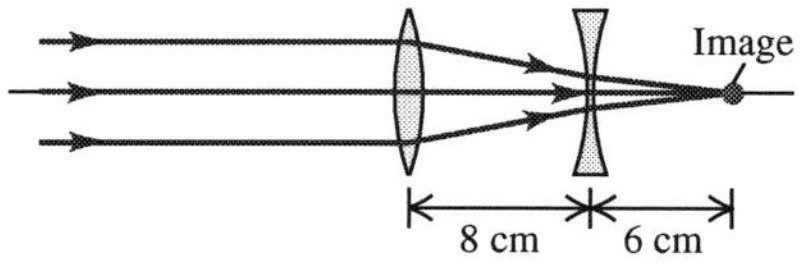

(c) A practical zoom lens

film, which acts as an object for the projection lens. The lens forms a real, enlarged, inverted image of the film on the projection screen. Because the image is inverted, the film goes through the projector upside down so that the image on the screen appears right-side up.

Example 34.12 Photographic exposures

A common telephoto lens for a 35-mm camera has a focal length of 200 mm and a range of f-stops from $f/5.6$ to $f/45$. (a) What is the corresponding range of aperture diameters? (b) What is the corresponding range of intensity of the image on the film?

SOLUTION

IDENTIFY: Part (a) of this problem uses the relationship among focal length, aperture diameter, and f-number for a lens. Part (b) uses the relationship between intensity and aperture diameter.

SET UP: We use Eq. (34.20) to relate the diameter D (the target variable) to the f-number and the focal length $f = 200$ mm. The intensity of the light reaching the film is proportional to D^2/f^2; since f is the same in each case, we conclude that the intensity in this case is proportional to D^2, the square of the aperture diameter.

EXECUTE: (a) From Eq. (34.20) the range of diameters is from

$$D = \frac{f}{f\text{-number}} = \frac{200\text{ mm}}{5.6} = 36\text{ mm}$$

to

$$D = \frac{200\text{ mm}}{45} = 4.4\text{ mm}$$

(b) Because the intensity is proportional to the square of the diameter, the ratio of the intensity at $f/5.6$ to the intensity at $f/45$ is

$$\left(\frac{36\text{ mm}}{4.4\text{ mm}}\right)^2 = \left(\frac{45}{5.6}\right)^2 = 65 \qquad (\text{about } 2^6)$$

EVALUATE: If the correct exposure time at $f/5.6$ is $\frac{1}{1000}$ s, then at $f/45$ it is $(65)(\frac{1}{1000}\text{ s}) = \frac{1}{15}$ s to compensate for the lower intensity. This illustrates a general rule: The smaller the aperture and the larger the f-number, the longer the required exposure time. Nevertheless, many photographers prefer to use small apertures so that only the central part of the lens is used to make the image. This minimizes aberrations that occur near the edges of the lens and gives the sharpest possible images.

Test Your Understanding of Section 34.5 When used with 35-mm film (image area 24 mm × 36 mm), a lens with $f = 50$ mm gives a 45° angle of view and is called a "normal lens." When used with a CCD array that measures 5 mm × 5 mm, this same lens is (i) a wide-angle lens; (ii) a normal lens; (iii) a telephoto lens.

34.6 The Eye

The optical behavior of the eye is similar to that of a camera. The essential parts of the human eye, considered as an optical system, are shown in Fig. 34.44a. The eye is nearly spherical and about 2.5 cm in diameter. The front portion is somewhat more sharply curved and is covered by a tough, transparent membrane called the

34.44 (a) The eye. (b) There are two types of light-sensitive cells on the retina. The rods are more sensitive to light than the cones, but only the cones are sensitive to differences in color. A typical human eye contains about 1.3×10^8 rods and about 7×10^6 cones.

(a) Diagram of the eye

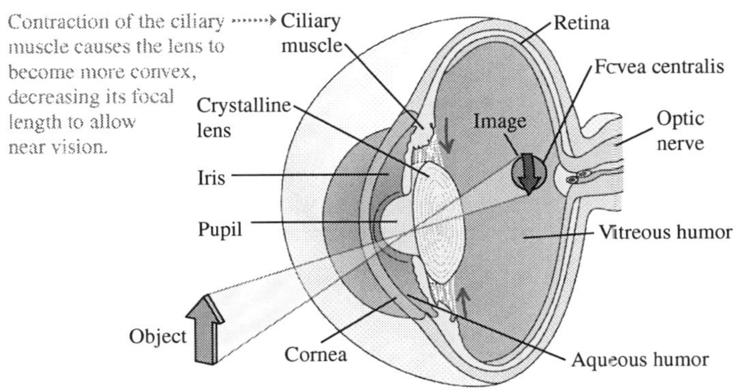

(b) Scanning electron micrograph showing retinal rods and cones in different colors

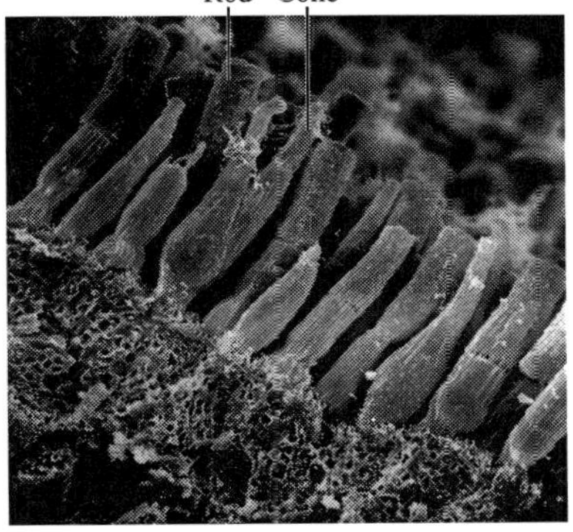

cornea. The region behind the cornea contains a liquid called the *aqueous humor.* Next comes the *crystalline lens,* a capsule containing a fibrous jelly, hard at the center and progressively softer at the outer portions. The crystalline lens is held in place by ligaments that attach it to the ciliary muscle, which encircles it. Behind the lens, the eye is filled with a thin watery jelly called the *vitreous humor.* The indexes of refraction of both the aqueous humor and the vitreous humor are about 1.336, nearly equal to that of water. The crystalline lens, while not homogeneous, has an average index of 1.437. This is not very different from the indexes of the aqueous and vitreous humors. As a result, most of the refraction of light entering the eye occurs at the outer surface of the cornea.

Refraction at the cornea and the surfaces of the lens produces a *real image* of the object being viewed. This image is formed on the light-sensitive *retina,* lining the rear inner surface of the eye. The retina plays the same role as the film in a camera. The *rods* and *cones* in the retina act like an array of miniature photocells (Fig. 34.44b); they sense the image and transmit it via the *optic nerve* to the brain. Vision is most acute in a small central region called the *fovea centralis,* about 0.25 mm in diameter.

In front of the lens is the *iris.* It contains an aperture with variable diameter called the *pupil,* which opens and closes to adapt to changing light intensity. The receptors of the retina also have intensity adaptation mechanisms.

For an object to be seen sharply, the image must be formed exactly at the location of the retina. The eye adjusts to different object distances s by changing the focal length f of its lens; the lens-to-retina distance, corresponding to s', does not change. (Contrast this with focusing a camera, in which the focal length is fixed and the lens-to-film distance is changed.) For the normal eye, an object at infinity is sharply focused when the ciliary muscle is relaxed. To permit sharp imaging on the retina of closer objects, the tension in the ciliary muscle surrounding the lens increases, the ciliary muscle contracts, the lens bulges, and the radii of curvature of its surfaces decrease; this decreases the focal length. This process is called *accommodation.*

The extremes of the range over which distinct vision is possible are known as the *far point* and the *near point* of the eye. The far point of a normal eye is at infinity. The position of the near point depends on the amount by which the ciliary muscle can increase the curvature of the crystalline lens. The range of accommodation gradually diminishes with age because the crystalline lens grows throughout a person's life (it is about 50% larger at age 60 than at age 20) and the ciliary muscles are less able to distort a larger lens. For this reason, the near point gradually recedes as one grows older. This recession of the near point is called *presbyopia.* Table 34.1 shows the approximate position of the near point for an average person at various ages. For example, an average person 50 years of age cannot focus on an object that is closer than about 40 cm.

Table 34.1 Receding of Near Point with Age

Age (years)	Near Point (cm)
10	7
20	10
30	14
40	22
50	40
60	200

34.45 Refractive errors for (a) a normal eye, (b) a myopic (nearsighted) eye, and (c) a hyperopic (farsighted) eye viewing a very distant object. The dashed blue curve indicates the required position of the retina.

(a) Normal eye

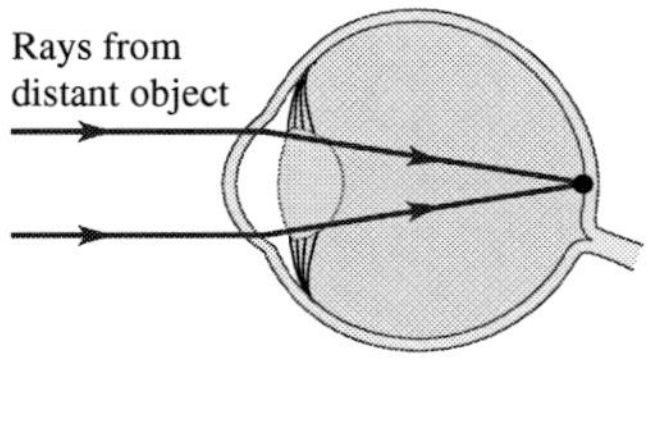

(b) Myopic (nearsighted) eye

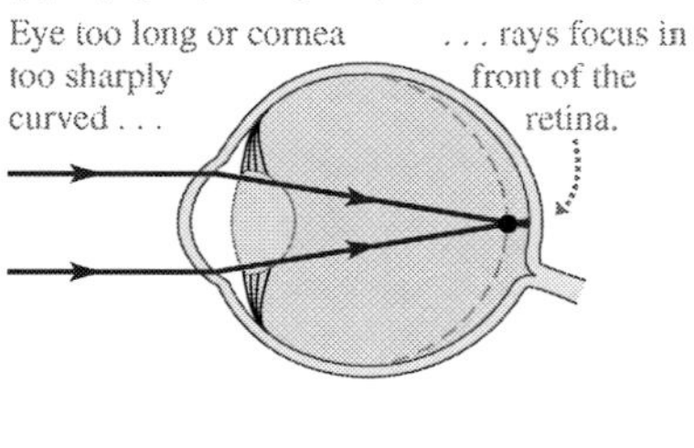

(c) Hyperopic (farsighted) eye

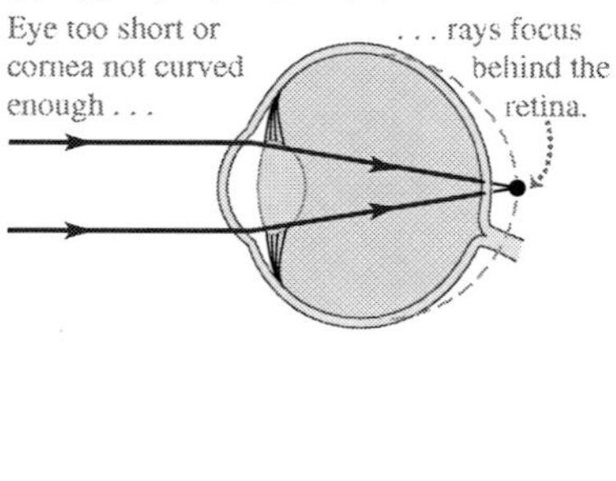

Defects of Vision

Several common defects of vision result from incorrect distance relationships in the eye. A normal eye forms an image on the retina of an object at infinity when the eye is relaxed (Fig. 34.45a). In the *myopic* (nearsighted) eye, the eyeball is too long from front to back in comparison with the radius of curvature of the cornea (or the cornea is too sharply curved), and rays from an object at infinity are focused in front of the retina (Fig. 34.45b). The most distant object for which an image can be formed on the retina is then nearer than infinity. In the *hyperopic* (farsighted) eye, the eyeball is too short or the cornea is not curved enough, and the image of an infinitely distant object is behind the retina (Fig. 34.45c). The myopic eye produces *too much* convergence in a parallel bundle of rays for an image to be formed on the retina; the hyperopic eye, *not enough* convergence.

All of these defects can be corrected by the use of corrective lenses (eyeglasses or contact lenses). The near point of either a presbyopic or a hyperopic

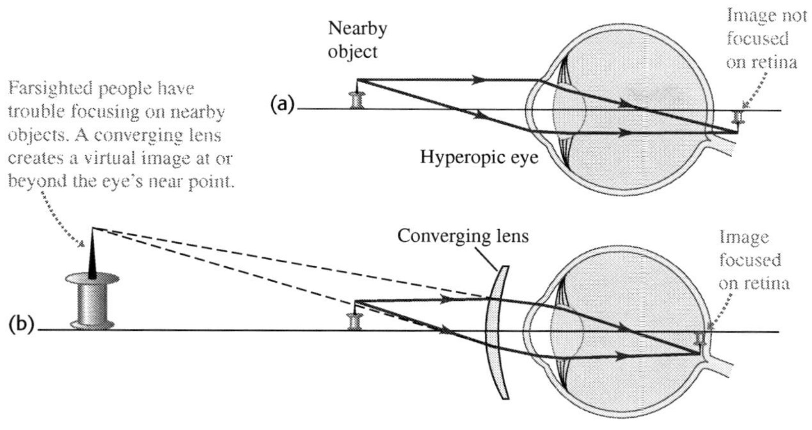

34.46 (a) An uncorrected hyperopic (farsighted) eye. (b) A positive (converging) lens gives the extra convergence needed for a hyperopic eye to focus the image on the retina.

34.47 (a) An uncorrected myopic (nearsighted) eye. (b) A negative (diverging) lens spreads the rays farther apart to compensate for the excessive convergence of the myopic eye.

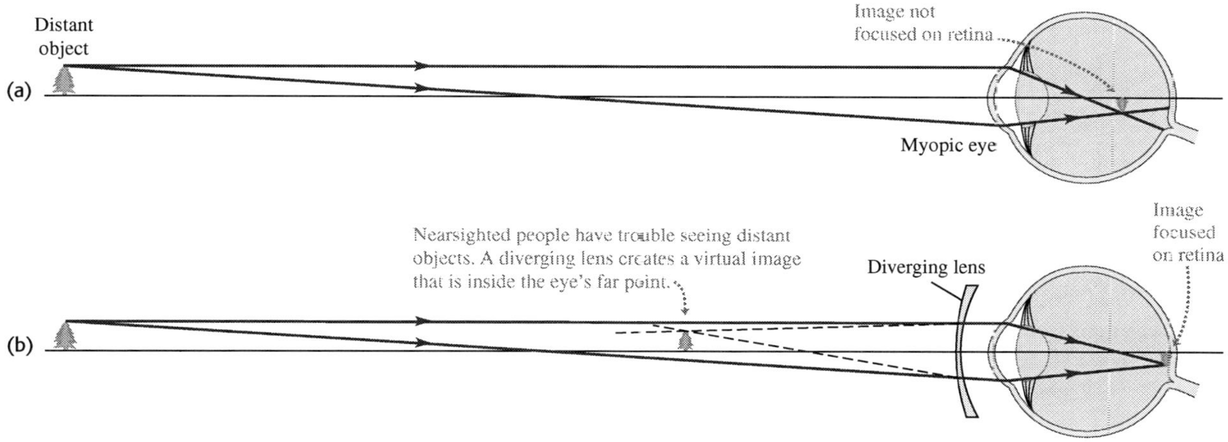

eye is *farther* from the eye than normal. To see clearly an object at normal reading distance (often assumed to be 25 cm), we need a lens that forms a virtual image of the object at or beyond the near point. This can be accomplished by a converging (positive) lens, as shown in Fig. 34.46. In effect the lens moves the object farther away from the eye to a point where a sharp retinal image can be formed. Similarly, correcting the myopic eye involves using a diverging (negative) lens to move the image closer to the eye than the actual object, as shown in Fig. 34.47.

Astigmatism is a different type of defect in which the surface of the cornea is not spherical but rather more sharply curved in one plane than in another. As a result, horizontal lines may be imaged in a different plane from vertical lines (Fig. 34.48a). Astigmatism may make it impossible, for example, to focus clearly on the horizontal and vertical bars of a window at the same time.

Astigmatism can be corrected by use of a lens with a *cylindrical* surface. For example, suppose the curvature of the cornea in a horizontal plane is correct to focus rays from infinity on the retina but the curvature in the vertical plane is too great to form a sharp retinal image. When a cylindrical lens with its axis horizontal is placed before the eye, the rays in a horizontal plane are unaffected, but the additional divergence of the rays in a vertical plane causes these to be sharply imaged on the retina (Fig. 34.48b).

Lenses for vision correction are usually described in terms of the **power,** defined as the reciprocal of the focal length expressed in meters. The unit of power is the **diopter.** Thus a lens with $f = 0.50$ m has a power of 2.0 diopters,

34.48 One type of astigmatism and how it is corrected.

(a) Vertical lines are imaged in front of the retina. (b) A cylindrical lens corrects for astigmatism.

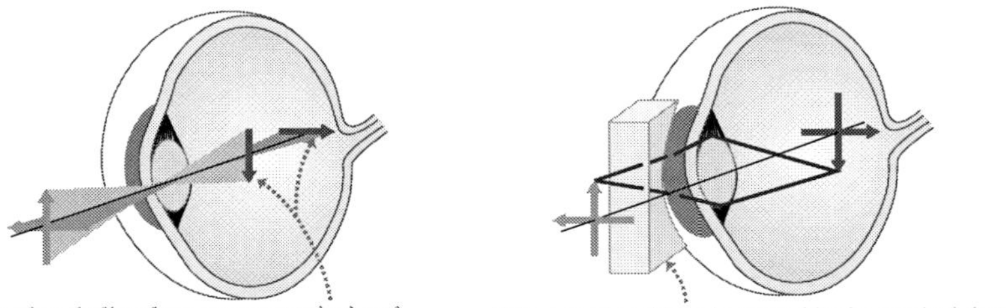

Shape of eyeball or lens causes vertical and horizontal elements to focus at different distances.

This cylindrical lens is curved in the vertical, but not the horizontal, direction; it changes the focal length of vertical elements.

$f = -0.25$ m corresponds to -4.0 diopters, and so on. The numbers on a prescription for glasses are usually powers expressed in diopters. When the correction involves both astigmatism and myopia or hyperopia, there are three numbers: one for the spherical power, one for the cylindrical power, and an angle to describe the orientation of the cylinder axis.

An alternative approach for correcting many defects of vision is to reshape the cornea. This is often done using a procedure called *laser-assisted in situ keratomileusis,* or LASIK. An incision is made into the cornea and a flap of outer corneal tissue is folded back. A pulsed ultraviolet laser with a beam only 50 μm wide (about $\frac{1}{200}$ the width of a human hair) is then used to vaporize away microscopic areas of the underlying tissue. The flap is then folded back into position, where it conforms to the new shape "carved" by the laser.

Example 34.13 Correcting for farsightedness

The near point of a certain hyperopic eye is 100 cm in front of the eye. To see clearly an object that is 25 cm in front of the eye, what contact lens is required?

SOLUTION

IDENTIFY: We want the lens to form a virtual image of the object at the near point of the eye, 100 cm from it. That is, when $s =$ 25 cm, we want s' to be 100 cm.

SET UP: Figure 34.49 shows the situation. We determine the required focal length of the contact lens using the object-image relationship for a thin lens, Eq. (34.16).

EXECUTE: From Eq. (34.16),

$$\frac{1}{f} = \frac{1}{s} + \frac{1}{s'} = \frac{1}{+25 \text{ cm}} + \frac{1}{-100 \text{ cm}}$$

$$f = +33 \text{ cm}$$

We need a converging lens with focal length $f = 33$ cm. The corresponding power is $1/(0.33 \text{ m})$, or $+3.0$ diopters.

EVALUATE: In this example we used a contact lens to correct hyperopia. Had we used eyeglasses, we would have had to account for the separation between the eye and the eyeglass lens, and a somewhat different power would have been required (see Example 34.14).

34.49 Using a contact lens to correct for farsightedness.

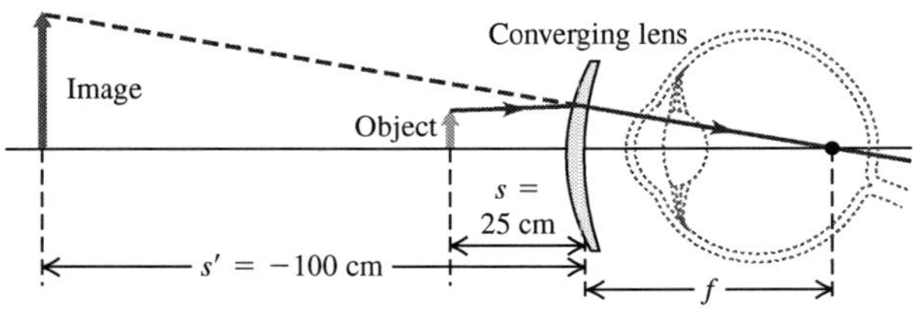

Example 34.14 Correcting for nearsightedness

The far point of a certain myopic eye is 50 cm in front of the eye. To see clearly an object at infinity, what eyeglass lens is required? Assume that the lens is worn 2 cm in front of the eye.

SOLUTION

IDENTIFY: The far point of a myopic eye is nearer than infinity. To see clearly objects beyond the far point, we need a lens that forms a virtual image of such objects no farther from the eye than the far point.

SET UP: Figure 34.50 shows the situation. Assume that the virtual image of the object at infinity is formed at the far point, 50 cm in front of the eye and 48 cm in front of the eyeglass lens. Then when $s = \infty$, s' -48 cm.

we want to be As in Example 34.13, we use the values of s and s' to calculate the required focal length.

EXECUTE: From Eq. (34.16),

$$\frac{1}{f} = \frac{1}{s} + \frac{1}{s'} = \frac{1}{\infty} + \frac{1}{-48\text{ cm}}$$

$$f = -48\text{ cm}$$

We need a *diverging* lens with focal length -48 cm $= -0.48$ m. The power is -2.1 diopter.

EVALUATE: If a *contact* lens were used instead, we would need $f = -50$ cm and a power of -2.0 diopters. Can you see why?

34.50 Using a contact lens to correct for nearsightedness.

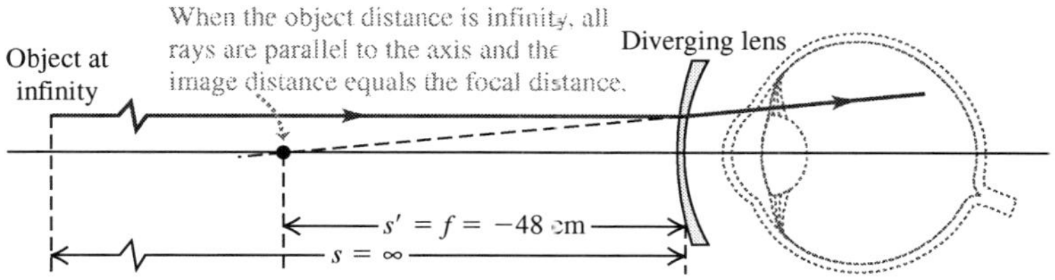

Test Your Understanding of Section 34.6 A certain eyeglass lens is thin at its center, even thinner at its top and bottom edges, and relatively thick at its left and right edges. What defects of vision is this lens intended to correct? (i) hyperopia for objects oriented both vertically and horizontally; (ii) myopia for objects oriented both vertically and horizontally; (iii) hyperopia for objects oriented vertically and myopia for objects oriented horizontally; (iv) hyperopia for objects oriented horizontally and myopia for objects oriented vertically.

34.7 The Magnifier

The apparent size of an object is determined by the size of its image on the retina. If the eye is unaided, this size depends on the *angle* θ subtended by the object at the eye, called its **angular size** (Fig. 34.51a).

To look closely at a small object, such as an insect or a crystal, you bring it close to your eye, making the subtended angle and the retinal image as large as possible. But your eye cannot focus sharply on objects that are closer than the near point, so the angular size of an object is greatest (that is, it subtends the largest possible viewing angle) when it is placed at the near point. In the following discussion we will assume an average viewer for whom the near point is 25 cm from the eye.

34.51 (a) The angular size θ is largest when the object is at the near point. (b) The magnifier gives a virtual image at infinity. This virtual image appears to the eye to be a real object subtending a larger angle θ' at the eye.

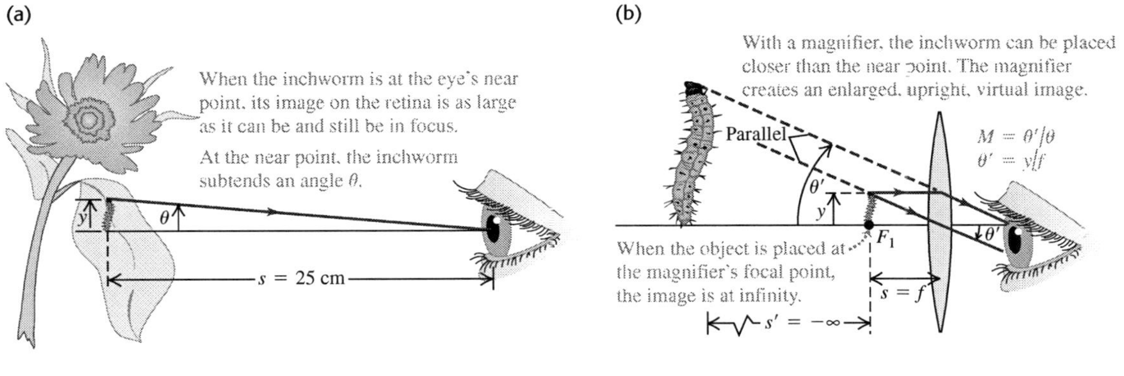

A converging lens can be used to form a virtual image that is larger and farther from the eye than the object itself, as shown in Fig. 34.51b. Then the object can be moved closer to the eye, and the angular size of the image may be substantially larger than the angular size of the object at 25 cm without the lens. A lens used in this way is called a **magnifier,** otherwise known as a *magnifying glass* or a *simple magnifier.* The virtual image is most comfortable to view when it is placed at infinity, so that the ciliary muscle of the eye is relaxed; this means that the object is placed at the focal point F_1 of the magnifier. In the following discussion we assume that this is done.

In Fig. 34.51a the object is at the near point, where it subtends an angle θ at the eye. In Fig. 34.51b a magnifier in front of the eye forms an image at infinity, and the angle subtended at the magnifier is θ'. The usefulness of the magnifier is given by the ratio of the angle θ' (with the magnifier) to the angle θ (without the magnifier). This ratio is called the **angular magnification** M:

$$M = \frac{\theta'}{\theta} \quad \text{(angular magnification)} \tag{34.21}$$

CAUTION Angular magnification vs. lateral magnification Don't confuse the *angular* magnification M with the *lateral* magnification m. Angular magnification is the ratio of the *angular* size of an image to the angular size of the corresponding object; lateral magnification refers to the ratio of the *height* of an image to the height of the corresponding object. For the situation shown in Fig. 34.51b, the angular magnification is about 3×, since the inchworm subtends an angle about three times larger than that in Fig. 34.51a; hence the inchworm will look about three times larger to the eye. The *lateral* magnification $m = -s'/s$ in Fig. 34.51b is *infinite* because the virtual image is at infinity, but that doesn't mean that the inchworm looks infinitely large through the magnifier! (That's why we didn't attempt to draw an infinitely large inchworm in Fig. 34.51b.) When dealing with a magnifier, M is useful but m is not.

To find the value of M, we first assume that the angles are small enough that each angle (in radians) is equal to its sine and its tangent. Using Fig. 34.451a and drawing the ray in Fig. 34.51b that passes undeviated through the center of the lens, we find that θ and θ' (in radians) are

$$\theta = \frac{y}{25\text{ cm}} \qquad \theta' = \frac{y}{f}$$

Combining these expressions with Eq. (34.21), we find

$$M = \frac{\theta'}{\theta} = \frac{y/f}{y/25\text{ cm}} = \frac{25\text{ cm}}{f} \quad \text{(angular magnification for a simple magnifier)} \tag{34.22}$$

It may seem that we can make the angular magnification as large as we like by decreasing the focal length f. In fact, the aberrations of a simple double-convex lens set a limit to M of about 3× to 4×. If these aberrations are corrected, the angular magnification may be made as great as 20×. When greater magnification than this is needed, we usually use a compound microscope, discussed in the next section.

Test Your Understanding of Section 34.7 You are examining a gem using a magnifier. If you change to a different magnifier with twice the focal length of the first one, (i) you will have to hold the object at twice the distance and the angular magnification will be twice as great; (ii) you will have to hold the object at twice the distance and the angular magnification will be $\frac{1}{2}$ as great; (iii) you will have to hold the object at $\frac{1}{2}$ the distance and the angular magnification will be twice as great; (iv) you will have to hold the object at $\frac{1}{2}$ the distance and the angular magnification will be $\frac{1}{2}$ as great.

34.8 Microscopes and Telescopes

Cameras, eyeglasses, and magnifiers use a single lens to form an image. Two important optical devices that use *two* lenses are the microscope and the telescope. In each device a primary lens, or *objective,* forms a real image, and a second lens, or *eyepiece,* is used as a magnifier to make an enlarged, virtual image.

ActivPhysics

15.12 Two-Lens System
15.3 The Telescope and Angular Magnification

Microscopes

When we need greater magnification than we can get with a simple magnifier, the instrument that we usually use is the **microscope,** sometimes called a *compound microscope.* The essential elements of a microscope are shown in Fig. 34.52a. To analyze this system, we use the principle that an image formed by one optical element such as a lens or mirror can serve as the object for a second element. We used this principle in Section 34.4 when we derived the thin-lens equation by repeated application of the single-surface refraction equation; we used this principle again in Example 34.11 (Section 34.4), in which the image formed by a lens was used as the object of a second lens.

The object O to be viewed is placed just beyond the first focal point F_1 of the **objective,** a converging lens that forms a real and enlarged image I (Fig. 34.52b). In a properly designed instrument this image lies just inside the first focal point F_1' of a second converging lens called the **eyepiece** or *ocular.* (The reason the image should lie just *inside* F_1' is left for you to discover; see Problem 34.108.) The eyepiece acts as a simple magnifier, as discussed in Section 34.7, and forms a final virtual image I' of I. The position of I' may be anywhere between the near

34.52 (a) Elements of a microscope. (b) The object O is placed just outside the first focal point of the objective (the distance s_1 has been exaggerated for clarity). (c) This microscope image shows single-celled organisms about 2×10^{-4} m (0.2 mm) across. Typical light microscopes can resolve features as small as 2×10^{-7} m, comparable to the wavelength of light.

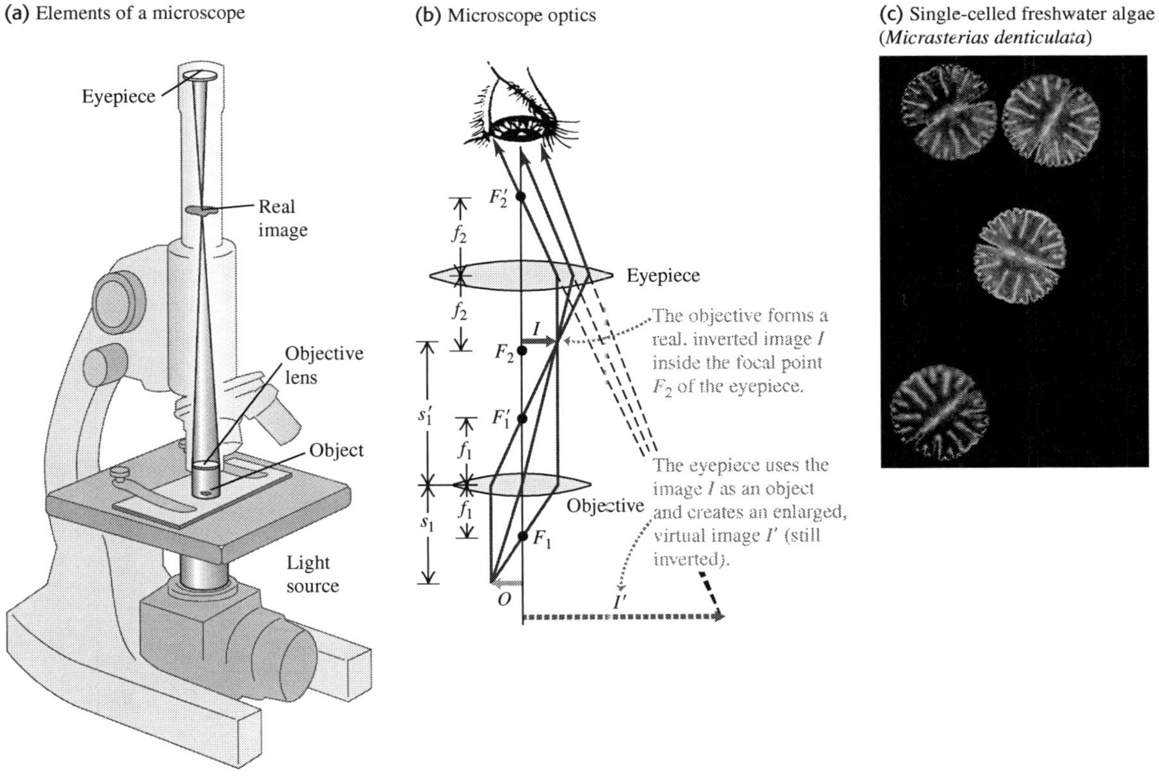

and far points of the eye. Both the objective and the eyepiece of an actual microscope are highly corrected compound lenses with several optical elements, but for simplicity we show them here as simple thin lenses.

As for a simple magnifier, what matters when viewing through a microscope is the *angular* magnification M. The overall angular magnification of the compound microscope is the product of two factors. The first factor is the *lateral* magnification m_1 of the objective, which determines the linear size of the real image I; the second factor is the *angular* magnification M_2 of the eyepiece, which relates the angular size of the virtual image seen through the eyepiece to the angular size that the real image I would have if you viewed it *without* the eyepiece. The first of these factors is given by

$$m_1 = -\frac{s_1'}{s_1} \qquad (34.23)$$

where s_1 and s_1' are the object and image distances, respectively, for the objective lens. Ordinarily, the object is very close to the focal point, and the resulting image distance s_1' is very great in comparison to the focal length f_1 of the objective lens. Thus s_1 is approximately equal to f_1, and we can write $m_1 = -s_1'/f_1$.

The real image I is close to the focal point F_1' of the eyepiece, so to find the eyepiece angular magnification, we can use Eq. (34.22): $M_2 = (25\text{ cm})/f_2$, where f_2 is the focal length of the eyepiece (considered as a simple lens). The overall angular magnification M of the compound microscope (apart from a negative sign, which is customarily ignored) is the product of the two magnifications:

$$M = m_1 M_2 = \frac{(25\text{ cm})s_1'}{f_1 f_2} \quad \text{(angular magnification for a microscope)} \qquad (34.24)$$

where s_1', f_1, and f_2 are measured in centimeters. The final image is inverted with respect to the object. Microscope manufacturers usually specify the values of m_1 and M_2 for microscope components rather than the focal lengths of the objective and eyepiece.

Equation (34.24) shows that the angular magnification of a microscope can be increased by using an objective of shorter focal length f_1, thereby increasing m_1 and the size of the real image I. Most optical microscopes have a rotating "turret" with three or more objectives of different focal lengths so that the same object can be viewed at different magnifications. The eyepiece should also have a short focal length f_2 to help to maximize the value of M.

To take a photograph using a microscope (called a *photomicrograph* or *micrograph*), the eyepiece is removed and a camera placed so that the real image I falls on the camera's CCD array or film. Figure 34.52c shows such a photograph. In this case what matters is the *lateral* magnification of the microscope as given by Eq. (34.23).

Telescopes

The optical system of a **telescope** is similar to that of a compound microscope. In both instruments the image formed by an objective is viewed through an eyepiece. The key difference is that the telescope is used to view large objects at large distances and the microscope is used to view small objects close at hand. Another difference is that many telescopes use a curved mirror, not a lens, as an objective.

Figure 34.53 shows an *astronomical telescope*. Because this telescope uses a lens as an objective, it is called a *refracting telescope* or *refractor*. The objective lens forms a real, reduced image I of the object. This image is the object for the eyepiece lens, which forms an enlarged, virtual image of I. Objects that are viewed with a telescope are usually so far away from the instrument that the first

34.53 Optical system of an astronomical refracting telescope.

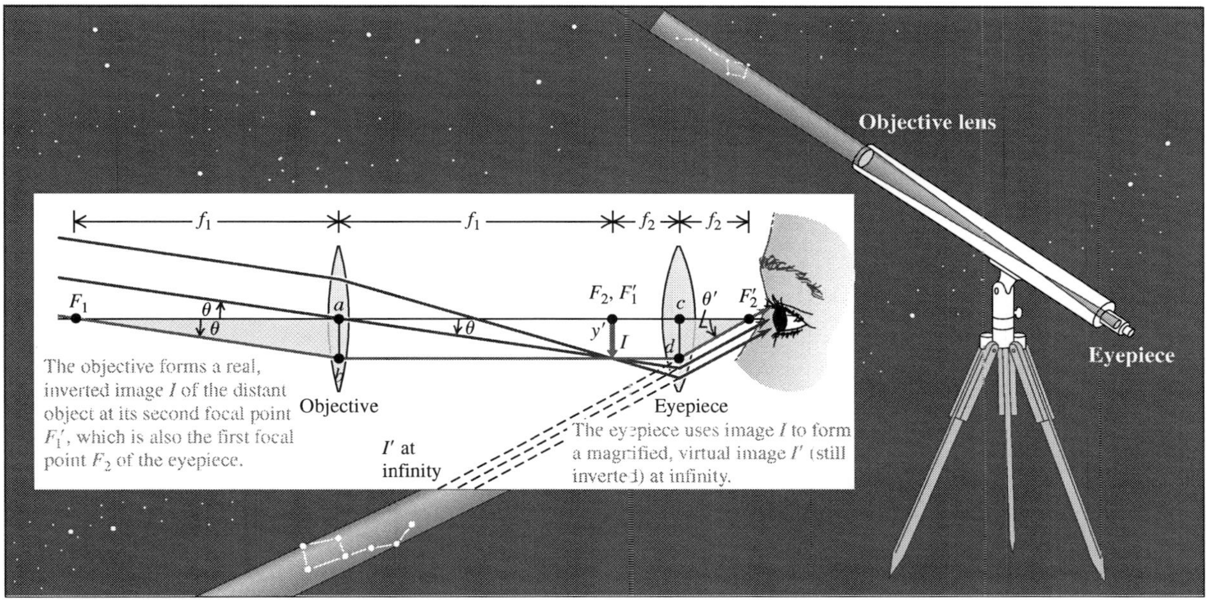

image I is formed very nearly at the second focal point of the objective lens. If the final image I' formed by the eyepiece is at infinity (for most comfortable viewing by a normal eye), the first image must also be at the first focal point of the eyepiece. The distance between objective and eyepiece, which is the length of the telescope, is therefore the *sum* of the focal lengths of objective and eyepiece, $f_1 + f_2$.

The angular magnification M of a telescope is defined as the ratio of the angle subtended at the eye by the final image I' to the angle subtended at the (unaided) eye by the object. We can express this ratio in terms of the focal lengths of objective and eyepiece. In Fig. 34.53 the ray passing through F_1, the first focal point of the objective, and through F_2', the second focal point of the eyepiece, is shown in red. The object (not shown) subtends an angle θ at the objective and would subtend essentially the same angle at the unaided eye. Also, since the observer's eye is placed just to the right of the focal point F_2', the angle subtended at the eye by the final image is very nearly equal to the angle θ'. Because bd is parallel to the optic axis, the distances ab and cd are equal to each other and also to the height y' of the real image I. Because the angles θ and θ' are small, they may be approximated by their tangents. From the right triangles F_1ab and $F_2'cd$,

$$\theta = \frac{-y'}{f_1} \qquad \theta' = \frac{y'}{f_2}$$

and the angular magnification M is

$$M = \frac{\theta'}{\theta} = -\frac{y'/f_2}{y'/f_1} = -\frac{f_1}{f_2} \qquad \text{(angular magnification for a telescope)} \qquad (34.25)$$

The angular magnification M of a telescope is equal to the ratio of the focal length of the objective to that of the eyepiece. The negative sign shows that the final image is inverted. Equation (34.25) shows that to achieve good angular magnification, a *telescope* should have a *long* objective focal length f_1. By contrast, Eq. (34.24) shows that a *microscope* should have a *short* objective focal length. However, a telescope objective with a long focal length should also have a large diameter D so that the f-number f_1/D will not be too large; as described in

Section 34.5, a large f-number means a dim, low-intensity image. Telescopes typically do not have interchangeable objectives; instead, the magnification is varied by using different eyepieces with different focal lengths f_2. Just as for a microscope, smaller values of f_2 give larger angular magnifications.

An inverted image is no particular disadvantage for astronomical observations. When we use a telescope or binoculars—essentially a pair of telescopes mounted side by side—to view objects on the earth, though, we want the image to be right-side up. In prism binoculars, this is accomplished by reflecting the light several times along the path from the objective to the eyepiece. The combined effect of the reflections is to flip the image both horizontally and vertically. Binoculars are usually described by two numbers separated by a multiplication sign, such as 7×50. The first number is the angular magnification M, and the second is the diameter of the objective lenses (in millimeters). The diameter helps to determine the light-gathering capacity of the objective lenses and thus the brightness of the image.

In the *reflecting telescope* (Fig. 34.54a) the objective lens is replaced by a concave mirror. In large telescopes this scheme has many advantages, both theoretical and practical. Mirrors are inherently free of chromatic aberrations (dependence of focal length on wavelength), and spherical aberrations (associated with the paraxial approximation) are easier to correct than with a lens. The reflecting surface is sometimes parabolic rather than spherical. The material of the mirror need not be transparent, and it can be made more rigid than a lens, which has to be supported only at its edges.

The largest reflecting telescopes in the world, the Keck telescopes atop Mauna Kea in Hawaii, each have an objective mirror of overall diameter 10 m made up of 36 separate hexagonal reflecting elements.

34.54 (a), (b), (c) Three designs for reflecting telescopes. (d) This photo shows the interior of the Gemini North telescope, which uses the design shown in (c). The objective mirror is 8 meters in diameter.

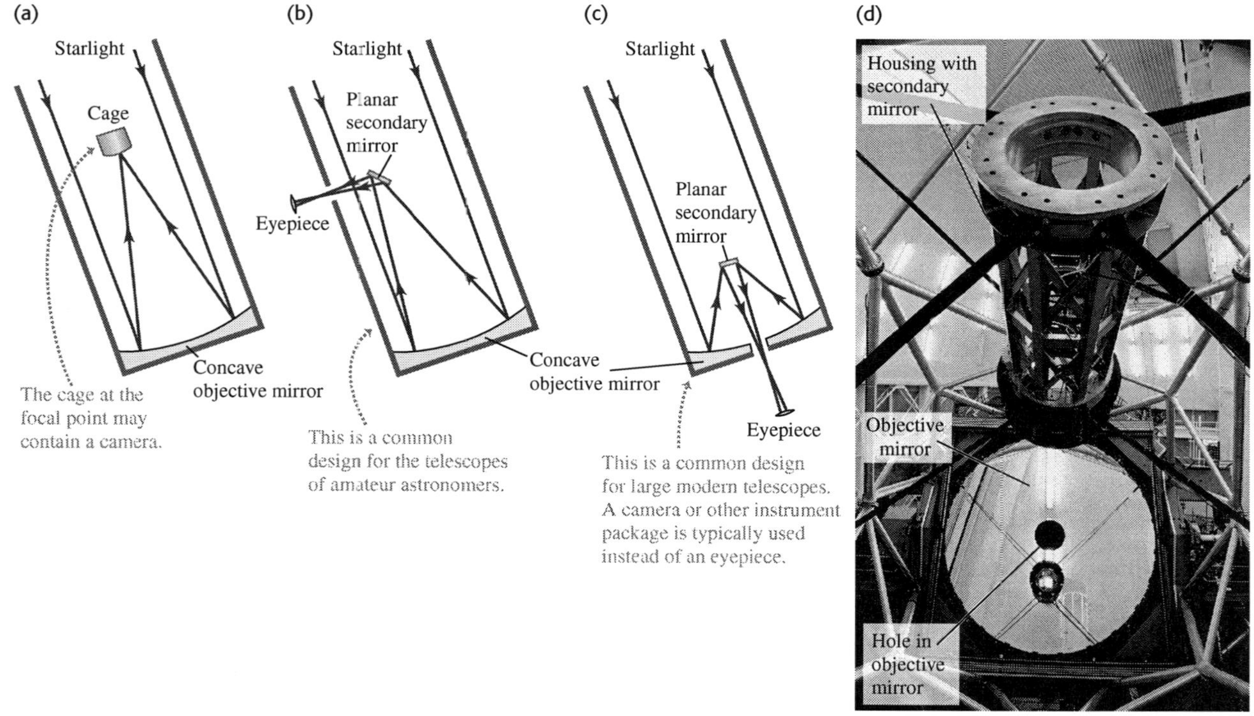

One challenge in designing reflecting telescopes is that the image is formed in front of the objective mirror, in a region traversed by incoming rays. Isaac Newton devised one solution to this problem. A flat secondary mirror oriented at 45° to the optic axis causes the image to be formed in a hole on the side of the telescope, where it can be magnified with an eyepiece (Fig. 34.54b). Another solution uses a secondary mirror that causes the focused light to pass through a hole in the objective mirror (Fig. 34.54c). Large research telescopes, as well as many amateur telescopes, use this design (Fig. 34.54d).

Like a microscope, when a telescope is used for photography the eyepiece is removed and a CCD array or photographic film is placed at the position of the real image formed by the objective. (Some long-focal-length "lenses" for photography are actually reflecting telescopes used in this way.) Most telescopes used for astronomical research are never used with an eyepiece.

Test Your Understanding of Section 34.8 Which gives a lateral magnification of greater absolute value: (i) the objective lens in a microscope (Fig. 34.52); (ii) the objective lens in a refracting telescope (Fig. 34.53); or (iii) not enough information is given to decide?

CHAPTER 34 SUMMARY

Reflection or refraction at a plane surface: When rays diverge from an object point P and are reflected or refracted, the directions of the outgoing rays are the same as though they had diverged from a point P' called the image point. If they actually converge at P' and diverge again beyond it, P' is a real image of P; if they only appear to have diverged from P', it is a virtual image. Images can be either erect or inverted.

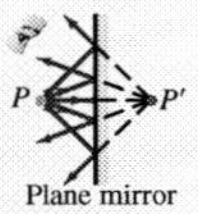

Lateral magnification: The lateral magnification m in any reflecting or refracting situation is defined as the ratio of image height y' to object height y. When m is positive, the image is erect; when m is negative, the image is inverted.

$$m = \frac{y'}{y} \qquad (34.2)$$

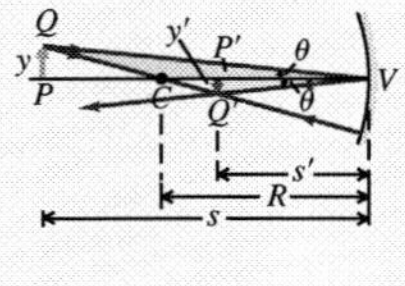

Focal point and focal length: The focal point of a mirror is the point where parallel rays converge after reflection from a concave mirror, or the point from which they appear to diverge after reflection from a convex mirror. Rays diverging from the focal point of a concave mirror are parallel after reflection; rays converging toward the focal point of a convex mirror are parallel after reflection. The distance from the focal point to the vertex is called the focal length, denoted as f. The focal points of a lens are defined similarly.

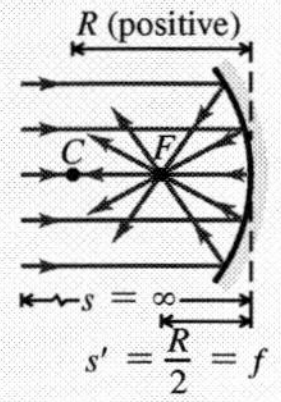

Relating object and image distances: The formulas for object distance s and image distance s' for plane and spherical mirrors and single refracting surfaces are summarized in the table. The equation for a plane surface can be obtained from the corresponding equation for a spherical surface by setting $R = \infty$. (See Examples 34.1–34.7.)

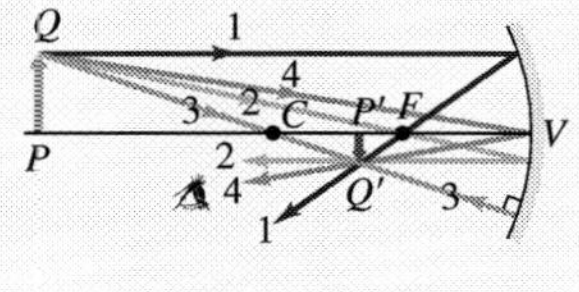

	Plane Mirror	Spherical Mirror	Plane Refracting Surface	Spherical Refracting Surface
Object and image distances	$\frac{1}{s} + \frac{1}{s'} = 0$	$\frac{1}{s} + \frac{1}{s'} = \frac{2}{R} = \frac{1}{f}$	$\frac{n_a}{s} + \frac{n_b}{s'} = 0$	$\frac{n_a}{s} + \frac{n_b}{s'} = \frac{n_b - n_a}{R}$
Lateral magnification	$m = -\frac{s'}{s} = 1$	$m = -\frac{s'}{s}$	$m = -\frac{n_a s'}{n_b s} = 1$	$m = -\frac{n_a s'}{n_b s}$

Object-image relationships derived in this chapter are valid only for rays close to and nearly parallel to the optic axis; these are called paraxial rays. Nonparaxial rays do not converge precisely to an image point. This effect is called spherical aberration.

Thin lenses: The object-image relationships, given by Eq. (34.16), is the same for a thin lens as for a spherical mirror. Equation (34.19), the lensmaker's equation, relates the focal length of a lens to its index of refraction and the radii of curvature of its surfaces. (See Examples 34.8–34.11.)

$$\frac{1}{s} + \frac{1}{s'} = \frac{1}{f} \qquad (34.16)$$

$$\frac{1}{f} = (n - 1)\left(\frac{1}{R_1} - \frac{1}{R_2}\right) \qquad (34.19)$$

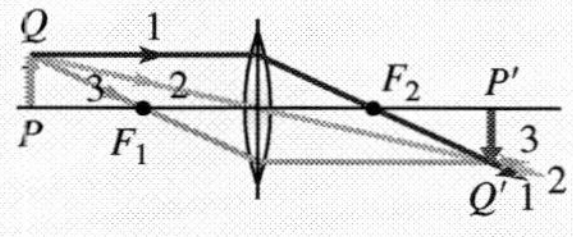

Sign rules: The following sign rules are used with all plane and spherical reflecting and refracting surfaces.

- $s > 0$ when the object is on the incoming side of the surface (a real object); $s < 0$ otherwise.
- $s' > 0$ when the image is on the outgoing side of the surface (a real image); $s' < 0$ otherwise.
- $R > 0$ when the center of curvature is on the outgoing side of the surface; $R < 0$ otherwise.
- $m > 0$ when the image is erect; $m < 0$ when inverted.

Cameras: A camera forms a real, inverted, reduced image of the object being photographed on a light-sensitive surface. The amount of light striking this surface is controlled by the shutter speed and the aperture. The intensity of this light is inversely proportional to the square of the *f*-number of the lens. (See Example 34.12.)

$$f\text{-number} = \frac{\text{Focal length}}{\text{Aperture diameter}} = \frac{f}{D} \quad (34.20)$$

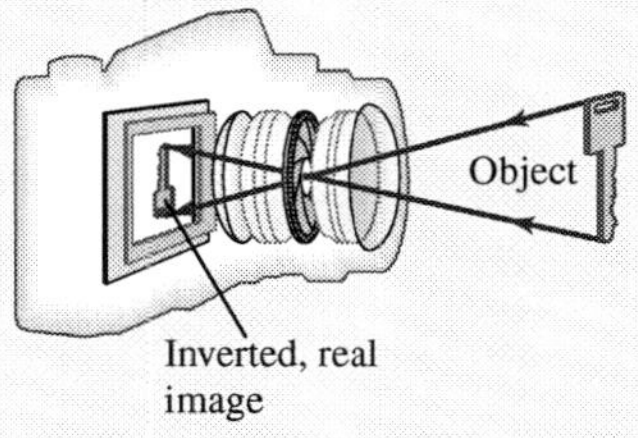

The eye: In the eye, refraction at the surface of the cornea forms a real image on the retina. Adjustment for various object distances is made by squeezing the lens, thereby making it bulge and decreasing its focal length. A nearsighted eye is too long for its lens; a farsighted eye is too short. The power of a corrective lens, in diopters, is the reciprocal of the focal length in meters. (See Examples 34.13 and 34.14.)

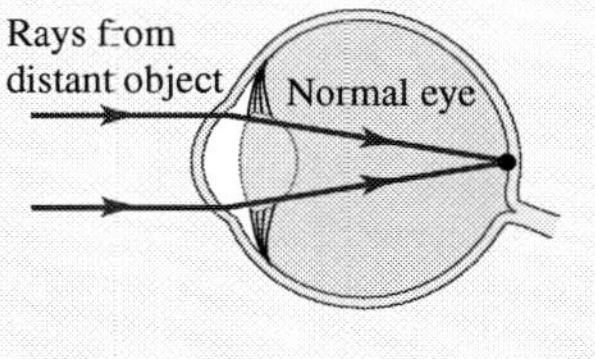

The simple magnifier: The simple magnifier creates a virtual image whose angular size θ' is larger than the angular size θ of the object itself at a distance of 25 cm, the nominal closest distance for comfortable viewing. The angular magnification M of a simple magnifier is the ratio of the angular size of the virtual image to that of the object at this distance.

$$M = \frac{\theta'}{\theta} = \frac{25\ \text{cm}}{f} \quad (34.22)$$

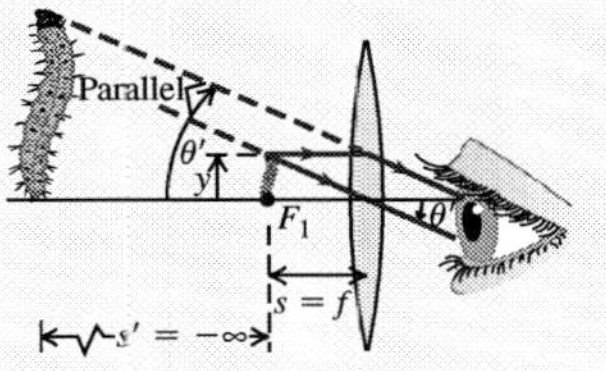

Microscopes and telescopes: In a compound microscope, the objective lens forms a first image in the barrel of the instrument, and the eyepiece forms a final virtual image, often at infinity, of the first image. The telescope operates on the same principle, but the object is far away. In a reflecting telescope, the objective lens is replaced by a concave mirror, which eliminates chromatic aberrations.

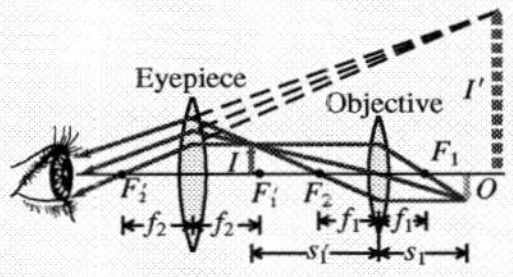

Key Terms

object, *1157*
point object, *1158*
extended object, *1158*
image, *1158*
virtual image, *1158*
real image, *1158*
object distance, *1159*
image distance, *1159*
lateral magnification, *1160*
erect image, *1160*
inverted image, *1160*
reversed image, *1160*
center of curvature, *1161*
vertex, *1161*
optic axis, *1161*
paraxial rays, *1162*
paraxial approximation, *1162*
spherical aberration, *1162*
focal point, *1163*
focal length, *1163*
principal rays, *1167*
thin lens, *1174*
converging lens, *1174*
diverging lens, *1176*
lensmaker's equation, *1177*
camera, *1182*
f-number, *1184*
power, *1187*
diopter, *1187*
angular size, *1189*
magnifier, *1190*
angular magnification, *1190*
microscope, *1191*
objective, *1191*
eyepiece, *1191*
telescope, *1192*

Answer to Chapter Opening Question ?

A magnifying lens (simple magnifier) produces a virtual image with a large angular size that is infinitely far away, so you can see it in sharp focus with your eyes relaxed. (A surgeon doing microsurgery would not appreciate having to strain his eyes while working.) The object should be at the focal point of the lens, so the object and lens are separated by one focal length.

Answers to Test Your Understanding Questions

34.1 Answer: (iv) When you are a distance s from the mirror, your image is a distance s on the other side of the mirror and the distance from you to your image is $2s$. As you move toward the mirror, the distance $2s$ changes at twice the rate of the distance s, so your image moves toward you at speed $2v$.

34.2 Answers: (a) concave, (b) (ii) A convex mirror always produces an erect image, but that image is smaller than the object (see

Fig. 34.16b). Hence a concave mirror must be used. The image will be erect and enlarged only if the distance from the object (your face) to the mirror is less than the focal length of the mirror, as in Fig. 34.20d.

34.3 Answer: no The sun is very far away, so the object distance is essentially infinite: $s = \infty$ and $1/s = 0$. Material a is air $(n_a = 1.00)$ and material b is water $(n_b = 1.33)$, so the image position s' is given by

$$\frac{n_a}{s} + \frac{n_b}{s'} = \frac{n_b - n_a}{R} \quad \text{or} \quad 0 + \frac{1.33}{s'} = \frac{1.33 - 1.00}{R}$$

$$s' = \frac{1.33}{0.33}R = 4.0R$$

The image would be formed 4.0 drop radii from the front surface of the drop. But since each drop is only a part of a complete sphere, the distance from the front to the back of the drop is less than $2R$. Thus the rays of sunlight never reach the image point, and the drops do not form an image of the sun on the leaf. While the rays are not focused to a point, they are nonetheless concentrated and can cause damage to the leaf.

34.4 Answers: A and C When rays A and D are extended backward, they pass through focal point F_2; thus, before they passed through the lens, they were parallel to the optic axis. The figures show that ray A emanated from point Q, but ray D did not. Ray B is parallel to the optic axis, so before it passed through the lens, it was directed toward focal point F_1. Hence it cannot have come from point Q. Ray C passes through the center of the lens and hence is not deflected by its passage; tracing the ray backward shows that it emanates from point Q.

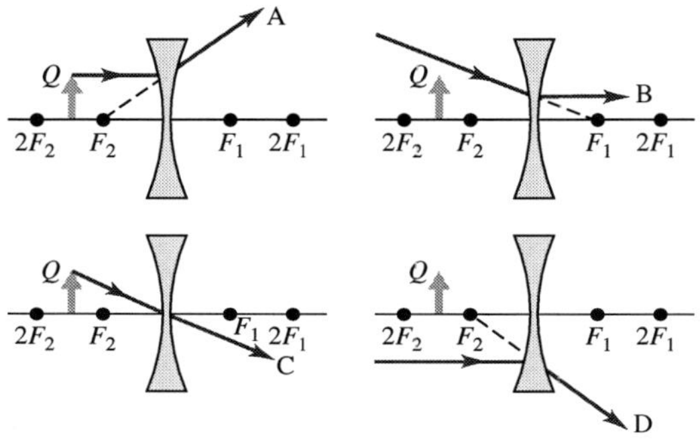

34.5 Answer: (iii) The smaller image area of the CCD array means that the angle of view is decreased for a given focal length. Individual objects make images of the same size in either case; when a smaller light-sensitive area is used, fewer images fit into the area and the field of view is narrower.

34.6 Answer: (iii) This lens is designed to correct for a type of astigmatism. Along the vertical axis, the lens is configured as a converging lens; along the horizontal axis, the lens is configured as a diverging lens. Hence the eye is hyperopic (see Fig. 34.46) for objects that are oriented vertically but myopic for objects that are oriented horizontally (see Fig. 34.47). Without correction, the eye focuses vertical objects behind the retina but horizontal objects in front of the retina.

34.7 Answer: (ii) The object must be held at the focal point, which is twice as far away if the focal length f is twice as great. Equation (24.22) shows that the angular magnification M is inversely proportional to f, so doubling the focal length makes $M\,\frac{1}{2}$ as great. To improve the magnification, you should use a magnifier with a *shorter* focal length.

34.8 Answer: (i) The objective lens of a microscope is designed to make enlarged images of small objects, so the absolute value of its lateral magnification m is greater than 1. By contrast, the objective lens of a refracting telescope is designed to make *reduced* images. For example, the moon is thousands of kilometers in diameter, but its image may fit on a CCD array a few centimeters across. Thus $|m|$ is much less than 1 for a refracting telescope. (In both cases m is negative because the objective makes an inverted image, which is why the question asks about the absolute value of m.)

PROBLEMS

For instructor-assigned homework, go to **www.masteringphysics.com**

Discussion Questions

Q34.1. A spherical mirror is cut in half horizontally. Will an image be formed by the bottom half of the mirror? If so, where will the image be formed?

Q34.2. For the situation shown in Fig. 34.3, is the image distance s' positive or negative? Is the image real or virtual? Explain your answers.

Q34.3. The laws of optics also apply to electromagnetic waves invisible to the eye. A satellite TV dish is used to detect radio waves coming from orbiting satellites. Why is a curved reflecting surface (a "dish") used? The dish is always concave, never convex; why? The actual radio receiver is placed on an arm and suspended in front of the dish. How far in front of the dish should it be placed?

Q34.4. Explain why the focal length of a *plane* mirror is infinite, and explain what it means for the focal point to be at infinity.

Q34.5. If a spherical mirror is immersed in water, does its focal length change? Explain.

Q34.6. For what range of object positions does a concave spherical mirror form a real image? What about a convex spherical mirror?

Q34.7. When a room has mirrors on two opposite walls, an infinite series of reflections can be seen. Discuss this phenomenon in terms of images. Why do the distant images appear fainter?

Q34.8. For a spherical mirror, if $s = f$, then $s' = \infty$, and the lateral magnification m is infinite. Does this make sense? If so, what does it mean?

Q34.9. You may have noticed a small convex mirror next to your bank's ATM. Why is this mirror convex, as opposed to flat or concave? What considerations determine its radius of curvature?

Q34.10. A student claims that she can start a fire on a sunny day using just the sun's rays and a concave mirror. How is this done? Is the concept of image relevant? Can she do the same thing with a convex mirror? Explain.

Q34.11. A person looks at his reflection in the concave side of a shiny spoon. Is it right side up or inverted? Does it matter how far his face is from the spoon? What if he looks in the convex side? (Try this yourself!)

Q34.12. In Example 34.4 (Section 34.2), there appears to be an ambiguity for the case $s = 10$ cm as to whether s' is $+\infty$ or $-\infty$ and whether the image is erect or inverted. How is this resolved? Or is it?

Q34.13. Suppose that in the situation of Example 34.7 of Section 34.3 (see Fig. 34.26) a vertical arrow 2.00 m tall is painted on the side of the pool beneath the water line. According to the calculations in the example, this arrow would appear to the person shown in Fig. 34.26 to be 1.50 m long. But the discussion following Eq. (34.13) states that the magnification for a plane refracting surface is $m = 1$, which suggests that the arrow would appear to the person to be 2.00 m long. How can you resolve this apparent contradiction?

Q34.14. The bottom of the passenger side mirror on your car notes, "Objects in mirror are closer than they appear." Is this true? Why?

Q34.15. How could you very quickly make an approximate measurement of the focal length of a converging lens? Could the same method be applied if you wished to use a diverging lens? Explain.

Q34.16. The focal length of a simple lens depends on the color (wavelength) of light passing through it. Why? Is it possible for a lens to have a positive focal length for some colors and negative for others? Explain.

Q34.17. When a converging lens is immersed in water, does its focal length increase or decrease in comparison with the value in air? Explain.

Q34.18. A spherical air bubble in water can function as a lens. Is it a converging or diverging lens? How is its focal length related to its radius?

Q34.19. Can an image formed by one reflecting or refracting surface serve as an object for a second reflection or refraction? Does it matter whether the first image is real or virtual? Explain.

Q34.20. If a piece of photographic film is placed at the location of a real image, the film will record the image. Can this be done with a virtual image? How might one record a virtual image?

Q34.21. According to the discussion in Section 34.2, light rays are reversible. Are the formulas in the table in this chapter's Summary still valid if object and image are interchanged? What does reversibility imply with respect to the *forms* of the various formulas?

Q34.22. You've entered a survival contest that will include building a crude telescope. You are given a large box of lenses. Which two lenses do you pick? How do you quickly identify them?

Q34.23. You can't see clearly underwater with the naked eye, but you *can* if you wear a face mask or goggles (with air between your eyes and the mask or goggles). Why is there a difference? Could you instead wear eyeglasses (with water between your eyes and the eyeglasses) in order to see underwater? If so, should the lenses be converging or diverging? Explain.

Q34.24. You take a lens and mask it so that light can pass through only the bottom half of the lens. How does the image formed by the masked lens compare to the image formed before masking?

Exercises

Section 34.1 Reflection and Refraction at a Plane Surface

34.1. A candle 4.85 cm tall is 39.2 cm to the left of a plane mirror. Where is the image formed by the mirror, and what is the height of this image?

34.2. The image of a tree just covers the length of a plane mirror 4.00 cm tall when the mirror is held 35.0 cm from the eye. The tree is 28.0 m from the mirror. What is its height?

34.3. As shown in Fig. 34.9, mirror 1 uses the image P_2' formed by mirror 2 as an object and forms an image of it. Show that this image is at point P_3' in the figure.

Section 34.2 Reflection at a Spherical Surface

34.4. A concave mirror has a radius of curvature of 34.0 cm. (a) What is its focal length? (b) If the mirror is immersed in water (refractive index 1.33), what is its focal length?

34.5. An object 0.600 cm tall is placed 16.5 cm to the left of the vertex of a concave spherical mirror having a radius of curvature of 22.0 cm. (a) Draw a principal-ray diagram showing the formation of the image. (b) Determine the position, size, orientation, and nature (real or virtual) of the image.

34.6. Repeat Exercise 34.5 for the case in which the mirror is convex.

34.7. The diameter of Mars is 6794 km, and its minimum distance from the earth is 5.58×10^7 km. When Mars is at this distance, find the diameter of the image of Mars formed by a spherical, concave, telescope mirror with a focal length of 1.75 m.

34.8. An object is 24.0 cm from the center of a silvered spherical glass Christmas tree ornament 6.00 cm in diameter. What are the position and magnification of its image?

34.9. A coin is placed next to the convex side of a thin spherical glass shell having a radius of curvature of 18.0 cm. An image of the 1.5-cm-tall coin is formed 6.00 cm behind the glass shell. Where is the coin located? Determine the size, orientation, and nature (real or virtual) of the image.

34.10. You hold a spherical salad bowl 90 cm in front of your face with the bottom of the bowl facing you. The salad bowl is made of polished metal with a 35-cm radius of curvature. (a) Where is the image of your 2.0-cm-tall nose located? (b) What are the image's size, orientation, and nature (real or virtual)?

34.11. (a) Show that Eq. (34.6) can be written as $s' = sf/(s - f)$ and hence that the lateral magnification, given by Eq. (34.7), can be expressed as $m = f/(f - s)$. (b) Use these formulas for s' and m to graph s' as a function of s for the case $f > 0$ (a concave mirror). (c) For what values of s is s' positive, so that the image is real? (d) For what values of s is s' negative, so that the image is virtual? (e) Where is the image if the object is just inside the focal point (s slightly less than f)? (f) Where is the image if the object is at infinity? (g) Where is the image if the object is next to the mirror $(s = 0)$? (h) Graph m as a function of s for the case of a concave mirror. (i) For which values of s is the image erect and larger than the object? (j) For what values of s is the image inverted? (k) For which values of s is the image smaller than the object? (l) What happens to the size of the image when the object is placed at the focal point?

34.12. Using the formulas for s' and m obtained in part (a) of Exercise 34.11, graph s' as a function of s, and graph m as a function of s, for the case $f < 0$ (a convex mirror), so that $f = -|f|$. (a) For which values of s is s' positive? (b) For what values of s is s' negative? (c) Where is the image if the object is at infinity? (d) Where is the image if the object is next to the mirror $(s = 0)$? For which values of s is the image (e) erect; (f) inverted; (g) larger than the object; (h) smaller than the object?

34.13. Dental Mirror. A dentist uses a curved mirror to view teeth on the upper side of the mouth. Suppose she wants an erect image with a magnification of 2.00 when the mirror is 1.25 cm from a tooth. (Treat this problem as though the object and image lie along a straight line.) (a) What kind of mirror (concave or convex) is needed? Use a ray diagram to decide, without performing any calculations. (b) What must be the focal length and radius of curvature of this mirror? (c) Draw a principal-ray diagram to check your answer in part (b).

34.14. A spherical, concave, shaving mirror has a radius of curvature of 32.0 cm. (a) What is the magnification of a person's face when it is 12.0 cm to the left of the vertex of the mirror? (b) Where is the image? Is the image real or virtual? (c) Draw a principal-ray diagram showing the formation of the image.

Section 34.3 Refraction at a Spherical Surface

34.15. A speck of dirt is embedded 3.50 cm below the surface of a sheet of ice $(n = 1.309)$. What is its apparent depth when viewed at normal incidence?

34.16. A tank whose bottom is a mirror is filled with water to a depth of 20.0 cm. A small fish floats motionless 7.0 cm under the surface of the water. (a) What is the apparent depth of the fish when viewed at normal incidence? (b) What is the apparent depth of the image of the fish when viewed at normal incidence?

34.17. A Spherical Fish Bowl. A small tropical fish is at the center of a water-filled, spherical fish bowl 28.0 cm in diameter. (a) Find the apparent position and magnification of the fish to an observer outside the bowl. The effect of the thin walls of the bowl may be ignored. (b) A friend advised the owner of the bowl to keep it out of direct sunlight to avoid blinding the fish, which might swim into the focal point of the parallel rays from the sun. Is the focal point actually within the bowl?

34.18. The left end of a long glass rod 6.00 cm in diameter has a convex hemispherical surface 3.00 cm in radius. The refractive index of the glass is 1.60. Determine the position of the image if an object is placed in air on the axis of the rod at the following distances to the left of the vertex of the curved end: (a) infinitely far, (b) 12.0 cm; (c) 2.00 cm.

34.19. The glass rod of Exercise 34.18 is immersed in oil $(n = 1.45)$. An object placed to the left of the rod on the rod's axis is to be imaged 1.20 m inside the rod. How far from the left end of the rod must the object be located to form the image?

34.20. The left end of a long glass rod 8.00 cm in diameter, with an index of refraction 1.60, is ground and polished to a convex hemispherical surface with a radius of 4.00 cm. An object in the form of an arrow 1.50 mm tall, at right angles to the axis of the rod, is located on the axis 24.0 cm to the left of the vertex of the convex surface. Find the position and height of the image of the arrow formed by paraxial rays incident on the convex surface. Is the image erect or inverted?

34.21. Repeat Exercise 34.20 for the case in which the end of the rod is ground to a *concave* hemispherical surface with radius 4.00 cm.

34.22. The glass rod of Exercise 34.21 is immersed in a liquid. An object 14.0 cm from the vertex of the left end of the rod and on its axis is imaged at a point 9.00 cm from the vertex inside the liquid. What is the index of refraction of the liquid?

Section 34.4 Thin Lenses

34.23. An insect 3.75 mm tall is placed 22.5 cm to the left of a thin planoconvex lens. The left surface of this lens is flat, the right surface has a radius of curvature of magnitude 13.0 cm, and the index of refraction of the lens material is 1.70. (a) Calculate the location and size of the image this lens forms of the insect. Is it real or virtual? Erect or inverted? (b) Repeat part (a) if the lens is reversed.

34.24. A lens forms an image of an object. The object is 16.0 cm from the lens. The image is 12.0 cm from the lens on the same side as the object. (a) What is the focal length of the lens? Is the lens converging or diverging? (b) If the object is 8.50 mm tall, how tall is the image? Is it erect or inverted? (c) Draw a principal-ray diagram.

34.25. A converging meniscus lens (see Fig. 34.32a) with a refractive index of 1.52 has spherical surfaces whose radii are 7.00 cm and 4.00 cm. What is the position of the image if an object is placed 24.0 cm to the left of the lens? What is the magnification?

34.26. A converging lens with a focal length of 90.0 cm forms an image of a 3.20-cm-tall real object that is to the left of the lens. The image is 4.50 cm tall and inverted. Where are the object and image located in relation to the lens? Is the image real or virtual?

34.27. A converging lens forms an image of an 8.00-mm-tall real object. The image is 12.0 cm to the left of the lens, 3.40 cm tall, and erect. What is the focal length of the lens? Where is the object located?

34.28. A photographic slide is to the left of a lens. The lens projects an image of the slide onto a wall 6.00 m to the right of the slide. The image is 80.0 times the size of the slide. (a) How far is the slide from the lens? (b) Is the image erect or inverted? (c) What is the focal length of the lens? (d) Is the lens converging or diverging?

34.29. A double-convex thin lens has surfaces with equal radii of curvature of magnitude 2.50 cm. Looking through this lens, you observe that it forms an image of a very distant tree at a distance of 1.87 cm from the lens. What is the index of refraction of the lens?

34.30. Six lenses in air are shown in Fig. 34.32. Each lens is made of a material with index of refraction $n > 1$. Considering each lens individually, imagine that light enters the lens from the left. Show that the three lenses shown in Fig. 34.32a have *positive* focal lengths and hence are *converging* lenses. In addition, show that the three lenses in Fig. 34.32b have *negative* focal lengths and hence are *diverging* lenses.

34.31. Exercises 34.11 and 34.12 deal with spherical mirrors. (a) Show that the equations for s' and m derived in part (a) of Exercise 34.11 also apply to a thin lens. (b) A concave mirror is used in Exercise 34.11. Repeat these exercises for a converging lens. Are there any differences in the results when the mirror is replaced by a lens? Explain. (c) A convex mirror is used in Exercise 34.12. Repeat these exercises for a diverging lens. Are there any differences in the results when the mirror is replaced by a lens? Explain.

34.32. A converging lens with a focal length of 12.0 cm forms a virtual image 8.00 mm tall, 17.0 cm to the right of the lens. Determine the position and size of the object. Is the image erect or inverted? Are the object and image on the same side or opposite sides of the lens? Draw a principal-ray diagram for this situation.

34.33. Repeat Exercise 34.32 for the case in which the lens is diverging, with a focal length of -48.0 cm.

34.34. An object is 16.0 cm to the left of a lens. The lens forms an image 36.0 cm to the right of the lens. (a) What is the focal length of the lens? Is the lens converging or diverging? (b) If the object is 8.00 mm tall, how tall is the image? Is it erect or inverted? (c) Draw a principal-ray diagram.

Section 34.5 Cameras

34.35. A camera lens has a focal length of 200 mm. How far from the lens should the subject for the photo be if the lens is 20.4 cm from the film?

34.36. When a camera is focused, the lens is moved away from or toward the film. If you take a picture of your friend, who is standing 3.90 m from the lens, using a camera with a lens with a 85-mm focal length, how far from the film is the lens? Will the whole image of your friend, who is 175 cm tall, fit on film that is 24×36 mm?

34.37. Figure 34.41 shows photographs of the same scene taken with the same camera with lenses of different focal length. If the object is 200 m from the lens, what is the magnitude of the lateral magnification for a lens of focal length (a) 28 mm; (b) 105 mm; (c) 300 mm?

34.38. A photographer takes a photograph of a Boeing 747 airliner (length 70.7 m) when it is flying directly overhead at an altitude of 9.50 km. The lens has a focal length of 5.00 m. How long is the image of the airliner on the film?

34.39. Choosing a Camera Lens. The picture size on ordinary 35-mm camera film is 24 mm $\times$ 36 mm. Focal lengths of lenses available for 35-mm cameras typically include 28, 35, 50 (the "normal" lens), 85, 100, 135, 200, and 300 mm, among others. Which of these lenses should be used to photograph the following objects, assuming that the object is to fill most of the picture area? (a) a building 240 m tall and 160 m wide at a distance of 600 m, and (b) a mobile home 9.6 m in length at a distance of 40.0 m.

34.40. Zoom Lens. Consider the simple model of the zoom lens shown in Fig. 34.43a. The converging lens has focal length $f_1 = 12$ cm, and the diverging lens has focal length $f_2 = -12$ cm. The lenses are separated by 4 cm as shown in Fig. 34.43a. (a) For a distant object, where is the image of the converging lens? (b) The image of the converging lens serves as the object for the diverging lens. What is the object distance for the diverging lens? (c) Where is the final image? Compare your answer to Fig. 34.43a. (d) Repeat parts (a), (b), and (c) for the situation shown in Fig. 34.43b, in which the lenses are separated by 8 cm.

34.41. A camera lens has a focal length of 180.0 mm and an aperture diameter of 16.36 mm. (a) What is the f-number of the lens? (b) If the correct exposure of a certain scene is $\frac{1}{30}$ s at $f/11$, what is the correct exposure at $f/2.8$?

34.42. Recall that the intensity of light reaching film in a camera is proportional to the effective area of the lens. Camera A has a lens with an aperture diameter of 8.00 mm. It photographs an object using the correct exposure time of $\frac{1}{30}$ s. What exposure time should be used with camera B in photographing the same object with the same film if this camera has a lens with an aperture diameter of 23.1 mm?

34.43. Photography. A 35-mm camera has a standard lens with focal length 50 mm and can focus on objects between 45 cm and infinity. (a) Is the lens for such a camera a concave or a convex lens? (b) The camera is focused by rotating the lens, which moves it on the camera body and changes its distance from the film. In what range of distances between the lens and the film plane must the lens move to focus properly over the 45 cm to infinity range?

34.44. You wish to project the image of a slide on a screen 9.00 m from the lens of a slide projector. (a) If the slide is placed 15.0 cm from the lens, what focal length lens is required? (b) If the dimensions of the picture on a 35-mm color slide are 24 mm $\times$ 36 mm, what is the minimum size of the projector screen required to accommodate the image?

Section 34.6 The Eye

34.45. (a) Where is the near point of an eye for which a contact lens with a power of +2.75 diopters is prescribed? (b) Where is the far point of an eye for which a contact lens with a power of -1.30 diopters is prescribed for distant vision?

34.46. Curvature of the Cornea. In a simplified model of the human eye, the aqueous and vitreous humors and the lens all have a refractive index of 1.40, and all the refraction occurs at the cornea, whose vertex is 2.60 cm from the retina. What should be the radius of curvature of the cornea such that the image of an object 40.0 cm from the cornea's vertex is focused on the retina?

34.47. Corrective Lenses. Determine the power of the corrective contact lenses required by (a) a hyperopic eye whose near point is at 60.0 cm and (b) a myopic eye whose far point is at 60.0 cm.

Section 34.7 The Magnifier

34.48. A thin lens with a focal length of 6.00 cm is used as a simple magnifier. (a) What angular magnification is obtainable with the lens if the object is at the focal point? (b) When an object is examined through the lens, how close can it be brought to the lens? Assume that the image viewed by the eye is at the near point, 25.0 cm from the eye, and that the lens is very close to the eye.

34.49. The focal length of a simple magnifier is 8.00 cm. Assume the magnifier is a thin lens placed very close to the eye. (a) How far in front of the magnifier should an object be placed if the image is formed at the observer's near point, 25.0 cm in front of her eye? (b) If the object is 1.00 mm high, what is the height of its image formed by the magnifier?

34.50. You want to view an insect 2.00 mm in length through a magnifier. If the insect is to be at the focal point of the magnifier, what focal length will give the image of the insect an angular size of 0.025 radian?

34.51. You are examining an ant with a magnifying lens that has focal length 5.00 cm. If the image of the ant appears 25.0 cm from the lens, how far is the ant from the lens? On which side of the lens is the image located?

Section 34.8 Microscopes and Telescopes

34.52. Resolution of a Microscope. The image formed by a microscope objective with a focal length of 5.00 mm is 160 mm from its second focal point. The eyepiece has a focal length of 25.0 mm. (a) What is the angular magnification of the microscope? (b) The unaided eye can distinguish two points at its near point as separate if they are about 0.10 mm apart. What is the minimum separation between two points that can be observed (or resolved) through this microscope?

34.53. The focal length of the eyepiece of a certain microscope is 18.0 mm. The focal length of the objective is 8.00 mm. The distance between objective and eyepiece is 19.7 cm. The final image formed by the eyepiece is at infinity. Treat all lenses as thin. (a) What is the distance from the objective to the object being viewed? (b) What is the magnitude of the linear magnification produced by the objective? (c) What is the overall angular magnification of the microscope?

34.54. A certain microscope is provided with objectives that have focal lengths of 16 mm, 4 mm, and 1.9 mm and with eyepieces that have angular magnifications of 5$\times$ and 10$\times$. Each objective forms an image 120 mm beyond its second focal point. Determine (a) the largest overall angular magnification obtainable and (b) the least overall angular magnification obtainable.

34.55. The Yerkes refracting telescope of the University of Chicago has an objective 1.02 m in diameter with an f-number of 19.0. (This is the largest-diameter refracting telescope in the world.) What is its focal length?

34.56. The eyepiece of a refracting telescope (see Fig. 34.53) has a focal length of 9.00 cm. The distance between objective and eyepiece is 1.80 m, and the final image is at infinity. What is the angular magnification of the telescope?

34.57. A telescope is constructed from two lenses with focal lengths of 95.0 cm and 15.0 cm, the 95.0-cm lens being used as the objective. Both the object being viewed and the final image are at infinity. (a) Find the angular magnification for the telescope. (b) Find the height of the image formed by the objective of a building 60.0 m tall, 3.00 km away. (c) What is the angular size of the final image as viewed by an eye very close to the eyepiece?

34.58. Saturn is viewed through the Lick Observatory refracting telescope (objective focal length 18 m). If the diameter of the image of Saturn produced by the objective is 1.7 mm, what angle does Saturn subtend from when viewed from earth?

34.59. A reflecting telescope (Fig. 34.55a) is to be made by using a spherical mirror with a radius of curvature of 1.30 m and an eyepiece with a focal length of 1.10 cm. The final image is at infinity. (a) What should the distance between the eyepiece and the mirror vertex be if the object is taken to be at infinity? (b) What will the angular magnification be?

Figure **34.55** Exercises 34.59 and 34.60 and Problem 34.112.

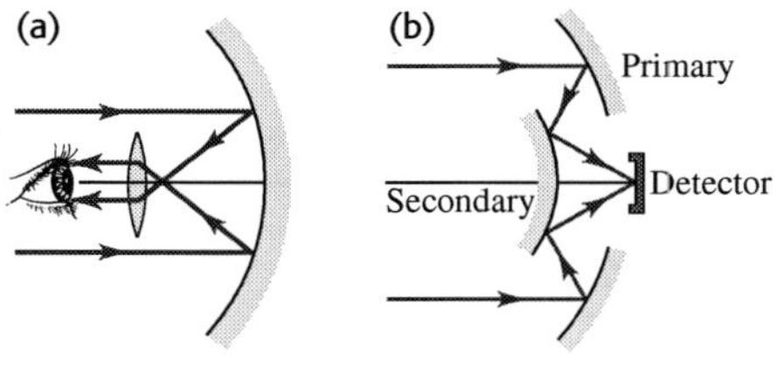

34.60. A Cassegrain telescope is a reflecting telescope that uses two mirrors, the secondary mirror focusing the image through a hole in the primary mirror (similar to that shown in Fig. 34.55b). You wish to focus the image of a distant galaxy onto the detector shown in the figure. If the primary mirror has a focal length of 2.5 m, the secondary mirror has a focal length of -1.5 m and the distance from the vertex of the primary mirror to the detector is 15 cm. What should be the distance between the vertices of the two mirrors?

Problems

34.61. If you run away from a plane mirror at 2.40 m/s, at what speed does your image move away from you?

34.62. An object is placed between two plane mirrors arranged at right angles to each other at a distance d_1 from the surface of one mirror and a distance d_2 from the other. (a) How many images are formed? Show the location of the images in a diagram. (b) Draw the paths of rays from the object to the eye of an observer.

34.63. What is the size of the smallest vertical plane mirror in which a woman of height h can see her full-length image?

34.64. A light bulb is 4.00 m from a wall. You are to use a concave mirror to project an image of the bulb on the wall, with the image 2.25 times the size of the object. How far should the mirror be from the wall? What should its radius of curvature be?

34.65. A concave mirror is to form an image of the filament of a headlight lamp on a screen 8.00 m from the mirror. The filament is 6.00 mm tall, and the image is to be 36.0 cm tall. (a) How far in front of the vertex of the mirror should the filament be placed? (b) What should be the radius of curvature of the mirror?

34.66. Rear-View Mirror. A mirror on the passenger side of your car is convex and has a radius of curvature with magnitude 18.0 cm. (a) Another car is seen in this side mirror and is 13.0 m behind the mirror. If this car is 1.5 m tall, what is the height of the image? (b) The mirror has a warning attached that objects viewed in it are closer than they appear. Why is this so?

34.67. Suppose the lamp filament shown in Example 34.1 (Section 34.2) is moved to a position 8.0 cm in front of the mirror. (a) Where is the image located now? Is it real or virtual? (b) What is the height of the image? Is it erect or inverted? (c) In Example 34.1, the filament is 10.0 cm in front of the mirror, and an image of the filament is formed on a wall 3.00 m from the mirror. If the filament is 8.0 cm from the mirror, can a wall be placed so that an image is formed on it? If so, where should the wall be placed? If not, why not?

34.68. Where must you place an object in front of a concave mirror with radius R so that the image is erect and $2\frac{1}{2}$ times the size of the object? Where is the image?

34.69. Virtual Object. If the light incident from the left onto a convex mirror does not diverge from an object point but instead converges toward a point at a (negative) distance s to the right of the mirror, this point is called a *virtual object.* (a) For a convex mirror having a radius of curvature of 24.0 cm, for what range of virtual-object positions is a real image formed? (b) What is the orientation of this real image? (c) Draw a principal-ray diagram showing the formation of such an image.

34.70. A layer of benzene ($n = 1.50$) 2.60 cm deep floats on water ($n = 1.33$) that is 6.50 cm deep. What is the apparent distance from the upper benzene surface to the bottom of the water layer when it is viewed at normal incidence?

34.71. Sketch the various possible thin lenses that can be obtained by combining two surfaces whose radii of curvature are 4.00 cm and 8.00 cm in absolute magnitude. Which are converging and which are diverging? Find the focal length of each if the surfaces are made of glass with index of refraction 1.60.

34.72. Figure 34.56 shows a small plant near a thin lens. The ray shown is one of the principal rays for the lens. Each square is 2.0 cm along the horizontal direction, but the vertical direction is not to the same scale. Use information from the diagram to answer the following questions: (a) Using only the ray shown, decide what type of lens (converging or diverging) this is. (b) What is the focal length of the lens? (c) Locate the image by drawing the other two principal rays. (d) Calculate where the image should be, and compare this result with the graphical solution in part (c).

Figure **34.56** Problem 34.72.

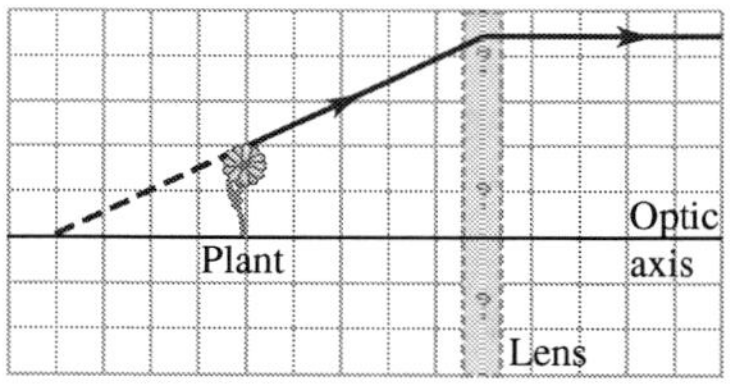

34.73. You are in your car driving on a highway at 25 m/s when you glance in the passenger side mirror (a convex mirror with radius of curvature 150 cm) and notice a truck approaching. If the image of the truck is approaching the vertex of the mirror at a speed of 1.5 m/s when the truck is 2.0 m away, what is the speed of the truck relative to the highway?

34.74. A microscope is focused on the upper surface of a glass plate. A second plate is then placed over the first. To focus on the bottom surface of the second plate, the microscope must be raised 0.780 mm. To focus on the upper surface, it must be raised another 2.50 mm. Find the index of refraction of the second plate.

34.75. Three-Dimensional Image. The *longitudinal* magnification is defined as $m' = ds'/ds$. It relates the longitudinal dimension of a small object to the longitudinal dimension of its image. (a) Show that for a spherical mirror, $m' = -m^2$. What is the significance of the fact that m' is *always* negative? (b) A wire frame in the form of a small cube 1.00 mm on a side is placed with its center on the axis of a concave mirror with radius of curvature 150.0 cm. The sides of the cube are all either parallel or perpendicular to the axis. The cube face toward the mirror is 200.0 cm to the left of the mirror vertex. Find (i) the location of the image of this face and of the opposite face of the cube; (ii) the lateral and longitudinal magnifications; (iii) the shape and dimensions of each of the six faces of the image.

34.76. Refer to Problem 34.75. Show that the longitudinal magnification m' for refraction at a spherical surface is given by

$$m' = -\frac{n_b}{n_a}m^2$$

34.77. Pinhole Camera. A pinhole camera is just a rectangular box with a tiny hole in one face. The film is on the face opposite this hole, and that is where the image is formed. The camera forms an image *without* a lens. (a) Make a clear ray diagram to show how a pinhole camera can form an image on the film without using a lens. (*Hint:* Put an object outside the hole, and then draw rays passing through the hole to the opposite side of the box.) (b) A certain pinhole camera is a box that is 25 cm square and 20.0 cm deep, with the hole in the middle of one of the 25 cm × 25 cm faces. If this camera is used to photograph a fierce chicken that is 18 cm high and 1.5 m in front of the camera, how large is the image of this bird on the film? What is the magnification of this camera?

34.78. A Glass Rod. Both ends of a glass rod with index of refraction 1.60 are ground and polished to convex hemispherical surfaces. The radius of curvature at the left end is 6.00 cm, and the radius of curvature at the right end is 12.0 cm. The length of the rod between vertices is 40.0 cm. The object for the surface at the left end is an arrow that lies 23.0 cm to the left of the vertex of this surface. The arrow is 1.50 mm tall and at right angles to the axis. (a) What constitutes the object for the surface at the right end of the rod? (b) What is the object distance for this surface? (c) Is the object for this surface real or virtual? (*Hint:* See Problem 34.69.) (d) What is the position of the final image? (e) Is the final image real or virtual? Is it erect or inverted with respect to the original object? (f) What is the height of the final image?

34.79. The rod in Problem 34.78 is shortened to a distance of 25.0 cm between its vertices; the curvatures of its ends remain the same. As in Problem 34.78, the object for the surface at the left end is an arrow that lies 23.0 cm to the left of the vertex of this surface. The arrow is 1.50 mm tall and at right angles to the axis. (a) What is the object distance for the surface at the right end of the rod? (b) Is the object for this surface real or virtual? (c) What is the position of the final image? (d) Is the final image real or virtual? Is it erect or inverted with respect to the original object? (e) What is the height of the final image?

34.80. Figure 34.57 shows an object and its image formed by a thin lens. (a) What is the focal length of the lens, and what type of lens (converging or diverging) is it? (b) What is the height of the image? Is it real or virtual?

Figure 34.57 Problem 34.80

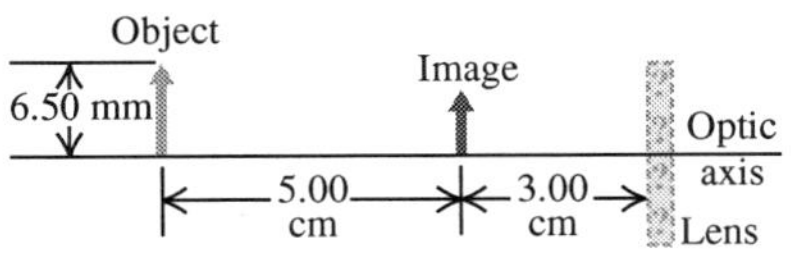

34.81. Figure 34.58 shows an object and its image formed by a thin lens. (a) What is the focal length of the lens, and what type of lens (converging or diverging) is it? (b) What is the height of the image? Is it real or virtual?

Figure 34.58 Problem 34.81

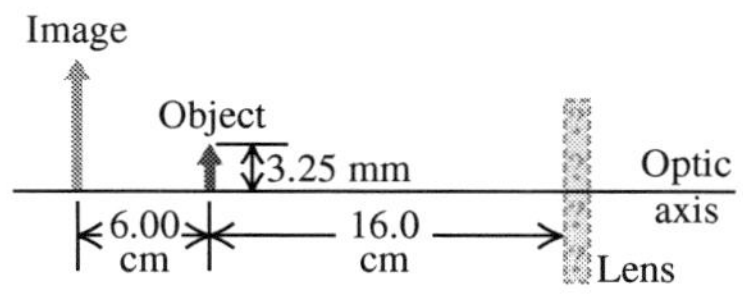

34.82. A transparent rod 30.0 cm long is cut flat at one end and rounded to a hemispherical surface of radius 10.0 cm at the other end. A small object is embedded within the rod along its axis and halfway between its ends, 15.0 cm from the flat end and 15.0 cm from the vertex of the curved end. When viewed from the flat end of the rod, the apparent depth of the object is 9.50 cm from the flat end. What is its apparent depth when viewed from the curved end?

34.83. A solid glass hemisphere of radius 12.0 cm and index of refraction $n = 1.50$ is placed with its flat face downward on a table. A parallel beam of light with a circular cross section 3.80 mm in diameter travels straight down and enters the hemisphere at the center of its curved surface. (a) What is the diameter of the circle of light formed on the table? (b) How does your result depend on the radius of the hemisphere?

34.84. A thick-walled wine goblet sitting on a table can be considered to be a hollow glass sphere with an outer radius of 4.00 cm and an inner radius of 3.40 cm. The index of refraction of the goblet glass is 1.50. (a) A beam of parallel light rays enters the side of the empty goblet along a horizontal radius. Where, if anywhere, will an image be formed? (b) The goblet is filled with white wine ($n = 1.37$). Where is the image formed?

34.85. Focus of the Eye. The cornea of the eye has a radius of curvature of approximately 0.50 cm, and the aqueous humor behind it has an index of refraction of 1.35. The thickness of the cornea itself is small enough that we shall neglect it. The depth of a typical human eye is around 25 mm. (a) What would have to be the radius of curvature of the cornea so that it alone would focus the image of a distant mountain on the retina, which is at the back of the eye opposite the cornea? (b) If the cornea focused the mountain correctly on the retina as described in part (a), would it also focus the text from a computer screen on the retina if that screen were 25 cm in front of the eye? If not, where would it focus that text: in front of or behind the retina? (c) Given that the cornea has a radius of curvature of about 5.0 mm, where does it actually focus the mountain? Is this in front of or behind the retina? Does this help you see why the eye needs help from a lens to complete the task of focusing?

34.86. A transparent rod 50.0 cm long and with a refractive index of 1.60 is cut flat at the right end and rounded to a hemispherical surface with a 15.0-cm radius at the left end. An object is placed on the axis of the rod 12.0 cm to the left of the vertex of the hemispherical end. (a) What is the position of the final image? (b) What is its magnification?

34.87. What should be the index of refraction of a transparent sphere in order for paraxial rays from an infinitely distant object to be brought to a focus at the vertex of the surface opposite the point of incidence?

34.88. A glass rod with a refractive index of 1.55 is ground and polished at both ends to hemispherical surfaces with radii of 6.00 cm. When an object is placed on the axis of the rod, 25.0 cm to the left of the left-hand end, the final image is formed 65.0 cm to the right of the right-hand end. What is the length of the rod measured between the vertices of the two hemispherical surfaces?

34.89. Two thin lenses with focal lengths of magnitude 15.0 cm, the first diverging and the second converging, are placed 12.00 cm apart. An object 4.00 mm tall is placed 5.00 cm to the left of the first (diverging) lens. (a) Where is the image formed by the first lens located? (b) How far from the object is the final image formed? (c) Is the final image real or virtual? (d) What is the height of the final image? Is the final image erect or inverted?

34.90. The radii of curvature of the surfaces of a thin converging meniscus lens are $R_1 = +12.0$ cm and $R_2 = +28.0$ cm. The index of refraction is 1.60. (a) Compute the position and size of the image of an object in the form of an arrow 5.00 mm tall, perpendicular to the lens axis, 45.0 cm to the left of the lens. (b) A second converging lens with the same focal length is placed 3.15 m to the right of the first. Find the position and size of the final image. Is the final image erect or inverted with respect to the original object? (c) Repeat part (b) except with the second lens 45.0 cm to the right of the first.

34.91. An object to the left of a lens is imaged by the lens on a screen 30.0 cm to the right of the lens. When the lens is moved 4.00 cm to the right, the screen must be moved 4.00 cm to the left to refocus the image. Determine the focal length of the lens.

34.92. For refraction at a spherical surface, the first focal length f is defined as the value of s corresponding to $s' = \infty$, as shown in Fig. 34.59a. The second focal length f' is defined as the value of s' when $s = \infty$, as shown in Fig. 34.59b. (a) Prove that $n_a/n_b = f/f'$. (b) Prove that the general relationship between object and image distance is

$$\frac{f}{s} + \frac{f'}{s'} = 1$$

Figure **34.59** Problem 34.92.

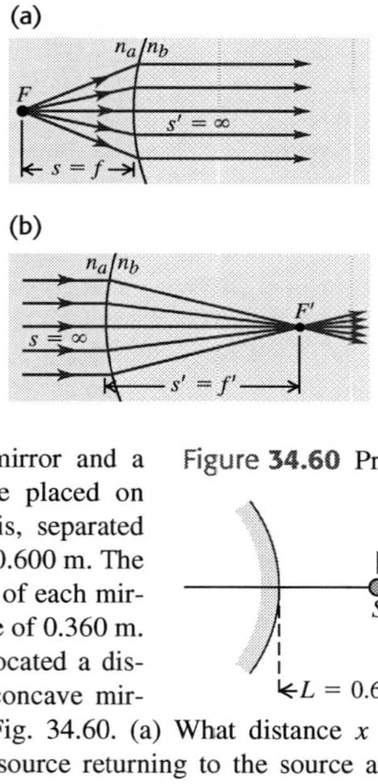

34.93. A convex mirror and a concave mirror are placed on the same optic axis, separated by a distance $L = 0.600$ m. The radius of curvature of each mirror has a magnitude of 0.360 m. A light source is located a distance x from the concave mirror, as shown in Fig. 34.60. (a) What distance x will result in the rays from the source returning to the source after reflecting first from the convex mirror and then from the concave mirror? (b) Repeat part (a), but now let the rays reflect first from the concave mirror and then from the convex one.

Figure **34.60** Problem 34.93.

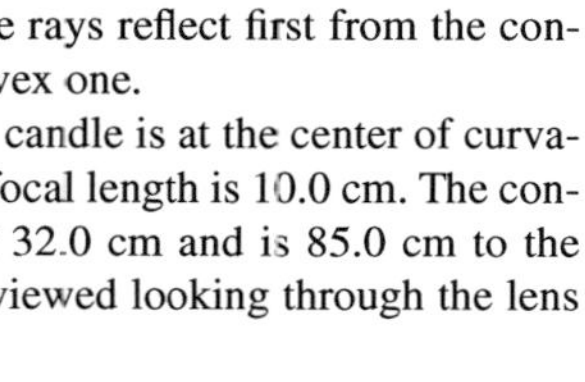

34.94. As shown in Fig. 34.61 the candle is at the center of curvature of the concave mirror, whose focal length is 10.0 cm. The converging lens has a focal length of 32.0 cm and is 85.0 cm to the right of the candle. The candle is viewed looking through the lens from the right. The lens forms two images of the candle. The first is formed by light passing directly through the lens. The second image is formed from the light that goes from the candle to the mirror, is reflected, and then passes through the lens. (a) For each of these two images, draw a principal-ray diagram that locates the image. (b) For each image, answer the following questions: (i) Where is the image? (ii) Is the image real or virtual? (iii) Is the image erect or inverted with respect to the original object?

Figure **34.61** Problem 34.94.

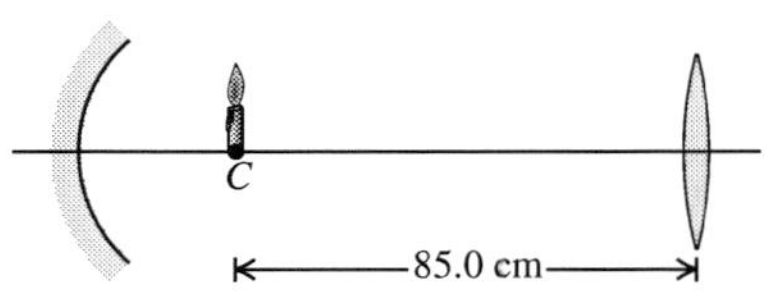

34.95. One end of a long glass rod is ground to a convex hemispherical shape. This glass has an index of refraction of 1.55. When a small leaf is placed 20.0 cm in front of the center of the hemisphere along the optic axis, an image is formed inside the glass 9.12 cm from the spherical surface. Where would the image be formed if the glass were now immersed in water (refractive index 1.33) but nothing else were changed?

34.96. Two Lenses in Contact. (a) Prove that when two thin lenses with focal lengths f_1 and f_2 are placed *in contact,* the focal length f of the combination is given by the relationship

$$\frac{1}{f} = \frac{1}{f_1} + \frac{1}{f_2}$$

(b) A converging meniscus lens (see Fig. 34.32a) has an index of refraction of 1.55 and radii of curvature for its surfaces of 4.50 cm and 9.00 cm. The concave surface is placed upward and filled with carbon tetrachloride (CCl_4), which has $n = 1.46$. What is the focal length of the CCl_4–glass combination?

34.97. Rays from a lens are converging toward a point image P located to the right of the lens. What thickness t of glass with index of refraction 1.60 must be interposed between the lens and P for the image to be formed at P', located 0.30 cm to the right of P? The locations of the piece of glass and of points P and P' are shown in Fig. 34.62.

Figure **34.62** Problem 34.97.

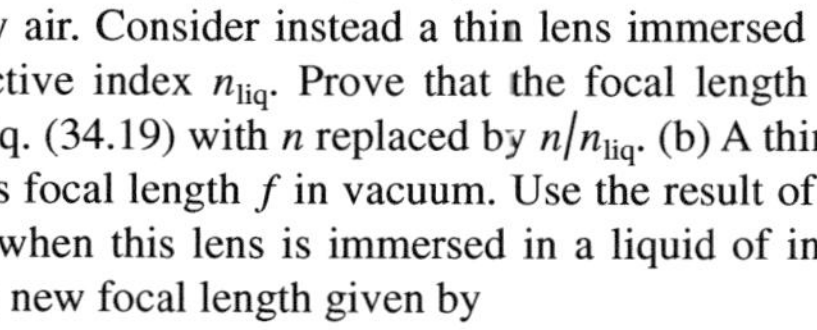

34.98. A Lens in a Liquid. A lens obeys Snell's law, bending light rays at each surface an amount determined by the index of refraction of the lens and the index of the medium in which the lens is located. (a) Equation (34.19) assumes that the lens is surrounded by air. Consider instead a thin lens immersed in a liquid with refractive index n_{liq}. Prove that the focal length f' is then given by Eq. (34.19) with n replaced by n/n_{liq}. (b) A thin lens with index n has focal length f in vacuum. Use the result of part (a) to show that when this lens is immersed in a liquid of index n_{liq}, it will have a new focal length given by

$$f' = \left[\frac{n_{liq}(n-1)}{n - n_{liq}}\right] f$$

34.99. When an object is placed at the proper distance to the left of a converging lens, the image is focused on a screen 30.0 cm to the right of the lens. A diverging lens is now placed 15.0 cm to the right of the converging lens, and it is found that the screen must be moved 19.2 cm farther to the right to obtain a sharp image. What is the focal length of the diverging lens?

34.100. A convex spherical mirror with a focal length of magnitude 24.0 cm is placed 20.0 cm to the left of a plane mirror. An object 0.250 cm tall is placed midway between the surface of the plane mirror and the vertex of the spherical mirror. The spherical mirror forms multiple images of the object. Where are the two images of the object formed by the spherical mirror that are closest to the spherical mirror, and how tall is each image?

34.101. A glass plate 3.50 cm thick, with an index of refraction of 1.55 and plane parallel faces, is held with its faces horizontal and its lower face 6.00 cm above a printed page. Find the position of the image of the page formed by rays making a small angle with the normal to the plate.

34.102. A symmetric, double-convex, thin lens made of glass with index of refraction 1.52 has a focal length in air of 40.0 cm. The lens is sealed into an opening in the left-hand end of a tank filled with water. At the right-hand end of the tank, opposite the lens, is a plane mirror 90.0 cm from the lens. The index of refraction of the water is $\frac{4}{3}$. (a) Find the position of the image formed by the lens–water–mirror system of a small object outside the tank on the lens axis and 70.0 cm to the left of the lens. (b) Is the image real or virtual? (c) Is it erect or inverted? (d) If the object has a height of 4.00 mm, what is the height of the image?

34.103. You have a camera with a 35.0-mm focal length lens and 36.0-mm-wide film. You wish to take a picture of a 12.0-m-long sailboat but find that the image of the boat fills only $\frac{1}{4}$ of the width of the film. (a) How far are you from the boat? (b) How much closer must the boat be to you for its image to fill the width of the film?

34.104. An object is placed 18.0 cm from a screen. (a) At what two points between object and screen may a converging lens with a 3.00-cm focal length be placed to obtain an image on the screen? (b) What is the magnification of the image for each position of the lens?

34.105. Three thin lenses, each with a focal length of 40.0 cm, are aligned on a common axis; adjacent lenses are separated by 52.0 cm. Find the position of the image of a small object on the axis, 80.0 cm to the left of the first lens.

34.106. A camera with a 90-mm-focal-length lens is focused on an object 1.30 m from the lens. To refocus on an object 6.50 m from the lens, by how much must the distance between the lens and the film be changed? To refocus on the more distant object, is the lens moved toward or away from the film?

34.107. The derivation of the expression for angular magnification, Eq. (34.22), assumed a near point of 25 cm. In fact, the near point changes with age as shown in Table 34.1. In order to achieve an angular magnification of 2.0×, what focal length should be used by a person of (a) age 10; (b) age 30; (c) age 60? (d) If the lens that gives $M = 2.0$ for a 10-year-old is used by a 60-year-old, what angular magnification will the older viewer obtain? (e) Does your answer in part (d) mean that older viewers are able to see more highly magnified images than younger viewers? Explain.

34.108. Angular Magnification. In deriving Eq. (34.22) for the angular magnification of a magnifier, we assumed that the object is placed at the focal point of the magnifier so that the virtual image is formed at infinity. Suppose instead that the object is placed so that the virtual image appears at an average viewer's near point of 25 cm, the closest point at which the viewer can bring an object into focus. (a) Where should the object be placed to achieve this? Give your answer in terms of the magnifier focal length f. (b) What angle θ' will an object of height y subtend at the position found in part (a)? (c) Find the angular magnification M with the object at the position found in part (a). The angle θ is the same as in Fig. 34.51a, since it refers to viewing the object *without* the magnifier. (d) For a convex lens with $f = +10.0$ cm, what is the value of M with the object at the position found in part (a)? How many times greater is M in this case than in the case where the image is formed at infinity? (e) In the description of a compound microscope in Section 34.8, it is stated that in a properly designed instrument, the real image formed by the objective lies *just inside* the first focal point F_1' of the eyepiece. What advantages are gained by having the image formed by the objective be just inside F_1', as opposed to precisely at F_1'? What happens if the image formed by the objective is *just outside* F_1'?

34.109. In one form of cataract surgery the person's natural lens, which has become cloudy, is replaced by an artificial lens. The refracting properties of the replacement lens can be chosen so that the person's eye focuses on distant objects. But there is no accommodation, and glasses or contact lenses are needed for close vision. What is the power, in diopters, of the corrective contact lenses that will enable a person who has had such surgery to focus on the page of a book at a distance of 24 cm?

34.110. A Nearsighted Eye. A certain very nearsighted person cannot focus on anything farther than 36.0 cm from the eye. Consider the simplified model of the eye described in Exercise 34.46. If the radius of curvature of the cornea is 0.75 cm when the eye is focusing on an object 36.0 cm from the cornea vertex and the indexes of refraction are as described in Exercise 34.46, what is the distance from the cornea vertex to the retina? What does this tell you about the shape of the nearsighted eye?

34.111. Focal Length of a Zoom Lens. Figure 34.63 shows a simple version of a zoom lens. The converging lens has focal length f_1, and the diverging lens has focal length $f_2 = -|f_2|$. The two lenses are separated by a variable distance d that is always less than f_1. Also, the magnitude of the focal length of the diverging lens satisfies the inequality $|f_2| > (f_1 - d)$. To determine the effective focal length of the combination lens, consider a bundle of parallel rays of radius r_0 entering the converging lens. (a) Show that the radius of the ray bundle decreases to $r_0' = r_0(f_1 - d)/f_1$ at the point that it enters the diverging lens. (b) Show that the final image I' is formed a distance $s_2' = |f_2|(f_1 - d)/(|f_2| - f_1 + d)$ to the right of the diverging lens. (c) If the rays that emerge from the diverging lens and reach the final image point are extended backward to the left of the diverging lens, they will eventually expand to the original radius r_0 at some point Q. The distance from the final image I' to the point Q is the *effective focal length* f of the lens combination; if the combination were replaced by a single lens of focal length f placed at Q, parallel rays would still be

Figure 34.63 Problem 34.111.

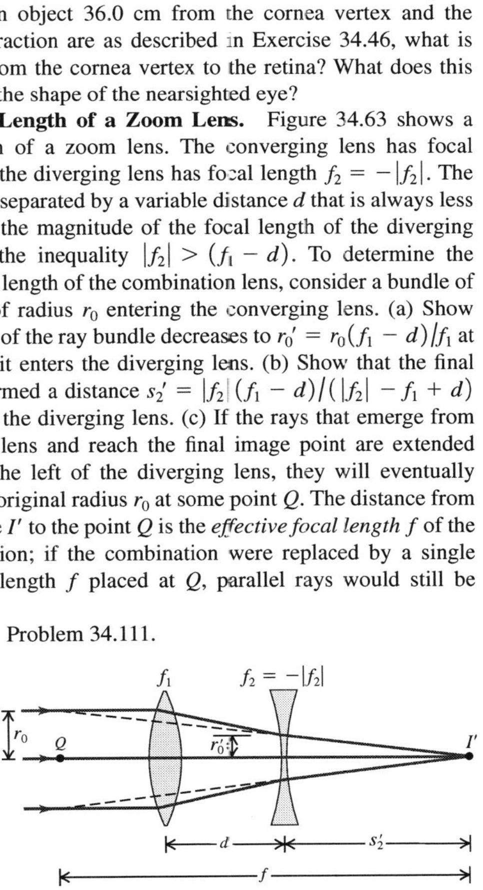

brought to a focus at I'. Show that the effective focal length is given by $f = f_1|f_2|/(|f_2| - f_1 + d)$. (d) If $f_1 = 12.0$ cm, $f_2 = -18.0$ cm, and the separation d is adjustable between 0 and 4.0 cm, find the maximum and minimum focal lengths of the combination. What value of d gives $f = 30.0$ cm?

34.112. A certain reflecting telescope, constructed as shown in Fig. 34.55a, has a spherical mirror with a radius of curvature of 96.0 cm and an eyepiece with a focal length of 1.20 cm. If the angular magnification has a magnitude of 36 and the object is at infinity, find the position of the eyepiece and the position and nature (real or virtual) of the final image. (*Note:* $|M|$ is *not* equal to $|f_1/f_2|$, so the image formed by the eyepiece is *not* at infinity.)

34.113. A microscope with an objective of focal length 8.00 mm and an eyepiece of focal length 7.50 cm is used to project an image on a screen 2.00 m from the eyepiece. Let the image distance of the objective be 18.0 cm. (a) What is the lateral magnification of the image? (b) What is the distance between the objective and the eyepiece?

34.114. The Galilean Telescope. Figure 34.64 is a diagram of a *Galilean telescope,* or *opera glass,* with both the object and its final image at infinity. The image I serves as a virtual object for the eyepiece. The final image is virtual and erect. (a) Prove that the angular magnification is $M = -f_1/f_2$. (b) A Galilean telescope is to be constructed with the same objective lens as in Exercise 34.57. What focal length should the eyepiece have if this telescope is to have the same magnitude of angular magnification as the one in Exercise 34.57? (c) Compare the lengths of the telescopes.

Figure **34.64** Problem 34.114.

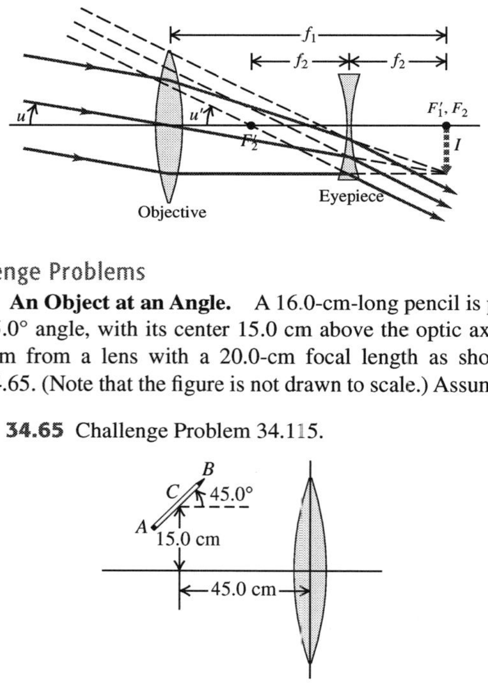

Challenge Problems

34.115. An Object at an Angle. A 16.0-cm-long pencil is placed at a 45.0° angle, with its center 15.0 cm above the optic axis and 45.0 cm from a lens with a 20.0-cm focal length as shown in Fig. 34.65. (Note that the figure is not drawn to scale.) Assume that the diameter of the lens is large enough for the paraxial approximation to be valid. (a) Where is the image of the pencil? (Give the location of the images of the points A, B, and C on the object, which are located at the eraser, point, and center of the pencil, respectively.) (b) What is the length of the image (that is, the distance between the images of points A and B)? (c) Show the orientation of the image in a sketch.

Figure **34.65** Challenge Problem 34.115.

34.116. *Spherical aberration* is a blurring of the image formed by a spherical mirror. It occurs because parallel rays striking the mirror far from the optic axis are focused at a different point than are rays near the axis. This problem is usually minimized by using only the center of a spherical mirror. (a) Show that for a spherical concave mirror, the focus moves toward the mirror as the parallel rays move toward the outer edge of the mirror. (*Hint:* Derive an analytic expression for the distance from the vertex to the focus of the ray for a particular parallel ray. This expression should be in terms of (i) the radius of curvature R of the mirror and (ii) the angle θ between the incident ray and a line connecting the center of curvature of the mirror with the point where the ray strikes the mirror.) (b) What value of θ produces a 2% change in the location of the focus, compared to the location for θ very close to zero?

34.117. (a) For a lens with focal length f, find the smallest distance possible between the object and its real image. (b) Graph the distance between the object and the real image as a function of the distance of the object from the lens. Does your graph agree with the result you found in part (a)?

34.118. Two mirrors are placed together as shown in Fig. 34.66. (a) Show that a point source in front of these mirrors and its two images lie on a circle. (b) Find the center of the circle. (c) In a diagram, show where an observer should stand so as to be able to see both images.

Figure **34.66** Challenge Problem 34.118.

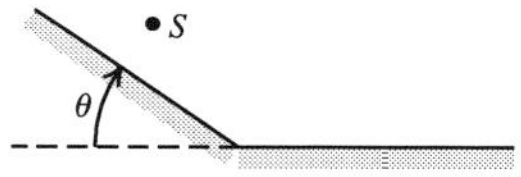

34.119. People with normal vision cannot focus their eyes underwater if they aren't wearing a face mask or goggles and there is water in contact with their eyes (see Discussion Question Q34.23). (a) Why not? (b) With the simplified model of the eye described in Exercise 34.46, what corrective lens (specified by focal length as measured in air) would be needed to enable a person underwater to focus an infinitely distant object? (Be careful—the focal length of a lens underwater is *not* the same as in air! See Problem 34.98. Assume that the corrective lens has a refractive index of 1.62 and that the lens is used in eyeglasses, not goggles, so there is water on both sides of the lens. Assume that the eyeglasses are 2.00 cm in front of the eye.)

INTERFERENCE 35

? Soapy water is colorless, but when blown into bubbles it shows vibrant colors. How does the thickness of the bubble walls determine the particular colors that appear?

LEARNING GOALS

By studying this chapter, you will learn:

- What happens when two waves combine, or interfere, in space.
- How to understand the interference pattern formed by the interference of two coherent light waves.
- How to calculate the intensity at various points in an interference pattern.
- How interference occurs when light reflects from the two surfaces of a thin film.
- How interference makes it possible to measure extremely small distances.

An ugly black oil spot on the pavement can become a thing of beauty after a rain, when the oil reflects a rainbow of colors. Multicolored reflections can also be seen from the surfaces of soap bubbles and compact discs. These familiar sights give us a hint that there are aspects of light that we haven't yet explored.

In our discussion of lenses, mirrors, and optical instruments we used the model of *geometric optics,* in which we represent light as *rays,* straight lines that are bent at a reflecting or refracting surface. But many aspects of the behavior of light *can't* be understood on the basis of rays. We have already learned that light is fundamentally a *wave,* and in some situations we have to consider its wave properties explicitly. If two or more light waves of the same frequency overlap at a point, the total effect depends on the *phases* of the waves as well as their amplitudes. The resulting patterns of light are a result of the *wave* nature of light and cannot be understood on the basis of rays. Optical effects that depend on the wave nature of light are grouped under the heading **physical optics.**

In this chapter we'll look at *interference* phenomena that occur when two waves combine. The colors seen in oil films and soap bubbles are a result of interference between light reflected from the front and back surfaces of a thin film of oil or soap solution. Effects that occur when *many* sources of waves are present are called *diffraction* phenomena; we'll study these in Chapter 36. In that chapter we'll see that diffraction effects occur whenever a wave passes through an aperture or around an obstacle. They are important in practical applications of physical optics such as diffraction gratings, x-ray diffraction, and holography.

While our primary concern is with light, interference and diffraction can occur with waves of *any* kind. As we go along, we'll point out applications to other types of waves such as sound and water waves.

35.1 Interference and Coherent Sources

As we discussed in Chapter 15, the term **interference** refers to any situation in which two or more waves overlap in space. When this occurs, the total wave at any point at any instant of time is governed by the **principle of superposition,** which we introduced in Section 15.6 in the context of waves on a string. This principle also applies to electromagnetic waves and is the most important principle in all of physical optics, so make sure you understand it well. The principle of superposition states:

When two or more waves overlap, the resultant displacement at any point and at any instant is found by adding the instantaneous displacements that would be produced at the point by the individual waves if each were present alone.

(In some special physical situations, such as electromagnetic waves propagating in a crystal, this principle may not apply. A discussion of these is beyond our scope.)

We use the term "displacement" in a general sense. With waves on the surface of a liquid, we mean the actual displacement of the surface above or below its normal level. With sound waves, the term refers to the excess or deficiency of pressure. For electromagnetic waves, we usually mean a specific component of electric or magnetic field.

Interference in Two or Three Dimensions

We have already discussed one important case of interference, in which two identical waves propagating in opposite directions combine to produce a *standing wave.* We saw this in Chapters 15 and 16 for transverse waves on a string and for longitudinal waves in a fluid filling a pipe; we described the same phenomenon for electromagnetic waves in Section 32.5. In all of these cases the waves propagated along only a single axis: along a string, along the length of a fluid-filled pipe, or along the propagation direction of an electromagnetic plane wave. But light waves can (and do) travel in *two* or *three* dimensions, as can any kind of wave that propagates in a two- or three-dimensional medium. In this section we'll see what happens when we combine waves that spread out in two or three dimensions from a pair of identical wave sources.

35.1 A "snapshot" of sinusoidal waves of frequency f and wavelength λ spreading out from source S_1 in all directions.

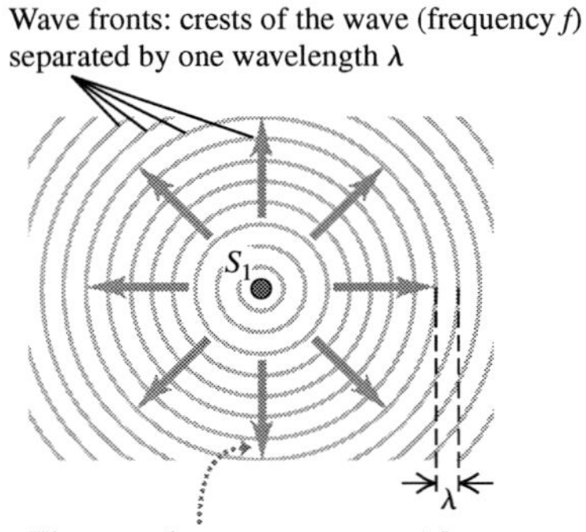

Interference effects are most easily seen when we combine *sinusoidal* waves with a single frequency f and wavelength λ. Figure 35.1 shows a "snapshot" or "freeze-frame" of a *single* source S_1 of sinusoidal waves and some of the wave fronts produced by this source. The figure shows only the wave fronts corresponding to wave *crests,* so the spacing between successive wave fronts is one wavelength. The material surrounding S_1 is uniform, so the wave speed is the same in all directions, and there is no refraction (and hence no bending of the wave fronts). If the waves are two-dimensional, like waves on the surface of a liquid, the circles in Fig. 35.1 represent circular wave fronts; if the waves propagate in three dimensions, the circles represent spherical wave fronts spreading away from S_1.

In optics, sinusoidal waves are characteristic of **monochromatic light** (light of a single color). While it's fairly easy to make water waves or sound waves of a single frequency, common sources of light *do not* emit monochromatic (single-frequency) light. For example, incandescent light bulbs and flames emit a continuous distribution of wavelengths. However, there are several ways to produce *approximately* monochromatic light. For example, some filters block all but a very narrow range of wavelengths. By far the most nearly monochromatic source that is available at present is the *laser.* An example is the helium–neon laser, which emits red light at 632.8 nm with a wavelength range of the order of ± 0.000001 nm, or about one part in 10^9. As we analyze interference and diffraction effects in this chapter and the next, we will assume that we are working with monochromatic waves (unless we explicitly state otherwise).

Constructive and Destructive Interference

Two identical sources of monochromatic waves, S_1 and S_2, are shown in Fig. 35.2a. The two sources produce waves of the same amplitude and the same wavelength λ. In addition, the two sources are permanently *in phase;* they vibrate in unison. They might be two synchronized agitators in a ripple tank, two loudspeakers driven by the same amplifier, two radio antennas powered by the same transmitter, or two small holes or slits in an opaque screen, illuminated by the same monochromatic light source. We will see that if there were not a constant phase relationship between the two sources, the phenomena we are about to discuss would not occur. Two monochromatic sources of the same frequency and with any definite, constant phase relationship (not necessarily in phase) are said to be **coherent.** We also use the term *coherent waves* (or, for light waves, *coherent light*) to refer to the waves emitted by two such sources.

If the waves emitted by the two coherent sources are *transverse,* like electromagnetic waves, then we will also assume that the wave disturbances produced by both sources have the same *polarization* (that is, they lie along the same line). For example, the sources S_1 and S_2 in Fig. 35.2a could be two radio antennas in the form of long rods oriented parallel to the z-axis (perpendicular to the plane of the figure); at any point in the xy-plane the waves produced by both antennas have $\vec{E}$ fields with only a z-component. Then we need only a single scalar function to describe each wave; this makes the analysis much easier.

We position the two sources of equal amplitude, equal wavelength, and (if the waves are transverse) the same polarization along the y-axis in Fig. 35.2a, equidistant from the origin. Consider a point a on the x-axis. From symmetry the two distances from S_1 to a and from S_2 to a are *equal;* waves from the two sources thus require equal times to travel to a. Hence waves that leave S_1 and S_2 in phase arrive at a in phase. The two waves add, and the total amplitude at a is *twice* the amplitude of each individual wave. This is true for *any* point on the x-axis.

Similarly, the distance from S_2 to point b is exactly two wavelengths *greater* than the distance from S_1 to b. A wave crest from S_1 arrives at b exactly two cycles earlier than a crest emitted at the same time from S_2, and again the two waves arrive in phase. As at point a, the total amplitude is the sum of the amplitudes of the waves from S_1 and S_2.

In general, when waves from two or more sources arrive at a point *in phase,* the amplitude of the resultant wave is the *sum* of the amplitudes of the individual waves; the individual waves reinforce each other. This is called **constructive interference** (Fig. 35.2b). Let the distance from S_1 to any point P be r_1, and let the distance from S_2 to P be r_2. For constructive interference to occur at P, the path difference $r_2 - r_1$ for the two sources must be an integral multiple of the wavelength λ:

$$r_2 - r_1 = m\lambda \qquad (m = 0, \pm 1, \pm 2, \pm 3, \ldots) \qquad \text{(constructive interference, sources in phase)} \qquad (35.1)$$

In Fig. 35.2a, points a and b satisfy Eq. (35.1) with $m = 0$ and $m = +2$, respectively.

Something different occurs at point c in Fig. 35.2a. At this point, the path difference $r_2 - r_1 = -2.50\lambda$, which is a *half-integral* number of wavelengths. Waves from the two sources arrive at point c exactly a half-cycle out of phase. A crest of one wave arrives at the same time as a crest in the opposite direction (a "trough") of the other wave (Fig. 35.2c). The resultant amplitude is the *difference* between the two individual amplitudes. If the individual amplitudes are equal, then the total amplitude is *zero!* This cancellation or partial cancellation of the

35.2 (a) A "snapshot" of sinusoidal waves spreading out from two coherent sources S_1 and S_2. Constructive interference occurs at point a (equidistant from the two sources) and (b) at point b. (c) Destructive interference occurs at point c.

(a) Two coherent wave sources separated by a distance 4λ

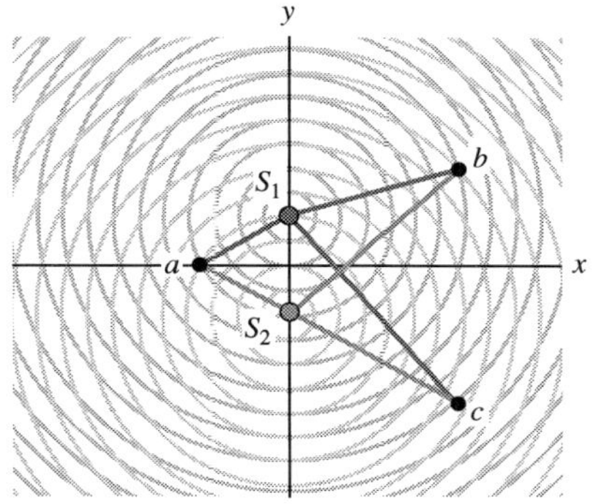

(b) Conditions for constructive interference: Waves interfere constructively if their path lengths differ by an integral number of wavelengths: $r_2 - r_1 = m\lambda$.

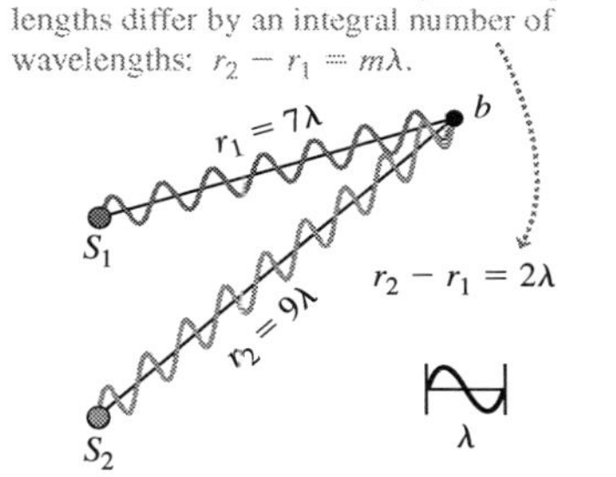

(c) Conditions for destructive interference: Waves interfere destructively if their path lengths differ by a half-integral number of wavelengths: $r_2 - r_1 = (m + \frac{1}{2})\lambda$.

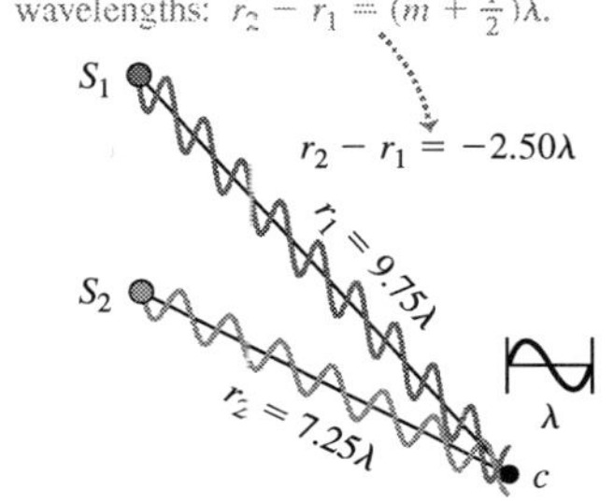

individual waves is called **destructive interference.** The condition for destructive interference in the situation shown in Fig. 35.2a is

$$r_2 - r_1 = \left(m + \frac{1}{2}\right)\lambda \quad (m = 0, \pm 1, \pm 2, \pm 3, \ldots) \quad \begin{array}{l}\text{(destructive}\\ \text{interference,}\\ \text{sources in phase)}\end{array} \quad (35.2)$$

The path difference at point c in Fig. 35.2a satisfies Eq. (35.2) with $m = -3$.

Figure 35.3 shows the same situation as in Fig. 35.2a, but with red curves that denote all points on which *constructive* interference occurs. On each curve, the path difference $r_2 - r_1$ is equal to an integer m times the wavelength, as in Eq. (35.1). These curves are called **antinodal curves.** They are directly analogous to *antinodes* in the standing-wave patterns described in Chapters 15 and 16 and Section 32.5. In a standing wave formed by interference between waves propagating in opposite directions, the antinodes are points at which the amplitude is maximum; likewise, the wave amplitude in the situation of Fig. 35.3 is maximum along the antinodal curves. Not shown in Fig. 35.3 are the **nodal curves,** which are the curves denoting points on which *destructive* interference occurs in accordance with Eq. (35.2); these are analogous to the *nodes* in a standing-wave pattern. A nodal curve lies between each two adjacent antinodal curves in Fig. 35.3; one such curve, corresponding to $r_2 - r_1 = -2.50\lambda$, passes through point c.

In some cases, such as two loudspeakers or two radio-transmitter antennas, the interference pattern is three-dimensional. Think of rotating the color curves of Fig. 35.3 around the y-axis; then maximum constructive interference occurs at all points on the resulting surfaces of revolution.

35.3 The same as Fig. 35.2a, but with red antinodal curves (curves of maximum amplitude) superimposed. All points on each curve satisfy Eq. (35.1) with the value of m shown. The nodal curves (not shown) lie between each adjacent pair of antinodal curves.

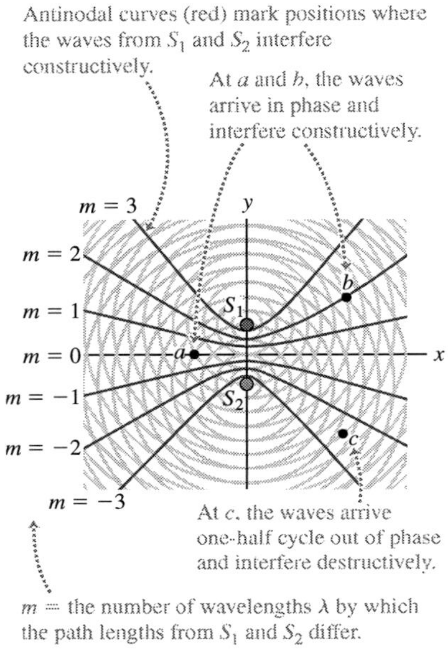

CAUTION **Interference patterns are not standing waves** The interference patterns in Figs. 35.2a and 35.3 are *not* standing waves, though they have some similarities to the standing-wave patterns described in Chapters 15 and 16 and Section 32.5. In a standing wave, the interference is between two waves propagating in opposite directions; a stationary pattern of antinodes and nodes appears, and there is *no* net energy flow in either direction (the energy in the wave is left "standing"). In the situations shown in Figs. 35.2a and 35.3, there is likewise a stationary pattern of antinodal and nodal curves, but there is a net flow of energy *outward* from the two sources. From the energy standpoint, all that interference does is to "channel" the energy flow so that it is greatest along the antinodal curves and least along the nodal curves. ▮

For Eqs. (35.1) and (35.2) to hold, the two sources must have the same wavelength and must *always* be in phase. These conditions are rather easy to satisfy for sound waves (see Example 16.15 in Section 16.6). But with *light* waves there is no practical way to achieve a constant phase relationship (coherence) with two independent sources. This is because of the way light is emitted. In ordinary light sources, atoms gain excess energy by thermal agitation or by impact with accelerated electrons. An atom that is "excited" in such a way begins to radiate energy and continues until it has lost all the energy it can, typically in a time of the order of 10^{-8} s. The many atoms in a source ordinarily radiate in an unsynchronized and random phase relationship, and the light that is emitted from *two* such sources has no definite phase relationship.

However, the light from a single source can be split so that parts of it emerge from two or more regions of space, forming two or more *secondary sources.* Then any random phase change in the source affects these secondary sources equally and does not change their *relative* phase.

The distinguishing feature of light from a *laser* is that the emission of light from many atoms is synchronized in frequency and phase. As a result, the random phase changes mentioned above occur much less frequently. Definite phase relationships are preserved over correspondingly much greater lengths in the beam, and laser light is much more coherent than ordinary light.

Test Your Understanding of Section 35.1 Consider a point in Fig. 35.3 on the positive y-axis above S_1. Does this point lie on (i) an antinodal curve; (ii) a nodal curve; or (iii) neither? (*Hint:* The distance between S_1 and S_2 is 4λ.)

35.2 Two-Source Interference of Light

The interference pattern produced by two coherent sources of *water* waves of the same wavelength can be readily seen in a ripple tank with a shallow layer of water (Fig. 35.4). This pattern is not directly visible when the interference is between *light* waves, since light traveling in a uniform medium cannot be seen. (A shaft of afternoon sunlight in a room is made visible by scattering from airborne dust particles.)

One of the earliest quantitative experiments to reveal the interference of light from two sources was performed in 1800 by the English scientist Thomas Young. We will refer back to this experiment several times in this and later chapters, so it's important to understand it in detail. Young's apparatus is shown in perspective in Fig. 35.5a. A light source (not shown) emits monochromatic light; however, this light is not suitable for use in an interference experiment because emissions from different parts of an ordinary source are not synchronized. To remedy this, the light is directed at a screen with a narrow slit S_0, 1 μm or so wide. The light emerging from the slit originated from only a small region of the light source; thus slit S_0 behaves more nearly like the idealized source shown in Fig. 35.1. (In modern versions of the experiment, a laser is used as a source of coherent light, and the slit S_0 isn't needed.) The light from slit S_0 falls on a screen with two other narrow slits S_1 and S_2, each 1 μm or so wide and a few tens or hundreds of micrometers apart. Cylindrical wave fronts spread out from slit S_0 and reach slits S_1 and S_2 *in phase* because they travel equal distances from S_0. The waves *emerging* from slits S_1 and S_2 are therefore also always in phase, so S_1 and S_2 are *coherent* sources. The interference of waves from S_1 and S_2 produces a pattern in space like that to the right of the sources in Figs. 35.2a and 35.3.

35.4 The concepts of constructive interference and destructive interference apply to these water waves as well as to light waves and sound waves.

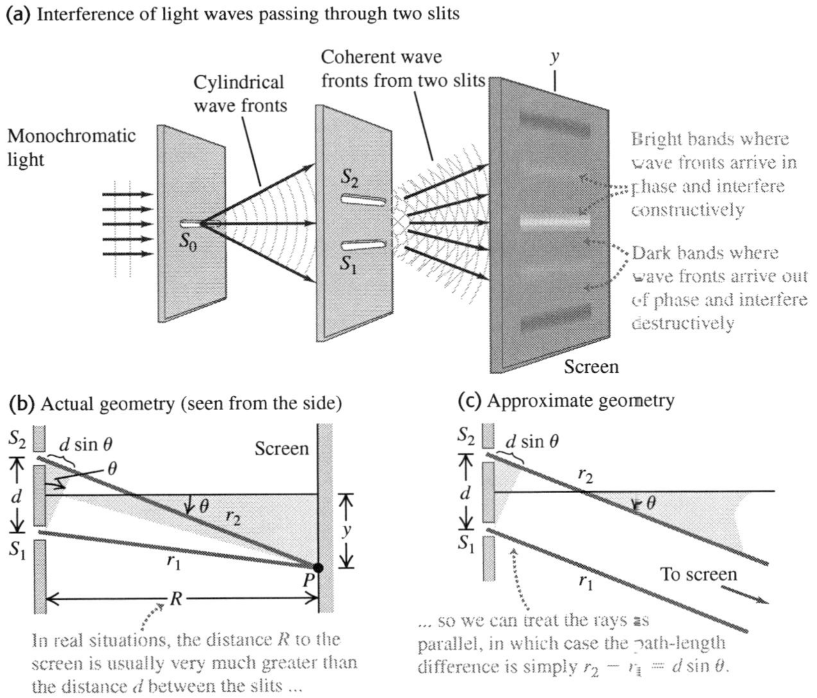

35.5 (a) Young's experiment to show interference of light passing through two slits. A pattern of bright and dark areas appears on the screen (see Fig. 35.6). (b) Geometrical analysis of Young's experiment. For the case shown, $r_2 > r_1$ and both y and θ are positive. If point P is on the other side of the screen's center, $r_2 < r_1$ and both y and θ are negative. (c) Approximate geometry when the distance R to the screen is much greater than the distance d between the slits.

ActivPhysics ONLINE

16.1 Two-Source Interference: Introduction

16.1 Two-Source Interference: Qualitative Questions

16.1 Two-Source Interference: Problems

To visualize the interference pattern, a screen is placed so that the light from S_1 and S_2 falls on it (Fig. 35.5b). The screen will be most brightly illuminated at points P, where the light waves from the slits interfere constructively, and will be darkest at points where the interference is destructive.

To simplify the analysis of Young's experiment, we assume that the distance R from the slits to the screen is so large in comparison to the distance d between the slits that the lines from S_1 and S_2 to P are very nearly parallel, as in Fig. 35.5c. This is usually the case for experiments with light; the slit separation is typically a few millimeters, while the screen may be a meter or more away. The difference in path length is then given by

$$r_2 - r_1 = d\sin\theta \qquad (35.3)$$

where θ is the angle between a line from slits to screen (shown in blue in Fig.35.5c) and the normal to the plane of the slits (shown as a thin black line).

Constructive and Destructive Two-Slit Interference

We found in Section 35.1 that constructive interference (reinforcement) occurs at points where the path difference is an integral number of wavelengths, $m\lambda$, where $m = 0, \pm1, \pm2, \pm3, \ldots$. So the bright regions on the screen in Fig. 35.5 occur at angles θ for which

$$d\sin\theta = m\lambda \qquad (m = 0, \pm1, \pm2, \ldots) \qquad \text{(constructive interference, two slits)} \qquad (35.4)$$

Similarly, destructive interference (cancellation) occurs, forming dark regions on the screen, at points for which the path difference is a half-integral number of wavelengths, $(m + \frac{1}{2})\lambda$:

$$d\sin\theta = \left(m + \frac{1}{2}\right)\lambda \qquad (m = 0, \pm1, \pm2, \ldots) \qquad \text{(destructive interference, two slits)} \qquad (35.5)$$

35.6 Photograph of interference fringes produced on a screen in Young's double-slit experiment.

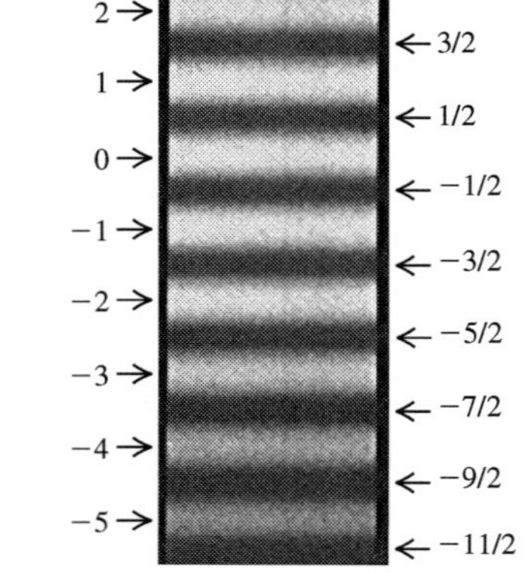

Thus the pattern on the screen of Figs. 35.5a and 35.5b is a succession of bright and dark bands, or **interference fringes,** parallel to the slits S_1 and S_2. A photograph of such a pattern is shown in Fig. 35.6. The center of the pattern is a bright band corresponding to $m = 0$ in Eq. (35.4); this point on the screen is equidistant from the two slits.

We can derive an expression for the positions of the centers of the bright bands on the screen. In Fig. 35.5b, y is measured from the center of the pattern, corresponding to the distance from the center of Fig. 35.6. Let y_m be the distance from the center of the pattern $(\theta = 0)$ to the center of the mth bright band. Let θ_m be the corresponding value of θ; then

$$y_m = R\tan\theta_m$$

In experiments such as this, the distances y_m are often much smaller than the distance R from the slits to the screen. Hence θ_m is very small, $\tan\theta_m$ is very nearly equal to $\sin\theta_m$, and

$$y_m = R\sin\theta_m$$

Combining this with Eq. (35.4), we find that *for small angles only,*

$$y_m = R\frac{m\lambda}{d} \qquad \text{(constructive interference in Young's experiment)} \qquad (35.6)$$

We can measure R and d, as well as the positions y_m of the bright fringes, so this experiment provides a direct measurement of the wavelength λ. Young's experiment was in fact the first direct measurement of wavelengths of light.

CAUTION **Equation (35.6) is for small angles only** While Eqs. (35.4) and (35.5) are valid at any angle, Eq. (35.6) is valid only for *small* angles. It can be used *only* if the distance R from slits to screen is much greater than the slit separation d and if R is much greater than the distance y_m from the center of the interference pattern to the mth bright fringe.

The distance between adjacent bright bands in the pattern is *inversely* proportional to the distance d between the slits. The closer together the slits are, the more the pattern spreads out. When the slits are far apart, the bands in the pattern are closer together.

While we have described the experiment that Young performed with visible light, the results given in Eqs. (35.4) and (35.5) are valid for *any* type of wave, provided that the resultant wave from two coherent sources is detected at a point that is far away in comparison to the separation d.

Example 35.1 Two-slit interference

In a two-slit interference experiment, the slits are 0.200 mm apart, and the screen is at a distance of 1.00 m. The third bright fringe (not counting the central bright fringe straight ahead from the slits) is found to be displaced 9.49 mm from the central fringe (Fig. 35.7). Find the wavelength of the light used.

SOLUTION

IDENTIFY: This problem asks us to determine the wavelength λ from the dimensions $d = 0.200$ mm (slit separation), $R = 1.00$ m (distance from slits to screen), and $y_m = 9.49$ mm (distance of the third bright fringe from the center of the pattern).

SET UP: The third bright fringe corresponds to $m = 3$ in Eqs. (35.4) and (35.6), as well as to the bright fringe labeled $m = 3$ in Fig. 35.6. To determine the value of the target variable λ, we may use Eq. (35.6) since $R = 1.00$ m is much greater than $d = 0.200$ mm or $y_3 = 9.49$ mm.

EXECUTE: Solving Eq. (35.6) for λ, we find

$$\lambda = \frac{y_m d}{mR} = \frac{(9.49 \times 10^{-3}\,\text{m})(0.200 \times 10^{-3}\,\text{m})}{(3)(1.00\,\text{m})}$$
$$= 633 \times 10^{-9}\,\text{m} = 633\,\text{nm}$$

35.7 Using a two-slit interference experiment to measure the wavelength of light.

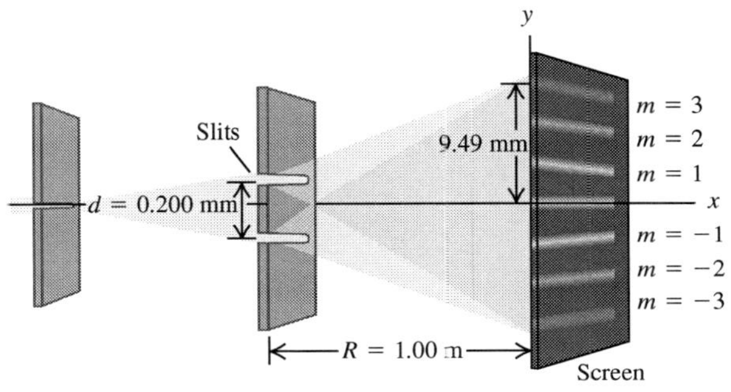

EVALUATE: This bright fringe could also correspond to $m = -3$; can you show that this gives the same result for λ?

Example 35.2 Broadcast pattern of a radio station

A radio station operating at a frequency of 1500 kHz $= 1.5 \times 10^6$ Hz (near the top end of the AM broadcast band) has two identical vertical dipole antennas spaced 400 m apart, oscillating in phase. At distances much greater than 400 m, in what directions is the intensity greatest in the resulting radiation pattern? (This is not just a hypothetical problem. It is often desirable to beam most of the radiated energy from a radio transmitter in particular directions rather than uniformly in all directions. Pairs or rows of antennas are often used to produce the desired radiation pattern.)

SOLUTION

IDENTIFY: The two antennas, shown in Fig. 35.8, correspond to sources S_1 and S_2 in Fig. 35.5. Hence we can apply the ideas of two-slit interference to this problem.

35.8 Two radio antennas broadcasting in phase. The purple arrows indicate the directions of maximum intensity. The waves that are emitted toward the lower half of the figure are not shown.

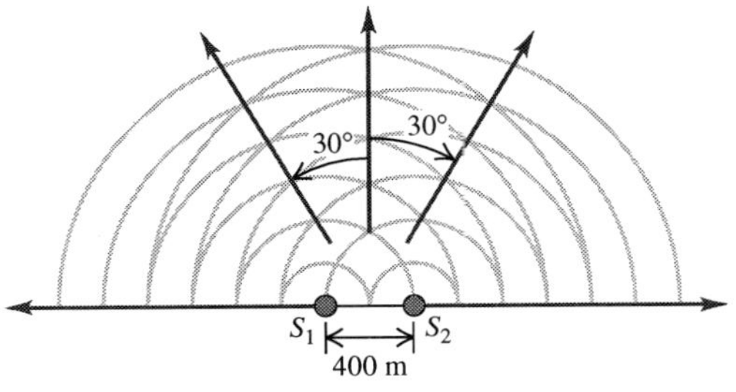

Continued

SET UP: Since the resultant wave is detected at distances much greater than $d = 400$ m, we may use Eq. (35.4) to give the directions of the intensity *maxima,* the values of θ for which the path difference is zero or a whole number of wavelengths.

EXECUTE: The wavelength is $\lambda = c/f = 200$ m. From Eq. (35.4) with $m = 0, \pm1$, and ±2, the intensity maxima are given by

$$\sin\theta = \frac{m\lambda}{d} = \frac{m(200\text{ m})}{400\text{ m}} = \frac{m}{2} \qquad \theta = 0, \pm30°, \pm90°$$

In this example, values of m greater than 2 or less than -2 give values of $\sin\theta$ greater than 1 or less than -1, which is impossible. There is *no* direction for which the path difference is three or more wavelengths. Thus values of m of ±3 and beyond have no physical meaning in this example.

EVALUATE: We can check our result by calculating the angles for *minimum* intensity (destructive interference). There should be one intensity minimum between each pair of intensity maxima, just as in the interference pattern shown in Fig. 35.6. The angles of the intensity minima are given by Eq. (35.5) with $m = -2, -1, 0$, and 1:

$$\sin\theta = \frac{(m + \frac{1}{2})\lambda}{d} = \frac{m + \frac{1}{2}}{2} \qquad \theta = \pm14.5°, \pm48.6°$$

(Other values of m have no physical significance in this example.) Note that these angles are intermediate between the angles for intensity maxima, as they should be. Note also that since the angles are not small, the angles for the minima are *not* exactly halfway between the angles for the maxima.

Test Your Understanding of Section 35.2 You shine a tunable laser (whose wavelength can be adjusted by turning a knob) on a pair of closely spaced slits. The light emerging from the two slits produces an interference pattern on a screen like that shown in Fig. 35.6. If you adjust the wavelength so that the laser light changes from red to blue, how will the spacing between bright fringes change? (i) The spacing increases; (ii) the spacing decreases; (iii) the spacing is unchanged; (iv) not enough information is given to decide.

35.3 Intensity in Interference Patterns

In Section 35.2 we found the positions of maximum and minimum intensity in a two-source interference pattern. Let's now see how to find the intensity at *any* point in the pattern. To do this, we have to combine the two sinusoidally varying fields (from the two sources) at a point P in the radiation pattern, taking proper account of the phase difference of the two waves at point P, which results from the path difference. The intensity is then proportional to the square of the resultant electric-field amplitude, as we learned in Chapter 32.

To calculate the intensity, we will assume that the two sinusoidal functions (corresponding to waves from the two sources) have equal amplitude E and that the $\vec{E}$ fields lie along the same line (have the same polarization). This assumes that the sources are identical and neglects the slight amplitude difference caused by the unequal path lengths (the amplitude decreases with increasing distance from the source). From Eq. (32.29), each source by itself would give an intensity $\frac{1}{2}\epsilon_0 cE^2$ at point P. If the two sources are in phase, then the waves that arrive at P differ in phase by an amount proportional to the difference in their path lengths, $(r_2 - r_1)$. If the phase angle between these arriving waves is ϕ, then we can use the following expressions for the two electric fields superposed at P:

$$E_1(t) = E\cos(\omega t + \phi)$$
$$E_2(t) = E\cos\omega t$$

Here is our program. The superposition of the two fields at P is a sinusoidal function with some amplitude E_P that depends on E and the phase difference ϕ. First we'll work on finding the amplitude E_P if E and ϕ are known. Then we'll find the intensity I of the resultant wave, which is proportional to E_P^2. Finally, we'll relate the phase difference ϕ to the path difference, which is determined by the geometry of the situation.

Amplitude in Two-Source Interference

To add the two sinusoidal functions with a phase difference, we use the same *phasor* representation that we used for simple harmonic motion (Section 13.2) and for voltages and currents in ac circuits (Section 31.1). We suggest that you review these sections so that phasors are fresh in your mind. Each sinusoidal function is represented by a rotating vector (phasor) whose projection on the horizontal axis at any instant represents the instantaneous value of the sinusoidal function.

In Fig. 35.9, E_1 is the horizontal component of the phasor representing the wave from source S_1, and E_2 is the horizontal component of the phasor for the wave from S_2. As shown in the diagram, both phasors have the same magnitude E, but E_1 is *ahead* of E_2 in phase by an angle ϕ. Both phasors rotate counterclockwise with constant angular speed ω, and the sum of the projections on the horizontal axis at any time gives the instantaneous value of the total E field at point P. Thus the amplitude E_P of the resultant sinusoidal wave at P is the magnitude of the dark red phasor in the diagram (labeled E_P); this is the *vector sum* of the other two phasors. To find E_P, we use the law of cosines and the trigonometric identity $\cos(\pi - \phi) = -\cos\phi$:

$$E_P^2 = E^2 + E^2 - 2E^2\cos(\pi - \phi)$$
$$= E^2 + E^2 + 2E^2\cos\phi$$

Then, using the identity $1 + \cos\phi = 2\cos^2(\phi/2)$, we obtain

$$E_P^2 = 2E^2(1 + \cos\phi) = 4E^2\cos^2\left(\frac{\phi}{2}\right)$$

$$E_P = 2E\left|\cos\frac{\phi}{2}\right| \quad \text{(amplitude in two-source interference)} \qquad (35.7)$$

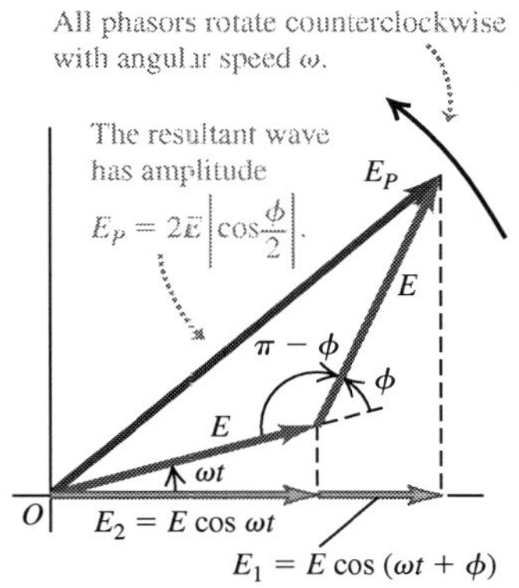

35.9 Phasor diagram for the superposition at a point P of two waves of equal amplitude E with a phase difference ϕ.

You can also obtain this result algebraically without using phasors (see Problem 35.48).

When the two waves are in phase, $\phi = 0$ and $E_P = 2E$. When they are exactly a half-cycle out of phase, $\phi = \pi$ rad $= 180°, \cos(\phi/2) = \cos(\pi/2) = 0$, and $E_P = 0$. Thus the superposition of two sinusoidal waves with the same frequency and amplitude but with a phase difference yields a sinusoidal wave with the same frequency and an amplitude between zero and twice the individual amplitudes, depending on the phase difference.

Intensity in Two-Source Interference

To obtain the intensity I at point P, we recall from Section 32.4 that I is equal to the average magnitude of the Poynting vector, S_{av}. For a sinusoidal wave with electric-field amplitude E_P, this is given by Eq. (32.29) with E_{max} replaced by E_P. Thus we can express the intensity in any of the following equivalent forms:

$$I = S_{av} = \frac{E_P^2}{2\mu_0 c} = \frac{1}{2}\sqrt{\frac{\epsilon_0}{\mu_0}}E_P^2 = \frac{1}{2}\epsilon_0 c E_P^2 \qquad (35.8)$$

The essential content of these expressions is that I is proportional to E_P^2. When we substitute Eq. (35.7) into the last expression in Eq. (35.8), we get

$$I = \frac{1}{2}\epsilon_0 c E_P^2 = 2\epsilon_0 c E^2 \cos^2\frac{\phi}{2} \qquad (35.9)$$

In particular, the *maximum* intensity I_0, which occurs at points where the phase difference is zero $(\phi = 0)$, is

$$I_0 = 2\epsilon_0 c E^2$$

Note that the maximum intensity I_0 is *four times* (not twice) as great as the intensity $\frac{1}{2}\epsilon_0 c E^2$ from each individual source.

Substituting the expression for I_0 into Eq. (35.9), we can express the intensity I at any point very simply in terms of the maximum intensity I_0:

$$I = I_0 \cos^2 \frac{\phi}{2} \quad \text{(intensity in two-source interference)} \qquad (35.10)$$

For some phase angles ϕ the intensity is I_0, four times as great as for an individual wave source, but for other phase angles the intensity is zero. If we average Eq. (35.10) over all possible phase differences, the result is $I_0/2 = \epsilon_0 c E^2$ (the average of $\cos^2(\phi/2)$ is $\frac{1}{2}$). This is just twice the intensity from each individual source, as we should expect. The total energy output from the two sources isn't changed by the interference effects, but the energy is redistributed (as we mentioned in Section 35.1).

Phase Difference and Path Difference

Our next task is to find how the phase difference ϕ between the two fields at point P is related to the geometry of the situation. We know that ϕ is proportional to the difference in path length from the two sources to point P. When the path difference is one wavelength, the phase difference is one cycle, and $\phi = 2\pi \text{ rad} = 360°$. When the path difference is $\lambda/2$, $\phi = \pi \text{ rad} = 180°$, and so on. That is, the ratio of the phase difference ϕ to 2π is equal to the ratio of the path difference $r_2 - r_1$ to λ:

$$\frac{\phi}{2\pi} = \frac{r_2 - r_1}{\lambda}$$

Thus a path difference $(r_2 - r_1)$ causes a phase difference given by

$$\phi = \frac{2\pi}{\lambda}(r_2 - r_1) = k(r_2 - r_1) \quad \text{(phase difference related to path difference)} \qquad (35.11)$$

where $k = 2\pi/\lambda$ is the *wave number* introduced in Section 15.3.

If the material in the space between the sources and P is anything other than vacuum, we must use the wavelength *in the material* in Eq. (35.11). If the material has index of refraction n, then

$$\lambda = \frac{\lambda_0}{n} \quad \text{and} \quad k = nk_0 \qquad (35.12)$$

where λ_0 and k_0 are the wavelength and the wave number, respectively, in vacuum.

Finally, if the point P is far away from the sources in comparison to their separation d, the path difference is given by Eq. (35.3):

$$r_2 - r_1 = d\sin\theta$$

Combining this with Eq. (35.11), we find

$$\phi = k(r_2 - r_1) = kd\sin\theta = \frac{2\pi d}{\lambda}\sin\theta \qquad (35.13)$$

When we substitute this into Eq. (35.10), we find

$$I = I_0 \cos^2\left(\frac{1}{2}kd\sin\theta\right) = I_0\cos^2\left(\frac{\pi d}{\lambda}\sin\theta\right) \quad \text{(intensity far from two sources)} \qquad (35.14)$$

The directions of *maximum* intensity occur when the cosine has the values ± 1—that is, when

$$\frac{\pi d}{\lambda}\sin\theta = m\pi \qquad (m = 0, \pm 1, \pm 2, \ldots)$$

35.10 Intensity distribution in the interference pattern from two identical slits.

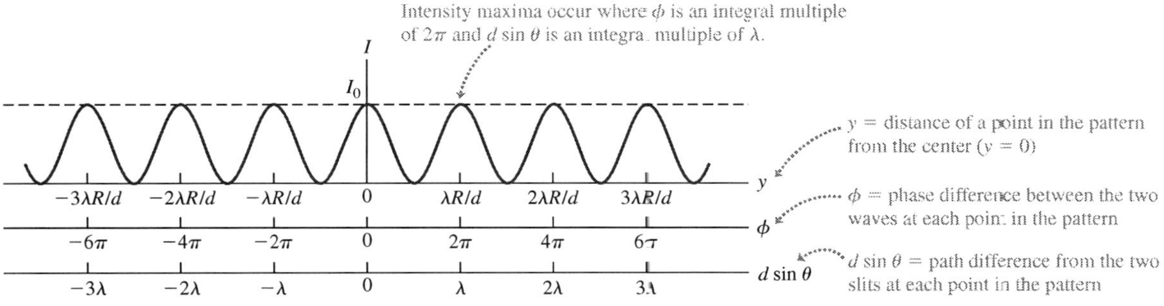

or

$$d\sin\theta = m\lambda$$

in agreement with Eq. (35.4). We leave it to you to show that Eq. (35.5) for the zero-intensity directions can also be derived from Eq. (35.14) (see Exercise 35.24).

As we noted in Section 35.2, in experiments with light we visualize the interference pattern due to two slits by using a screen placed at a distance R from the slits. We can describe positions on the screen with the coordinate y; the positions of the bright fringes are given by Eq. (35.6), where ordinarily $y \ll R$. In that case, $\sin\theta$ is approximately equal to y/R, and we obtain the following expressions for the intensity at *any* point on the screen as a function of y:

$$I = I_0\cos^2\left(\frac{kdy}{2R}\right) = I_0\cos^2\left(\frac{\pi dy}{\lambda R}\right) \quad \text{(intensity in two-slit interference)} \quad (35.15)$$

Figure 35.10 shows a graph of Eq. (35.15); we can compare this with the photographically recorded pattern of Fig. 35.6. The peaks in Fig. 35.10 all have the same intensity, while those in Fig. 35.6 fade off as we go away from the center. We'll explore the reasons for this variation in peak intensity in Chapter 36.

Example 35.3 A directional transmitting antenna array

Suppose the two identical radio antennas in Fig. 35.8 are moved to be only 10.0 m apart and the frequency of the radiated waves is increased to $f = 60.0$ MHz. The intensity at a distance of 700 m in the $+x$-direction (corresponding to $\theta = 0$ in Fig. 35.5) is $I_0 = 0.020\ \text{W/m}^2$. (a) What is the intensity in the direction $\theta = 4.0°$? (b) In what direction near $\theta = 0$ is the intensity $I_0/2$? (c) In what directions is the intensity zero?

SOLUTION

IDENTIFY: This problem involves the intensity distribution as a function of *direction*—that is, as a function of angle. (In other problems we are concerned with the intensity as a function of *position* on a screen, as in the interference pattern shown in Fig. 35.6.)

SET UP: Because the 700-m distance from the antennas to the point where the intensity is measured is much greater than the distance between the antennas ($d = 10.0$ m), the amplitudes of the waves from the two antennas are very nearly equal. Hence we can use Eq. (35.14) to relate intensity I and angle θ.

EXECUTE: To use Eq. (35.14), we must first find the wavelength λ using the relationship $c = \lambda f$:

$$\lambda = \frac{c}{f} = \frac{3.00 \times 10^8\ \text{m/s}}{60.0 \times 10^6\ \text{s}^{-1}} = 5.00\ \text{m}$$

The spacing $d = 10.0$ m between the antennas is just twice the wavelength. Equation (35.14) then becomes

$$\begin{aligned} I &= I_0\cos^2\left(\frac{\pi d}{\lambda}\sin\theta\right) \\ &= (0.020\ \text{W/m}^2)\cos^2\left[\frac{\pi(10.0\ \text{m})}{5.00\ \text{m}}\sin\theta\right] \\ &= (0.020\ \text{W/m}^2)\cos^2[(2.00\pi\ \text{rad})\sin\theta] \end{aligned}$$

(a) When $\theta = 4.0°$,

$$\begin{aligned} I &= (0.020\ \text{W/m}^2)\cos^2[(2.00\pi\ \text{rad})\sin 4.0°] \\ &= 0.016\ \text{W/m}^2 \end{aligned}$$

Continued

This is about 82% of the intensity at $\theta = 0$.

(b) The intensity I equals $I_0/2$ when the cosine in Eq. (35.14) has the value $\pm 1/\sqrt{2}$. This occurs when $2.00\pi \sin\theta = \pm\pi/4$ rad, so that $\sin\theta = \pm(1/8.00) = \pm 0.125$ and $\theta = \pm 7.2°$.

(c) The intensity is zero when $\cos[(2.00\pi \text{ rad})\sin\theta] = 0$. This occurs when $2.00\pi \sin\theta = \pm\pi/2, \pm 3\pi/2, \pm 5\pi/2, \ldots$, or $\sin\theta = \pm 0.250, \pm 0.750, \pm 1.25, \ldots$. Values of $\sin\theta$ greater than 1 have no meaning, and we find

$$\theta = \pm 14.5°, \pm 48.6°$$

EVALUATE: The condition in part (b) that $I = I_0/2$, so that $(2.00\pi \text{ rad})\sin\theta = \pm\pi/4$ rad, is also satisfied when $\sin\theta = \pm 0.375, \pm 0.625$, or ± 0.875 so that $\theta = \pm 22.0°, \pm 38.7°$, or $\pm 61.0°$. (Can you verify this?) It would be incorrect to include these angles in the solution, however, because the problem asked for the angle *near* $\theta = 0$ at which $I = I_0/2$. These additional values of θ aren't the ones we're looking for.

Test Your Understanding of Section 35.3 A two-slit interference experiment uses coherent light of wavelength 5.00×10^{-7} m. Rank the following points in the interference pattern according to the intensity at each point, from highest to lowest. (i) a point that is closer to one slit than the other by 4.00×10^{-7} m; (ii) a point where the light waves received from the two slits are out of phase by 4.00 rad; (iii) a point that is closer to one slit than the other by 7.50×10^{-7} m; (iv) a point where the light waves received by the two slits are out of phase by 2.00 rad.

35.4 Interference in Thin Films

35.11 (a) A diagram and (b) a photograph showing interference of light reflected from a thin film.

(a) Interference between rays reflected from the two surfaces of a thin film

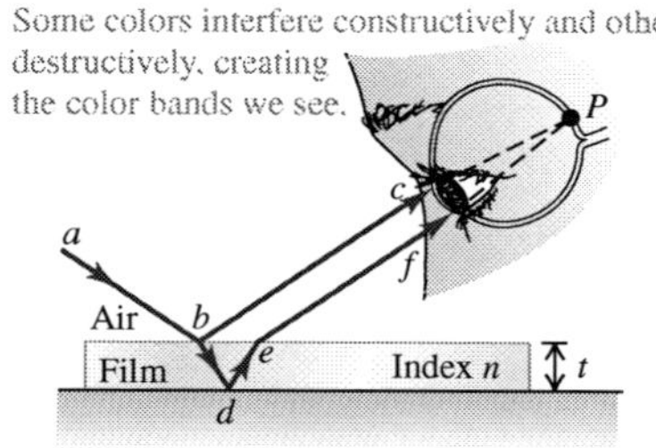

(b) The rainbow fringes of an oil slick on water

You often see bright bands of color when light reflects from a thin layer of oil floating on water or from a soap bubble (see the photograph that opens this chapter). These are the results of interference. Light waves are reflected from the front and back surfaces of such thin films, and constructive interference between the two reflected waves (with different path lengths) occurs in different places for different wavelengths. Figure 35.11a shows the situation. Light shining on the upper surface of a thin film with thickness t is partly reflected at the upper surface (path *abc*). Light *transmitted* through the upper surface is partly reflected at the lower surface (path *abdef*). The two reflected waves come together at point P on the retina of the eye. Depending on the phase relationship, they may interfere constructively or destructively. Different colors have different wavelengths, so the interference may be constructive for some colors and destructive for others. That's why we see colored rings or fringes in Fig. 35.11b (which shows a thin film of oil floating on water) and in the photograph that opens this chapter (which shows thin films of soap solution that make up the bubble walls). The complex shapes of the colored rings in each photograph result from variations in the thickness of the film.

Thin-Film Interference and Phase Shifts During Reflection

Let's look at a simplified situation in which *monochromatic* light reflects from two nearly parallel surfaces at nearly normal incidence. Figure 35.12 shows two plates of glass separated by a thin wedge, or film, of air. We want to consider interference between the two light waves reflected from the surfaces adjacent to the air wedge, as shown. (Reflections also occur at the top surface of the upper plate and the bottom surface of the lower plate; to keep our discussion simple, we won't include these.) The situation is the same as in Fig. 35.11a except that the film (wedge) thickness is not uniform. The path difference between the two waves is just twice the thickness t of the air wedge at each point. At points where $2t$ is an integer number of wavelengths, we expect to see constructive interference and a bright area; where it is a half-integer number of wavelengths, we expect to see destructive interference and a dark area. Along the line where the plates are in contact, there is practically *no* path difference, and we expect a bright area.

When we carry out the experiment, the bright and dark fringes appear, but they are interchanged! Along the line where the plates are in contact, we find a *dark* fringe, not a bright one. This suggests that one or the other of the reflected waves has undergone a half-cycle phase shift during its reflection. In that case the two waves that are reflected at the line of contact are a half-cycle out of phase even though they have the same path length.

In fact, this phase shift can be predicted from Maxwell's equations and the electromagnetic nature of light. The details of the derivation are beyond our scope, but here is the result. Suppose a light wave with electric-field amplitude E_i is traveling in an optical material with index of refraction n_a. It strikes, at normal incidence, an interface with another optical material with index n_b. The amplitude E_r of the wave reflected from the interface is proportional to the amplitude E_i of the incident wave and is given by

$$E_r = \frac{n_a - n_b}{n_a + n_b} E_i \quad \text{(normal incidence)} \tag{35.16}$$

This result shows that the incident and reflected amplitudes have the same sign when n_a is larger than n_b and opposite sign when n_b is larger than n_a. We can distinguish three cases, as shown in Fig. 35.13:

Figure 35.13a: When $n_a > n_b$, light travels more slowly in the first material than in the second. In this case, E_r and E_i have the same sign, and the phase shift of the reflected wave relative to the incident wave is zero. This is analogous to reflection of a transverse mechanical wave on a heavy rope at a point where it is tied to a lighter rope or a ring that can move vertically without friction.

Figure 35.13b: When $n_a = n_b$, the amplitude E_r of the reflected wave is zero. The incident light wave can't "see" the interface, and there is *no* reflected wave.

Figure 35.13c: When $n_a < n_b$, light travels more slowly in the second material than in the first. In this case, E_r and E_i have opposite signs, and the phase shift of the reflected wave relative to the incident wave is π rad (180° or a half-cycle). This is analogous to reflection (with inversion) of a transverse mechanical wave on a light rope at a point where it is tied to a heavier rope or a rigid support.

Let's compare with the situation of Fig. 35.12. For the wave reflected from the upper surface of the air wedge, n_a (glass) is greater than n_b, so this wave has zero phase shift. For the wave reflected from the lower surface, n_a (air) is less than n_b (glass), so this wave has a half-cycle phase shift. Waves that are reflected from the line of contact have no path difference to give additional phase shifts, and they interfere destructively; this is what we observe. You can use the above

35.12 Interference between light waves reflected from the two sides of an air wedge separating two glass plates. The angles and the thickness of the air wedge have been exaggerated for clarity; in the text we assume that the light strikes the upper plate at normal incidence and that the distances h and t are much less than l.

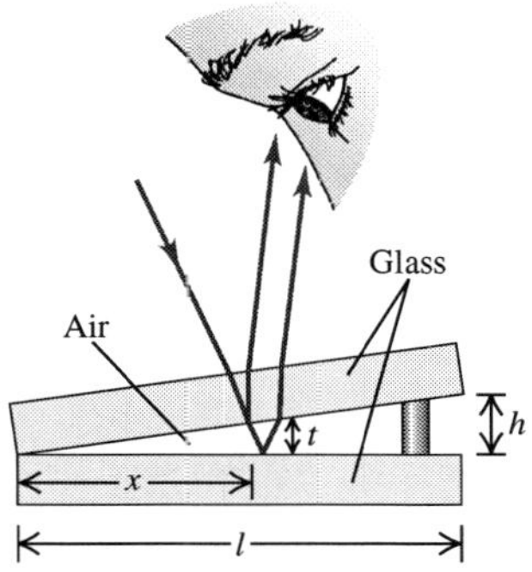

35.13 Upper figures: electromagnetic waves striking an interface between optical materials at normal incidence (shown as a small angle for clarity). Lower figures: mechanical wave pulses on ropes.

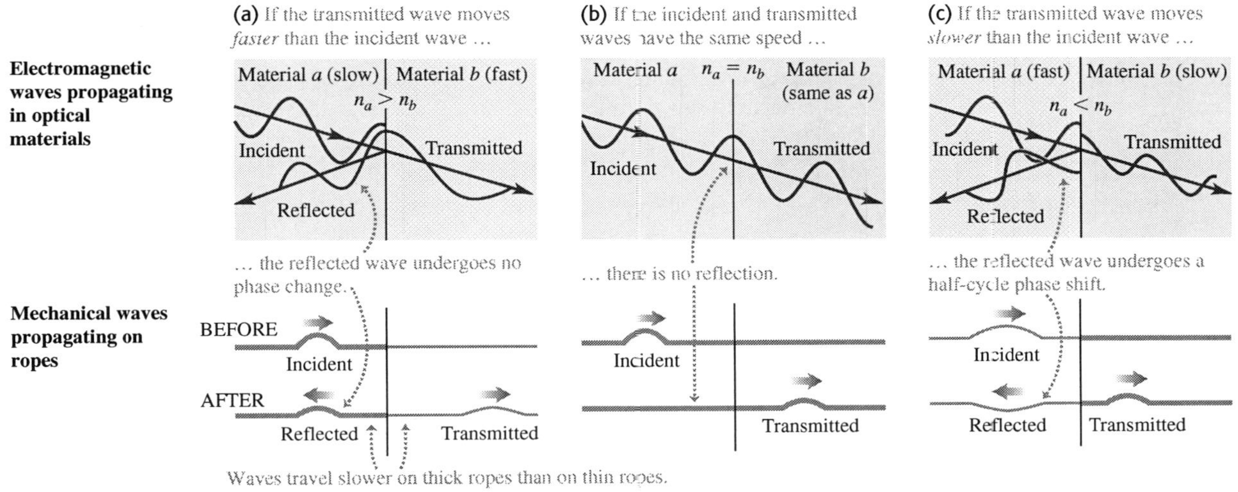

principle to show that for normal incidence, the wave reflected at point b in Fig. 35.11a is shifted by a half-cycle, while the wave reflected at d is not (if there is air below the film).

We can summarize this discussion mathematically. If the film has thickness t, the light is at normal incidence and has wavelength λ in the film; if neither or both of the reflected waves from the two surfaces have a half-cycle reflection phase shift, the conditions for constructive and destructive interference are

$$2t = m\lambda \quad (m = 0, 1, 2, \dots) \quad \text{(constructive reflection from thin film, no relative phase shift)} \tag{35.17a}$$

$$2t = \left(m + \frac{1}{2}\right)\lambda \quad (m = 0, 1, 2, \dots) \quad \text{(destructive reflection from thin film, no relative phase shift)} \tag{35.17b}$$

If *one* of the two waves has a half-cycle reflection phase shift, the conditions for constructive and destructive interference are reversed:

$$2t = \left(m + \frac{1}{2}\right)\lambda \quad (m = 0, 1, 2, \dots) \quad \text{(constructive reflection from thin film, half-cycle relative phase shift)} \tag{35.18a}$$

$$2t = m\lambda \quad (m = 0, 1, 2, \dots) \quad \text{(destructive reflection from thin film, half-cycle relative phase shift)} \tag{35.18b}$$

Thin and Thick Films

You may wonder why we have emphasized *thin* films in our discussion. We have done so because of a principle we introduced in Section 35.1: In order for two waves to cause a steady interference pattern, the waves must be *coherent,* with a definite and constant phase relationship. However, the sun and light bulbs emit light in a stream of short bursts, each of which is only a few micrometers long (1 micrometer $= 1\ \mu m = 10^{-6}$ m). If light reflects from the two surfaces of a thin film, the two reflected waves are part of the same burst (Fig. 35.14a). Hence these waves are coherent and interference occurs as we have described. If the film is too thick, however, the two reflected waves will belong to different bursts (Fig. 35.14b). There is no definite phase relationship between different light bursts, so the two waves are incoherent and there is no fixed interference pattern. That's why you see interference colors in light reflected from an oil slick a few micrometers thick (Fig. 35.11b), but you do *not* see such colors in the light reflected from a pane of window glass with a thickness of a few millimeters (a thousand times greater).

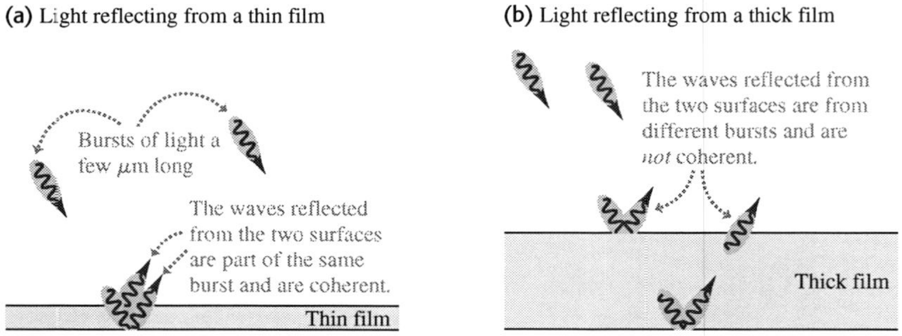

35.14 (a) Light reflecting from a thin film produces a steady interference pattern, but (b) light reflecting from a thick film does not.

Problem-Solving Strategy 35.1 Interference in Thin Films

IDENTIFY *the relevant concepts:* Problems with thin films involve interference of two waves, one reflected from the film's front surface and one reflected from the back surface. Typically you will be asked to relate the wavelength, the index of refraction of the film, and the dimensions of the film.

SET UP *the problem* using the following steps:

1. Make a drawing showing the geometry of the thin film. Your drawing should also depict the materials that adjoin the film; their properties determine whether one or both of the reflected waves have a half-cycle phase shift.
2. Determine the target variable.

EXECUTE *the solution* as follows:

1. Apply the rule for phase changes to each reflected wave. There is a half-cycle phase shift when $n_b > n_a$, and none when $n_b < n_a$.
2. If neither reflected wave undergoes a phase shift, or if both reflected waves do, you can apply Eqs. (35.17). If only one of the reflected waves undergoes a phase shift, you must use Eqs. (35.18).
3. Solve the resulting interference equation for the target variable. If the film consists of anything other than vacuum, be sure to use the wavelength of light *in the film* in your calculations. If the film is anything except vacuum, this is smaller than the wavelength in vacuum by a factor of n. (For air, $n = 1.000$ to four-figure precision.)
4. If you are asked about the wave that is transmitted through the film, keep in mind that *minimum* intensity in the *reflected* wave corresponds to *maximum transmitted* intensity, and vice versa.

EVALUATE *your answer:* You can interpret your results by examining what would happen if the wavelength were changed or if the film had a different thickness.

Example 35.4 Thin-film interference I

Suppose the two glass plates in Fig. 35.12 are two microscope slides 10.0 cm long. At one end they are in contact; at the other end they are separated by a piece of paper 0.0200 mm thick. What is the spacing of the interference fringes seen by reflection? Is the fringe at the line of contact bright or dark? Assume monochromatic light with a wavelength in air of $\lambda = \lambda_0 = 500$ nm.

SOLUTION

IDENTIFY: We'll consider only interference between the light reflected from the upper and lower surfaces of the air wedge between the slides. The glass plate is a millimeter or so thick, so we can ignore interference between the light reflected from the upper and lower surfaces of this plate (see Fig. 35.14b).

SET UP: Figure 35.15 depicts the situation. Light travels more slowly in the glass of the microscope slides than it does in air. Hence the wave reflected from the upper surface of the air wedge has no phase shift (see Fig. 35.13a), while the wave reflected from the lower surface has a half-cycle phase shift (see Fig. 35.13c).

35.15 Our sketch for this problem.

EXECUTE: Since only one of the reflected waves undergoes a phase shift, the condition for *destructive* interference (a dark fringe) is Eq. (35.18b):

$$2t = m\lambda_0 \qquad (m = 0, 1, 2, \dots)$$

From similar triangles in Fig. 35.15 the thickness t of the air wedge at each point is proportional to the distance x from the line of contact:

$$\frac{t}{x} = \frac{h}{l}$$

Combining this with Eq. (35.18b), we find

$$\frac{2xh}{l} = m\lambda_0$$

$$x = m\frac{l\lambda_0}{2h} = m\frac{(0.100\text{ m})(500 \times 10^{-9}\text{ m})}{(2)(0.0200 \times 10^{-3}\text{ m})} = m(1.25\text{ mm})$$

Successive dark fringes, corresponding to successive integer values of m, are spaced 1.25 mm apart. Substituting $m = 0$ into this equation gives $x = 0$, corresponding to the line of contact between the two slides (at the left-hand side of Fig. 35.15). Hence there is a dark fringe at the line of contact.

EVALUATE: Our result shows that the fringe spacing is proportional to the wavelength of the light used; the fringes would be farther apart with red light (larger λ_0) than with blue light (smaller λ_0). If we use white light, the reflected light at any point is a mixture of wavelengths for which constructive interference occurs; the wavelengths that interfere destructively are weak or absent in the reflected light. (This same effect explains the colors seen when an oil film on water is illuminated by white light, as in Fig. 35.11b).

Example 35.5 Thin-film interference II

In Example 35.4, suppose the glass plates have $n = 1.52$ and the space between plates contains water $(n = 1.33)$ instead of air. What happens now?

SOLUTION

IDENTIFY: The index of refraction of the water film is still less than that of the glass on either side of the film, so the phase shifts are the same as in Example 35.4. The only difference is that the wavelength in water is different than in air.

SET UP: Once again we use Eq. (35.18b) to find the positions of the dark fringes. The wavelengths λ in water is related to the wavelength λ_0 in air (essentially vacuum) by Eq. (33.5), $\lambda = \lambda_0/n$.

EXECUTE: In the film of water $(n = 1.33)$, the wavelength is

$$\lambda = \frac{\lambda_0}{n} = \frac{500\text{ nm}}{1.33} = 376\text{ nm}$$

When we replace λ_0 by λ in the expression from Example 35.4 for the position x of the mth dark fringe, we find that the fringe spacing is reduced by the same factor of 1.33 and is equal to 0.940 mm. Note that there is still a dark fringe at the line of contact.

EVALUATE: Can you see that to return to the same fringe spacing as in Example 35.4, the dimension h in Fig. 35.15 would have to be reduced to $(0.0200\text{ mm})/1.33 = 0.0150$ mm? This shows that what matters in thin-film interference is the *ratio* between the wavelength and the thickness of the film. [You can see this by considering Eqs. (35.17a) and (35.17b).]

Example 35.6 Thin-film interference III

Suppose the upper of the two plates in Example 35.4 is a plastic material with $n = 1.40$, the wedge is filled with a silicone grease having $n = 1.50$, and the bottom plate is a dense flint glass with $n = 1.60$. What happens now?

SOLUTION

IDENTIFY: The geometry is still as shown in Fig. 35.15, but now half-cycle phase shifts occur at *both* surfaces of the wedge of grease (see Fig. 35.13c).

SET UP: Figure 35.16 shows the situation. Since there is a half-cycle phase shift at both surfaces, there is no *relative* phase shift and we must use Eq. (35.17b) to find the positions of the dark fringes.

EXECUTE: The value of λ to use in Eq. (35.17b) is the wavelength in the silicone grease: $\lambda = \lambda_0/n = (500\text{ nm})/1.50 = 333$ nm. You can readily show that the fringe spacing is 0.833 mm. Note that the two reflected waves from the line of contact are in phase (they both undergo the same phase shift), so the line of contact is at a *bright* fringe.

35.16 Our sketch for this problem.

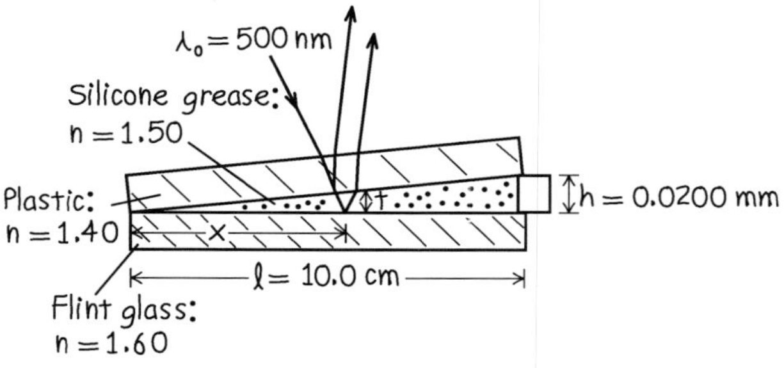

EVALUATE: What would happen if you carefully removed the upper microscope slide so that the grease wedge retained its shape? There would still be half-cycle phase changes at the upper and lower surfaces of the wedge, so the pattern of fringes would be the same as with the upper slide present.

Newton's Rings

Figure 35.17a shows the convex surface of a lens in contact with a plane glass plate. A thin film of air is formed between the two surfaces. When you view the setup with monochromatic light, you see circular interference fringes (Fig. 35.17b). These were studied by Newton and are called **Newton's rings.**

We can use interference fringes to compare the surfaces of two optical parts by placing the two in contact and observing the interference fringes. Figure 35.18 is a photograph made during the grinding of a telescope objective lens. The lower, larger-diameter, thicker disk is the correctly shaped master, and the smaller, upper disk is the lens under test. The "contour lines" are Newton's interference fringes; each one indicates an additional distance between the specimen and the master of one half-wavelength. At 10 lines from the center spot the distance between the two surfaces is five wavelengths, or about 0.003 mm. This

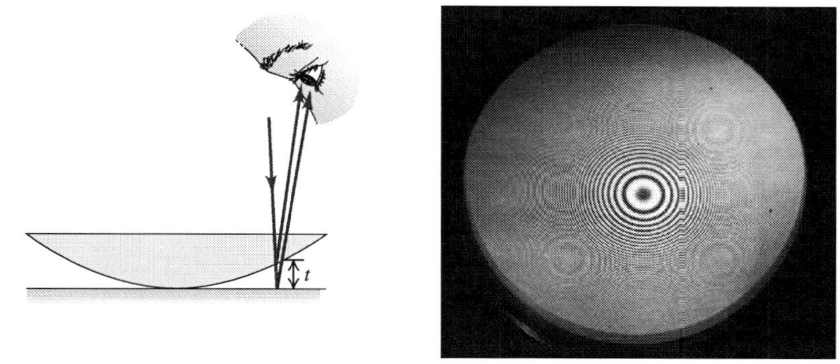

35.17 (a) Air film between a convex lens and a plane surface. The thickness of the film t increases from zero as we move out from the center, giving (b) a series of alternating dark and bright rings for monochromatic light.

isn't very good; high-quality lenses are routinely ground with a precision of less than one wavelength. The surface of the primary mirror of the Hubble Space Telescope was ground to a precision of better than $\frac{1}{50}$ wavelength. Unfortunately, it was ground to incorrect specifications, creating one of the most precise errors in the history of optical technology (see Section 34.2).

35.18 The surface of a telescope objective lens under inspection during manufacture.

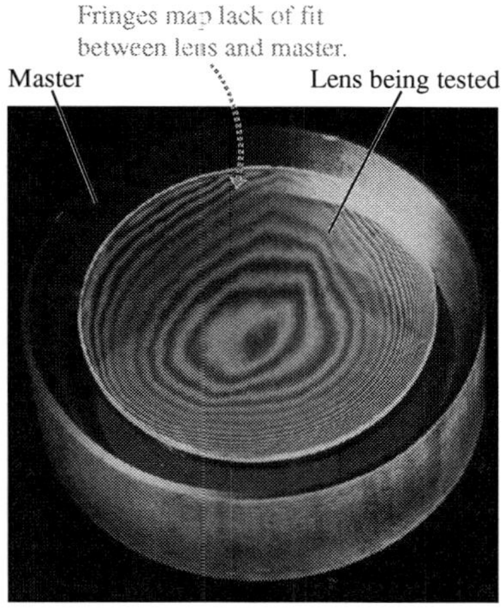

Nonreflective and Reflective Coatings

Nonreflective coatings for lens surfaces make use of thin-film interference. A thin layer or film of hard transparent material with an index of refraction smaller than that of the glass is deposited on the lens surface, as in Fig. 35.19. Light is reflected from both surfaces of the layer. In both reflections the light is reflected from a medium of greater index than that in which it is traveling, so the same phase change occurs in both reflections. If the film thickness is a quarter (one-fourth) of the wavelength *in the film* (assuming normal incidence), the total path difference is a half-wavelength. Light reflected from the first surface is then a half-cycle out of phase with light reflected from the second, and there is destructive interference.

The thickness of the nonreflective coating can be a quarter-wavelength for only one particular wavelength. This is usually chosen in the central yellow-green portion of the spectrum $(\lambda = 550\text{ nm})$, where the eye is most sensitive. Then there is somewhat more reflection at both longer (red) and shorter (blue) wavelengths, and the reflected light has a purple hue. The overall reflection from a lens or prism surface can be reduced in this way from 4–5% to less than 1%. This treatment is particularly important in eliminating stray reflected light in highly corrected photographic lenses with many individual pieces of glass and many air-glass surfaces. It also increases the net amount of light that is *transmitted* through the lens, since light that is not reflected will be transmitted. The same principle is used to minimize reflection from silicon photovoltaic solar cells $(n = 3.5)$ by use of a thin surface layer of silicon monoxide $(SiO, n = 1.45)$; this helps to increase the amount of light that actually reaches the solar cells.

If a quarter-wavelength thickness of a material with an index of refraction *greater* than that of glass is deposited on glass, then the reflectivity is *increased,* and the deposited material is called a **reflective coating.** In this case there is a half-cycle phase shift at the air–film interface but none at the film–glass interface, and reflections from the two sides of the film interfere constructively. For example, a coating with refractive index 2.5 causes 38% of the incident energy to be reflected, compared with 4% or so with no coating. By use of multiple-layer coatings, it is possible to achieve nearly 100% transmission or reflection for particular wavelengths. Some practical applications of these coatings are for color separation in color television cameras and for infrared "heat reflectors" in motion-picture projectors, solar cells, and astronauts' visors. Reflective coatings occur in nature on the scales of herring and other silvery fish; this gives these fish their characteristic shiny appearance (see Problem 35.56).

35.19 A nonreflective coating has an index of refraction intermediate between those of glass and air.

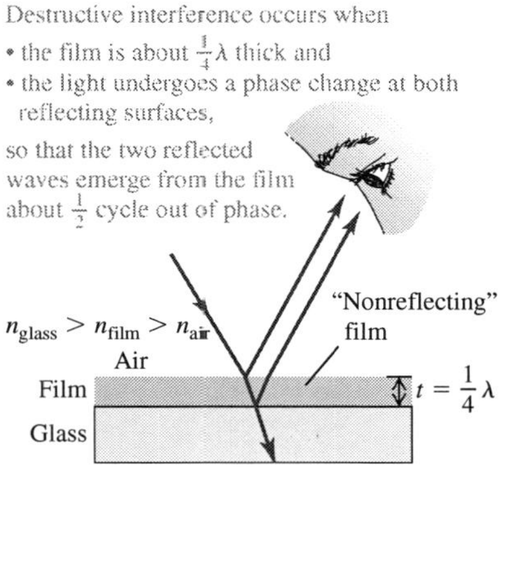

Example 35.7 A nonreflective coating

A commonly used lens coating material is magnesium fluoride, MgF_2, with $n = 1.38$. What thickness should a nonreflective coating have for 550-nm light if it is applied to glass with $n = 1.52$?

SOLUTION

IDENTIFY: This coating is of the sort depicted in Fig. 35.19.

SET UP: The thickness must be one-quarter of the wavelength in the coating.

EXECUTE: The wavelength of yellow-green light in air is $\lambda_0 = 550$ nm, so its wavelength in the MgF_2 coating is

$$\lambda = \frac{\lambda_0}{n} = \frac{550 \text{ nm}}{1.38} = 400 \text{ nm}$$

To be a nonreflective film, the coating should have a thickness of one-quarter λ, or 100 nm. This is a very thin film, no more than a few hundred molecules thick.

EVALUATE: Note that such a coating becomes *reflective* if its thickness is equal to one-*half* of a wavelength; then light reflected from the lower surface of the coating travels one wavelength farther than light reflected from the upper surface, so the two waves are in phase and interfere constructively. This occurs for light with a wavelength in MgF_2 of 200 nm and a wavelength in air of $(200 \text{ nm})(1.38) = 276$ nm. This is an ultraviolet wavelength (see Section 32.1), so designers of optical lenses with nonreflective coatings need not worry about enhanced reflection of this kind.

Test Your Understanding of Section 35.4 A thin layer of benzene $(n = 1.501)$ ties on top of a sheet of fluorite $(n = 1.434)$. It is illuminated from above with light whose wavelength in benzene is 400 nm. Which of the following possible thicknesses of the benzene layer will maximize the brightness of the reflected light? (i) 100 nm; (ii) 200 nm; (iii) 300 nm; (iv) 400 nm.

35.5 The Michelson Interferometer

An important experimental device that uses interference is the **Michelson interferometer.** In the late 19th century, it helped to provide one of the key experimental underpinnings of the theory of relativity. More recently, Michelson interferometers have been used to make precise measurements of wavelengths and of very small distances, such as the minute changes in thickness of an axon when a nerve impulse propagates along its length. Like the Young two-slit experiment, a Michelson interferometer takes monochromatic light from a single source and divides it into two waves that follow different paths. In Young's experiment, this is done by sending part of the light through one slit and part through another; in a Michelson interferometer a device called a *beam splitter* is used. Interference occurs in both experiments when the two light waves are recombined.

How a Michelson Interferometer Works

The principal components of a Michelson interferometer are shown schematically in Fig. 35.20. A ray of light from a monochromatic source A strikes the beam splitter C, which is a glass plate with a thin coating of silver on its right side. Part of the light (ray 1) passes through the silvered surface and the compensator plate D and is reflected from mirror M_1. It then returns through D and is reflected from the silvered surface of C to the observer. The remainder of the light (ray 2) is reflected from the silvered surface at point P to the mirror M_2 and back through C to the observer's eye. The purpose of the compensator plate D is to ensure that rays 1 and 2 pass through the same thickness of glass; plate D is cut from the same piece of glass as plate C, so their thicknesses are identical to within a fraction of a wavelength.

The whole apparatus in Fig. 35.20 is mounted on a very rigid frame, and the position of mirror M_2 can be adjusted with a fine, very accurate micrometer screw. If the distances L_1 and L_2 are exactly equal and the mirrors M_1 and M_2 are exactly at right angles, the virtual image of M_1 formed by reflection at the silvered

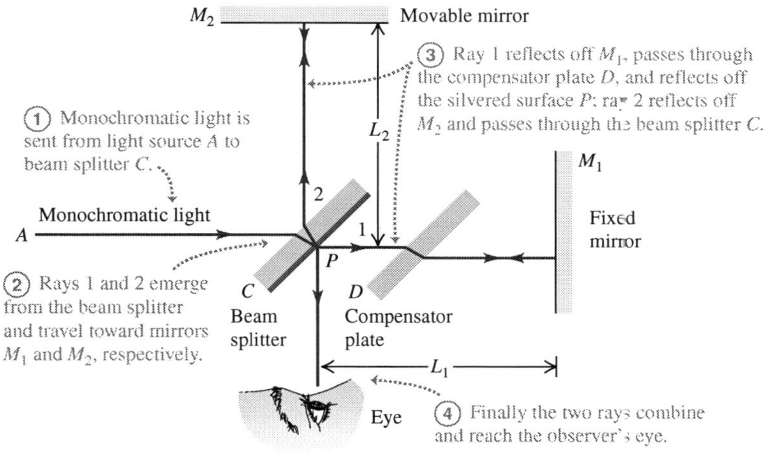

35.20 A schematic Michelson interferometer. The observer sees an interference pattern that results from the difference in path lengths for rays 1 and 2.

surface of plate C coincides with mirror M_2. If L_1 and L_2 are *not* exactly equal, the image of M_1 is displaced slightly from M_2; and if the mirrors are not exactly perpendicular, the image of M_1 makes a slight angle with M_2. Then the mirror M_2 and the virtual image of M_1 play the same roles as the two surfaces of a wedge-shaped thin film (see Section 35.4), and light reflected from these surfaces forms the same sort of interference fringes.

Suppose the angle between mirror M_2 and the virtual image of M_1 is just large enough that five or six vertical fringes are present in the field of view. If we now move the mirror M_2 slowly either backward or forward a distance $\lambda/2$, the difference in path length between rays 1 and 2 changes by λ, and each fringe moves to the left or right a distance equal to the fringe spacing. If we observe the fringe positions through a telescope with a crosshair eyepiece and m fringes cross the crosshairs when we move the mirror a distance y, then

$$y = m\frac{\lambda}{2} \qquad \text{or} \qquad \lambda = \frac{2y}{m} \tag{35.19}$$

If m is several thousand, the distance y is large enough that it can be measured with good accuracy, and we can obtain an accurate value for the wavelength λ. Alternatively, if the wavelength is known, a distance y can be measured by simply counting fringes when M_2 is moved by this distance. In this way, distances that are comparable to a wavelength of light can be measured with relative ease.

The Michelson-Morley Experiment

The original application of the Michelson interferometer was to the historic **Michelson-Morley experiment.** Before the electromagnetic theory of light and Einstein's special theory of relativity became established, most physicists believed that the propagation of light waves occurred in a medium called the **ether,** which was believed to permeate all space. In 1887 the American scientists Albert Michelson and Edward Morley used the Michelson interferometer in an attempt to detect the motion of the earth through the ether. Suppose the interferometer in Fig. 35.20 is moving from left to right relative to the ether. According to the ether theory, this would lead to changes in the speed of light in the portions of the path shown as horizontal lines in the figure. There would be fringe shifts relative to the positions that the fringes would have if the instrument were at rest in the ether. Then when the entire instrument was rotated 90°, the other portions of the paths would be similarly affected, giving a fringe shift in the opposite direction.

Michelson and Morley expected that the motion of the earth through the ether would cause a fringe shift of about four-tenths of a fringe when the instrument was rotated. The shift that was actually observed was less than a hundredth of a fringe and, within the limits of experimental uncertainty, appeared to be exactly zero. Despite its orbital motion around the sun, the earth appeared to be *at rest* relative to the ether. This negative result baffled physicists until Einstein developed the special theory of relativity in 1905. Einstein postulated that the speed of a light wave in vacuum has the same magnitude c relative to *all* inertial reference frames, no matter what their velocity may be relative to each other. The presumed ether then plays no role, and the concept of an ether has been abandoned.

The theory of relativity is a well-established cornerstone of modern physics, and we will study it in detail in Chapter 37. In retrospect, the Michelson-Morley experiment gives strong experimental support to the special theory of relativity, and it is often called the most significant "negative-result" experiment ever performed.

Test Your Understanding of Section 35.5 You are observing the pattern of fringes in a Michelson interferometer like that shown in Fig. 35.20. If you change the index of refraction (but not the thickness) of the compensator plate, will the pattern change?

CHAPTER 35 SUMMARY

Interference and coherent sources: Monochromatic light is light with a single frequency. Coherence is a definite, unchanging phase relationship between two waves. The overlap of waves from two coherent sources of monochromatic light forms an interference pattern. The principle of superposition states that the total wave disturbance at any point is the sum of the disturbances from the separate waves.

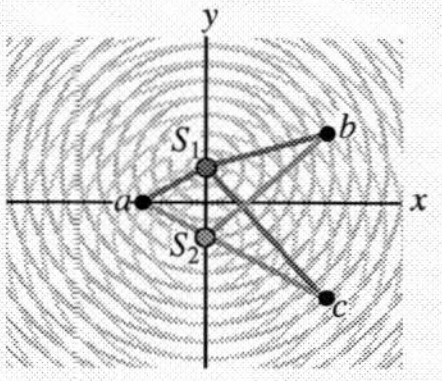

Two-source interference of light: When two sources are in phase, constructive interference occurs at points where the difference in path length from the two sources is zero or an integer number of wavelengths; destructive interference occurs at points where the path difference is a half-integer number of wavelengths. If two sources separated by a distance d are both very far from a point P, and the line from the sources to P makes an angle θ with the line perpendicular to the line of the sources, then the condition for constructive interference at P is Eq. (35.4). The condition for destructive interference is Eq. (35.5). When θ is very small, the position y_m of the mth bright fringe on a screen located a distance R from the sources is given by Eq. (35.6). (See Examples 35.1 and 35.2.)

$$d\sin\theta = m\lambda \quad (m = 0, \pm1, \pm2, \dots) \tag{35.4}$$
(constructive interference)

$$d\sin\theta = \left(m + \frac{1}{2}\right)\lambda \quad (m = 0, \pm1, \pm2, \dots) \tag{35.5}$$
(destructive interference)

$$y_m = R\frac{m\lambda}{d} \tag{35.6}$$
(bright fringes)

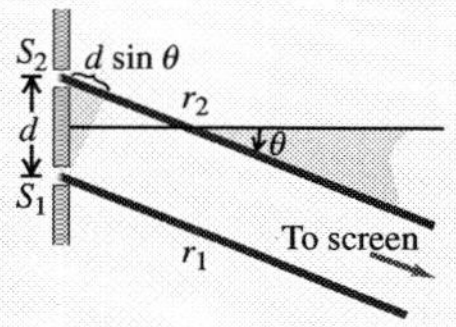

Intensity in interference patterns: When two sinusoidal waves with equal amplitude E and phase difference ϕ are superimposed, the resultant amplitude E_P and intensity I are given by Eqs. (35.7) and (35.10), respectively. If the two sources emit in phase, the phase difference ϕ at a point P (located a distance r_1 from source 1 and a distance r_2 from source 2) is directly proportional to the difference in path length $r_2 - r_1$. (See Example 35.3.)

$$E_P = 2E\left|\cos\frac{\phi}{2}\right| \tag{35.7}$$

$$I = I_0\cos^2\frac{\phi}{2} \tag{35.10}$$

$$\phi = \frac{2\pi}{\lambda}(r_2 - r_1) = k(r_2 - r_1) \tag{35.11}$$

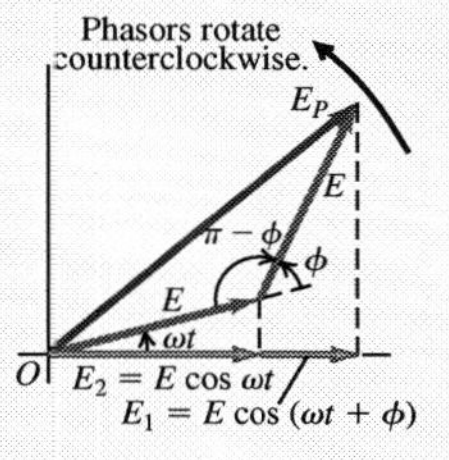

Interference in thin films: When light is reflected from both sides of a thin film of thickness t and no phase shift occurs at either surface, constructive interference between the reflected waves occurs when $2t$ is equal to an integral number of wavelengths. If a half-cycle phase shift occurs at one surface, this is the condition for destructive interference. A half-cycle phase shift occurs during reflection whenever the index of refraction in the second material is greater than that in the first. (See Examples 35.4–35.7.)

$$2t = m\lambda \quad (m = 0, 1, 2, \dots) \tag{35.17a}$$
(constructive reflection from thin film, no relative phase shift)

$$2t = \left(m + \frac{1}{2}\right)\lambda \quad (m = 0, 1, 2, \dots) \tag{35.17b}$$
(destructive reflection from thin film, no relative phase shift)

$$2t = \left(m + \frac{1}{2}\right)\lambda \quad (m = 0, 1, 2, \dots) \tag{35.18a}$$
(constructive reflection from thin film, half-cycle relative phase shift)

$$2t = m\lambda \quad (m = 0, 1, 2, \dots) \tag{35.18b}$$
(destructive reflection from thin film, half-cycle relative phase shift)

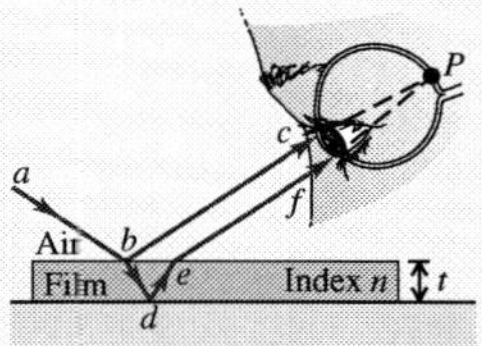

Michelson interferometer: The Michelson interferometer uses a monochromatic light source and can be used for high-precision measurements of wavelengths. Its original purpose was to detect motion of the earth relative to a hypothetical ether, the supposed medium for electromagnetic waves. The ether has never been detected, and the concept has been abandoned; the speed of light is the same relative to all observers. This is part of the foundation of the special theory of relativity.

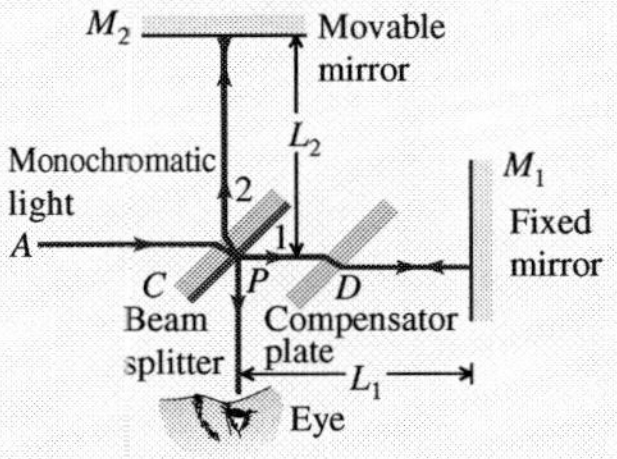

Key Terms

physical optics, *1207*
interference, *1208*
principle of superposition, *1208*
monochromatic light, *1208*
coherent, *1209*
constructive interference, *1209*
destructive interference, *1210*
antinodal curves, *1210*
nodal curves, *1210*
interference fringes, *1212*
Newton's rings, *1222*
nonreflective coating, *1223*
reflective coating, *1223*
Michelson interferometer, *1224*
Michelson-Morley experiment, *1225*
ether, *1225*

Answer to Chapter Opening Question ?

The colors appear due to constructive interference between light waves reflected from the outer and inner surfaces of the soap bubble. The thickness of the bubble walls at each point determines the wavelength of light for which the most constructive interference occurs and hence the color that appears the brightest at that point (see Example 35.4 in Section 35.4).

Answers to Test Your Understanding Questions

35.1 Answer: (i) At any point P on the positive y-axis above S_1, the distance r_2 from S_2 to P is greater than the distance r_1 from S_1 to P by 4λ. This corresponds to $m = 4$ in Eq. (35.1), the equation for constructive interference. Hence all such points make up an antinodal curve.
35.2 Answer: (ii) Blue light has a shorter wavelength than red light (see Section 32.1). Equation (35.6) tells us that the distance y_m from the center of the pattern to the mth bright fringe is proportional to the wavelength λ. Hence all of the fringes will move toward the center of the pattern as the wavelength decreases, and the spacing between fringes will decrease.
35.3 Answer: (i), (iv), (ii), (iii) In cases (i) and (iii) we are given the wavelength λ and path difference $d\sin\theta$. Hence we use Eq. (35.14), $I = I_0\cos^2[(\pi d\sin\theta)/\lambda]$. In parts (ii) and (iii) we are given the phase difference ϕ and we use Eq. (35.10), $I = I_0\cos^2(\phi/2)$. We find:
(i) $I = I_0\cos^2[\pi(4.00 \times 10^{-7}\text{ m})/(5.00 \times 10^{-7}\text{ m})] = I_0\cos^2(0.800\pi\text{ rad}) = 0.655I_0$;
(ii) $I = I_0\cos^2[(4.00\text{ rad})/2] = I_0\cos^2(2.00\text{ rad}) = 0.173I_0$;
(iii) $I = I_0\cos^2[\pi(7.50 \times 10^{-7}\text{ m})/(5.00 \times 10^{-7}\text{ m})] = I_0\cos^2(1.50\pi\text{ rad}) = 0$;
(iv) $I = I_0\cos^2[(2.00\text{ rad})/2] = I_0\cos^2(1.00\text{ rad}) = 0.292I_0$.
35.4 Answers: (i) and (iii) Benzene has a larger index of refraction than air, so light that reflects off the upper surface of the benzene undergoes a half-cycle phase shift. Fluorite has a *smaller* index of refraction than benzene, so light that reflects off the benzene–fluorite interface does not undergo a phase shift. Hence the equation for constructive reflection is Eq. (35.18a), $2t = (m + \frac{1}{2})\lambda$, which we can rewrite as $t = (m + \frac{1}{2})\lambda/2 = (m + \frac{1}{2})(400\text{ mm})/2 = 100\text{ nm}, 300\text{ nm}, 500\text{ nm}, \ldots$.
35.5 Answer: yes Changing the index of refraction changes the wavelength of the light inside the compensator plate, and so changes the nuimber of wavelengths within the thickness of the plate. Hence this has the same effect as changing the distance L_1 from the beam splitter to mirror M_1, which would change the interference pattern.

PROBLEMS

For instructor-assigned homework, go to **www.masteringphysics.com**

Discussion Questions

Q35.1. A two-slit interference experiment is set up, and the fringes are displayed on a screen. Then the whole apparatus is immersed in the nearest swimming pool. How does the fringe pattern change?
Q35.2. Could an experiment similar to Young's two-slit experiment be performed with sound? How might this be carried out? Does it matter that sound waves are longitudinal and electromagnetic waves are transverse? Explain.
Q35.3. Monochromatic coherent light passing through two thin slits is viewed on a distant screen. Are the bright fringes equally spaced on the screen? If so, why? If not, which ones are closest to being equally spaced?
Q35.4. In a two-slit interference pattern on a distant screen, are the bright fringes midway between the dark fringes? Is this ever a good approximation?
Q35.5. Would the headlights of a distant car form a two-source interference pattern? If so, how might it be observed? If not, why not?

Q35.6. The two sources S_1 and S_2 shown in Fig. 35.3 emit waves of the same wavelength λ and are in phase with each other. Suppose S_1 is a weaker source, so that the waves emitted by S_1 have half the amplitude of the waves emitted by S_2. How would this affect the positions of the antinodal lines and nodal lines? Would there be total reinforcement at points on the antinodal curves? Would there be total cancellation at points on the nodal curves? Explain your answers.

Q35.7. Could the Young two-slit interference experiment be performed with gamma rays? If not, why not? If so, discuss differences in the experimental design compared to the experiment with visible light.

Q35.8. Coherent red light illuminates two narrow slits that are 25 cm apart. Will a two-slit interference pattern be observed when the light from the slits falls on a screen? Explain.

Q35.9. Coherent light with wavelength λ falls on two narrow slits separated by a distance d. If d is less than some minimum value, no dark fringes are observed. Explain. In terms of λ, what is this minimum value of d?

Q35.10. A fellow student, who values memorizing equations above understanding them, combines Eqs. (35.4) and (35.13) to "prove" that ϕ can *only* equal $2\pi m$. How would you explain to this student that ϕ can have values other than $2\pi m$?

Q35.11. If the monochromatic light shown in Fig. 35.5a were replaced by white light, would a two-slit interference pattern be seen on the screen? Explain.

Q35.12. In using the superposition principle to calculate intensities in interference patterns, could you add the intensities of the waves instead of their amplitudes? Explain.

Q35.13. A glass windowpane with a thin film of water on it reflects less than when it is perfectly dry. Why?

Q35.14. A *very* thin soap film $(n = 1.33)$, whose thickness is much less than a wavelength of visible light, looks black; it appears to reflect no light at all. Why? By contrast, an equally thin layer of soapy water $(n = 1.33)$ on glass $(n = 1.50)$ appears quite shiny. Why is there a difference?

Q35.15. Interference can occur in thin films. Why is it important that the films be *thin?* Why don't you get these effects with a relatively *thick* film? Where should you put the dividing line between "thin" and "thick"? Explain your reasoning.

Q35.16. If we shine white light on an air wedge like that shown in Fig. 35.12, the colors that are weak in the light *reflected* from any point along the wedge are strong in the light *transmitted* through the wedge. Explain why this should be so.

Q35.17. Monochromatic light is directed at normal incidence on a thin film. There is destructive interference for the reflected light, so the intensity of the reflected light is very low. What happened to the energy of the incident light?

Q35.18. When a thin oil film spreads out on a puddle of water, the thinnest part of the film looks dark in the resulting interference pattern. What does this tell you about the relative magnitudes of the refractive indexes of oil and water?

Exercises

Section 35.1 Interference and Coherent Sources

35.1. Two coherent sources A and B of radio waves are 5.00 m apart. Each source emits waves with wavelength 6.00 m. Consider points along the line between the two sources. At what distances, if any, from A is the interference (a) constructive and (b) destructive?

35.2. Radio Interference. Two radio antennas A and B radiate in phase. Antenna B is 120 m to the right of antenna A. Consider point Q along the extension of the line connecting the antennas, a horizontal distance of 40 m to the right of antenna B. The frequency, and hence the wavelength, of the emitted waves can be varied. (a) What is the longest wavelength for which there will be destructive interference at point Q? (b) What is the longest wavelength for which there will be constructive interference at point Q?

35.3. A radio transmitting station operating at a frequency of 120 MHz has two identical antennas that radiate in phase. Antenna B is 9.00 m to the right of antenna A. Consider point P between the antennas and along the line connecting them, a horizontal distance x to the right of antenna A. For what values of x will constructive interference occur at point P?

35.4. Two light sources can be adjusted to emit monochromatic light of any visible wavelength. The two sources are coherent, 2.04 μm apart, and in line with an observer, so that one source is 2.04 μm farther from the observer than the other. (a) For what visible wavelengths (400 to 700 nm) will the observer see the brightest light, owing to constructive interference? (b) How would your answers to part (a) be affected if the two sources were not in line with the observer, but were still arranged so that one source is 2.04 μm farther away from the observer than the other? (c) For what visible wavelengths will there be *destructive* interference at the location of the observer?

35.5. Two speakers, emitting identical sound waves of wavelength 2.0 m in phase with each other, and an observer are located as shown in Fig. 35.21. (a) At the observer's location, what is the path difference for waves from the two speakers? (b) Will the sound waves interfere constructively or destructively at the observer's location—or something in between constructive and destructive? (c) Suppose the observer now increases her distance from the speakers to 17.0 m, staying directly in front of the same speaker as initially. Answer the questions of parts (a) and (b) for this new situation.

Figure **35.21**
Exercise 35.5

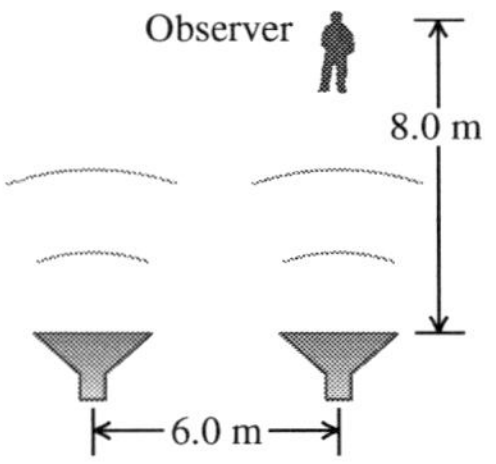

35.6. Figure 35.3 shows the wave pattern produced by two identical, coherent sources emitting waves with wavelength λ and separated by a distance $d = 4\lambda$. (a) Explain why the positive y-axis above S_1 constitutes an antinodal curve with $m = +4$ and why the negative y-axis below S_2 constitutes an antinodal curve with $m = -4$. (b) Draw the wave pattern produced when the separation between the sources is reduced to 3λ. In your drawing, sketch all antinodal curves—that is, the curves on which $r_2 - r_1 = m\lambda$. Label each curve by its value of m. (c) In general, what determines the maximum (most positive) and minimum (most negative) values of the integer m that labels the antinodal lines? (d) Suppose the separation between the sources is increased to $7\frac{1}{2}\lambda$. How many antinodal curves will there be? To what values of m do they correspond? Explain your reasoning. (You should not have to make a drawing to answer these questions.)

35.7. Consider Fig. 35.3, which could represent interference between water waves in a ripple tank. Pick at least three points on the antinodal curve labeled "$m = 3$," and make measurements from the figure to show that Eq. (35.1) is indeed satisfied. Explain what measurements you made and how you measured the wavelength λ.

Section 35.2 Two-Source Interference of Light

35.8. Young's experiment is performed with light from excited helium atoms $(\lambda = 502\text{ nm})$. Fringes are measured carefully on a screen 1.20 m away from the double slit, and the center of the 20th fringe (not counting the central bright fringe) is found to be 10.6 mm from the center of the central bright fringe. What is the separation of the two slits?

35.9. Two slits spaced 0.450 mm apart are placed 75.0 cm from a screen. What is the distance between the second and third dark lines of the interference pattern on the screen when the slits are illuminated with coherent light with a wavelength of 500 nm?

35.10. Coherent light with wavelength 450 nm falls on a double slit. On a screen 1.80 m away, the distance between dark fringes is 4.20 mm. What is the separation of the slits?

35.11. Coherent light from a sodium-vapor lamp is passed through a filter that blocks everything except light of a single wavelength. It then falls on two slits separated by 0.460 mm. In the resulting interference pattern on a screen 2.20 m away, adjacent bright fringes are separated by 2.82 mm. What is the wavelength?

35.12. Coherent light with wavelength 400 nm passes through two very narrow slits that are separated by 0.200 mm and the interference pattern is observed on a screen 4.00 m from the slits. (a) What is the width (in mm) of the central interference maximum? (b) What is the width of the first-order bright fringe?

35.13. Two very narrow slits are spaced 1.80 μm apart and are placed 35.0 cm from a screen. What is the distance between the first and second dark lines of the interference pattern when the slits are illuminated with coherent light with $\lambda = 550$ nm? (*Hint:* The angle θ in Eq. (35.5) is *not* small.)

35.14. Coherent light that contains two wavelengths, 660 nm (red) and 470 nm (blue), passes through two narrow slits separated by 0.300 mm, and the interference pattern is observed on a screen 5.00 m from the slits. What is the distance on the screen between the first-order bright fringes for the two wavelengths?

35.15. Coherent light with wavelength 600 nm passes through two very narrow slits and the interference pattern is observed on a screen 3.00 m from the slits. The first-order bright fringe is at 4.84 mm from the center of the central bright fringe. For what wavelength of light will the first-order dark fringe be observed at this same point on the screen?

35.16. Coherent light of frequency 6.32×10^{14} Hz passes through two thin slits and falls on a screen 85.0 cm away. You observe that the third bright fringe occurs at ± 3.11 cm on either side of the central bright fringe. (a) How far apart are the two slits? (b) At what distance from the central bright fringe will the third dark fringe occur?

35.17. Two thin parallel slits that are 0.0116 mm apart are illuminated by a laser beam of wavelength 585 nm. (a) On a very large distant screen, what is the *total* number of bright fringes (those indicating complete constructive interference), including the central fringe and those on both sides of it? Solve this problem without calculating all the angles! (*Hint:* What is the largest that $\sin\theta$ can be? What does this tell you is the largest value of m?) (b) At what angle, relative to the original direction of the beam, will the fringe that is most distant from the central bright fringe occur?

35.18. An FM radio station has a frequency of 107.9 MHz and uses two identical antennas mounted at the same elevation, 12.0 m apart. The antennas radiate in phase. The resulting radiation pattern has a maximum intensity along a horizontal line perpendicular to the line joining the antennas and midway between them. Assume that the intensity is observed at distances from the antennas that are much greater than 12.0 m. (a) At which other angles (measured from the line of maximum intensity) is the intensity maximum? (b) At which angles is it zero?

Section 35.3 Intensity in Interference Patterns

35.19. In a two-slit interference pattern, the intensity at the peak of the central maximum is I_0. (a) At a point in the pattern where the phase difference between the waves from the two slits is 60.0°, what is the intensity? (b) What is the path difference for 480-nm light from the two slits at a point where the phase angle is 60.0°?

35.20. Coherent sources A and B emit electromagnetic waves with wavelength 2.00 cm. Point P is 4.86 m from A and 5.24 m from B. What is the phase difference at P between these two waves?

35.21. Coherent light with wavelength 500 nm passes through narrow slits separated by 0.340 mm. At a distance from the slits large compared to their separation, what is the phase difference (in radians) in the light from the two slits at an angle of 23.0° from the centerline?

35.22. GPS Transmission. The GPS (Global Positioning System) satellites are approximately 5.18 m across and transmit two low-power signals, one of which is at 1575.42 MHz (in the UHF band). In a series of laboratory tests on the satellite, you put two 1575.42-MHz UHF transmitters at opposite ends of the satellite. These broadcast in phase uniformly in all directions. You measure the intensity at points on a circle that is several hundred meters in radius and centered on the satellite. You measure angles on this circle relative to a point that lies along the centerline of the satellite (that is, the perpendicular bisector of a line which extends from one transmitter to the other). At this point on the circle, the measured intensity is 2.00 W/m^2. (a) At how many other angles in the range $0° < \theta < 90°$ is the intensity also 2.00 W/m^2? (b) Find the four smallest angles in the range $0° < \theta < 90°$ for which the intensity is 2.00 W/m^2. (c) What is the intensity at a point on the circle at an angle of 4.65° from the centerline?

35.23. Two slits spaced 0.260 mm apart are placed 0.700 m from a screen and illuminated by coherent light with a wavelength of 660 nm. The intensity at the center of the central maximum ($\theta = 0°$) is I_0. (a) What is the distance on the screen from the center of the central maximum to the first minimum? (b) What is the distance on the screen from the center of the central maximum to the point where the intensity has fallen to $I_0/2$?

35.24. Show that Eq. (35.14) gives zero-intensity directions that agree with Eq. (35.5).

35.25. Points A and B are 56.0 m apart along an east-west line. At each of these points, a radio transmitter is emitting a 12.5-MHz signal horizontally. These transmitters are in phase with other and emit their beams uniformly in a horizontal plane. A receiver is taken 0.500 km north of the AB line and initially placed at point C, directly opposite the midpoint of AB. The receiver can be moved only along an east-west direction but, due to its limited sensitivity, it must always remain within a range so that the intensity of the signal it receives from the transmitter is no less than $\frac{1}{4}$ of its maximum value. How far from point C (along an east-west line) can the receiver be moved and always be able to pick up the signal?

35.26. Consider two antennas separated by 9.00 m that radiate in phase at 120 MHz, as described in Exercise 35.3. A receiver placed 150 m from both antennas measures an intensity I_0. The receiver is moved so that it is 1.8 m closer to one antenna than to the other. (a) What is the phase difference ϕ between the two radio waves produced by this path difference? (b) In terms of I_0, what is the intensity measured by the receiver at its new position?

Section 35.4 Interference in Thin Films

35.27. What is the thinnest film of a coating with $n = 1.42$ on glass ($n = 1.52$) for which destructive interference of the red component (650 nm) of an incident white light beam in air can take place by reflection?

35.28. Nonglare Glass. When viewing a piece of art that is behind glass, one often is affected by the light that is reflected off the front of the glass (called *glare*), which can make it difficult to see the art clearly. One solution is to coat the outer surface of the glass with a film to cancel part of the glare. (a) If the glass has a refractive index of 1.62 and you use TiO_2, which has an index of refraction of 2.62, as the coating, what is the minimum film thickness that will cancel light of wavelength 505 nm? (b) If this coating is too thin to stand up to wear, what other thickness would also work? Find only the three thinnest ones.

35.29. Two rectangular pieces of plane glass are laid one upon the other on a table. A thin strip of paper is placed between them at one edge so that a very thin wedge of air is formed. The plates are illuminated at normal incidence by 546-nm light from a mercury-vapor lamp. Interference fringes are formed, with 15.0 fringes per centimeter. Find the angle of the wedge.

35.30. A plate of glass 9.00 cm long is placed in contact with a second plate and is held at a small angle with it by a metal strip 0.0800 mm thick placed under one end. The space between the plates is filled with air. The glass is illuminated from above with light having a wavelength in air of 656 nm. How many interference fringes are observed per centimeter in the reflected light?

35.31. A uniform film of TiO_2, 1036 nm thick and having index of refraction 2.62, is spread uniformly over the surface of crown glass of refractive index 1.52. Light of wavelength 520.0 nm falls at normal incidence onto the film from air. You want to increase the thickness of this film so that the reflected light cancels. (a) What is the *minimum* thickness of TiO_2 that you must *add* so the reflected light cancels as desired? (b) After you make the adjustment in part (a), what is the path difference between the light reflected off the top of the film and the light that cancels it after traveling through the film? Express your answer in (i) nanometers and (ii) wavelengths of the light in the TiO_2 film.

35.32. A plastic film with index of refraction 1.85 is put on the surface of a car window to increase the reflectivity and thus to keep the interior of the car cooler. The window glass has index of refraction 1.52. (a) What minimum thickness is required if light with wavelength 550 nm in air reflected from the two sides of the film is to interfere constructively? (b) It is found to be difficult to manufacture and install coatings as thin as calculated in part (a). What is the next greatest thickness for which there will also be constructive interference?

35.33. The walls of a soap bubble have about the same index of refraction as that of plain water, $n = 1.33$. There is air both inside and outside the bubble. (a) What wavelength (in air) of visible light is most strongly reflected from a point on a soap bubble where its wall is 290 nm thick? To what color does this correspond (see Fig. 32.4 and Table 32.1)? (b) Repeat part (a) for a wall thickness of 340 nm.

35.34. Light with wavelength 648 nm in air is incident perpendicularly from air on a film 8.76 μm thick and with refractive index 1.35. Part of the light is reflected from the first surface of the film, and part enters the film and is reflected back at the second surface, where the film is again in contact with air. (a) How many waves are contained along the path of this second part of the light in its round trip through the film? (b) What is the phase difference between these two parts of the light as they leave the film?

35.35. Compact Disc Player. A compact disc (CD) is read from the bottom by a semiconductor laser with wavelength 790 nm passing through a plastic substrate of refractive index 1.8. When the beam encounters a pit, part of the beam is reflected from the pit and part from the flat region between the pits, so these two beams interfere with each other (Fig. 35.22). What must the minimum pit depth be so that the part of the beam reflected from a pit cancels the part of the beam reflected from the flat region? (It is this cancellation that allows the player to recognize the beginning and end of a pit. For a fuller explanation of the physics behind CD technology, see the article "The Compact Disc Digital Audio System," by Thomas D. Rossing, in the December 1987 issue of *The Physics Teacher*.)

Figure 35.22 Exercise 35.35.

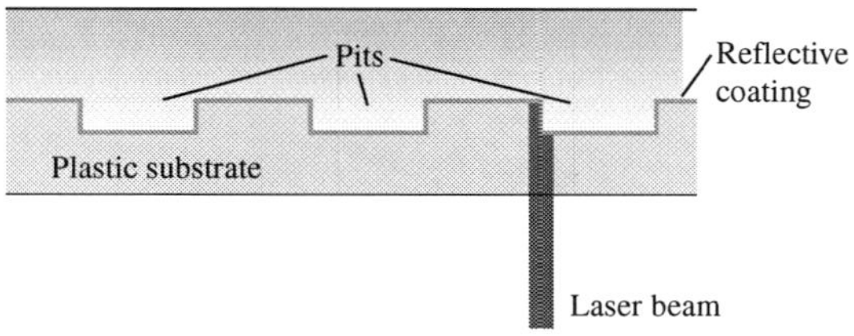

35.36. What is the thinnest soap film (excluding the case of zero thickness) that appears black when illuminated with light with wavelength 480 nm? The index of refraction of the film is 1.33, and there is air on both sides of the film.

Section 35.5 The Michelson Interferometer

35.37. How far must the mirror M_2 (see Fig. 35.20) of the Michelson interferometer be moved so that 1800 fringes of He-Ne laser light $(\lambda = 633 \text{ nm})$ move across a line in the field of view?

35.38. Jan first uses a Michelson interferometer with the 606-nm light from a krypton-86 lamp. He displaces the movable mirror away from him, counting 818 fringes moving across a line in his field of view. Then Linda replaces the krypton lamp with filtered 502-nm light from a helium lamp and displaces the movable mirror toward her. She also counts 818 fringes, but they move across the line in her field of view opposite to the direction they moved for Jan. Assume that both Jan and Linda counted to 818 correctly. (a) What distance did each person move the mirror? (b) What is the resultant displacement of the mirror?

Problems

35.39. The radius of curvature of the convex surface of a planoconvex lens is 95.2 cm. The lens is placed convex side down on a perfectly flat glass plate that is illuminated from above with red light having a wavelength of 580 nm. Find the diameter of the second bright ring in the interference pattern.

35.40. Newton's rings can be seen when a planoconvex lens is placed on a flat glass surface (see Problem 35.39). For a particular lens with an index of refraction of $n = 1.50$ and a glass plate with an index of $n = 1.80$, the diameter of the third bright ring is 0.350 mm. If water $(n = 1.33)$ now fills the space between the lens and the plate, what is the new diameter of this ring?

35.41. Suppose you illuminate two thin slits by monochromatic coherent light in air and find that they produce their first interference *minima* at $\pm 35.20°$ on either side of the central bright spot. You then immerse these slits in a transparent liquid and illuminate them with the same light. Now you find that the first minima occur at $\pm 19.46°$ instead. What is the index of refraction of this liquid?

35.42. A very thin sheet of brass contains two thin parallel slits. When a laser beam shines on these slits at normal incidence and room temperature (20.0°C), the first interference dark fringes occur at ±32.5° from the original direction of the laser beam when viewed from some distance. If this sheet is now slowly heated up to 135°C, by how many degrees do these dark fringes change position? Do they move closer together or get farther apart? See Table 17.1 for pertinent information, and ignore any effects that might occur due to change in the thickness of the slits. (*Hint:* Since thermal expansion normally produces very small changes in length, you can use differentials to find the change in the angle.)

35.43. Two speakers, 2.50 m apart, are driven by the same audio oscillator so that each one produces a sound consisting of *two* distinct frequencies, 0.900 kHz and 1.20 kHz. The speed of sound in the room is 344 m/s. Find all the angles relative to the usual center line in front of (and far from) the speakers at which *both* frequencies interfere constructively.

35.44. Two radio antennas radiating in phase are located at points A and B, 200 m apart (Fig. 35.23). The radio waves have a frequency of 5.80 MHz. A radio receiver is moved out from point B along a line perpendicular to the line connecting A and B (line BC shown in Fig. 35.23). At what distances from B will there be *destructive* interference? (*Note:* The distance of the receiver from the sources is not large in comparison to the separation of the sources, so Eq. (35.5) does not apply.)

Figure 35.23 Problem 35.44.

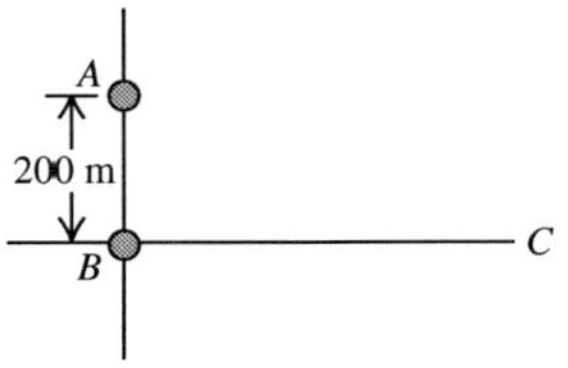

35.45. One round face of a 3.25-m, solid, cylindrical plastic pipe is covered with a thin black coating that completely blocks light. The opposite face is covered with a fluorescent coating that glows when it is struck by light. Two straight, thin, parallel scratches, 0.225 mm apart, are made in the center of the black face. When laser light of wavelength 632.8 nm shines through the slits perpendicular to the black face, you find that the central bright fringe on the opposite face is 5.82 mm wide, measured between the dark fringes that border it on either side. What is the index of refraction of the plastic?

35.46. A uniform thin film of material of refractive index 1.40 coats a glass plate of refractive index 1.55. This film has the proper thickness to cancel normally incident light of wavelength 525 nm that strikes the film surface from air, but it is somewhat greater than the minimum thickness to achieve this cancellation. As time goes by, the film wears away at a steady rate of 4.20 nm per year. What is the minimum number of years before the reflected light of this wavelength is now enhanced instead of cancelled?

35.47. (a) In Fig. 35.3, suppose source S_2 is *not* in phase with S_1, but instead is *out* of phase by $\frac{1}{2}$ cycle. In this situation, Eq. (35.1) is the condition for *destructive* interference, and Eq. (35.2) is the condition for *constructive* interference. Explain why this is so. (b) Suppose S_2 *leads* S_1 by a phase angle ϕ; that is, if the displacement of source S_1 is given by $x_1(t) = A\cos\omega t$, then the displacement of source S_2 is $x_2(t) = A\cos(\omega t + \phi)$. (In the situation of part (a), $\phi = \pi$.) Find expressions for the values of the path difference $r_2 - r_1$ that correspond to constructive interference and to destructive interference.

35.48. The electric fields received at point P from two identical, coherent wave sources are $E_1(t) = E\cos(\omega t + \phi)$ and $E_2(t) = E\cos\omega t$. (a) Use one of the trigonometric identities in Appendix B to show that the resultant wave is $E_P(t) = 2E\cos(\phi/2)\cos(\omega t + \phi/2)$. (b) Show that the amplitude of this resultant wave is given by Eq. (35.7). (c) Use the result of part (a) to show that at an interference maximum, the amplitude of the resultant wave is in phase with the original waves $E_1(t)$ and $E_2(t)$. (d) Use the result of part (a) to show that near an interference minimum, the resultant wave is approximately $\frac{1}{4}$ cycle out of phase with either of the original waves. (e) Show that the *instantaneous* Poynting vector at point P has magnitude $S = 4\epsilon_0 cE^2\cos^2(\phi/2)\cos^2(\omega t + \phi/2)$ and that the *time-averaged* Poynting vector is given by Eq. (35.9).

35.49. Let the two sources S_1 and S_2 shown in Fig. 35.3 be located at $y = d$ and $y = -d$, respectively. (a) Rewrite Eq. (35.1) in terms of the x- and y-coordinates of a point P in Fig. 35.3 at which constructive interference occurs. (b) Your expression in part (a) is the equation for the antinodal curves shown in Fig. 35.3. Show that these curves are hyperbolas. (*Hint:* You may want to review the definition of a hyperbola in analytic geometry.) (c) Repeat part (a) for Eq. (35.2), which describes points at which *destructive* interference occurs, and show that the *nodal* curves (not shown in Fig. 35.3) are also hyperbolas.

35.50. Consider a two-slit interference experiment in which the two slits are of different widths. As measured on a distant screen, the amplitude of the wave from the first slit is E, while the amplitude of the wave from the second slit is $2E$. (a) Show that the intensity at any point in the interference pattern is

$$I = I_0\left(\frac{5}{9} + \frac{4}{9}\cos\phi\right)$$

where ϕ is the phase difference between the two waves as measured at a particular point on the screen and I_0 is the maximum intensity in the pattern. (b) Graph I versus ϕ (like Fig. 35.10). What is the minimum value of the intensity, and for which values of ϕ does it occur?

35.51. A thin uniform film of refractive index 1.750 is placed on a sheet of glass of refractive index 1.50. At room temperature (20.0°C), this film is just thick enough for light with wavelength 582.4 nm reflected off the top of the film to be cancelled by light reflected from the top of the glass. After the glass is placed in an oven and slowly heated to 170°C, you find that the film cancels reflected light with wavelength 588.5 nm. What is the coefficient of linear expansion of the film? (Ignore any changes in the refractive index of the film due to the temperature change.)

35.52. Red light with wavelength 700 nm is passed through a two-slit apparatus. At the same time, monochromatic visible light with another wavelength passes through the same apparatus. As a result, most of the pattern that appears on the screen is a mixture of two colors; however, the center of the third bright fringe ($m = 3$) of the red light appears pure red, with none of the other color. What are the possible wavelengths of the second type of visible light? Do you need to know the slit spacing to answer this question? Why or why not?

35.53. Consider a two-slit interference pattern, for which the intensity distribution is given by Eq. (35.14). Let θ_m be the angular position of the mth bright fringe, where the intensity is I_0. Assume that θ_m is small, so that $\sin\theta_m \cong \theta_m$. Let θ_m^+ and θ_m^- be the two angles on either side of θ_m for which $I = \frac{1}{2}I_0$. The quantity $\Delta\theta_m = |\theta_m^+ - \theta_m^-|$ is the half-width of the mth fringe. Calculate $\Delta\theta_m$. How does $\Delta\theta_m$ depend on m?

35.54. White light reflects at normal incidence from the top and bottom surfaces of a glass plate ($n = 1.52$). There is air above and below the plate. Constructive interference is observed for light whose wavelength in air is 477.0 nm. What is the thickness of the plate if the next longer wavelength for which there is constructive interference is 540.6 nm?

35.55. A source S of monochromatic light and a detector D are both located in air a distance h above a horizontal plane sheet of glass, and are separated by a horizontal distance x. Waves reaching D directly from S interfere with waves that reflect off the glass. The distance x is small compared to h so that the reflection is at close to normal incidence. (a) Show that the condition for constructive interference is $\sqrt{x^2 + 4h^2} - x = (m + \frac{1}{2})\lambda$, and the condition for destructive interference is $\sqrt{x^2 + 4h^2} - x = m\lambda$. (*Hint:* Take into account the phase change on reflection.) (b) Let $h = 24$ cm and $x = 14$ cm. What is the longest wavelength for which there will be constructive interference?

35.56. Reflective Coatings and Herring. Herring and related fish have a brilliant silvery appearance that camouflages them while they are swimming in a sunlit ocean. The silveriness is due to *platelets* attached to the surfaces of these fish. Each platelet is made up of several alternating layers of crystalline guanine $(n = 1.80)$ and of cytoplasm ($n = 1.333$, the same as water), with a guanine layer on the outside in contact with the surrounding water (Fig. 35.24). In one typical platelet, the guanine layers are 74 nm thick and the cytoplasm layers are 100 nm thick. (a) For light striking the platelet surface at normal incidence, for which vacuum wavelengths of visible light will all of the reflections R_1, R_2, R_3, R_4, and R_5, shown in Fig. 35.24, be approximately in phase? If white light is shone on this platelet, what color will be most strongly reflected (see Fig. 32.4)? The surface of a herring has very many platelets side by side with layers of different thickness, so that *all* visible wavelengths are reflected. (b) Explain why such a "stack" of layers is more reflective than a single layer of guanine with cytoplasm underneath. (A stack of five guanine layers separated by cytoplasm layers reflects more than 80% of incident light at the wavelength for which it is "tuned.") (c) The color that is most strongly reflected from a platelet depends on the angle at which it is viewed. Explain why this should be so. (You can see these changes in color by examining a herring from different angles. Most of the platelets on these fish are oriented in the same way, so that they are vertical when the fish is swimming.)

Figure **35.24** Problem 35.56.

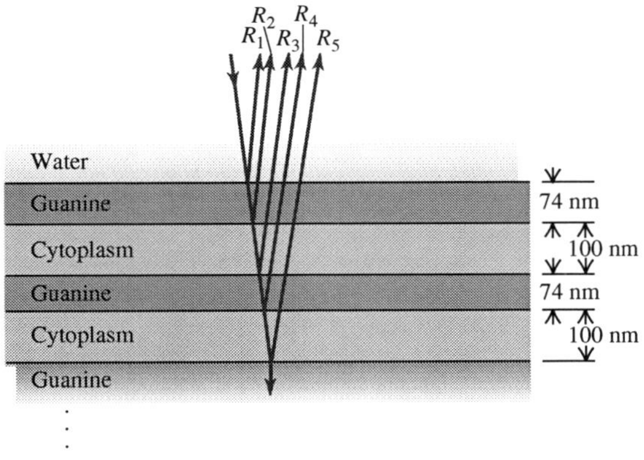

35.57. Two thin parallel slits are made in an opaque sheet of film. When a monochromatic beam of light is shone through them at normal incidence, the first bright fringes in the transmitted light occur in air at $\pm 18.0°$ with the original direction of the light beam on a distant screen when the apparatus is in air. When the apparatus is immersed in a liquid, the same bright fringes now occur at $\pm 12.6°$. Find the index of refraction of the liquid.

35.58. An oil tanker spills a large amount of oil $(n = 1.45)$ into the sea $(n = 1.33)$. (a) If you look down onto the oil spill from overhead, what predominant wavelength of light do you see at a point where the oil is 380 nm thick? What color is the light? (*Hint:* See Table 32.1.) (b) In the water under the slick, what visible wavelength (as measured in air) is predominant in the transmitted light at the same place in the slick as in part (a)?

35.59. In a Young's two-slit experiment a piece of glass with an index of refraction n and a thickness L is placed in front of the upper slit. (a) Describe qualitatively what happens to the interference pattern. (b) Derive an expression for the intensity I of the light at points on a screen as a function of n, L, and θ. Here θ is the usual angle measured from the center of the two slits. That is, determine the equation analogous to Eq. (35.14) but that also involves L and n for the glass plate. (c) From your result in part (b) derive an expression for the values of θ that locate the maxima in the interference pattern [that is, derive an equation analogous to Eq. (35.4)].

35.60. After a laser beam passes through two thin parallel slits, the first completely dark fringes occur at $\pm 15.0°$ with the original direction of the beam, as viewed on a screen far from the slits. (a) What is the ratio of the distance between the slits to the wavelength of the light illuminating the slits? (b) What is the smallest angle, relative to the original direction of the laser beam, at which the intensity of the light is $\frac{1}{10}$ the maximum intensity on the screen?

Challenge Problems

35.61. The index of refraction of a glass rod is 1.48 at $T = 20.0°C$ and varies linearly with temperature, with a coefficient of $2.50 \times 10^{-5}/C°$. The coefficient of linear expansion of the glass is $5.00 \times 10^{-6}/C°$. At 20.0°C the length of the rod is 3.00 cm. A Michelson interferometer has this glass rod in one arm, and the rod is being heated so that its temperature increases at a rate of $5.00\ C°/min$. The light source has wavelength $\lambda = 589$ nm, and the rod initially is at $T = 20.0°C$. How many fringes cross the field of view each minute?

35.62. Figure 35.25 shows an interferometer known as *Fresnel's biprism*. The magnitude of the prism angle A is extremely small. (a) If S_0 is a very narrow source slit, show that the separation of the two virtual coherent sources S_1 and S_2 is given by $d = 2aA(n - 1)$, where n is the index of refraction of the material of the prism. (b) Calculate the spacing of the fringes of green light with wavelength 500 nm on a screen 2.00 m from the biprism. Take $a = 0.200$ m, $A = 3.50$ mrad, and $n = 1.50$.

Figure **35.25** Challenge Problem 35.62.

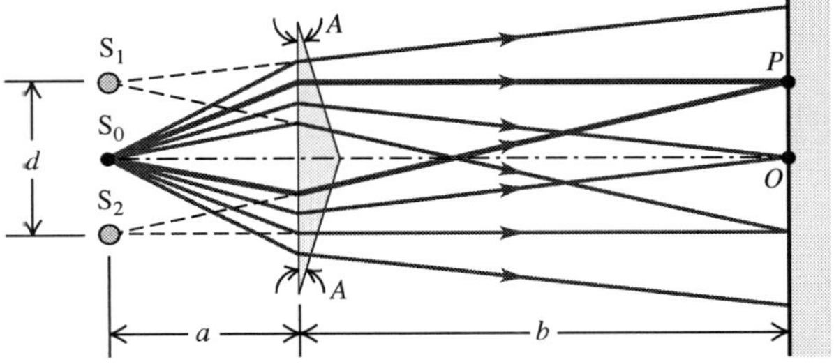

36 DIFFRACTION

LEARNING GOALS

By studying this chapter, you will learn:

- What happens when coherent light shines on an object with an edge or aperture.
- How to understand the diffraction pattern formed when coherent light passes through a narrow slit.
- How to calculate the intensity at various points in a single-slit diffraction pattern.
- What happens when coherent light shines on an array of narrow, closely spaced slits.
- How scientists use diffraction gratings for precise measurements of wavelength.
- How x-ray diffraction reveals the arrangement of atoms in a crystal.
- How diffraction sets limits on the smallest details that can be seen with a telescope.

? The laser used to read a compact disc (CD) has a wavelength of 780 nm, while the laser used to read a DVD has a wavelength of 650 nm. How does this make it possible for a DVD to hold more information than a CD?

Everyone is used to the idea that sound bends around corners. If sound didn't behave this way, you couldn't hear a police siren that's out of sight around a corner or the speech of a person whose back is turned to you. What may surprise you (and certainly surprised many scientists of the early 19th century) is that *light* can bend around corners as well. When light from a point source falls on a straightedge and casts a shadow, the edge of the shadow is never perfectly sharp. Some light appears in the area that we expect to be in the shadow, and we find alternating bright and dark fringes in the illuminated area. In general, light emerging from apertures doesn't behave precisely according to the predictions of the straight-line ray model of geometric optics.

The reason for these effects is that light, like sound, has wave characteristics. In Chapter 35 we studied the interference patterns that can arise when two light waves are combined. In this chapter we'll investigate interference effects due to combining *many* light waves. Such effects are referred to as *diffraction.* We'll find that the behavior of waves after they pass through an aperture is an example of diffraction; each infinitesimal part of the aperture acts as a source of waves, and the resulting pattern of light and dark is a result of interference among the waves emanating from these sources.

Light emerging from arrays of apertures also forms patterns whose character depends on the color of the light and the size and spacing of the apertures. Examples of this effect include the colors of iridescent butterflies and the "rainbow" you see reflected from the surface of a compact disc. We'll explore similar effects with x rays that are used to study the atomic structure of solids and liquids. Finally, we'll look at the physics of a *hologram,* a special kind of interference pattern recorded on photographic film and reproduced. When properly illuminated, it forms a three-dimensional image of the original object.

36.1 Fresnel and Fraunhofer Diffraction

According to geometric optics, when an opaque object is placed between a point light source and a screen, as in Fig. 36.1, the shadow of the object forms a perfectly sharp line. No light at all strikes the screen at points within the shadow, and the area outside the shadow is illuminated nearly uniformly. But as we saw in Chapter 35, the *wave* nature of light causes effects that can't be understood with the simple model of geometric optics. An important class of such effects occurs when light strikes a barrier that has an aperture or an edge. The interference patterns formed in such a situation are grouped under the heading **diffraction.**

An example of diffraction is shown in Fig. 36.2. The photograph in Fig. 36.2a was made by placing a razor blade halfway between a pinhole, illuminated by monochromatic light, and a photographic film. The film recorded the shadow cast by the blade. Figure 36.2b is an enlargement of a region near the shadow of the right edge of the blade. The position of the *geometric* shadow line is indicated by arrows. The area outside the geometric shadow is bordered by alternating bright and dark bands. There is some light in the shadow region, although this is not very visible in the photograph. The first bright band in Fig. 36.2b, just to the right of the geometric shadow, is considerably brighter than in the region of uniform illumination to the extreme right. This simple experiment gives us some idea of the richness and complexity of what might seem to be a simple idea, the casting of a shadow by an opaque object.

We don't often observe diffraction patterns such as Fig. 36.2 in everyday life because most ordinary light sources are not monochromatic and are not point sources. If we use a white frosted light bulb instead of a point source in Fig. 36.1, each wavelength of the light from every point of the bulb forms its own diffraction pattern, but the patterns overlap to such an extent that we can't see any individual pattern.

36.1 A point source of light illuminates a straightedge.

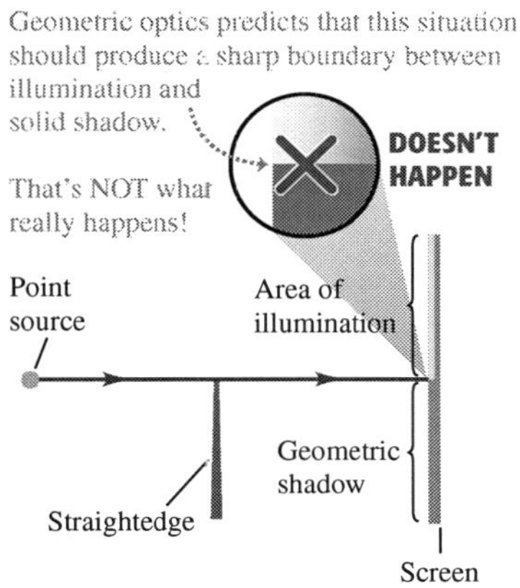

Diffraction and Huygens's Principle

Diffraction patterns can be analyzed by use of Huygens's principle (see Section 33.7). Let's review that principle briefly. Every point of a wave front can be considered the source of secondary wavelets that spread out in all directions with a speed equal to the speed of propagation of the wave. The position of the wave front at any later time is the *envelope* of the secondary waves at that time. To find the resultant displacement at any point, we combine all the individual displacements

36.2 An example of diffraction.

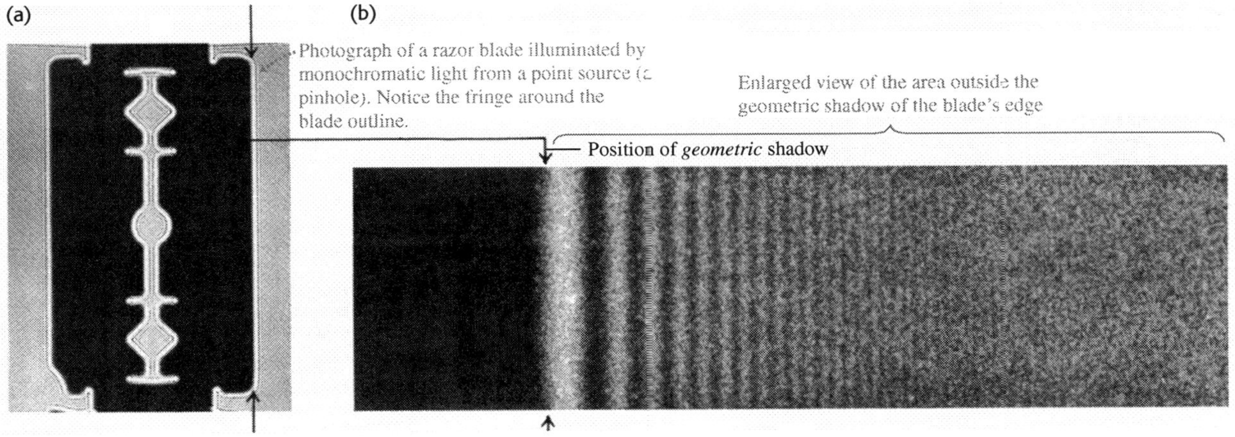

produced by these secondary waves, using the superposition principle and taking into account their amplitudes and relative phases.

In Fig. 36.1, both the point source and the screen are relatively close to the obstacle forming the diffraction pattern. This situation is described as *near-field diffraction* or **Fresnel diffraction,** pronounced "Freh-nell" (after the French scientist Augustin Jean Fresnel, 1788–1827). If the source, obstacle, and screen are far enough away that all lines from the source to the obstacle can be considered parallel and all lines from the obstacle to a point in the pattern can be considered parallel, the phenomenon is called *far-field diffraction* or **Fraunhofer diffraction** (after the German physicist Joseph von Fraunhofer, 1787–1826). We will restrict the following discussion to Fraunhofer diffraction, which is usually simpler to analyze in detail than Fresnel diffraction.

Diffraction is sometimes described as "the bending of light around an obstacle." But the process that causes diffraction is present in the propagation of *every* wave. When part of the wave is cut off by some obstacle, we observe diffraction effects that result from interference of the remaining parts of the wave fronts. Optical instruments typically use only a limited portion of a wave; for example, a telescope uses only the part of a wave that is admitted by its objective lens or mirror. Thus diffraction plays a role in nearly all optical phenomena.

Finally, we emphasize that there is no fundamental distinction between *interference* and *diffraction.* In Chapter 35 we used the term *interference* for effects involving waves from a small number of sources, usually two. *Diffraction* usually involves a *continuous* distribution of Huygens's wavelets across the area of an aperture, or a very large number of sources or apertures. But both categories of phenomena are governed by the same basic physics of superposition and Huygens's principle.

Test Your Understanding of Section 36.1 Can *sound* waves undergo diffraction around an edge?

36.2 Diffraction from a Single Slit

ActivPhysics ONLINE
16.6 Single-Slit Diffraction

In this section we'll discuss the diffraction pattern formed by plane-wave (parallel-ray) monochromatic light when it emerges from a long, narrow slit, as shown in Fig. 36.3. We call the narrow dimension the *width,* even though in this figure it is a vertical dimension.

According to geometric optics, the transmitted beam should have the same cross section as the slit, as in Fig. 36.3a. What is *actually* observed is the pattern shown in Fig. 36.3b. The beam spreads out vertically after passing through the

36.3 (a) The "shadow" of a horizontal slit as incorrectly predicted by geometric optics. (b) A horizontal slit actually produces a diffraction pattern. The slit width has been greatly exaggerated.

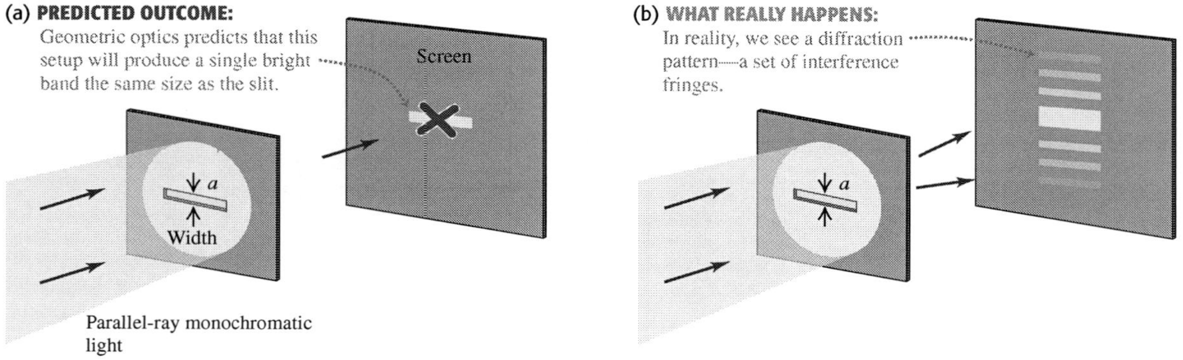

slit. The diffraction pattern consists of a central bright band, which may be much broader than the width of the slit, bordered by alternating dark and bright bands with rapidly decreasing intensity. About 85% of the power in the transmitted beam is in the central bright band, whose width is found to be *inversely* proportional to the width of the slit. In general, the smaller the width of the slit, the broader the entire diffraction pattern. (The *horizontal* spreading of the beam in Fig. 36.3b is negligible because the horizontal dimension of the slit is relatively large.) You can easily observe a similar diffraction pattern by looking at a point source, such as a distant street light, through a narrow slit formed between your two thumbs held in front of your eye; the retina of your eye corresponds to the screen.

Single-Slit Diffraction: Locating the Dark Fringes

Figure 36.4 shows a side view of the same setup; the long sides of the slit are perpendicular to the figure, and plane waves are incident on the slit from the left. According to Huygens's principle, each element of area of the slit opening can be considered as a source of secondary waves. In particular, imagine dividing the slit into several narrow strips of equal width, parallel to the long edges and perpendicular to the page. Two such strips are shown in Fig. 36.4a. Cylindrical secondary wavelets, shown in cross section, spread out from each strip.

In Fig. 36.4b a screen is placed to the right of the slit. We can calculate the resultant intensity at a point P on the screen by adding the contributions from the individual wavelets, taking proper account of their various phases and amplitudes. It's easiest to do this calculation if we assume that the screen is far enough away that all the rays from various parts of the slit to a particular point P on the screen are parallel, as in Fig. 36.4c. An equivalent situation is Fig. 36.4d, in which the rays to the lens are parallel and the lens forms a reduced image of the same pattern that would be formed on an infinitely distant screen without the lens. We might expect that the various light paths through the lens would introduce additional phase shifts, but in fact it can be shown that all the paths have *equal* phase shifts, so this is not a problem.

The situation of Fig. 36.4b is Fresnel diffraction; those in Figs. 36.4c and 36.4d, where the outgoing rays are considered parallel, are Fraunhofer diffraction. We can derive quite simply the most important characteristics of the Fraunhofer diffraction pattern from a single slit. First consider two narrow strips, one just below the top edge of the drawing of the slit and one at its center, shown in end view in Fig. 36.5. The difference in path length to point P is $(a/2)\sin\theta$, where a is the slit width and θ is the angle between the perpendicular to the slit and a line from the center of the slit to P. Suppose this path difference happens to be equal to $\lambda/2$; then light from these two strips arrives at point P with a half-cycle phase difference, and cancellation occurs.

36.4 Diffraction by a single rectangular slit. The long sides of the slit are perpendicular to the figure.

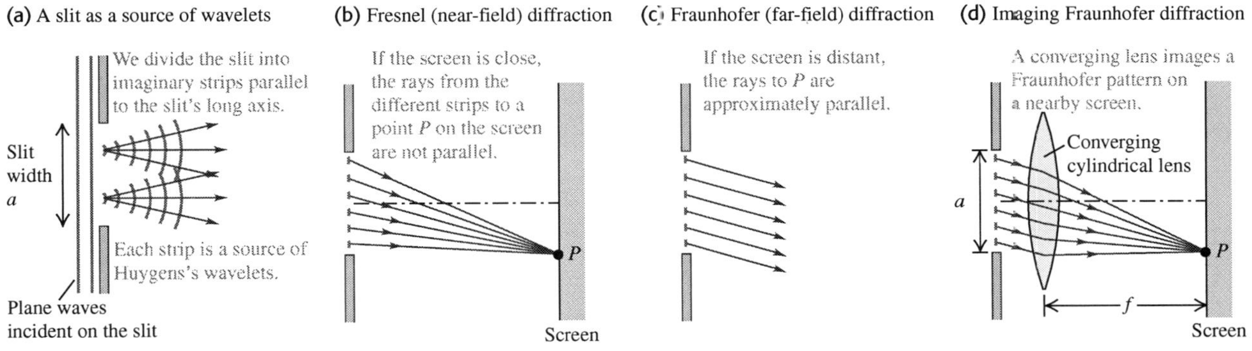

36.5 Side view of a horizontal slit. When the distance x to the screen is much greater than the slit width a, the rays from a distance $a/2$ apart may be considered parallel.

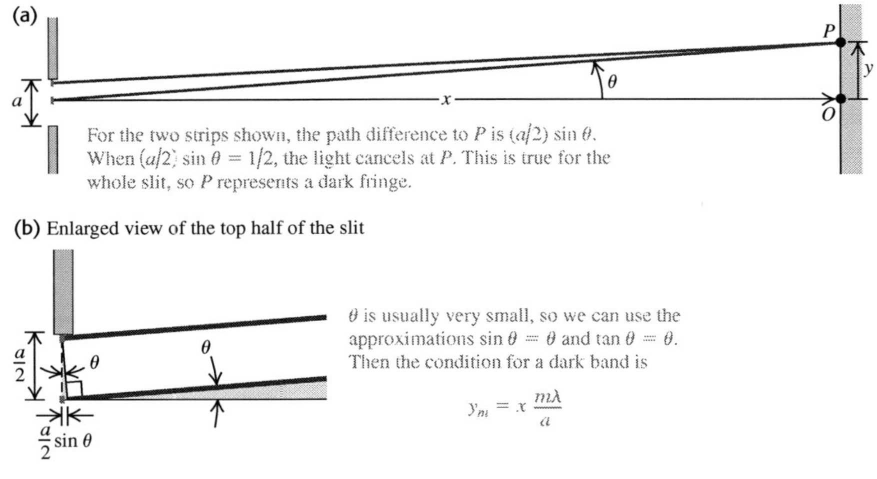

Similarly, light from two strips immediately *below* the two in the figure also arrives at P a half-cycle out of phase. In fact, the light from *every* strip in the top half of the slit cancels out the light from a corresponding strip in the bottom half. The result is complete cancellation at P for the combined light from the entire slit, giving a dark fringe in the interference pattern. That is, a dark fringe occurs whenever

$$\frac{a}{2}\sin\theta = \pm\frac{\lambda}{2} \quad \text{or} \quad \sin\theta = \pm\frac{\lambda}{a} \tag{36.1}$$

The plus-or-minus ($\pm$) sign in Eq. (36.1) says that there are symmetrical dark fringes above and below point O in Fig. 36.5a. The upper fringe ($\theta > 0$) occurs at a point P where light from the bottom half of the slit travels $\lambda/2$ farther to P than does light from the top half; the lower fringe ($\theta < 0$) occurs where light from the *top* half travels $\lambda/2$ farther than light from the *bottom* half.

We may also divide the screen into quarters, sixths, and so on, and use the above argument to show that a dark fringe occurs whenever $\sin\theta = \pm 2\lambda/a$, $\pm 3\lambda/a$, and so on. Thus the condition for a *dark* fringe is

$$\sin\theta = \frac{m\lambda}{a} \quad (m = \pm 1, \pm 2, \pm 3, \dots) \quad \text{(dark fringes in single-slit diffraction)} \tag{36.2}$$

For example, if the slit width is equal to ten wavelengths ($a = 10\lambda$), dark fringes occur at $\sin\theta = \pm\frac{1}{10}, \pm\frac{2}{10}, \pm\frac{3}{10}, \dots$. Between the dark fringes are bright fringes. We also note that $\sin\theta = 0$ corresponds to a *bright* band; in this case, light from the entire slit arrives at P in phase. Thus it would be wrong to put $m = 0$ in Eq. (36.2). The central bright fringe is wider than the other bright fringes, as Fig. 36.3 shows. In the small-angle approximation that we will use below, it is exactly *twice* as wide.

With light, the wavelength λ is of the order of 500 nm $= 5 \times 10^{-7}$ m. This is often much smaller than the slit width a; a typical slit width is 10^{-2} cm $= 10^{-4}$ m. Therefore the values of θ in Eq. (36.2) are often so small that the approximation $\sin\theta \approx \theta$ (where θ is in radians) is a very good one. In that case we can rewrite this equation as

$$\theta = \frac{m\lambda}{a} \quad (m = \pm 1, \pm 2, \pm 3, \dots) \quad \text{(for small angles } \theta\text{)}$$

where θ is in *radians*. Also, if the distance from slit to screen is x, as in Fig. 36.5a, and the vertical distance of the mth dark band from the center of the pattern is y_m, then $\tan\theta = y_m/x$. For small θ we may also approximate $\tan\theta$ by θ (in radians), and we then find

$$y_m = x\frac{m\lambda}{a} \qquad (\text{for } y_m \ll x) \tag{36.3}$$

Figure 36.6 is a photograph of a single-slit diffraction pattern with the $m = \pm 1, \pm 2$, and ± 3 minima labeled.

CAUTION **Single-slit diffraction vs. two-slit interference** Equation (36.3) has the same form as the equation for the two-slit pattern, Eq. (35.6), except that in Eq. (36.3) we use x rather than R for the distance to the screen. But Eq. (36.3) gives the positions of the *dark* fringes in a *single-slit* pattern rather than the *bright* fringes in a *double-slit* pattern. Also, $m = 0$ in Eq. (36.2) is *not* a dark fringe. Be careful!

36.6 Photograph of the Fraunhofer diffraction pattern of a single horizontal slit.

Example 36.1 Single-slit diffraction

You pass 633-nm laser light through a narrow slit and observe the diffraction pattern on a screen 6.0 m away. You find that the distance on the screen between the centers of the first minima outside the central bright fringe is 32 mm (Fig. 36.7). How wide is the slit?

SOLUTION

IDENTIFY: This problem involves the relationship between the dark fringes in a single-slit diffraction pattern and the width of the slit (our target variable).

36.7 A single-slit diffraction experiment.

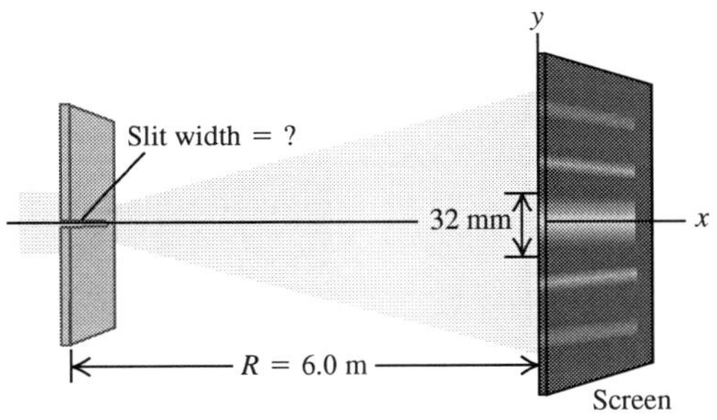

SET UP: The distances between points on the screen are much smaller than the distance from the slit to the screen, so the angle θ shown in Fig. 36.5a is very small. Hence we can use the approximate relationship of Eq. (36.3) to solve for the slit width a (the target variable).

EXECUTE: The first minimum corresponds to $m = 1$ in Eq. (36.3). The distance y_1 from the central maximum to the first minimum on either side is half the distance between the two first minima, so $y_1 = (32\text{ mm})/2$. Substituting these values and solving for a, we find

$$a = \frac{x\lambda}{y_1} = \frac{(6.0\text{ m})(633 \times 10^{-9}\text{ m})}{(32 \times 10^{-3}\text{ m})/2}$$
$$= 2.4 \times 10^{-4}\text{ m} = 0.24\text{ mm}$$

EVALUATE: The angle θ is small only if the wavelength is small compared to the slit width. Since $\lambda = 633\text{ nm} = 6.33 \times 10^{-7}\text{ m}$ and we have found $a = 0.24\text{ mm} = 2.4 \times 10^{-4}\text{ m}$, our result is consistent with this: The wavelength is $(6.33 \times 10^{-7}\text{ m})/(2.4 \times 10^{-4}\text{ m}) = 0.0026$ as large as the slit width.

Can you show that the distance between the *second* minima on the two sides is $2(32\text{ mm}) = 64$ mm, and so on?

Test Your Understanding of Section 36.2 Rank the following single-slit diffraction experiments in order of the size of the angle from the center of the diffraction pattern to the first dark fringe, from largest to smallest (i) wavelength 400 nm, slit width 0.20 mm; (ii) wavelength 600 nm, slit width 0.20 mm; (iii) wavelength 400 nm, slit width 0.30 mm; (iv) wavelength 600 nm, slit width 0.30 mm.

36.3 Intensity in the Single-Slit Pattern

We can derive an expression for the intensity distribution for the single-slit diffraction pattern by the same phasor-addition method that we used in Section 35.3 to obtain Eqs. (35.10) and (35.14) for the two-slit interference pattern. We again imagine a plane wave front at the slit subdivided into a large number of strips. We superpose the contributions of the Huygens wavelets from all the strips at a point P on a distant screen at an angle θ from the normal to the slit plane

36.8 Using phasor diagrams to find the amplitude of the $\vec{E}$ field in single-slit diffraction. Each phasor represents the $\vec{E}$ field from a single strip within the slit.

(a)

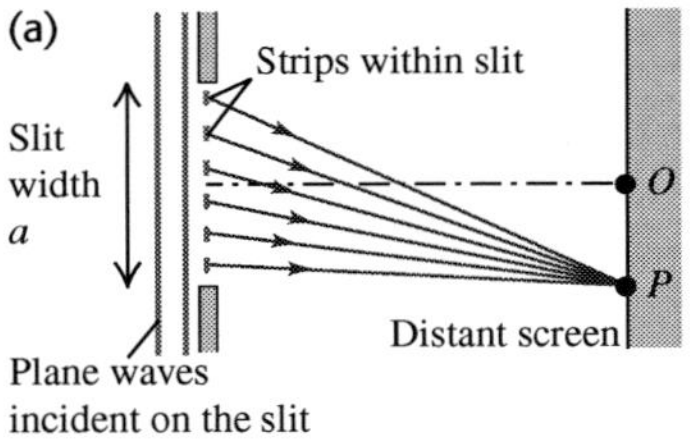

(b) At the center of the diffraction pattern (point O), the phasors from all strips within the slit are in phase.

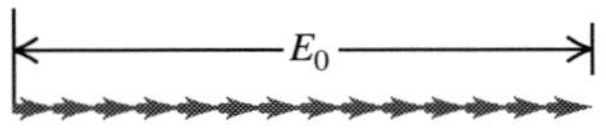

(c) Phasor diagram at a point slightly off the center of the pattern; β = total phase difference between the first and last phasors.

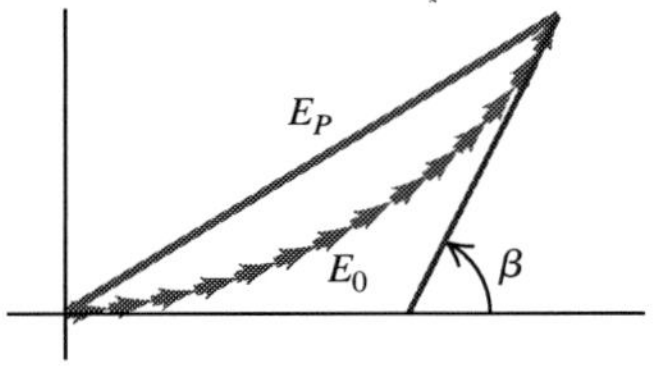

(d) As in (c), but in the limit that the slit is subdivided into infinitely many strips

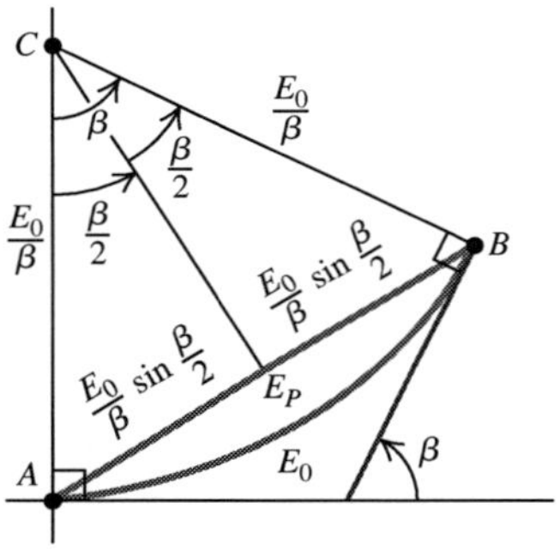

(Fig. 36.8a). To do this, we use a phasor to represent the sinusoidally varying $\vec{E}$ field from each individual strip. The magnitude of the vector sum of the phasors at each point P is the amplitude E_P of the total $\vec{E}$ field at that point. The intensity at P is proportional to E_P^2.

At the point O shown in Figure 36.8a, corresponding to the center of the pattern where $\theta = 0$, there are negligible path differences for $x \gg a$; the phasors are all essentially *in phase* (that is, have the same direction). In Fig. 36.8b we draw the phasors at time $t = 0$ and denote the resultant amplitude at O by E_0. In this illustration we have divided the slit into 14 strips.

Now consider wavelets arriving from different strips at point P in Fig. 36.8a, at an angle θ from point O. Because of the differences in path length, there are now phase differences between wavelets coming from adjacent strips; the corresponding phasor diagram is shown in Fig. 36.8c. The vector sum of the phasors is now part of the perimeter of a many-sided polygon, and E_P, the amplitude of the resultant electric field at P, is the *chord.* The angle β is the total phase difference between the wave from the top strip of Fig. 36.8a and the wave from the bottom strip; that is, β is the phase of the wave received at P from the top strip with respect to the wave received at P from the bottom strip.

We may imagine dividing the slit into narrower and narrower strips. In the limit that there is an infinite number of infinitesimally narrow strips, the curved trail of phasors becomes an *arc of a circle* (Fig. 36.8d), with arc length equal to the length E_0 in Fig. 36.8b. The center C of this arc is found by constructing perpendiculars at A and B. From the relationship among arc length, radius, and angle, the radius of the arc is E_0/β; the amplitude E_P of the resultant electric field at P is equal to the chord AB, which is $2(E_0/\beta)\sin(\beta/2)$. (Note that β *must* be in radians!) We then have

$$E_P = E_0\frac{\sin(\beta/2)}{\beta/2} \quad \text{(amplitude in single-slit diffraction)} \qquad (36.4)$$

The intensity at each point on the screen is proportional to the square of the amplitude given by Eq. (36.4). If I_0 is the intensity in the straight-ahead direction where $\theta = 0$ and $\beta = 0$, then the intensity I at any point is

$$I = I_0\left[\frac{\sin(\beta/2)}{\beta/2}\right]^2 \quad \text{(intensity in single-slit diffraction)} \qquad (36.5)$$

We can express the phase difference β in terms of geometric quantities, as we did for the two-slit pattern. From Eq. (35.11) the phase difference is $2\pi/\lambda$ times the path difference. Figure 36.5 shows that the path difference between the ray from the top of the slit and the ray from the middle of the slit is $(a/2)\sin\theta$. The path difference between the rays from the top of the slit and the bottom of the slit is twice this, so

$$\beta = \frac{2\pi}{\lambda}a\sin\theta \qquad (36.6)$$

and Eq. (36.5) becomes

$$I = I_0\left\{\frac{\sin[\pi a(\sin\theta)/\lambda]}{\pi a(\sin\theta)/\lambda}\right\}^2 \quad \text{(intensity in single-slit diffraction)} \qquad (36.7)$$

This equation expresses the intensity directly in terms of the angle θ. In many calculations it is easier first to calculate the phase angle β, using Eq. (36.6), and then to use Eq. (36.5).

Equation (36.7) is plotted in Fig. 36.9a. Note that the central intensity peak is much larger than any of the others. This means that most of the power in the wave remains within an angle θ from the perpendicular to the slit, where $\sin\theta = \lambda/a$ (the first diffraction minimum). You can see this easily in Fig. 36.9b, which is a photograph of water waves undergoing single-slit diffraction. Note also that the peak intensities in Fig. 36.9a decrease rapidly as we go away from the center of the pattern. (Compare Fig. 36.6, which shows a single-slit diffraction pattern for light.)

36.9 **(a)** Intensity versus angle in single-slit diffraction. The values of m label intensity minima given by Eq. (36.8). Most of the wave power goes into the central intensity peak (between the $m = 1$ and $m = -1$ intensity minima). **(b)** These water waves passing through a small aperture behave exactly like light waves in single-slit diffraction. Only the diffracted waves within the central intensity peak are visible; the waves at larger angles are too faint to see.

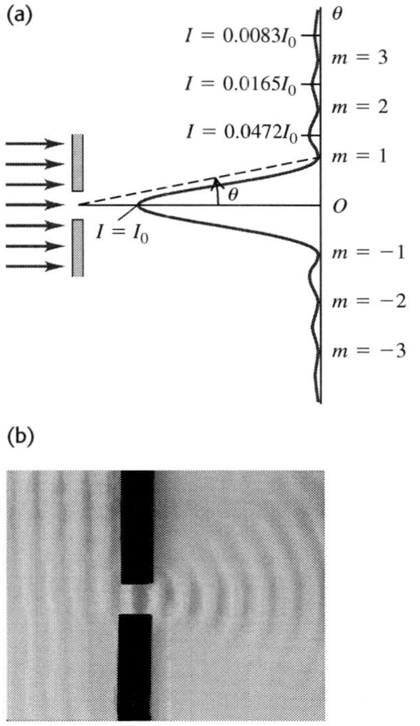

The dark fringes in the pattern are the places where $I = 0$. These occur at points for which the numerator of Eq. (36.5) is zero so that β is a multiple of 2π. From Eq. (36.6) this corresponds to

$$\frac{a\sin\theta}{\lambda} = m \qquad (m = \pm 1, \pm 2, \ldots)$$

$$\sin\theta = \frac{m\lambda}{a} \qquad (m = \pm 1, \pm 2, \ldots) \tag{36.8}$$

This agrees with our previous result, Eq. (36.2). Note again that $\beta = 0$ (corresponding to $\theta = 0$) is *not* a minimum. Equation (36.5) is indeterminate at $\beta = 0$, but we can evaluate the limit as $\beta \rightarrow 0$ using L'Hôpital's rule. We find that at $\beta = 0$, $I = I_0$, as we should expect.

Intensity Maxima in the Single-Slit Pattern

We can also use Eq. (36.5) to calculate the positions of the peaks, or *intensity maxima,* and the intensities at these peaks. This is not quite as simple as it may appear. We might expect the peaks to occur where the sine function reaches the value ± 1—namely, where $\beta = \pm\pi, \pm 3\pi, \pm 5\pi$, or in general,

$$\beta \approx \pm(2m + 1)\pi \qquad (m = 0, 1, 2, \ldots) \tag{36.9}$$

This is *approximately* correct, but because of the factor $(\beta/2)^2$ in the denominator of Eq. (36.5), the maxima don't occur precisely at these points. When we take the derivative of Eq. (36.5) with respect to β and set it equal to zero to try to find the maxima and minima, we get a transcendental equation that has to be solved numerically. In fact there is *no* maximum near $\beta = \pm\pi$. The first maxima on either side of the central maximum, near $\beta = \pm 3\pi$, actually occur at $\pm 2.860\pi$. The second side maxima, near $\beta = \pm 5\pi$, are actually at $\pm 4.918\pi$, and so on. The error in Eq. (36.9) vanishes in the limit of large m—that is, for intensity maxima far from the center of the pattern.

To find the intensities at the side maxima, we substitute these values of β back into Eq. (36.5). Using the approximate expression in Eq. (36.9), we get

$$I_m \approx \frac{I_0}{\left(m + \frac{1}{2}\right)^2 \pi^2} \tag{36.10}$$

where I_m is the intensity of the mth side maximum and I_0 is the intensity of the central maximum. Equation (36.10) gives the series of intensities

$$0.0450I_0 \qquad 0.0162I_0 \qquad 0.0083I_0$$

and so on. As we have pointed out, this equation is only approximately correct. The actual intensities of the side maxima turn out to be

$$0.0472I_0 \qquad 0.0165I_0 \qquad 0.0083I_0 \qquad \ldots$$

Note that the intensities of the side maxima decrease very rapidly, as Fig. 36.9a also shows. Even the first side maxima have less than 5% of the intensity of the central maximum.

Width of the Single-Slit Pattern

For small angles the angular spread of the diffraction pattern is inversely proportional to the slit width a or, more precisely, to the ratio of a to the wavelength λ. Figure 36.10 shows graphs of intensity I as a function of the angle θ for three values of the ratio a/λ.

36.10 The single-slit diffraction pattern depends on the ratio of the slit width a to the wavelength λ.

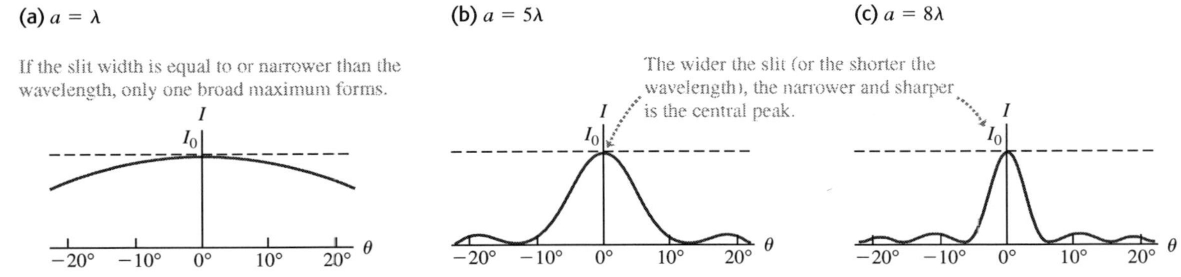

With light waves, the wavelength λ is often much smaller than the slit width a, and the values of θ in Eqs. (36.6) and (36.7) are so small that the approximation $\sin\theta = \theta$ is very good. With this approximation the position θ_1 of the first minimum beside the central maximum, corresponding to $\beta/2 = \pi$, is, from Eq. (36.7),

$$\theta_1 = \frac{\lambda}{a} \tag{36.11}$$

This characterizes the width (angular spread) of the central maximum, and we see that it is *inversely* proportional to the slit width a. When the small-angle approximation is valid, the central maximum is exactly twice as wide as each side maximum. When a is of the order of a centimeter or more, θ_1 is so small that we can consider practically all the light to be concentrated at the geometrical focus. But when a is less than λ, the central maximum spreads over 180°, and the fringe pattern is not seen at all.

It's important to keep in mind that diffraction occurs for *all* kinds of waves, not just light. Sound waves undergo diffraction when they pass through a slit or aperture such as an ordinary doorway. The sound waves used in speech have wavelengths of about a meter or greater, and a typical doorway is less than 1 m wide; in this situation, a is less than λ, and the central intensity maximum extends over 180°. This is why the sounds coming through an open doorway can easily be heard by an eavesdropper hiding out of sight around the corner. In the same way, sound waves can bend around the head of an instructor who faces the blackboard while lecturing (Fig. 36.11). By contrast, there is essentially no diffraction of visible light through such a doorway because the width a is very much greater than the wavelength λ (of order 5×10^{-7} m). You can *hear* around corners because typical sound waves have relatively long wavelengths; you cannot *see* around corners because the wavelength of visible light is very short.

36.11 The sound waves used in speech have a long wavelength (about 1 m) and can easily bend around this instructor's head. By contrast, light waves have very short wavelengths and undergo very little diffraction. Hence you can't *see* around his head!

Example 36.2 Single-slit diffraction: Intensity I

(a) In a single-slit diffraction pattern, what is the intensity at a point where the total phase difference between wavelets from the top and bottom of the slit is 66 rad? (b) If this point is 7.0° away from the central maximum, how many wavelengths wide is the slit?

SOLUTION

IDENTIFY: This problem asks us to find the intensity at a point in a single-slit diffraction pattern where there is a specified phase difference between waves coming from the two edges of the slit (Fig. 36.8a). It also asks us to relate phase difference, slit width, wavelength, and the θ shown in Fig. 36.9a.

SET UP: The total phase difference between wavelets from the two edges of the slit is the quantity we called β in Fig. 36.8d. Given $\beta = 66$ rad, we use Eq. (36.5) to find the intensity I at the point in question, and we use Eq. (36.6) to find the slit width a in terms of the wavelength λ.

EXECUTE: (a) Since $\beta = 66$ rad, $\beta/2 = 33$ rad and Eq. (36.5) becomes

$$I = I_0\left[\frac{\sin(33\text{ rad})}{33\text{ rad}}\right]^2 = (9.2 \times 10^{-4})I_0$$

(b) We solve Eq. (36.6) for a:

$$a = \frac{\beta\lambda}{2\pi\sin\theta} = \frac{(66\text{ rad})\lambda}{(2\pi\text{ rad})\sin 7.0°} = 86\lambda$$

For example, for 550-nm light, the slit width a is $(86)(550\text{ nm}) = 4.7 \times 10^{-5}\text{ m} = 0.047\text{ mm}$, or roughly $\frac{1}{20}$ mm.

EVALUATE: To what point in the diffraction pattern does this value of β correspond? To find out, note that $\beta = 66\text{ rad} = 21\pi$. Comparing to Eq. (36.9) shows that this is approximately equal to the value of β at the *tenth* side maximum, well beyond the range shown in Fig. 36.9a (which shows only the first three side maxima). The intensity is very much less than the intensity I_0 at the central maximum. (The *actual* position of this maximum is at $\beta = 65.91\text{ rad} = 20.98\pi$, or approximately midway between the minima at $\beta = 20\pi$ and $\beta = 22\pi$.)

Example 36.3 Single-slit diffraction: Intensity II

In the experiment described in Example 36.1 (Section 36.2), what is the intensity at a point on the screen 3.0 mm from the center of the pattern? The intensity at the center of the pattern is I_0.

SOLUTION

IDENTIFY: This is similar to Example 36.2, except that we are not given the value of the phase difference β at the point in question.

SET UP: We use geometry to determine the angle θ for our point and then use Eq. (36.7) to calculate the intensity I (our target variable).

EXECUTE: Referring to Fig. 36.5a, we have $y = 3.0\text{ mm}$ and $x = 6.0\text{ m}$, so $\tan\theta = y/x = (3.0 \times 10^{-3}\text{ m})/(6.0\text{ m}) = 5.0 \times 10^{-4}$; since this is so small, the values of $\tan\theta$, $\sin\theta$, and θ (in radians) are all nearly the same. Then, using Eq. (36.7), we have

$$\frac{\pi a\sin\theta}{\lambda} = \frac{\pi(2.4 \times 10^{-4}\text{ m})(5.0 \times 10^{-4})}{6.33 \times 10^{-7}\text{ m}} = 0.60$$

$$I = I_0\left(\frac{\sin 0.60}{0.60}\right)^2 = 0.89I_0$$

EVALUATE: Examining Fig. 36.9a shows that an intensity this large can occur only within the central intensity maximum. This checks out; from Example 36.1, the first intensity minimum ($m = 1$ in Fig. 36.9a) is $(32\text{ mm})/2 = 16\text{ mm}$ from the center of the pattern, so the point in question here does, indeed, lie within the central maximum.

Test Your Understanding of Section 36.3 Coherent electromagnetic radiation is sent through a slit of width 0.0100 mm. For which of the following wavelengths will there be *no* points in the diffraction pattern where the intensity is zero? (i) blue light of wavelength 500 nm; (ii) infrared light of wavelength 10.6 μm; (iii) microwaves of wavelength 1.00 mm; (iv) ultraviolet light of wavelength 50.0 nm.

36.4 Multiple Slits

In Sections 35.2 and 35.3 we analyzed interference from two point sources or from two very narrow slits; in this analysis we ignored effects due to the finite (that is, nonzero) slit width. In Sections 36.2 and 36.3 we considered the diffraction effects that occur when light passes through a single slit of finite width. Additional interesting effects occur when we have two slits with finite width or when there are several very narrow slits.

Two Slits of Finite Width

Let's take another look at the two-slit pattern in the more realistic case in which the slits have finite width. If the slits are narrow in comparison to the wavelength, we can assume that light from each slit spreads out uniformly in all directions to the right of the slit. We used this assumption in Section 35.3 to calculate the interference pattern described by Eq. (35.10) or (35.15), consisting of a series of equally spaced, equally intense maxima. However, when the slits have finite width, the peaks in the two-slit interference pattern are modulated by the single-slit diffraction pattern characteristic of the width of each slit.

36.12 Finding the intensity pattern for two slits of finite width.

(a) Single-slit diffraction pattern for a slit width a

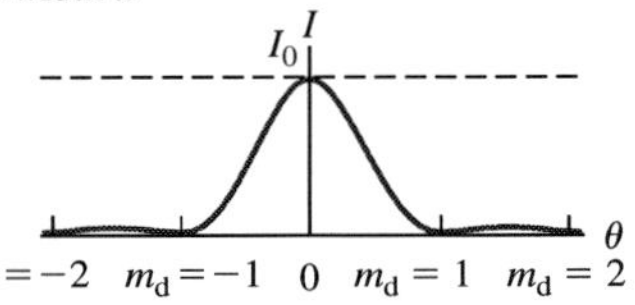

(b) Two-slit interference pattern for narrow slits whose separation d is four times the width of the slit in **(a)**

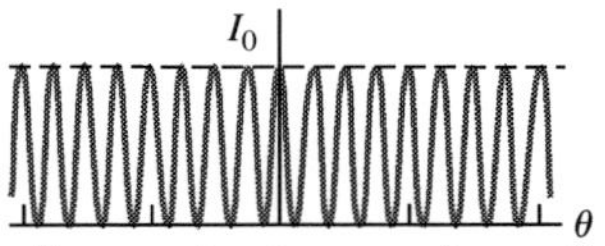

(c) Calculated intensity pattern for two slits of width a and separation $d = 4a$, including both interference and diffraction effects

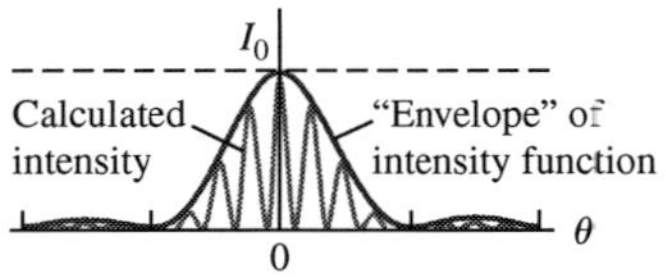

(d) Actual photograph of the pattern calculated in **(c)**

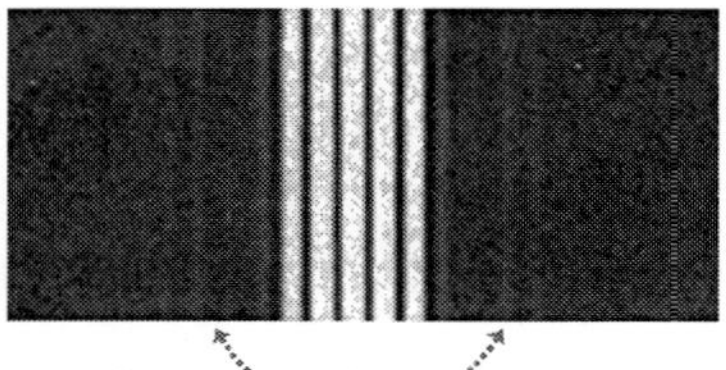

For $d = 4a$, every fourth interference maximum at the sides ($m_i = \pm 4, \pm 8, \ldots$) is missing.

36.13 Multiple-slit diffraction. Here a lens is used to give a Fraunhofer pattern on a nearby screen, as in Fig. 36.4d.

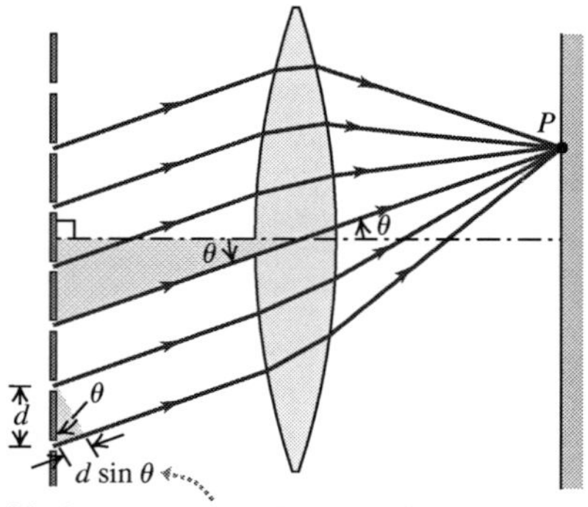

Maxima occur where the path difference for adjacent slits is a whole number of wavelengths: $d \sin\theta = m\lambda$.

Figure 36.12a shows the intensity in a single-slit diffraction pattern with slit width a. The *diffraction minima* are labeled by the integer $m_d = \pm 1, \pm 2, \ldots$ ("d" for "diffraction"). Figure 36.12b shows the pattern formed by two very narrow slits with distance d between slits, where d is four times as great as the single-slit width a in Fig. 36.12a; that is, $d = 4a$. The *interference maxima* are labeled by the integer $m_i = 0, \pm 1, \pm 2, \ldots$ ("i" for "interference"). We note that the spacing between adjacent minima in the single-slit pattern is four times as great as in the two-slit pattern. Now suppose we widen each of the narrow slits to the same width a as that of the single slit in Fig. 36.12a. Figure 36.12c shows the pattern from two slits with width a, separated by a distance (between centers) $d = 4a$. The effect of the finite width of the slits is to superimpose the two patterns—that is, to multiply the two intensities at each point. The two-slit peaks are in the same positions as before, but their intensities are modulated by the single-slit pattern, which acts as an "envelope" for the intensity function. The expression for the intensity shown in Fig. 36.12c is proportional to the product of the two-slit and single-slit expressions, Eqs. (35.10) and (36.5):

$$I = I_0 \cos^2\frac{\phi}{2}\left[\frac{\sin(\beta/2)}{\beta/2}\right]^2 \qquad \text{(two slits of finite width)} \qquad (36.12)$$

where, as before,

$$\phi = \frac{2\pi d}{\lambda}\sin\theta \qquad \beta = \frac{2\pi a}{\lambda}\sin\theta$$

Note that in Fig. 36.12c, every fourth interference maximum at the sides is *missing* because these interference maxima $(m_i = \pm 4, \pm 8, \ldots)$ coincide with diffraction minima $(m_d = \pm 1, \pm 2, \ldots)$. This can also be seen in Fig. 36.12d, which is a photograph of an actual pattern with $d = 4a$. You should be able to convince yourself that there will be "missing" maxima whenever d is an integer multiple of a.

Figures 36.12c and 36.12d show that as you move away from the central bright maximum of the two-slit pattern, the intensity of the maxima decreases. This is a result of the single-slit modulating pattern shown in Fig. 36.12a; mathematically, the decrease in intensity arises from the factor $(\beta/2)^2$ in the denominator of Eq. (36.12). This decrease in intensity can also be seen in Fig. 35.6 (Section 35.2). The narrower the slits, the broader the single-slit pattern (as in Fig. 36.10) and the slower the decrease in intensity from one interference maximum to the next.

Shall we call the pattern in Fig. 36.12d *interference* or *diffraction?* It's really both, since it results from superposition of waves coming from various parts of the two apertures. There is no truly fundamental distinction between interference and diffraction.

Several Slits

Next let's consider patterns produced by *several* very narrow slits. As we will see, systems of narrow slits are of tremendous practical importance in *spectroscopy,* the determination of the particular wavelengths of light coming from a source. Assume that each slit is narrow in comparison to the wavelength, so its diffraction pattern spreads out nearly uniformly. Figure 36.13 shows an array of eight narrow slits, with distance d between adjacent slits. Constructive interference occurs for rays at angle θ to the normal that arrive at point P with a path difference between adjacent slits equal to an integer number of wavelengths,

$$d\sin\theta = m\lambda \qquad (m = 0, \pm 1, \pm 2, \ldots)$$

This means that reinforcement occurs when the phase difference ϕ at P for light from adjacent slits is an integer multiple of 2π. That is, the maxima in the pattern occur at the *same* positions as for *two* slits with the same spacing. To this extent the pattern resembles the two-slit pattern.

But what happens *between* the maxima? In the two-slit pattern, there is exactly one intensity minimum located midway between each pair of maxima, corresponding to angles for which the phase difference between waves from the two sources is π, 3π, 5π, and so on. In the eight-slit pattern these are also minima because the light from adjacent slits cancels out in pairs, corresponding to the phasor diagram in Fig. 36.14a. But these are not the only minima in the eight-slit pattern. For example, when the phase difference ϕ from adjacent sources is $\pi/4$, the phasor diagram is as shown in Fig. 36.14b; the total (resultant) phasor is zero, and the intensity is zero. When $\phi = \pi/2$, we get the phasor diagram of Fig. 36.14c, and again both the total phasor and the intensity are zero. More generally, the intensity with eight slits is zero whenever ϕ is an integer multiple of $\pi/4$, *except* when ϕ is a multiple of 2π. Thus there are seven minima for every maximum.

Detailed calculation shows that the eight-slit pattern is as shown in Fig. 36.15b. The large maxima, called *principal maxima,* are in the same positions as for the two-slit pattern of Fig. 36.15a but are much narrower. If the phase difference ϕ between adjacent slits is slightly different from a multiple of 2π, the waves from slits 1 and 2 will be only a little out of phase; however, the phase difference between slits 1 and 3 will be greater, that between slits 1 and 4 will be greater still, and so on. This leads to a partial cancellation for angles that are only slightly different from the angle for a maximum, giving the narrow maxima in Fig. 36.15b. The maxima are even narrower with 16 slits (Fig. 36.15c).

You should show that when there are N slits, there are $(N - 1)$ minima between each pair of principal maxima and a minimum occurs whenever ϕ is an integral multiple of $2\pi/N$ (except when ϕ is an integral multiple of 2π, which gives a principal maximum). There are small *secondary* intensity maxima between the minima; these become smaller in comparison to the principal maxima as N increases. The greater the value of N, the narrower the principal maxima become. From an energy standpoint the total power in the entire pattern is proportional to N. The height of each principal maximum is proportional to N^2, so from energy conservation the width of each principal maximum must be proportional to $1/N$. As we will see in the next section, the narrowness of the principal maxima in a multiple-slit pattern is of great practical importance in physics and astronomy.

36.14 Phasor diagrams for light passing through eight narrow slits. Intensity maxima occur when the phase difference $\phi = 0, 2\pi, 4\pi, \ldots$. Between the maxima at $\phi = 0$ and $\phi = 2\pi$ are seven minima, corresponding to $\phi = \pi/4, \pi/2, 3\pi/4, \pi, 5\pi/4, 3\pi/2$, and $7\pi/4$. Can you draw phasor diagrams for the other minima?

(a) Phasor diagram for $\phi = \pi$

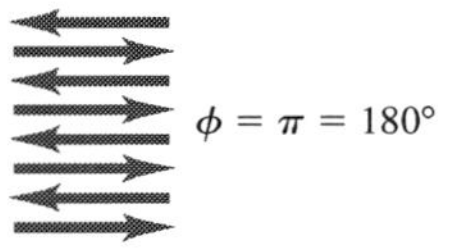

(b) Phasor diagram for $\phi = \frac{\pi}{4}$

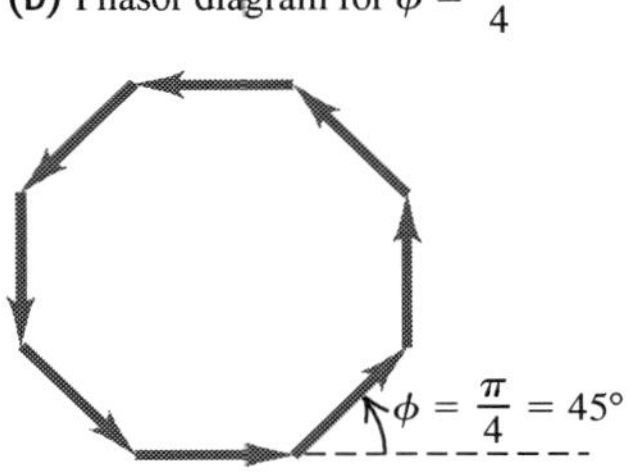

(c) Phasor diagram for $\phi = \frac{\pi}{2}$

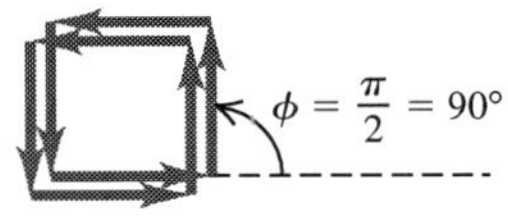

36.15 Interference patterns for N equally spaced, very narrow slits. (a) Two slits. (b) Eight slits. (c) Sixteen slits. The vertical scales are different for each graph; I_0 is the maximum intensity for a single slit and the maximum intensity for N slits is N^2I_0. The width of each peak is proportional to $1/N$.

(a) $N = 2$: two slits produce one minimum between adjacent maxima.

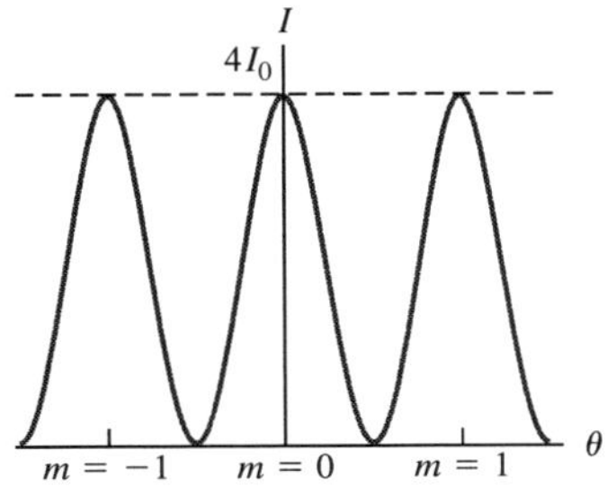

(b) $N = 8$: eight slits produce taller, narrower maxima in the same locations, separated by seven minima.

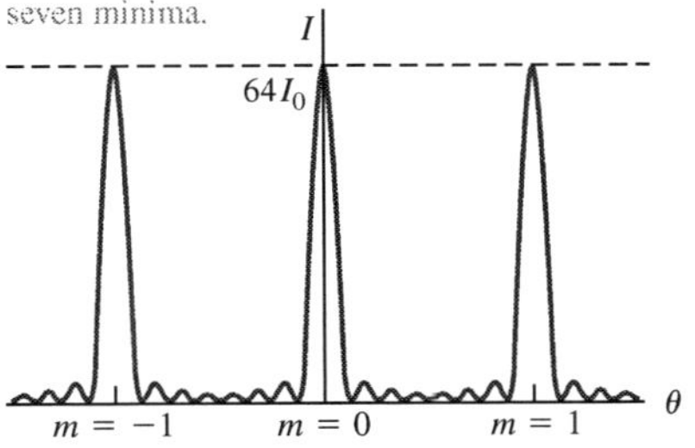

(c) $N = 16$: with 16 slits, the maxima are even taller and narrower, with more intervening minima.

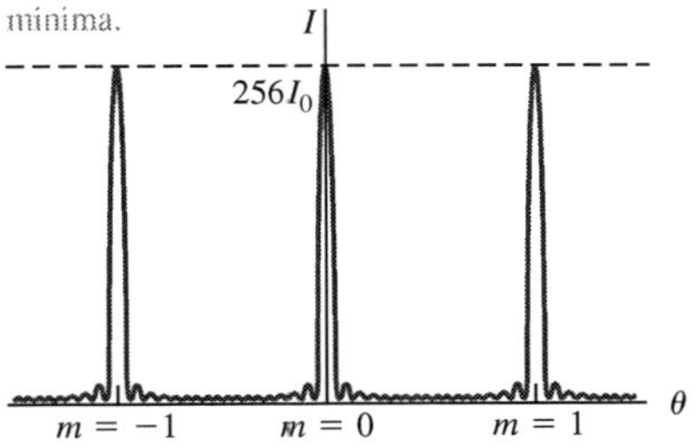

Test Your Understanding of Section 36.4 Suppose two slits, each of width a, are separated by a distance $d = 2.5a$. Are there any missing maxima in the interference pattern produced by these slits? If so, which are missing? If not, why not?

36.16 A portion of a transmission diffraction grating. The separation between the centers of adjacent slits is d.

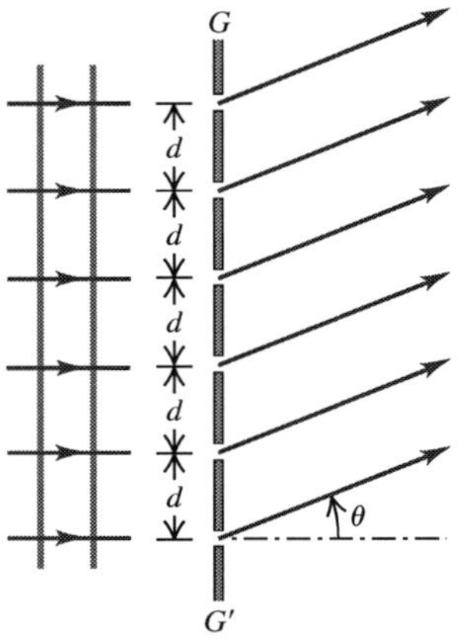

ActivPhysics Online

16.4 The Grating: Introduction and Questions

16.5 The Grating: Problems

36.17 The millions of microscopic scales in the wings of the tropical butterfly *Morpho peleides* act as a reflection grating. When viewed at the right angle, these scales strongly reflect blue light. This may be a defense mechanism: The flashes of light from the flapping wings of a *Morpho* could momentarily dazzle predators such as lizards and birds.

36.5 The Diffraction Grating

We have just seen that increasing the number of slits in an interference experiment (while keeping the spacing of adjacent slits constant) gives interference patterns in which the maxima are in the same positions, but progressively narrower, than with two slits. Because these maxima are so narrow, their angular position, and hence the wavelength, can be measured to very high precision. As we will see, this effect has many important applications.

An array of a large number of parallel slits, all with the same width a and spaced equal distances d between centers, is called a **diffraction grating.** The first one was constructed by Fraunhofer using fine wires. Gratings can be made by using a diamond point to scratch many equally spaced grooves on a glass or metal surface, or by photographic reduction of a pattern of black and white stripes on paper. For a grating, what we have been calling *slits* are often called *rulings* or *lines.*

In Fig. 36.16, GG' is a cross section of a *transmission grating;* the slits are perpendicular to the plane of the page, and an interference pattern is formed by the light that is transmitted through the slits. The diagram shows only six slits; an actual grating may contain several thousand. The spacing d between centers of adjacent slits is called the *grating spacing.* A plane monochromatic wave is incident normally on the grating from the left side. We assume far-field (Fraunhofer) conditions; that is, the pattern is formed on a screen that is far enough away that all rays emerging from the grating and going to a particular point on the screen can be considered to be parallel.

We found in Section 36.4 that the principal intensity maxima with multiple slits occur in the same directions as for the two-slit pattern. These are the directions for which the path difference for adjacent slits is an integer number of wavelengths. So the positions of the maxima are once again given by

$$d\sin\theta = m\lambda \qquad (m = 0, \pm1, \pm2, \pm3, \dots) \qquad \text{(intensity maxima, multiple slits)} \qquad (36.13)$$

The intensity patterns for two, eight, and 16 slits displayed in Fig. 36.15 show the progressive increase in sharpness of the maxima as the number of slits increases.

When a grating containing hundreds or thousands of slits is illuminated by a beam of parallel rays of monochromatic light, the pattern is a series of very sharp lines at angles determined by Eq. (36.13). The $m = \pm1$ lines are called the *first-order lines,* the $m = \pm2$ lines the *second-order lines,* and so on. If the grating is illuminated by white light with a continuous distribution of wavelengths, each value of m corresponds to a continuous spectrum in the pattern. The angle for each wavelength is determined by Eq. (36.13); for a given value of m, long wavelengths (the red end of the spectrum) lie at larger angles (that is, are deviated more from the straight-ahead direction) than do the shorter wavelengths at the violet end of the spectrum.

As Eq. (36.13) shows, the sines of the deviation angles of the maxima are proportional to the ratio λ/d. For substantial deviation to occur, the grating spacing d should be of the same order of magnitude as the wavelength λ. Gratings for use with visible light (λ from 400 to 700 nm) usually have about 1000 slits per millimeter; the value of d is the *reciprocal* of the number of slits per unit length, so d is of the order of $\frac{1}{1000}$ mm $= 1000$ nm.

In a *reflection grating,* the array of equally spaced slits shown in Fig. 36.16 is replaced by an array of equally spaced ridges or grooves on a reflective screen. The reflected light has maximum intensity at angles where the phase difference between light waves reflected from adjacent ridges or grooves is an integral multiple of 2π. If light of wavelength λ is incident normally on a reflection grating with a spacing d between adjacent ridges or grooves, the *reflected* angles at which intensity maxima occur are given by Eq. (36.13). The iridescent colors of certain butterflies arise from microscopic ridges on the butterfly's wings that form a reflection grating (Fig 36.17). When the wings are viewed from different angles, corresponding to varying θ in Eq. (36.13), the wavelength and color that are predominantly reflected to the viewer's eye vary as well.

The rainbow-colored reflections that you see from the surface of a compact disc are a reflection-grating effect (Fig. 36.18). The "grooves" are tiny pits 0.1 μm deep in the surface of the disc, with a uniform radial spacing of $d = 1.60\ \mu\text{m} = 1600$ nm. Information is coded on the CD by varying the *length* of the pits; the reflection-grating aspect of the disc is merely an aesthetic side benefit.

36.18 Microscopic pits on the surface of this compact disc act as a reflection grating, splitting white light into its component colors.

Example 36.4 Width of a grating spectrum

The wavelengths of the visible spectrum are approximately 400 nm (violet) to 700 nm (red). (a) Find the angular width of the first-order visible spectrum produced by a plane grating with 600 slits per millimeter when white light falls normally on the grating. (b) Do the first-order and second-order spectra overlap? What about the second-order and third-order spectra? Do your answers depend on the grating spacing?

SOLUTION

IDENTIFY: The first-, second-, and third-order spectra correspond to $m = 1, 2,$ and 3 in Eq. (36.13). This problem asks us to look at the angles spanned by the visible spectrum in each of these orders.

SET UP: We use Eq. (36.13) with $m = 1$ to find the angular deviation θ for 400-nm violet light and 700-nm red light in the first-order spectrum. The difference between these is the angular width of the first-order spectrum, our target variable in part (a). Using the same technique for $m = 2$ and $m = 3$ tells us the maximum and minimum angular deviation for these orders.

EXECUTE: (a) The grating spacing d is

$$d = \frac{1}{600\ \text{slits/mm}} = 1.67 \times 10^{-6}\ \text{m}$$

From Eq. (36.13), with $m = 1$, the angular deviation θ_v of the violet light (400 nm or 400×10^{-9} m) is

$$\sin\theta_v = \frac{400 \times 10^{-9}\ \text{m}}{1.67 \times 10^{-6}\ \text{m}} = 0.240$$

$$\theta_v = 13.9°$$

The angular deviation θ_r of the red light (700 nm) is

$$\sin\theta_r = \frac{700 \times 10^{-9}\ \text{m}}{1.67 \times 10^{-6}\ \text{m}} = 0.419$$

$$\theta_r = 24.8°$$

So the angular width of the first-order visible spectrum is

$$24.8° - 13.9° = 10.9°$$

(b) From Eq. (36.13), with a grating spacing of d the angular deviation θ_{vm} of the 400-nm violet light in the mth-order spectrum is given by

$$\sin\theta_{vm} = \frac{m(400 \times 10^{-9}\ \text{m})}{d}$$
$$= \frac{4.00 \times 10^{-7}\ \text{m}}{d} \quad (m = 1)$$
$$= \frac{8.00 \times 10^{-7}\ \text{m}}{d} \quad (m = 2)$$
$$= \frac{1.20 \times 10^{-6}\ \text{m}}{d} \quad (m = 3)$$

Similarly, the angular deviation θ_{rm} of the 700-nm red light in the mth-order spectrum is given by

$$\sin\theta_{rm} = \frac{m(700 \times 10^{-9}\ \text{m})}{d}$$
$$= \frac{7.00 \times 10^{-7}\ \text{m}}{d} \quad (m = 1)$$
$$= \frac{1.40 \times 10^{-6}\ \text{m}}{d} \quad (m = 2)$$
$$= \frac{2.10 \times 10^{-6}\ \text{m}}{d} \quad (m = 3)$$

The greater the value of $\sin\theta$, the greater the value of θ (for angles between zero and 90°). Hence our results show that for any value of the grating spacing d, the largest angle (at the red end) of the $m = 1$ spectrum is always less than the smallest angle (at the violet end) of the $m = 2$ spectrum, so the first and second orders *never* overlap. By contrast, the largest (red) angle of the $m = 2$ spectrum is always greater than the smallest (violet) angle of the $m = 3$ spectrum, so the second and third orders *always* overlap.

EVALUATE: The fundamental reason the first-order and second-order visible spectra don't overlap is that the human eye is sensitive to only a narrow range of wavelengths. Can you show that if the eye could detect wavelengths from 400 nm to 900 nm (in the near-infrared range), the first and second orders *would* overlap?

36.19 **(a)** A visible-light photograph of the sun. **(b)** Sunlight is dispersed into a spectrum by a diffraction grating. Specific wavelengths are absorbed as sunlight passes through the sun's atmosphere, leaving dark lines in the spectrum.

(a)

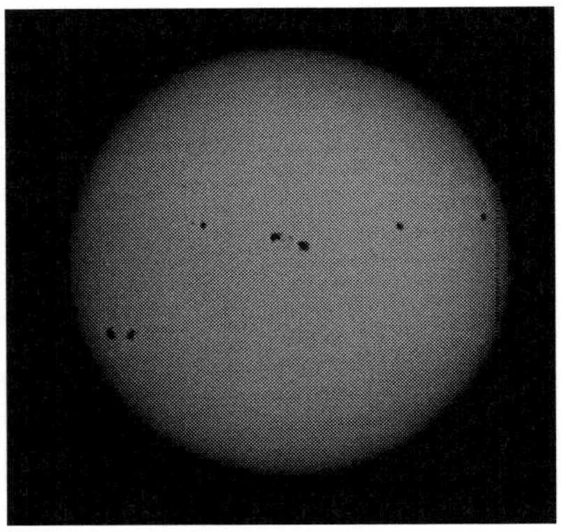

(b)

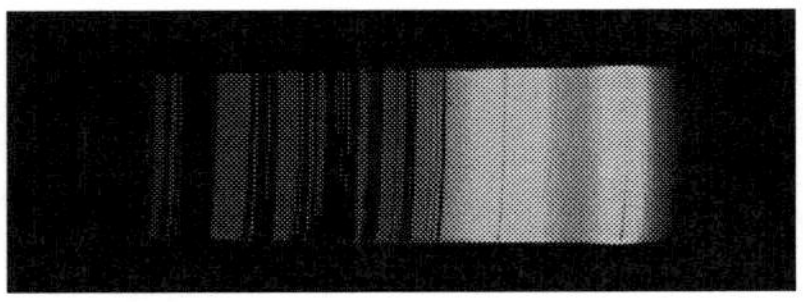

Grating Spectrographs

Diffraction gratings are widely used to measure the spectrum of light emitted by a source, a process called *spectroscopy* or *spectrometry.* Light incident on a grating of known spacing is dispersed into a spectrum. The angles of deviation of the maxima are then measured, and Eq. (36.13) is used to compute the wavelength. With a grating that has many slits, very sharp maxima are produced, and the angle of deviation (and hence the wavelength) can be measured very precisely.

An important application of this technique is to astronomy. As light generated within the sun passes through the sun's atmosphere, certain wavelengths are selectively absorbed. The result is that the spectrum of sunlight produced by a diffraction grating has dark *absorption lines* (Fig. 36.19). Experiments in the laboratory show that different types of atoms and ions absorb light at different wavelengths. By comparing these laboratory results with the wavelengths of absorption lines in the spectrum of sunlight, astronomers can deduce the chemical composition of the sun's atmosphere. The same technique is used to make chemical assays of galaxies that are millions of light-years away.

Figure 36.20 shows one design for a *grating spectrograph* used in astronomy. A transmission grating is used in the figure; in other setups, a reflection grating is used. In older designs a prism was used rather than a grating, and a spectrum was formed by dispersion (see Section 33.4) rather than diffraction. However, there is no simple relationship between wavelength and angle of deviation for a prism, prisms absorb some of the light that passes through them, and they are less effective for many nonvisible wavelengths that are important in astronomy. For these and other reasons, gratings are preferred in precision applications.

Resolution of a Grating Spectrograph

In spectroscopy it is often important to distinguish slightly differing wavelengths. The minimum wavelength difference $\Delta\lambda$ that can be distinguished by a spectrograph is described by the **chromatic resolving power** R, defined as

$$R = \frac{\lambda}{\Delta\lambda} \quad \text{(chromatic resolving power)} \qquad (36.14)$$

36.20 A schematic diagram of a diffraction-grating spectrograph for use in astronomy. Note that the light does not strike the grating normal to its surface, so the intensity maxima are given by a somewhat different expression than Eq. (36.13). (See Problem 36.66).

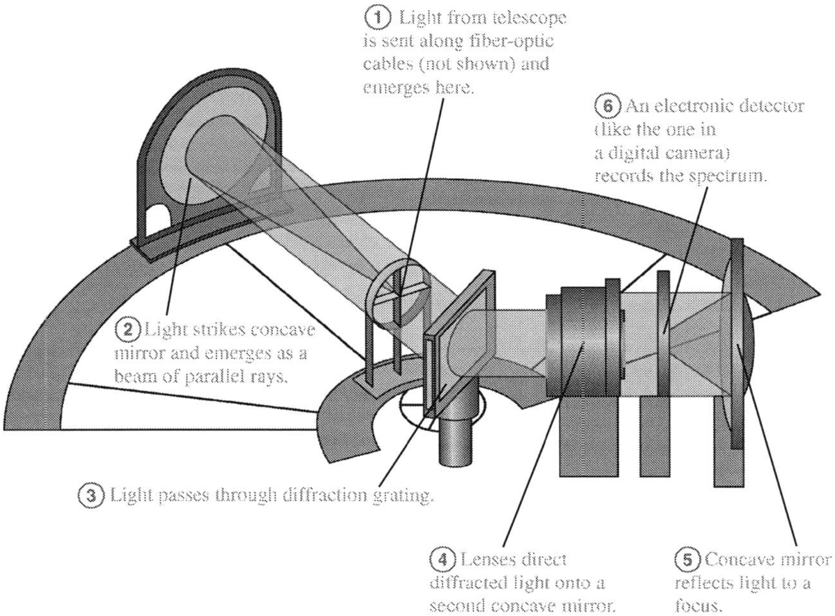

As an example, when sodium atoms are heated, they emit strongly at the yellow wavelengths 589.00 nm and 589.59 nm. A spectrograph that can barely distinguish these two lines in the spectrum of sodium light (called the *sodium doublet*) has a chromatic resolving power $R = (589.00\ \text{nm})/(0.59\ \text{nm}) = 1000$. (You can see these wavelengths when boiling water on a gas range. If the water boils over onto the flame, dissolved sodium from table salt emits a burst of yellow light.)

We can derive an expression for the resolving power of a diffraction grating used in a spectrograph. Two different wavelengths give diffraction maxima at slightly different angles. As a reasonable (though arbitrary) criterion, let's assume that we can distinguish them as two separate peaks if the maximum of one coincides with the first minimum of the other.

From our discussion in Section 36.4 the mth-order maximum occurs when the phase difference ϕ for adjacent slits is $\phi = 2\pi m$. The first minimum beside that maximum occurs when $\phi = 2\pi m + 2\pi/N$, where N is the number of slits. The phase difference is also given by $\phi = (2\pi d \sin\theta)/\lambda$, so the angular interval $d\theta$ corresponding to a small increment $d\phi$ in the phase shift can be obtained from the differential of this equation:

$$d\phi = \frac{2\pi d \cos\theta\, d\theta}{\lambda}$$

When $d\phi = 2\pi/N$, this corresponds to the angular interval $d\theta$ between a maximum and the first adjacent minimum. Thus $d\theta$ is given by

$$\frac{2\pi}{N} = \frac{2\pi d \cos\theta\, d\theta}{\lambda} \qquad \text{or} \qquad d\cos\theta\, d\theta = \frac{\lambda}{N}$$

CAUTION **Watch out for different uses of the symbol *d*** Don't confuse the spacing d with the differential "d" in the angular interval $d\theta$ or in the phase shift increment $d\phi$!

Now we need to find the angular spacing $d\theta$ between maxima for two slightly different wavelengths. This is easy; we have $d\sin\theta = m\lambda$, so the differential of this equation gives

$$d\cos\theta\, d\theta = m\, d\lambda$$

According to our criterion, the limit or resolution is reached when these two angular spacings are equal. Equating the two expressions for the quantity $(d\cos\theta\, d\theta)$, we find

$$\frac{\lambda}{N} = m\, d\lambda \qquad \text{and} \qquad \frac{\lambda}{d\lambda} = Nm$$

If $\Delta\lambda$ is small, we can replace $d\lambda$ by $\Delta\lambda$, and the resolving power R is given simply by

$$R = \frac{\lambda}{\Delta\lambda} = Nm \qquad (36.15)$$

The greater the number of slits N, the better the resolution; also, the higher the order m of the diffraction-pattern maximum that we use, the better the resolution.

Test Your Understanding of Section 36.5 What minimum number of slits would be required in a grating to resolve the sodium doublet in the fourth order? (i) 250; (ii) 400; (iii) 1000; (iv) 4000.

36.6 X-Ray Diffraction

X rays were discovered by Wilhelm Röntgen (1845–1923) in 1895, and early experiments suggested that they were electromagnetic waves with wavelengths of the order of 10^{-10} m. At about the same time, the idea began to emerge that in a crystalline solid the atoms are arranged in a regular repeating pattern, with spacing between adjacent atoms also of the order of 10^{-10} m. Putting these two ideas together, Max von Laue (1879–1960) proposed in 1912 that a crystal might serve as a kind of three-dimensional diffraction grating for x rays. That is, a beam of x rays might be scattered (that is, absorbed and re-emitted) by the individual atoms in a crystal, and the scattered waves might interfere just like waves from a diffraction grating.

The first **x-ray diffraction** experiments were performed in 1912 by Friederich, Knipping, and von Laue, using the experimental setup sketched in Fig. 36.21a. The scattered x rays *did* form an interference pattern, which they recorded on photographic film. Figure 36.21b is a photograph of such a pattern. These experiments verified that x rays *are* waves, or at least have wavelike properties, and also that the atoms in a crystal *are* arranged in a regular pattern (Fig. 36.22). Since that time, x-ray diffraction has proved to be an invaluable research tool, both for measuring x-ray wavelengths and for studying the structure of crystals and complex molecules.

A Simple Model of X-Ray Diffraction

To better understand x-ray diffraction, we consider first a two-dimensional scattering situation, as shown in Fig. 36.23a, in which a plane wave is incident on a rectangular array of scattering centers. The situation might be a ripple tank with an array of small posts, 3-cm microwaves striking an array of small conducting spheres, or x rays incident on an array of atoms. In the case of electromagnetic waves, the wave induces an oscillating electric dipole moment in each scatterer. These dipoles act like little antennas, emitting scattered waves. The resulting interference pattern is the superposition of all these scattered waves. The situation is different from that with a diffraction grating, in which the waves from all the slits are emitted *in phase* (for a plane wave at normal incidence). Here the scattered waves are *not* all in phase because their distances from the *source* are different. To compute the interference pattern, we have to consider the *total* path differences for the scattered waves, including the distances from source to scatterer and from scatterer to observer.

As Fig. 36.23b shows, the path length from source to observer is the same for all the scatterers in a single row if the two angles θ_a and θ_r are equal. Scattered radia-

36.21 (a) An x-ray diffraction experiment. (b) Diffraction pattern (or *Laue pattern*) formed by directing a beam of x rays at a thin section of quartz crystal.

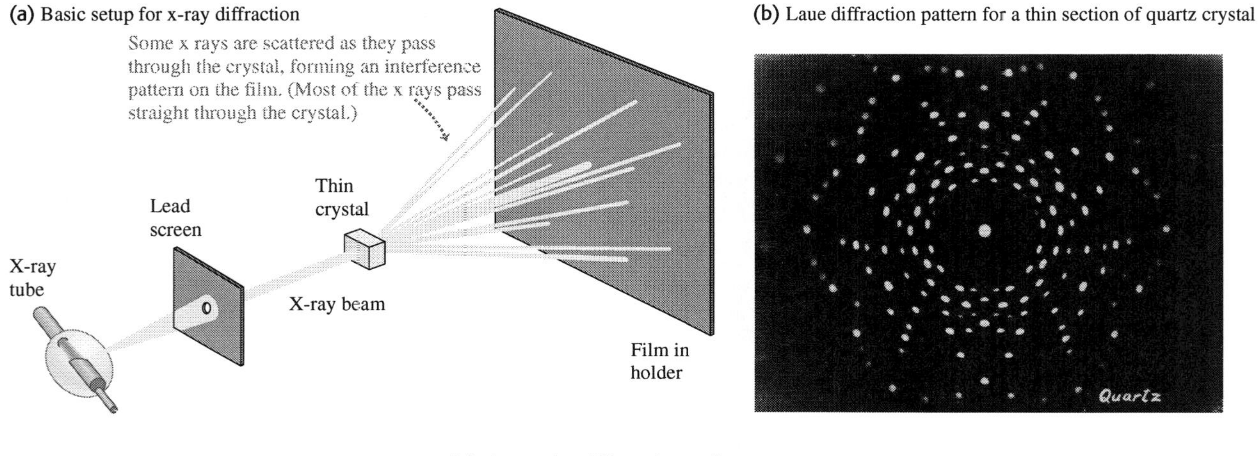

tion from *adjacent* rows is *also* in phase if the path difference for adjacent rows is an integer number of wavelengths. Figure 36.23c shows that this path difference is $2d\sin\theta$, where θ is the common value of θ_a and θ_r. Therefore the conditions for radiation from the *entire array* to reach the observer in phase are (1) the angle of incidence must equal the angle of scattering and (2) the path difference for adjacent rows must equal $m\lambda$, where m is an integer. We can express the second condition as

$$2d\sin\theta = m\lambda \qquad (m = 1, 2, 3, \ldots) \qquad \text{(Bragg condition for constructive interference from an array)} \qquad (36.16)$$

CAUTION **Scattering from an array** In Eq. (36.16) the angle θ is measured with respect to the *surface* of the crystal, rather than with respect to the *normal* to the plane of an array of slits or a grating. Also, note that the path difference in Eq. (36.16) is $2d\sin\theta$, not $d\sin\theta$ as in Eq. (36.13) for a diffraction grating.

In directions for which Eq. (36.16) is satisfied, we see a strong maximum in the interference pattern. We can describe this interference in terms of *reflections* of the wave from the horizontal rows of scatterers in Fig. 36.23a. Strong reflection (constructive interference) occurs at angles such that the incident and scattered angles are equal and Eq. (36.16) is satisfied. Since $\sin\theta$ can never be greater than 1, Eq. (36.16) says that to have constructive interference the quantity $m\lambda$ must be less than $2d$ and so λ must be less than $2d/m$. For example, the value of d in an NaCl crystal (Fig. 36.22) is only 0.282 nm. Hence to have the mth-order maximum present in the diffraction pattern, λ must be less than $2(0.282\text{ nm})/m$; that is, $\lambda < 0.564$ nm for $m = 1$, $\lambda < 0.282$ nm for $m = 2$, $\lambda < 0.188$ nm for $m = 3$, and so on. These are all x-ray wavelengths (see Fig. 32.4), which is why x rays are used for studying crystal structure.

We can extend this discussion to a three-dimensional array by considering *planes* of scatterers instead of *rows*. Figure 36.24 shows two different sets of parallel planes that pass through all the scatterers. Waves from all the scatterers in a

36.22 Model of the arrangement of ions in a crystal of NaCl (table salt). The spacing of adjacent atoms is 0.282 nm. (The electron clouds of the atoms actually overlap slightly.)

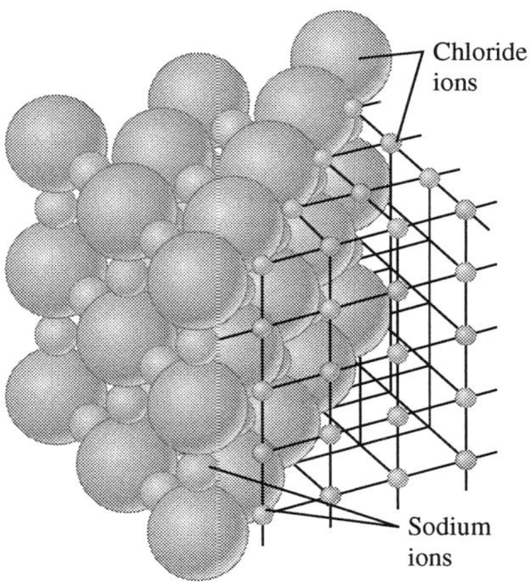

36.23 A two-dimensional model of scattering from a rectangular array. Note that the angles in (b) are measured from the *surface* of the array, not from its normal.

(a) Scattering of waves from a rectangular array

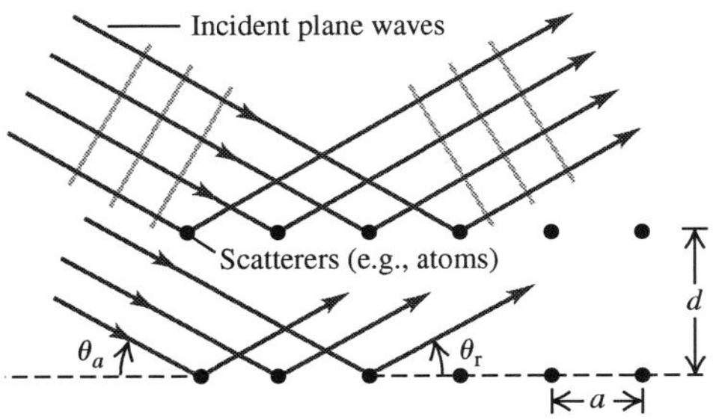

(b) Scattering from adjacent atoms in a row
Interference from adjacent atoms in a row is constructive when the path lengths $a\cos\theta_a$ and $a\cos\theta_r$ are equal, so that the angle of incidence θ_a equals the angle of reflection (scattering) θ_r.

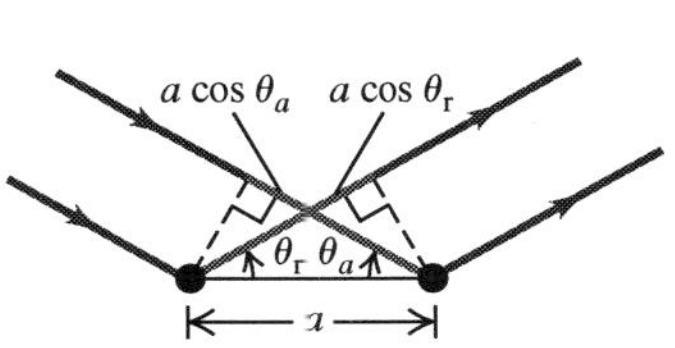

(c) Scattering from atoms in adjacent rows
Interference from atoms in adjacent rows is constructive when the path difference $2d\sin\theta$ is an integral number of wavelengths, as in Eq. (36.16).

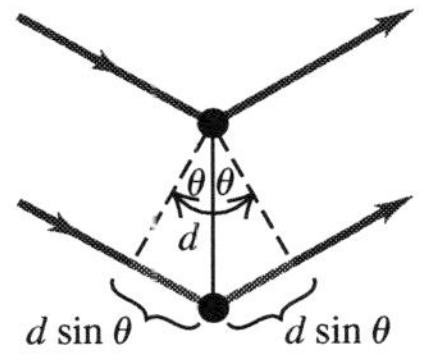

36.24 A cubic crystal and two different families of crystal planes. There are also three sets of planes parallel to the cube faces, with spacing a.

(a) Spacing of planes is $d = a/\sqrt{2}$.

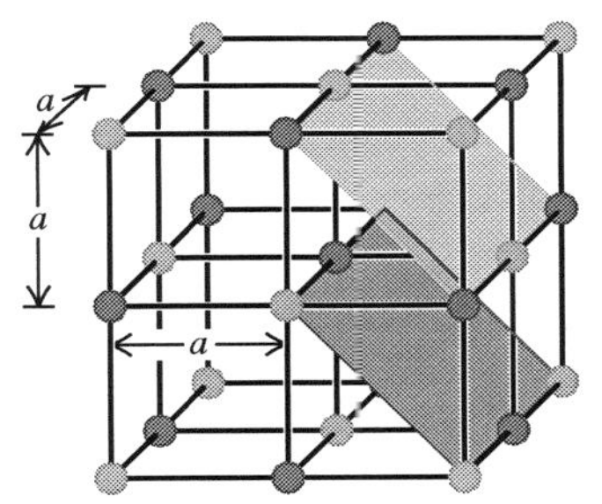

(b) Spacing of planes is $d = a/\sqrt{3}$.

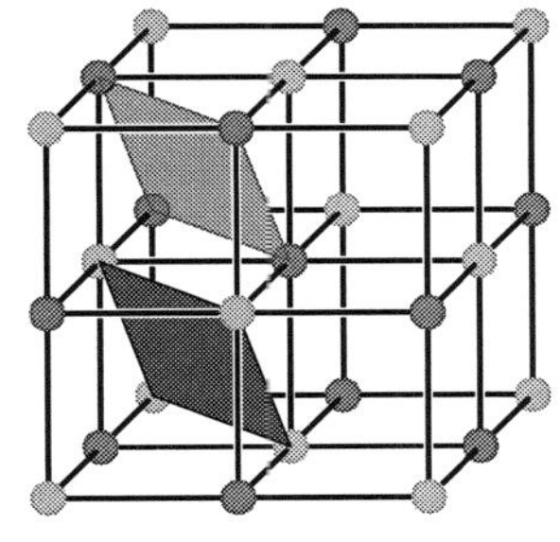

given plane interfere constructively if the angles of incidence and scattering are equal. There is also constructive interference between planes when Eq. (36.16) is satisfied, where d is now the distance between adjacent planes. Because there are many different sets of parallel planes, there are also many values of d and many sets of angles that give constructive interference for the whole crystal lattice. This phenomenon is called **Bragg reflection,** and Eq. (36.16) is called the **Bragg condition,** in honor of Sir William Bragg and his son Laurence Bragg, two pioneers in x-ray analysis.

CAUTION **Bragg *reflection* is really Bragg *interference*** While we are using the term *reflection,* remember that we are dealing with an *interference* effect. In fact, the reflections from various planes are closely analogous to interference effects in thin films (see Section 35.4).

As Fig. 36.21b shows, in x-ray diffraction there is nearly complete cancellation in all but certain very specific directions in which constructive interference occurs and forms bright spots. Such a pattern is usually called an x-ray *diffraction* pattern, although *interference* pattern might be more appropriate.

We can determine the wavelength of x rays by examining the diffraction pattern for a crystal of known structure and known spacing between atoms, just as we determined wavelengths of visible light by measuring patterns from slits or gratings. (The spacing between atoms in simple crystals of known structure, such as sodium chloride, can be found from the density of the crystal and Avogadro's number.) Then, once we know the x-ray wavelength, we can use x-ray diffraction to explore the structure and determine the spacing between atoms in crystals with unknown structure.

X-ray diffraction is by far the most important experimental tool in the investigation of crystal structure of solids. X-ray diffraction also plays an important role in studies of the structures of liquids and of organic molecules. It has been one of the chief experimental techniques in working out the double-helix structure of DNA (Fig. 36.25) and subsequent advances in molecular genetics.

36.25 The British scientist Rosalind Franklin made this groundbreaking x-ray diffraction image of DNA in 1953. The dark bands arranged in a cross provided the first evidence of the helical structure of the DNA molecule.

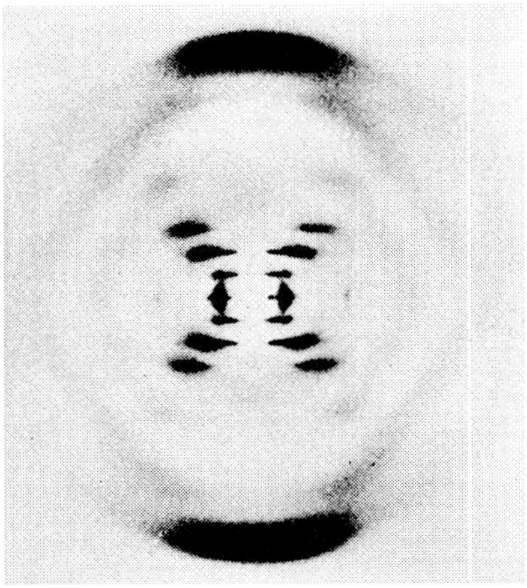

Example 36.5 X-ray diffraction

You direct a beam of x rays with wavelength 0.154 nm at certain planes of a silicon crystal. As you increase the angle of incidence from zero, you find the first strong interference maximum from these planes when the beam makes an angle of 34.5° with the planes. (a) How far apart are the planes? (b) Will you find other interference maxima from these planes at larger angles?

SOLUTION

IDENTIFY: This problem involves Bragg reflection of x rays from the planes of a crystal.

SET UP: In part (a) we use the Bragg condition, Eq. (36.16), to relate the wavelength λ and the angle θ for the $m = 1$ interference maximum (both of which are given) to the spacing d between planes (which is the target variable). Given the value of d, we use the Bragg condition again in part (b) to find the values of θ for interference maxima corresponding to other values of m.

EXECUTE: (a) We solve the Bragg equation, Eq. (36.16), for d and set $m = 1$:

$$d = \frac{m\lambda}{2\sin\theta} = \frac{(1)(0.154\text{ nm})}{2\sin 34.5°} = 0.136\text{ nm}$$

This is the distance between adjacent planes.

(b) To calculate other angles, we solve Eq. (36.16) for $\sin\theta$:

$$\sin\theta = \frac{m\lambda}{2d} = m\frac{0.154\text{ nm}}{2(0.136\text{ nm})} = m(0.566)$$

Values of m of 2 or greater give values of $\sin\theta$ greater than unity, which is impossible. Hence there are *no* other angles for interference maxima for this particular set of crystal planes.

EVALUATE: Our result in part (b) shows that there *would* be a second interference maximum if the quantity $\lambda/2d$ were equal to 0.500 or less. This would be the case if the wavelength of the x rays were less than $2d = 0.272$ nm. How short would the wavelength need to be to have *three* interference maxima?

Test Your Understanding of Section 36.6 You are doing an x-ray diffraction experiment with a crystal in which the atomic planes are 0.200 nm apart. You are using x rays of wavelength 0.100 nm. Will the fifth-order maximum be present in the diffraction pattern?

36.7 Circular Apertures and Resolving Power

We have studied in detail the diffraction patterns formed by long, thin slits or arrays of slits. But an aperture of *any* shape forms a diffraction pattern. The diffraction pattern formed by a *circular* aperture is of special interest because of its role in limiting how well an optical instrument can resolve fine details. In principle, we could compute the intensity at any point P in the diffraction pattern by dividing the area of the aperture into small elements, finding the resulting wave amplitude and phase at P, and then integrating over the aperture area to find the resultant amplitude and intensity at P. In practice, the integration cannot be carried out in terms of elementary functions. We will simply *describe* the pattern and quote a few relevant numbers.

ActivPhysics Online
16.7 Circular Hole Diffraction
16.8 Resolving Power

The diffraction pattern formed by a circular aperture consists of a central bright spot surrounded by a series of bright and dark rings, as shown in Fig. 36.26. We can describe the pattern in terms of the angle θ, representing the angular radius of each ring. If the aperture diameter is D and the wavelength is λ, the angular radius θ_1 of the first *dark* ring is given by

$$\sin\theta_1 = 1.22\frac{\lambda}{D} \quad \text{(diffraction by a circular aperture)} \tag{36.17}$$

The angular radii of the next two dark rings are given by

$$\sin\theta_2 = 2.23\frac{\lambda}{D} \qquad \sin\theta_3 = 3.24\frac{\lambda}{D} \tag{36.18}$$

Between these are bright rings with angular radii given by

$$\sin\theta = 1.63\frac{\lambda}{D}, \quad 2.68\frac{\lambda}{D}, \quad 3.70\frac{\lambda}{D} \tag{36.19}$$

and so on. The central bright spot is called the **Airy disk,** in honor of Sir George Airy (1801–1892), Astronomer Royal of England, who first derived the expression for the intensity in the pattern. The angular radius of the Airy disk is that of the first dark ring, given by Eq. (36.17).

The intensities in the bright rings drop off very quickly with increasing angle. When D is much larger than the wavelength λ, the usual case for optical instruments, the peak intensity in the first ring is only 1.7% of the value at the center of the Airy disk, and the peak intensity of the second ring is only 0.4%. Most (85%) of the light energy falls within the Airy disk. Figure 36.27 shows a diffraction pattern from a circular aperture.

36.26 Diffraction pattern formed by a circular aperture of diameter D. The pattern consists of a central bright spot and alternating dark and bright rings. The angular radius θ_1 of the first dark ring is shown. (This diagram is not drawn to scale.)

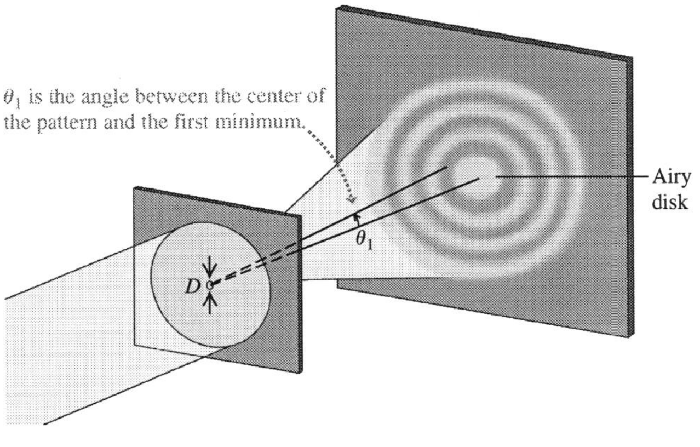

36.27 Photograph of the diffraction pattern formed by a circular aperture.

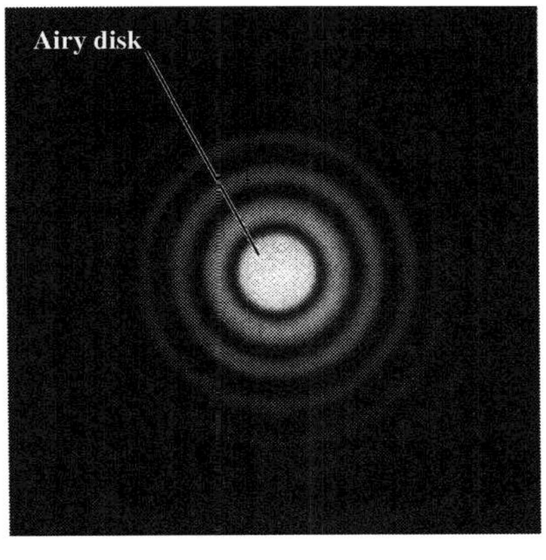

36.28 Diffraction patterns of four very small ("point") sources of light. The photographs were made with a circular aperture in front of the lens. (a) The aperture is so small that the patterns of sources 3 and 4 overlap and are barely resolved by Rayleigh's criterion. Increasing the size of the aperture decreases the size of the diffraction patterns, as shown in (b) and (c).

(a) Small aperture

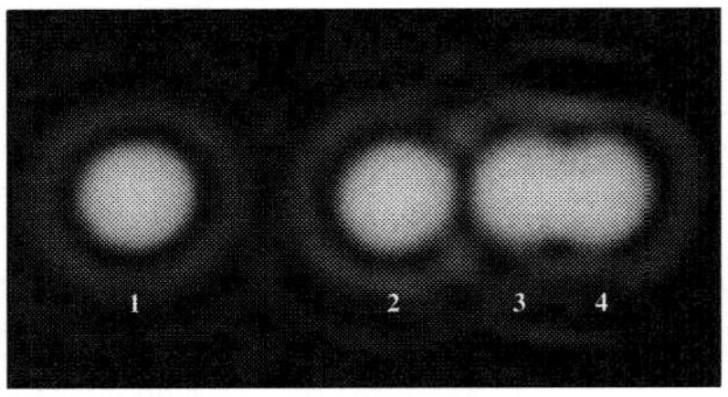

(b) Medium aperture

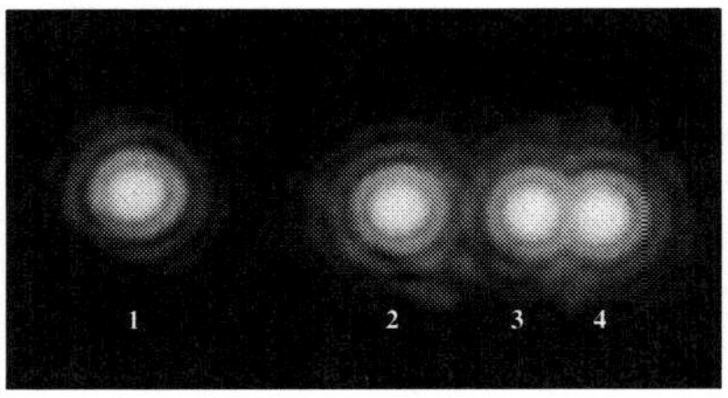

(c) Large aperture

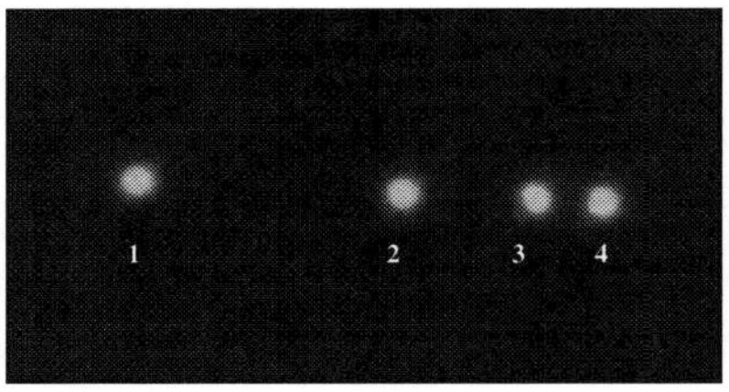

Diffraction and Image Formation

Diffraction has far-reaching implications for image formation by lenses and mirrors. In our study of optical instruments in Chapter 34 we assumed that a lens with focal length f focuses a parallel beam (plane wave) to a *point* at a distance f from the lens. This assumption ignored diffraction effects. We now see that what we get is not a point but the diffraction pattern just described. If we have two point objects, their images are not two points but two diffraction patterns. When the objects are close together, their diffraction patterns overlap; if they are close enough, their patterns overlap almost completely and cannot be distinguished. The effect is shown in Fig. 36.28, which presents the patterns for four very small "point" sources of light. In Fig. 36.28a the image of the left-hand source is well separated from the others, but the images of the middle and right-hand sources have merged. In Fig. 36.28b, with a larger aperture diameter and hence smaller Airy disks, the middle and right-hand images are better resolved. In Fig. 36.28c, with a still larger aperture, they are well resolved.

A widely used criterion for resolution of two point objects, proposed by the English physicist Lord Rayleigh (1842–1919) and called **Rayleigh's criterion,** is that the objects are just barely resolved (that is, distinguishable) if the center of one diffraction pattern coincides with the first minimum of the other. In that case the angular separation of the image centers is given by Eq. (36.17). The angular separation of the *objects* is the same as that of the *images* made by a telescope, microscope, or other optical device. So two point objects are barely resolved, according to Rayleigh's criterion, when their angular separation is given by Eq. (36.17).

The minimum separation of two objects that can just be resolved by an optical instrument is called the **limit of resolution** of the instrument. The smaller the limit of resolution, the greater the *resolution,* or **resolving power,** of the instrument. Diffraction sets the ultimate limits on resolution of lenses. *Geometric* optics may make it seem that we can make images as large as we like. Eventually, though, we always reach a point at which the image becomes larger but does not gain in detail. The images in Fig. 36.28 would not become sharper with further enlargement.

CAUTION **Resolving power vs. chromatic resolving power** Be careful not to confuse the resolving power of an optical instrument with the *chromatic* resolving power of a grating (described in Section 36.5). Resolving power refers to the ability to distinguish the images of objects that appear close to each other, when looking either through an optical instrument or at a photograph made with the instrument. Chromatic resolving power describes how well different wavelengths can be distinguished in a spectrum formed by a diffraction grating.

Rayleigh's criterion combined with Eq. (36.17) shows that resolution (resolving power) improves with larger diameter; it also improves with shorter wavelengths. Ultraviolet microscopes have higher resolution than visible-light microscopes. In electron microscopes the resolution is limited by the wavelengths associated with the electrons, which have wavelike aspects (to be discussed further in Chapter 39). These wavelengths can be made 100,000 times smaller than wavelengths of visible light, with a corresponding gain in resolution. Resolving power also explains the difference in storage capacity between compact discs (CDs) and digital video discs (DVDs). Information is stored in both of these in a series of tiny pits. In order not to lose information in the scanning process, the scanning optics must be able to resolve two adjacent pits so that they do not seem to blend into a single pit (see sources 3 and 4 in Fig. 36.28). The red laser used in a DVD player has a shorter wavelength (650 nm) and hence better resolving power than the infrared laser in a CD player (780 nm). Hence pits can be spaced closer together in a DVD than in a CD, and more information can be stored on a disc of the same size (4.7 gigabytes on a DVD versus 700 megabytes, or 0.7 gigabyte, on a CD). The latest disc storage technology uses a blue-violet laser of 405-nm wavelength; this makes it possible to use an even smaller pit spacing and hence store even more data (15 to 25 gigabytes) on a disc of the same size as a CD or DVD.

Diffraction is an important consideration for satellite "dishes," parabolic reflectors designed to receive satellite transmission. Satellite dishes have to be able to pick up transmissions from two satellites that are only a few degrees apart, transmitting at the same frequency; the need to resolve two such transmissions determines the minimum diameter of the dish. As higher frequencies are used, the needed diameter decreases. For example, when two satellites 5.0° apart broadcast 7.5-cm microwaves, the minimum dish diameter to resolve them (by Rayleigh's criterion) is about 1.0 m.

One reason for building very large telescopes is to increase the aperture diameter and thus minimize diffraction effects. The effective diameter of a telescope can be increased by using arrays of smaller telescopes. The Very Large Array (VLA) is a collection of 27 radio telescopes that can be spread out in a Y-shaped arrangement 36 km across (Fig. 36.29a). Hence the effective aperture diameter is 36 km, giving the VLA a limit of resolution of less than 3×10^{-7} rad. This is comparable, in the optical realm, to being able to read the bottom line of an eye chart 7 km away! Such an arrangement is called a *radio interferometer* because it makes use of the phase differences between the signals received in different telescopes. The same principle can also be used to improve the resolution of visible-light telescopes (Fig. 36.29b).

36.29 By simultaneously observing the same object with widely separated telescopes, astronomers can obtain far better resolving power than with a single telescope.

(a) **Radio interferometry.** The Very Large Array 80 km west of Socorro, New Mexico, consists of 27 radio dishes that can be moved on tracks; at their greatest separation, their resolution equals that of a single dish 36 km across.

(b) **Optical interferometry.** The four 8.2-m telescopes of the European Southern Observatory's Very Large Telescope in Cerro Paranal, Chile, can be combined optically in pairs. Functioning together, the outer two telescopes have the resolution of a single telescope 130 m across.

Example 36.6 Resolving power of a camera lens

A camera lens with focal length $f = 50$ mm and maximum aperture $f/2$ forms an image of an object 9.0 m away. (a) If the resolution is limited by diffraction, what is the minimum distance between two points on the object that are barely resolved, and what is the corresponding distance between image points? (b) How does the situation change if the lens is "stopped down" to $f/16$? Assume that $\lambda = 500$ nm in both cases.

SOLUTION

IDENTIFY: This example uses ideas from this section as well as Sections 34.4 (in which we discussed image formation by a lens) and 34.5 (in which the idea of *f*-number was introduced).

SET UP: From Eq. (34.20) the *f*-number of a lens is its focal length f divided by the aperture diameter D. We use the information provided to determine D and then use Eq. (36.17) to find the angular separation θ between two barely resolved points on the object. We then use the geometry of image formation by a lens (see Section 34.4) to determine the distance between those points and the distance between the corresponding image points.

EXECUTE: (a) The aperture diameter is $D = f/(f\text{-number}) = (50\text{ mm})/2 = 25\text{ mm} = 25 \times 10^{-3}\text{ m}$. From Eq. (36.17) the

Continued

angular separation θ of two object points that are barely resolved is given by

$$\theta \approx \sin\theta = 1.22\frac{\lambda}{D} = 1.22\frac{500 \times 10^{-9}\ \text{m}}{25 \times 10^{-3}\ \text{m}}$$
$$= 2.4 \times 10^{-5}\ \text{rad}$$

Let y be the separation of the object points, and let y' be the separation of the corresponding image points. We know from our thin-lens analysis in Section 34.4 that, apart from sign, $y/s = y'/s'$. Thus the angular separations of the object points and the corresponding image points are both equal to θ. Because the object distance s is much greater than the focal length $f = 50$ mm, the image distance s' is approximately equal to f. Thus

$$\frac{y}{9.0\ \text{m}} = 2.4 \times 10^{-5} \qquad y = 2.2 \times 10^{-4}\ \text{m} = 0.22\ \text{mm}$$
$$\frac{y'}{50\ \text{mm}} = 2.4 \times 10^{-5} \qquad y' = 1.2 \times 10^{-3}\ \text{mm}$$
$$= 0.0012\ \text{mm} \approx \frac{1}{800}\ \text{mm}$$

(b) The aperture diameter is now $(50\ \text{mm})/16$, or one-eighth as large as before. The angular separation between barely resolved points is eight times as great, and the values of y and y' are also eight times as great as before:

$$y = 1.8\ \text{mm} \qquad y' = 0.0096\ \text{mm} = \frac{1}{100}\ \text{mm}$$

Only the best camera lenses can approach this resolving power.

EVALUATE: Many photographers use the smallest possible aperture for maximum sharpness, since lens aberrations cause light rays that are far from the optic axis to converge to a different image point than do rays near the axis. Photographers should be aware that, as this example shows, diffraction effects become more significant at small apertures. One cause of fuzzy images has to be balanced against another.

Test Your Understanding of Section 36.7 You have been asked to compare four different proposals for telescopes to be placed in orbit, above the blurring effects of the earth's atmosphere. Rank the proposed telescopes in order of their ability to resolve small details, from best to worst. (i) a radio telescope 100 m in diameter observing at a wavelength of 21 cm; (ii) an optical telescope 2.0 m in diameter observing at a wavelength of 500 nm; (iii) an ultraviolet telescope 1.0 m in diameter observing at a wavelength of 100 nm; (iv) an infrared telescope 2.0 m in diameter observing at a wavelength of 10 μm.

*36.8 Holography

Holography is a technique for recording and reproducing an image of an object through the use of interference effects. Unlike the two-dimensional images recorded by an ordinary photograph or television system, a holographic image is truly three-dimensional. Such an image can be viewed from different directions to reveal different sides and from various distances to reveal changing perspective. If you had never seen a hologram, you wouldn't believe it was possible!

Figure 36.30a shows the basic procedure for making a hologram. We illuminate the object to be holographed with monochromatic light, and we place a photographic film so that it is struck by scattered light from the object and also by direct light from the source. In practice, the light source must be a laser, for reasons we will discuss later. Interference between the direct and scattered light leads to the formation and recording of a complex interference pattern on the film.

To form the images, we simply project light through the developed film, as shown in Fig. 36.30b. Two images are formed: a virtual image on the side of the film nearer the source and a real image on the opposite side.

Holography and Interference Patterns

A complete analysis of holography is beyond our scope, but we can gain some insight into the process by looking at how a single point is holographed and imaged. Consider the interference pattern that is formed on a sheet of photographic negative film by the superposition of an incident plane wave and a spherical wave, as shown in Fig. 36.31a. The spherical wave originates at a point

36.30 (a) A hologram is the record on film of the interference pattern formed with light from the coherent source and light scattered from the object. (b) Images are formed when light is projected through the hologram. The observer sees the virtual image formed behind the hologram.

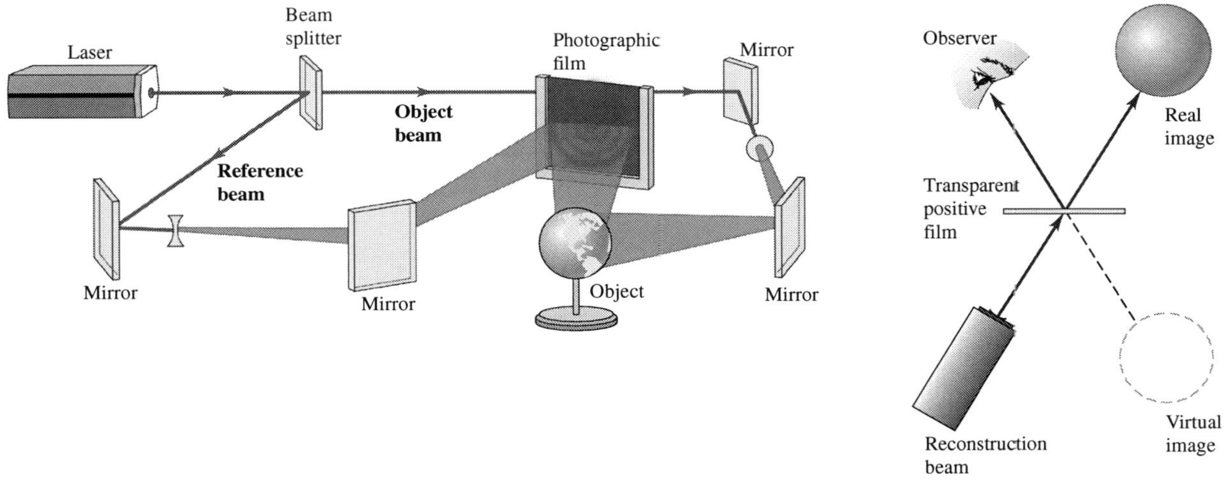

36.31 (a) Constructive interference of the plane and spherical waves occurs in the plane of the film at every point Q for which the distance b_m from P is greater than the distance b_0 from P to O by an integral number of wavelengths $m\lambda$. For the point Q shown, $m = 2$. (b) When a plane wave strikes a transparent positive print of the developed film, the diffracted wave consists of a wave converging to P' and then diverging again and a diverging wave that appears to originate at P. These waves form the real and virtual images, respectively.

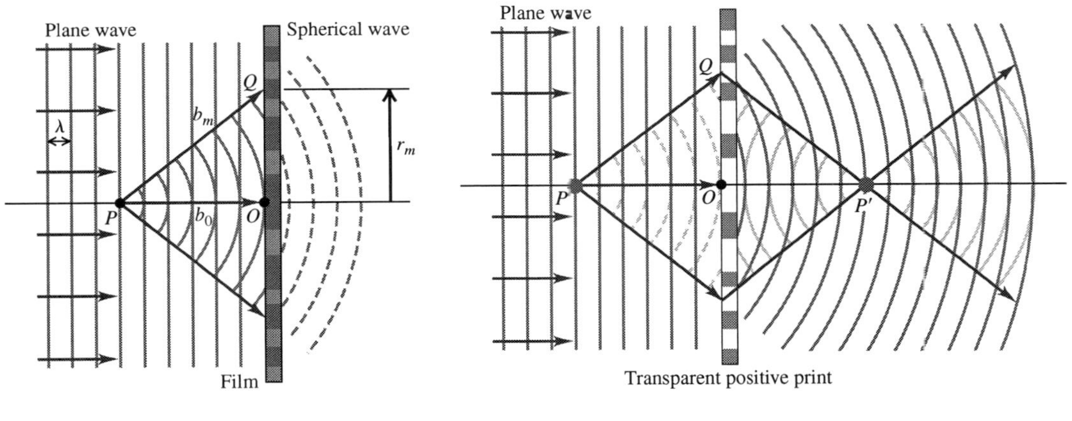

source P at a distance b_0 from the film; P may in fact be a small object that scatters part of the incident plane wave. We assume that the two waves are monochromatic and coherent and that the phase relationship is such that constructive interference occurs at point O on the diagram. Then constructive interference will *also* occur at any point Q on the film that is farther from P than O is by an integer number of wavelengths. That is, if $b_m - b_0 = m\lambda$, where m is an integer, then constructive interference occurs. The points where this condition is satisfied form circles on the film centered at O, with radii r_m given by

$$b_m - b_0 = \sqrt{b_0^2 + r_m^2} - b_0 = m\lambda \qquad (m = 1, 2, 3, \dots) \qquad (36.20)$$

Solving this for r_m^2, we find

$$r_m^2 = \lambda(2mb_0 + m^2\lambda)$$

36.32 Two views of the same hologram seen from different angles.

Ordinarily, b_0 is very much larger than λ, so we neglect the second term in parentheses and obtain

$$r_m = \sqrt{2m\lambda b_0} \qquad (m = 1, 2, 3, \ldots) \tag{36.21}$$

The interference pattern consists of a series of concentric bright circular fringes with radii given by Eq. (36.21). Between these bright fringes are dark fringes.

Now we develop the film and make a transparent positive print, so the bright-fringe areas have the greatest transparency on the film. Then we illuminate it with monochromatic plane-wave light of the same wavelength λ that we used initially. In Fig. 36.31b, consider a point P' at a distance b_0 along the axis from the film. The centers of successive bright fringes differ in their distances from P' by an integer number of wavelengths, and therefore a strong *maximum* in the diffracted wave occurs at P'. That is, light converges to P' and then diverges from it on the opposite side. Therefore P' is a *real image* of point P.

This is not the entire diffracted wave, however. The interference of the wavelets that spread out from all the transparent areas forms a second spherical wave that is diverging rather than converging. When this wave is traced back behind the film in Fig. 36.31b, it appears to be spreading out from point P. Thus the total diffracted wave from the hologram is a superposition of a spherical wave converging to form a real image at P' and a spherical wave that diverges as though it had come from the virtual image point P.

Because of the principle of superposition for waves, what is true for the imaging of a single point is also true for the imaging of any number of points. The film records the superposed interference pattern from the various points, and when light is projected through the film, the various image points are reproduced simultaneously. Thus the images of an extended object can be recorded and reproduced just as for a single point object. Figure 36.32 shows photographs of a holographic image from two different angles, showing the changing perspective in this three-dimensional image.

In making a hologram, we have to overcome two practical problems. First, the light used must be *coherent* over distances that are large in comparison to the dimensions of the object and its distance from the film. Ordinary light sources *do not* satisfy this requirement, for reasons that we discussed in Section 35.1. Therefore laser light is essential for making a hologram. (Ordinary white light can be used for *viewing* certain types of hologram, such as those used on credit cards.) Second, extreme mechanical stability is needed. If any relative motion of source, object, or film occurs during exposure, even by as much as a quarter of a wavelength, the interference pattern on the film is blurred enough to prevent satisfactory image formation. These obstacles are not insurmountable, however, and holography has become important in research, entertainment, and a wide variety of technological applications.

CHAPTER 36 SUMMARY

Fresnel and Fraunhofer diffraction: Diffraction occurs when light passes through an aperture or around an edge. When the source and the observer are so far away from the obstructing surface that the outgoing rays can be considered parallel, it is called Fraunhofer diffraction. When the source or the observer is relatively close to the obstructing surface, it is Fresnel diffraction.

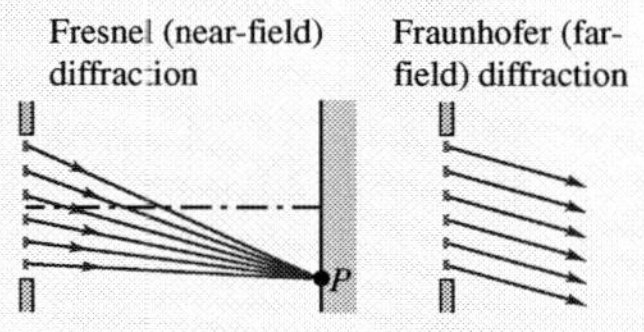

Single-slit diffraction: Monochromatic light sent through a narrow slit of width a produces a diffraction pattern on a distant screen. Equation (36.2) gives the condition for destructive interference (a dark fringe) at a point P in the pattern at angle θ. Equation (36.7) gives the intensity in the pattern as a function of θ. (See Examples 36.1–36.3.)

$$\sin\theta = \frac{m\lambda}{a} \quad (m = \pm 1, \pm 2, \ldots) \tag{36.2}$$

$$I = I_0\left\{\frac{\sin[\pi a(\sin\theta)/\lambda]}{\pi a(\sin\theta)/\lambda}\right\}^2 \tag{36.7}$$

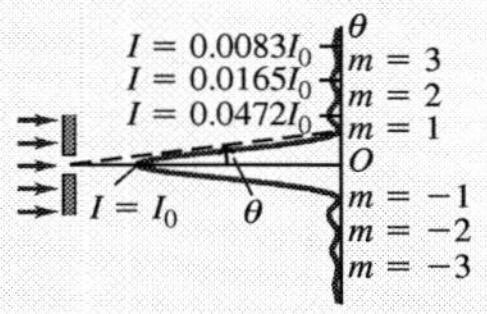

Diffraction gratings: A diffraction grating consists of a large number of thin parallel slits, spaced a distance d apart. The condition for maximum intensity in the interference pattern is the same as for the two-source pattern, but the maxima for the grating are very sharp and narrow. (See Example 36.4.)

$$d\sin\theta = m\lambda$$
$$(m = 0, \pm 1, \pm 2, \pm 3, \ldots) \tag{36.13}$$

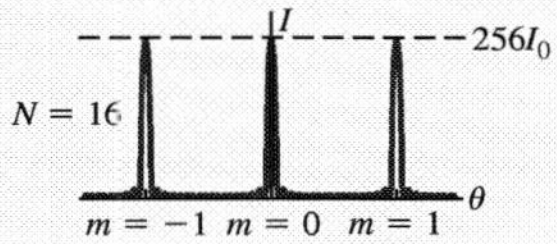

X-ray diffraction: A crystal serves as a three-dimensional diffraction grating for x rays with wavelengths of the same order of magnitude as the spacing between atoms in the crystal. For a set of crystal planes spaced a distance d apart, constructive interference occurs when the angles of incidence and scattering (measured from the crystal planes) are equal and when the Bragg condition [Eq. (36.16)] is satisfied. (See Example 36.5.)

$$2d\sin\theta = m\lambda \quad (m = 1, 2, 3, \ldots) \tag{36.16}$$

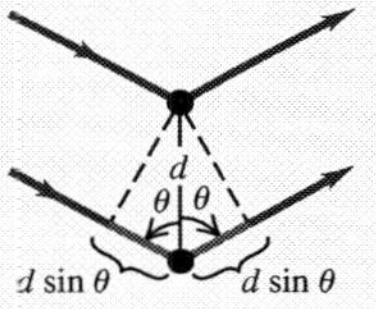

Circular apertures and resolving power: The diffraction pattern from a circular aperture of diameter D consists of a central bright spot, called the Airy disk, and a series of concentric dark and bright rings. Equation (36.17) gives the angular radius θ_1 of the first dark ring, equal to the angular size of the Airy disk. Diffraction sets the ultimate limit on resolution (image sharpness) of optical instruments. According to Rayleigh's criterion, two point objects are just barely resolved when their angular separation θ is given by Eq. (36.17). (See Example 36.6.)

$$\sin\theta_1 = 1.22\frac{\lambda}{D} \tag{36.17}$$

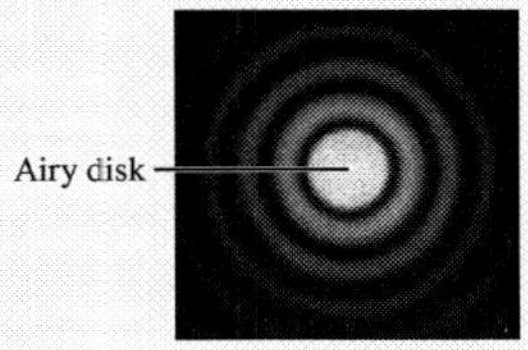

Key Terms

Answer to Chapter Opening Question ?

The shorter wavelength of a DVD scanning laser gives it superior resolving power, so information can be more tightly packed onto a DVD than a CD. See Section 36.7 for details.

Answers to Test Your Understanding Questions

36.1 Answer: yes When you hear the voice of someone standing around a corner, you are hearing sound waves that underwent diffraction. If there were no diffraction of sound, you could hear sounds only from objects that were in plain view.

36.2 Answers: (ii), (i) and (iv) (tie), (iii) The angle θ of the first dark fringe is given by Eq. (36.2) with $m = 1$, or $\sin\theta = \lambda/a$. The larger the value of the ratio λ/a, the larger the value of $\sin\theta$ and hence the value of θ. The ratio λ/a in each case is (i) $(400\text{ nm})/(0.20\text{ mm}) = (4.0 \times 10^{-7}\text{ m})/(2.0 \times 10^{-4}\text{ m}) = 2.0 \times 10^{-3}$; (ii) $(600\text{ nm})/(0.20\text{ mm}) = (6.0 \times 10^{-7}\text{ m})/(2.0 \times 10^{-4}\text{ m}) = 3.0 \times 10^{-3}$; (iii) $(400\text{ nm})/(0.30\text{ mm}) = (4.0 \times 10^{-7}\text{ m})/(3.0 \times 10^{-4}\text{ m}) = 1.3 \times 10^{-3}$; (iv) $(600\text{ nm})/(0.30\text{ mm}) = (6.0 \times 10^{-7}\text{ m})/(3.0 \times 10^{-4}\text{ m}) = 2.0 \times 10^{-3}$.

36.3 Answers: (ii) and (iii) If the slit width a is less than the wavelength λ, there are no points in the diffraction pattern at which the intensity is zero (see Fig. 36.10a). The slit width is $0.0100\text{ mm} = 1.00 \times 10^{-5}\text{ m}$, so this condition is satisfied for (ii) $(\lambda = 10.6\ \mu\text{m} = 10.6 \times 10^{-5}\text{ m})$ and (iii) $(\lambda = 1.00\text{ mm} = 1.00 \times 10^{-3}\text{ m})$ but not for (i) $(\lambda = 500\text{ nm} = 500 \times 10^{-7})$ or (iv) $(\lambda = 50.0\text{ nm} = 5.00 \times 10^{-8}\text{ m})$.

36.4 Answers: yes; $m_i = \pm 5, \pm 10, \ldots$ A "missing maximum" satisfies both $d\sin\theta = m_i\lambda$ (the condition for an interference maximum) and $a\sin\theta = m_d\lambda$ (the condition for a diffraction minimum). Substituting $d = 2.5a$, we can combine these two conditions into the relationship $m_i = 2.5m_d$. This is satisfied for $m_i = \pm 5$ and $m_d = \pm 2$ (the fifth interference maximum is missing because it coincides with the second diffraction minimum), $m_i = \pm 10$ and $m_d = \pm 4$ (the tenth interference maximum is missing because it coincides with the fourth diffraction minimum), and so on.

36.5 Answer: (i) As described in the text, the resolving power needed is $R = Nm = 1000$. In the first order $(m = 1)$ we need $N = 1000$ slits, but in the fourth order $(m = 4)$ we need only $N = R/m = 1000/4 = 250$ slits. (These numbers are only approximate because of the arbitrary nature of our criterion for resolution and because real gratings always have slight imperfections in the shapes and spacings of the slits.)

36.6 Answer: no The angular position of the mth maximum is given by Eq. (36.16), $2d\sin\theta = m\lambda$. With $d = 0.200$ nm, $\lambda = 0.100$ nm, and $m = 5$, this gives $\sin\theta = m\lambda/2d = (5)(0.100\text{ nm})/(2)(0.200\text{ nm}) = 1.25$. Since the sine function can never be greater than 1, this means that there is no solution to this equation and the $m = 5$ maximum does not appear.

36.7 Answer: (iii), (ii), (iv), (i) Rayleigh's criterion combined with Eq. (36.17) shows that the smaller the value of the ratio λ/D, the better the resolving power of a telescope of diameter D. For the four telescopes, this ratio is equal to (i) $(21\text{ cm})/(100\text{ m}) = (0.21\text{ m})/(100\text{ m}) = 2.1 \times 10^{-3}$; (ii) $(500\text{ nm})/(2.0\text{ m}) = (5.0 \times 10^{-7}\text{ m})/(2.0\text{ m}) = 2.5 \times 10^{-7}$; (iii) $(100\text{ nm})/(1.0\text{ m}) = (1.0 \times 10^{-7}\text{ m})/(1.0\text{ m}) = 1.0 \times 10^{-7}$; (iv) $(10\ \mu\text{m})/(2.0\text{ m}) = (1.0 \times 10^{-5}\text{ m})/(2.0\text{ m}) = 5.0 \times 10^{-6}$.

PROBLEMS

For instructor-assigned homework, go to **www.masteringphysics.com**

Discussion Questions

Q36.1. Why can we readily observe diffraction effects for sound waves and water waves, but not for light? Is this because light travels so much faster than these other waves? Explain.

Q36.2. What is the difference between Fresnel and Fraunhofer diffraction? Are they different *physical* processes? Explain.

Q36.3. You use a lens of diameter D and light of wavelength λ and frequency f to form an image of two closely spaced and distant objects. Which of the following will increase the resolving power? (a) Use a lens with a smaller diameter; (b) use light of higher frequency; (c) use light of longer wavelength. In each case justify your answer.

Q36.4. Light of wavelength λ and frequency f passes through a single slit of width a. The diffraction pattern is observed on a screen a distance x from the slit. Which of the following will *decrease* the width of the central maximum? (a) Decrease the slit width; (b) decrease the frequency f of the light; (c) decrease the wavelength λ of the light; (d) decrease the distance x of the screen from the slit. In each case justify your answer.

Q36.5. In a diffraction experiment with waves of wavelength λ, there will be *no* intensity minima (that is, no dark fringes) if the slit width is small enough. What is the maximum slit width for which this occurs? Explain your answer.

Q36.6. The predominant sound waves used in human speech have wavelengths in the range from 1.0 to 3.0 meters. Using the ideas of diffraction, explain how it is possible to hear a person's voice even when he is facing away from you.

Q36.7. In single-slit diffraction, what is $\sin(\beta/2)$ when $\theta = 0$? In view of your answer, why is the single-slit intensity *not* equal to zero at the center?

Q36.8. A rainbow ordinarily shows a range of colors (see Section 33.4). But if the water droplets that form the rainbow are small enough, the rainbow will appear white. Explain why, using diffraction ideas. How small do you think the raindrops would have to be for this to occur?

Q36.9. Some loudspeaker horns for outdoor concerts (at which the entire audience is seated on the ground) are wider vertically than horizontally. Use diffraction ideas to explain why this is more efficient at spreading the sound uniformly over the audience than either a square speaker horn or a horn that is wider horizontally than vertically. Would this still be the case if the audience were seated at different elevations, as in an amphitheater? Why or why not?

Q36.10. Figure 31.12 (Section 31.2) shows a loudspeaker system. Low-frequency sounds are produced by the *woofer*, which is a speaker with large diameter; the *tweeter*, a speaker with smaller diameter, produces high-frequency sounds. Use diffraction ideas to explain why the tweeter is more effective for distributing high-frequency sounds uniformly over a room than is the woofer.

Q36.11. Information is stored on an audio compact disc, CD-ROM, or DVD disc in a series of pits on the disc. These pits are scanned by a laser beam. An important limitation on the amount of information that can be stored on such a disc is the width of the laser beam. Explain why this should be, and explain how using a shorter-wavelength laser allows more information to be stored on a disc of the same size.

Q36.12. With which color of light can the Hubble Space Telescope see finer detail in a distant astronomical object: red, blue, or ultraviolet? Explain your answer.

Q36.13. A typical telescope used by amateur astronomers has a mirror 20 cm in diameter. With such a telescope (and a filter to cut the intensity of sunlight to a safe level for viewing), fine details can be seen on the surface of the sun. Explain why a *radio* telescope would have to be *much* larger to "see" comparable details on the sun.

Q36.14. Could x-ray diffraction effects with crystals be observed by using visible light instead of x rays? Why or why not?

Q36.15. Why is a diffraction grating better than a two-slit setup for measuring wavelengths of light?

Q36.16. One sometimes sees rows of evenly spaced radio antenna towers. A student remarked that these act like diffraction gratings. What did she mean? Why would one want them to act like a diffraction grating?

Q36.17. If a hologram is made using 600-nm light and then viewed with 500-nm light, how will the images look compared to those observed when viewed with 600-nm light? Explain.

Q36.18. A hologram is made using 600-nm light and then viewed by using white light from an incandescent bulb. What will be seen? Explain.

Q36.19. Ordinary photographic film reverses black and white, in the sense that the most brightly illuminated areas become blackest upon development (hence the term *negative*). Suppose a hologram negative is viewed directly, without making a positive transparency. How will the resulting images differ from those obtained with the positive? Explain.

Exercises

Section 36.2 Diffraction from a Single Slit

36.1. Monochromatic light from a distant source is incident on a slit 0.750 mm wide. On a screen 2.00 m away, the distance from the central maximum of the diffraction pattern to the first minimum is measured to be 1.35 mm. Calculate the wavelength of the light.

36.2. Parallel rays of green mercury light with a wavelength of 546 nm pass through a slit covering a lens with a focal length of 60.0 cm. In the focal plane of the lens the distance from the central maximum to the first minimum is 10.2 mm. What is the width of the slit?

36.3. Light of wavelength 585 nm falls on a slit 0.0666 mm wide. (a) On a very large distant screen, how many *totally* dark fringes (indicating complete cancellation) will there be, including both sides of the central bright spot? Solve this problem *without* calculating all the angles! (*Hint:* What is the largest that $\sin\theta$ can be? What does this tell you is the largest that m can be?) (b) At what angle will the dark fringe that is most distant from the central bright fringe occur?

36.4. Light of wavelength 633 nm from a distant source is incident on a slit 0.750 mm wide, and the resulting diffraction pattern is observed on a screen 3.50 m away. What is the distance between the two dark fringes on either side of the central bright fringe?

36.5. Diffraction occurs for all types of waves, including sound waves. High-frequency sound from a distant source with wavelength 9.00 cm passes through a narrow slit 12.0 cm wide. A microphone is placed 40.0 cm directly in front of the center of the slit, corresponding to point O in Fig. 36.5a. The microphone is then moved in a direction perpendicular to the line from the center of the slit to point O. At what distances from O will the intensity detected by the microphone be zero?

36.6. Tsunami! On December 26, 2004, a violent magnitude-9.1 earthquake occurred off the coast of Sumatra. This quake triggered a huge tsunami (similar to a tidal wave) that killed more than 150,000 people. Scientists observing the wave on the open ocean measured the time between crests to be 1.0 h and the speed of the wave to be 800 km/h. Computer models of the evolution of this enormous wave showed that it bent around the continents and spread to all the oceans of the earth. When the wave reached the gaps between continents, it diffracted between them as through a slit. (a) What was the wavelength of this tsunami? (b) The distance between the southern tip of Africa and northern Antarctica is about 4500 km, while the distance between the southern end of Australia and Antarctica is about 3700 km. As an approximation, we can model this wave's behavior by using Fraunhofer diffraction. Find the smallest angle away from the central maximum for which the waves would cancel after going through each of these continental gaps.

36.7. A series of parallel linear water wave fronts are traveling directly toward the shore at 15.0 cm/s on an otherwise placid lake. A long concrete barrier that runs parallel to the shore at a distance of 3.20 m away has a hole in it. You count the wave crests and observe that 75.0 of them pass by each minute, and you also observe that no waves reach the shore at ± 61.3 cm from the point directly opposite the hole, but waves do reach the shore everywhere within this distance. (a) How wide is the hole in the barrier? (b) At what other angles do you find no waves hitting the shore?

36.8. Monochromatic light of wavelength 580 nm passes through a single slit and the diffraction pattern is observed on a screen. Both the source and screen are far enough from the slit for Fraunhofer diffraction to apply. (a) If the first diffraction minima are at $\pm 90.0°$, so the central maximum completely fills the screen, what is the width of the slit? (b) For the width of the slit as calculated in part (a), what is the ratio of the intensity at $\theta = 45.0°$ to the intensity at $\theta = 0$?

36.9. Doorway Diffraction. Sound of frequency 1250 Hz leaves a room through a 1.00-m-wide doorway (see Exercise 36.5). At which angles relative to the centerline perpendicular to the doorway will someone outside the room hear no sound? Use 344 m/s for the speed of sound in air and assume that the source and listener are both far enough from the doorway for Fraunhofer diffraction to apply. You can ignore effects of reflections.

36.10. Light waves, for which the electric field is given by $E_y(x, t) = E_{max}\sin[(1.20 \times 10^7\ \text{m}^{-1})x - \omega t]$, pass through a slit and produce the first dark bands at $\pm 28.6°$ from the center of the diffraction pattern. (a) What is the frequency of this light? (b) How wide is the slit? (c) At which angles will other dark bands occur?

36.11. Parallel rays of light with wavelength 620 nm pass through a slit covering a lens with a focal length of 40.0 cm. The diffraction pattern is observed in the focal plane of the lens, and the distance from the center of the central maximum to the first minimum is 36.5 cm. What is the width of the slit? (*Note:* The angle that locates the first minimum is *not* small.)

36.12. Monochromatic electromagnetic radiation with wavelength λ from a distant source passes through a slit. The diffraction pattern is observed on a screen 2.50 m from the slit. If the width of the central maximum is 6.00 mm, what is the slit width a if the wavelength is (a) 500 nm (visible light); (b) 50.0 μm (infrared radiation); (c) 0.500 nm (x rays)?

36.13. Red light of wavelength 633 nm from a helium–neon laser passes through a slit 0.350 mm wide. The diffraction pattern is observed on a screen 3.00 m away. Define the width of a bright fringe as the distance between the minima on either side. (a) What is the width of the central bright fringe? (b) What is the width of the first bright fringe on either side of the central one?

Section 36.3 Intensity in the Single-Slit Pattern

36.14. Monochromatic light of wavelength $\lambda = 620$ nm from a distant source passes through a slit 0.450 mm wide. The diffraction pattern is observed on a screen 3.00 m from the slit. In terms of the intensity I_0 at the peak of the central maximum, what is the intensity of the light at the screen the following distances from the center of the central maximum: (a) 1.00 mm; (b) 3.00 mm; (c) 5.00 mm?

36.15. A slit 0.240 mm wide is illuminated by parallel light rays of wavelength 540 nm. The diffraction pattern is observed on a screen that is 3.00 m from the slit. The intensity at the center of the central maximum ($\theta = 0°$) is $6.00 \times 10^{-6}\ \text{W/m}^2$. (a) What is the distance on the screen from the center of the central maximum to the first minimum? (b) What is the intensity at a point on the screen midway between the center of the central maximum and the first minimum?

36.16. Laser light of wavelength 632.8 nm falls normally on a slit that is 0.0250 mm wide. The transmitted light is viewed on a distant screen where the intensity at the center of the central bright fringe is 8.50 W/m^2. (a) Find the maximum number of totally dark fringes on the screen, assuming the screen is large enough to show them all. (b) At what angle does the dark fringe that is most distant from the center occur? (c) What is the maximum intensity of the bright fringe that occurs immediately before the dark fringe in part (b)? Approximate the angle at which this fringe occurs by assuming it is midway between the angles to the dark fringes on either side of it.

36.17. A single-slit diffraction pattern is formed by monochromatic electromagnetic radiation from a distant source passing through a slit 0.105 mm wide. At the point in the pattern 3.25° from the center of the central maximum, the total phase difference between wavelets from the top and bottom of the slit is 56.0 rad. (a) What is the wavelength of the radiation? (b) What is the intensity at this point, if the intensity at the center of the central maximum is I_0?

36.18. Consider a single-slit diffraction experiment in which the amplitude of the wave at point O in Fig. 36.5a is E_0. For each of the following cases, draw a phasor diagram like that in Fig. 36.8c and determine *graphically* the amplitude of the wave at the point in question. (*Hint:* Use Eq. (36.6) to determine the value of β for each case.) Compute the intensity and compare to Eq. (36.5). (a) $\sin\theta = \lambda/2a$; (b) $\sin\theta = \lambda/a$; (c) $\sin\theta = 3\lambda/2a$.

36.19. Public Radio station KXPR-FM in Sacramento broadcasts at 88.9 MHz. The radio waves pass between two tall skyscrapers that are 15.0 m apart along their closest walls. (a) At what horizontal angles, relative to the original direction of the waves, will a distant antenna not receive any signal from this station? (b) If the maximum intensity is 3.50 W/m^2 at the antenna, what is the intensity at $\pm 5.00°$ from the center of the central maximum at the distant antenna?

Section 36.4 Multiple Slits

36.20. Diffraction and Interference Combined. Consider the interference pattern produced by two parallel slits of width a and separation d, in which $d = 3a$. The slits are illuminated by normally incident light of wavelength λ. (a) First we ignore diffraction effects due to the slit width. At what angles θ from the central maximum will the next four maxima in the two-slit interference pattern occur? Your answer will be in terms of d and λ. (b) Now we include the effects of diffraction. If the intensity at $\theta = 0$ is I_0, what is the intensity at each of the angles in part (a)? (c) Which double-slit interference maxima are missing in the pattern? (d) Compare your results to those illustrated in Fig. 36.12c. In what ways is your result different?

36.21. Number of Fringes in a Diffraction Maximum. In Fig. 36.12c the central diffraction maximum contains exactly seven interference fringes, and in this case $d/a = 4$. (a) What must the ratio d/a be if the central maximum contains exactly five fringes? (b) In the case considered in part (a), how many fringes are contained within the first diffraction maximum on one side of the central maximum?

36.22. An interference pattern is produced by eight parallel and equally spaced, narrow slits. There is an interference minimum when the phase difference ϕ between light from adjacent slits is $\pi/4$. The phasor diagram is given in Fig. 36.14b. For which pairs of slits is there totally destructive interference?

36.23. An interference pattern is produced by light of wavelength 580 nm from a distant source incident on two identical parallel slits separated by a distance (between centers) of 0.530 mm. (a) If the slits are very narrow, what would be the angular positions of the first-order and second-order, two-slit, interference maxima? (b) Let the slits have width 0.320 mm. In terms of the intensity I_0 at the center of the central maximum, what is the intensity at each of the angular positions in part (a)?

36.24. Monochromatic light illuminates a pair of thin parallel slits at normal incidence, producing an interference pattern on a distant screen. The width of each slit is $\frac{1}{7}$ the center-to-center distance between the slits. (a) Which interference maxima are missing in the pattern on the screen? (b) Does the answer to part (a) depend on the wavelength of the light used? Does the location of the missing maxima depend on the wavelength?

36.25. An interference pattern is produced by four parallel and equally spaced, narrow slits. By drawing appropriate phasor diagrams, show that there is an interference minimum when the phase difference ϕ from adjacent slits is (a) $\pi/2$; (b) π; (c) $3\pi/2$. In each case, for which pairs of slits is there totally destructive interference?

36.26. A diffraction experiment involving two thin parallel slits yields the pattern of closely spaced bright and dark fringes shown in Fig. 36.33. Only the central portion of the pattern is shown in the figure. The bright spots are equally spaced at 1.53 mm center to center (except for the missing spots) on a screen 2.50 m from the slits. The light source was a He-Ne laser producing a wavelength of 632.8 nm. (a) How far apart are the two slits? (b) How wide is each one?

Figure **36.33** Exercise 36.26

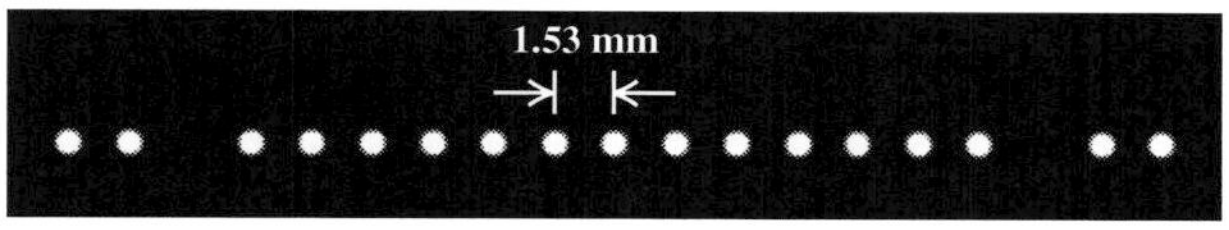

36.27. Laser light of wavelength 500.0 nm illuminates two identical slits, producing an interference pattern on a screen 90.0 cm from the slits. The bright bands are 1.00 cm apart, and the third bright bands on either side of the central maximum are missing in the pattern. Find the width and the separation of the two slits.

Section 36.5 The Diffraction Grating

36.28. Monochromatic light is at normal incidence on a plane transmission grating. The first-order maximum in the interference pattern is at an angle of 8.94°. What is the angular position of the fourth-order maximum?

36.29. If a diffraction grating produces its third-order bright band at an angle of 78.4° for light of wavelength 681 nm, find (a) the number of slits per centimeter for the grating and (b) the angular location of the first-order and second-order bright bands. (c) Will there be a fourth-order bright band? Explain.

36.30. If a diffraction grating produces a third-order bright spot for red light (of wavelength 700 nm) at 65.0° from the central maximum, at what angle will the second-order bright spot be for violet light (of wavelength 400 nm)?

36.31. Visible light passes through a diffraction grating that has 900 slits/cm, and the interference pattern is observed on a screen that is 2.50 m from the grating. (a) Is the angular position of the first-order spectrum small enough for $\sin\theta \approx \theta$ to be a good approximation? (b) In the first-order spectrum, the maxima for two different wavelengths are separated on the screen by 3.00 mm. What is the difference in these wavelengths?

36.32. The wavelength range of the visible spectrum is approximately 400–700 nm. White light falls at normal incidence on a diffraction grating that has 350 slits/mm. Find the angular width of the visible spectrum in (a) the first order and (b) the third order. (*Note:* An advantage of working in higher orders is the greater angular spread and better resolution. A disadvantage is the overlapping of different orders, as shown in Example 36.4.)

36.33. Measuring Wavelengths with a CD. A laser beam of wavelength $\lambda = 632.8$ nm shines at normal incidence on the reflective side of a compact disc. The tracks of tiny pits in which information is coded onto the CD are 1.60 μm apart. For what angles of reflection (measured from the normal) will the intensity of light be maximum?

36.34. (a) What is the wavelength of light that is deviated in the first order through an angle of 13.5° by a transmission grating having 5000 slits/cm? (b) What is the second-order deviation of this wavelength? Assume normal incidence.

36.35. Plane monochromatic waves with wavelength 520 nm are incident normally on a plane transmission grating having 350 slits/mm. Find the angles of deviation in the first, second, and third orders.

36.36. Identifying Isotopes by Spectra. Different isotopes of the same element emit light at slightly different wavelengths. A wavelength in the emission spectrum of a hydrogen atom is 656.45 nm; for deuterium, the corresponding wavelength is 656.27 nm. (a) What minimum number of slits is required to resolve these two wavelengths in second order? (b) If the grating has 500.00 slits/mm, find the angles and angular separation of these two wavelengths in the second order.

36.37. A typical laboratory diffraction grating has 5.00×10^3 lines/cm, and these lines are contained in a 3.50-cm width of grating. (a) What is the chromatic resolving power of such a grating in the first order? (b) Could this grating resolve the lines of the sodium doublet (see Section 36.5) in the first order? (c) While doing spectral analysis of a star, you are using this grating in the *second* order to resolve spectral lines that are very close to the 587.8002-nm spectral line of iron. (i) For wavelengths longer than the iron line, what is the shortest wavelength you could distinguish from the iron line? (ii) For wavelengths shorter than the iron line, what is the longest wavelength you could distinguish from the iron line? (iii) What is the range of wavelengths you could *not* distinguish from the iron line?

36.38. The light from an iron arc includes many different wavelengths. Two of these are at $\lambda = 587.9782$ nm and $\lambda = 587.8002$ nm. You wish to resolve these spectral lines in first order using a grating 1.20 cm in length. What minimum number of slits per centimeter must the grating have?

Section 36.6 X-Ray Diffraction

36.39. X rays of wavelength 0.0850 nm are scattered from the atoms of a crystal. The second-order maximum in the Bragg reflection occurs when the angle θ in Fig. 36.23 is 21.5°. What is the spacing between adjacent atomic planes in the crystal?

36.40. If the planes of a crystal are 3.50 Å (1 Å $= 10^{-10}$ m $=$ 1 Ångstrom unit) apart, (a) what wavelength of electromagnetic waves is needed so that the first strong interference maximum in the Bragg reflection occurs when the waves strike the planes at an angle of 15.0°, and in what part of the electromagnetic spectrum do these waves lie? (See Fig. 32.4.) (b) At what other angles will strong interference maxima occur?

Section 36.7 Circular Apertures and Resolving Power

36.41. Due to blurring caused by atmospheric distortion, the best resolution that can be obtained by a normal, earth-based, visible-light telescope is about 0.3 arcsecond (there are 60 arcminutes in a degree and 60 arcseconds in an arcminute). (a) Using Rayleigh's criterion, calculate the diameter of an earth-based telescope that gives this resolution with 550-nm light. (b) Increasing the telescope diameter beyond the value found in part (a) will increase the light-gathering power of the telescope, allowing more distant and dimmer astronomical objects to be studied, but it will not improve the resolution. In what ways are the Keck telescopes (each of 10-m diameter) atop Mauna Kea in Hawaii superior to the Hale Telescope (5-m diameter) on Palomar Mountain in California? In what ways are they *not* superior? Explain.

36.42. If you can read the bottom row of your doctor's eye chart, your eye has a resolving power of 1 arcminute, equal to $\frac{1}{60}$ degree. If this resolving power is diffraction limited, to what effective diameter of your eye's optical system does this correspond? Use Rayleigh's criterion and assume $\lambda = 550$ nm.

36.43. Two satellites at an altitude of 1200 km are separated by 28 km. If they broadcast 3.6-cm microwaves, what minimum receiving-dish diameter is needed to resolve (by Rayleigh's criterion) the two transmissions?

36.44. The Very Long Baseline Array can resolve (by Rayleigh's criterion) signals from sources separated by 1.0×10^{-8} rad. If the effective diameter of the receiver is 8000 km, what is the wavelength of these signals?

36.45. Monochromatic light with wavelength 620 nm passes through a circular aperture with diameter 7.4 μm. The resulting diffraction pattern is observed on a screen that is 4.5 m from the aperture. What is the diameter of the Airy disk on the screen?

36.46. Photography. A wildlife photographer uses a moderate telephoto lens of focal length 135 mm and maximum aperture $f/4.00$ to photograph a bear that is 11.5 m away. Assume the wavelength is 550 nm. (a) What is the width of the smallest feature on the bear that this lens can resolve if it is opened to its maximum aperture? (b) If, to gain depth of field, the photographer stops the lens down to $f/22.0$, what would be the width of the smallest resolvable feature on the bear?

36.47. Observing Jupiter. You are asked to design a space telescope for earth orbit. When Jupiter is 5.93×10^8 km away (its closest approach to the earth), the telescope is to resolve, by Rayleigh's criterion, features on Jupiter that are 250 km apart. What minimum-diameter mirror is required? Assume a wavelength of 500 nm.

36.48. A converging lens 7.20 cm in diameter has a focal length of 300 mm. If the resolution is diffraction limited, how far away can an object be if points on it 4.00 mm apart are to be resolved (according to Rayleigh's criterion)? Use $\lambda = 550$ nm.

36.49. Hubble Versus Arecibo. The Hubble Space Telescope has an aperture of 2.4 m and focuses visible light (400–700 nm). The Arecibo radio telescope in Puerto Rico is 305 m (1000 ft) in diameter (it is built in a mountain valley) and focuses radio waves of wavelength 75 cm. (a) Under optimal viewing conditions, what is the smallest crater that each of these telescopes could resolve on our moon? (b) If the Hubble Space Telescope were to be converted to surveillance use, what is the highest orbit above the surface of the earth it could have and still be able to resolve the license plate (not the letters, just the plate) of a car on the ground? Assume optimal viewing conditions, so that the resolution is diffraction limited.

36.50. Searching for Starspots. The Hale Telescope on Palomar Mountain in California has a mirror 200 in. (5.08 m) in diameter and it focuses visible light. Given that a large sunspot is about 10,000 mi in diameter, what is the most distant star on which this telescope could resolve a sunspot to see whether other stars have them? (Assume optimal viewing conditions, so that the resolution is diffraction limited.) Are there any stars this close to us, besides our sun?

36.51. Searching for Planets. The Keck Telescopes, on Mauna Kea, Hawaii have a 10.0-m-diameter mirror. Could these telescopes resolve Jupiter-sized planets about our nearest star, Alpha Centauri, which is 4.28 light-years away?

Problems

36.52. Suppose the entire apparatus (slit, screen, and space in between) in Exercise 36.4 is immersed in water ($n = 1.33$). Then what is the distance between the two dark fringes?

36.53. Consider a single-slit diffraction pattern. The center of the central maximum, where the intensity is I_0, is located at $\theta = 0$. (a) Let θ_+ and θ_- be the two angles on either side of $\theta = 0$ for which $I = \frac{1}{2}I_0$. $\Delta\theta = |\theta_+ - \theta_-|$ is called the *full width at half maximum,* or *FWHM,* of the central diffraction maximum. Solve for $\Delta\theta$ when the ratio between slit width a and wavelength λ is (i) $a/\lambda = 2$; (ii) $a/\lambda = 5$; (iii) $a/\lambda = 10$. (*Hint:* Your equation for θ_+ or θ_- cannot be solved analytically. You must use trial and error or solve it graphically.) (b) The width of the central maximum can alternatively be defined as $2\theta_0$, where θ_0 is the angle that locates the minimum on one side of the central maximum. Calculate $2\theta_0$ for each case considered in part (a), and compare to $\Delta\theta$.

36.54. A loudspeaker having a diaphragm that vibrates at 1250 Hz is traveling at 80.0 m/s directly toward a pair of holes in a very large wall in a region for which the speed of sound is 344 m/s. You observe that the sound coming through the openings first cancels at $\pm 12.7°$ with respect to the original direction of the speaker when observed far from the wall. (a) How far apart are the two openings? (b) At what angles would the sound first cancel if the source stopped moving?

36.55. Measuring Refractive Index. A thin slit illuminated by light of frequency f produces its first dark band at $\pm 38.2°$ in air. When the entire apparatus (slit, screen, and space in between) is immersed in an unknown transparent liquid, the slit's first dark bands occur instead at $\pm 17.4°$. Find the refractive index of the liquid.

36.56. Grating Design. Your boss asks you to design a diffraction grating that will disperse the first-order visible spectrum through an angular range of 15.0° (see Example 36.4 in Section 36.5). (a) What must the number of slits per centimeter be for this grating? (b) At what angles will the first-order visible spectrum begin and end?

36.57. A slit 0.360 mm wide is illuminated by parallel rays of light that have a wavelength of 540 nm. The diffraction pattern is observed on a screen that is 1.20 m from the slit. The intensity at the center of the central maximum ($\theta = 0°$) is I_0. (a) What is the distance on the screen from the center of the central maximum to the first minimum? (b) What is the distance on the screen from the center of the central maximum to the point where the intensity has fallen to $I_0/2$? (See Problem 36.53, part (a), for a hint about how to solve for the phase angle β.)

36.58. The intensity of light in the Fraunhofer diffraction pattern of a single slit is

$$I = I_0\left(\frac{\sin\gamma}{\gamma}\right)^2$$

where

$$\gamma = \frac{\pi a \sin\theta}{\lambda}$$

(a) Show that the equation for the values of γ at which I is a maximum is $\tan\gamma = \gamma$. (b) Determine the three smallest positive values of γ that are solutions of this equation. (*Hint:* You can use a trial-and-error procedure. Guess a value of γ and adjust your guess to bring $\tan\gamma$ closer to γ. A graphical solution of the equation is very helpful in locating the solutions approximately, to get good initial guesses.)

36.59. Angular Width of a Principal Maximum. Consider N evenly spaced, narrow slits. Use the small-angle approximation $\sin\theta = \theta$ (for θ in radians) to prove the following: For an intensity maximum that occurs at an angle θ, the intensity minima immediately adjacent to this maximum are at angles $\theta + \lambda/Nd$ and $\theta - \lambda/Nd$, so that the angular width of the principal maximum is $2\lambda/Nd$. This is proportional to $1/N$, as we concluded in Section 36.4 on the basis of energy conservation.

36.60. The Expanding Universe. A cosmologist who is studying the light from a galaxy has identified the spectrum of hydrogen but finds that the wavelengths are somewhat shifted from those found in the laboratory. In the lab, the H_α line has a wavelength of 656.3 nm. The cosmologist is using a transmission diffraction grating having 5758 lines/cm in the first order and finds that the first bright fringe for the H_α line occurs at $\pm 23.41°$ from the central spot. How fast is the galaxy moving? Express your answer in m/s and as a percentage of the speed of light. Is it moving toward us or away from us? (*Hint:* See Section 16.8.)

36.61. Phasor Diagram for Eight Slits. An interference pattern is produced by eight equally spaced, narrow slits. Figure 36.14 shows phasor diagrams for the cases in which the phase difference ϕ between light from adjacent slits is $\phi = \pi$, $\phi = \pi/4$, and $\phi = \pi/2$. Each of these cases gives an intensity minimum. The caption for Fig. 36.14 also claims that minima occur for $\phi = 3\pi/4$, $\phi = 5\pi/4$, $\phi = 3\pi/2$, and $\phi = 7\pi/4$. (a) Draw the phasor diagram for each of these four cases, and explain why each diagram proves that there is in fact a minimum. (*Note:* You may find it helpful to use a different colored pencil for each slit!) (b) For each of the four cases $\phi = 3\pi/4$, $\phi = 5\pi/4$, $\phi = 3\pi/2$, and $\phi = 7\pi/4$, for which pairs of slits is there totally destructive interference?

36.62. X-Ray Diffraction of Salt. X rays with a wavelength of 0.125 nm are scattered from a cubic array (of a sodium chloride crystal), for which the spacing of adjacent atoms is $a = 0.282$ nm. (a) If diffraction from planes parallel to a cube face is considered, at what angles θ of the incoming beam relative to the crystal planes will maxima be observed? (b) Repeat part (a) for diffraction produced by the planes shown in Fig. 36.24a, which are separated by $a/\sqrt{2}$.

36.63. At the end of Section 36.4, the following statements were made about an array of N slits. Explain, using phasor diagrams, why each statement is true. (a) A minimum occurs whenever ϕ is an integral multiple of $2\pi/N$, except when ϕ is an integral multiple of 2π (which gives a principal maximum). (b) There are $(N - 1)$ minima between each pair of principal maxima.

36.64. In Eq. (36.12), consider the case in which $d = a$. In a sketch, show that in this case the two slits reduce to a single slit with width $2a$. Then show that Eq. (36.12) reduces to Eq. (36.5) with slit width $2a$.

36.65. What is the longest wavelength that can be observed in the third order for a transmission grating having 6500 slits/cm? Assume normal incidence.

36.66. (a) Figure 36.16 shows plane waves of light incident *normally* on a diffraction grating. If instead the light strikes the grating at an angle of incidence θ' (measured from the normal), show that the condition for an intensity maximum is *not* Eq. (36.13), but rather

$$d(\sin\theta + \sin\theta') = m\lambda \qquad (m = 0, \pm 1, \pm 2, \pm 3, \dots)$$

(b) For the grating described in Example (Section 36.5), with 600 slits/mm, find the angles of the maxima corresponding to $m = 0$, 1, and -1 with red light ($\lambda = 650$ nm) for the cases $\theta' = 0$ (normal incidence) and $\theta' = 20.0°$.

36.67. A diffraction grating has 650 slits/mm. What is the highest order that contains the entire visible spectrum? (The wavelength range of the visible spectrum is approximately 400–700 nm.)

36.68. *Quasars,* an abbreviation for *quasi-stellar radio sources,* are distant objects that look like stars through a telescope but that emit far more electromagnetic radiation than an entire normal galaxy of stars. An example is the bright object below and to the left of center in Fig. 36.34; the other elongated objects in this image are normal galaxies. The leading model for the structure of a quasar is a galaxy with a supermassive black hole at its center. In this model, the radiation is emitted by interstellar gas and dust within the galaxy as this material falls toward the black hole. The radiation is thought to emanate from a region just a few light-years in diameter. (The diffuse glow surrounding the bright quasar shown in Fig. 36.34 is thought to be this quasar's host galaxy.) To investigate this model of quasars and to study other exotic astronomical objects, the Russian Space Agency plans to place a radio telescope in an orbit that extends to 77,000 km from the earth. When the signals from this telescope are combined with signals from the ground-based telescopes of the VLBA, the resolution will be that of a single radio telescope 77,000 km in diameter. What is the size of the smallest detail that this arrangement could resolve in quasar 3C 405, which is 7.2×10^8 light-years from earth, using radio waves at a frequency of 1665 MHz? (*Hint:* Use Rayleigh's criterion.) Give your answer in light-years and in kilometers.

Figure 36.34 Problem 36.68

36.69. Phased-Array Radar. In one common type of radar installation, a rotating antenna sweeps a radio beam around the sky. But in a *phased-array* radar system, the antennas remain stationary and the beam is swept electronically. To see how this is done, consider an array of N antennas that are arranged along the horizontal x-axis at $x = 0, \pm d, \pm 2d, \dots, \pm(N - 1)d/2$. (The number N is odd.) Each antenna emits radiation uniformly in all directions in the horizontal xy-plane. The antennas all emit radiation coherently, with the same amplitude E_0 and the same wavelength λ. The relative phase δ of the emission from adjacent antennas can be varied, however. If the antenna at $x = 0$ emits a signal that is given by $E_0 \cos\omega t$, as measured at a point next to the antenna, the antenna at $x = d$ emits a signal given by $E_0\cos(\omega t + \delta)$, as measured at a point next to that antenna. The corresponding quantity for the antenna at $x = -d$ is $E_0\cos(\omega t - \delta)$; for the antennas at $x = \pm 2d$, it is $E_0\cos(\omega t \pm 2\delta)$; and so on. (a) If $\delta = 0$, the interference pattern at a distance from the antennas is large compared to d and has a principal maximum at $\theta = 0$ (that is, in the $+y$-direction, perpendicular to the line of the antennas). Show that if $d < \lambda$, this is the *only* principal interference maximum in the angular range $-90° < \theta < 90°$. Hence this principal maximum describes a beam emitted in the direction $\theta = 0$. As described in Section 36.4, if N is large, the beam will have a large intensity and be quite narrow. (b) If $\delta \neq 0$, show that the principal intensity maximum described in part (a) is located at

$$\theta = \arcsin\left(\frac{\delta\lambda}{2\pi d}\right)$$

where δ is measured in radians. Thus, by varying δ from positive to negative values and back again, which can easily be done electronically, the beam can be made to sweep back and forth around $\theta = 0$. (c) A weather radar unit to be installed on an airplane emits radio waves at 8800 MHz. The unit uses 15 antennas in an array 28.0 cm long (from the antenna at one end of the array to the antenna at the other end). What must the maximum and minimum values of δ be (that is, the most positive and most negative values) if the radar beam is to sweep 45° to the left or right of the airplane's direction of flight? Give your answer in radians.

36.70. Underwater Photography. An underwater camera has a lens of focal length 35.0 mm and a maximum aperture of $f/2.80$. The film it uses has an emulsion that is sensitive to light of frequency 6.00×10^{14} Hz. If the photographer takes a picture of an object 2.75 m in front of the camera with the lens wide open, what is the width of the smallest resolvable detail on the subject if the object is (a) a fish underwater with the camera in the water and (b) a person on the beach, with the camera out of the water?

36.71. An astronaut in orbit can just resolve two point sources on the earth that are 75.0 m apart. Assume that the resolution is diffraction limited, and use Rayleigh's criterion. What is the astronaut's altitude above the earth? Treat her eye as a circular aperture with a diameter of 4.00 mm (the diameter of her pupil), and take the wavelength of the light to be 500 nm.

36.72. Observing Planets Beyond Our Solar System. NASA is considering a project called *Planet Imager* that would give astronomers the ability to see details on planets orbiting other stars. Using the same principle as the Very Large Array (see Section 36.7), *Planet Imager* will use an array of infrared telescopes spread over thousands of kilometers of space. (Visible light would give even better resolution. Unfortunately, at visible wavelengths, stars are so bright that a planet would be lost in the glare. This is less of a problem at infrared wavelengths.) (a) If *Planet Imager* has an effective diameter of 6000 km and observes infrared radiation at a wavelength of 10 μm, what is the greatest distance at which it would be able to observe details as small as 250 km across (about the size of the greater Los Angeles area) on a planet? Give your answer in light-years (see Appendix E). (*Hint:* Use Rayleigh's criterion.) (b) For comparison, consider the resolution of a single infrared telescope in space that has a diameter of 1.0 m and that observes 10-μm radiation. What is the size of the smallest details that such a telescope could resolve at the distance of the nearest star to the sun, Proxima Centauri, which is 4.22 light-years distant? How does this compare to the diameter of the earth $(1.27 \times 10^4 \text{ km})$? To the average distance from the earth to the sun $(1.50 \times 10^8 \text{ km})$? Would a single telescope of this kind be able to detect the presence of a planet like the earth, in an orbit the size of the earth's orbit, around *any* other star? Explain. (c) Suppose *Planet Imager* is used to observe a planet orbiting the star 70 Virginis, which is 59 light-years from our solar system. A planet (though not an earthlike one) has in fact been detected orbiting this star, not by imaging it directly but by observing the slight "wobble" of the star as both it and the planet orbit their common center of mass. What is the size of the smallest details that *Planet Imager* could hope to resolve on the planet of 70 Virginis? How does this compare to the diameter of the planet, assumed to be comparable to that of Jupiter $(1.38 \times 10^5 \text{ km})$? (Although the planet of 70 Virginis is thought to be at least 6.6 times more massive than Jupiter, its radius is probably not too different from that of Jupiter. The reason is that such large planets are thought to be composed primarily of gases, not rocky material, and hence can be greatly compressed by the mutual gravitational attraction of different parts of the planet.)

Challenge Problems

36.73. It is possible to calculate the intensity in the single-slit Fraunhofer diffraction pattern *without* using the phasor method of Section 36.3. Let y' represent the position of a point within the slit of width a in Fig. 36.5a, with $y' = 0$ at the center of the slit so that the slit extends from $y' = -a/2$ to $y' = a/2$. We imagine dividing the slit up into infinitesimal strips of width dy', each of which acts as a source of secondary wavelets. (a) The amplitude of the total wave at the point O on the distant screen in Fig. 36.5a is E_0. Explain why the amplitude of the wavelet from each infinitesimal strip within the slit is $E_0(dy'/a)$, so that the electric field of the wavelet a distance x from the infinitesimal strip is $dE = E_0(dy'/a)\sin(kx - \omega t)$. (b) Explain why the wavelet from each strip as detected at point P in Fig. 36.5a can be expressed as

$$dE = E_0\frac{dy'}{a}\sin[k(D - y'\sin\theta) - \omega t]$$

where D is the distance from the center of the slit to point P and $k = 2\pi/\lambda$. (c) By integrating the contributions dE from all parts of the slit, show that the total wave detected at point P is

$$\begin{aligned} E &= E_0\sin(kD - \omega t)\frac{\sin[ka(\sin\theta)/2]}{ka(\sin\theta)/2} \\ &= E_0\sin(kD - \omega t)\frac{\sin[\pi a(\sin\theta)/\lambda]}{\pi a(\sin\theta)/\lambda} \end{aligned}$$

(The trigonometric identities in Appendix B will be useful.) Show that at $\theta = 0$, corresponding to point O in Fig. 36.5a, the wave is $E = E_0\sin(Kd - \omega t)$ and has amplitude E_0, as stated in part (a). (d) Use the result of part (c) to show that if the intensity at point O is I_0, then the intensity at a point P is given by Eq. (36.7).

36.74. Intensity Pattern of N Slits. (a) Consider an arrangement of N slits with a distance d between adjacent slits. The slits emit coherently and in phase at wavelength λ. Show that at a time t, the electric field at a distant point P is

$$\begin{aligned} E_P(t) &= E_0\cos(kR - \omega t) + E_0\cos(kR - \omega t + \phi) \\ &\quad + E_0\cos(kR - \omega t + 2\phi) + \cdots \\ &\quad + E_0\cos(kR - \omega t + (N - 1)\phi) \end{aligned}$$

where E_0 is the amplitude at P of the electric field due to an individual slit, $\phi = (2\pi d\sin\theta)/\lambda$, θ is the angle of the rays reaching P (as measured from the perpendicular bisector of the slit arrangement), and R is the distance from P to the most distant slit. In this problem, assume that R is much larger than d. (b) To carry out the sum in part (a), it is convenient to use the complex-number relationship

$$e^{iz} = \cos z + i \sin z$$

where $i = \sqrt{-1}$. In this expression, $\cos z$ is the *real part* of the complex number e^{iz}, and $\sin z$ is its *imaginary part*. Show that the electric field $E_P(t)$ is equal to the real part of the complex quantity

$$\sum_{n=0}^{N-1} E_0 e^{i(kR - \omega t + n\phi)}$$

(c) Using the properties of the exponential function that $e^A e^B = e^{(A+B)}$ and $(e^A)^n = e^{nA}$, show that the sum in part (b) can be written as

$$E_0\left(\frac{e^{iN\phi} - 1}{e^{i\phi} - 1}\right)e^{i(kR-\omega t)} = E_0\left(\frac{e^{iN\phi/2} - e^{-iN\phi/2}}{e^{i\phi/2} - e^{-i\phi/2}}\right)e^{i[kR-\omega t+(N-1)\phi/2]}$$

Then, using the relationship $e^{iz} = \cos z + i \sin z$, show that the (real) electric field at point P is

$$E_p(t) = \left[E_0 \frac{\sin(N\phi/2)}{\sin(\phi/2)}\right]\cos[kR - \omega t + (N-1)\phi/2]$$

The quantity in the first square brackets in this expression is the amplitude of the electric field at P. (d) Use the result for the electric-field amplitude in part (c) to show that the intensity at an angle θ is

$$I = I_0\left[\frac{\sin(N\phi/2)}{\sin(\phi/2)}\right]^2$$

where I_0 is the maximum intensity for an individual slit. (e) Check the result in part (d) for the case $N = 2$. It will help to recall that $\sin 2A = 2 \sin A \cos A$. Explain why your result differs from Eq. (35.10), the expression for the intensity in two-source interference, by a factor of 4. (*Hint:* Is I_0 defined in the same way in both expressions?)

36.75. Intensity Pattern of N Slits, Continued. Part (d) of Challenge Problem 36.74 gives an expression for the intensity in the interference pattern of N identical slits. Use this result to verify the following statements. (a) The maximum intensity in the pattern is $N^2 I_0$. (b) The principal maximum at the center of the pattern extends from $\phi = -2\pi/N$ to $\phi = 2\pi/N$, so its width is inversely proportional to $1/N$. (c) A minimum occurs whenever ϕ is an integral multiple of $2\pi/N$, except when ϕ is an integral multiple of 2π (which gives a principal maximum). (d) There are $(N - 1)$ minima between each pair of principal maxima. (e) Halfway between two principal maxima, the intensity can be no greater than I_0; that is, it can be no greater than $1/N^2$ times the intensity at a principal maximum.

PHOTONS, ELECTRONS, AND ATOMS

38

? Although the glowing gas cloud called the Lagoon Nebula is more than 5000 light-years away, we can tell that it contains mostly hydrogen because of its red color. This red light has a wavelength of 656.3 nm, which is emitted by hydrogen and no other substance. What happens within a hydrogen atom to emit light of this wavelength?

LEARNING GOALS

By studying this chapter, you will learn:

- How experiments involving line spectra, the photoelectric effect, and x rays pointed the way to a radical reinterpretation of the nature of light.
- How Einstein's photon picture of light explains the photoelectric effect.
- How the spectrum of light emitted by atomic hydrogen reveals the existence of atomic energy levels.
- How physicists discovered the atomic nucleus.
- How Bohr's model of electron orbits explained the spectra of hydrogen and hydrogenlike atoms.
- The operating principles of a laser.
- How experiments with x rays helped confirm the photon picture of light.
- How the photon picture explains the spectrum of light emitted by a hot, opaque object.
- How we can reconcile the wave and particle aspects of light.

In Chapter 32 we saw how Maxwell, Hertz, and others established firmly that light is an electromagnetic wave. Interference, diffraction, and polarization, discussed in Chapters 35 and 36, further demonstrate this *wave nature* of light.

When we look more closely at the emission, absorption, and scattering of electromagnetic radiation, however, we discover a completely different aspect of light. We find that the energy of an electromagnetic wave is *quantized;* it is emitted and absorbed in particle-like packages of definite energy, called *photons* or *quanta.* The energy of a single photon is proportional to the frequency of the radiation.

The internal energy of atoms is also quantized. For a given kind of individual atom the energy can't have just any value; only discrete values called *energy levels* are possible.

The basic ideas of photons and energy levels take us a long way toward understanding a wide variety of otherwise puzzling observations. Among these are the unique sets of wavelengths emitted and absorbed by gaseous elements, the emission of electrons from an illuminated surface, the operation of lasers, and the production and scattering of x rays. Our studies of photons and energy levels will take us to the threshold of *quantum mechanics,* which involves some radical changes in our views of the nature of electromagnetic radiation and of matter itself.

38.1 Emission and Absorption of Light

How is light produced? In Chapter 32 we discussed how Heinrich Hertz produced electromagnetic waves using oscillations in a resonant *L-C* circuit similar to those we studied in Chapter 30. He used frequencies of the order of 10^8 Hz, but visible light has frequencies of the order of 10^{15} Hz, far higher than any frequency that can be attained with conventional electronic circuits. At the end of the 19th century, some physicists speculated that waves in this frequency range might be produced by oscillating electric charges within individual atoms. However, their speculations were unable to give explanations for some key experimental data. Among the great challenges facing physicists in 1900 were how to explain *line spectra,* the *photoelectric effect,* and the nature of x rays. We'll describe each of these in turn.

Line Spectra

We can use a prism or a diffraction grating to separate the various wavelengths in a beam of light into a spectrum. If the light source is a hot solid (such as a light-bulb filament) or liquid, the spectrum is *continuous;* light of all wavelengths is present (Fig. 38.1a). But if the source is a gas carrying an electric discharge (as in a neon sign) or a volatile salt heated in a flame (as when table salt is thrown into a campfire), only a few colors appear, in the form of isolated sharp parallel lines (Fig. 38.1b). (Each "line" is an image of the spectrograph slit, deviated through an angle that depends on the wavelength of the light forming that image; see Section 36.5.) A spectrum of this sort is called a **line spectrum.** Each line corresponds to a definite wavelength and frequency.

It was discovered early in the 19th century that each element in its gaseous state has a unique set of wavelengths in its line spectrum. The spectrum of hydrogen always contains a certain set of wavelengths; sodium produces a different set, iron still another, and so on. Scientists find the use of spectra to identify elements and compounds to be an invaluable tool (see the photograph that opens this chapter). For instance, astronomers have detected the spectra from more than 100 different molecules in interstellar space, including some that are not found naturally on earth. The characteristic spectrum of an atom was presumably related to its internal structure, but attempts to understand this relationship solely on the basis of *classical* mechanics and electrodynamics—the physics summarized in Newton's three laws and Maxwell's four equations—were not successful.

Photoelectric Effect

ActivPhysics Online 17.3 Photoelectric Effect

There were also mysteries associated with *absorption* of light. In 1887, during his electromagnetic-wave experiments, Hertz discovered the *photoelectric effect.* When light struck a metal surface, some electrons near the surface absorbed enough energy to overcome the attraction of the positive ions in the metal and

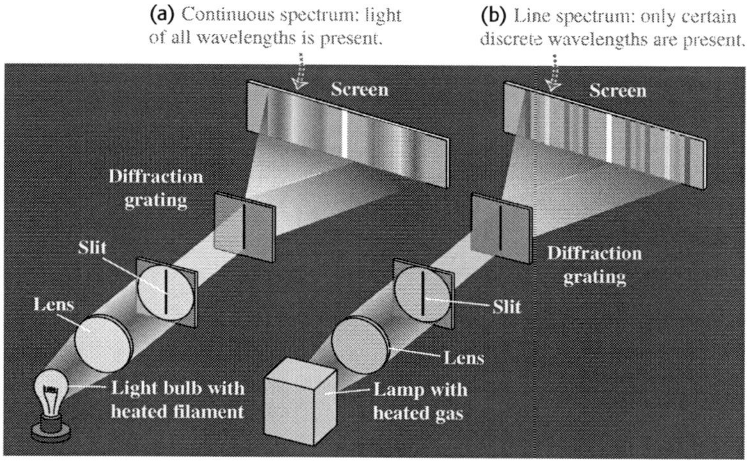

38.1 (a) Continuous spectrum produced by a glowing light bulb filament. (b) Line spectrum emitted by a lamp containing a heated gas.

escape into the surrounding space. Detailed investigation of this effect revealed some puzzling features that couldn't be understood on the basis of classical optics. We will discuss these in the next section.

X Rays

Other unsolved problems in the emission and absorption of radiation centered on the production and scattering of *x rays,* discovered in 1895. These rays were produced in high-voltage electric discharge tubes, but no one understood how or why they were produced or what determined their wavelengths (which are much shorter those of visible light). Even worse, when x rays collided with matter, the scattered rays sometimes had longer wavelengths than the original ray. This is analogous to a beam of blue light striking a mirror and reflecting back as red!

Photons and Energy Levels

All these phenomena (and several others) pointed forcefully to the conclusion that classical optics, successful though it was in explaining lenses, mirrors, interference, and polarization, had its limitations. We now understand that all these phenomena result from the *quantum* nature of radiation. Electromagnetic radiation, along with its *wave* nature, has properties resembling those of *particles.* In particular, the energy in an electromagnetic wave is always emitted and absorbed in packages called *photons* or *quanta,* with energy proportional to the frequency of the radiation.

The two common threads that are woven through this chapter are the quantization of electromagnetic radiation and the existence of discrete energy levels in atoms. In the remainder of this chapter we will show how these two concepts contribute to understanding the phenomena mentioned above. We are not yet ready for a comprehensive theory of atomic structure; that will come in Chapters 40 and 41. But we'll look at the Bohr model of the hydrogen atom, one attempt to predict atomic energy levels on the basis of atomic structure.

38.2 (a) A night vision scope makes use of the photoelectric effect. Photons entering the scope strike a plate, ejecting electrons that pass through a thin disk in which there are millions of tiny channels. The current through each channel is amplified electronically and then directed toward a screen that glows when hit by electrons. (b) The image formed on the screen, which is a combination of these millions of glowing spots, is thousands of times brighter than the naked-eye view.

(a)

(b)

38.2 The Photoelectric Effect

The **photoelectric effect** is the emission of electrons when light strikes a surface. This effect has numerous practical applications (Fig. 38.2). To escape from the surface, the electron must absorb enough energy from the incident radiation to overcome the attraction of positive ions in the material of the surface. This attraction causes a potential-energy barrier that normally confines the electrons inside the material. Think of this barrier as being like a rounded curb separating a flat street from a raised sidewalk. The curb will keep a slowly moving soccer ball in the street. But if the ball is kicked hard enough, it can roll up onto the sidewalk, with the work done against the gravitational attraction (the gain in gravitational potential energy) equal to its loss in kinetic energy.

Threshold Frequency and Stopping Potential

The photoelectric effect was first observed in 1887 by Heinrich Hertz, who noticed that a spark would jump more readily between two electrically charged spheres when their surfaces were illuminated by the light from another spark. Light shining on the surfaces somehow facilitated the escape of what we now know to be electrons. This idea in itself was not revolutionary. The existence of the surface potential-energy barrier was already known. In 1883 Thomas Edison had discovered *thermionic emission,* in which the escape energy is supplied by

38.3 Demonstrating the photoelectric effect.

(a) Light causes the cathode to emit electrons, which are pushed toward the anode by the electric-field force.

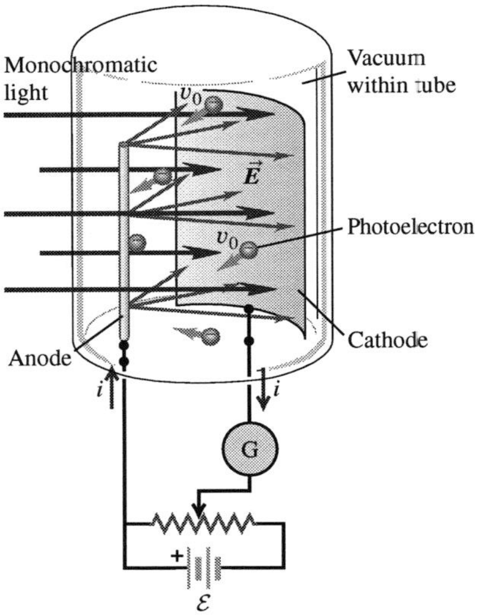

(b) Overhead view with $\vec{E}$ field reversed. Even when the direction of $\vec{E}$ field is reversed so that the electric-field force points away from the anode, some electrons still reach the anode ...

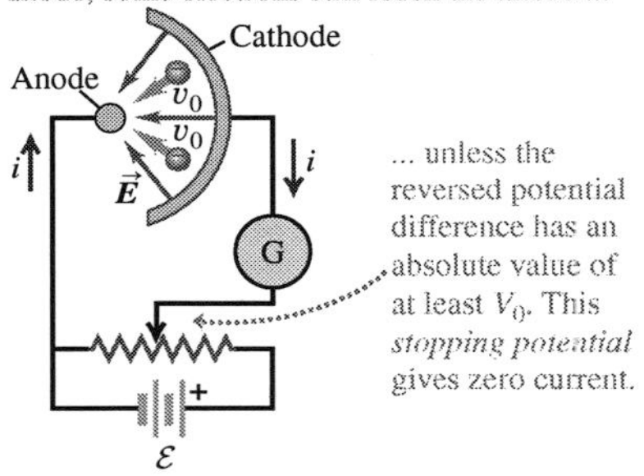

38.4 Photocurrent i for a constant light frequency f as a function of the potential V_{AC} of the anode with respect to the cathode.

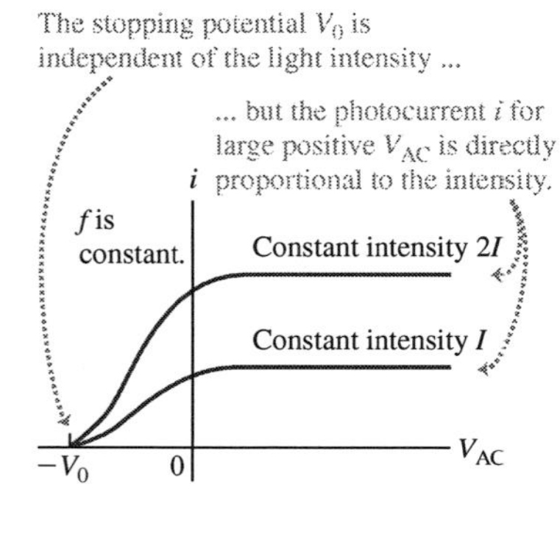

heating the material to a very high temperature, liberating electrons by a process analogous to boiling a liquid. The *minimum* amount of energy an individual electron has to gain to escape from a particular surface is called the **work function** for that surface, denoted by ϕ. However, the surfaces that Hertz used were not at the high temperatures needed for thermionic emission.

The photoelectric effect was investigated in detail by the German physicists Wilhelm Hallwachs and Philipp Lenard during the years 1886–1900; their results were quite unexpected. We will describe their work in terms of a more modern phototube (Fig. 38.3). Two conducting electrodes, the anode and the cathode, are enclosed in an evacuated glass tube. The battery or other source of potential difference creates an electric field in the direction from anode to cathode in Fig. 38.3a. Light (indicated by the magenta arrows) falling on the surface of the cathode causes a current in the external circuit; the current is measured by the galvanometer (G). Hallwachs and Lenard studied how this *photocurrent* varies with voltage and with the frequency and intensity of the light.

After the discovery of the electron in 1897, it became clear that the light causes electrons to be emitted from the cathode. Because of their negative charge $-e$, the emitted *photoelectrons* are then pushed toward the anode by the electric field. A high vacuum with residual pressure of 0.01 Pa (10^{-7} atm) or less is needed to minimize collisions of the electrons with gas molecules.

Hallwachs and Lenard found that when monochromatic light fell on the cathode, *no* photoelectrons at all were emitted unless the frequency of the light was greater than some minimum value called the **threshold frequency.** This minimum frequency depends on the material of the cathode. For most metals the threshold frequency is in the ultraviolet (corresponding to wavelengths λ between 200 and 300 nm), but for potassium and cesium oxides it is in the visible spectrum (λ between 400 and 700 nm).

When the frequency f is *greater* than the threshold frequency, some electrons are emitted from the cathode with substantial initial speeds. This can be shown by reversing the polarity of the battery (Fig. 38.3b) so that the electric-field force on the electrons is back toward the cathode. If the magnitude of the field is not too great, the highest-energy emitted electrons still reach the anode and there is still a current. We can determine the *maximum* kinetic energy of the emitted electrons by making the potential of the anode relative to the cathode, V_{AC}, just negative enough so that the current stops. This occurs for $V_{AC} = -V_0$, where V_0 is called the **stopping potential.** As an electron moves from the cathode to the anode, the potential decreases by V_0 and negative work $-eV_0$ is done on the (negatively charged) electron; the most energetic electron leaves the cathode with kinetic energy $K_{max} = \frac{1}{2}mv_{max}{}^2$ and has zero kinetic energy at the anode. Using the work–energy theorem, we have

$$W_{tot} = -eV_0 = \Delta K = 0 - K_{max}$$

$$K_{max} = \frac{1}{2}mv_{max}{}^2 = eV_0 \qquad \text{(maximum kinetic energy of photoelectrons)} \qquad (38.1)$$

Hence by measuring the stopping potential V_0, we can determine the maximum kinetic energy with which electrons leave the cathode. (We are ignoring any effects due to differences in the materials of the cathode and anode.)

Figure 38.4 shows graphs of photocurrent as a function of potential difference V_{AC} for light of constant frequency and two different intensities. When V_{AC} is sufficiently large and positive, the curves level off, showing that *all* the emitted electrons are being collected by the anode. The reverse potential difference $-V_0$ needed to reduce the current to zero is shown.

If the intensity of light is increased while its frequency is kept the same, the current levels off at a higher value, showing that more electrons are being emitted per time. But the stopping potential V_0 is found to be the same.

Figure 38.5 shows current as a function of potential difference for two different frequencies. We see that when the frequency of the incident monochromatic light is increased, the stopping potential V_0 increases. In fact, V_0 turns out to be a linear function of the frequency f.

38.5 Photocurrent i for two different light frequencies f_1 and f_2 as functions of the potential V_{AC} of the anode with respect to the cathode.

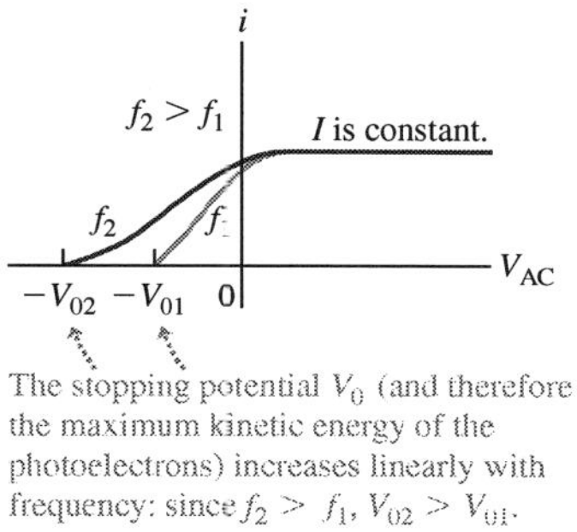

Einstein's Photon Explanation

These results are hard to understand on the basis of classical physics. When the intensity (average energy per unit area per unit time) increases, electrons should be able to gain more energy, increasing the stopping potential V_0. But V_0 was found *not* to depend on intensity. Also, classical physics offers no explanation for the threshold frequency. In Section 32.4 we found that the intensity of an electromagnetic wave such as light does not depend on frequency, so an electron should be able to acquire its needed escape energy from light of any frequency. Thus there should not be a threshold frequency f_0. Finally, we would expect it to take a while for an electron to collect enough energy from extremely faint light. But experiment also shows that electrons are emitted as soon as *any* light with $f \geq f_0$ hits the surface.

The correct analysis of the photoelectric effect was developed by Albert Einstein in 1905. Building on an assumption made five years earlier by Max Planck (to be discussed in Section 38.8), Einstein postulated that a beam of light consists of small packages of energy called **photons** or *quanta*. The energy E of a photon is equal to a constant h times its frequency f. From $f = c/\lambda$ for electromagnetic waves in vacuum we have

$$E = hf = \frac{hc}{\lambda} \quad \text{(energy of a photon)} \qquad (38.2)$$

where h is a universal constant called **Planck's constant.** The numerical value of this constant, to the accuracy known at present, is

$$h = 6.6260693(11) \times 10^{-34}\ \text{J} \cdot \text{s}$$

A photon arriving at the surface is absorbed by an electron. This energy transfer is an all-or-nothing process, in contrast to the continuous transfer of energy in the classical theory; the electron gets all the photon's energy or none at all. If this energy is greater than the work function ϕ, the electron may escape from the surface. Greater intensity at a particular frequency means a proportionally greater number of photons per second absorbed, and thus a proportionally greater number of electrons emitted per second and the proportionally greater current seen in Fig. 38.4.

Recall that ϕ is the *minimum* energy needed to remove an electron from the surface. Einstein applied conservation of energy to find that the *maximum* kinetic energy $K_{max} = \frac{1}{2}mv_{max}^2$ for an emitted electron is the energy hf gained from a photon minus the work function ϕ:

$$K_{max} = \frac{1}{2}mv_{max}^2 = hf - \phi \qquad (38.3)$$

Substituting $K_{max} = eV_0$ from Eq. (38.1), we find

$$eV_0 = hf - \phi \quad \text{(photoelectric effect)} \qquad (38.4)$$

We can measure the stopping potential V_0 for each of several values of frequency f for a given cathode material. A graph of V_0 as a function of f turns out to be a straight line, verifying Eq. (38.4), and from such a graph we can determine both the work function ϕ for the material and the value of the quantity h/e. (We do this graphical analysis in Example 38.3.) After the electron charge $-e$ was measured by Robert Millikan in 1909, Planck's constant h could also be determined from these measurements.

Electron energies and work functions are usually expressed in electron volts (eV), defined in Section 23.2. To four significant figures,

$$1 \text{ eV} = 1.602 \times 10^{-19} \text{ J}$$

To this accuracy, Planck's constant is

$$h = 6.626 \times 10^{-34} \text{ J} \cdot \text{s} = 4.136 \times 10^{-15} \text{ eV} \cdot \text{s}$$

Table 38.1 Work Functions of Several Elements

Element	Work Function (eV)
Aluminum	4.3
Carbon	5.0
Copper	4.7
Gold	5.1
Nickel	5.1
Silicon	4.8
Silver	4.3
Sodium	2.7

Table 38.1 lists some typical work functions of elements. These values are approximate because they are very sensitive to surface impurities. The greater the work function, the higher the minimum frequency needed to emit photoelectrons.

Photon Momentum

We have discussed photons mostly in the context of light. However, the quantization concept applies to *all* regions of the electromagnetic spectrum, including radio waves, x rays, and so on. A photon of any electromagnetic radiation with frequency f and wavelength λ has energy E given by Eq. (38.2). Furthermore, according to the special theory of relativity, every particle that has energy must also have momentum, even if it has no rest mass. Photons have zero rest mass. As we saw in Eq. (37.40), a photon with energy E has momentum with magnitude p given by $E = pc$. Thus the wavelength λ of a photon and the magnitude of its momentum p are related simply by

$$p = \frac{E}{c} = \frac{hf}{c} = \frac{h}{\lambda} \quad \text{(momentum of a photon)} \qquad (38.5)$$

The direction of the photon's momentum is simply the direction in which the electromagnetic wave is moving.

Problem-Solving Strategy 38.1 Photons

MP

IDENTIFY *the relevant concepts:* The energy and momentum of an individual photon are proportional to the frequency and inversely proportional to the wavelength. These concepts, along with the idea of stopping potential in the photoelectric effect, allow you to solve almost any problem involving photons.

SET UP *the problem:* Decide what your target variable is. It could be the photon's wavelength λ, frequency f, energy E, or momentum p. If the problem involves the photoelectric effect, the target variable could be the maximum kinetic energy of photoelectrons K_{max}, the stopping potential V_0, or the work function ϕ.

EXECUTE *the solution* as follows:

1. Use Eqs. (38.2) and (38.5) to relate the energy and momentum of a photon to its wavelength and frequency. If the problem involves the photoelectric effect, use Eqs. (38.1), (38.3), and (38.4) to relate the photon frequency, stopping potential, work function, and maximum photoelectron kinetic energy.
2. The electron volt is an important and convenient unit. We used it in Chapter 37, and we will use it even more in this and the next three chapters. One electron volt (eV) is the amount of kinetic energy gained by an electron when it moves freely through an increase of potential of one volt: $1 \text{ eV} = 1.602 \times 10^{-19}$ J. If the photon energy E is given in electron volts, use $h = 4.136 \times 10^{-15}$ eV · s; if E is in joules, use $h = 6.626 \times 10^{-34}$ J · s.

EVALUATE *your answer:* In problems involving photons, the numbers are in such unfamiliar ranges that common sense may not help you identify whether your calculation is off by a large factor. It helps to remember that a visible-light photon with $\lambda = 600$ nm and $f = 5 \times 10^{14}$ Hz has an energy E of about 2 eV, or about 3×10^{-19} J.

Example 38.1 FM radio photons

Radio station WQED in Pittsburgh broadcasts at 89.3 MHz with a radiated power of 43.0 kW. (a) What is the magnitude of the momentum of each photon? (b) How many photons does WQED emit each second?

SOLUTION

IDENTIFY: This problem involves the ideas of (a) photon momentum and (b) photon energy.

SET UP: We are given the frequency of the radiation, so we can use Eq. (38.5) directly to find the magnitude of each photon's momentum in part (a). In part (b), note that "radiated power" means the energy emitted per second and that Eq. (38.2) gives the energy per photon. We can combine these to calculate the number of photons emitted per second.

EXECUTE: (a) From Eq. (38.5), the magnitude of momentum of each photon is

$$p = \frac{hf}{c} = \frac{(6.626 \times 10^{-34}\ \text{J} \cdot \text{s})(89.3 \times 10^{6}\ \text{Hz})}{3.00 \times 10^{8}\ \text{m/s}}$$
$$= 1.97 \times 10^{-34}\ \text{kg} \cdot \text{m/s}$$

(Recall that $1\ \text{J} = 1\ \text{kg} \cdot \text{m}^2/\text{s}^2$ and that $1\ \text{Hz} = 1\ \text{s}^{-1}$.)

(b) From Eq. (38.2), the energy of each photon is

$$E = pc = (1.97 \times 10^{-34}\ \text{kg} \cdot \text{m/s})(3.00 \times 10^{8}\ \text{m/s})$$
$$= 5.92 \times 10^{-26}\ \text{J}$$

The station emits 43.0×10^3 joules each second, so the rate at which photons are emitted is

$$\frac{43.0 \times 10^{3}\ \text{J/s}}{5.92 \times 10^{-26}\ \text{J/photon}} = 7.26 \times 10^{29}\ \text{photons/s}$$

EVALUATE: The result in part (a) is very small, about the magnitude of the momentum an electron would have if it crawled along at a speed of one meter per hour. The photon energy E calculated in part (b) is also very small, equal to 3.69×10^{-7} eV. A visible-light photon has about 10^7 times as much energy. This makes sense: The energy of a photon is proportional to its frequency, and the frequencies of visible light are about 10^7 times greater than those used in FM radio. (You can also check the value of the photon energy by using $E = hf$.)

Our result in part (b) shows that a huge number of photons leave the station each second, each of which has an infinitesimal amount of energy. Hence the discreteness of the photons isn't noticed, and the radiated energy appears to be a continuous flow.

Example 38.2 A photoelectric-effect experiment

While conducting a photoelectric-effect experiment with light of a certain frequency, you find that a reverse potential difference of 1.25 V is required to reduce the current to zero. Find (a) the maximum kinetic energy and (b) the maximum speed of the emitted photoelectrons.

SOLUTION

IDENTIFY: This problem uses the relationship between the maximum kinetic energy K_{max} of a electron in the photoelectric effect and the associated stopping potential V_0.

SET UP: The value of 1.25 V is the stopping potential V_0. We can find the maximum photoelectron kinetic energy K_{max} by using Eq. (38.1); given this value, we can find the maximum photoelectron speed.

EXECUTE: (a) From Eq. (38.1),

$$K_{max} = eV_0 = (1.60 \times 10^{-19}\ \text{C})(1.25\ \text{V}) = 2.00 \times 10^{-19}\ \text{J}$$

Recall that $1\ \text{V} = 1\ \text{J/C}$. In terms of electron volts,

$$K_{max} = eV_0 = e(1.25\ \text{V}) = 1.25\ \text{eV}$$

since the electron volt (eV) is the magnitude of the electron charge e times one volt (1 V).

(b) From $K_{max} = \frac{1}{2}mv_{max}{}^2$ we get

$$v_{max} = \sqrt{\frac{2K_{max}}{m}} = \sqrt{\frac{2(2.00 \times 10^{-19}\ \text{J})}{9.11 \times 10^{-31}\ \text{kg}}}$$
$$= 6.63 \times 10^{5}\ \text{m/s}$$

EVALUATE: The value of V_{max} is about $\frac{1}{500}$ the speed of light c, so we are justified in using the nonrelativistic expression for kinetic energy. An equivalent justification is that the electron's 1.25-eV kinetic energy is much lower than its rest energy $mc^2 =$ 0.511 MeV.

Example 38.3 Determining ϕ and h experimentally

For a certain cathode material in a photoelectric-effect experiment, you measure a stopping potential of 1.0 V for light of wavelength 600 nm, 2.0 V for 400 nm, and 3.0 V for 300 nm. Determine the work function for this material and the value of Planck's constant.

SOLUTION

IDENTIFY: This example uses the relationship among stopping potential V_0, frequency f, and work function ϕ in the photoelectric effect.

SET UP: According to Eq. (38.4), a graph of stopping potential V_0 versus frequency f should be a straight line. Such a graph is completely determined by its slope and the value at which it intercepts the vertical axis; we will use these to determine the values of the target variables ϕ and h.

EXECUTE: We rewrite Eq. (38.4) as

$$V_0 = \frac{h}{e}f - \frac{\phi}{e}$$

Continued

In this form we see that the slope of the line is h/e and the intercept on the vertical axis (corresponding to $f = 0$) is at $-\phi/e$. The frequencies, obtained from $f = c/\lambda$ and $c = 3.00 \times 10^8$ m/s, are 0.50×10^{15} Hz, 0.75×10^{15} Hz, and 1.0×10^{15} Hz, respectively. The graph is shown in Fig. 38.6. From it we find

$$-\frac{\phi}{e} = \text{vertical intercept} = -1.0\text{ V}$$

$$\phi = 1.0\text{ eV} = 1.6 \times 10^{-19}\text{ J}$$

and

$$\text{Slope} = \frac{\Delta V_0}{\Delta f} = \frac{3.0\text{ V} - (-1.0\text{ V})}{1.00 \times 10^{15}\text{ s}^{-1} - 0} = 4.0 \times 10^{-15}\text{ J}\cdot\text{s/C}$$

$$h = \text{slope} \times e = (4.0 \times 10^{-15}\text{ J}\cdot\text{s/C})(1.60 \times 10^{-19}\text{ C})$$
$$= 6.4 \times 10^{-34}\text{ J}\cdot\text{s}$$

EVALUATE: This experimental value differs by about 3% from the accepted value. The small value of ϕ tells us that the cathode surface is not composed solely of one of the elements in Table 38.1.

38.6 Stopping potential as a function of frequency. For a different cathode material having a different work function, the line would be displaced up or down but would have the same slope, equal to h/e within experimental error.

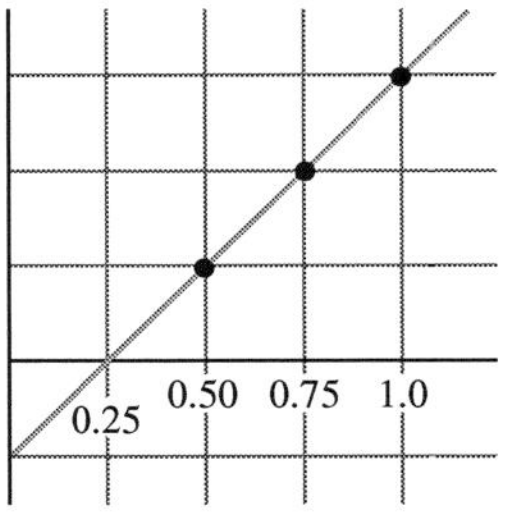

Test Your Understanding of Section 38.2 Silicon films become better electrical conductors when illuminated by photons with energies of 1.14 eV or greater, an effect called *photoconductivity.* Which of the following wavelengths of electromagnetic radiation can cause photoconductivity in silicon films? (i) ultraviolet light with $\lambda = 300$ nm; (ii) red light with wavelength $\lambda = 600$ nm; (iii) infrared light with $\lambda = 1200$ nm.

38.3 Atomic Line Spectra and Energy Levels

ActivPhysics Online
18.2 Spectroscopy

The origin of *line spectra,* described in Section 38.1, can be understood in general terms on the basis of two central ideas. One is the photon concept; the other is the concept of *energy levels* of atoms. These two ideas were combined by the Danish physicist Niels Bohr in 1913.

Photon Emission by Atoms

Bohr's hypothesis represented a bold breakaway from 19th-century ideas. His reasoning went like this. The line spectrum of an element results from the emission of photons with specific energies from the atoms of that element. During the emission of a photon, the internal energy of the atom changes by an amount equal to the energy of the photon. Therefore, said Bohr, each atom must be able to exist with only certain specific values of internal energy. Each atom has a set of possible **energy levels.** An atom can have an amount of internal energy equal to any one of these levels, but it *cannot* have an energy *intermediate* between two levels. All isolated atoms of a given element have the same set of energy levels, but atoms of different elements have different sets. In electric discharge tubes atoms are raised, or *excited,* to higher energy levels mainly through inelastic collisions with electrons.

38.7 An atom emitting a photon.

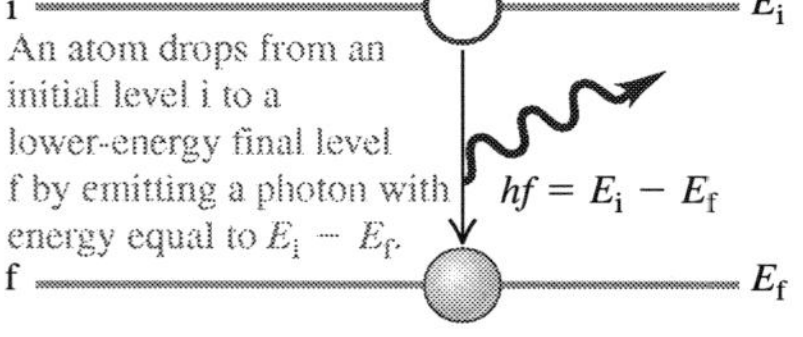

According to Bohr, an atom can make a *transition* from one energy level to a lower level by emitting a photon with energy equal to the energy *difference* between the initial and final levels (Fig. 38.7). If E_i is the initial energy of the atom before such a transition, E_f is its final energy after the transition, and the photon's energy is $hf = hc/\lambda$, then conservation of energy gives

$$hf = \frac{hc}{\lambda} = E_i - E_f \quad \text{(energy of emitted photon)} \qquad (38.6)$$

For example, when a krypton atom emits a photon of orange light with wavelength $\lambda = 606$ nm, the corresponding photon energy is

$$E = \frac{hc}{\lambda} = \frac{(6.63 \times 10^{-34}\ \text{J}\cdot\text{s})(3.00 \times 10^{8}\ \text{m/s})}{606 \times 10^{-9}\ \text{m}}$$
$$= 3.28 \times 10^{-19}\ \text{J} = 2.05\ \text{eV}$$

This photon is emitted during a transition like that in Fig. 38.7 between two levels of the atom that differ in energy by 2.05 eV.

The Hydrogen Spectrum

By 1913 the spectrum of hydrogen, the simplest and least massive atom, had been studied intensively. In an electric discharge tube, atomic hydrogen emits the series of lines shown in Fig. 38.8. The visible line with longest wavelength, or lowest frequency, is in the red and is called H_α; the next line, in the blue-green, is called H_β; and so on. In 1885 the Swiss teacher Johann Balmer (1825–1898) found (by trial and error) a formula that gives the wavelengths of these lines, which are now called the *Balmer series.* We may write Balmer's formula as

$$\frac{1}{\lambda} = R\left(\frac{1}{2^2} - \frac{1}{n^2}\right) \qquad (38.7)$$

where λ is the wavelength, R is a constant called the **Rydberg constant** (chosen to make Eq. (38.7) fit the measured wavelengths), and n may have the integer values 3, 4, 5, ... If λ is in meters, the numerical value of R is

$$R = 1.097 \times 10^{7}\ \text{m}^{-1}$$

Letting $n = 3$ in Eq. (38.7), we obtain the wavelength of the H_α line:

$$\frac{1}{\lambda} = (1.097 \times 10^{7}\ \text{m}^{-1})\left(\frac{1}{4} - \frac{1}{9}\right) \qquad \text{or} \qquad \lambda = 656.3\ \text{nm}$$

For $n = 4$ we obtain the wavelength of the H_β line, and so on. For $n = \infty$ we obtain the *smallest* wavelength in the series, $\lambda = 364.6$ nm.

Balmer's formula has a very direct relationship to Bohr's hypothesis about energy levels. Using the relationship $E = hc/\lambda$, we can find the *photon energies* corresponding to the wavelengths of the Balmer series. Multiplying Eq. (38.7) by hc, we find

$$E = \frac{hc}{\lambda} = hcR\left(\frac{1}{2^2} - \frac{1}{n^2}\right) = \frac{hcR}{2^2} - \frac{hcR}{n^2} \qquad (38.8)$$

Equations (38.6) and (38.8) for the photon's energy agree most simply if we identify $-hcR/n^2$ as the initial energy E_i of the atom and $-hcR/2^2$ as its final energy E_f, in a transition in which a photon with energy $E_i - E_f$ is emitted. The energies of the levels are negative because we choose the potential energy to be zero when the electron and nucleus are infinitely far apart. The Balmer (and other) series

38.8 The Balmer series of spectral lines for atomic hydrogen.

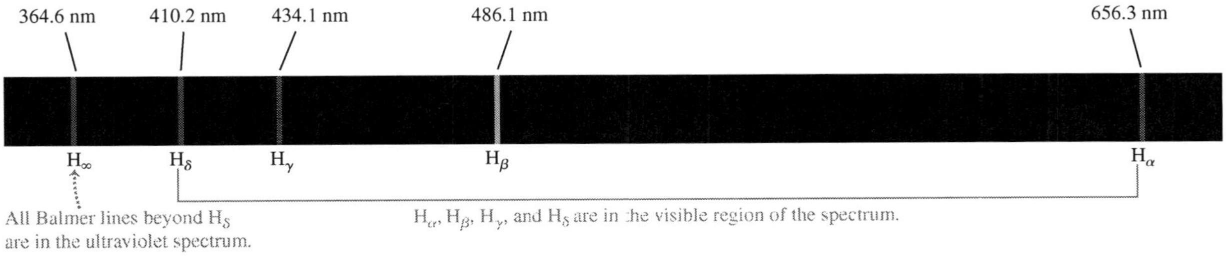

suggest that the hydrogen atom has a series of energy levels, which we may call E_n, given by

$$E_n = -\frac{hcR}{n^2} \quad (n = 1, 2, 3, 4, \dots) \quad \text{(energy levels of the hydrogen atom)} \tag{38.9}$$

Each wavelength in the Balmer series corresponds to a transition from a level with n equal to 3 or greater to the level with $n = 2$.

The numerical value of the product hcR is

$$\begin{aligned} hcR &= (6.626 \times 10^{-34}\ \text{J} \cdot \text{s})(2.998 \times 10^8\ \text{m/s})(1.097 \times 10^7\ \text{m}^{-1}) \\ &= 2.179 \times 10^{-18}\ \text{J} = 13.60\ \text{eV} \end{aligned}$$

Thus the magnitudes of the energy levels given by Eq. (38.9) are -13.60 eV, -3.40 eV, -1.51 eV, -0.85 eV, and so on.

Other spectral series for hydrogen have been discovered. These are known, after their discoverers, as the Lyman, Paschen, Brackett, and Pfund series. Their wavelengths can be represented by formulas similar to Balmer's formula:

$$\textit{Lyman series} \quad \frac{1}{\lambda} = R\left(\frac{1}{1^2} - \frac{1}{n^2}\right) \quad (n = 2, 3, 4, \dots)$$

$$\textit{Paschen series} \quad \frac{1}{\lambda} = R\left(\frac{1}{3^2} - \frac{1}{n^2}\right) \quad (n = 4, 5, 6, \dots)$$

$$\textit{Brackett series} \quad \frac{1}{\lambda} = R\left(\frac{1}{4^2} - \frac{1}{n^2}\right) \quad (n = 5, 6, 7, \dots)$$

$$\textit{Pfund series} \quad \frac{1}{\lambda} = R\left(\frac{1}{5^2} - \frac{1}{n^2}\right) \quad (n = 6, 7, 8, \dots)$$

The Lyman series is in the ultraviolet, and the Paschen, Brackett, and Pfund series are in the infrared. We see that the Balmer series fits into the scheme between the Lyman and Paschen series.

Thus *all* the spectral series of hydrogen can be understood on the basis of Bohr's picture of transitions from one energy level (and corresponding electron orbit) to another, with the energy levels given by Eq. (38.9) with $n = 1, 2, 3, \dots$. For the Lyman series the final level is always $n = 1$; for the Paschen series it is $n = 3$; and so on. The relationship of the various spectral series to the energy levels and to electron orbits is shown in Fig. 38.9. Taken together, these spectral series give very strong support to Bohr's picture of energy levels in atoms.

CAUTION **Producing a line spectrum** The lines of a spectrum, such as the hydrogen spectrum shown in Fig. 38.8, are *not* all produced by a single atom. The sample of hydrogen gas that produced the spectrum in Fig. 38.8 contained a large number of atoms; these were excited in an electric discharge tube to various energy levels. The spectrum of the gas shows the light emitted from all the different transitions that occurred in different atoms of the sample.

We haven't yet discussed any way to *predict* what the energy levels for a particular atom should be. Neither have we shown how to derive Eq. (38.9) from fundamental theory or to relate the Rydberg constant R to other fundamental constants. We'll return to these problems later.

The Franck-Hertz Experiment

In 1914, James Franck and Gustav Hertz found even more direct experimental evidence for the existence of atomic energy levels. Franck and Hertz studied the motion of electrons through mercury vapor under the action of an electric field.

38.9 Two ways to represent the energy levels of the hydrogen atom and transitions between them.

(a) "Permitted" orbits of an electron in the Bohr model of a hydrogen atom (not to scale). Arrows indicate the transitions responsible for some of the lines of various series.

(b) Energy-level diagram for hydrogen, showing some transitions corresponding to the various series

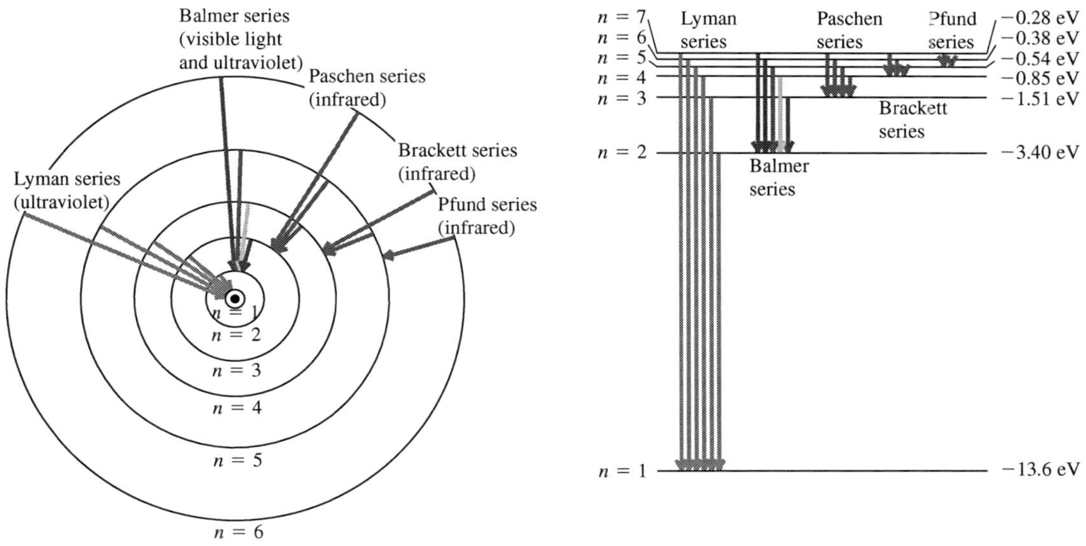

They found that when the electron kinetic energy was 4.9 eV or greater, the vapor emitted ultraviolet light of wavelength 0.25 μm. Suppose mercury atoms have an energy level 4.9 eV above the lowest energy level. An atom can be raised to this level by collision with an electron; it later decays back to the lowest energy level by emitting a photon. According to Eq. (38.2), the wavelength of the photon should be

$$\lambda = \frac{hc}{E} = \frac{(4.136 \times 10^{-15}\ \text{eV} \cdot \text{s})(3.00 \times 10^{8}\ \text{m/s})}{4.9\ \text{eV}}$$
$$= 2.5 \times 10^{-7}\ \text{m} = 0.25\ \mu\text{m}$$

This is equal to the measured wavelength, confirming the existence of this energy level of the mercury atom. Similar experiments with other atoms yield the same kind of evidence for atomic energy levels.

Energy Levels

Only a few atoms and ions (such as hydrogen, singly ionized helium, doubly ionized lithium) have spectra whose wavelengths can be represented by a simple formula such as Balmer's. But it is *always* possible to analyze the more complicated spectra of other elements in terms of transitions among various energy levels and to deduce the numerical values of these levels from the measured spectrum wavelengths.

Every atom has a lowest energy level that includes the *minimum* internal energy state that the atom can have. This is called the *ground-state level,* or **ground level,** and all higher levels are called **excited levels.** A photon corresponding to a particular spectrum line is emitted when an atom makes a transition from a state in an excited level to a state in a lower excited level or the ground level.

Some energy levels for sodium are shown in Fig. 38.10a (next page). You may have noticed the yellow-orange light emitted by sodium vapor street lights. Sodium atoms emit this characteristic yellow-orange light with wavelengths 589.0 and 589.6 nm when they make transitions from the two closely spaced levels labeled *lowest excited levels* to the ground level. A standard test for the presence of

38.10 (a) Energy levels of the sodium atom relative to the ground level. Numbers on the lines between levels are wavelengths. The column labels, such as $^2S_{1/2}$, refer to some quantum states of the valence electron (to be discussed in Chapter 41). (b) When a sodium compound is placed in a flame, sodium atoms are excited into the lowest excited levels. As they drop back to the ground level, the atoms emit photons of yellow-orange light with wavelengths 589.0 and 589.6 nm.

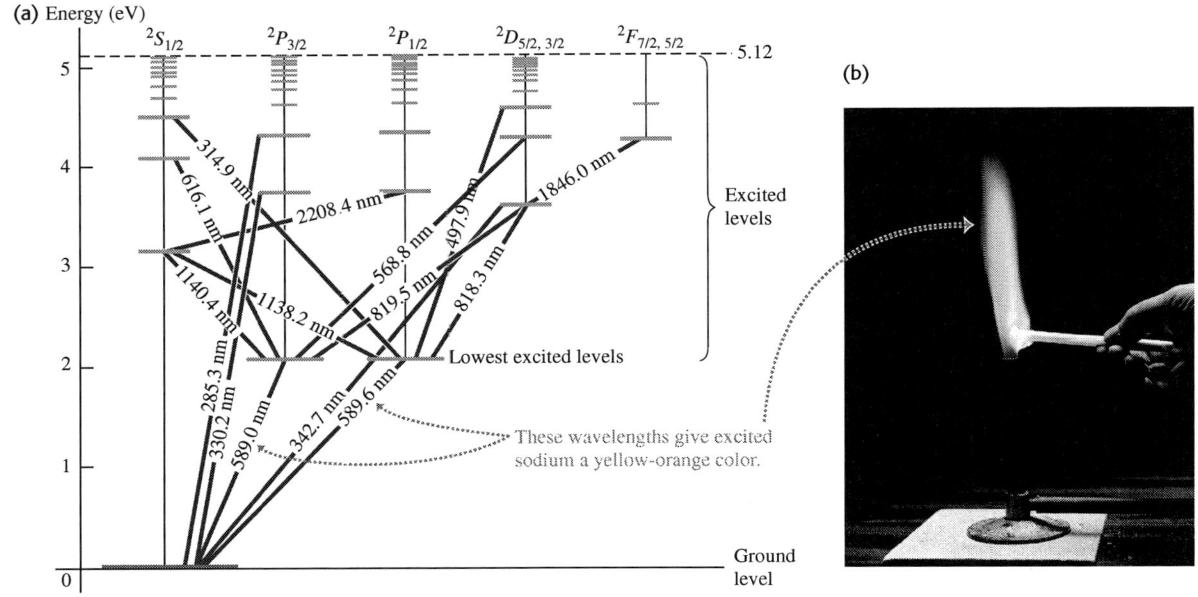

sodium compounds is to look for this yellow-orange light from a sample placed in a flame (Fig. 38.10b).

38.11 An atom absorbing a photon. (Compare with Fig. 38.7.)

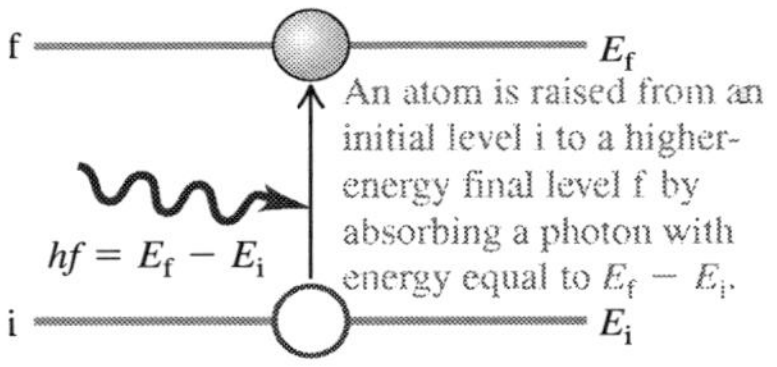

38.12 The absorption spectrum of the sun. (You should read the spectrum from left to right in each strip and then downward from one strip to the next, like words in a printed paragraph.) The relatively cool solar atmosphere selectively absorbs photons from the sun's glowing surface, producing dark absorption lines that indicate what kinds of atoms are present in the solar atmosphere.

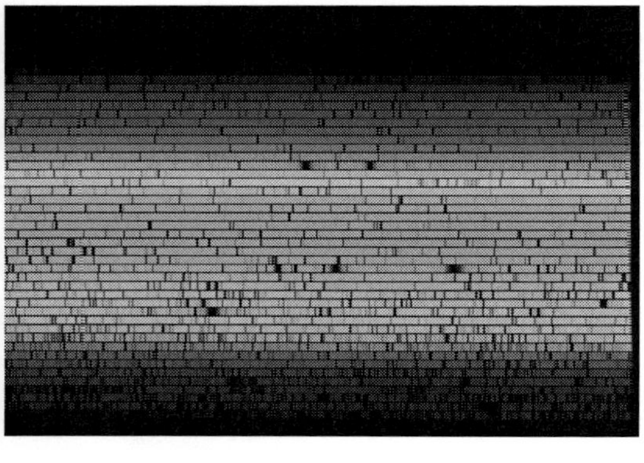

Photon Absorption

A sodium atom in the ground level can also *absorb* a photon with wavelength 589.0 or 589.6 nm. To demonstrate this process, we pass a beam of light from a sodium-vapor lamp through a bulb containing sodium vapor. The atoms in the vapor absorb the 589.0-nm or 589.6-nm photons from the beam, reaching the lowest excited levels; after a short time they return to the ground level, emitting photons in all directions and causing the sodium vapor to glow with the characteristic yellow light. The average time spent in an excited level is called the lifetime of the level; for the lowest excited levels of the sodium atom, the *lifetime* is about 1.6×10^{-8} s.

More generally, a photon *emitted* when an atom makes a transition from an excited level to a lower level can also be *absorbed* by a similar atom that is initially in the lower level (Fig. 38.11). If we pass white (continuous-spectrum) light through a gas and look at the *transmitted* light with a spectrometer, we find a series of dark lines corresponding to the wavelengths that have been absorbed, as shown in Fig. 38.12. This is called an **absorption spectrum.**

A related phenomenon is *fluorescence.* An atom absorbs a photon (often in the ultraviolet region) to reach an excited level and then drops back to the ground level in steps, emitting two or more photons with lower energy and longer wavelength. For example, the electric discharge in a fluorescent tube causes the mercury vapor in the tube to emit ultraviolet radiation. This radiation is absorbed by the atoms of the coating on the inside of the tube. The coating atoms then re-emit light in the longer-wavelength visible portion of the spectrum. Fluorescent lamps are more efficient than incandescent lamps in converting electrical energy to visible light because they do not waste as much energy producing (invisible) infrared photons.

The Bohr hypothesis established the relationship of wavelengths to energy levels, but it provided no general principles for *predicting* the energy levels of a particular atom. Bohr provided a partial analysis for the hydrogen atom; we will discuss this in Section 38.5. A more general understanding of atomic structure and energy levels rests on the concepts of *quantum mechanics,* which we will introduce in Chapters 39 and 40. Quantum mechanics provides all the principles needed to calculate energy levels from fundamental theory. Unfortunately, for many-electron atoms the calculations are so complex that they can be carried out only approximately.

Example 38.4 Emission and absorption spectra

A hypothetical atom has three energy levels: the ground level and 1.00 eV and 3.00 eV above the ground level. (a) Find the frequencies and wavelengths of the spectral lines that this atom can emit when excited. (b) What wavelengths can be absorbed by this atom if it is initially in its ground level?

SOLUTION

IDENTIFY: We use the idea that energy is conserved when a photon is emitted or absorbed.

SET UP: In each transition the photon energy is equal to the difference between the energies of the higher level and the lower level involved in the transition.

EXECUTE: (a) Figure 38.13a shows an energy-level diagram. The possible photon energies, corresponding to the transitions shown, are 1.00 eV, 2.00 eV, and 3.00 eV. For 1.00 eV we have, from Eq. (38.2),

$$f = \frac{E}{h} = \frac{1.00\text{ eV}}{4.136 \times 10^{-15}\text{ eV}\cdot\text{s}} = 2.42 \times 10^{14}\text{ Hz}$$

For 2.00 eV and 3.00 eV, $f = 4.84 \times 10^{14}$ Hz and 7.25×10^{14} Hz, respectively. We can find the wavelengths using $\lambda = c/f$. For 1.00 eV,

$$\lambda = \frac{c}{f} = \frac{3.00 \times 10^8\text{ m/s}}{2.42 \times 10^{14}\text{ Hz}} = 1.24 \times 10^{-6}\text{ m} = 1240\text{ nm}$$

in the infrared region of the spectrum. For 2.00 eV and 3.00 eV the wavelengths are 620 nm (red) and 414 nm (violet), respectively (Fig. 38.13b).

(b) From the atom's ground level, only a 1.00-eV or 3.00-eV photon can be absorbed; a 2.00-eV photon cannot be because there is no energy level 2.00 eV above the ground level. As we have calculated, the wavelengths of 1.00-eV and 3.00-eV photons are 1240 nm and 414 nm, respectively. Passing light from a hot solid through a cool gas of these atoms would result in a continuous spectrum with dark absorption lines at 1240 nm and 414 nm.

EVALUATE: Note that if a gas of these atoms were at a high enough temperature, collisions would excite a number of atoms into the 1.00-eV energy level. Such excited atoms *can* absorb 2.00-eV photons, and an absorption line at 620 nm would appear in the spectrum. Thus the observed spectrum of a given substance depends on its energy levels and its temperature.

38.13 (a) Energy-level diagram, showing the possible transitions for emission from excited levels and for absorption from the ground level. (b) Emission spectrum of this hypothetical atom.

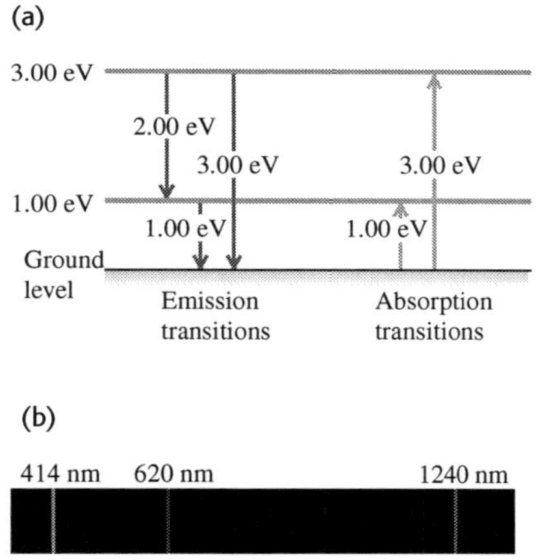

Test Your Understanding of Section 38.3 Would it be possible to excite the hypothetical atom in Example 38.4 from the ground level to the 1.00-eV level with a 1.50-eV photon?

38.4 The Nuclear Atom

Before we can make further progress in relating the energy levels of an atom to its internal structure, we need to have a better idea of what the inside of an atom is like. We know that atoms are much smaller than the wavelengths of visible light, so there is no hope of actually *seeing* an atom using that light. But we can still describe how the mass and electric charge are distributed throughout the volume of the atom.

Here's where things stood in 1910. J. J. Thomson had discovered the electron and measured its charge-to-mass ratio (e/m) in 1897; by 1909, Millikan had completed his first measurements of the electron charge $-e$. These and other experiments showed that almost all the mass of an atom had to be associated with the *positive* charge, not with the electrons. It was also known that the overall size of atoms is of the order of 10^{-10} m and that all atoms except hydrogen contain more than one electron. What was *not* known was how the mass and charge were distributed within the atom. Thomson had proposed a model in which the atom consisted of a sphere of positive charge, of the order of 10^{-10} m in diameter, with the electrons embedded in it like raisins in a more or less spherical muffin.

Rutherford's Exploration of the Atom

38.14 Born in New Zealand, Ernest Rutherford (1871–1937) spent his professional life in England and Canada. Before carrying out the experiments that established the existence of atomic nuclei, he shared (with Frederick Soddy) the 1908 Nobel Prize in chemistry for showing that radioactivity results from the disintegration of atoms.

ActivPhysics Online

19.1 Particle Scattering

The first experiments designed to probe the interior structure of the atom were the **Rutherford scattering experiments,** carried out in 1910–1911 by Ernest Rutherford (Fig. 38.14) and two of his students, Hans Geiger and Ernest Marsden, at the University of Manchester in England. These experiments consisted of projecting charged particles at thin foils of the elements under study and observing the deflections of the particles. The particle accelerators that are now in common use in laboratories had not yet been invented, and Rutherford's projectiles were *alpha particles* emitted from naturally radioactive elements. Alpha particles are identical with the nuclei of most helium atoms: two protons and two neutrons bound together. They are ejected from unstable nuclei with speeds of the order of 10^7 m/s, and they can travel several centimeters through air or 0.1 mm or so through solid matter before they are brought to rest by collisions.

Rutherford's experimental setup is shown schematically in Fig. 38.15. A radioactive substance at the left emits alpha particles. Thick lead screens stop all particles except those in a narrow beam defined by small holes. The beam then passes through a target consisting of a thin gold, silver, or copper foil and strikes screens coated with zinc sulfide, similar in principle to the screen of a TV picture tube. A momentary flash, or *scintillation,* can be seen on the screen whenever it is struck by an alpha particle. Rutherford and his students counted the numbers of particles deflected through various angles.

Think of the atoms of the target material as being packed together like marbles in a box. An alpha particle can pass through a thin sheet of metal foil, so the alpha particle must be able actually to pass through the interiors of atoms. The *total* electric charge of the atom is zero, so outside the atom there is little force on the alpha particle. Within the atom there are electrical forces caused by the electrons and by the positive charge. But the mass of an alpha particle is about 7300 times that of an electron. Momentum considerations show that the alpha particle can be scattered only a very small amount by its interaction with the much lighter electrons. It's like throwing a pebble through a swarm of mosquitoes; the mosquitoes don't deflect the pebble very much. Only interactions with the *positive* charge, which is tied to almost all of the mass of the atom, can deflect the alpha particle appreciably.

In the Thomson model, the positive charge and the negative electrons are distributed through the whole atom. Hence the electric field inside the atom should

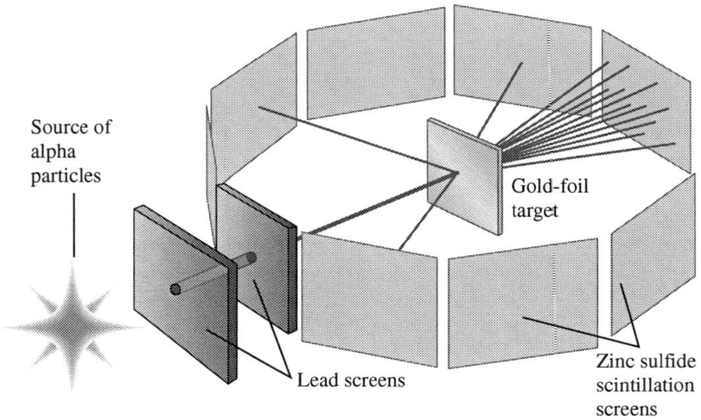

38.15 The scattering of alpha particles by a thin metal foil. The source of alpha particles is a radioactive element such as radium. The two lead screens with small holes form a narrow beam of alpha particles, which are scattered by the gold foil. The directions of the scattered particles are determined from the flashes on the scintillation screens.

be quite small, and the electric force on an alpha particle that enters the atom should be quite weak. The maximum deflection to be expected is then only a few degrees (Fig. 38.16a).

The results were very different from this and were totally unexpected. Some alpha particles were scattered by nearly 180°—that is, almost straight backward (Fig. 38.16b). Rutherford later wrote:

> It was quite the most incredible event that ever happened to me in my life. It was almost as incredible as if you had fired a 15-inch shell at a piece of tissue paper and it came back and hit you.

Back to the drawing board! Suppose the positive charge, instead of being distributed through a sphere with atomic dimensions (of the order of 10^{-10} m), is all concentrated in a much *smaller* space. Then it would act like a point charge down to much smaller distances. The maximum electric field repelling the alpha particle would be much larger, and the amazing large-angle scattering would be possible. Rutherford called this concentration of positive charge the **nucleus.** He again computed the numbers of particles expected to be scattered through various angles. Within the accuracy of his experiments, the computed and measured results agreed, down to distances of the order of 10^{-14} m. His experiments therefore established that the atom does have a nucleus, a very small, very dense structure, no larger than 10^{-14} m in diameter. The nucleus occupies only about 10^{-12} of the total volume of the atom or less, but it contains *all* the positive charge and at least 99.95% of the total mass of the atom.

Figure 38.17 shows a computer simulation of the scattering of 5.0-MeV alpha particles from a gold nucleus of radius 7.0×10^{-15} m (the actual value) and from a nucleus with a hypothetical radius ten times this great. In the second case there is *no* large-angle scattering. The presence of large-angle scattering in Rutherford's experiments thus attested to the small size of the nucleus.

38.16 **(a)** Thomson's model of the atom was ruled out by Rutherford's scattering experiments. **(b)** To account for his experimental results, Rutherford developed a nuclear model of the atom.

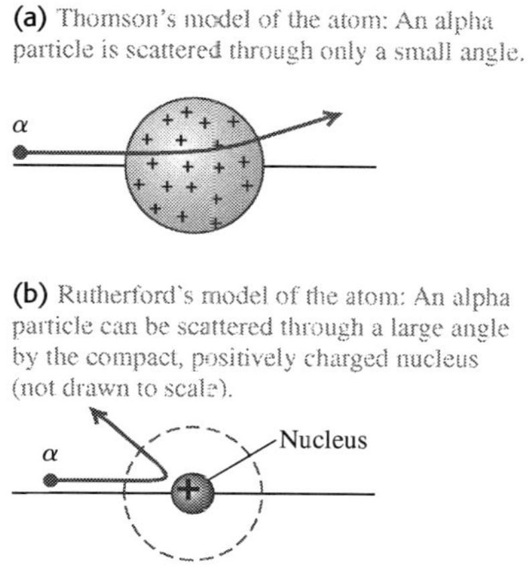

38.17 Computer simulation of scattering of 5.0-MeV alpha particles from a gold nucleus. Each curve shows a possible alpha-particle trajectory. **(a)** The scattering curves match Rutherford's experimental data if a radius of 7.0×10^{-15} m is assumed for a gold nucleus. **(b)** A model with a much larger radius does not match the data.

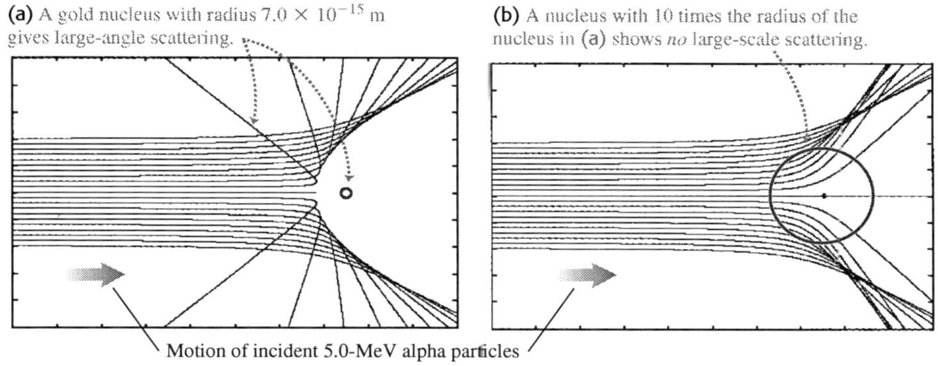

Example 38.5 A Rutherford experiment

An alpha particle is aimed directly at a gold nucleus. An alpha particle has two protons and a charge of $2e = 2(1.60 \times 10^{-19}\,\text{C})$, while a gold nucleus has 79 protons and a charge of $79e = 79(1.60 \times 10^{-19}\,\text{C})$. What minimum initial kinetic energy must the alpha particle have to approach within 5.0×10^{-14} m of the center of the gold nucleus? Assume that the gold nucleus, which has about 50 times the rest mass of an alpha particle, remains at rest.

SOLUTION

IDENTIFY: As the alpha particle approaches the gold nucleus, it slows down due to the repulsive electric force that acts on it. This is a conservative force, so the total mechanical energy (kinetic energy of the alpha particle plus electric potential energy of the system) is conserved.

Continued

SET UP: Let point 1 be very far from the gold nucleus (the initial position of the alpha particle), and let point 2 be 5.0×10^{-14} m from the center of the gold nucleus. Our target variable is the kinetic energy of the alpha particle at point 1 that allows it to reach point 2. To find this we use the law of conservation of energy and Eq. (23.9) for electric potential energy, $U = qq_0/4\pi\epsilon_0 r$.

EXECUTE: The kinetic energy at point 1 is K_1 (the target variable), and the kinetic energy at point 2 (where the alpha particle comes momentarily to rest) is $K_2 = 0$. At point 1 the alpha particle and gold nucleus are so far apart that we can regard their separation r as infinite, so from Eq. (23.9) the electric potential energy at point 1 is $U_1 = 0$. At point 2, the potential is

$$U_2 = \frac{1}{4\pi\epsilon_0}\frac{qq_0}{r}$$
$$= (9.0 \times 10^9\ \text{N}\cdot\text{m}^2/\text{C}^2)\frac{(2)(79)(1.60 \times 10^{-19}\ \text{C})^2}{5.0 \times 10^{-14}\ \text{m}}$$
$$= 7.3 \times 10^{-13}\ \text{J} = 4.6 \times 10^6\ \text{eV} = 4.6\ \text{MeV}$$

By energy conservation $K_1 + U_1 = K_2 + U_2$, so $K_1 = K_2 + U_2 - U_1 = 0 + 4.6\ \text{MeV} - 0 = 4.6\ \text{MeV}$. Thus, to approach to 5.0×10^{-14} m, the alpha particle must have 4.6 MeV when it is far away from the nucleus; to approach even closer, even greater kinetic energy is needed. In fact, alpha particles emitted from naturally occurring radioactive elements typically have energies in the range 4 to 6 MeV. For example, the common isotope of radium, ^{226}Ra, emits an alpha particle with energy 4.78 MeV.

EVALUATE: Was it valid to assume that the gold nucleus remains at rest? To find out, note that when the alpha particle stops momentarily, all of its initial momentum has been transferred to the gold nucleus. An alpha particle has a mass $m_\alpha = 6.64 \times 10^{-27}$ kg; if its initial kinetic energy $K_1 = \frac{1}{2}mv_1^2$ is 7.3×10^{-13} J, you can show that its initial speed is $v_1 = 1.5 \times 10^7$ m/s and its initial momentum is $p_1 = m_\alpha v_1 = 9.8 \times 10^{-20}\ \text{kg}\cdot\text{m/s}$. A gold nucleus (mass $m_{\text{Au}} = 3.27 \times 10^{-25}$ kg) with this much momentum has a much slower speed $v_{\text{Au}} = 3.0 \times 10^5$ m/s and a kinetic energy $K_{\text{Au}} = \frac{1}{2}mv_{\text{Au}}^2 = 1.5 \times 10^{-14}\ \text{J} = 0.092$ MeV. This *recoil kinetic energy* of the gold nucleus is a small fraction of the total energy in this situation, so we are justified in neglecting it.

Test Your Understanding of Section 38.4 Suppose you repeated Rutherford's scattering experiment using a thin sheet of solid hydrogen in place of the gold foil. (Hydrogen is a solid at temperatures below 14.0 K.) Would you expect to observe large-angle scattering of alpha particles?

38.5 The Bohr Model

ActivPhysics ONLINE
18.1 The Bohr Model

At the same time (1913) that Bohr established the relationship between spectral wavelengths and energy levels, he also proposed a model of the hydrogen atom. He developed his ideas while working in Rutherford's laboratory. Using this model, now known as the **Bohr model,** he was able to *calculate* the energy levels of hydrogen and obtain agreement with values determined from spectra.

Rutherford's discovery of the atomic nucleus raised a serious question. What kept the negatively charged electrons at relatively large distances ($\sim 10^{-10}$ m) away from the very small ($\sim 10^{-14}$ m), positively charged nucleus despite their electrostatic attraction? Rutherford suggested that perhaps the electrons *revolve* in orbits about the nucleus, just as the planets revolve around the sun.

But according to classical electromagnetic theory, any accelerating electric charge (either oscillating or revolving) radiates electromagnetic waves. An example is radiation from an oscillating point charge, shown in Fig. 32.2 (Section 32.1). The energy of an orbiting electron should therefore decrease continuously, its orbit should become smaller and smaller, and it should spiral rapidly into the nucleus (Fig. 38.18). Even worse, according to classical theory the *frequency* of the electromagnetic waves emitted should equal the frequency of revolution. As the electrons radiated energy, their angular speeds would change continuously, and they would emit a *continuous* spectrum (a mixture of all frequencies), not the *line* spectrum actually observed.

38.18 Classical physics predicts that an orbiting electron should continuously radiate electromagnetic waves and spiral into the nucleus. Because the electron's angular speed varies with its radius, classical physics predicts that the frequency of the emitted radiation should change continuously.

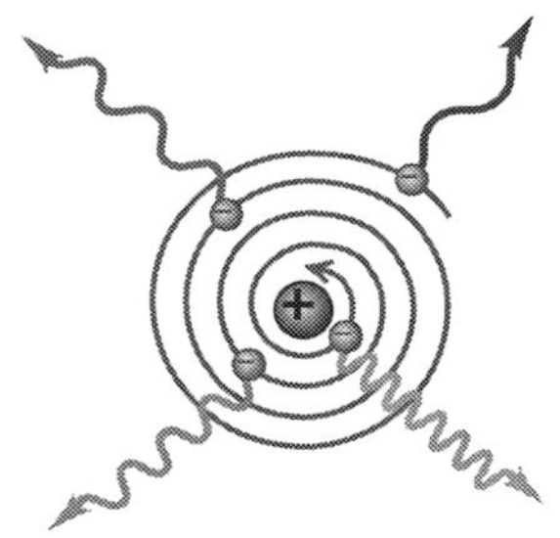

Stable Electron Orbits

To solve this problem, Bohr made a revolutionary proposal. He postulated that an electron in an atom can move around the nucleus in certain circular *stable orbits, without* emitting radiation, contrary to the predictions of classical electromagnetic theory. According to Bohr, there is a definite energy associated with each stable orbit, and an atom radiates energy only when it makes a transition from one of these orbits to another. The energy is radiated in the form of a photon with energy and frequency given by Eq. (38.6), $hf = E_i - E_f$.

As a result of a rather complicated argument that related the angular frequency of the light emitted to the angular speeds of the electron in highly excited energy levels, Bohr found that the magnitude of the electron's angular momentum is *quantized,* that this magnitude for the electron must be an integral multiple of $h/2\pi$. (Note that because $1\ \text{J} = 1\ \text{kg}\cdot\text{m}^2/\text{s}^2$, the SI units of Planck's constant h, $\text{J}\cdot\text{s}$, are the same as the SI units of angular momentum, usually written as $\text{kg}\cdot\text{m}^2/\text{s}$.) From Section 10.5, Eq. (10.28), the magnitude of the angular momentum is $L = mvr$ for a particle with mass m moving with speed v in a circle of radius r (thus with $\phi = 90°$). So Bohr's argument led to

$$L = mvr = n\frac{h}{2\pi}$$

where $n = 1, 2, 3, \ldots$. Each value of n corresponds to a permitted value of the orbit radius, which we denote from now on by r_n, and a corresponding speed v_n. The value of n for each orbit is called the **principal quantum number** for the orbit. With this notation the above equation becomes

$$L_n = mv_nr_n = n\frac{h}{2\pi} \quad \text{(quantization of angular momentum)} \tag{38.10}$$

Now let's consider a model of the hydrogen atom (Fig. 38.19) that is Newtonian in spirit but incorporates this quantization assumption. This atom consists of a single electron with mass m and charge $-e$, revolving around a single proton with charge $+e$. The proton is nearly 2000 times as massive as the electron, so we have been assuming that the proton does not move. We learned in Section 5.4 that when a particle with mass m moves with speed v_n in a circular orbit with radius r_n, its radially inward acceleration is v_n^2/r_n. According to Newton's second law, a radially inward net force with magnitude $F = mv_n^2/r_n$ is needed to cause this acceleration. We discussed in Section 12.4 how the gravitational attraction provided that force for satellite orbits. In hydrogen the force F is provided by the electrical attraction between the positive proton and the negative electron. From Coulomb's law (Eq. 21.2),

$$F = \frac{1}{4\pi\epsilon_0}\frac{e^2}{r_n^2}$$

so Newton's second law states that

$$\frac{1}{4\pi\epsilon_0}\frac{e^2}{r_n^2} = \frac{mv_n^2}{r_n} \tag{38.11}$$

38.19 Bohr model of the hydrogen atom.

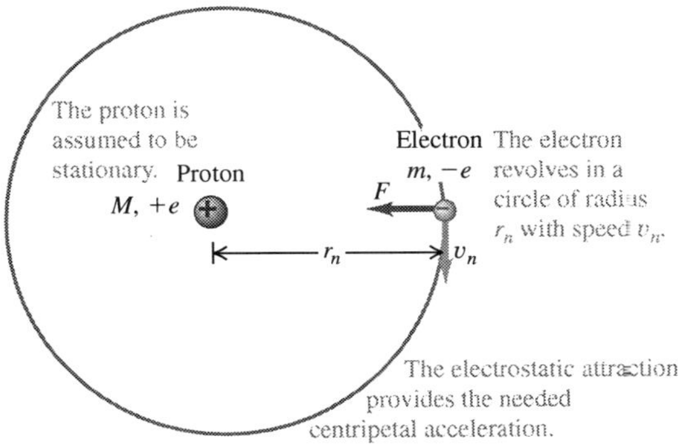

When we solve Eqs. (38.10) and (38.11) simultaneously for r_n and v_n, we get

$$r_n = \epsilon_0 \frac{n^2 h^2}{\pi m e^2} \quad \text{(orbit radii in the Bohr model)} \qquad (38.12)$$

$$v_n = \frac{1}{\epsilon_0} \frac{e^2}{2nh} \quad \text{(orbital speeds in the Bohr model)} \qquad (38.13)$$

Equation (38.12) shows that the orbit radius r_n is proportional to n^2; the smallest orbit radius corresponds to $n = 1$. We'll denote this minimum radius, called the *Bohr radius,* as a_0:

$$a_0 = \epsilon_0 \frac{h^2}{\pi m e^2} \qquad (38.14)$$

Then Eq. (38.12) can be written as

$$r_n = n^2 a_0 \qquad (38.15)$$

The permitted orbits have radii a_0, $4a_0$, $9a_0$, and so on.

The numerical values of the quantities on the right side of Eq. (38.14) are given in Appendix F. Using these values, we find that the radius a_0 of the smallest Bohr orbit is

$$\begin{aligned} a_0 &= \frac{(8.854 \times 10^{-12}\ \text{C}^2/\text{N} \cdot \text{m}^2)(6.626 \times 10^{-34}\ \text{J} \cdot \text{s})^2}{(3.142)(9.109 \times 10^{-31}\ \text{kg})(1.602 \times 10^{-19}\ \text{C})^2} \\ &= 5.29 \times 10^{-11}\ \text{m} \end{aligned}$$

This result, giving an atomic diameter of about 10^{-10} m = 0.1 nm, is consistent with atomic dimensions estimated by other methods.

We can use Eq. (38.13) to find the orbital *speed* of the electron. We leave this calculation to you (see Exercise 38.27); the result is that for the $n = 1$ state, $v_1 = 2.19 \times 10^6$ m/s. This, the greatest possible speed of the electron in the hydrogen atom, is less than 1% of the speed of light, showing that relativistic considerations aren't significant.

Hydrogen Energy Levels in the Bohr Model

We can use Eqs. (38.13) and (38.12) to find the kinetic and potential energies K_n and U_n when the electron is in the orbit with quantum number n:

$$K_n = \frac{1}{2} m v_n^2 = \frac{1}{\epsilon_0^2} \frac{m e^4}{8 n^2 h^2} \qquad (38.16)$$

$$U_n = -\frac{1}{4\pi\epsilon_0} \frac{e^2}{r_n} = -\frac{1}{\epsilon_0^2} \frac{m e^4}{4 n^2 h^2} \qquad (38.17)$$

The total energy E_n is the sum of the kinetic and potential energies:

$$E_n = K_n + U_n = -\frac{1}{\epsilon_0^2} \frac{m e^4}{8 n^2 h^2} \qquad (38.18)$$

The potential energy has a negative sign because we have taken the potential energy to be zero when the electron is infinitely far from the nucleus. We are interested only in energy *differences,* so the reference position doesn't matter.

The orbits and energy levels are depicted in Fig. 38.9 (Section 38.3). The possible energy levels of the atom are labeled by values of the quantum number n. For each value of n there are corresponding values of orbit radius r_n, speed v_n, angular momentum $L_n = nh/2\pi$, and total energy E_n. The energy of the atom is least when $n = 1$ and E_n has its most negative value. This is the *ground level* of the atom; it is the level with the smallest orbit, with radius a_0. For $n = 2, 3, \ldots,$

the absolute value of E_n is smaller and the energy is progressively larger (less negative). The orbit radius increases as n^2, as shown by Eqs. (38.12) and (38.15), while the speed decreases as $1/n$, as shown by Eq. (38.13).

Comparing the expression for E_n in Eq. (38.18) with Eq. (38.9) (deduced from the measured spectrum of hydrogen), we see that they agree only if the coefficients are equal:

$$hcR = \frac{1}{\epsilon_0^2}\frac{me^4}{8h^2} \quad \text{or} \quad R = \frac{me^4}{8\epsilon_0^2 h^3 c} \tag{38.19}$$

This equation therefore shows us how to *calculate* the value of the Rydberg constant from the fundamental physical constants m, c, e, h, and ϵ_0, all of which can be determined quite independently of the Bohr theory. When we substitute the numerical values of these quantities, we obtain the value $R = 1.097 \times 10^7\ \text{m}^{-1}$. To four significant figures, this is the value determined from wavelength measurements. This agreement provides very strong and direct confirmation of Bohr's theory. You should substitute numerical values into Eq. (38.19) and compute the value of R to confirm these statements.

The *ionization energy* of the hydrogen atom is the energy required to remove the electron completely. Ionization corresponds to a transition from the ground level $(n = 1)$ to an infinitely large orbit radius $(n = \infty)$ and thus equals $-E_1$. Substituting the constants from Appendix F into Eq. (38.18) gives an ionization energy of 13.606 eV. The ionization energy can also be measured directly; the result is 13.60 eV. These two values agree within 0.1%.

Example 38.6 Exploring the Bohr model

Find the kinetic, potential, and total energies of the hydrogen atom in the first excited level, and find the wavelength of the photon emitted in the transition from the first excited level to the ground level.

SOLUTION

IDENTIFY: This problem uses the ideas of the Bohr model discussed in this section.

SET UP: We use Eqs. (38.16), (38.17), and (38.18) to find the energies of the atom. To find the photon wavelength λ we use Eq. (38.6), which relates the energy and wavelength of an emitted photon to the energies of the initial and final levels of the atom.

EXECUTE: We could evaluate Eqs. (38.16), (38.17), and (38.18) for the nth level by substituting the values of m, e, ϵ_0, and h. But we can simplify the calculation by comparing with Eq. (38.19), which shows that the constant that appears in Eqs. (38.14), (38.15), and (38.16) is equal to hcR and is experimentally equal to 13.60 eV:

$$\frac{me^4}{8\epsilon_0^2 h^2} = hcR = 13.60\ \text{eV}$$

Using this expression, we can rewrite Eqs. (38.16), (38.17), and (38.18) as

$$K_n = \frac{13.60\ \text{eV}}{n^2} \qquad U_n = \frac{-27.20\ \text{eV}}{n^2} \qquad E_n = \frac{-13.60\ \text{eV}}{n^2}$$

The first excited level is the $n = 2$ level, which has $K_2 = 3.40$ eV, $U_2 = -6.80$ eV, and $E_2 = -3.40$ eV.

The ground level is the $n = 1$ level, for which the energy is $E_1 = -13.60$ eV. The energy of the emitted photon is $E_2 - E_1 = -3.40\ \text{eV} - (-13.60\ \text{eV}) = 10.20$ eV. The photon energy equals hc/λ, so we find

$$\lambda = \frac{hc}{E_2 - E_1} = \frac{(4.136 \times 10^{-15}\ \text{eV}\cdot\text{s})(3.00 \times 10^8\ \text{m/s})}{10.20\ \text{eV}}$$
$$= 1.22 \times 10^{-7}\ \text{m} = 122\ \text{nm}$$

This is the wavelength of the "Lyman alpha" line, the longest-wavelength line in the Lyman series of ultraviolet lines in the hydrogen spectrum (see Fig. 38.9).

EVALUATE: Our results show that the total mechanical energy is negative and equal to one-half the potential energy. Remarkably, we found the same energy relationship for satellite orbits in Section 12.4. The reason for the similarity between the two situations is that both the electrostatic force and the gravitational force are inversely proportional to $1/r^2$.

Nuclear Motion and the Reduced Mass of an Atom

The values of the Rydberg constant R and the ionization energy of hydrogen predicted by Bohr's analysis are within 0.1% of the measured values. The agreement would be even better if we had not assumed that the nucleus (a proton) remains at rest. Rather, the proton and electron both revolve in circular orbits about their common *center of mass* (see Section 8.5), as shown in Fig. 38.20. It turns out that we can take this motion into account very simply by using in Bohr's equations not the electron rest mass m but a quantity called the **reduced mass** m_r of the system. For a system composed of two bodies of masses m_1 and m_2, the reduced mass is defined as

$$m_r = \frac{m_1 m_2}{m_1 + m_2} \tag{38.20}$$

For ordinary hydrogen we let m_1 equal m and m_2 equal the proton mass, $m_p = 1836.2m$. Thus the proton-electron system of ordinary hydrogen has a reduced mass of

$$m_r = \frac{m(1836.2m)}{m + 1836.2m} = 0.99946m$$

When this value is used instead of the electron mass m in the Bohr equations, the predicted values agree very well with the measured values.

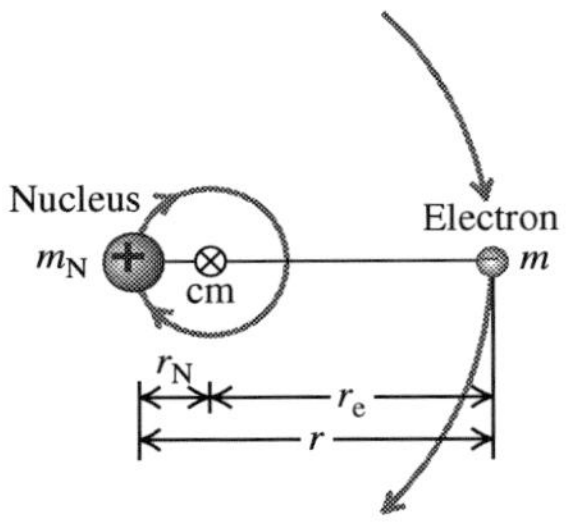

38.20 The nucleus and the electron revolve around their common center of mass. The distance r_N has been exaggerated for clarity; for ordinary hydrogen it actually equals $r_e/1836.2$.

In an atom of deuterium, also called *heavy hydrogen,* the nucleus is not a single proton but a proton and a neutron bound together to form a composite body called the *deuteron.* The reduced mass of the deuterium atom turns out to be $0.99973m$. Equations (38.6) and (38.18) (with m replaced by m_r) show that all energies and frequencies are proportional to m_r, while all wavelengths are inversely proportional to m_r. Thus the spectral wavelengths of deuterium should be those of hydrogen divided by $(0.99973m)/(0.99946m) = 1.00027$. This is a small effect but well within the precision of modern spectrometers. This small wavelength shift led the American scientist Harold Urey to the discovery of deuterium in 1932, an achievement that earned him the 1934 Nobel Prize in chemistry.

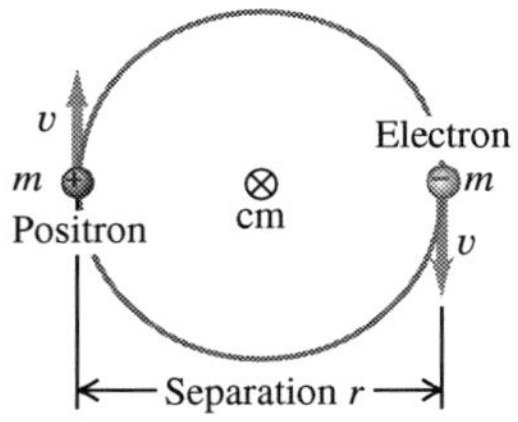

38.21 Applying the Bohr model to positronium. The electron and the positron revolve about their common center of mass, which is located midway between them because they have equal mass. Their separation is twice either orbit radius.

The concept of reduced mass has other applications. A *positron* has the same rest mass as an electron but a charge $+e$. A positronium "atom" (Fig. 38.21) consists of an electron and a positron, each with mass m, in orbit around their common center of mass. This structure lasts only about 10^{-6} s before the two particles annihilate one another and disappear, but this is enough time to study the positronium spectrum. The reduced mass is $m/2$, so the energy levels and photon frequencies have exactly half the values for the simple Bohr model with infinite proton mass. The existence of positronium atoms was confirmed by observation of the corresponding spectrum lines.

Hydrogenlike Atoms

We can extend the Bohr model to other one-electron atoms, such as singly ionized helium (He^+), doubly ionized lithium (Li^{2+}), and so on. Such atoms are called *hydrogenlike* atoms. In such atoms, the nuclear charge is not e but Ze, where Z is the *atomic number,* equal to the number of protons in the nucleus. The effect in the previous analysis is to replace e^2 everywhere by Ze^2. In particular, the orbit radii r_n given by Eq. (38.12) become smaller by a factor of Z, and the energy levels E_n given by Eq. (38.18) are multiplied by Z^2. We invite you to verify these statements. The reduced-mass correction in these cases is even less than 0.1% because the nuclei are more massive than the single proton of ordinary hydrogen. Figure 38.22 compares the energy levels for H and for He^+, which has $Z = 2$.

Atoms of the *alkali* metals have one electron outside a core consisting of the nucleus and the inner electrons, with net core charge $+e$. These atoms are approximately hydrogenlike, especially in excited levels. Even trapped electrons and electron vacancies in semiconducting solids act somewhat like hydrogen atoms.

Although the Bohr model predicted the energy levels of the hydrogen atom correctly, it raised as many questions as it answered. It combined elements of

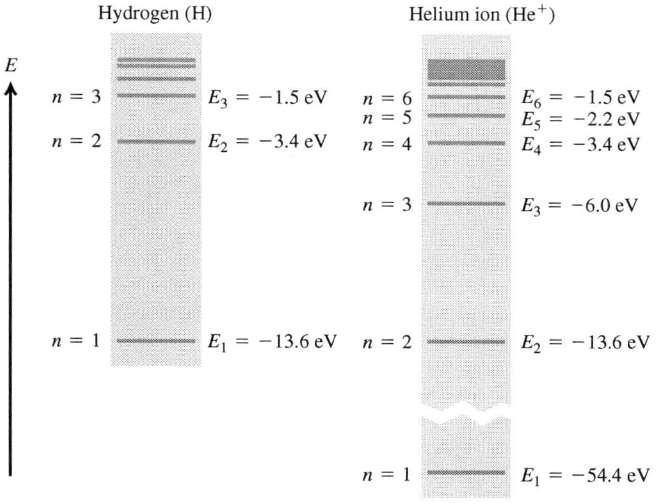

38.22 Energy levels of H and He^+. The energy expression, Eq. (38.18), is multiplied by $Z^2 = 4$ for He^+, so the energy of an He^+ ion with a given n is almost exactly four times that of an H atom with the same n. (There are small differences of the order of 0.05% because the reduced masses are slightly different.)

38.23 Three processes in which an atom interacts with electromagnetic waves. In part **(a)**, A denotes an atom in its ground level; in parts **(b)** and **(c)**, A* denotes an atom in an excited level.

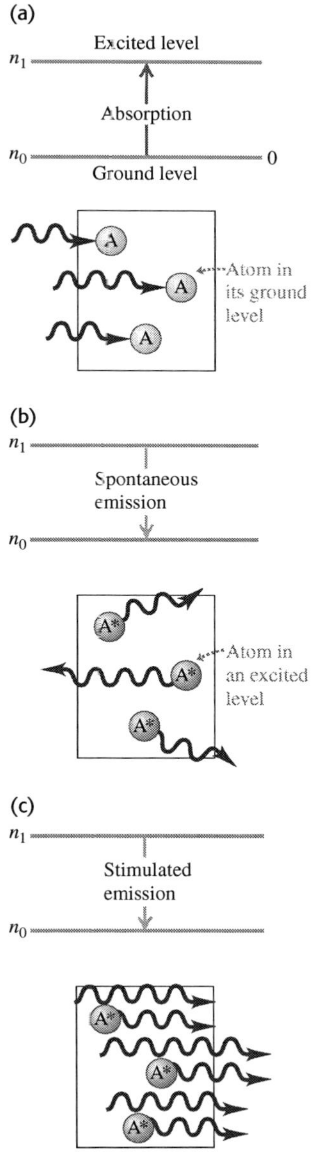

classical physics with new postulates that were inconsistent with classical ideas. The model provided no insight into what happens during a transition from one orbit to another; the angular speeds of the electron motion were not in general the angular frequencies of the emitted radiation, a result that is contrary to classical electrodynamics. Attempts to extend the model to atoms with two or more electrons were not successful. An electron moving in one of Bohr's circular orbits forms a current loop and should produce a magnetic moment. However, a hydrogen atom in its ground level has no magnetic moment due to orbital motion. In Chapters 39 and 40 we will find that an even more radical departure from classical concepts was needed before the understanding of atomic structure could progress further.

Test Your Understanding of Section 38.5 Consider the wavelengths in the spectrum of light emitted by excited He^+ ions. Are any of these wavelengths nearly equal to any of the wavelengths emitted by excited H atoms?

38.6 The Laser

The **laser** is a light source that produces a beam of highly coherent and very nearly monochromatic light as a result of cooperative emission from many atoms. The name "laser" is an acronym for "light amplification by stimulated emission of radiation." We can understand the principles of laser operation on the basis of photons and atomic energy levels. During the discussion we'll also introduce two new concepts: *stimulated emission* and *population inversion.*

Spontaneous and Stimulated Emission

If an atom has an excited level an energy E above the ground level, the atom in its ground level can absorb a photon with frequency f given by $E = hf$. This process is shown schematically in Fig. 38.23a, which shows a gas in a transparent container. Three atoms A each absorb a photon, reaching an excited level and being denoted as A*. Some time later, the excited atoms return to the ground level by each emitting a photon with the same frequency as the one originally absorbed. This process is called *spontaneous emission;* the direction and phase of the emitted photons are random (Fig. 38.23b).

In **stimulated emission** (Fig. 38.23c), each incident photon encounters a previously excited atom. A kind of resonance effect induces each atom to emit a second photon with the same frequency, direction, phase, and polarization as the incident photon, which is not changed by the process. For each atom there is one photon before a stimulated emission and two photons after—thus the name *light amplification.* Because the two photons have the same phase, they emerge together as *coherent* radiation. The laser makes use of stimulated emission to produce a beam consisting of a large number of such coherent photons.

To discuss stimulated emission from atoms in excited levels, we need to know something about how many atoms are in each of the various energy levels. First, we need to make the distinction between the terms *energy level* and *state.* A system may have more than one way to attain a given energy level; each different way is a different **state.** For instance, there are two ways of putting an ideal unstretched spring in a given energy level. Remembering that $U = \frac{1}{2}kx^2$, we could compress the spring by $x = -b$ or we could stretch it by $x = +b$ to get the same $U = \frac{1}{2}kb^2$. The Bohr model had only one state in each energy level, but we will find in Section 41.4 that the hydrogen atom actually has two states in its -13.60-eV ground level, eight states in its -3.40-eV first exited level, and so on.

The Maxwell-Boltzmann distribution function (see Section 18.5) determines the number of atoms in a given state in a gas. The function tells us that when the gas is in thermal equilibrium at absolute temperature T, the number n_i of atoms in a state with energy E_i equals $Ae^{-E_i/kT}$, where k is Boltzmann's constant and A is another constant determined by the total number of atoms in the gas. (In Section 18.5, E was the kinetic energy $\frac{1}{2}mv^2$ of a gas molecule; here we're talking about the internal energy of an atom.) Because of the negative exponent, fewer atoms are in higher-energy states, as we should expect. If E_g is a ground-state energy and E_{ex} is the energy of an excited state, then the ratio of numbers of atoms in the two states is

$$\frac{n_{ex}}{n_g} = \frac{Ae^{-E_{ex}/kT}}{Ae^{-E_g/kT}} = e^{-(E_{ex}-E_g)/kT} \tag{38.21}$$

For example, suppose $E_{ex} - E_g = 2.0\text{ eV} = 3.2 \times 10^{-19}\text{ J}$, the energy of a 620 nm visible-light photon. At $T = 3000\text{ K}$ (the temperature of the filament in an incandescent light bulb),

$$\frac{E_{ex} - E_g}{kT} = \frac{3.2 \times 10^{-19}\text{ J}}{(1.38 \times 10^{-23}\text{ J/K})(3000\text{ K})} = 7.73$$

and

$$e^{-(E_{ex}-E_g)/kT} = e^{-7.73} = 0.00044$$

That is, the fraction of atoms in a state 2.0 eV above a ground state is extremely small, even at this high temperature. The point is that at any reasonable temperature there aren't enough atoms in excited states for any appreciable amount of stimulated emission from these states to occur. Rather, absorption is much more probable.

Enhancing Stimulated Emission: Population Inversions

We could try to enhance the number of atoms in excited states by sending through the container a beam of radiation with frequency $f = E/h$ corresponding to the energy difference $E = E_{ex} - E_g$. Some of the atoms absorb photons of energy E and are raised to the excited state, and the population ratio n_{ex}/n_g momentarily increases. But because n_g is originally so much larger than n_{ex}, an enormously intense beam of light would be required to momentarily increase n_{ex} to a value comparable to n_g. The rate at which energy is *absorbed* from the beam by the n_g ground-state atoms far exceeds the rate at which energy is added to the beam by stimulated emission from the relatively rare (n_{ex}) excited atoms.

We need to create a nonequilibrium situation in which the number of atoms in a higher-energy state is greater than the number in a lower-energy state. Such a situation is called a **population inversion.** Then the rate of energy radiation by stimulated emission can *exceed* the rate of absorption, and the system acts as a net *source* of radiation with photon energy E. Furthermore, because the photons are the result of stimulated emission, they all have the same frequency, phase, polarization, and direction of propagation. The resulting radiation is therefore very much more coherent than light from ordinary sources, in which the emissions of individual atoms are not coordinated. This coherent emission is exactly what happens in a laser.

The necessary population inversion can be achieved in a variety of ways. A familiar example is the helium-neon laser, a common and inexpensive laser that is available in many undergraduate laboratories. A mixture of helium and neon, typically at a pressure of the order of 10^2 Pa $(10^{-3}$ atm$)$, is sealed in a glass enclosure provided with two electrodes. When a sufficiently high voltage is applied, an electric discharge occurs. Collisions between ionized atoms and electrons carrying the discharge current excite atoms to various energy states.

Figure 38.24a shows an energy-level diagram for the system. The labels for the various energy levels, such as 1*s*, 3*p*, and 5*s*, refer to states of the electrons in the levels. (We'll discuss this notation in Chapter 41; we don't need detailed knowledge of the nature of the states to understand laser action.) Because of restrictions imposed by conservation of angular momentum, a helium atom with an electron excited to a 2*s* state cannot return to a ground $(1s)$ state by emitting a 20.61-eV photon, as you might expect it to do. Such a state, in which single-photon emission is impossible, has an unusually long lifetime and is called a **metastable state.** Helium atoms "pile up" in the metastable 2*s* states, creating a population inversion relative to the ground states.

However, excited helium atoms *can* lose energy by energy-exchange collisions with neon atoms that are initially in a ground state. A helium atom, excited to a 2*s* state 20.61 eV above its ground states and possessing a little additional kinetic energy, can collide with a neon atom in a ground state, exciting the *neon* atom to a 5*s* excited state at 20.66 eV and dropping the *helium* atom back to a ground state. Thus we have the necessary mechanism for a population inversion in neon, with more neon atoms per 5*s* state than per 3*p* state.

38.24 (a) Energy-level diagram for a helium-neon laser. (b) Operation of a helium-neon laser. The light emitted by electrons returning to lower-energy states is reflected between mirrors, so it continues to stimulate emission of more coherent light. One mirror is partially transmitting and allows the high-intensity light beam to escape.

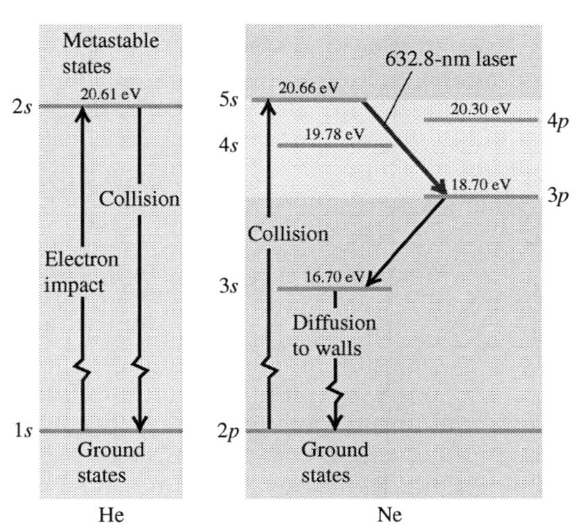

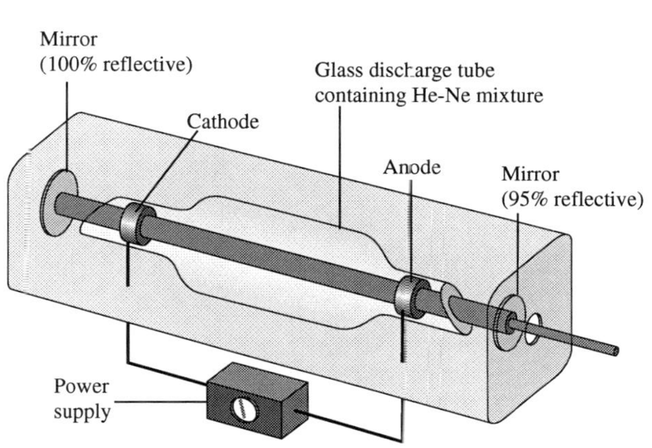

Stimulated emissions during transitions from a $5s$ to a $3p$ state then result in the emission of highly coherent light at 632.8 nm, as shown in Fig. 38.24a. In practice the beam is sent back and forth through the gas many times by a pair of parallel mirrors (Fig. 38.24b), so as to stimulate emission from as many excited atoms as possible. One of the mirrors is partially transparent, so a portion of the beam emerges as an external beam. The net result of all these processes is a beam of light that can be quite intense, has parallel rays, is highly monochromatic, and is spatially *coherent* at all points within a given cross section.

Other types of lasers use different processes to achieve the necessary population inversion. In a semiconductor laser the inversion is obtained by driving electrons and holes to a *p-n* junction (to be discussed in Section 42.7) with a steady electric field. In one type of *chemical* laser a chemical reaction forms molecules in metastable excited states. In a carbon dioxide gas dynamic laser the population inversion results from rapid expansion of the gas. A *maser* (acronym for "microwave amplification by stimulated emission of radiation") operates on the basis of population inversions in molecules, using closely spaced energy levels. The corresponding emitted radiation is in the microwave range. Maser action even occurs naturally in interstellar molecular clouds.

Lasers have found a wide variety of practical applications. A high-intensity laser beam can drill a very small hole in a diamond for use as a die in drawing very small-diameter wire. Surveyors often use lasers, especially in situations requiring great precision such as drilling a long tunnel from both ends; the laser beam has parallel rays that can travel long distances without appreciable spreading.

Lasers are widely used in medicine. A laser with a very narrow intense beam can be used in the treatment of a detached retina; a short burst of radiation damages a small area of the retina, and the resulting scar tissue "welds" the retina back to the membrane from which it has become detached (see Fig. 33.2). Laser beams are used in surgery; blood vessels cut by the beam tend to seal themselves off, making it easier to control bleeding. Lasers are also used for selective destruction of tissue, as in the removal of tumors.

Test Your Understanding of Section 38.6 An ordinary neon light fixture (like those used in advertising signs) emits red light with the same 632.8-nm wavelength as a helium-neon laser. The light emitted by a neon light fixture is (i) spontaneous emission; (ii) stimulated emission; (iii) both spontaneous and stimulated emission.

38.7 X-Ray Production and Scattering

The production and scattering of x rays provide additional examples of the quantum nature of electromagnetic radiation. X rays are produced when rapidly moving electrons that have been accelerated through a potential difference of the order of 10^3 to 10^6 V strike a metal target. They were first produced in 1895 by Wilhelm Röntgen (1845–1923), using an apparatus similar in principle to the setup shown in Fig. 38.25. Electrons are "boiled off" from the heated cathode by thermionic emission and are accelerated toward the anode (the target) by a large potential difference V_{AC}. The bulb is evacuated (residual pressure 10^{-7} atm or less), so the electrons can travel from the cathode to the anode without colliding with air molecules. When V_{AC} is a few thousand volts or more, a very penetrating radiation is emitted from the anode surface.

38.25 An apparatus used to produce x rays, similar to Röntgen's 1895 apparatus.

Electrons are emitted thermionically from the heated cathode and are accelerated toward the anode; when they strike it, x rays are produced.

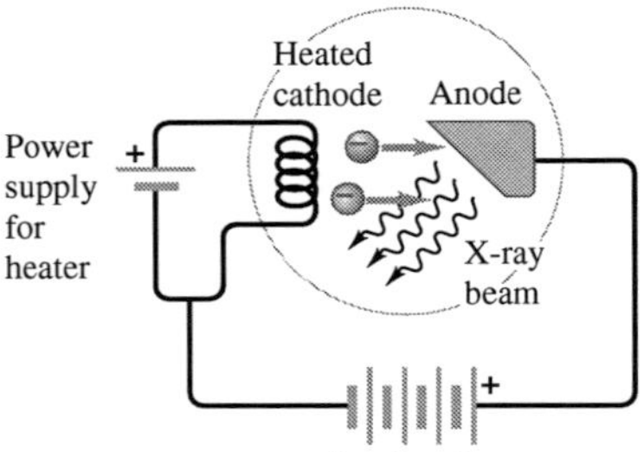

X-Ray Photons

Because they are emitted by accelerated charges, it is clear that x rays are electromagnetic waves. Like light, x rays are governed by quantum relationships in their interaction with matter. Thus we can talk about x-ray photons or quanta, and the energy of an x-ray photon is related to its frequency and wavelength in the same way as for photons of light, $E = hf = hc/\lambda$. Typical x-ray wavelengths are 0.001 to 1 nm $(10^{-12}$ to 10^{-9} m$)$. X-ray wavelengths can be measured quite precisely by crystal diffraction techniques, which we discussed in Section 36.6.

X-ray emission is the inverse of the photoelectric effect. In photoelectric emission there is a transformation of the energy of a photon into the kinetic energy of an electron; in x-ray production there is a transformation of the kinetic energy of an electron into the energy of a photon. The energy relationships are similar. In x-ray production we usually neglect the work function of the target and the initial kinetic energy of the "boiled-off" electrons because they are very small in comparison to the other energies.

Two distinct processes are involved in x-ray emission. In the first process, some electrons are slowed down or stopped by the target, and part or all of their kinetic energy is converted directly to a continuous spectrum of photons, including x rays. This process is called *bremsstrahlung* (German for "braking radiation"). Classical physics is completely unable to explain why the x rays that are emitted in a bremsstrahlung process have a maximum frequency f_{max} and a corresponding minimum wavelength λ_{min}, much less predict their values. With quantum concepts it is easy. An electron has charge $-e$ and gains kinetic energy eV_{AC} when accelerated through a potential increase V_{AC}. The most energetic photon (highest frequency and shortest wavelength) is produced when all the electron's kinetic energy goes to produce one photon; that is,

$$eV_{AC} = hf_{max} = \frac{hc}{\lambda_{min}} \quad \text{(bremsstrahlung)} \qquad (38.22)$$

Note that the maximum frequency and minimum wavelength in the bremsstrahlung process do *not* depend on the target material.

The second process gives peaks in the x-ray spectrum at characteristic frequencies and wavelengths that *do* depend on the target material. Other electrons, if they have enough kinetic energy, can transfer that energy partly or completely to individual atoms within the target. These atoms are left in excited levels; when they decay back to their ground levels, they may emit x-ray photons. Since each element has a unique set of atomic energy levels, each also has a characteristic x-ray spectrum. The energy levels associated with x rays are rather different in character from those associated with visible spectra. They involve vacancies in the inner electron configurations of complex atoms. The energy differences between these levels can be hundreds or thousands of electron volts, rather than a few electron volts as is typical for optical spectra. We will return to energy levels and spectra associated with x rays in Section 41.5.

Example 38.7 Producing x rays

Electrons in an x-ray tube are accelerated by a potential difference of 10.0 kV. If an electron produces one photon on impact with the target, what is the minimum wavelength of the resulting x rays? Answer using both SI units and electron volts.

SOLUTION

IDENTIFY: To produce an x-ray photon with minimum wavelength and hence maximum energy, all of the kinetic energy of an electron must go into producing a single x-ray photon.

SET UP: We use Eq. (38.22) to determine the wavelength in this situation.

EXECUTE: From Eq. (38.22), using SI units we find

$$\lambda_{min} = \frac{hc}{eV_{AC}} = \frac{(6.626 \times 10^{-34}\ \text{J}\cdot\text{s})(3.00 \times 10^{8}\ \text{m/s})}{(1.602 \times 10^{-19}\ \text{C})(10.0 \times 10^{3}\ \text{V})}$$
$$= 1.24 \times 10^{-10}\ \text{m} = 0.124\ \text{nm}$$

Using electron volts, we have

$$\lambda_{min} = \frac{hc}{eV_{AC}} = \frac{(4.136 \times 10^{-15}\ \text{eV}\cdot\text{s})(3.00 \times 10^{8}\ \text{m/s})}{e(10.0 \times 10^{3}\ \text{V})}$$
$$= 1.24 \times 10^{-10}\ \text{m} = 0.124\ \text{nm}$$

Continued

Note that the "e" in the unit eV is canceled by the "e" for the magnitude of the electron charge, because the electron volt (eV) is the magnitude of the electron charge e times one volt (1 V).

EVALUATE: To check our result, recall from Example 38.6 (Section 38.5) that a photon of energy 10.2 eV has a wavelength of 122 nm. In the present example the electron energy, and therefore the energy of the x-ray photon, is 10.0×10^3 eV = 10.0 keV (about 10^3 times greater than in Example 38.6) and the wavelength is about 10^{-3} times as great as in Example 38.6. This makes sense, since wavelength and photon energy are inversely proportional.

ActivPhysics Online
17.4 Compton Scattering

Compton Scattering

A phenomenon called **Compton scattering,** first explained in 1923 by the American physicist Arthur H. Compton, provides additional direct confirmation of the quantum nature of x rays. When x rays strike matter, some of the radiation is *scattered,* just as visible light falling on a rough surface undergoes diffuse reflection. Compton and others discovered that some of the scattered radiation has smaller frequency (longer wavelength) than the incident radiation and that the change in wavelength depends on the angle through which the radiation is scattered. Specifically, if the scattered radiation emerges at an angle ϕ with respect to the incident direction (Fig. 38.26) and if λ and λ' are the wavelengths of the incident and scattered radiation, respectively, we find that

$$\lambda' - \lambda = \frac{h}{mc}(1 - \cos\phi) \qquad \text{(Compton scattering)} \qquad (38.23)$$

where m is the electron rest mass. The quantity h/mc that appears in Eq. (38.23) has units of length. Its numerical value is

$$\frac{h}{mc} = \frac{6.626 \times 10^{-34}\ \text{J} \cdot \text{s}}{(9.109 \times 10^{-31}\ \text{kg})(2.998 \times 10^{8}\ \text{m/s})} = 2.426 \times 10^{-12}\ \text{m}$$

We cannot explain Compton scattering using classical electromagnetic theory, which predicts that the scattered wave has the same wavelength as the incident wave. However the quantum theory provides a beautifully clear explanation. We imagine the scattering process as a collision of two *particles,* the incident photon and an electron that is initially at rest, as in Fig. 38.27a. The incident photon disappears, giving part of its energy and momentum to the electron, which recoils as a result of this impact. The remainder goes to a new scattered photon that therefore has less energy, smaller frequency, and longer wavelength than the incident one (Fig. 38.27b).

We can derive Eq. (38.23) from the principles of conservation of energy and momentum. We outline the derivation below; we invite you to fill in the details (see Exercise 38.41). The electron recoil energy may be in the relativistic range, so we have to use the relativistic energy–momentum relationships, Eqs. (37.39) and (37.40). The incident photon has momentum $\vec{p}$, with magnitude p and energy pc. The scattered photon has momentum $\vec{p}\,'$, with magnitude p' and energy $p'c$. The electron is initially at rest, so its initial momentum is zero and its initial energy is its rest energy mc^2. The final electron momentum $\vec{P}_e$ has magnitude P_e, and the final electron energy is given by $E_e^2 = (mc^2)^2 + (P_e c)^2$. Then energy conservation gives us the relationship

$$pc + mc^2 = p'c + E_e$$

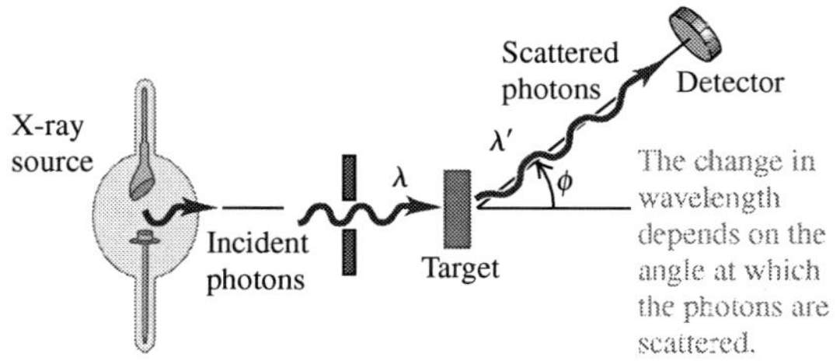

38.26 A Compton-effect experiment.

38.27 Schematic diagram of Compton scattering.

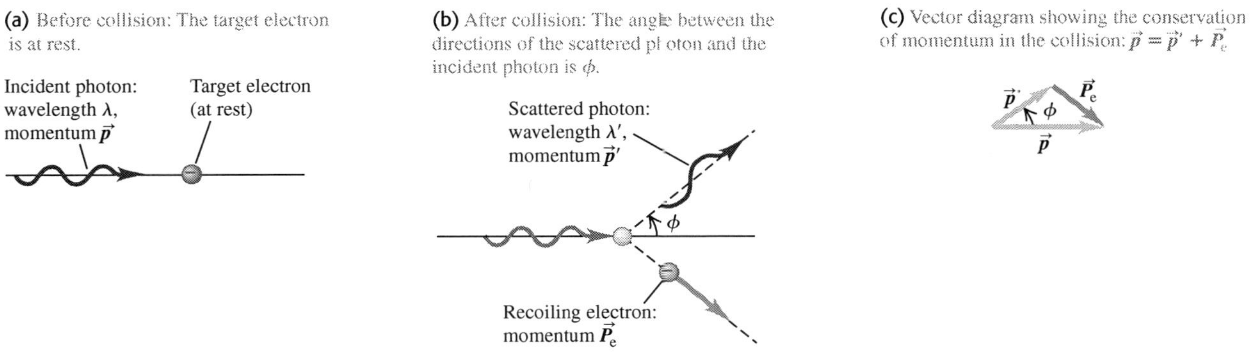

Rearranging, we find

$$(pc - p'c + mc^2)^2 = E_e^2 = (mc^2)^2 + (P_e c)^2 \qquad (38.24)$$

We can eliminate the electron momentum $\vec{P}_e$ from Eq. (38.24) by using momentum conservation (Fig. 38.27c):

$$\vec{p} = \vec{p}\,' + \vec{P}_e \quad \text{or} \quad \vec{P}_e = \vec{p} - \vec{p}\,' \qquad (38.25)$$

By taking the scalar product of each side of Eq. (38.25) with itself or by using the law of cosines with the vector diagram in Fig. 38.27c, we find

$$P_e^2 = p^2 + p'^2 - 2pp'\cos\phi \qquad (38.26)$$

We now substitute this expression for P_e^2 into Eq. (38.24) and multiply out the left side. We divide out a common factor c^2; several terms cancel, and when the resulting equation is divided through by (pp'), the result is

$$\frac{mc}{p'} - \frac{mc}{p} = 1 - \cos\phi \qquad (38.27)$$

Finally, we substitute $p' = h/\lambda'$ and $p = h/\lambda$, then multiply by h/mc to obtain Eq. (38.23).

When the wavelengths of x rays scattered at a certain angle are measured, the curve of intensity per unit wavelength as a function of wavelength has two peaks (Fig. 38.28). The longer-wavelength peak represents Compton scattering. The shorter-wavelength peak, labeled λ_0, is at the wavelength of the incident x rays and corresponds to x-ray scattering from tightly bound electrons. In such scattering processes the entire atom must recoil, so the m in Eq. (38.23) is the mass of the entire atom rather than of a single electron. The resulting wavelength shifts are negligible.

38.28 Intensity as a function of wavelength for photons scattered at an angle of 135° in a Compton scattering experiment.

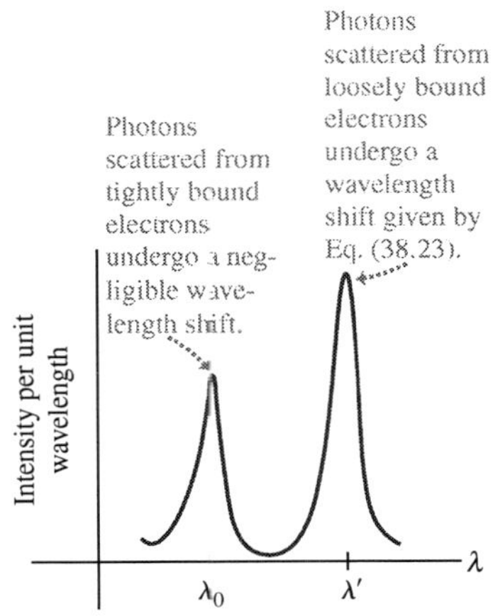

Example 38.8 Compton scattering

You use the x-ray photons in Example 38.7 (with $\lambda = 0.124$ nm) in a Compton scattering experiment. (a) At what angle is the wavelength of the scattered x rays 1.0% longer than that of the incident x rays? (b) At what angle is it 0.050% longer?

SOLUTION

IDENTIFY: This problem uses the relationship between scattering angle and wavelength shift in the Compton effect.

SET UP: In each case our target variable is the angle ϕ shown in Fig. 38.27b. We solve for it using Eq. (38.23).

EXECUTE: (a) In Eq. (38.23) we want $\Delta\lambda = \lambda' - \lambda$ to be 1.0% of 0.124 nm. That is, $\Delta\lambda = 0.00124\ \text{nm} = 1.24 \times 10^{-12}$ m. Using the value $h/mc = 2.426 \times 10^{-12}$ m, we find

$$\Delta\lambda = \frac{h}{mc}(1 - \cos\phi)$$

$$\cos\phi = 1 - \frac{\Delta\lambda}{h/mc} = 1 - \frac{1.24 \times 10^{-12}\ \text{m}}{2.426 \times 10^{-12}\ \text{m}} = 0.4889$$

$$\phi = 60.7°$$

Continued

(b) For $\Delta\lambda$ to be 0.050% of 0.124 nm, or 6.2×10^{-14} m,

$$\cos\phi = 1 - \frac{6.2 \times 10^{-14}\ \text{m}}{2.426 \times 10^{-12}\ \text{m}} = 0.9744$$

$$\phi = 13.0°$$

EVALUATE: Comparing our results for parts (a) and (b) shows that smaller angles give smaller wavelength shifts. Thus in a grazing collision the photon energy loss and the electron recoil energy are smaller than when the scattering angle is larger. This is just what we would expect for an elastic collision, whether between a photon and an electron or between two billiard balls.

Applications of X Rays

X rays have many practical applications in medicine and industry. Because x-ray photons are of such high energy, they can penetrate several centimeters of solid matter and so can be used to visualize the interiors of materials that are opaque to ordinary light, such as broken bones or defects in structural steel. The object to be visualized is placed between an x-ray source and a large sheet of photographic film; the darkening of the film is proportional to the radiation exposure. A crack or air bubble allows greater transmission and shows as a dark area. Bones appear lighter than the surrounding soft tissue because they contain greater proportions of elements with higher atomic number (and greater absorption); in the soft tissue the light elements carbon, hydrogen, and oxygen predominate.

A widely used and vastly improved x-ray technique is *computed tomography;* the corresponding instrument is called a *CT scanner.* The x-ray source produces a thin, fan-shaped beam that is detected on the opposite side of the subject by an array of several hundred detectors in a line. Each detector measures absorption along a thin line through the subject. The entire apparatus is rotated around the subject in the plane of the beam during a few seconds. The changing photon-counting rates of the detectors are recorded digitally; a computer processes this information and reconstructs a picture of absorption over an entire cross section of the subject (Fig. 38.29). Differences as small as 1% can be detected with CT scans, and tumors and other anomalies that are much too small to be seen with older x-ray techniques can be detected.

38.29 This radiologist is operating a CT scanner (seen through the window) from a separate room to avoid repeated exposure to x rays.

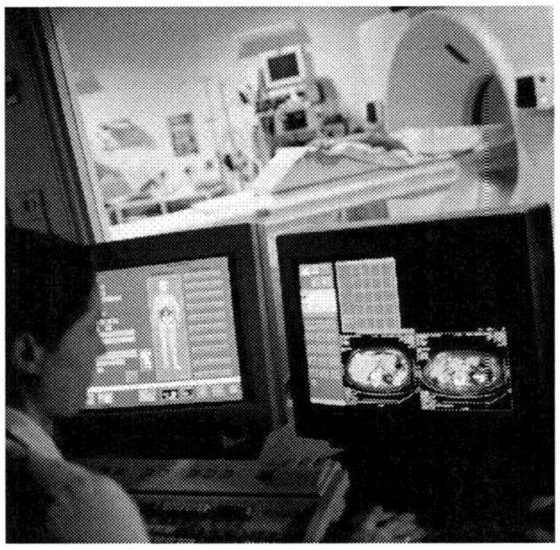

X rays cause damage to living tissues. As x-ray photons are absorbed in tissues, their energy breaks molecular bonds and creates highly reactive free radicals (such as neutral H and OH), which in turn can disturb the molecular structure of proteins and especially genetic material. Young and rapidly growing cells are particularly susceptible; thus x rays are useful for selective destruction of cancer cells. Conversely, however, a cell may be damaged by radiation but survive, continue dividing, and produce generations of defective cells; thus x rays can *cause* cancer.

Even when the organism itself shows no apparent damage, excessive radiation exposure can cause changes in the organism's reproductive system that will affect its offspring. A careful assessment of the balance between risks and benefits of radiation exposure is essential in each individual case.

Test Your Understanding of Section 38.7 If you used visible-light photons in the experiment shown in Fig. 38.26, would the photons undergo a wavelength shift in the scattering? If so, is it possible to detect the shift with the human eye?

38.8 Continuous Spectra

Line spectra are emitted by matter in the gaseous state, in which the atoms are so far apart that interactions between them are negligible and each atom behaves as an isolated system. Hot matter in condensed states (solid or liquid) nearly always emits radiation with a *continuous* distribution of wavelengths rather than a line spectrum. An ideal surface that absorbs all wavelengths of electromagnetic radiation incident upon it is also the best possible emitter of electromagnetic radiation at any wavelength. Such an ideal surface is called a *blackbody,* and the continuous-spectrum radiation that it emits is called **blackbody radiation.** By 1900 this radiation had been studied extensively, and two characteristics had been established.

First, the total intensity I (the average rate of radiation of energy per unit surface area or average power per area) emitted from the surface of an ideal radiator is proportional to the fourth power of the absolute temperature (Fig. 38.30). We studied this relationship in Section 17.7 during our study of heat transfer mechanisms. This total intensity I emitted at absolute temperature T is given by the **Stefan-Boltzmann law:**

$$I = \sigma T^4 \qquad \text{(Stefan-Boltzmann law for a blackbody)} \qquad (38.28)$$

where σ is a fundamental physical constant called the *Stefan-Boltzmann constant.* In SI units,

$$\sigma = 5.670400(40) \times 10^{-8} \frac{\text{W}}{\text{m}^2 \cdot \text{K}^4}$$

Second, the intensity is not uniformly distributed over all wavelengths. Its distribution can be measured and described by the intensity per wavelength interval $I(\lambda)$, called the *spectral emittance.* Thus $I(\lambda)\,d\lambda$ is the intensity corresponding to wavelengths in the interval from λ to $\lambda + d\lambda$. The *total* intensity I, given by Eq. (38.28), is the *integral* of the distribution function $I(\lambda)$ over all wavelengths, which equals the area under the $I(\lambda)$ versus λ curve:

$$I = \int_0^\infty I(\lambda)\,d\lambda \qquad (38.29)$$

CAUTION **Spectral emittance vs. intensity** Although we use the symbol $I(\lambda)$ for spectral emittance, keep in mind that spectral emittance is *not* the same thing as intensity I. Intensity is power per unit area, with units W/m^2; spectral emittance is power per unit area per unit wavelength interval, with units W/m^3.

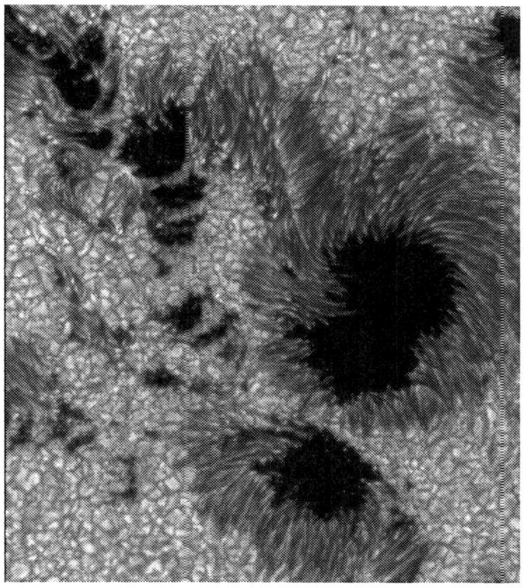

38.30 This close-up view of the sun's surface shows two dark sunspots. Their temperature is about 4000 K, while the surrounding solar material is at $T =$ 5800 K. From the Stefan-Boltzmann law, the intensity from a given area of sunspot is only $(4000\text{ K}/5800\text{ K})^4 = 0.23$ as great as the intensity from the same area of the surrounding material—which is why sunspots appear dark.

Measured spectral emittances $I(\lambda)$ for three different temperatures are shown in Fig. 38.31. Each has a peak wavelength λ_m at which the emitted intensity per wavelength interval is largest. Experiment shows that λ_m is inversely proportional to T, so their product is constant. This result is called the **Wien displacement law.** The experimental value of the constant is $2.90 \times 10^{-3}\text{ m} \cdot \text{K}$:

$$\lambda_m T = 2.90 \times 10^{-3}\text{ m} \cdot \text{K} \qquad \text{(Wien displacement law)} \qquad (38.30)$$

As the temperature rises, the peak of $I(\lambda)$ becomes higher and shifts to shorter wavelengths. A body that glows yellow is hotter and brighter than one that glows red; yellow light has shorter wavelengths than red light. Finally, experiments show that the *shape* of the distribution function is the same for all temperatures; we can make a curve for one temperature fit any other temperature by simply changing the scales on the graph.

38.31 Spectral emittance $I(\lambda)$ for radiation from a blackbody at three different temperatures.

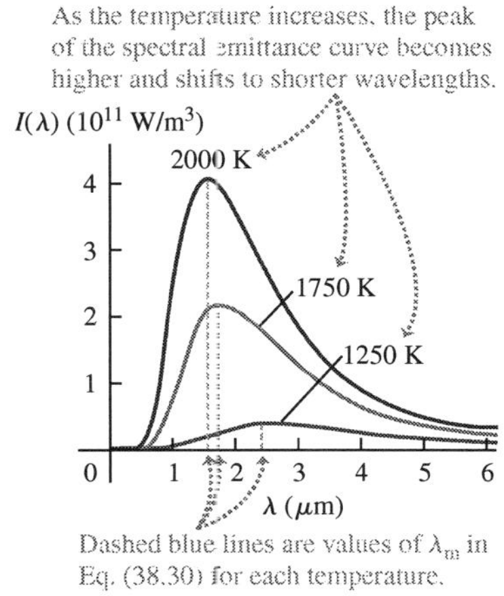

Rayleigh and the "Ultraviolet Catastrophe"

During the last decade of the 19th century, many attempts were made to *derive* these empirical results from basic principles. In one attempt, the English physicist Lord Rayleigh considered light enclosed in a rectangular box with perfectly reflecting sides. Such a box has a series of possible *normal modes* for electromagnetic waves, as discussed in Section 32.5. It seemed reasonable to assume that the distribution of energy among the various modes was given by the equipartition principle (see Section 18.4), which had been successfully used in the analysis of heat capacities. A small hole in the box would behave as an ideal blackbody radiator.

Including both the electric- and magnetic-field energies, Rayleigh assumed that the total energy of each normal mode was equal to kT. Then by computing the *number* of normal modes corresponding to a wavelength interval $d\lambda$, Rayleigh could predict the distribution of wavelengths in the radiation within the box. Finally, he could compute the intensity distribution $I(\lambda)$ of the radiation emerging from a small hole in the box. His result was quite simple:

$$I(\lambda) = \frac{2\pi ckT}{\lambda^4} \tag{38.31}$$

At large wavelengths this formula agrees quite well with the experimental results shown in Fig. 38.31, but there is serious disagreement at small wavelengths. The experimental curve falls toward zero at small λ, but Rayleigh's curve goes the opposite direction, approaching infinity as $1/\lambda^4$, a result that was called in Rayleigh's time the "ultraviolet catastrophe." Even worse, the integral of Eq. (38.31) over all λ is infinite, indicating an infinitely large *total* radiated intensity. Clearly, something is wrong.

Planck and the Quantum Hypothesis

Finally, in 1900, the German physicist Max Planck succeeded in deriving a function, now called the **Planck radiation law,** that agreed very well with experimental intensity distribution curves. In his derivation he made what seemed at the time to be a crazy assumption. Planck assumed that electromagnetic oscillators (electrons) in the walls of Rayleigh's box vibrating at a frequency f could have only certain values of energy equal to nhf, where $n = 0, 1, 2, 3, \ldots$ and h turned out to be the constant that now bears Planck's name. These oscillators were in equilibrium with the electromagnetic waves in the box. His assumption gave quantized energy levels and was in sharp contrast to Rayleigh's point of view, which was that each normal mode could have any amount of energy.

Planck was not comfortable with this *quantum hypothesis;* he regarded it as a calculational trick rather than a fundamental principle. In a letter to a friend, he called it "an act of desperation" into which he was forced because "a theoretical explanation had to be found at any cost, whatever the price." But five years later, Einstein identified the energy change hf between levels as the energy of a photon to explain the photoelectric effect (see Section 38.2), and other evidence quickly mounted. By 1915 there was little doubt about the validity of the quantum concept and the existence of photons. By discussing atomic spectra *before* continuous spectra, we have departed from the historical order of things. The credit for inventing the concept of quantization of energy levels goes to Planck, even though he didn't believe it at first.

We won't go into the details of Planck's derivation of the intensity distribution. Here is his result:

$$I(\lambda) = \frac{2\pi hc^2}{\lambda^5(e^{hc/\lambda kT} - 1)} \quad \text{(Planck radiation law)} \tag{38.32}$$

where h is Planck's constant, c is the speed of light, k is Boltzmann's constant, T is the *absolute* temperature, and λ is the wavelength. This function turns out to agree well with experimental intensity curves such as those in Fig. 38.31.

The Planck radiation law also contains the Wien displacement law and the Stefan-Boltzmann law as consequences. To derive the Wien law, we take the derivative of Eq. (38.32) and set it equal to zero to find the value of λ at which

$I(\lambda)$ is maximum. We leave the details as a problem (see Exercise 38.47); the result is

$$\lambda_m = \frac{hc}{4.965kT} \tag{38.33}$$

To obtain this result, you have to solve the equation

$$5 - x = 5e^{-x} \tag{38.34}$$

The root of this equation, found by trial and error or more sophisticated means, is 4.965 to four significant figures. You should evaluate the constant $hc/4.965k$ and show that it agrees with the experimental value of 2.90×10^{-3} m · K given in Eq. (38.30).

We can obtain the Stefan-Boltzmann law for a blackbody by integrating Eq. (38.32) over all λ to find the *total* radiated intensity (see Problem 38.77). This is not a simple integral; the result is

$$I = \int_0^\infty I(\lambda)\, d\lambda = \frac{2\pi^5 k^4}{15c^2h^3}T^4 = \sigma T^4 \tag{38.35}$$

in agreement with Eq. (38.28). This result also shows that the constant σ in that law can be expressed as a combination of other fundamental constants:

$$\sigma = \frac{2\pi^5 k^4}{15c^2h^3} \tag{38.36}$$

You should substitute the values of k, c, and h from Appendix F and verify that you obtain the Stefan-Boltzmann constant $\sigma = 5.6704 \times 10^{-8}\ \text{W/m}^2 \cdot \text{K}^4$.

The general form of Eq. (38.33) is what we should expect from kinetic theory. If photon energies are typically of the order of kT, as suggested by the equipartition theorem, then for a typical photon we would expect

$$E \approx kT \approx \frac{hc}{\lambda} \quad \text{and} \quad \lambda \approx \frac{hc}{kT} \tag{38.37}$$

Indeed, a photon with wavelength given by Eq. (38.33) has an energy $E = 4.965kT$.

The Planck radiation law, Eq. (38.32), looks so different from the unsuccessful Rayleigh expression, Eq. (38.31), that it may seem unlikely that they would agree at large values of λ. But when λ is large, the exponent in the denominator of Eq. (38.32) is very small. We can then use the approximation $e^x \approx 1 + x$ (for $x \ll 1$). You should verify that when this is done, the result approaches Eq. (38.31), showing that the two expressions do agree in the limit of very large λ. We also note that the Rayleigh expression does not contain h. At very long wavelengths and correspondingly very small photon energies, quantum effects become unimportant.

Example 38.9 Light from the sun

The surface of the sun has a temperature of approximately 5800 K. To a good approximation we may treat it as a blackbody. (a) What is the peak-intensity wavelength λ_m? (b) What is the total radiated power per unit area of surface?

SOLUTION

IDENTIFY: This problem involves the Wien displacement law (which relates the temperature T of a blackbody and its peak-intensity wavelength λ_m) and the Stefan-Boltzmann law (which relates the blackbody temperature and the radiated power per area I).

SET UP: We use Eq. (38.30) to determine λ_m and Eq. (38.28) to calculate I.

EXECUTE: (a) From Eq. (38.30),

$$\lambda_m = \frac{2.90 \times 10^{-3}\ \text{m} \cdot \text{K}}{T} = \frac{2.90 \times 10^{-3}\ \text{m} \cdot \text{K}}{5800\ \text{K}}$$
$$= 0.500 \times 10^{-6}\ \text{m} = 500\ \text{nm}$$

Continued

(b) From Eq. (38.28),

$$I = \sigma T^4 = (5.67 \times 10^{-8}\ \mathrm{W/m^2 \cdot K^4})(5800\ \mathrm{K})^4$$
$$= 6.42 \times 10^7\ \mathrm{W/m^2} = 64.2\ \mathrm{MW/m^2}$$

EVALUATE: The wavelength found in part (b) is near the middle of the visible spectrum. This is not a surprising result: The human eye evolved to take maximum advantage of natural light.

The enormous value of I found in part (b) is the intensity at the *surface* of the sun. When the radiated power reaches the earth, the intensity drops to about 1.4 $\mathrm{kW/m^2}$ because the power spreads out over the much larger area of a sphere with a radius equal to that of the earth's orbit.

Example 38.10 A slice of sunlight

Find the power per unit area radiated from the surface of the sun in the wavelength range 600.0 to 605.0 nm.

SOLUTION

IDENTIFY: This question concerns the power emitted by a blackbody over a narrow range of wavelengths, and so involves the spectral emittance (intensity distribution) $I(\lambda)$.

SET UP: For an exact result we should integrate Eq. (38.32) between the limits 600.0 and 605.0 nm, finding the area under the $I(\lambda)$ curve between these limits. This integral can't be evaluated in terms of familiar functions, so we *approximate* the area by the height of the curve at the median wavelength $\lambda = 602.5$ nm multiplied by the width of the interval ($\Delta\lambda = 5.0$ nm).

EXECUTE: To obtain the height of the $I(\lambda)$ curve at $\lambda = 602.5\ \mathrm{nm} = 6.025 \times 10^{-7}$ m, we first evaluate the quantity $hc/\lambda kT$ for this value of λ and for $T = 5800$ K (the temperature of the solar surface; see Example 38.9). We then substitute the result into Eq. (38.32):

$$\frac{hc}{\lambda kT} = \frac{(6.626 \times 10^{-34}\ \mathrm{J \cdot s})(2.998 \times 10^8\ \mathrm{m/s})}{(6.025 \times 10^{-7}\ \mathrm{m})(1.381 \times 10^{-23}\ \mathrm{J/K})(5800\ \mathrm{K})}$$
$$= 4.116$$

$$I(\lambda) = \frac{2\pi(6.626 \times 10^{-34}\ \mathrm{J \cdot s})(2.998 \times 10^8\ \mathrm{m/s})^2}{(6.025 \times 10^{-7}\ \mathrm{m})^5(e^{4.116} - 1)}$$
$$= 7.81 \times 10^{13}\ \mathrm{W/m^3}$$

The intensity in the 5.0-nm range from 600.0 to 605.0 nm is then approximately

$$I(\lambda)\Delta\lambda = (7.81 \times 10^{13}\ \mathrm{W/m^3})(5.0 \times 10^{-9}\ \mathrm{m})$$
$$= 3.9 \times 10^5\ \mathrm{W/m^2} = 0.39\ \mathrm{MW/m^2}$$

EVALUATE: We see from this result and from part (b) of Example 38.9 (in which we found the power radiated per unit area at *all* wavelengths) that about 0.6% of the total power radiated from the sun is in the wavelength range from 600 to 605 nm. Had we found that more power is radiated in this range than the total calculated in Example 38.9, something would have been very wrong with our calculations!

Test Your Understanding of Section 38.8 (a) Does a blackbody at 2000 K emit x rays? (b) Does it emit radio waves?

38.9 Wave–Particle Duality

We have studied many examples of the behavior of light and other electromagnetic radiation. Some, including the interference and diffraction effects described in Chapters 35 and 36, demonstrate conclusively the *wave* nature of light. Others, the subject of the present chapter, point with equal force to the *particle* nature of light. At first glance these two aspects seem to be in direct conflict. How can light be a wave and a particle at the same time?

We can find the answer to this apparent wave–particle conflict in the **principle of complementarity,** first stated by Niels Bohr in 1928. The wave descriptions and the particle descriptions are complementary. That is, we need both descriptions to complete our model of nature, but we will never need to use both descriptions at the same time to describe a single part of an occurrence.

Diffraction and Interference in the Photon Picture

Let's start by considering again the diffraction pattern for a single slit, which we analyzed in Sections 36.2 and 36.3. Instead of recording the pattern on photographic film, we use a detector called a *photomultiplier* that can actually detect individual photons. Using the setup shown in Fig. 38.32, we place the photomultiplier at various positions for equal time intervals, count photons at each position, and plot out the intensity distribution.

We find that, on average, the distribution of photons agrees with our predictions from Section 36.3. At points corresponding to the maxima of the pattern, we count many photons; at minimum points, we count almost none, and so on. The graph of the counts at various points gives us the same diffraction pattern that we predicted with Eq. (36.7).

But suppose we now reduce the intensity to the point at which only a few photons per second pass through the slit. With only a few photons we can't expect to get the smooth diffraction curve that we found with very large numbers. In fact, there is no way to predict exactly where any individual photon will go. To reconcile the wave and particle aspects of this pattern, we have to regard the pattern as a *statistical* distribution that tells us how many photons, on average, go various places, or the *probability* for an individual photon to land in each of several places. But we can't predict exactly where an individual photon will go.

Now let's take a brief look at a quantum interpretation of a *two-slit* optical interference pattern, which we studied in Section 35.2. We can again trace out the pattern using a photomultiplier and a counter. We reduce the light intensity to a level of a few photons per second (Fig. 38.33). Again we can't predict exactly where an individual photon will go; the interference pattern is a *statistical distribution.*

How does the principle of complementarity apply to these diffraction and interference experiments? The wave description, not the particle description, explains the single- and double-slit patterns. But the particle description, not the wave description, explains why the photomultiplier measures the pattern to be built up by discrete packages of energy. The two descriptions complete our understanding of the results.

In these experiments, light goes through a slit or slits, and photons are detected at various positions on a screen. Suppose we ask the following question: "We just detected a photon at a certain position; when the photon went through the slit, how did it know which way to go?" The problem with the question is that it is asked in terms of a *particle* description. The wave nature of light, not the particle nature, determines the distribution of photons. Asking that question is trying to force a particle description (the photon being somehow directed which way to go) on a wave phenomenon (the formation of the pattern).

38.32 Single-slit diffraction pattern observed with a movable photomultiplier. The curve shows the intensity distribution predicted by the wave picture. The photon distribution is shown by the numbers of photons counted at various positions.

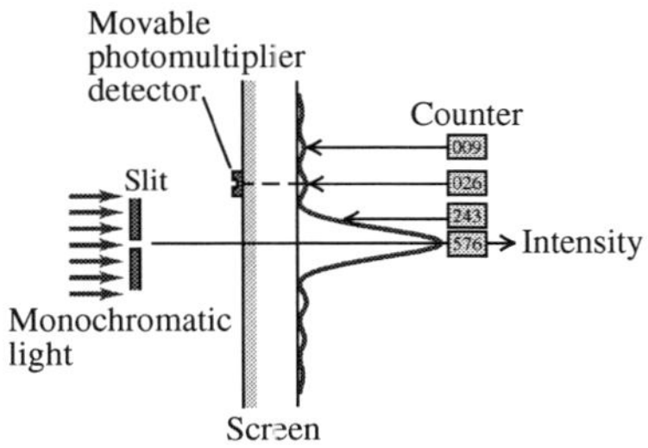

38.33 These images record the positions where individual photons in a two-slit interference experiment strike the screen. As more photons reach the screen, a recognizable interference pattern appears.

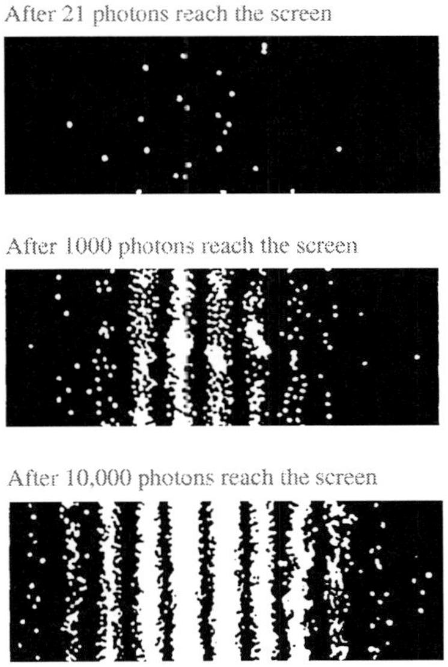

Quantum Electrodynamics

We know from experiments that sending electromagnetic waves through slits gives interference patterns and that these patterns are built up by individual photons. But if we didn't know the experimental results, how would we predict when to apply the wave description and when to apply the particle description? We need a theory that includes both these descriptions and predicts as well as explains both types of behavior. Such a comprehensive theory is called *quantum electrodynamics* (QED). In this theory the concept of energy levels of an atomic system is extended to electromagnetic fields. Just as an atom exists only in certain definite energy states, so the electromagnetic field has certain well-defined energy states, corresponding to the presence of various numbers of photons with assorted energies, momenta, and polarizations. QED came to full flower 50 years after Planck's quantum hypothesis of 1900 gave quantum mechanics its conceptual birth.

In the following chapters we will find that particles such as electrons also have a dual wave–particle personality. One of the great achievements of quantum mechanics has been to reconcile these apparently incompatible aspects of behavior of photons, electrons, and other constituents of matter.

CHAPTER 38 SUMMARY

Photons: Electromagnetic radiation behaves as both waves and particles. The energy in an electromagnetic wave is carried in units called photons. The energy E of one photon is proportional to the wave frequency f and inversely proportional to the wavelength λ, and is proportional to a universal quantity h called Planck's constant. The momentum of a photon has magnitude E/c. (See Example 38.1.)

$$E = hf = \frac{hc}{\lambda} \qquad (38.2)$$

$$p = \frac{E}{c} = \frac{hf}{c} = \frac{h}{\lambda} \qquad (38.5)$$

The photoelectric effect: In the photoelectric effect, a surface can eject an electron by absorbing a photon whose energy hf is greater than or equal to the work function ϕ of the material. The stopping potential V_0 is the voltage required to stop a current of ejected electrons from reaching an anode. (See Examples 38.2 and 38.3.)

$$eV_0 = hf - \phi \qquad (38.4)$$

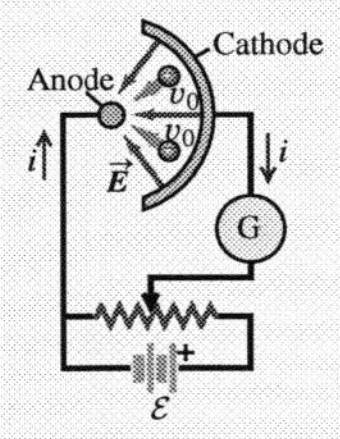

Atomic line spectra and energy levels: When an atom makes a transition from an energy level E_i to a lower level E_f, the energy of the emitted photon is equal to $E_i - E_f$. The energy levels of the hydrogen atom are given by Eq. (38.9), where R is the Rydberg constant. All of the observed spectral series of hydrogen can be understood in terms of these levels. (See Example 38.4.)

$$hf = \frac{hc}{\lambda} = E_i - E_f \qquad (38.6)$$

$$E_n = -\frac{hcR}{n^2} = -\frac{13.60\text{ eV}}{n^2} \quad (n = 1, 2, 3, \ldots) \qquad (38.9)$$

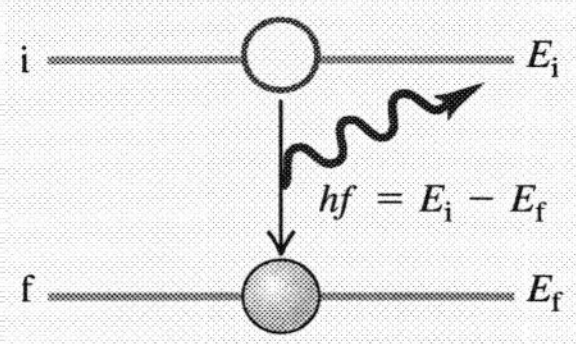

The nuclear atom: The Rutherford scattering experiments show that at the center of an atom is a dense nucleus, much smaller than the overall size of the atom but containing all of the positive charge and most of the mass. (See Example 38.5.)

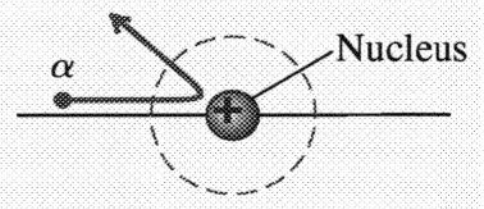

The Bohr model: In the Bohr model of the hydrogen atom, the permitted values of angular momentum are integral multiples of $h/2\pi$. The integer multiplier n is called the principal quantum number for the level. The orbital radii are proportional to n^2 and the orbital speeds are proportional to $1/n$. (See Example 38.6.)

$$L_n = mv_n r_n = n\frac{h}{2\pi} \quad n = 1, 2, 3, \ldots \qquad (38.10)$$

$$r_n = \epsilon_0 \frac{n^2 h^2}{\pi m e^2} = n^2 a_0 = n^2(5.29 \times 10^{-11}\text{ m}) \qquad (38.12)$$

$$v_n = \frac{1}{\epsilon_0}\frac{e^2}{2nh} = \frac{2.19 \times 10^6\text{ m/s}}{n} \qquad (38.13)$$

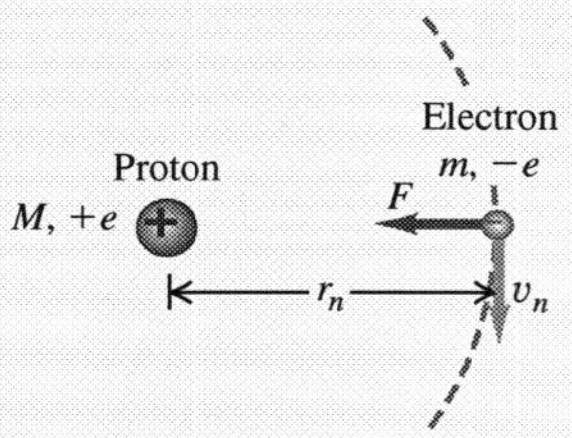

The laser: The laser operates on the principle of stimulated emission, by which many photons with identical wavelength and phase are emitted. Laser operation requires a nonequilibrium condition called a population inversion, in which more atoms are in a higher-energy state than are in a lower-energy state.

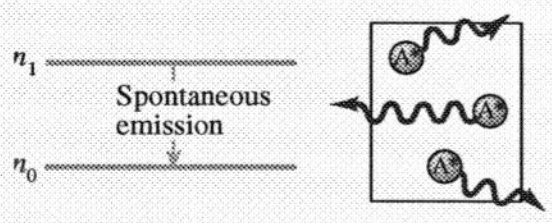

X-ray production and scattering: X rays can be produced by electron impact on a target. If electrons are accelerated through a potential increase V_{AC}, the maximum frequency and minimum wavelength that they can produce are given by Eq. (38.22). (See Example 38.7.) Compton scattering is scattering of x-ray photons by electrons. For free electrons (mass m), the wavelengths of incident and scattered photons are related to the photon scattering angle ϕ by Eq. (38.23). (See Example 38.8.)

$$eV_{AC} = hf_{max} = \frac{hc}{\lambda_{min}} \quad (38.22)$$

$$\lambda' - \lambda = \frac{h}{mc}(1 - \cos\phi) \quad (38.23)$$

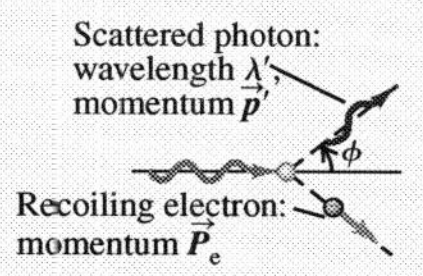

Blackbody radiation: The total radiated intensity (average power radiated per area) from a blackbody surface is proportional to the fourth power of the absolute temperature T. The quantity $\sigma = 5.67 \times 10^{-8}\ \mathrm{W/m^2 \cdot K^4}$ is called the Stefan-Boltzmann constant. The wavelength λ_m at which a blackbody radiates most strongly is inversely proportional to T. The Planck radiation law gives the spectral emittance $I(\lambda)$ (intensity per wavelength interval in blackbody radiation). (See Examples 38.9 and 38.10.)

$$I = \sigma T^4 \quad (38.28)$$
(Stefan-Boltzmann law)

$$\lambda_m T = 2.90 \times 10^{-3}\ \mathrm{m \cdot K} \quad (38.30)$$
(Wien displacement law)

$$I(\lambda) = \frac{2\pi hc^2}{\lambda^5(e^{hc/\lambda kT} - 1)} \quad (38.32)$$
(Planck radiation law)

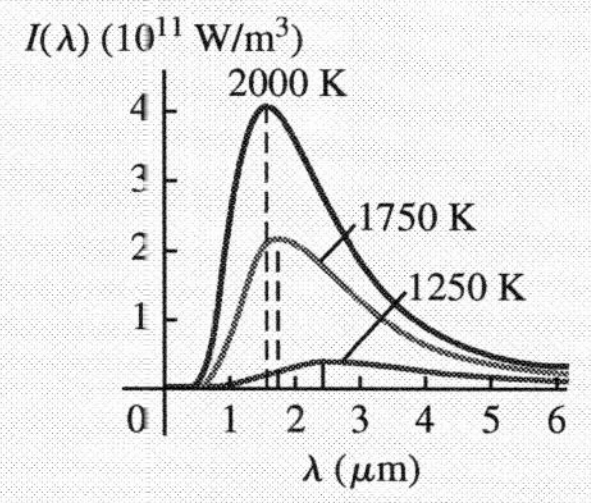

Key Terms

line spectrum, *1308*
photoelectric effect, *1309*
work function, *1310*
threshold frequency, *1310*
stopping potential, *1310*
photon, *1311*
Planck's constant, *1311*
energy level, *1314*
Rydberg constant, *1315*
ground level, *1317*
excited level, *1317*
absorption spectrum, *1318*
Rutherford scattering experiments, *1320*
nucleus, *1321*
Bohr model, *1322*
principal quantum number, *1323*
reduced mass, *1326*
laser, *1327*
stimulated emission, *1327*
state, *1328*
population inversion, *1329*
metastable state, *1329*
Compton scattering, *1332*
blackbody radiation, *1335*
Stefan-Boltzmann law, *1335*
Wien displacement law, *1335*
Planck radiation law, *1336*
principle of complementarity, *1338*

Answer to Chapter Opening Question ?

A photon of wavelength 656.3 nm (the H_α line in Fig. 38.8) is emitted when the electron in a hydrogen atom drops from the $n = 3$ energy level to the $n = 2$ level. The fact that we observe this tells us that the hydrogen gas in the Lagoon Nebula is very highly excited. (If the gas were not excited, all of the atoms would be in the lowest energy level and could not emit photons.) The gas is excited by high-energy ultraviolet photons coming from hot, luminous stars within the Lagoon Nebula.

Answers to Test Your Understanding Questions

38.2 Answers: (i) and (ii) From Eq. (38.2), a photon of energy $E = 1.14\ \mathrm{eV}$ has wavelength $\lambda = hc/E = (4.136 \times 10^{-15}\ \mathrm{eV}) \cdot (3.00 \times 10^8\ \mathrm{m/s})/(1.14\ \mathrm{eV}) = 1.09 \times 10^{-6}\ \mathrm{m} = 1090\ \mathrm{nm}$. This is in the infrared part of the spectrum. Since wavelength is inversely proportional to photon energy, the *minimum* photon energy of 1.14 eV corresponds to the *maximum* wavelength that causes photoconductivity in silicon. Thus the wavelength must be 1090 nm or less.

38.3 Answer: no For the atom to absorb a photon, the photon energy must be precisely equal to the energy difference between the initial and final energy levels of the atom (see Fig. 38.11).

38.4 Answer: no The nucleus of a hydrogen atom is a proton (mass 1.67×10^{-27} kg), which has only about one-fourth of the mass of an alpha particle $(6.64 \times 10^{-27}\ \mathrm{kg})$. Because the alpha particle is more massive, it won't bounce back in even a head-on collision with a proton. It's like a bowling ball colliding with a ping-pong ball at rest (see Fig. 8.22b). Thus there would be no large-angle scattering in this case. In Rutherford's experiment, by contrast, there *was* large-angle scattering because a gold nucleus is more massive than an alpha particle. The analogy there is a ping-pong ball hitting a bowling ball at rest (see Fig. 8.22a).

38.5 Answer: yes Figure 38.22 shows that many (though not all) of the energy levels of He^+ are the same as those of H. Hence photons emitted during transitions between corresponding pairs of levels in He^+ and H have the same energy E and the same wavelength $\lambda = hc/E$. For example, a H atom that drops from the $n = 2$ level to the $n = 1$ level emits a photon of energy 10.20 eV and wavelength 122 nm (see Example 38.6); a He^+ ion emits a photon of the same energy and wavelength when it drops from the $n = 4$ level to the $n = 2$ level.

38.6 Answer: (i) In a neon light fixture, a large potential difference is applied between the ends of a neon-filled glass tube. This ionizes some of the neon atoms, allowing a current of electrons to flow through the gas. Some of the neon atoms are struck by fast-moving electrons, exciting the atoms to the 5s level shown in Fig. 38.24a. From this level the atoms transition spontaneously to the $3p$ level and emit 632.8-nm photons in the process. The photons escape out of the sides of the glass tube. This process is *spontaneous* emission, as depicted in Fig. 38.23b. No population inversion occurs and the photons are not trapped by mirrors as shown in Fig. 38.24b, so there is no stimulated emission. Hence there is no laser action.

38.7 Answer: yes, no Equation (38.23) shows that the wavelength shift $\Delta\lambda = \lambda' - \lambda$ depends only on the photon scattering angle ϕ, not on the wavelength of the incident photon. So a visible-light photon scattered through an angle ϕ undergoes the same wavelength shift as an x-ray photon. Equation (38.23) also shows that this shift is of the order of $h/mc = 2.426 \times 10^{-12}$ m = 0.002426 nm. This is a few percent of the wavelength of x rays (see Example 38.8), so the effect is noticeable in x-ray scattering. However, h/mc is a tiny fraction of the wavelength of visible light (between 400 and 700nm). The human eye cannot distinguish such minuscule differences in wavelength (that is, differences in color).

38.8 Answer: (a) yes, (b) yes The Planck radiation law, Eq. (38.32), shows that an ideal blackbody emits radiation at *all* wavelengths: The spectral emittance $I(\lambda)$ is equal to zero only for $\lambda = 0$ and in the limit $\lambda \to \infty$. So a blackbody at 2000 K does indeed emit both x rays and radio waves. However, Fig. 38.31 shows that the spectral emittance for this temperature is very low for wavelengths much shorter than 1 μm (including x rays) and for wavelengths much longer than a few μm (including radio waves). Hence such a blackbody emits very little in the way of x rays or radio waves.

PROBLEMS

For instructor-assigned homework, go to **www.masteringphysics.com**

Discussion Questions

Q38.1. In what ways do photons resemble other particles such as electrons? In what ways do they differ? Do photons have mass? Do they have electric charge? Can they be accelerated? What mechanical properties do they have?

Q38.2. There is a certain probability that a single electron may simultaneously absorb *two* identical photons from a high-intensity laser. How would such an occurrence affect the threshold frequency and the equations of Section 38.2? Explain.

Q38.3. According to the photon model, light carries its energy in packets called quanta or photons. Why then don't we see a series of flashes when we look at things?

Q38.4. Would you expect effects due to the photon nature of light to be generally more important at the low-frequency end of the electromagnetic spectrum (radio waves) or at the high-frequency end (x rays and gamma rays)? Why?

Q38.5. During the photoelectric effect, light knocks electrons out of metals. So why don't the metals in your home lose their electrons when you turn on the lights?

Q38.6. Most black-and-white photographic film (with the exception of some special-purpose films) is less sensitive to red light than blue light and has almost no sensitivity to infrared. How can these properties be understood on the basis of photons?

Q38.7. Human skin is relatively insensitive to visible light, but ultraviolet radiation can cause severe burns. Does this have anything to do with photon energies? Explain.

Q38.8. Explain why Fig. 38.4 shows that most photoelectrons have kinetic energies less than $hf - \phi$, and also explain how these smaller kinetic energies occur.

Q38.9. Figure 38.5 shows that in a photoelectric-effect experiment, the photocurrent i for large positive values of V_{AC} has the same value no matter what the light frequency f (provided that f is higher than the threshold frequency f_0). Explain why.

Q38.10. In an experiment involving the photoelectric effect, if the intensity of the incident light (having frequency higher than the threshold frequency) is reduced by a factor of 10 without changing anything else, which (if any) of the following statements about this process will be true? (a) The number of photoelectrons will most likely be reduced by a factor of 10. (b) The maximum kinetic energy of the ejected photoelectrons will most likely be reduced by a factor of 10. (c) The maximum speed of the ejected photoelectrons will most likely be reduced by a factor of 10. (d) The maximum speed of the ejected photoelectrons will most likely be reduced by a factor of $\sqrt{10}$. (e) The time for the first photoelectron to be ejected will be increased by a factor of 10.

Q38.11. The materials called *phosphors* that coat the inside of a fluorescent lamp convert ultraviolet radiation (from the mercury-vapor discharge inside the tube) into visible light. Could one also make a phosphor that converts visible light to ultraviolet? Explain.

Q38.12. In a photoelectric-effect experiment, which of the following will increase the maximum kinetic energy of the photoelectrons; (a) Use light of greater intensity; (b) use light of higher frequency; (c) use light of longer wavelength; (d) use a metal surface with a larger work function. In each case justify your answer.

Q38.13. Galaxies tend to be strong emitters of Lyman-α photons (from the $n = 2$ to $n = 1$ transition in atomic hydrogen). But the intergalactic medium—the very thin gas between the galaxies—tends to *absorb* Lyman-α photons. What can you infer from these observations about the temperature in these two environments? Explain.

Q38.14. A doubly ionized lithium atom (Li^{++}) is one that has had two of its three electrons removed. The energy levels of the remaining single-electron ion are closely related to those of the hydrogen atom. The nuclear charge for lithium is $+3e$ instead of just $+e$. How are the energy levels related to those of hydrogen? How is the *radius* of the ion in the ground level related to that of the hydrogen atom? Explain.

Q38.15. The emission of a photon by an isolated atom is a recoil process in which momentum is conserved. Thus Eq. (38.6) should include a recoil kinetic energy K_r for the atom. Why is this energy negligible in that equation?

Q38.16. How might the energy levels of an atom be measured directly—that is, without recourse to analysis of spectra?

Q38.17. Elements in the gaseous state emit line spectra with well-defined wavelengths. But hot solid bodies always emit a continuous spectrum—that is, a continuous smear of wavelengths. Can you account for this difference?

Q38.18. As a body is heated to a very high temperature and becomes self-luminous, the apparent color of the emitted radiation shifts from red to yellow and finally to blue as the temperature increases. Why does the color shift? What other changes in the character of the radiation occur?

Q38.19. The peak-intensity wavelength of red dwarf stars, which have surface temperatures around 3000 K, is about 1000 nm, which is beyond the visible spectrum. So why are we able to see these stars, and why do they appear red?

Q38.20. A photon of frequency f undergoes Compton scattering from an electron at rest and scatters through an angle ϕ. The frequency of the scattered photon is f'. How is f' related to f? Does your answer depend on ϕ? Explain.

Q38.21. Can Compton scattering occur with protons as well as electrons? For example, suppose a beam of x rays is directed at a target of liquid hydrogen. (Recall that the nucleus of hydrogen consists of a single proton.) Compared to Compton scattering with electrons, what similarities and differences would you expect? Explain.

Q38.22. Why must engineers and scientists shield against x-ray production in high-voltage equipment?

Exercises

Section 38.2 The Photoelectric Effect

38.1. The graph in Figure 38.34 shows the stopping potential as a function of the frequency of the incident light falling on a metal surface, (a) Find the photoelectric work function for this metal. (b) What value of Planck's constant does the graph yield? (c) Why does the graph *not* extend below the x-axis? (d) If a different metal were used, what characteristics of the graph would you expect to be the same and which ones to be different?

Figure **38.34** Exercise 38.1.

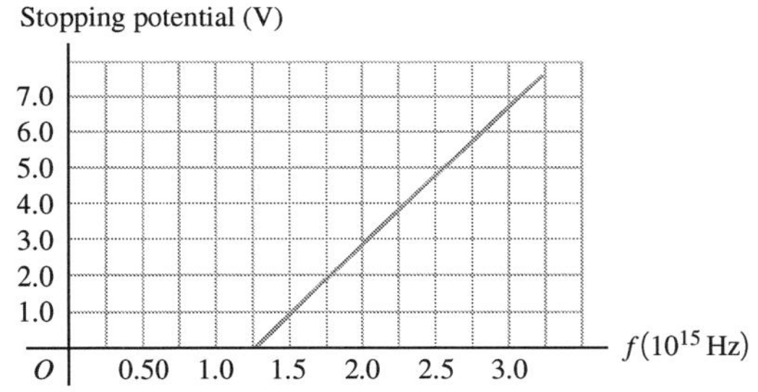

38.2. Response of the Eye. The human eye is most sensitive to green light of wavelength 505 nm. Experiments have found that when people are kept in a dark room until their eyes adapt to the darkness, a *single* photon of green light will trigger receptor cells in the rods of the retina. (a) What is the frequency of this photon? (b) How much energy (in joules and electron volts) does it deliver to the receptor cells? (c) to appreciate what a small amount of energy this is, calculate how fast a typical bacterium of mass 9.5×10^{-12} g would move if it had that much energy.

38.3. A photon of green light has a wavelength of 520 nm. Find the photon's frequency, magnitude of momentum, and energy. Express the energy in both joules and electron volts.

38.4. A laser used to weld detached retinas emits light with a wavelength of 652 nm in pulses that are 20.0 ms in duration. The average power during each pulse is 0.600 W. (a) How much energy is in each pulse in joules? In electron volts? (b) What is the energy of one photon in joules? In electron volts? (c) How many photons are in each pulse?

38.5. An excited nucleus emits a gamma-ray photon with an energy of 2.45 MeV. (a) What is the photon frequency? (b) What is the photon wavelength? (c) How does the wavelength compare with a typical nuclear diameter of 10^{-14} m?

38.6. The photoelectric threshold wavelength of a tungsten surface is 272 nm. Calculate the maximum kinetic energy of the electrons ejected from this tungsten surface by ultraviolet radiation of frequency 1.45×10^{15} Hz. Express the answer in electron volts.

38.7. A clean nickel surface is exposed to light of wavelength 235 nm. What is the maximum speed of the photoelectrons emitted from this surface? Use Table 38.1.

38.8. What would the minimum work function for a metal have to be for visible light (400–700 mn) to eject photoelectrons?

38.9. A 75-W light source consumes 75 W of electrical power. Assume all this energy goes into emitted light of wavelength 600 nm. (a) Calculate the frequency of the emitted light. (b) How many photons per second does the source emit? (c) Are the answers to parts (a) and (b) the same? Is the frequency of the light the same thing as the number of photons emitted per second? Explain.

38.10. In a photoelectric-effect experiment, the maximum kinetic energy of the ejected photoelectrons is measured for various wavelengths of the incident light. Figure 38.35 shows a graph of this maximum kinetic energy, K_{max}, as a function of the wavelength λ of the light falling on the surface of the metal. What are (a) the threshold frequency and (b) the work function (in electron volts) for this metal? (c) Data from experiments like this are often graphed showing K_{max} as a function of $1/\lambda$. Make a *qualitative* (no numbers) sketch of what this graph would look like. Identify the threshold wavelength (λ_0) on your sketch. What advantages are there to graphing the data this way?

Figure **38.35** Exercise 38.10.

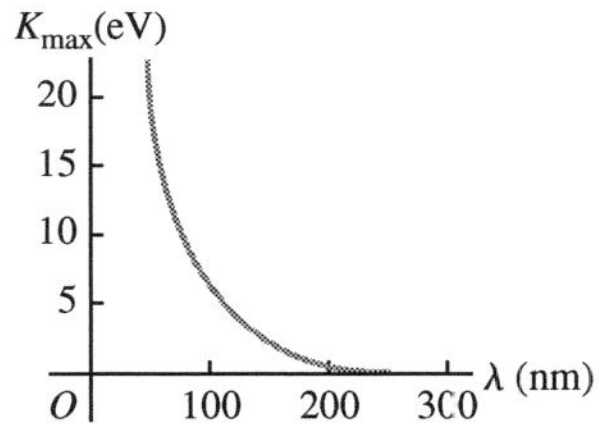

38.11. (a) A proton is moving at a speed much slower than the speed of light. It has kinetic energy K_1 and momentum p_1. If the momentum of the proton is doubled, so $p_2 = 2p_1$, how is its new kinetic energy K_2 related to K_1? (b) A photon with energy E_1 has momentum p_1. If another photon has momentum p_2 that is twice p_1, how is the energy E_2 of the second photon related to E_1?

38.12. The photoelectric work function of potassium is 2.3 eV. If light having a wavelength of 250 nm falls on potassium, find (a) the stopping potential in volts; (b) the kinetic energy in electron volts of the most energetic electrons ejected; (c) the speed of these electrons.

38.13. When ultraviolet light with a wavelength of 254 nm falls on a clean copper surface, the stopping potential necessary to stop emission of photoelectrons is 0.181 V. (a) What is the photoelectric threshold wavelength for this copper surface? (b) What is the work function for this surface, and how does your calculated value compare with that given in Table 38.1?

38.14. A photon has momentum of magnitude 8.24×10^{-28} kg · m/s. (a) What is the energy of this photon? Give your answer in joules and in electron volts. (b) What is the wavelength of this photon? In what region of the electromagnetic spectrum does it lie?

Section 38.3 Atomic Line Spectra and Energy Levels

38.15. Use Balmer's formula to calculate (a) the wavelength, (b) the frequency, and (c) the photon energy for the H_γ line of the Balmer series for hydrogen.

38.16. Find the longest and shortest wavelengths in the Lyman and Paschen series for hydrogen. In what region of the electromagnetic spectrum does each series lie?

38.17. (a) An atom initially in an energy level with $E = -6.52$ eV absorbs a photon that has wavelength 860 nm. What is the internal energy of the atom after it absorbs the photon? (b) An atom initially in an energy level with $E = -2.68$ eV emits a photon that has wavelength 420 nm. What is the internal energy of the atom after it emits the photon?

38.18. The energy-level scheme for the hypothetical one-electron element Searsium is shown in Fig. 38.36. The potential energy is taken to be zero for an electron at an infinite distance from the nucleus. (a) How much energy (in electron volts) does it take to ionize an electron from the ground level? (b) An 18-eV photon is absorbed by a Searsium atom in its ground level. As the atom returns to its ground level, what possible energies can the emitted photons have? Assume that there can be transitions between all pairs of levels. (c) What will happen if a photon with an energy of 8 eV strikes a Searsium atom in its ground level? Why? (d) Photons emitted in the Searsium transitions $n = 3 \rightarrow n = 2$ and $n = 3 \rightarrow n = 1$ will eject photoelectrons from an unknown metal, but the photon emitted from the transition $n = 4 \rightarrow n = 3$ will not. What are the limits (maximum and minimum possible values) of the work function of the metal?

Figure **38.36** Exercise 38.18.

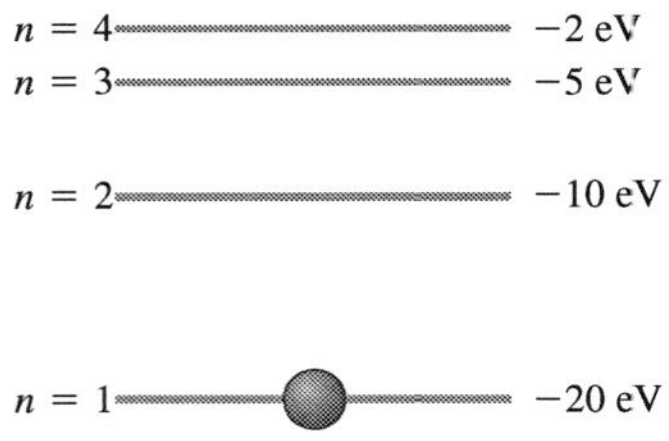

38.19. In a set of experiments on a hypothetical one-electron atom, you measure the wavelengths of the photons emitted from transitions ending in the ground state $(n = 1)$, as shown in the energy-level diagram in Fig. 38.37. You also observe that it takes 17.50 eV to ionize this atom. (a) What is the energy of the atom in each of the levels $(n = 1, n = 2$, etc.) shown in the figure? (b) If an electron made a transition from the $n = 4$ to the $n = 2$ level what wavelength of light would it emit?

Figure **38.37** Exercise 38.19.

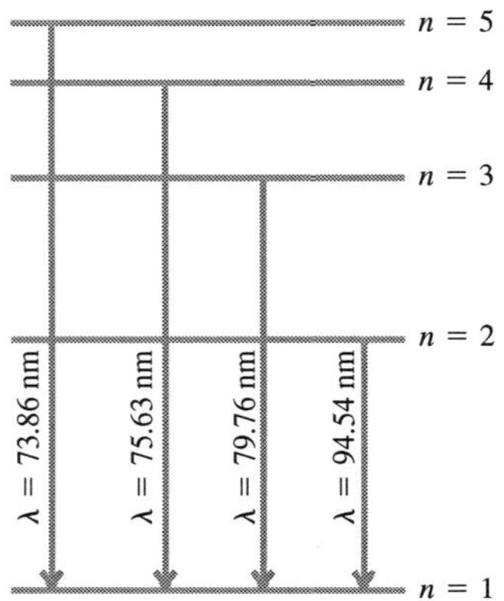

Section 38.4 The Nuclear Atom

38.20. A 4.78-MeV alpha particle from a ^{226}Ra decay makes a head-on collision with a uranium nucleus. A uranium nucleus has 92 protons. (a) What is the distance of closest approach of the alpha particle to the center of the nucleus? Assume that the uranium nucleus remains at rest and that the distance of closest approach is much greater than the radius of the uranium nucleus. (b) What is the force on the alpha particle at the instant when it is at the distance of closest approach?

38.21. A beam of alpha particles is incident on a target of lead. A particular alpha particle comes in "head-on" to a particular lead nucleus and stops 6.50×10^{-14} m away from the center of the nucleus. (This point is well outside the nucleus.) Assume that the lead nucleus, which has 82 protons, remains at rest. The mass of the alpha particle is 6.64×10^{-27} kg. (a) Calculate the electrostatic potential energy at the instant that the alpha particle stops. Express your result in joules and in MeV. (b) What initial kinetic energy (in joules and in MeV) did the alpha particle have? (c) What was the initial speed of the alpha particle?

Section 38.5 The Bohr Model

38.22. (a) What is the angular momentum L of the electron in a hydrogen atom, with respect to an origin at the nucleus, when the atom is in its lowest energy level? (b) Repeat part (a) for the ground level of He^+. Compare to the answer in part (a).

38.23. A hydrogen atom is in a state with energy -1.51 eV. In the Bohr model, what is the angular momentum of the electron in the atom, with respect to an axis at the nucleus?

38.24. A hydrogen atom initially in the ground level absorbs a photon, which excites it to the $n = 4$ level. Determine the wavelength and frequency of the photon.

38.25. A triply ionized beryllium ion, Be^{3+} (a beryllium atom with three electrons removed), behaves very much like a hydrogen atom except that the nuclear charge is four times as great. (a) What is the ground-level energy of Be^{3+}? How does this compare to the ground-level energy of the hydrogen atom? (b) What is the ionization energy of Be^{3+}? How does this compare to the ionization energy of the hydrogen atom? (c) For the hydrogen atom the wavelength of the photon emitted in the $n = 2$ to $n = 1$ transition is 122 nm (see Example 38.6). What is the wavelength of the photon emitted when a Be^{3+} ion undergoes this transition? (d) For a given value of n, how does the radius of an orbit in Be^{3+} compare to that for hydrogen?

38.26. A hydrogen atom undergoes a transition from the $n = 5$ to the $n = 2$ state. (a) What are the energy and wavelength of the photon that is emitted? (b) If the angular momentum is conserved and if the Bohr model is used to describe the atom, what must the angular momentum be of the photon that is emitted? (As we will see in Chapter 41, the modern quantum-mechanical description of the hydrogen atom gives a different result.)

38.27. (a) Using the Bohr model, calculate the speed of the electron in a hydrogen atom in the $n = 1$, 2 and 3 levels. (b) Calculate the orbital period in each of these levels. (c) The average lifetime of the first excited level of a hydrogen atom is 1.0×10^{-8} s. In the Bohr model, how many orbits does an electron in the $n = 2$ level complete before returning to the ground level?

38.28. (a) Show that, as n gets very large, the energy levels of the hydrogen atom get closer and closer together in energy. (b) Do the radii of these energy levels also get closer together?

Section 38.6 The Laser

38.29. How many photons per second are emitted by a 7.50-mW CO_2 laser that has a wavelength of 10.6 μm?

38.30. PRK Surgery. Photorefractive keratectomy (PRK) is a laser-based surgical procedure that corrects near- and farsightedness by removing part on the lens of the eye to change its curvature and hence focal length. This procedure can remove layers 0.25 μm thick using pulses lasting 12.0 ns from a laser beam of wavelength 193 nm. Low-intensity beams can be used because each individual photon has enough energy to break the covalent bonds of the tissue. (a) In what part of the electromagnetic spectrum does this light lie? (b) What is the energy of a single photon? (c) If a 1.50-mW beam is used, how many photons are delivered to the lens in each pulse?

38.31. A large number of neon atoms are in thermal equilibrium. What is the ratio of the number of atoms in a 5s state to the number in a 3p state at (a) 300 K; (b) 600 K; (c) 1200 K? The energies of these states are shown in Fig. 38.24a. (d) At any of these temperatures, the rate at which a neon gas will spontaneously emit 632.8-nm radiation is quite low. Explain why.

38.32. Figure 38.10 a shows the energy levels of the sodium atom. The two lowest excited levels are shown in columns labeled $^2P_{3/2}$ and $^2P_{1/2}$. Find the ratio of the number of atoms in a $^2P_{3/2}$ state to the number in a $^2P_{1/2}$ state for a sodium gas in thermal equilibrium at 500 K. In which state are more atoms found?

Section 38.7 X-Ray Production and Scattering

38.33. Protons are accelerated from rest by a potential difference of 4.00 kV and strike a metal target. If a proton produces one photon on impact, what is the minimum wavelength of the resulting x rays? How does your answer compare to the minimum wavelength if 4.00-keV electrons are used instead? Why do x-ray tubes use electrons rather than protons to produce x rays?

38.34. (a) What is the minimum potential difference between the filament and the target of an x-ray tube if the tube is to produce x rays with a wavelength of 0.150 nm? (b) What is the shortest wavelength produced in an x-ray tube operated at 30.0 kV?

38.35. X Rays from Television Screens. Accelerating voltages in cathode-ray-tube (CRT) TVs are about 25.0 kV. What are (a) the highest frequency and (b) the shortest wavelength (in nm) of the x rays that such a TV screen could produce? (c) What assumptions did you need to make? (CRT televisions contain shielding to absorb these x rays.)

38.36. X rays are produced in a tube operating at 18.0 kV. After emerging from the tube, x rays with the minimum wavelength produced strike a target and are Compton-scattered through an angle of 45.0°. (a) What is the original x-ray wavelength? (b) What is the wavelength of the scattered x rays? (c) What is the energy of the scattered x rays (in electron volts)?

38.37. X rays with initial wavelength 0.0665 nm undergo Compton scattering. What is the longest wavelength found in the scattered x rays? At which scattering angle is this wavelength observed?

38.38. A beam of x rays with wavelength 0.0500 nm is Compton-scattered by the electrons in a sample. At what angle from the incident beam should you look to find x rays with a wavelength of (a) 0.0542 nm; (b) 0.0521 nm; (c) 0.0500 nm?

38.39. If a photon of wavelength 0.04250 nm strikes a free electron and is scattered at an angle of 35.0° from its original direction, find (a) the change in the wavelength of this photon; (b) the wavelength of the scattered light; (c) the change in energy of the photon (is it a loss or a gain?); (d) the energy gained by the electron.

38.40. A photon scatters in the backward direction $(\theta = 180°)$ from a free proton that is initially at rest. What must the wavelength of the incident photon be if it is to undergo a 10.0% change in wavelength as a result of the scattering?

38.41. Complete the derivation of the Compton-scattering formula, Eq. (38.23), following the outline given in Eqs. (38.24) through (38.27).

Section 38.8 Continuous Spectra

38.42. Determine λ_m, the wavelength at the peak of the Planck distribution, and the corresponding frequency f, at these temperatures: (a) 3.00 K; (b) 300 K; (c) 3000 K.

38.43. A 100-W incandescent light bulb has a cylindrical tungsten filament 30.0 cm long, 0.40 mm in diameter, and with an emissivity of 0.26. (a) What is the temperature of the filament? (b) For what wavelength does the spectral emittance of the bulb peak? (c) Incandescent light bulbs are not very efficient sources of visible light. Explain why this is so.

38.44. The shortest visible wavelength is about 400 nm. What is the temperature of an ideal radiator whose spectral emittance peaks at this wavelength?

38.45. Radiation has been detected from space that is characteristic of an ideal radiator at $T = 2.728$ K. (This radiation is a relic of the Big Bang at the beginning of the universe.) For this temperature, at what wavelength does the Planck distribution peak? In what part of the electromagnetic spectrum is this wavelength?

38.46. Two stars, both of which behave like ideal blackbodies, radiate the same total energy per second. The cooler one has a surface temperature T and 3.0 times the diameter of the hotter star. (a) What is the temperature of the hotter star in terms of T? (b) What is the ratio of the peak-intensity wavelength of the hot star to the peak-intensity wavelength of the cool star?

38.47. (a) Show that the maximum in the Planck distribution, Eq. (38.32), occurs at a wavelength λ_m given by $\lambda_m = hc/4.965kT$ (Eq. 38.33). As discussed in the text, 4.965 is the root of Eq. (38.34). (b) Evaluate the constants in the expression derived in part (a) to show that $\lambda_m T$ has the numerical value given in the Wien displacement law, Eq. (38.30).

38.48. Sirius B. The brightest star in the sky is Sirius, the Dog Star. It is actually a binary system of two stars, the smaller one (Sirius B) being a white dwarf. Spectral analysis of Sirius B indicates that its surface temperature is 24,000 K and that it radiates energy at a total rate of 1.0×10^{25} W. Assume that it behaves like an ideal blackbody. (a) What is the total radiated intensity of Sirius B? (b) What is the peak-intensity wavelength? Is this wavelength visible to humans? (c) What is the radius of Sirius B? Express your answer in kilometers and as a fraction of our sun's radius. (d) Which star radiates more *total* energy per second, the hot Sirius B or the (relatively) cool sun with a surface temperature of 5800 K? To find out, calculate the ratio of the total power radiated by our sun to the power radiated by Sirius B.

38.49. Show that for large values of λ the Planck distribution, Eq. (38.32), agrees with the Rayleigh distribution, Eq. (38.31).

38.50. Blue Supergiants. A typical blue supergiant star (the type that explode and leave behind black holes) has a surface temperature of 30,000 K and a visual luminosity 100,000 times that of our sun. Our sun radiates at the rate of 3.86×10^{26} W. (Visual luminosity is the total power radiated at visible wavelengths.) (a) Assuming that this star behaves like an ideal blackbody, what is the principal wavelength it radiates? Is this light visible? Use your answer to explain why these stars are blue. (b) If we assume that the power radiated by the star is also 100,000 times that of our sun, what is the radius of this star? Compare its size to that of our sun, which has a radius of 6.96×10^5 km. (c) Is it really correct to say that the visual luminosity is proportional to the total power radiated? Explain.

Problems

38.51. Exposing Photographic Film. The light-sensitive compound on most photographic films is silver bromide, AgBr. A film is "exposed" when the light energy absorbed dissociates this molecule into its atoms. (The actual process is more complex, but the quantitative result does not differ greatly.) The energy of dissociation of AgBr is 1.00×10^5 J/mol. For a photon that is just able to dissociate a molecule of silver bromide, find (a) the photon energy in electron volts; (b) the wavelength of the photon; (c) the frequency of the photon. (d) What is the energy in electron volts of a photon having a frequency of 100 MHz? (e) Light from a firefly can expose photographic film, but the radiation from an FM station broadcasting 50,000 W at 100 MHz cannot. Explain why this is so.

38.52. An atom with mass m emits a photon of wavelength λ. (a) What is the recoil speed of the atom? (b) What is the kinetic energy K of the recoiling atom? (c) Find the ratio K/E, where E is the energy of the emitted photon. If this ratio is much less than unity, the recoil of the atom can be neglected in the emission process. Is the recoil of the atom more important for small or large atomic masses? For long or short wavelengths? (d) Calculate K (in electron volts) and K/E for a hydrogen atom (mass 1.67×10^{-27} kg) that emits an ultraviolet photon of energy 10.2 eV. Is recoil an important consideration in this emission process?

38.53. When a certain photoelectric surface is illuminated with light of different wavelengths, the following stopping potentials are observed:

Wavelength (nm)	Stopping potential (V)
366	1.48
405	1.15
436	0.93
492	0.62
546	0.36
579	0.24

Plot the stopping potential on the vertical axis against the frequency of the light on the horizontal axis. Determine (a) the threshold frequency; (b) the threshold wavelength; (c) the photoelectric work function of the material (in electron volts); (d) the value of Planck's constant h (assuming that the value of e is known).

38.54. (a) If the average frequency emitted by a 200-W light bulb is 5.00×10^{14} Hz, and 10.0% of the input power is emitted as visible light, approximately how many visible-light photons are emitted per second? (b) At what distance would this correspond to 1.00×10^{11} visible-light photons per square centimeter per second if the light is emitted uniformly in all directions?

38.55. (a) The wavelength of light incident on a metal surface is reduced from λ_1 to λ_2. (Both λ_1 and λ_2 are less than the threshold wavelength for the surface.) When the wavelength is reduced in this way, what is the change in the stopping potential for photoelectrons emitted from the surface? (b) Evaluate the change in stopping potential for $\lambda_1 = 295$ nm and $\lambda_2 = 265$ nm.

38.56. A 2.50-W beam of light of wavelength 124 nm falls on a metal surface. You observe that the maximum kinetic energy of the ejected electrons is 4.16 eV. Assume that each photon in the beam ejects a photoelectron. (a) What is the work function (in electron volts) of this metal? (b) How many photoelectrons are ejected each second from this metal? (c) If the power of the light beam, but not its wavelength, were reduced by half, what would be the answer to part (b)? (d) If the wavelength of the beam, but not its power, were reduced by half, what would be the answer to part (b)?

38.57. Removing Vascular Lesions. A pulsed dye laser emits light of wavelength 585 nm in 450-μs pulses. Because this wavelength is strongly absorbed by the hemoglobin in the blood, the method is especially effective for removing various types of blemishes due to blood, such as port-wine–colored birth-marks. To get a reasonable estimate of the power required for such laser surgery, we can model the blood as having the same specific heat and heat of vaporization as water $(4190 \text{ J/kg} \cdot \text{K}, 2.256 \times 10^6 \text{ J/kg})$. Suppose that each pulse must remove 2.0 μg of blood by evaporating it, starting at 33°C. (a) How much energy must each pulse deliver to the blemish? (b) What must be the power output of this laser? (c) How many photons does each pulse deliver to the blemish?

38.58. The photoelectric work functions for particular samples of certain metals are as follows: cesium, 2.1 eV; copper, 4.7 eV; potassium, 2.3 eV; and zinc, 4.3 eV. (a) What is the threshold wavelength for each metal surface? (b) Which of these metals could *not* emit photoelectrons when irradiated with visible light (400–700 nm)?

38.59. The negative muon has a charge equal to that of an electron but a mass that is 207 times as great. Consider a hydrogenlike atom consisting of a proton and a muon. (a) What is the reduced mass of the atom? (b) What is the ground-level energy (in electron volts)? (c) What is the wavelength of the radiation emitted in the transition from the $n = 2$ level to the $n = 1$ level?

38.60. An incident x-ray photon is scattered from a free electron that is initially at rest. The photon is scattered straight back at an angle of 180° from its initial direction. The wavelength of the scattered photon is 0.0830 nm. (a) What is the wavelength of the incident photon? (b) What is the magnitude of the momentum of the electron after the collision? (c) What is the kinetic energy of the electron after the collision?

38.61. An incident x-ray photon of wavelength 0.0900 nm is scattered in the backward direction from a free electron that is initially at rest. (a) What is the magnitude of the momentum of the scattered photon? (b) What is the kinetic energy of the electron after the photon is scattered?

38.62. Bohr Orbits of a Satellite. A 20.0-kg satellite circles the earth once every 2.00 h in an orbit having a radius of 8060 km. (a) Assuming that Bohr's angular-momentum result $(L = nh/2\pi)$ applies to satellites just as it does to an electron in the hydrogen atom, find the quantum number n of the orbit of the satellite. (b) Show from Bohr's angular momentum result and Newton's law of gravitation that the radius of an earth-satellite orbit is directly proportional to the square of the quantum number, $r = kn^2$, where k is the constant of proportionality. (c) Using the result from part (b), find the distance between the orbit of the satellite in this problem and its next "allowed" orbit. (Calculate a numerical value.) (d) Comment on the possibility of observing the separation of the two adjacent orbits. (e) Do quantized and classical orbits correspond for this satellite? Which is the "correct" method for calculating the orbits?

38.63. (a) What is the smallest amount of energy in electron volts that must be given to a hydrogen atom initially in its ground level so that it can emit the H_α line in the Balmer series? (b) How many different possibilities of spectral-line emissions are there for this atom when the electron starts in the $n = 3$ level and eventually ends up in the ground level? Calculate the wavelength of the emitted photon in each case.

38.64. A large number of hydrogen atoms are in thermal equilibrium. Let n_2/n_1 be the ratio of the number of atoms in an $n = 2$ excited state to the number of atoms in an $n = 1$ ground state. At what temperature is n_2/n_1 equal to (a) 10^{-12}; (b) 10^{-8}; (c) 10^{-4}? (d) Like the sun, other stars have continuous spectra with dark absorption lines (see Fig. 38.12). The absorption takes place in the star's atmosphere, which in all stars is composed primarily of hydrogen. Explain why the Balmer absorption lines are relatively weak in stars with low atmospheric temperatures such as the sun (atmosphere temperature 5800 K) but strong in stars with higher atmospheric temperatures.

38.65. A sample of hydrogen atoms is irradiated with light with wavelength 85.5 nm, and electrons are observed leaving the gas. (a) If each hydrogen atom were initially in its ground level, what would be the maximum kinetic energy in electron volts of these photoelectrons? (b) A few electrons are detected with energies as much as 10.2 eV greater than the maximum kinetic energy calculated in part (a). How can this be?

38.66. Light from an ideal spherical blackbody 15.0 cm in diameter is analyzed using a diffraction grating having 3850 lines/cm. When you shine this light through the grating, you observe that the peak-intensity wavelength forms a first-order bright fringe at $\pm 11.6°$ from the central bright fringe. (a) What is the temperature of the blackbody? (b) How long will it take this sphere to radiate 12.0 MJ of energy?

38.67. The Red Giant Betelgeuse. The red giant Betelgeuse has a surface temperature of 3000 K and is 600 times the diameter of our sun. (If our sun were that large, we would be inside it!) Assume that it radiates like an ideal blackbody. (a) If Betelgeuse were to radiate all of its energy at the peak-intensity wavelength, how many photons per second would it radiate? (b) Find the ratio of the power radiated by Betelgeuse to the power radiated by our sun (at 5800 K).

38.68. An ideal spherical blackbody 24.0 cm in diameter is maintained at 225°C by an internal electrical heater and is immersed in a very large open-faced tank of water that is kept boiling by the energy radiated by the sphere. You can neglect any heat transferred by conduction and convection. Consult Table 17.4 as needed. (a) At what rate, in g/s, is water evaporating from the tank? (b) If a physics-wise thermophile organism living in the hot water is observing this process, what will it measure for the peak-intensity (i) wavelength and (ii) frequency of the electromagnetic waves emitted by the sphere?

38.69. What must be the temperature of an ideal blackbody so that photons of its radiated light having the peak-intensity wavelength can excite the electron in the Bohr-model hydrogen atom from the ground state to the third excited state?

38.70. An x-ray tube is operating at voltage V and current I. (a) If only a fraction p of the electric power supplied is converted into x rays, at what rate is energy being delivered to the target? (b) If the target has mass m and specific heat capacity c (in J/kg · K), at what average rate would its temperature rise if there were no thermal losses? (c) Evaluate your results from parts (a) and (b) for an x-ray tube operating at 18.0 kV and 60.0 mA that converts 1.0% of the electric power into x rays. Assume that the 0.250-kg target is made of lead $(c = 130 \text{ J/kg} \cdot \text{K})$. (d) What must the physical properties of a practical target material be? What would be some suitable target elements?

38.71. When a photon is emitted by an atom, the atom must recoil to conserve momentum. This means that the photon and the recoiling atom share the transition energy. (a) For an atom with mass m, calculate the correction $\Delta\lambda$ due to recoil to the wavelength of an emitted photon. Let λ be the wavelength of the photon if recoil is not taken into consideration. (*Hint:* The correction is very small, as Problem 38.52 suggests, so $|\Delta\lambda|/\lambda \ll 1$. Use this fact to obtain an approximate but very accurate expression for $\Delta\lambda$.) (b) Evaluate the correction for a hydrogen atom in which an electron in the nth level returns to the ground level. How does the answer depend on n?

38.72. (a) Derive an expression for the total shift in photon wavelength after two successive Compton scatterings from electrons at rest. The photon is scattered by an angle θ_1 in the first scattering and by θ_2 in the second. (b) In general, is the total shift in wavelength produced by two successive scatterings of an angle $\theta/2$ the same as by a single scattering of θ? If not, are there any specific values of θ, other than $\theta = 0°$, for which the total shifts are the same? (c) Use the result of part (a) to calculate the total wavelength shift produced by two successive Compton scatterings of 30.0° each. Express your answer in terms of h/mc. (d) What is the wavelength shift produced by a single Compton scattering of 60.0°? Compare to the answer in part (c).

38.73. Nuclear fusion reactions at the center of the sun produce gamma-ray photons with energies of order 1 MeV (10^6 eV). By contrast, what we see emanating from the sun's surface are visible-light photons with wavelengths of order 500 nm. A simple model that explains this difference in wavelength is that a photon undergoes Compton scattering many times—in fact, about 10^{26} times, as suggested by models of the solar interior—as it travels from the center of the sun to its surface. (a) Estimate the increase in wavelength of a photon in an average Compton-scattering event. (b) Find the angle in degrees through which the photon is scattered in the scattering event described in part (a). (*Hint:* A useful approximation is $\cos\phi \approx 1 - \phi^2/2$, which is valid for $\phi \ll 1$. Note that ϕ is in radians in this expression.) (c) It is estimated that a photon takes about 10^6 years to travel from the core to the surface of the sun. Find the average distance that light can travel within the interior of the sun without being scattered. (This distance is roughly equivalent to how far you could see if you were inside the sun and could survive the extreme temperatures there. As your answer shows, the interior of the sun is *very* opaque.)

38.74. An x-ray photon is scattered from a free electron (mass m) at rest. The wavelength of the scattered photon is λ', and the final speed of the struck electron is v. (a) What was the initial wavelength λ of the photon? Express your answer in terms of λ', v, and m. (*Hint:* Use the relativistic expression for the electron kinetic energy.) (b) Through what angle ϕ is the photon scattered? Express your answer in terms of λ, λ', and m. (c) Evaluate your results in parts (a) and (b) for a wavelength of 5.10×10^{-3} nm for the scattered photon and a final electron speed of 1.80×10^8 m/s. Give ϕ in degrees.

38.75. A photon with wavelength 0.1100 nm collides with a free electron that is initially at rest. After the collision the wavelength is 0.1132 nm. (a) What is the kinetic energy of the electron after the collision? What is its speed? (b) If the electron is suddenly stopped (for example, in a solid target), all of its kinetic energy is used to create a photon. What is the wavelength of this photon?

38.76. (a) Calculate the maximum increase in photon wavelength that can occur during Compton scattering. (b) What is the energy (in electron volts) of the lowest-energy x-ray photon for which Compton scattering could result in doubling the original wavelength?

38.77. (a) Write the Planck distribution law in terms of the frequency f rather than the wavelength λ, to obtain $I(f)$. (b) Show that

$$\int_0^\infty I(\lambda)\,d\lambda = \frac{2\pi^5 k^4}{15c^2h^3}T^4$$

where $I(\lambda)$ is the Planck distribution formula of Eq. (38.32). (*Hint:* Change the integration variable from λ to f.) You will need to use the following tabulated integral:

$$\int_0^\infty \frac{x^3}{e^{\alpha x} - 1}\,dx = \frac{1}{240}\left(\frac{2\pi}{\alpha}\right)^4$$

(c) The result of (b) is I and has the form of the Stefan-Boltzmann law, $I = \sigma T^4$ (Eq. 38.28). Evaluate the constants in (b) to show that σ has the value given in Section 38.8.

38.78. An Ideal Blackbody. A large cavity with a very small hole and maintained at a temperature T is a good approximation to an ideal radiator or blackbody. Radiation can pass into or out of the cavity only through the hole. The cavity is a perfect absorber, since any radiation incident on the hole becomes trapped inside the cavity. Such a cavity at 200°C has a hole with area 4.00 mm^2. How long does it take for the cavity to radiate 100 J of energy through the hole?

Challenge Problems

38.79. (a) Show that in the Bohr model, the frequency of revolution of an electron in its circular orbit around a stationary hydrogen nucleus is $f = me^4/4\epsilon_0^2 n^3 h^3$. (b) In classical physics, the frequency of revolution of the electron is equal to the frequency of the radiation that it emits. Show that when n is very large, the frequency of revolution does indeed equal the radiated frequency calculated from Eq. (38.6) for a transition from $n_1 = n + 1$ to $n_2 = n$. (This illustrates Bohr's *correspondence principle,* which is often used as a check on quantum calculations. When n is small, quantum physics gives results that are very different from those of classical physics. When n is large, the differences are not significant, and the two methods then "correspond." In fact, when Bohr first tackled the hydrogen atom problem, he sought to determine f as a function of n such that it would correspond to classical results for large n.)

38.80. Consider a beam of monochromatic light with intensity I incident on a perfectly absorbing surface oriented perpendicular to the beam. Use the photon concept to show that the radiation pressure exerted by the light on the surface is given by I/c.

38.81. Consider Compton scattering of a photon by a *moving* electron. Before the collision the photon has wavelength λ and is moving in the $+x$-direction, and the electron is moving in the $-x$-direction with total energy E (including its rest energy mc^2). The photon and electron collide head-on. After the collision, both are moving in the $-x$-direction (that is, the photon has been scattered by 180°). (a) Derive an expression for the wavelength λ' of the scattered photon. Show that if $E \gg mc^2$, where m is the rest mass of the electron, your result reduces to

$$\lambda' = \frac{hc}{E}\left(1 + \frac{m^2c^4\lambda}{4hcE}\right)$$

(b) A beam of infrared radiation from a CO_2 laser ($\lambda = 10.6\ \mu m$) collides head-on with a beam of electrons, each of total energy $E = 10.0$ GeV ($1\text{ GeV} = 10^9$ eV). Calculate the wavelength λ' of the scattered photons, assuming a 180° scattering angle. (c) What kind of scattered photons are these (infrared, microwave, ultraviolet, etc.)? Can you think of an application of this effect?

THE WAVE NATURE OF PARTICLES

39

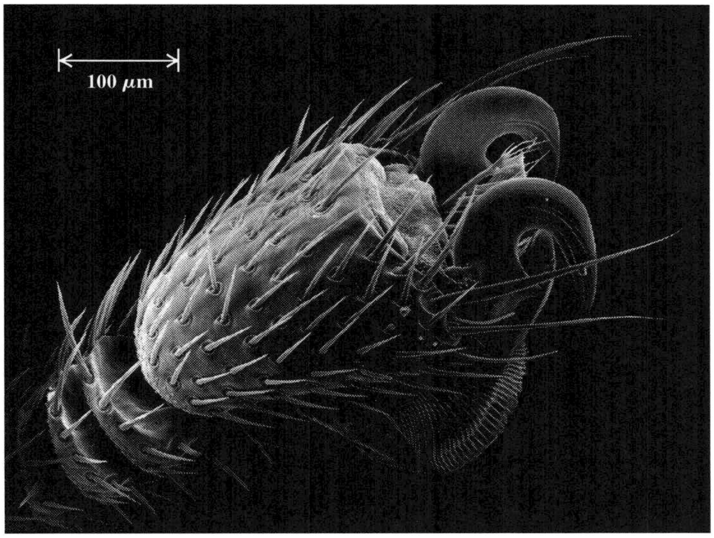

? This image of the foot of a bat fly *(Streblidae)* was made using a beam of electrons rather than a light beam. What properties of electrons make them useful for imaging such fine details?

LEARNING GOALS

By studying this chapter, you will learn:

- De Broglie's proposal that electrons, protons, and other particles can behave like waves.
- How electron diffraction experiments provided evidence for de Broglie's ideas.
- How the Heisenberg uncertainty principle imposes fundamental limits on what can be measured.
- How electron microscopes can provide much higher magnification than visible-light microscopes.
- About the wave functions that describe the behavior of particles and the Schrödinger equation that these functions must satisfy.

In preceding chapters we've seen one aspect of nature's wave–particle duality: Light and other electromagnetic radiation sometimes act like waves and sometimes like particles. Interference and diffraction demonstrate wave behavior, while emission and absorption of photons demonstrate the particle behavior.

A complete theory should also be able to *predict,* on theoretical grounds, the energy levels of any particular atom. The 1913 Bohr model of the hydrogen atom was a step in this direction. But it combined classical principles with new ideas that were inconsistent with classical theory, and it raised as many questions as it answered. More drastic departures from classical concepts were needed.

A successful drastic departure is *quantum mechanics,* a theory that began to emerge in the 1920s. Besides waves that sometimes act like particles, quantum mechanics extends the concept of wave–particle duality to include particles that sometimes show *wavelike* behavior. In these situations a particle is modeled as an inherently spread-out entity that can't be described as a point with a perfectly definite position and velocity.

In this chapter we will explore the wavelike nature of matter and some of its applications. We'll also introduce the *Schrödinger equation,* which is as fundamental to quantum mechanics as Newton's laws are to mechanics or as Maxwell's equations are to electromagnetism. In Chapter 40 we'll see how to use this equation to find the possible energy levels of a quantum-mechanical system.

Quantum mechanics is the key to understanding atoms and molecules, including their structure, spectra, chemical behavior, and many other properties. It has the happy effect of restoring unity and symmetry to our description of both particles and radiation. We will learn about quantum mechanics and use its results throughout the remainder of this book.

39.1 De Broglie Waves

A major advance in the understanding of atomic structure began in 1924 with a proposition made by a French physicist and nobleman, Prince Louis de Broglie (pronounced "de broy"). His reasoning, freely paraphrased, went like this: Nature loves symmetry. Light is dualistic in nature, behaving in some situations like waves and in others like particles. If nature is symmetric, this duality should also hold for matter. Electrons and protons, which we usually think of as *particles,* may in some situations behave like *waves.*

If a particle acts like a wave, it should have a wavelength and a frequency. De Broglie (Fig. 39.1) postulated that a free particle with rest mass m, moving with nonrelativistic speed v, should have a wavelength λ related to its momentum $p = mv$ in exactly the same way as for a photon, as expressed by Eq. (38.5): $\lambda = h/p$. The **de Broglie wavelength** of a particle is then

39.1 Louis-Victor de Broglie, the seventh Duke de Broglie (1892–1987), broke with family tradition by choosing a career in physics rather than as a diplomat. His revolutionary proposal that particles have wave characteristics—for which de Broglie won the 1929 Nobel Prize in physics—was published in his doctoral thesis.

$$\lambda = \frac{h}{p} = \frac{h}{mv} \quad \text{(de Broglie wavelength of a particle)} \qquad (39.1)$$

where h is Planck's constant. If the particle's speed is an appreciable fraction of the speed of light c, we use Eq. (37.27) to replace mv in Eq. (39.1) with $\gamma mv = mv/\sqrt{1 - v^2/c^2}$.The frequency f, according to de Broglie, is also related to the particle's energy E in the same way as for a photon, namely,

$$E = hf \qquad (39.2)$$

Thus the relationships of wavelength to momentum and of frequency to energy, in de Broglie's hypothesis, are exactly the same for particles as for photons.

CAUTION **Not all photon equations apply to particles with mass** The relationship $E = hf$ must be carefully applied to particles with nonzero rest mass, such as electrons and protons. Unlike a photon, they do *not* travel at speed c, so neither the equation $f = c/\lambda$ nor $E = pc$ applies to them!

To appreciate the enormous significance of de Broglie's proposal, we have to realize that at the time there was no direct experimental evidence that particles have wave characteristics. It is one thing to suggest a new hypothesis to explain experimental observations; it is quite another to propose such a radical departure from established concepts on theoretical grounds alone. But it was clear that a radical idea was needed. The dual nature of electromagnetic radiation had led to adoption of the photon concept, also a radical idea. The limited success in understanding atomic structure indicated that a similar revolution was needed in the mechanics of particles.

De Broglie's hypothesis was the beginning of that revolution. Within a few years after 1924 it was developed by Heisenberg, Schrödinger, Dirac, Born, and many others into a detailed theory called **quantum mechanics.** This development was well under way even before direct experimental evidence for the wave properties of particles was found.

Quantum mechanics involves sweeping revisions of our fundamental concepts of the description of matter. A particle is not a geometric point but an entity that is spread out in space. The spatial distribution of a particle is defined by a function called a **wave function,** which is closely analogous to the wave functions we used for mechanical waves in Chapter 15 and for electromagnetic waves in Chapter 32. The wave function for a *free* particle with definite energy has a recurring wave pattern with definite wavelength and frequency. The wave and particle aspects are not inconsistent; the *principle of complementarity,* which we discussed in Section 38.9, tells us that we need both the particle model and the wave model for a complete description of nature.

The Bohr Model and de Broglie Waves

In the Bohr model, we pictured the energy levels of the hydrogen atom in terms of definite electron orbits, as shown in Fig. 38.9. This is an oversimplification and should not be taken literally. But the most important idea in Bohr's theory was the existence of discrete energy levels and their relationship to the frequencies of emitted photons. The new quantum mechanics still assigns only certain allowed energy states to an atom, but with a more general description of the electron motion in terms of wave functions. In the hydrogen atom the energy levels predicted by quantum mechanics turn out to be the same as those given by Bohr's theory. In more complicated atoms, for which the Bohr theory does not work, the quantum-mechanical picture is in excellent agreement with observation.

The de Broglie wave hypothesis has an interesting relationship to the Bohr model. We can use Eq. (39.1) to obtain the Bohr quantum condition that the angular momentum $L = mvr$ must be an integer multiple of Planck's constant h. The method is analogous to determining the normal-mode frequencies of standing waves. We discussed this problem in Sections 15.8 and 32.5; the central idea was to satisfy the **boundary conditions** for the waves. For example, for standing waves on a string that is fixed at both ends, the ends are always nodes, and there will be additional nodes along the string for all but the fundamental mode. For the boundary conditions to be satisfied, the total length of the string must equal some *integral* number of half-wavelengths.

A standing wave on a string transmits no energy, and electrons in Bohr's orbits radiate no energy. So think of an electron as a standing wave fitted around a circle in one of the Bohr orbits. For the wave to "come out even" and join onto itself smoothly, the circumference of this circle must include some *whole number* of wavelengths, as suggested by Fig. 39.2. For an orbit with radius r and circumference $2\pi r$, we must have $2\pi r = n\lambda$, where $n = 1, 2, 3, \ldots$. According to the de Broglie relationship, Eq. (39.1), the wavelength λ of a particle with rest mass m, moving with nonrelativistic speed v, is $\lambda = h/mv$. Combining $2\pi r = n\lambda$ and $\lambda = h/mv$, we find $2\pi r = nh/mv$, or

$$mvr = n\frac{h}{2\pi} \qquad (39.3)$$

We recognize Eq. (39.3) as being identical to Eq. (38.10), Bohr's result that the magnitude of the angular momentum $L = mvr$ must equal an integer n times $h/2\pi$. Thus a wave-mechanical picture leads naturally to the quantization of the electron's angular momentum.

To be sure, the idea of fitting a standing wave around a circular orbit is a rather vague notion. But the agreement of Eq. (39.3) with Bohr's result is much too remarkable to be a coincidence. It strongly suggests that the wave properties of electrons do indeed have something to do with atomic structure.

Later we will learn how wave functions for specific systems are determined by solution of a wave equation called the Schrödinger equation. Boundary conditions play a central role in finding solutions of this equation and thus in determining possible energy levels, values of angular momentum, and other properties.

39.2 Diagrams showing the idea of fitting a standing wave around a circular orbit. For the wave to join onto itself smoothly, the circumference of the orbit must be an integral number n of wavelengths. Examples are shown for $n = 2$, $n = 3$, and $n = 4$.

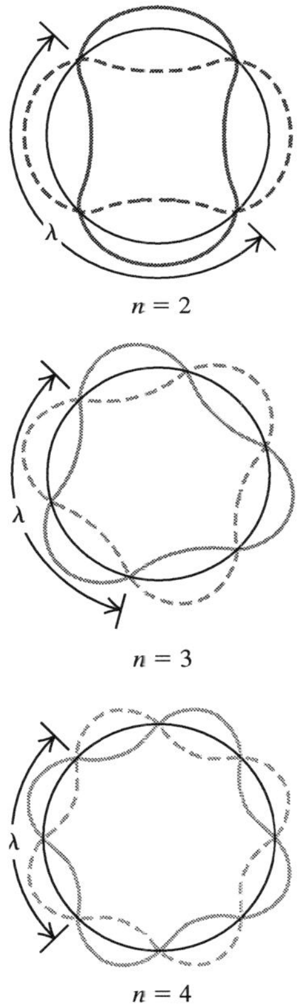

Problem-Solving Strategy 39.1 Particles and Waves

IDENTIFY *the relevant concepts:* Jarring as it may seem, both particles and light have wavelike properties as well as particlelike properties. The wavelength of a particle is inversely proportional to the momentum, and the frequency is proportional to the energy. These concepts will help you solve many problems involving the wave character of matter.

SET UP *the problem:* As always, determine the target variable and decide which equations you will use to calculate it.

EXECUTE *the solution* as follows:

1. Use Eq. (39.1) to relate the momentum p to the wavelength λ, and use Eq. (39.2) to relate the energy E to the frequency f.

Continued

2. Nonrelativistic kinetic energy may be expressed as either $K = \frac{1}{2}mv^2$ or (because $p = mv$) $K = p^2/2m$. The latter form is often useful in calculations involving the de Broglie wavelength.
3. Be careful to use consistent units. Lengths, such as wavelengths, are always in meters if you use the other quantities consistently in SI units. If you want a length in nanometers or some other units, don't forget to convert. As in Chapter 38, energies may be expressed in either joules or electron volts. Depending on your choice, you can use either $h = 6.626 \times 10^{-34}\ \text{J}\cdot\text{s}$ or $h = 4.136 \times 10^{-15}\ \text{eV}\cdot\text{s}$.

EVALUATE *your answer:* To check your numerical results, it helps to remember some typical orders of magnitude for quantities on the atomic scale:

Size of an atom: 10^{-10} m
Mass of an atom: 10^{-26} kg
Mass of an electron: 10^{-30} kg
Energy magnitude of an atomic state: 1 to 10 eV (10^{-19} to 10^{-18} J) (but some interaction energies are much smaller)
Speed of an electron in the Bohr model of a hydrogen atom: 10^6 m/s
Electron charge magnitude: 10^{-19} C
kT at room temperature: $\frac{1}{40}$ eV

You may want to add items to this list. These approximate values will also help you in Chapter 43, in which we will deal with magnitudes that are characteristic of *nuclear* structure rather than atomic structure; these are often different by factors of 10^4 to 10^6.

Example 39.1 Energy of a thermal neutron

Find the speed and kinetic energy of a neutron ($m = 1.675 \times 10^{-27}$ kg) that has a de Broglie wavelength $\lambda = 0.200$ nm, approximately the atomic spacing in many crystals. Compare the energy with the average translational kinetic energy of a gas molecule at room temperature ($T = 20°\text{C} = 293$ K).

SOLUTION

IDENTIFY: This problem uses several relationships: between the wavelength and speed of a particle, between particle speed and kinetic energy, and between the temperature of a gas and the average kinetic energy of a gas molecule.

SET UP: Our target variables are the speed v and kinetic energy K. We determine v using Eq. (39.1), and then calculate the kinetic energy using $K = \frac{1}{2}mv^2$. To find the average kinetic energy of a molecule at temperature T, we use the results of our discussion of kinetic theory in Section 18.3.

EXECUTE: From Eq. (39.1),

$$v = \frac{h}{\lambda m} = \frac{6.626 \times 10^{-34}\ \text{J}\cdot\text{s}}{(0.200 \times 10^{-9}\ \text{m})(1.675 \times 10^{-27}\ \text{kg})}$$
$$= 1.98 \times 10^3\ \text{m/s}$$

The kinetic energy is

$$K = \frac{1}{2}mv^2 = \frac{1}{2}(1.675 \times 10^{-27}\ \text{kg})(1.98 \times 10^3\ \text{m/s})^2$$
$$= 3.28 \times 10^{-21}\ \text{J} = 0.0204\ \text{eV}$$

The average translational kinetic energy of a molecule of an ideal gas is given by Eq. (18.16):

$$\frac{1}{2}m(v^2)_{\text{av}} = \frac{3}{2}kT = \frac{3}{2}(1.38 \times 10^{-23}\ \text{J/K})(293\ \text{K})$$
$$= 6.07 \times 10^{-21}\ \text{J} = 0.0397\ \text{eV}$$

The two energies are comparable in magnitude. In fact, a neutron with kinetic energy in this range is called a *thermal neutron.* Diffraction of thermal neutrons, which we'll discuss in the next section, is used to study crystal and molecular structure in the same way as x-ray diffraction. Neutron diffraction has proved to be especially useful in the study of large organic molecules.

EVALUATE: Note that the calculated neutron speed is much less than the speed of light. This justifies our use of the nonrelativistic form of Eq. (39.1).

Test Your Understanding of Section 39.1 A proton has a slightly smaller mass than a neutron. Compared to the neutron described in Example 39.1, would a proton of the same wavelength have (i) more kinetic energy; (ii) less kinetic energy; or (iii) the same kinetic energy?

39.2 Electron Diffraction

De Broglie's wave hypothesis, radical though it seemed, almost immediately received experimental confirmation. The first direct evidence involved a diffraction experiment with electrons that was analogous to the x-ray diffraction experiments that we described in Section 36.6. In those experiments, atoms in a crystal act as a three-dimensional diffraction grating for x rays. An x-ray beam is strongly reflected when it strikes a crystal at an angle that gives constructive interference among the waves scattered from the various atoms in the crystal. These interference effects demonstrate the *wave* nature of x rays.

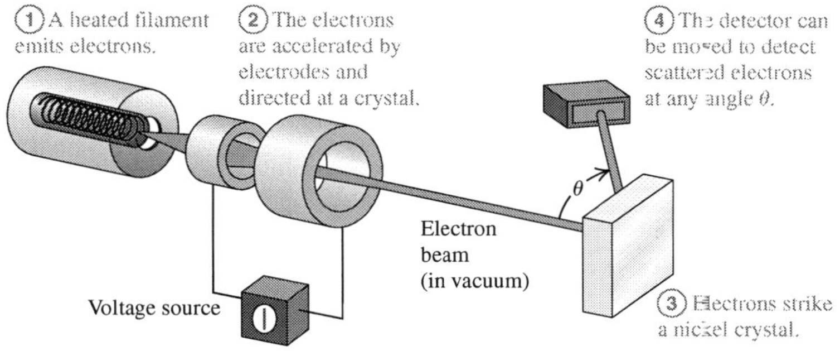

39.3 An apparatus similar to that used by Davisson and Germer to discover electron diffraction.

Observing the Wave Nature of Electrons

In 1927, Clinton Davisson and Lester Germer, working at the Bell Telephone Laboratories, were studying the surface of a piece of nickel by directing a beam of *electrons* at the surface and observing how many electrons bounced off at various angles. Figure 39.3 shows an experimental setup like theirs. The specimen was *polycrystalline:* Like many ordinary metals, it consisted of many microscopic crystals bonded together with random orientations. The experimenters expected that even the smoothest surface attainable would still look rough to an electron and that the electron beam would be diffusely reflected, with a smooth distribution of intensity as a function of the angle θ.

During the experiment an accident occurred that permitted air to enter the vacuum chamber, and an oxide film formed on the metal surface. To remove this film, Davisson and Germer baked the specimen in a high-temperature oven, almost hot enough to melt it. Unknown to them, this had the effect of creating large single-crystal regions with crystal planes that were continuous over the width of the electron beam.

When the observations were repeated, the results were quite different. Strong maxima in the intensity of the reflected electron beam occurred at specific angles (Fig. 39.4a), in contrast to the smooth variation of intensity with angle that Davisson and Germer had observed before the accident. The angular positions of the maxima depended on the accelerating voltage V_{ba} used to produce the electron beam. Davisson and Germer were familiar with de Broglie's hypothesis, and they noticed the similarity of this behavior to x-ray diffraction. This was not the effect they had been looking for, but they immediately recognized that the electron beam was being *diffracted.* They had discovered a very direct experimental confirmation of the wave hypothesis.

Davisson and Germer could determine the speeds of the electrons from the accelerating voltage, so they could compute the de Broglie wavelength from Eq. (39.1). The electrons were scattered primarily by the planes of atoms at the surface of the crystal. Atoms in a surface plane are arranged in rows, with a distance d that can be measured by x-ray diffraction techniques. These rows act like a reflecting diffraction grating; the angles at which strong reflection occurs are the same as for a grating with center-to-center distance d between its slits (Fig. 39.4b). From Eq. (36.13) the angles of maximum reflection are given by

$$d\sin\theta = m\lambda \qquad (m = 1, 2, 3, \dots) \tag{39.4}$$

where θ is the angle shown in Fig. 39.3. The angles predicted by this equation, using the de Broglie wavelength, were found to agree with the observed values (Fig. 39.4a). Thus the accidental discovery of **electron diffraction** was the first direct evidence confirming de Broglie's hypothesis.

The de Broglie wavelength of a nonrelativistic particle is $\lambda = h/p = h/mv$. We can also express λ in terms of the particle's kinetic energy. For example,

ActivPhysics ONLINE

17.5 Electron Interference

39.4 (a)A graph of intensity of the scattered electron beam shown in Fig. 39.3 as a function of the scattering angle θ.
(b) Constructive interference of electron waves scattered from two adjacent atoms occurs when $d\sin\theta = m\lambda$. In the case shown here, $\theta = 50°$ and $m = 1$.

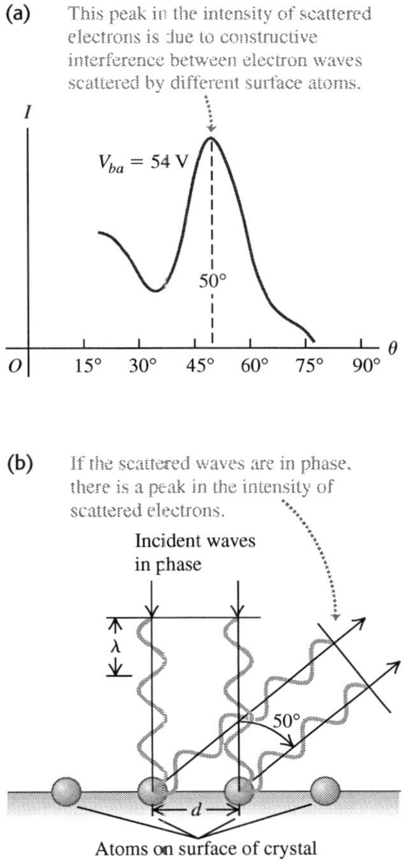

consider an electron freely accelerated from rest at point a to point b through a potential increase $V_b - V_a = V_{ba}$. The work done on the electron eV_{ba} equals its kinetic energy K. Using $K = p^2/2m$, we have

$$eV_{ba} = \frac{p^2}{2m} \qquad p = \sqrt{2meV_{ba}}$$

and the de Broglie wavelength of the electron is

$$\lambda = \frac{h}{p} = \frac{h}{\sqrt{2meV_{ba}}} \qquad \text{(de Broglie wavelength of an electron)} \quad (39.5)$$

Example 39.2 An electron-diffraction experiment

In a particular electron-diffraction experiment using an accelerating voltage of 54 V, an intensity maximum occurs when the angle θ in Fig. 39.3 is 50° (see Fig. 39.4a). The initial kinetic energy of the electrons is negligible. The rows of atoms have been found by x-ray diffraction to have a separation $d = 2.15 \times 10^{-10}$ m = 0.215 nm. Find the electron wavelength.

SOLUTION

IDENTIFY: This problem uses the relationships among accelerating voltage, electron wavelength, spacing between atoms, and angle of maximum reflection for an electron-diffraction experiment.

SET UP: We are given the accelerating voltage $V_{ba} = 54$ V, so we can determine the wavelength λ using Eq. (39.5). We are also given the atomic spacing d and the angle θ for an intensity maximum, so we can also determine λ using Eq. (39.4).

EXECUTE: From Eq. (39.5), the electron wavelength is

$$\lambda = \frac{6.626 \times 10^{-34}\ \text{J} \cdot \text{s}}{\sqrt{2(9.109 \times 10^{-31}\ \text{kg})(1.602 \times 10^{-19}\ \text{C})(54\ \text{V})}} = 1.7 \times 10^{-10}\ \text{m}$$

Alternatively, using Eq. (39.4) and assuming $m = 1$,

$$\lambda = d\sin\theta = (2.15 \times 10^{-10}\ \text{m})\sin 50° = 1.7 \times 10^{-10}\ \text{m}$$

EVALUATE: The two numbers agree within the accuracy of the experimental results, which gives us an excellent check on our calculations. Note that this electron wavelength is less than the spacing between the atoms.

Diffraction of Electrons and Other Particles

In 1928, just a year after the Davisson-Germer discovery, the English physicist G. P. Thomson carried out electron-diffraction experiments using a thin polycrystalline metallic foil as a target. Debye and Sherrer had used a similar technique several years earlier to study x-ray diffraction from polycrystalline specimens. Because of the random orientations of the individual microscopic crystals in his foil, the diffraction pattern consisted of intensity maxima forming rings around the direction of the incident beam. Thomson's results again confirmed the de Broglie relationship. Figure 39.5 shows both x-ray and electron diffraction patterns for a polycrystalline aluminum foil. (G. P. Thomson was the son of J. J. Thomson, who 31 years earlier had performed the definitive experiment to establish the *particle* nature of electrons.)

Additional experiments were soon carried out in many laboratories. In Germany, Estermann and Stern demonstrated diffraction of alpha particles. Diffraction experiments have also been performed with various ions and low-energy neutrons (see Example 39.1). Thus the wave nature of particles, so strange in 1924, became firmly established in the years that followed.

39.5 X-ray and electron diffraction. The upper half of the photo shows the diffraction pattern for 71-pm x rays passing through aluminum foil. The lower half, with a different scale, shows the diffraction pattern for 600-eV electrons from aluminum.

Top: x-ray diffraction

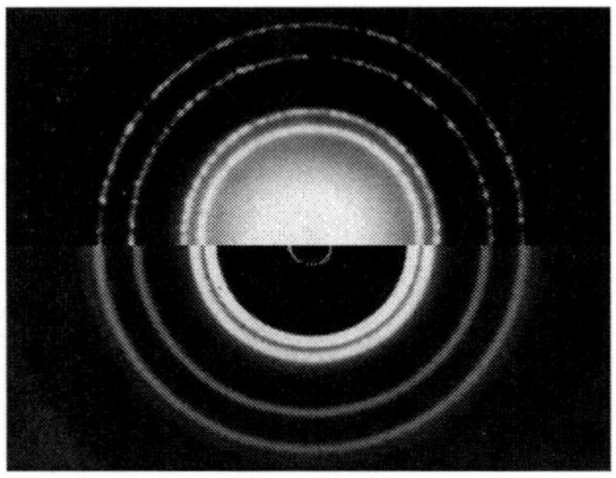

Bottom: electron diffraction

Test Your Understanding of Section 39.2 Example 39.2 shows that to give electrons a wavelength of 1.7×10^{-10} m, they must be accelerated from rest through a voltage of 54 V and so acquire a kinetic energy of 54 eV. Does a photon of this same energy also have a wavelength of 1.7×10^{-10} m?

39.3 Probability and Uncertainty

The discovery of the dual wave–particle nature of matter has forced us to reevaluate the kinematic language we use to describe the position and motion of a particle. In classical Newtonian mechanics we think of a particle as a point. We can describe its location and state of motion at any instant with three spatial coordinates and three components of velocity. But in general such a specific description is not possible. When we look on a small enough scale, there are fundamental limitations on the precision with which we can determine the position and velocity of a particle. Many aspects of a particle's behavior can be stated only in terms of *probabilities*.

ActivPhysics ONLINE
17.6 Uncertainty Principle

Single-Slit Diffraction

To try to get some insight into the nature of the problem, let's review the single-slit diffraction of light described in Section 36.2. Suppose the wavelength λ is much less than the slit width a. Then most (85%) of the light in the diffraction pattern is concentrated in the central maximum, bounded on either side by the first intensity minimum. We use θ_1 to denote the angle between the central maximum and the first minimum. Using Eq. (36.2) with $m = 1$, we find that θ_1 is given by $\sin\theta_1 = \lambda/a$. Since we assume $\lambda \ll a$, it follows that θ_1 is very small, $\sin\theta_1$ is very nearly equal to θ_1 (in radians), and

$$\theta_1 = \frac{\lambda}{a} \tag{39.6}$$

Now we perform the same experiment again, but using a beam of *electrons* instead of a beam of monochromatic light (Fig. 39.6). We have to do the experiment in vacuum (10^{-7} atm or less) so that the electrons don't bounce off air molecules. We can produce the electron beam with a setup that is similar in principle to the electron gun in a cathode-ray tube. This produces a narrow beam of electrons that all have very nearly the same direction and speed and therefore also the same de Broglie wavelength.

The result of this experiment, recorded on photographic film or by use of more sophisticated detectors, is a diffraction pattern identical to the one shown in Fig. 39.6. This pattern gives us additional direct evidence of the *wave* nature of electrons. About 85% of the electrons strike the film within the central maximum; the remainder strike the film within the subsidiary maxima on both sides.

If we believe that electrons are waves, the wave behavior in this experiment isn't surprising. But if we try to interpret it in terms of *particles*, we run into very serious problems. First, the electrons don't all follow the same path, even though they all have the same initial state of motion. In fact, we can't predict the exact

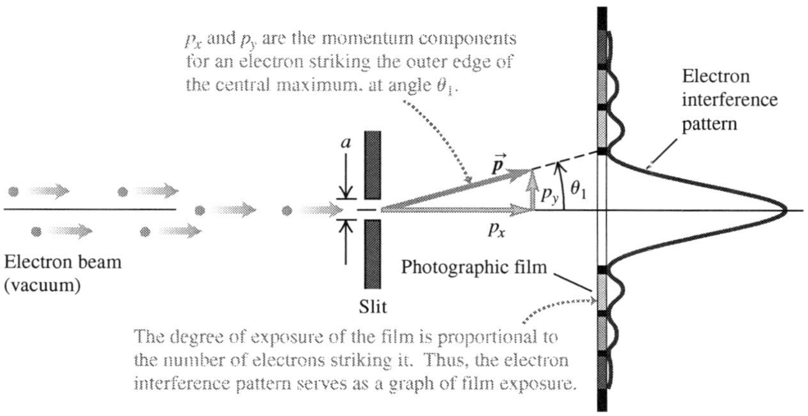

39.6 An electron-diffraction experiment.

trajectory of any individual electron from knowledge of its initial state. The best we can do is to say that *most* of the electrons go to a certain region, *fewer* go to other regions, and so on. That is, we can describe only the *probability* that an individual electron will strike each of various areas on the film. This fundamental indeterminacy has no counterpart in Newtonian mechanics, in which the motion of a particle can always be well predicted if we know the initial position and motion with sufficient accuracy.

Second, there are fundamental *uncertainties* in both the position and the momentum of an individual particle, and these two uncertainties are related inseparably. To clarify this point, let's go back to Fig. 39.6. An electron that strikes the film at the outer edge of the central maximum, at angle θ_1, must have a component of momentum p_y in the y-direction, as well as a component p_x in the x-direction, despite the fact that initially the beam was directed along the x-axis. From the geometry of the situation the two components are related by $p_y/p_x = \tan\theta_1$. Since θ_1 is small, we may use the approximation $\tan\theta_1 = \theta_1$, and

$$p_y = p_x\theta_1 \tag{39.7}$$

Substituting Eq. (39.6), $\theta_1 = \lambda/a$, we have

$$p_y = p_x\frac{\lambda}{a} \tag{39.8}$$

For the 85% of the electrons that strike the film within the central maximum (that is, at angles between $-\lambda/a$ and $+\lambda/a$), we see that the y-component of momentum is spread out over a range from $-p_x\lambda/a$ to $+p_x\lambda/a$. Now let's consider *all* the electrons that pass through the slit and strike the film. Again, they may hit above or below the center of the pattern, so their component p_y may be positive or negative. However the symmetry of the diffraction pattern shows us the average value $(p_y)_{av} = 0$. There will be an *uncertainty* Δp_y in the y-component of momentum at least as great as $p_x\lambda/a$. That is,

$$\Delta p_y \geq p_x\frac{\lambda}{a} \tag{39.9}$$

The narrower the slit width a, the broader is the diffraction pattern and the greater is the uncertainty in the y-component of momentum p_y.

The electron wavelength λ is related to the momentum $p_x = mv_x$ by the de Broglie relationship Eq. (39.1), which we can rewrite as $\lambda = h/p_x$. Using this relationship in Eq. (39.9) and simplifying, we find

$$\begin{aligned}\Delta p_y &\geq p_x\frac{h}{p_x a} = \frac{h}{a}\\ \Delta p_y a &\geq h\end{aligned} \tag{39.10}$$

What does this result mean? The slit width a represents an uncertainty in the y-component of the *position* of an electron as it passes through the slit. We don't know exactly *where* in the slit each particle passes through. So both the y-position and the y-component of momentum have uncertainties, and the two uncertainties are related by Eq. (39.10). We can reduce the *momentum* uncertainty Δp_y only by reducing the width of the diffraction pattern. To do this, we have to increase the slit width a, which increases the *position* uncertainty. Conversely, when we *decrease* the position uncertainty by narrowing the slit, the diffraction pattern broadens and the corresponding momentum uncertainty *increases.*

You may protest that it doesn't seem to be consistent with common sense for a particle not to have a definite position and momentum. We reply that what we call *common sense* is based on familiarity gained through experience. Our usual experience includes very little contact with the microscopic behavior of particles. Sometimes we have to accept conclusions that violate our intuition when we are dealing with areas that are far removed from everyday experience.

The Uncertainty Principle

In more general discussions of uncertainty relationships, the uncertainty of a quantity is usually described in terms of the statistical concept of *standard deviation,* which is a measure of the spread or dispersion of a set of numbers around their average value. Suppose we now begin to describe uncertainties in this way [neither Δp_y nor a in Eq. (39.10) is a standard deviation]. If a coordinate x has an uncertainty Δx and if the corresponding momentum component p_x has an uncertainty Δp_x, then those standard-deviation uncertainties are found to be related in general by the inequality

$$\Delta x \Delta p_x \geq \hbar \quad \text{(Heisenberg uncertainty principle for position and momentum)} \tag{39.11}$$

In this expression the quantity $\hbar$ (pronounced "h-bar") is Planck's constant divided by 2π:

$$\hbar = \frac{h}{2\pi} = 1.05457168(18) \times 10^{-34}\ \text{J} \cdot \text{s} \tag{39.12}$$

We will use this quantity frequently to avoid writing a lot of factors of 2π in later equations.

Equation (39.11) is one form of the **Heisenberg uncertainty principle,** first discovered by the German physicist Werner Heisenberg (1901–1976). It states that, in general, neither the position nor the momentum of a particle can be determined with arbitrarily great precision, as classical physics would predict. Instead, the uncertainties in the two quantities play complementary roles, as we have described. Figure 39.7 shows the relationship between the two uncertainties.

39.7 The Heisenberg uncertainty principle for position and momentum components. It is impossible for the product $\Delta x \Delta p_x$ to be less than $\hbar = h/2\pi$.

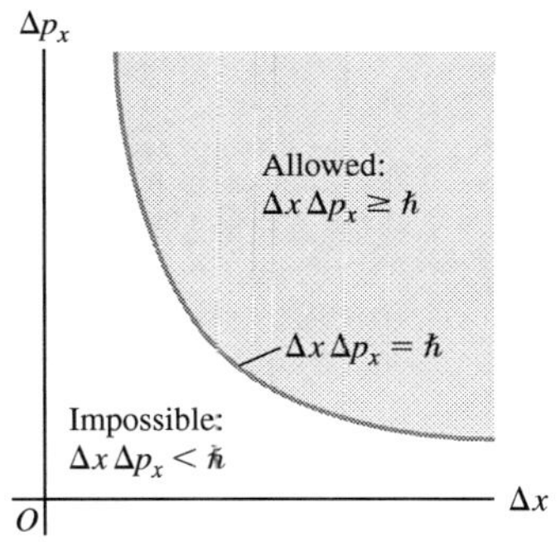

It is tempting to suppose that we could get greater precision by using more sophisticated detectors of position and momentum. This turns out not to be possible. To detect a particle, the detector must *interact* with it, and this interaction unavoidably changes the state of motion of the particle, introducing uncertainty about its original state. For example, if we were to bounce shorter-wavelength photons off a particle to better locate its position, the larger photon momentum h/λ would make the particle recoil more, giving us greater uncertainty in its momentum. A more detailed analysis of such hypothetical experiments shows that the uncertainties we have described are fundamental and intrinsic. They *cannot* be circumvented *even in principle* by any experimental technique, no matter how sophisticated.

There is nothing special about the x-axis. In a three-dimensional situation with coordinates (x, y, z) there is an uncertainty relationship for each coordinate and its corresponding momentum component: $\Delta x \Delta p_x \geq \hbar$, $\Delta y \Delta p_y \geq \hbar$, and $\Delta z \Delta p_z \geq \hbar$. However, the uncertainty in one coordinate is *not* related to the uncertainty in a different component of momentum. For example, Δx is not related directly to Δp_y.

For a particle moving along a radius, we can replace x in Eq. (39.11) with r, giving $\Delta r \Delta p_r \geq \hbar$. In the Bohr model, an electron moves in a circle of exact radius r, giving $\Delta r = 0$ and $\Delta p_r = 0$. Thus the Bohr model violates the Heisenberg uncertainty principle. We'll give a more correct description of atomic structure in Chapter 41; happily, it turns out that the energy-level predictions of the Bohr model *are* correct.

39.8 The longer the lifetime Δt of a state, the smaller is its spread in energy (shown by the width of the energy levels).

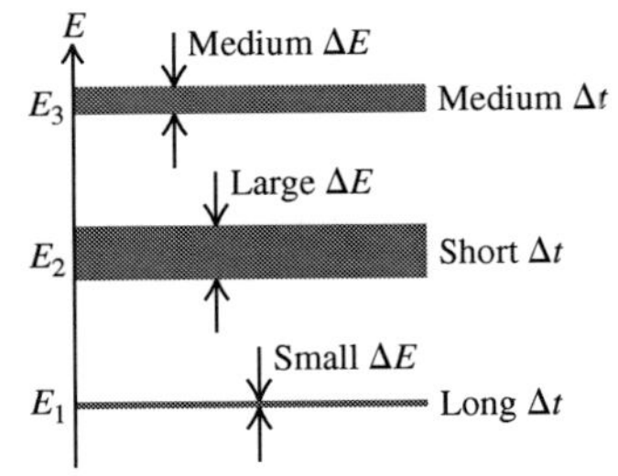

Uncertainty in Energy

There is also an uncertainty principle for *energy.* It turns out that the energy of a system also has inherent uncertainty. The uncertainty ΔE depends on the *time interval* Δt during which the system remains in the given state. The relationship is

$$\Delta E \Delta t \geq \hbar \quad \text{(Heisenberg uncertainty principle for energy and time interval)} \tag{39.13}$$

A system that remains in a metastable state for a very long time (large Δt) can have a very well-defined energy (small ΔE), but if it remains in a state for only a short time (small Δt), the uncertainty in energy must be correspondingly greater (large ΔE). Figure 39.8 illustrates this idea.

Example 39.3 The uncertainty principle: momentum and position

An electron is confined within a region of width 1.0×10^{-10} m. (a) Estimate the minimum uncertainty in the x-component of the electron's momentum. (b) If the electron has momentum with magnitude equal to the uncertainty found in part (a), what is its kinetic energy? Express the result in joules and in electron volts.

SOLUTION

IDENTIFY: This problem uses the Heisenberg uncertainty principle for position and momentum, as well as the relationship between a particle's momentum and its kinetic energy.

SET UP: We know that the electron is *somewhere* within the region to which it is confined. Since we don't know exactly where within this region the electron is, the width of the region is the position uncertainty: $\Delta x = 1.0 \times 10^{-10}$ m. (You could argue that the uncertainty should actually be half the width of the region. All we want is an estimate, however, so we won't worry about factors of 2.) We then find the momentum uncertainty Δp_x using Eq. (39.11), and we find the kinetic energy using $K = p^2/2m$.

EXECUTE: (a) For a given value of Δx, there is minimum uncertainty in momentum when the product $\Delta x \Delta p_x$ in Eq. (39.11) is equal to $\hbar$. Using the value of $\hbar$ from Eq. (39.12),

$$(\Delta p_x)_{\min} = \frac{\hbar}{\Delta x} = \frac{1.055 \times 10^{-34}\ \text{J} \cdot \text{s}}{1.0 \times 10^{-10}\ \text{m}} = 1.1 \times 10^{-24}\ \text{kg} \cdot \text{m/s}$$

(b) An electron with this magnitude of momentum has kinetic energy

$$K = \frac{p^2}{2m} = \frac{(1.1 \times 10^{-24}\ \text{kg} \cdot \text{m/s})^2}{2(9.11 \times 10^{-31}\ \text{kg})}$$
$$= 6.1 \times 10^{-19}\ \text{J} = 3.8\ \text{eV}$$

EVALUATE: The region is roughly the same width as an atom, and the energy is of the same order of magnitude as typical electron energies in atoms. This is a very rough calculation, but it is reassuring that the energy has a reasonable order of magnitude.

In a type of radioactive decay called *beta-minus decay,* an electron is emitted from a nucleus. Although it seems reasonable to assume that the electron was confined within the nucleus before the decay, that would require the position of the electron to be known to within an uncertainty of $\Delta x = 10^{-14}$ m or so. But this would give the confined electron a value of Δp_x that is 10^4 times greater than for the electron in this example and an expected kinetic energy so large that we would need to use relativistic equations to calculate it. This high energy is one reason to believe that there are no electrons confined in nuclei. In beta-minus decay the electron is actually *produced* within the nucleus when a neutron changes into a proton.

Example 39.4 The uncertainty principle: energy and time

A sodium atom is in one of the states labeled "Lowest excited levels" in Fig. 38.10 (Section 38.3). It remains in that state for an average time of 1.6×10^{-8} s before it makes a transition back to a ground state, emitting a photon with wavelength 589.0 nm and energy 2.105 eV. What is the uncertainty in energy of that excited state? What is the wavelength spread of the corresponding spectrum line?

SOLUTION

IDENTIFY: This problem uses the Heisenberg uncertainty principle for energy and time interval, as well as the relationship between the energy and wavelength of a photon.

SET UP: The average time that the atom spends in this excited state is equal to Δt in Eq. (39.13). We find the uncertainty in energy by replacing the $\geq$ sign in Eq. (39.13) with an equals sign and solving for ΔE.

EXECUTE: From Eq. (39.13),

$$\Delta E = \frac{\hbar}{\Delta t} = \frac{1.055 \times 10^{-34}\ \text{J} \cdot \text{s}}{1.6 \times 10^{-8}\ \text{s}}$$
$$= 6.6 \times 10^{-27}\ \text{J} = 4.1 \times 10^{-8}\ \text{eV}$$

The atom remains an indefinitely long time in the ground state, so there is *no* fundamental uncertainty there. The fractional uncertainty of the photon energy is

$$\frac{4.1 \times 10^{-8}\ \text{eV}}{2.105\ \text{eV}} = 1.95 \times 10^{-8}$$

The corresponding spread in wavelength, or "width," of the spectrum line is approximately

$$\Delta\lambda = (1.95 \times 10^{-8})(589.0\ \text{nm}) = 0.000011\ \text{nm}$$

EVALUATE: The irreducible uncertainty $\Delta\lambda$ that we have calculated is called the *natural line width* of this particular spectral line. Though very small, it is within the limits of resolution of present-day spectrometers. Ordinarily, the natural line width is much smaller than the line width from other causes such as the Doppler effect and collisions among the rapidly moving atoms.

Two-Slit Interference

Now let's take a brief look at a quantum interpretation of a *two-slit* optical interference pattern. We studied these patterns in detail for light in Sections 35.2 and 35.3, and in Section 38.9 we discussed their interpretation in terms of the probability that photons strike various regions of the screen where the pattern is formed.

It is natural to ask what happens when we do a two-slit interference experiment with electrons. The answer is: exactly the same thing we saw in Section 38.9 with photons! We can again use photographic film (Fig. 39.9) or electronic detectors to trace out the interference pattern, as we did with photons. The principle of complementarity, introduced in Section 38.9, again tells us that we cannot simultaneously attempt to apply the wave model and the particle model to describe a single part of this experiment. Thus we *cannot* predict exactly where in the pattern (a wave phenomenon) any individual electron (a particle) will land. We can't even ask which slit an individual electron passed through in building up the two-slit interference pattern. If we do determine which slits that the electrons pass through by scattering photons off them, the electrons recoil, and the two-slit interference pattern is not built up.

39.9 **(a)** Formation of an interference pattern for electrons incident on two slits, **(b)** after 28, 1000, and 10,000 electrons.

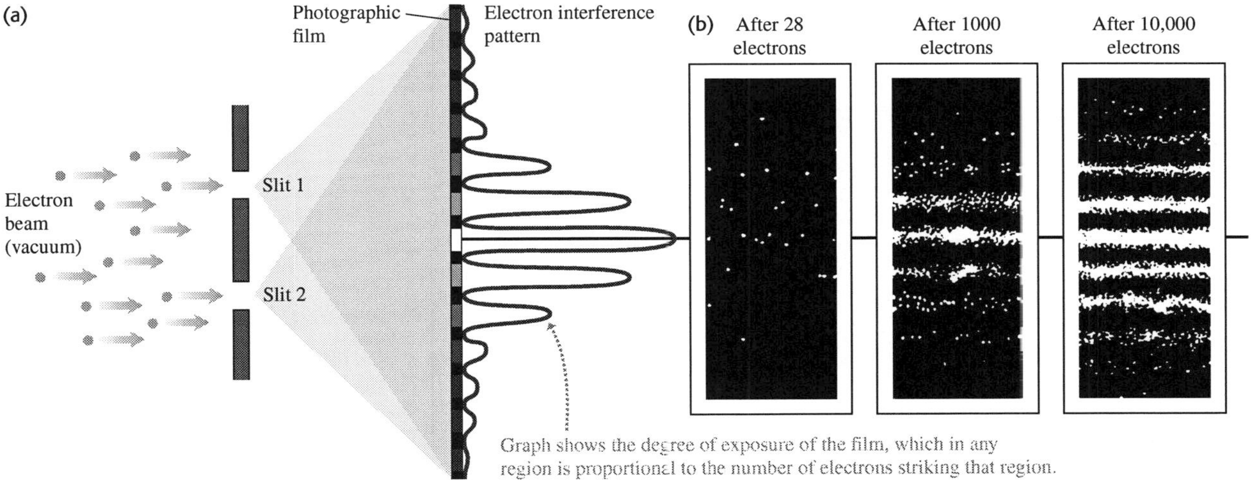

Test Your Understanding of Section 39.3 Rank the following situations according to the uncertainty in x-momentum, from largest to smallest. The mass of the proton is 1836 times the mass of the electron. (i) an electron whose x-coordinate is known to within 2×10^{-15} m; (ii) an electron whose x-coordinate is known to within 4×10^{-15} m; (iii) a proton whose x-coordinate is known to within 2×10^{-15} m; (iv) a proton whose x-coordinate is known to within 4×10^{-15} m.

39.4 The Electron Microscope

The **electron microscope** offers an important and interesting example of the interplay of wave and particle properties of electrons. An electron beam can be used to form an image of an object in much the same way as a light beam. A ray of light can be bent by reflection or refraction, and an electron trajectory can be bent by an electric or magnetic field. Rays of light diverging from a point on an object can be brought to convergence by a converging lens or concave mirror, and electrons diverging from a small region can be brought to convergence by electric and/or magnetic fields.

The analogy between light rays and electrons goes deeper. The *ray* model of geometric optics is an approximate representation of the more general *wave* model. Geometric optics (ray optics) is valid whenever interference and diffraction effects can be neglected. Similarly, the model of an electron as a point particle following a line trajectory is an approximate description of the actual behavior of the electron; this model is useful when we can neglect effects associated with the wave nature of electrons.

How is an electron microscope superior to an optical microscope? The *resolution* of an optical microscope is limited by diffraction effects, as we discussed in Section 36.7. Using wavelengths around 500 nm, an optical microscope can't resolve objects smaller than a few hundred nanometers, no matter how carefully its lenses are made. The resolution of an electron microscope is similarly limited by the wavelengths of the electrons, but these wavelengths may be many thousands of times *smaller* than wavelengths of visible light. As a result, the useful magnification of an electron microscope can be thousands of times as great as that of an optical microscope. ?

Note that the ability of the electron microscope to form a magnified image *does not* depend on the wave properties of electrons. Within the limitations of the Heisenberg uncertainty principle, we can compute the electron trajectories by treating them as classical charged particles under the action of electric- and magnetic-field forces (in analogy to ray optics). Only when we talk about *resolution* do the wave properties become important.

Example 39.5 An electron microscope

The nonrelativistic electron beam in an electron microscope is formed by a setup similar to the electron gun used in the Davisson-Germer experiment (Fig. 39.3). What accelerating voltage is needed to produce electrons with wavelength 10 pm = 0.010 nm (roughly 50,000 times smaller than typical visible-light wavelengths)? The initial kinetic energy of the electrons is negligible.

SOLUTION

IDENTIFY: Since this situation is similar to the Davisson-Germer experiment, we can use all of the concepts that we developed to understand that experiment.

SET UP: The accelerating voltage (our target variable) is the quantity V_{ba} in Eq. (39.5). We are given the values of all the other quantities in that equation, so we can solve for V_{ba}.

EXECUTE: Solving Eq. (39.5) for V_{ba} and inserting the appropriate numbers, we find

$$V_{ba} = \frac{h^2}{2me\lambda^2}$$
$$= \frac{(6.626 \times 10^{-34}\ \mathrm{J \cdot s})^2}{2(9.109 \times 10^{-31}\ \mathrm{kg})(1.602 \times 10^{-19}\ \mathrm{C})(10 \times 10^{-12}\ \mathrm{m})^2}$$
$$= 1.5 \times 10^4\ \mathrm{V} = 15{,}000\ \mathrm{V}$$

EVALUATE: This is approximately equal to the accelerating voltage for the electron beam in a TV picture tube. This example shows, incidentally, that the sharpness of a TV picture is *not* limited by electron diffraction effects. Also note that this 15-kV voltage will increase the kinetic energy of the electrons from a relatively small value to 15 keV. Since electrons have a rest energy of 0.511 MeV = 511 keV, we can accurately describe these 15-keV electrons as nonrelativistic.

The Transmission Electron Microscope

Except within their electron guns, most practical electron microscopes use magnetic fields rather than electric fields as "lenses" for focusing the beam. A common setup for *transmission electron microscopes* includes three such lenses in a compound-microscope arrangement, as shown in Fig. 39.10a. Electrons are emitted from a hot cathode and accelerated by a potential difference, typically 10 to 100 kV. The electrons pass through a condensing lens and are formed into a parallel beam before passing through the specimen or object to be viewed. The specimen to be viewed is very thin, typically 10 to 100 nm, so the electrons are not slowed appreciably as they pass through. The objective lens then forms an intermediate image of this object, and the projection lens produces a final real image of that image. The objective and projection lenses play the roles of the objective and eyepiece lenses, respectively, of a compound optical microscope (see Section 34.8). The final image is recorded on photographic film or projected onto a fluorescent screen for viewing or photographing. The entire apparatus, including the specimen, must be enclosed in a vacuum container, just as with the cathode-ray tube; otherwise, electrons would scatter off air molecules and muddle the image. Figure 39.10b shows an image made with a transmission electron microscope.

We might think that when the electron wavelength is 0.01 nm (as in Example 39.5), the resolution would also be about 0.01 nm. In fact, it is seldom better than 0.5 nm, in part because the focal length of a magnetic lens depends on the electron speed, which is never exactly the same for all electrons in the beam.

39.10 (a) Schematic diagram of a transmission electron microscope (TEM). (b) This TEM image shows a skin cell in the process of dividing into two daughter cells. False color has been used to show the genetic material (dark green).

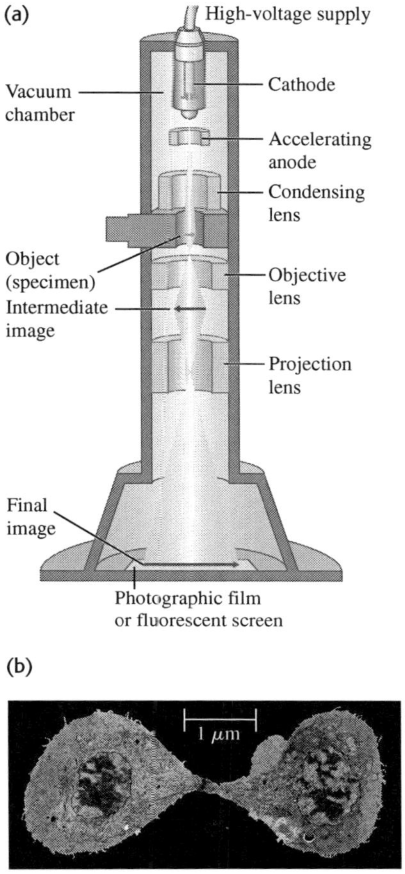

The Scanning Electron Microscope

An important variation is the *scanning electron microscope.* The electron beam is focused to a very fine line and is swept across the specimen, just as the electron beam in a TV picture tube traces out the picture. As the beam scans the specimen, electrons are knocked off and are collected by a collecting anode that is kept at a potential a few hundred volts positive with respect to the specimen. The current in the collecting anode is amplified and used to modulate the electron beam in a cathode-ray tube, which is swept in synchronization with the microscope beam. Thus the cathode-ray tube traces out a greatly magnified image of the specimen. This scheme has several advantages. The specimen can be thick because the beam does not need to pass through it. Also, the knock-off electron production depends on the *angle* at which the beam strikes the surface. Thus scanning electron micrographs have an appearance that is much more three-dimensional than conventional visible-light micrographs. The resolution is typically of the order of 10 nm, still much finer than the best optical microscopes. The image that opens this chapter was made with a scanning electron microscope.

39.5 Wave Functions and the Schrödinger Equation

We have now seen persuasive evidence that on an atomic or subatomic scale, a particle such as an electron cannot be described simply as a point. Instead we use a *wave function* to describe the state of a particle. Let's describe more specifically the kinematic language we must use to replace the classical scheme of describing a particle by its coordinates and velocity components.

Our new scheme for describing the state of a particle has a lot in common with the language of classical wave motion. In Chapter 15, we described transverse waves on a string by specifying the position of each point in the string at each instant of time by means of a *wave function* (see Section 15.3). If y represents the displacement from equilibrium, at time t, of a point on the string at a distance x from the origin, then the function $y(x, t)$ represents the displacement of any point x at any time t. Once we know the wave function for a particular wave motion, we know everything there is to know about the motion. We can find the position and velocity of any point on the string at any time, and so on (Fig. 39.11). We worked out specific forms for these functions for *sinusoidal* waves, in which each particle undergoes simple harmonic motion.

39.11 These children are talking over a cup-and-string "telephone." The displacement of the string is completely described by a wave function $y(x, t)$. In an analogous way, a particle is completely described by a quantum-mechanical wave function $\Psi(x, y, z, t)$.

We followed a similar pattern for sound waves in Chapter 16. The wave function $p(x, t)$ for a wave traveling along the x-direction represented the pressure variation at any point x at any time t. We used this language once more in Section 32.3, where we used *two* wave functions to describe the electric and magnetic fields of *electromagnetic* waves at any point in space at any time.

Thus it is natural to use a wave function as the central element of our new language. The symbol customarily used for this wave function is Ψ or ψ. In general, Ψ is a function of all the space coordinates and time, whereas ψ is a function of the space coordinates only—*not* of time. Just as the wave function $y(x, t)$ for mechanical waves on a string provides a complete description of the motion, so the wave function $\Psi(x, y, z, t)$ for a particle contains all the information that can be known about the particle. The mathematical theory of quantum mechanics describes how to use $\Psi(x, y, z, t)$ to determine the average values of the particle's position, velocity, momentum, energy, and angular momentum. The required techniques are beyond the scope of this discussion, but they are well established and well supported by experimental results.

CAUTION **Particle waves vs. mechanical waves** Unlike for mechanical waves on a string or sound waves in air, the wave function for a particle is *not* a mechanical wave that needs some material medium in order to propagate. The wave function describes the particle, but we cannot define the function itself in terms of anything material. We can only describe how it is related to physically observable effects.

Interpretation of the Wave Function

The wave function describes the distribution of a particle in space, just as the wave functions for an electromagnetic wave describe the distribution of the electric and magnetic fields. When we worked out interference and diffraction patterns in Chapters 35 and 36, we found that the intensity I of the radiation at any point in a pattern is proportional to the square of the electric-field magnitude—that is, to E^2. In the photon interpretation of interference and diffraction (see Section 38.9), the intensity at each point is proportional to the number of photons striking around that point or, alternatively, to the *probability* that any individual photon will strike around the point. Thus the square of the electric-field magnitude at each point is proportional to the probability of finding a photon around that point.

In exactly the same way, the square of the wave function of a particle at each point tells us about the probability of finding the particle around that point. More precisely, we should say the square of the *absolute value* of the wave function, $|\Psi|^2$. This is necessary because, as we'll see later, Ψ may be a *complex* quantity with real and imaginary parts. (The imaginary part of the function is a real function multiplied by the imaginary number $i = \sqrt{-1}$.)

For a particle moving in three dimensions, the quantity $|\Psi(x, y, z, t)|^2\, dV$ is the probability that the particle will be found at time t within a volume dV around the point (x, y, z). The particle is most likely to be found in regions where $|\Psi|^2$ is large, and so on. This interpretation, first made by the German physicist Max Born (Fig. 39.12), requires that the wave function Ψ be *normalized.* That is, the integral of $|\Psi|^2\, dV$ over all space must equal exactly 1. In other words, the probability is exactly 1, or 100%, that the particle is *somewhere* in the universe.

CAUTION **Interpreting $|\Psi|^2$** Note that $|\Psi(x, y, z, t)|^2$ itself is *not* a probability. Rather, $|\Psi(x, y, z, t)|^2\, dV$ is the probability of finding the particle within a volume dV around the point (x, y, z) at time t. If the volume is made smaller, it becomes less likely that the particle will be found within that volume, so the probability decreases. A better name for $|\Psi(x, y, z, t)|^2$ is the **probability distribution function,** since it describes how the probability of finding the particle at different locations is distributed over space.

39.12 In 1926, the German physicist Max Born (1882–1970) devised the interpretation that $|\Psi|^2\, dV$ is the probability of finding a particle within a given volume dV. He also coined the term "quantum mechanics" (in the original German, *Quantenmechanik*). For his contributions, Born shared (with Walther Bothe) the 1954 Nobel Prize in physics.

Stationary States

In general, the value of $|\Psi(x, y, z, t)|^2$ at a particular point varies with time. This makes sense: As an electron in a television tube flies from the cathode toward the screen, the place it is most likely to be found changes with time. But if the particle is in a state of a definite energy, such as an atomic electron in an atom in a definite energy level, the value of $|\Psi|^2$ at each point is *independent* of time. Because the particle's probability distribution in such a state doesn't change with time, a state with a definite energy is called a **stationary state.** Such states are of tremendous importance in quantum mechanics. For example, for each definite-energy, stationary state in a hydrogen atom, there is a specific wave function. It is possible for an atom to be in a state that is not identical to any one of these stationary-state wave functions and that does not have a definite energy. Nevertheless, the wave function for any state can always be written as a combination of stationary-state wave functions. We saw something similar in Section 15.8: Any standing wave on a string, no matter how complex, can be expressed as a combination of the normal-mode vibrations of the string. We will devote most of Chapter 40 to finding the stationary-state wave functions for different physical situations.

Why *is* $|\Psi|^2$ independent of time if the particle is in a state of definite energy? That is, why are such states stationary? To answer this question, we first note the following result from quantum mechanics: For a particle in a state of definite energy E, the time-dependent wave function $\Psi(x, y, z, t)$ can be written as a product of a time-*independent* function $\psi(x, y, z)$ and a simple function of time:

$$\Psi(x, y, z, t) = \psi(x, y, z)e^{-iEt/\hbar} \quad \text{(time-dependent wave function for a stationary state)} \tag{39.14}$$

The exponential function in Eq. (39.14) is defined by *Euler's formula,* which states that for any angle θ,

$$e^{i\theta} = \cos\theta + i\sin\theta \quad \text{and} \quad e^{-i\theta} = \cos\theta - i\sin\theta \tag{39.15}$$

Thus Eq. (39.14) shows that the wave function for any stationary state is a *complex* function.

Let's now look at the probability distribution function $|\Psi|^2$ obtained from Eq. (39.14). Note that $|\Psi|^2$ is the product of Ψ and its *complex conjugate* Ψ^*. To find the complex conjugate of a complex number, simply replace all i with $-i$. For example, the complex conjugate of $c = a + ib$, where a and b are real, is $c^* = a - ib$, so $|c|^2 = c^*c = (a + ib)(a - ib) = a^2 + b^2$ (recall that $i^2 = -1$). The complex conjugate of Eq. (39.14) is

$$\Psi^*(x, y, z, t) = \psi^*(x, y, z)e^{+iEt/\hbar} \tag{39.16}$$

Hence

$$\begin{aligned} |\Psi(x, y, z, t)|^2 &= \Psi^*(x, y, z, t)\Psi(x, y, z, t) \\ &= \psi^*(x, y, z)\psi(x, y, z)e^{+iEt/\hbar}e^{-iEt/\hbar} \\ &= \psi^*(x, y, z)\psi(x, y, z)e^0 = |\psi(x, y, z)|^2 \end{aligned} \tag{39.17}$$

Since $|\psi(x, y, z)|^2$ does not depend on time, Eq. (39.17) shows that the same must be true for the probability distribution function $|\Psi(x, y, z, t)|^2$. This justifies the term "stationary state" for a state of definite energy.

The Schrödinger Equation

We have stressed the importance of stationary states in describing quantum-mechanical systems. As Eq. (39.14) shows, to describe a stationary state we must know its spatial wave function $\psi(x, y, z)$ and its energy E. To determine these

39.13 Erwin Schrödinger (1887–1961) developed the equation that bears his name in 1926, an accomplishment for which he shared (with the British physicist P. A. M. Dirac) the 1933 Nobel Prize in physics. In later life he did research into the philosophy and history of science.

values we use a tool developed in 1926 by the Austrian physicist Erwin Schrödinger (Fig. 39.13) and known as the *Schrödinger equation.* The Schrödinger equation plays the same central role in quantum mechanics as do Newton's laws in mechanics and Maxwell's equations in electromagnetism. Our understanding of every quantum-mechanical system, including atoms, molecules, atomic nuclei, and electrons in solids, is based on the solutions to this equation for that system.

We can't *derive* the Schrödinger equation from other principles; it is a new principle of its own. But we can show how it is related to the de Broglie relationships, and we can make it seem plausible.

The simplest form of the Schrödinger equation is for a particle of mass m that moves in one dimension only, parallel to the x-axis, so that the spatial wave function ψ is a function only of x. We assume that the particle moves in the presence of a conservative force that has only an x-component, so there is a corresponding potential energy $U(x)$. The **Schrödinger equation** for such a particle with a definite energy E is

$$-\frac{\hbar^2}{2m}\frac{d^2\psi(x)}{dx^2} + U(x)\psi(x) = E\psi(x) \quad \text{(one-dimensional Schrödinger equation)} \tag{39.18}$$

In this equation E is a constant. How do we know that this equation is correct? Because it works. Predictions made using this equation agree with experimental results. In Chapter 40 we'll apply Eq. (39.18) to several physical situations, each with a different form of the function $U(x)$.

Wave Equation for a Free Particle

As an example, let's consider a *free* particle that experiences no force at all. If there is no force, $U(x)$ is independent of x; for simplicity we choose $U(x) = 0$. If such a free particle is moving in the $+x$-direction with momentum of magnitude p, its kinetic energy (and hence its total energy) is $E = p^2/2m$. Such a particle is in a state of definite energy (a stationary state).

From the de Broglie equations, the particle has a definite wavelength $\lambda = h/p$ and a definite frequency $f = E/h$. By analogy to the traveling mechanical waves we studied in Section 15.3, we write the wave function for the particle as

$$\Psi(x, t) = A\cos(kx - \omega t) + B\sin(kx - \omega t) \tag{39.19}$$

where A and B are constants. As we did for mechanical waves in Chapter 15 and electromagnetic waves in Chapter 32, we use the wave number $k = 2\pi/\lambda$ and the angular frequency $\omega = 2\pi f$ in Eq. (39.19). These are related to the momentum and energy by

$$k = \frac{2\pi}{\lambda} = \frac{2\pi}{h}\frac{h}{\lambda} = \frac{p}{\hbar} \qquad \omega = 2\pi f = \frac{2\pi}{h}hf = \frac{E}{\hbar} \tag{39.20}$$

The wave function in Eq. (39.19) doesn't seem to be of the stationary-state form given in Eq. (39.14). We can cast it into this form, however, if we let $B = iA$. Then

$$\begin{aligned}\Psi(x, t) &= A\cos(kx - \omega t) + iA\sin(kx - \omega t)\\ &= A[\cos(kx - \omega t) + i\sin(kx - \omega t)]\\ &= Ae^{i(kx-\omega t)} = Ae^{ikx}e^{-i\omega t}\end{aligned} \tag{39.21}$$

In Eq. (39.21) we have again used Euler's formula, which states that for any angle θ,

$$e^{i\theta} = \cos\theta + i\sin\theta \quad \text{and} \quad e^{-i\theta} = \cos\theta - i\sin\theta \tag{39.22}$$

Comparing Eq. (39.21) and Eq. (39.14), we see that our wave function does indeed represent a stationary state with energy $E = \hbar\omega$ and spatial wave function $\psi(x) = Ae^{ikx}$. Figure 39.14 graphs the real and imaginary parts of this function.

39.14 The spatial wave function $\psi(x) = Ae^{ikx}$ for a free particle of definite momentum $p = \hbar k$ is a *complex* function: It has both a real part and an imaginary part.

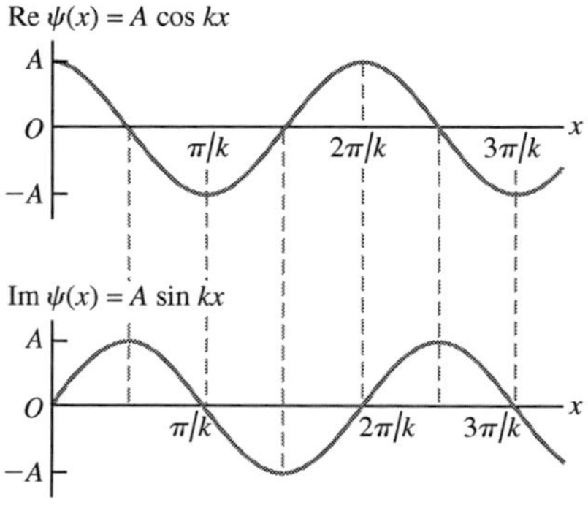

If we substitute $\psi(x)$ into the left-hand side of the Schrödinger equation, Eq. (39.18), we obtain

$$-\frac{\hbar^2}{2m}\frac{d^2(Ae^{ikx})}{dx^2} + (0)(Ae^{ikx}) = -\frac{\hbar^2}{2m}(ik)^2 Ae^{ikx} = \frac{\hbar^2 k^2}{2m}Ae^{ikx}$$
$$= \frac{p^2}{2m}\psi(x) \qquad (39.23)$$

(We used the result that the second derivative of e^{ax}, where a is a constant, is a^2e^{ax}.) The right-hand side of Eq. (39.18) is $E\psi(x)$. Since $E = p^2/2m$ from Eq. (39.18), we see from Eq. (39.23) that the free-particle wave function in Eq. (39.21) satisfies the Schrödinger equation.

If k is positive in Eq. (39.21), the wave function represents a free particle moving in the positive x-direction. If k is negative, the momentum and hence the motion are in the negative x-direction. (With a negative value of k, the wavelength is $\lambda = 2\pi/|k|$.)

For a free particle, there is no restriction on the value of p and hence no restriction on the value of the energy $E = p^2/2m$. We will find in Chapter 40 that if $U(x)$ is not a constant, then solutions of the Schrödinger equation are possible only for certain values of E. These values represent the allowed *energy levels* of the system described by $U(x)$. This discovery is of the utmost importance. Before the development of the Schrödinger equation, there was no way to predict energy levels from any fundamental theory except the Bohr model, which had very limited success.

There is also a version of the Schrödinger equation that includes time dependence. It is needed to study states that are *not* stationary and for which the probability distribution function $|\Psi(x, y, z, t)|^2$ depends on time. However, we don't need that version to calculate the energy levels and wave functions of stationary states. If we do need a time-dependent wave function for a stationary state of energy E, we simply use Eq. (39.14). The time-dependent Schrödinger equation is essential for studying the details of *transitions* between states. In more advanced courses you may study the quantum mechanics of time-dependent phenomena such as photon emission and absorption and the lifetimes of states.

Example 39.6 The Schrödinger equation

Consider the wave function $\psi(x) = A_1e^{ikx} + A_2e^{-ikx}$, where k is positive. Is this a valid stationary-state wave function for a free particle? What is the energy?

SOLUTION

IDENTIFY: A valid stationary-state wave function for a free particle must satisfy the Schrödinger equation, Eq. (39.18), with $U(x) = 0$.

SET UP: To test the given function $\psi(x)$, we simply substitute it into the left-hand side of the equation. If the result is a constant times $\psi(x)$, then the wave function is indeed a solution and the constant is equal to the energy E.

EXECUTE: Substituting $\psi(x) = A_1e^{ikx} + A_2e^{-ikx}$ and $U(x) = 0$ into Eq. (39.18), we obtain

$$-\frac{\hbar^2}{2m}\frac{d^2\psi(x)}{dx^2} = -\frac{\hbar^2}{2m}\frac{d^2(A_1e^{ikx} + A_2e^{-ikx})}{dx^2}$$
$$= -\frac{\hbar^2}{2m}[(ik)^2A_1e^{ikx} + (-ik)^2A_2e^{ikx}]$$
$$= \frac{\hbar^2k^2}{2m}(A_1e^{ikx} + A_2e^{-ikx}) = \frac{\hbar^2k^2}{2m}\psi(x)$$

We see that the left-hand side of the Schrödinger equation is equal to a constant times $\psi(x)$, so this $\psi(x)$ is indeed a valid stationary-state wave function for a free particle. The constant is just equal to the energy: $E = \hbar^2k^2/2m$.

EVALUATE: Note that $\psi(x)$ is a *superposition* of two different wave functions: one for a particle with a given magnitude of momentum $p = \hbar k$ moving in the positive x-direction and one for a particle with the same magnitude of momentum moving in the negative x-direction. While the combined wave function represents a stationary state with a definite energy, this state does *not* have a definite momentum. We will see in Chapter 40 that such a wave function can represent a *standing wave,* and we will explore the situations in which such standing matter waves can arise.

Wave Packets

ActivPhysics ONLINE
17.7 Wave Packets

The wave function for a free particle given by Eq. (39.21) has a definite momentum p in the x-direction. For such a state, there is *no* uncertainty in momentum: $\Delta p_x = 0$. The Heisenberg uncertainty principle, Eq. (39.11), says that $\Delta x \Delta p_x \geq \hbar$. If Δp_x is zero, then Δx must be infinite. The price we pay for knowing the particle's momentum precisely is that we have no idea *where* the particle is! We can show this by calculating the probability distribution function $|\Psi(x, t)|^2$:

$$|\Psi(x,t)|^2 = \Psi^*(x,t)\Psi(x,t) = (A^*e^{-ikx}e^{+i\omega t})(Ae^{ikx}e^{-i\omega t}) \\ = A^*Ae^0 = |A|^2$$

The probability distribution function doesn't depend on time, as we would expect because this is a stationary state (a state of definite energy). It also doesn't depend on position, which says that we are equally likely to find the particle *anywhere* in space! Mathematically, this situation occurs because the spatial wave function $\psi(x) = Ae^{ikx} = A\cos kx + iA\sin kx$ is a sinusoidal function that extends all the way from $x = -\infty$ to $x = +\infty$ with the same amplitude A. (This also means that the wave function can't be normalized: The integral of $|\Psi(x, t)|^2$ over all space would be infinite for any value of A.)

In practical situations, we always have *some* idea where a particle is. To describe this situation, we need a wave function that is more localized in space. We can create one by superimposing two or more sinusoidal functions, as we did in Example 39.6. In that example we added two wave functions with opposite wave numbers k and $-k$; to make a localized wave function, we must make a different choice.

To keep things simple, we'll imagine wave functions that depend on only one spatial coordinate (x), and we'll consider them at one instant of time (say, $t = 0$). Our wave functions are then functions only of x, so we'll denote them as ψ. For example, let's add together two different stationary-state wave functions for a free particle as given by Eq. (39.21), but with slightly different values k_1 and k_2 of the wave number. At $t = 0$ the time factors $e^{-i\omega_1 t}$ and $e^{-i\omega_2 t}$ are both equal to $e^0 = 1$, so the wave function at $t = 0$ is

$$\begin{aligned} \psi(x) &= A_1e^{ik_1x} + A_2e^{ik_2x} \\ &= [A_1\cos(k_1x) + iA_1\sin(k_1x)] + [A_2\cos(k_2x) + iA_2\sin(k_2x)] \\ &= [A_1\cos(k_1x) + A_2\cos(k_2x)] + i[A_1\sin(k_1x) + A_2\sin(k_2x)] \end{aligned} \quad (39.24)$$

We have used Euler's formula, Eq. (39.22), to separate the real and imaginary parts of $\psi(x)$.

Figure 39.15a is a graph of the real parts of the individual wave functions for the case $A_2 = -A_1$; Fig. 39.15b graphs the real part of the combined wave function $\psi(x)$ given by Eq. (39.24). We saw something very similar to Fig. 39.15b in our discussion of beats in Section 16.7: When we superimposed two sinusoidal waves with slightly different frequencies (Fig. 16.24), the result was a wave with

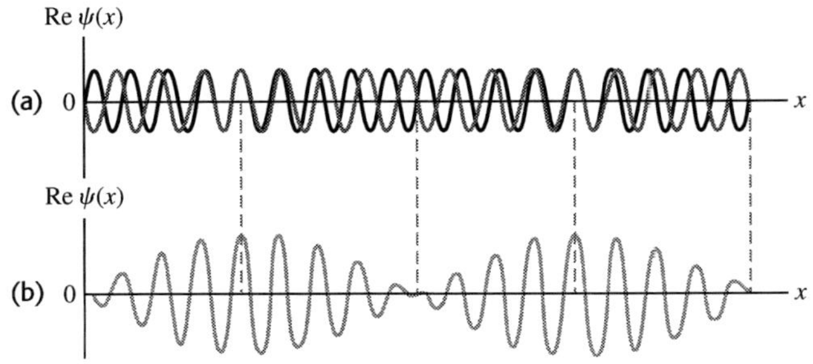

39.15 (a) The real parts of two sinusoidal waves with slightly different wave numbers, shown at one instant of time. (b) The superposition of these waves has a wave number equal to the average of the two individual wave numbers. The amplitude varies, giving the total wave a lumpy character not possessed by either individual wave.

a lumpy character that the individual waves did not possess. In the same way, a particle represented by the lumpy wave function given in Eq. (39.24) is more likely to be found in some regions than in others; that is, it is *localized.* However, the particle's momentum no longer has a definite value because we began with two different wave numbers. This agrees with the Heisenberg uncertainty principle: By decreasing the uncertainty in the particle's position, we have increased the uncertainty in its momentum.

It's not hard to imagine superposing two additional sinusoidal waves with different wave numbers and amplitudes so as to reinforce alternate lumps in Fig. 39.15b and cancel out the in-between ones. Finally, if we superpose waves with a very large number of different wave numbers, we can construct a wave with only one lump (Fig. 39.16). Then, finally, we have something that begins to look like both a particle and a wave. It is a particle in the sense that it is localized in space; if we look from a distance, it may look like a point. But it also has a periodic structure that is characteristic of a wave.

Such a wave pulse is called a **wave packet.** We can represent such a superposition by an expression such as

$$\psi(x) = \int_{-\infty}^{\infty} A(k)e^{ikx}\,dk \qquad (39.25)$$

This integral represents a superposition of a very large number of waves, each with a different wave number and each with an amplitude $A(k)$ that depends on k.

There is an important relationship between the two functions $\psi(x)$ and $A(k)$. This relationship is shown qualitatively in Fig. 39.17. If the function $A(k)$ is sharply peaked, as in Fig. 39.17a, we are superposing only a narrow range of wave numbers. The resulting wave pulse is then relatively broad (Fig. 39.17b). But if we use a wider range of wave numbers, so that the function $A(k)$ is broader (Fig. 39.17c), then the wave pulse is more narrowly localized (Fig. 39.17d).

What we are seeing is the uncertainty principle in action. A narrow range of k means a narrow range of $p_x = \hbar k$ and thus a small Δp_x; the result is a relatively large Δx. A broad range of k corresponds to a large Δp_x, and the resulting Δx is smaller. Thus we see that the uncertainty principle $\Delta x \Delta p_x \geq \hbar$ is an inevitable consequence of the de Broglie relationship and the properties of integrals such as Eq. (39.25). These integrals are called *Fourier integrals;* they are a generalization of the concept of Fourier series, which we mentioned in Section 15.8. In both cases we are representing a complex wave form as a superposition of sinusoidal functions. With Fourier series we use a sequence of frequencies (or values of k) that are integer multiples of some basic value, while with Fourier integrals we superpose functions with a *continuous* distribution of values of k.

Test Your Understanding of Section 39.5 Does a wave packet given by Eq. (39.25) represent a stationary state?

39.16 Superimposing a large number of sinusoidal waves with different wave numbers and appropriate amplitudes can produce a wave pulse with a wavelength $\lambda_{av} = 2\pi/k_{av}$ that is localized in a region of space of length Δx. This localized pulse has aspects of both particle and wave.

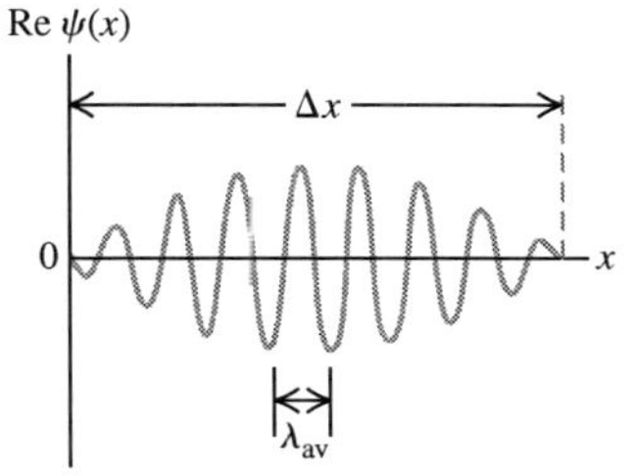

39.17 How varying the function $A(k)$ in the wave-packet expression, Eq. (39.25), changes the character of the wave function $\psi(x)$.

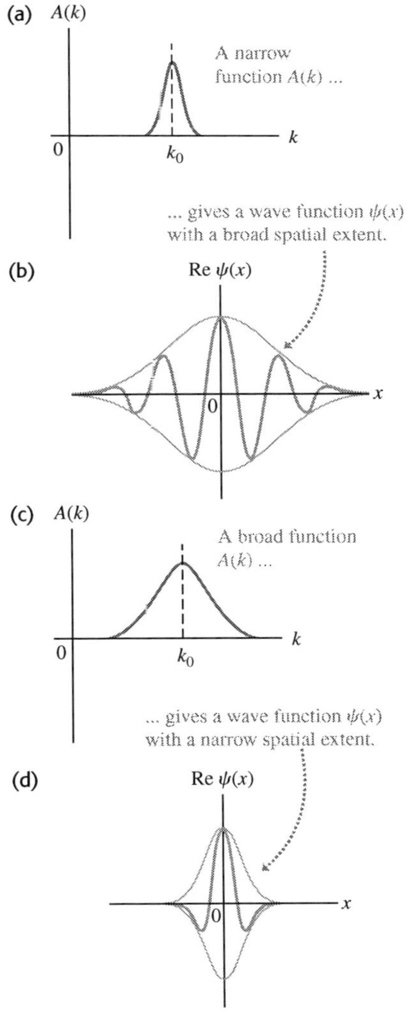

CHAPTER 39 SUMMARY

De Broglie waves: Electrons and other particles have wave properties. The wavelength of the wave depends on the particle's momentum in the same way as for photons. The state of a particle is described not by its coordinates and velocity components but rather by a wave function, which in general is a function of the three space coordinates and time. (See Example 39.1.)

$$\lambda = \frac{h}{p} = \frac{h}{mv} \quad (39.1)$$

$$E = hf \quad (39.2)$$

Electron diffraction and electron microscopes: Diffraction of an electron beam from the surface of a metallic crystal provided the first direct confirmation of the wave nature of particles. If a nonrelativistic electron has been accelerated from rest through a potential difference V_{ba}, its wavelength is given by Eq. (39.5). Electron microscopes use the very small wavelengths of fast-moving electrons to make images with resolution thousands of times finer than is possible with visible-light microscopes. (See Examples 39.2 and 39.5.)

$$\lambda = \frac{h}{p} = \frac{h}{\sqrt{2meV_{ba}}} \quad (39.5)$$

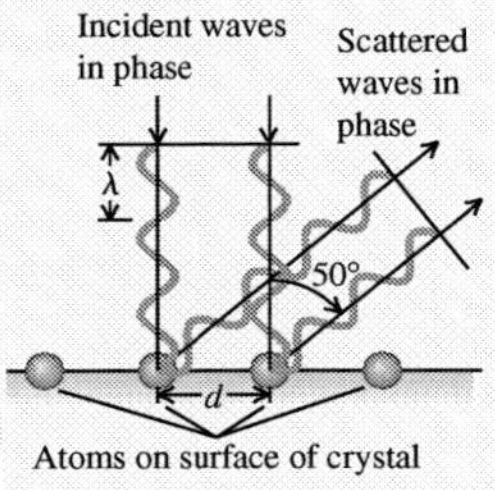

The Heisenberg uncertainty principle: It is impossible to make precise determinations of a coordinate of a particle and of the corresponding momentum component at the same time. The precision of such measurements for the x-components is limited by the Heisenberg uncertainty principle, Eq. (39.11); there are corresponding relationships for the y- and z-components. The uncertainty ΔE in the energy of a state that is occupied for a time Δt is given by Eq. (39.13). In these expressions, $\hbar = h/2\pi$. (See Examples 39.3 and 39.4.)

$$\Delta x \Delta p_x \geq \hbar \quad (39.11)$$
(Heisenberg uncertainty principle for position and momentum)

$$\Delta E \Delta t \geq \hbar \quad (39.13)$$
(Heisenberg uncertainty principle for energy and time)

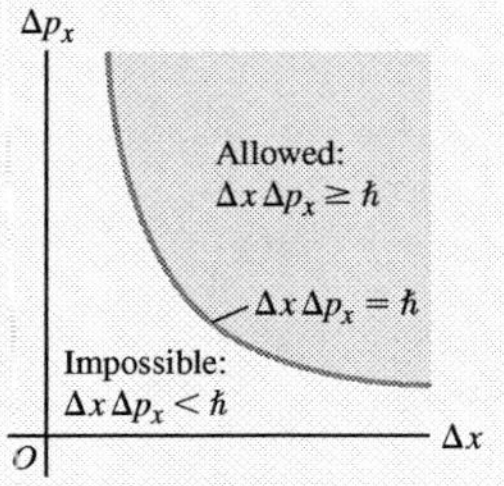

Wave functions: The wave function $\Psi(x, y, z, t)$ for a particle contains all of the information about that particle. The quantity $|\Psi(x, y, z, t)|^2$, called the probability distribution function, determines the relative probability of finding a particle near a given position at a given time. If the particle is in a state of definite energy, called a stationary state, $\Psi(x, y, z, t)$ is a product of a function ψ that depends only on spatial coordinates and a function $e^{-iEt/\hbar}$ that depends only on time. For a stationary state, the probability distribution function is independent of time.

$$\Psi(x, y, z, t) = \psi(x, y, z)\, e^{-iEt/\hbar} \quad (39.14)$$
(stationary state)

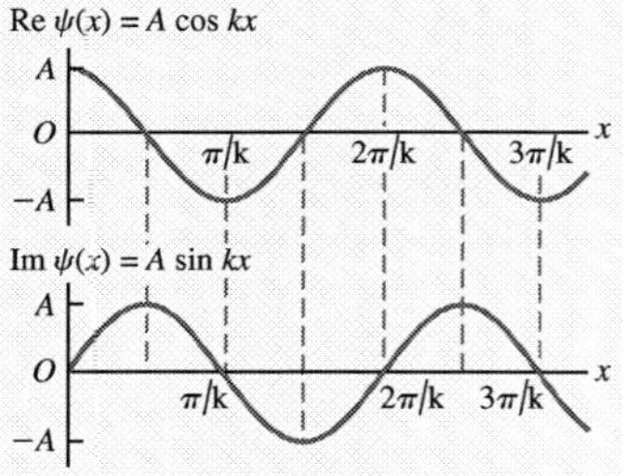

The Schrödinger equation: For a particle that moves in one dimension in the presence of a potential-energy function $U(x)$, the wave function for a stationary state of energy E satisfies the Schrödinger equation. More complex wave functions can be made by superposing stationary-state wave functions. These can represent particles that are localized in a certain region and still have wave properties, giving them both particle and wave aspects. (See Example 39.6.)

$$-\frac{\hbar^2}{2m}\frac{d^2\psi(x)}{dx^2} + U(x)\psi(x) = E\psi(x)$$
(one-dimensional Schrödinger equation) (39.18)

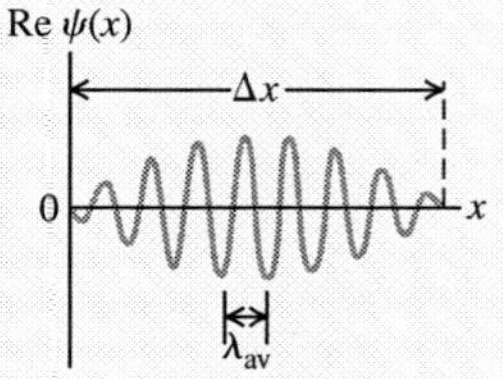

Key Terms

de Broglie wavelength, *1350*
quantum mechanics, *1350*
wave function, *1350*
boundary condition, *1351*
electron diffraction, *1353*
Heisenberg uncertainty principle, *1357*
electron microscope, *1360*
probability distribution function, *1362*
stationary state, *1363*
Schrödinger equation, *1364*
wave packet, *1367*

Answer to Chapter Opening Question ?

The smallest detail visible in an image is comparable to the wavelength used to make the image. Electrons can easily be given a large momentum p and hence a short wavelength $\lambda = h/p$, and so can be used to resolve extremely fine details. (See Section 39.4.)

Answers to Test Your Understanding Questions

39.1 Answer: (i) From Example 39.1, the speed of a particle is $v = h/\lambda m$ and the kinetic energy is $K = \frac{1}{2}mv^2 = (m/2)(h/\lambda m)^2 = h^2/2\lambda^2 m$. This shows that for a given wavelength, the kinetic energy is inversely proportional to the mass. Hence the proton, with a smaller mass, has more kinetic energy than the neutron.
39.2 Answer: no The energy of a photon is $E = hf$, and the frequency of a photon is $f = c/\lambda$. Hence $E = hc/\lambda$ and $\lambda = hc/E = (4.136 \times 10^{-15}\ \text{eV}\cdot\text{s}) \times (2.998 \times 10^8\ \text{m/s})/(54\ \text{eV}) = 2.3 \times 10^{-8}$ m. This is more than 100 times greater than the wavelength of an electron of the same energy. While both photons and electrons have wavelike properties, they have different relationships between their energy and momentum and hence between their frequency and wavelength.
39.3 Answer: (i) and (iii) (tie), (ii) and (iv) (tie) According to the Heisenberg uncertainty principle, the smaller the uncertainty Δx in the x-coordinate, the greater the uncertainty Δp_x in the x-momentum. The relationship between Δx and Δp_x does not depend on the mass of the particle, and so is the same for a proton as for an electron.
39.5 Answer: no Equation (39.25) represents a superposition of wave functions with different values of wave number k and hence different values of energy $E = \hbar^2 k^2/2m$. The state that this combined wave function represents is not a state of definite energy, and therefore not a stationary state. Another way to see this is to note that the time-dependent wave function $\Psi(x, t)$ would include a factor $e^{-iEt/\hbar}$ inside the integral in Eq. (39.25), with a different value of E for each value of k. This wave function therefore has a very complicated time dependence, and the probability distribution function $|\Psi(x, t)|^2$ does depend on time.

PROBLEMS

For instructor-assigned homework, go to **www.masteringphysics.com**

Discussion Questions

Q39.1. In attempting to reconcile the wave and particle models of light, some people have suggested that the photon rides up and down on the crests and troughs of the electromagnetic wave. What things are *wrong* with this description?
Q39.2. If a proton and an electron have the same speed, which has the longer de Broglie wavelength? Explain.
Q39.3. If a proton and an electron have the same kinetic energy, which has the longer de Broglie wavelength? Explain.
Q39.4. Does a photon have a de Broglie wavelength? If so, how is it related to the wavelength of the associated electromagnetic wave? Explain.
Q39.5. When an electron beam goes through a very small hole, it produces a diffraction pattern on a screen, just like that of light. Does this mean that an electron spreads out as it goes through the hole? What does this pattern mean?
Q39.6. You have been asked to design a magnet system to steer a beam of 54-eV electrons like those described in Example 39.2 (Section 39.2). The goal is to be able to direct the electron beam to a specific target location with an accuracy of ± 1.0 mm. In your design, do you need to take the wave nature of electrons into account? Explain.
Q39.7. The upper half of the electron diffraction pattern shown in Fig. 39.6 is the mirror image of the lower half of the pattern. Would it be correct to say that the upper half of the pattern is caused by electrons that pass through the upper half of the slit? Explain.
Q39.8. A particular electron in the experimental setup shown in Fig. 39.6 lands on the photographic film a distance x above the center of the pattern. Given the value of x, is it possible to calculate the precise trajectory that the electron followed? Explain.
Q39.9. Does the uncertainty principle have anything to do with marksmanship? That is, is the accuracy with which a bullet can be aimed at a target limited by the uncertainty principle? Explain.
Q39.10. Suppose a two-slit interference experiment is carried out using an electron beam. Would the same interference pattern result if one slit at a time is uncovered instead of both at once? If not, why not? Doesn't each electron go through one slit or the other? Or does every electron go through both slits? Discuss the latter possibility in light of the principle of complementarity.
Q39.11. Equation (39.13) states that the energy of a system can have uncertainty. Does this mean that the principle of conservation of energy is no longer valid? Explain.
Q39.12. Laser light results from transitions from long-lived metastable states. Why is it more monochromatic than ordinary light?
Q39.13. Could an electron-diffraction experiment be carried out using three or four slits? Using a grating with many slits? What sort of results would you expect with a grating? Would the uncertainty principle be violated? Explain.
Q39.14. As the lower half of Fig. 39.5 shows, the diffraction pattern made by electrons that pass through aluminum foil is a series of concentric rings. But if the aluminum foil is replaced by a single crystal of aluminum, only certain points on these rings appear in the pattern. Explain.
Q39.15. Why can an electron microscope have greater magnification than an ordinary microscope?

Q39.16. If quantum mechanics replaces the language of Newtonian mechanics, why don't we have to use wave functions to describe the motion of macroscopic bodies such as baseballs and cars?

Q39.17. A student remarks that the relationship of ray optics to the more general wave picture is analogous to the relationship of Newtonian mechanics, with well-defined particle trajectories, to quantum mechanics. Comment on this remark.

Q39.18. As Eq. (39.14) indicates, the time-dependent wave function for a stationary state is a complex number having a real part and an imaginary part. How can this function have any physical meaning, since part of it is *imaginary?*

Q39.19. When you check the air pressure in a tire, a little air always escapes; the process of making the measurement changes the quantity being measured. Think of other examples of measurements that change or disturb the quantity being measured.

Q39.20. Why must the wave function of a particle be normalized?

Q39.21. If a particle is in a stationary state, does that mean that the particle is not moving? If a particle moves in empty space with constant momentum $\vec{p}$ and hence constant energy $E = p^2/2m$, is it in a stationary state? Explain your answers.

Q39.22. Some lasers emit light in pulses that are only 10^{-12} s in duration. The length of such a pulse is $(3 \times 10^8 \text{ m/s})(10^{-12} \text{ s}) = 3 \times 10^{-4} \text{ m} = 0.3$ mm. Can pulsed laser light be as monochromatic as light from a laser that emits a steady, continuous beam? Explain.

Exercises

Section 39.1 De Broglie Waves

39.1. (a) An electron moves with a speed of 4.70×10^6 m/s. What is its de Broglie wavelength? (b) A proton moves with the same speed. Determine its de Broglie wavelength.

39.2. For crystal diffraction experiments (discussed in Section 39.2), wavelengths on the order of 0.20 nm are often appropriate. Find the energy in electron volts for a particle with this wavelength if the particle is (a) a photon; (b) an electron; (c) an alpha particle $(m = 6.64 \times 10^{-27} \text{ kg})$.

39.3. An electron has a de Broglie wavelength of 2.80×10^{-10} m. Determine (a) the magnitude of its momentum and (b) its kinetic energy (in joules and in electron volts).

39.4. Wavelength of an Alpha Particle. An alpha particle $(m = 6.64 \times 10^{-27} \text{ kg})$ emitted in the radioactive decay of uranium-238 has an energy of 4.20 MeV. What is its de Broglie wavelength?

39.5. In the Bohr model of the hydrogen atom, what is the de Broglie wavelength for the electron when it is in (a) the $n = 1$ level and (b) the $n = 4$ level? In each case, compare the de Broglie wavelength to the circumference $2\pi r_n$ of the orbit.

39.6. (a) A nonrelativistic free particle with mass m has kinetic energy K. Derive an expression for the de Broglie wavelength of the particle in terms of m and K. (b) What is the de Broglie wavelength of an 800-eV electron?

39.7. Why Don't *We* Diffract? (a) Calculate the de Broglie wavelength of a typical person walking through a doorway. Make reasonable approximations for the necessary quantities. (b) Will the person in part (a) exhibit wave-like behavior when walking through the "single slit" of a doorway? Why?

39.8. What is the de Broglie wavelength for an electron with speed (a) $v = 0.480c$ and (b) $v = 0.960c$? (*Hint:* Use the correct relativistic expression for linear momentum if necessary.)

39.9. (a) If a photon and an electron each have the same energy of 20.0 eV, find the wavelength of each. (b) If a photon and an electron each have the same wavelength of 250 nm, find the energy of each. (c) You want to study an organic molecule that is about 250 nm long using either a photon or an electron microscope. Approximately what wavelength should you use, and which probe, the electron or the photon, is likely to damage the molecule the least?

39.10. Hydrogen gas (H_2) is at 0°C. The mass of a hydrogen *atom* is 1.67×10^{-27} kg. (a) What is the average de Broglie wavelength of the hydrogen molecules? (b) How fast would an electron have to move to have the same de Broglie wavelength as the hydrogen? Do we need to consider relativity for this electron? (c) What would be the energy of a photon having the same wavelength as the H_2 molecules and the electrons? Compare it to the kinetic energy of the hydrogen molecule in part (a) and the electron in part (b).

39.11. Wavelength of a Bullet. Calculate the de Broglie wavelength of a 5.00-g bullet that is moving at 340 m/s. Will the bullet exhibit wavelike properties?

Section 39.2 Electron Diffraction

39.12. Through what potential difference must electrons be accelerated so they will have (a) the same wavelength as an x ray of wavelength 0.150 nm and (b) the same energy as the x ray in part (a)?

39.13. (a) Approximately how fast should an electron move so it has a wavelength that makes it useful to measure the distance between adjacent atoms in typical crystals (about 0.10 nm)? (b) What is the kinetic energy of the electron in part (a)? (c) What would be the energy of a photon of the same wavelength as the electron in part (b)? (d) Which would make a more effective probe of small-scale structures, electrons or photons? Why?

39.14. A beam of electrons is accelerated from rest through a potential difference of 0.100 kV and then passes through a thin slit. The diffracted beam shows its first diffraction minima at ±11.5° from the original direction of the beam when viewed far from the slit. (a) Do we need to use relativity formulas? How do you know? (b) How wide is the slit?

39.15. A beam of neutrons that all have the same energy scatters from the atoms that have a spacing of 0.0910 nm in the surface plane of a crystal. The $m = 1$ intensity maximum occurs when the angle θ in Fig. 39.3 is 28.6°. What is the kinetic energy (in electron volts) of each neutron in the beam?

39.16. A beam of 188-eV electrons is directed at normal incidence onto a crystal surface as shown in Fig. 39.4b. The $m = 2$ intensity maximum occurs at an angle $\theta = 60.6°$. (a) What is the spacing between adjacent atoms on the surface? (b) At what other angle or angles is there an intensity maximum? (c) For what electron energy (in electron volts) would the $m = 1$ intensity maximum occur at $\theta = 60.6°$? For this energy, is there an $m = 2$ intensity maximum? Explain.

39.17. A CD-ROM is used instead of a crystal in an electron-diffraction experiment like that shown in Fig. 39.3. The surface of the CD-ROM has tracks of tiny pits with a uniform spacing of 1.60 μm. (a) If the speed of the electrons is 1.26×10^4/s, at which values of θ will the $m = 1$ and $m = 2$ intensity maxima appear? (b) The scattered electrons in these maxima strike at normal incidence a piece of photographic film that is 50.0 cm from the CD-ROM. What is the spacing on the film between these maxima?

Section 39.3 Probability and Uncertainty

39.18. A pesky 1.5-mg mosquito is annoying you as you attempt to study physics in your room, which is 5.0 m wide and 2.5 m high. You decide to swat the bothersome insect as it flies toward you, but you need to estimate its speed to make a successful hit. (a) What is the maximum uncertainty in the horizontal position of the mosquito? (b) What limit does the Heisenberg uncertainty principle place on your ability to know the horizontal velocity of this mosquito? Is this limitation a serious impediment to your attempt to swat it?

39.19. By extremely careful measurement, you determine the x-coordinate of a car's center of mass with an uncertainty of only 1.00 μm. The car has a mass of 1200 kg. (a) What is the minimum uncertainty in the x-component of the velocity of the car's center of mass as prescribed by the Heisenberg uncertainty principle? (b) Does the uncertainty principle impose a practical limit on our ability to make simultaneous measurements of the positions and velocities of ordinary objects like cars, books, and people? Explain.

39.20. A 10.0-g marble is gently placed on a horizontal tabletop that is 1.75 m wide. (a) What is the maximum uncertainty in the horizontal position of the marble? (b) According to the Heisenberg uncertainty principle, what is the minimum uncertainty in the horizontal velocity of the marble? (c) In light of your answer to part (b), what is the longest time the marble could remain on the table? Compare this time to the age of the universe, which is approximately 14 billion years. (*Hint:* Can you know that the horizontal velocity of the marble is *exactly* zero?)

39.21. A scientist has devised a new method of isolating individual particles. He claims that this method enables him to detect simultaneously the position of a particle along an axis with a standard deviation of 0.12 nm and its momentum component along this axis with a standard deviation of $3.0 \times 10^{-25}\ \text{kg}\cdot\text{m/s}$. Use the Heisenberg uncertainty principle to evaluate the validity of this claim.

39.22. (a) The x-coordinate of an electron is measured with an uncertainty of 0.20 mm. What is the x-component of the electron's velocity, v_x, if the minimum percentage uncertainty in a simultaneous measurement of v_x is 1.0%? (b) Repeat part (a) for a proton.

39.23. An atom in a metastable state has a lifetime of 5.2 ms. What is the uncertainty in energy of the metastable state?

39.24. (a) The uncertainty in the y-component of a proton's position is 2.0×10^{-12} m. What is the minimum uncertainty in a simultaneous measurement of the y-component of the proton's velocity? (b) The uncertainty in the z-component of an electron's velocity is 0.250 m/s. What is the minimum uncertainty in a simultaneous measurement of the z-coordinate of the electron?

39.25. The ψ ("psi") particle has a rest energy of 3097 MeV $(1\ \text{MeV} = 10^6\ \text{eV})$. The ψ particle is unstable with a lifetime of 7.6×10^{-21} s. Estimate the uncertainty in rest energy of the ψ particle. Express your answer in MeV and as a fraction of the rest energy of the particle.

39.26. Particle Lifetime. The unstable W^+ particle has a rest energy of 80.41 GeV $(1\ \text{GeV} = 10^9\ \text{eV})$ and an uncertainty in rest energy of 2.06 GeV. Estimate the lifetime of the W^+ particle.

Section 39.4 The Electron Microscope

39.27. You want to study a biological specimen by means of a wavelength of 10.0 nm, and you have a choice of using electromagnetic waves or an electron microscope. (a) Calculate the ratio of the energy of a 10.0-nm-wavelength photon to the kinetic energy of a 10.0-nm-wavelength electron. (b) In view of your answer to part (a), which would be less damaging to the specimen you are studying: photons or electrons?

39.28. (a) In an electron microscope, what accelerating voltage is needed to produce electrons with wavelength 0.0600 nm? (b) If protons are used instead of electrons, what accelerating voltage is needed to produce protons with wavelength 0.0600 nm? (*Hint:* In each case the initial kinetic energy is negligible.)

Section 39.5 Wave Functions and the Schrödinger Equation

39.29. Consider a wave function given by $\psi(x) = A\sin kx$, where $k = 2\pi/\lambda$ and A is a real constant. (a) For what values of x is there the highest probability of finding the particle described by this wave function? Explain. (b) For which values of x is the probability *zero?* Explain.

39.30. Compute $|\Psi|^2$ for $\Psi = \psi \sin \omega t$, where ψ is time independent and ω is a real constant. Is this a wave function for a stationary state? Why or why not?

39.31. Normalization of the Wave Function. Consider a particle moving in one dimension, which we shall call the x-axis. (a) What does it mean for the wave function of this particle to be *normalized?* (b) Is the wave function $\psi(x) = e^{ax}$, where a is a positive real number, normalized? Could this be a valid wave function? (c) If the particle described by the wave function $\psi(x) = Ae^{bx}$, where A and b are positive real numbers, is confined to the range $x \geq 0$, determine A (including its units) so that the wave function is normalized.

39.32. A particle is described by a wave function $\psi(x) = Ae^{-\alpha x^2}$, where A and α are real, positive constants. If the value of α is increased, what effect does this have on (a) the particle's uncertainty in position and (b) the particle's uncertainty in momentum? Explain your answers.

39.33. Consider the complex-valued function $f(x, y) = (x - iy)/(x + iy)$. Calculate $|f|^2$.

39.34. Particle A is described by the wave function $\psi(x, y, z)$. Particle B is described by the wave function $\psi(x, y, z)e^{i\phi}$, where ϕ is a real constant. How does the probability of finding particle A within a volume dV around a certain point in space compare with the probability of finding particle B within this same volume?

39.35. A particle moving in one dimension (the x-axis) is described by the wave function

$$\psi(x) = \begin{cases} Ae^{-bx}, & \text{for } x \geq 0 \\ Ae^{bx}, & \text{for } x < 0 \end{cases}$$

where $b = 2.00\ \text{m}^{-1}$, $A > 0$, and the $+x$-axis points toward the right. (a) Determine A so that the wave function is normalized. (b) Sketch the graph of the wave function. (c) Find the probability of finding this particle in each of the following regions: (i) within 50.0 cm of the origin, (ii) on the left side of the origin (can you first guess the answer by looking at the graph of the wave function?), (iii) between $x = 0.500$ m and $x = 1.00$ m.

39.36. Linear Combinations of Wave Functions. Let ψ_1 and ψ_2 be two solutions of Eq. (39.18) with the same energy E. Show that $\psi = B\psi_1 + C\psi_2$ is also a solution with energy E, for any values of the constants B and C.

39.37. Let ψ_1 and ψ_2 be two solutions of Eq. (39.18) with energies E_1 and E_2, respectively, where $E_1 \neq E_2$. Is $\psi = A\psi_1 + B\psi_2$, where A and B are nonzero constants, a solution to Eq. (39.18)? Explain your answer.

Problems

39.38. A beam of 40-eV electrons traveling in the $+x$-direction passes through a slit that is parallel to the y-axis and 5.0 μm wide. The diffraction pattern is recorded on a screen 2.5 m from the slit. (a) What is the de Broglie wavelength of the electrons? (b) How much time does it take the electrons to travel from the slit to the screen? (c) Use the width of the central diffraction pattern to calculate the uncertainty in the y-component of momentum of an electron just after it has passed through the slit. (d) Use the result of part (c) and the Heisenberg uncertainty principle (Eq. 39.11 for y) to estimate the minimum uncertainty in the y-coordinate of an electron just after it has passed through the slit. Compare your result to the width of the slit.

39.39. (a) What is the energy of a photon that has wavelength 0.10 μm? (b) Through approximately what potential difference must electrons be accelerated so that they will exhibit wave nature in passing through a pinhole 0.10 μm in diameter? What is the speed of these electrons? (c) If protons rather than electrons were used, through what potential difference would protons have to be accelerated so they would exhibit wave nature in passing through this pinhole? What would be the speed of these protons?

39.40. Electrons go through a single slit 150 nm wide and strike a screen 24.0 cm away. You find that at angles of $\pm 20.0°$ from the center of the diffraction pattern, no electrons hit the screen but electrons hit at all points closer to the center. (a) How fast were these electrons moving when they went through the slit? (b) What will be the next larger angles at which no electrons hit the screen?

39.41. A beam of electrons is accelerated from rest and then passes through a pair of identical thin slits that are 1.25 nm apart. You observe that the first double-slit interference dark fringe occurs at $\pm 18.0°$ from the original direction of the beam when viewed on a distant screen. (a) Are these electrons relativistic? How do you know? (b) Through what potential difference were the electrons accelerated?

39.42. A beam of protons and a beam of alpha particles (of mass 6.64×10^{-27} kg and charge $+2e$) are accelerated from rest through the same potential difference and pass through identical circular holes in a very thin, opaque film. When viewed far from the hole, the diffracted proton beam forms its first dark ring at 15° with respect to its original direction. When viewed similarly, at what angle will the alpha particle form its first dark ring?

39.43. An electron beam and a photon beam pass through identical slits. On a distant screen, the first dark fringe occurs at the same angle for both of the beams. The electron speeds are much slower than that of light. (a) Express the energy of a photon in terms of the kinetic energy K of one of the electrons. (b) Which is greater, the energy of a photon or the kinetic energy of an electron?

39.44. Coherent light is passed through two narrow slits whose separation is 40.0 μm. The second-order bright fringe in the interference pattern is located at an angle of 0.0300 rad. If electrons are used instead of light, what must the kinetic energy (in electron volts) of the electrons be if they are to produce an interference pattern for which the second-order maximum is also at 0.0300 rad?

39.45. What is the de Broglie wavelength of a red blood cell, with mass 1.00×10^{-11} g, that is moving with a speed of 0.400 cm/s? Do we need to be concerned with the wave nature of the blood cells when we describe the flow of blood in the body?

39.46. High-speed electrons are used to probe the interior structure of the atomic nucleus. For such electrons the expression $\lambda = h/p$ still holds, but we must use the relativistic expression for momentum, $p = mv/\sqrt{1 - v^2/c^2}$. (a) Show that the speed of an electron that has de Broglie wavelength λ is

$$v = \frac{c}{\sqrt{1 + (mc\lambda/h)^2}}$$

(b) The quantity h/mc equals 2.426×10^{-12} m. (As we saw in Section 38.7, this same quantity appears in Eq. (38.23), the expression for Compton scattering of photons by electrons.) If λ is small compared to h/mc, the denominator in the expression found in part (a) is close to unity and the speed v is very close to c. In this case it is convenient to write $v = (1 - \Delta)c$ and express the speed of the electron in terms of Δ rather than v. Find an expression for Δ valid when $\lambda \ll h/mc$. [*Hint:* Use the binomial expansion $(1 + z)^n = 1 + nz + n(n - 1)z^2/2 + \cdots$, valid for the case $|z| < 1$. (c) How fast must an electron move for its de Broglie wavelength to be 1.00×10^{-15} m, comparable to the size of a proton? Express your answer in the form $v = (1 - \Delta)c$, and state the value of Δ.

39.47. (a) What is the de Broglie wavelength of an electron accelerated from rest through a potential increase of 125 V? (b) What is the de Broglie wavelength of an alpha particle $(q = +2e,$ $m = 6.64 \times 10^{-27}$ kg$)$ accelerated from rest through a potential drop of 125 V?

39.48. Suppose that the uncertainty of position of an electron is equal to the radius of the $n = 1$ Bohr orbit for hydrogen. Calculate the simultaneous minimum uncertainty of the corresponding momentum component, and compare this with the magnitude of the momentum of the electron in the $n = 1$ Bohr orbit. Discuss your results.

39.49. (a) A particle with mass m has kinetic energy equal to three times its rest energy. What is the de Broglie wavelength of this particle? (*Hint:* You must use the relativistic expressions for momentum and kinetic energy: $E^2 = (pc)^2 + (mc^2)^2$ and $K = E - mc^2$.) (b) Determine the numerical value of the kinetic energy (in MeV) and the wavelength (in meters) if the particle in part (a) is (i) an electron and (ii) a proton.

39.50. Proton Energy in a Nucleus. The radii of atomic nuclei are of the order of 5.0×10^{-15} m. (a) Estimate the minimum uncertainty in the momentum of a proton if it is confined within a nucleus. (b) Take this uncertainty in momentum to be an estimate of the magnitude of the momentum. Use the relativistic relationship between energy and momentum, Eq. (37.39), to obtain an estimate of the kinetic energy of a proton confined within a nucleus. (c) For a proton to remain bound within a nucleus, what must the magnitude of the (negative) potential energy for a proton be within the nucleus? Give your answer in eV and in MeV. Compare to the potential energy for an electron in a hydrogen atom, which has a magnitude of a few tens of eV. (This shows why the interaction that binds the nucleus together is called the "strong nuclear force.")

39.51. Electron Energy in a Nucleus. The radii of atomic nuclei are of the order of 5.0×10^{-15} m. (a) Estimate the minimum uncertainty in the momentum of an electron if it is confined within a nucleus. (b) Take this uncertainty in momentum to be an estimate of the magnitude of the momentum. Use the relativistic relationship between energy and momentum, Eq. (37.39), to obtain an estimate of the kinetic energy of an electron confined within a nucleus. (c) Compare the energy calculated in part (b) to the magnitude of the Coulomb potential energy of a proton and an electron separated by 5.0×10^{-15} m. On the basis of your result, could there be electrons within the nucleus? (*Note:* It is interesting to compare this result to that of Problem 39.50.)

39.52. In a TV picture tube the accelerating voltage is 15.0 kV, and the electron beam passes through an aperture 0.50 mm in diameter to a screen 0.300 m away. (a) Calculate the uncertainty in the component of the electron's velocity perpendicular to the line between aperture and screen. (b) What is the uncertainty in position of the point where the electrons strike the screen? (c) Does this uncertainty affect the clarity of the picture significantly? (Use nonrelativistic expressions for the motion of the electrons. This is fairly accurate and is certainly adequate for obtaining an estimate of uncertainty effects.)

39.53. The neutral pion (π^0) is an unstable particle produced in high-energy particle collisions. Its mass is about 264 times that of the electron, and it exists for an average lifetime of 8.4×10^{-17} s before decaying into two gamma-ray photons. Using the relationship $E = mc^2$ between rest mass and energy, find the uncertainty in the mass of the particle and express it as a fraction of the mass.

39.54. Quantum Effects in Daily Life? A 1.25-mg insect flies through a 4.00-mm-diameter hole in an ordinary window screen. The thickness of the screen is 0.500 mm. (a) What should be the approximate wavelength and speed of the insect for her to show wave behavior as she goes through the hole? (b) At the speed found in part (a), how long would it take the insect to pass through the 0.500-mm thickness of the hole in the screen? Compare this time to the age of the universe (about 14 billion years). Would you expect to see "insect diffraction" in daily life?

39.55. Doorway Diffraction. If your wavelength were 1.0 m, you would undergo considerable diffraction in moving through a doorway. (a) What must your speed be for you to have this wavelength? (Assume that your mass is 60.0 kg.) (b) At the speed calculated in part (a), how many years would it take you to move 0.80 m (one step)? Will you notice diffraction effects as you walk through doorways?

39.56. Atomic Spectra Uncertainties. A certain atom has an energy level 2.58 eV above the ground level. Once excited to this level, the atom remains in this level for 1.64×10^{-7} s (on average) before emitting a photon and returning to the ground level. (a) What is the energy of the photon (in electron volts)? What is its wavelength (in nanometers)? (b) What is the smallest possible uncertainty in energy of the photon? Give your answer in electron volts. (c) Show that $|\Delta E/E| = |\Delta\lambda/\lambda|$ if $|\Delta\lambda/\lambda| \ll 1$. Use this to calculate the magnitude of the smallest possible uncertainty in the wavelength of the photon. Give your answer in nanometers.

39.57. You intend to use an electron microscope to study the structure of some crystals. For accurate resolution, you want the electron wavelength to be 1.00 nm. (a) Are these electrons relativistic? How do you know? (b) What accelerating potential is needed? (c) What is the kinetic energy of the electrons you are using? To see if it is great enough to damage the crystals you are studying, compare it to the potential energy of a typical NaCl molecule, which is about 6.0 eV. (d) If you decided to use electromagnetic waves as you probe, what energy should their photons have to provide the same resolution as the electrons? Would this energy damage the crystal?

39.58. For x rays with wavelength 0.0300 nm, the $m = 1$ intensity maximum for a crystal occurs when the angle θ in Fig. 36.23c is 35.8°. At what angle θ does the $m = 1$ maximum occur when a beam of 4.50-keV electrons is used instead? Assume that the electrons also scatter from the atoms in the surface plane of this same crystal.

39.59. Electron diffraction can also take place when there is interference between electron waves that scatter from atoms on the surface of a crystal and waves that scatter from atoms in the next plane below the surface, a distance d from the surface (see Fig. 36.23c). (a) Find an equation for the angles θ at which there is an intensity maximum for electron waves of wavelength λ. (b) The spacing between crystal planes in a certain metal is 0.091 nm. If 71.0-eV electrons are used, find the angle at which there is an intensity maximum due to interference between scattered waves from adjacent crystal planes. The angle is measured as shown in Fig. 36.23c. (c) The actual angle of the intensity maximum is slightly different from your result in part (b). The reason is the work function ϕ of the metal (see Section 38.2), which changes the electron potential energy by $-e\phi$ when it moves from vacuum into the metal. If the effect of the work function is taken into account, is the angle of the intensity maximum larger or smaller than the value found in part (b)? Explain.

39.60. Zero-Point Energy. Consider a particle with mass m moving in a potential $U = \frac{1}{2}kx^2$, as in a mass-spring system. The total energy of the particle is $E = p^2/2m + \frac{1}{2}kx^2$. Assume that p and x are approximately related by the Heisenberg uncertainty principle, $px \approx h$. (a) Calculate the minimum possible value of the energy E, and the value of x that gives this minimum E. This lowest possible energy, which is not zero, is called the *zero-point energy*. (b) For the x calculated in part (a), what is the ratio of the kinetic to the potential energy of the particle?

39.61. A particle with mass m moves in a potential $U(x) = A|x|$, where A is a positive constant. In a simplified picture, quarks (the constituents of protons, neutrons, and other particles, as will be described in Chapter 44) have a potential energy of interaction of approximately this form, where x represents the separation between a pair of quarks. Because $U(x) \to \infty$ as $x \to \infty$, it's not possible to separate quarks from each other (a phenomenon called *quark confinement*). (a) Classically, what is the force acting on this particle as a function of x? (b) Using the uncertainty principle as in Problem 39.60, determine approximately the zero-point energy of the particle.

39.62. The discussion in Section 39.5 shows that the wave function $\Psi = \psi e^{-i\omega t}$ is a stationary state, where ψ is time independent and ω is a real (not complex) constant. Consider the wave function $\Psi = \psi_1 e^{-i\omega_1 t} + \psi_2 e^{-i\omega_2 t}$, where ψ_1 and ψ_2 are different time-independent functions and ω_1 and ω_2 are different real constants. Assume that ψ_1 and ψ_2 are real-valued functions, so that $\psi_1{}^* = \psi_1$ and $\psi_2{}^* = \psi_2$. Is this Ψ a wave function for a stationary state? Why or why not?

39.63. The Time-Dependent Schrödinger Equation. Equation (39.18) is the time-independent Schrödinger equation in one dimension. The time-dependent Schrödinger equation is

$$-\frac{\hbar^2}{2m}\frac{\partial^2\Psi(x,t)}{\partial x^2} + U(x)\Psi(x,t) = i\hbar\frac{\partial\Psi(x,t)}{\partial t}$$

If $\psi(x)$ is a solution to Eq. (39.18) with energy E, show that the time-dependent function $\Psi(x,t) = \psi(x)e^{-i\omega t}$ is a solution to the time-dependent Schrödinger equation if ω is chosen appropriately. What is the value of ω that makes Ψ a solution?

39.64. Time-Dependent Wave Function for a Free Particle. One example of a time-dependent wave function is that of a free particle [one for which $U(x) = 0$ for all x] of energy E and x-component of momentum p. From the de Broglie relationships (see Section 39.1), such a particle has associated with it a frequency $f = E/h$ and a wavelength $\lambda = h/p$. A reasonable first guess for the time-dependent wave function for such a particle is $\Psi(x,t) = A\cos(kx - \omega t)$, where A is a constant, $\omega = 2\pi f$ is the angular frequency, and $k = 2\pi/\lambda$ is the wave number. This is the same function we used to describe a mechanical wave [see Eq. (15.7)] or an electromagnetic wave propagating in the x-direction [see Eq. (32.16)]. (a) Show that $\omega = E\hbar, k = p/\hbar$, and $\omega = \hbar k^2/2m$. (*Hint:* The energy is purely kinetic, so $E = p^2/2m$.) (b) To check this guess for the time-dependent wave function, substitute $\Psi(x,t) = A\cos(kx - \omega t)$ into the time-dependent Schrödinger equation (see Problem 39.63) with $U(x) = 0$ (so the particle is free). Show that this guess for $\Psi(x,t)$ does *not* satisfy this equation and so is not a suitable wave function for a free particle. (c) Use the procedure described in part (b) to show that a second guess, $\Psi(x,t) = A\sin(kx - \omega t)$, is also not a suitable wave function for a free particle. (d) Consider a combination of the functions proposed in parts (b) and (c):

$$\Psi(x, t) = A\cos(kx - \omega t) + B\sin(kx - \omega t)$$

By using the procedure described in part (b), show that this wave function is a solution to the time-dependent Schrödinger equation with $U(x) = 0$, but only if $B = iA$. (*Hint:* To satisfy the time-dependent Schrödinger equation for all x and t, the coefficients of $\cos(kx - \omega t)$ on both sides of the equation must be equal. The same is true for the coefficients of $\sin(kx - \omega t)$ on both sides of the equation.) This is an example of the general result that time-dependent wave functions always have both a real part and an imaginary part.

39.65. Imagine another universe in which the value of Planck's constant is 0.0663 J · s, but in which the physical laws and all other physical constants are the same as in our universe. In this universe, two physics students are playing catch. They are 12 m apart, and one throws a 0.25-kg ball directly toward the other with a speed of 6.0 m/s. (a) What is the uncertainty in the ball's horizontal momentum, in a direction perpendicular to that in which it is being thrown, if the student throwing the ball knows that it is located within a cube with volume 125 cm^3 at the time she throws it? (b) By what horizontal distance could the ball miss the second student?

39.66. A particle is described by the normalized wave function $\psi(x, y, z) = Axe^{-\alpha x^2}e^{-\beta y^2}e^{-\gamma z^2}$, where A, α, β, and γ are all real, positive constants. The probability that the particle will be found in the infinitesimal volume $dx\,dy\,dz$ centered at the point (x_0, y_0, z_0) is $|\psi(x_0, y_0, z_0)|^2\,dx\,dy\,dz$. (a) At what value of x_0 is the particle most likely to be found? (b) Are there values of x_0 for which the probability of the particle being found is zero? If so, at what x_0?

39.67. A particle is described by the normalized wave function $\psi(x, y, z) = Ae^{-\alpha(x^2+y^2+z^2)}$, where A and α are real, positive constants. (a) Determine the probability of finding the particle at a distance between r and $r + dr$ from the origin. (*Hint:* See Problem 39.66. Consider a spherical shell centered on the origin with inner radius r and thickness dr.) (b) For what value of r does the probability in part (a) have its maximum value? Is this the same value of r for which $|\psi(x, y, z)|^2$ is a maximum? Explain any differences.

39.68. Consider the wave packet defined by

$$\psi(x) = \int_0^\infty B(k)\cos kx\,dk$$

Let $B(k) = e^{-\alpha^2k^2}$. (a) The function $B(k)$ has its maximum value at $k = 0$. Let k_h be the value of k at which $B(k)$ has fallen to half its maximum value, and define the width of $B(k)$ as $w_k = k_h$. In terms of α, what is w_k? (b) Use integral tables to evaluate the integral that gives $\psi(x)$. For what value of x is $\psi(x)$ maximum? (c) Define the width of $\psi(x)$ as $w_x = x_h$, where x_h is the positive value of x where $\psi(x)$ has fallen to half its maximum value. Calculate w_x in terms of α. (d) The momentum p is equal to $hk/2\pi$, so the width of B in momentum is $w_p = hw_k/2\pi$. Calculate the product w_pw_x and compare to the Heisenberg uncertainty principle.

39.69. (a) Using the integral in Problem 39.68, determine the wave function $\psi(x)$ for a function $B(k)$ given by

$$B(k) = \begin{cases} 0 & k < 0 \\ 1/k_0, & 0 \le k \le k_0 \\ 0, & k > k_0 \end{cases}$$

This represents an equal combination of all wave numbers between 0 and k_0. Thus $\psi(x)$ represents a particle with average wave number $k_0/2$, with a total spread or uncertainty in wave number of k_0. We will call this spread the *width* w_k of $B(k)$, so $w_k = k_0$. (b) Graph $B(k)$ versus k and $\psi(x)$ versus x for the case $k_0 = 2\pi/L$, where L is a length. Locate the point where $\psi(x)$ has its maximum value and label this point on your graph. Locate the two points closest to this maximum (one on each side of it) where $\psi(x) = 0$, and define the distance along the x-axis between these two points as w_x, the width of $\psi(x)$. Indicate the distance w_x on your graph. What is the value of w_x if $k_0 = 2\pi/L$? (c) Repeat part (b) for the case $k_0 = \pi/L$. (d) The momentum p is equal to $hk/2\pi$, so the width of B in momentum is $w_p = hw_k/2\pi$. Calculate the product w_pw_x for each of the cases $k_0 = 2\pi/L$ and $k_0 = \pi/L$. Discuss your results in light of the Heisenberg uncertainty principle.

Challenge Problems

39.70. The wave nature of particles results in the quantum-mechanical situation that a particle confined in a box can assume only wavelengths that result in standing waves in the box, with nodes at the box walls. (a) Show that an electron confined in a one-dimensional box of length L will have energy levels given by

$$E_n = \frac{n^2h^2}{8mL^2}$$

(*Hint:* Recall that the relationship between the de Broglie wavelength and the speed of a nonrelativistic particle is $mv = h/\lambda$. The energy of the particle is $\frac{1}{2}mv^2$.) (b) If a hydrogen atom is modeled as a one-dimensional box with length equal to the Bohr radius, what is the energy (in electron volts) of the lowest energy level of the electron?

39.71. You have entered a contest in which the contestants drop a marble with mass 20.0 g from the roof of a building onto a small target 25.0 m below. From uncertainty considerations, what is the typical distance by which you will miss the target, given that you aim with the highest possible precision? (*Hint:* The uncertainty Δx_f in the x-coordinate of the marble when it reaches the ground comes in part from the uncertainty Δx_i in the x-coordinate initially and in part from the initial uncertainty in v_x. The latter gives rise to an uncertainty Δv_x in the horizontal motion of the marble as it falls. The values of Δx_i and Δv_x are related by the uncertainty principle. A small Δx_i gives rise to a large Δv_x, and vice versa. Find the value of Δx_i that gives the smallest total uncertainty in x at the ground. Ignore any effects of air resistance.)

QUANTUM MECHANICS 40

? Each of these containers holds a solution of microscopic semiconductor particles, each of which acts like a nanometer-sized "atom." The particles glow when exposed to ultraviolet light. Why do the smallest particles (on the left) glow blue, while the largest particles (on the right) glow red?

LEARNING GOALS

By studying this chapter, you will learn:

- How to calculate the wave functions and energy levels for a particle confined to a box.
- How to analyze the quantum-mechanical behavior of a particle in a potential well.
- How quantum mechanics makes it possible for particles to go where Newtonian mechanics says they cannot.
- How to use quantum mechanics to analyze a harmonic oscillator.
- How to extend quantum-mechanical calculations to three-dimensional problems.

In Chapter 39 we found that particles sometimes behave like waves and that they can be described by wave functions. Now we're ready for a systematic analysis of particles that are trapped in *bound states* (such as electrons in atoms), including finding their possible wave functions and energy levels.

Our analysis involves finding solutions of the Schrödinger equation, which we introduced in Chapter 39. For any quantum-mechanical system, the wave functions for any state of definite energy must be solutions of the Schrödinger equation for that system. As we will see, solving the Schrödinger equation automatically gives the possible energy levels for the system. We'll discuss several simple one-dimensional applications of the Schrödinger equation.

Besides energies, solving the Schrödinger equation gives us the probabilities of finding a particle in various regions. One surprising result is that there is a nonzero probability that microscopic particles will pass through thin barriers, even though such a process is forbidden by Newtonian mechanics.

Finally, we'll generalize the Schrödinger equation to three dimensions. This will pave the way for describing the wave functions for the hydrogen atom in Chapter 41. The hydrogen-atom wave functions in turn form the foundation for our analysis of more complex atoms, of the periodic table of the elements, of x-ray energy levels, and of other properties of atoms.

40.1 Particle in a Box

In this chapter we want to learn how to use the Schrödinger equation to determine the possible energy levels and the corresponding wave functions for various systems. As in Chapter 39, we refer to the states of definite energy as *stationary states*. For simplicity, we begin with problems in which the particle can move

ActivPhysics Online
20.2 Particle in a Box

only along the x-axis. From Section 39.5, the Schrödinger equation for such one-dimensional problems is

$$-\frac{\hbar^2}{2m}\frac{d^2\psi(x)}{dx^2} + U(x)\psi(x) = E\psi(x) \qquad \text{(one-dimensional Schrödinger equation)} \tag{40.1}$$

This equation is really a statement that the kinetic energy [represented by the term $-(\hbar^2/2m)d^2\psi(x)/dx^2$] plus the potential energy [represented by the term $U(x)\psi(x)$] is equal to the total energy [represented by the term $E\psi(x)$]. The fundamental problem is then the following: For a given potential energy function $U(x)$, what are the possible stationary-state wave functions $\psi(x)$, and what are the corresponding energies E?

In Section 39.5 we solved this problem for the case $U(x) = 0$, corresponding to a *free* particle. The allowed wave functions and corresponding energies are

$$\psi(x) = Ae^{ikx} \qquad E = \frac{\hbar^2k^2}{2m} \qquad \text{(free particle)} \tag{40.2}$$

The wave number k is equal to $2\pi/\lambda$, where λ is the wavelength. We found that k can have any real value, so the energy E of a free particle can have any value from zero to infinity. Furthermore, the particle can be found with equal probability at any value of x from $-\infty$ to $+\infty$.

Now let's look at a simple model in which a particle is *bound* so that it cannot escape to infinity, but rather is confined to a restricted region of space. Our system consists of a particle confined between two rigid walls separated by a distance L (Fig. 40.1). The motion is purely one dimensional, with the particle moving along the x-axis only and the walls at $x = 0$ and $x = L$. The potential energy corresponding to the rigid walls is infinite, so the particle cannot escape; between the walls, the potential energy is zero. This situation is often described as a **"particle in a box."** This model might represent an electron that is free to move within a long, straight molecule or along a very thin wire. The potential-energy function is graphed in Fig. 40.2.

40.1 The Newtonian view of a particle in a box.

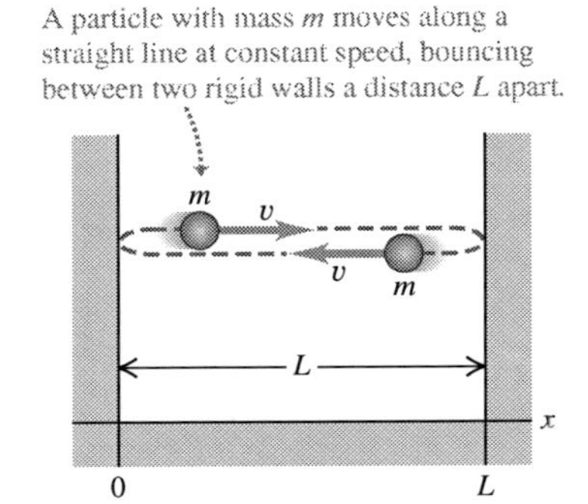

40.2 The potential-energy function for a particle in a box.

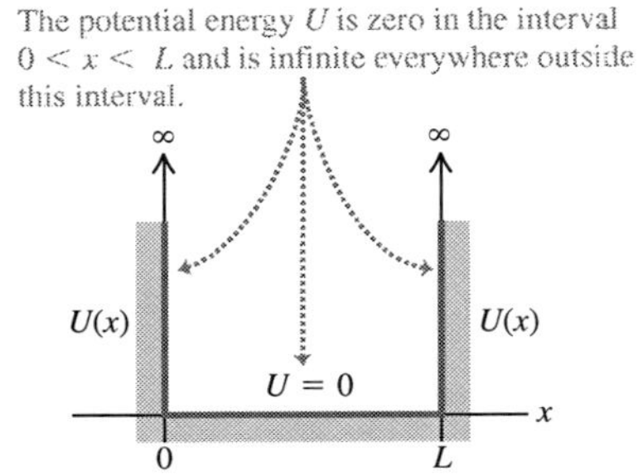

Wave Function for a Particle in a Box

To solve the Schrödinger equation for this system, we begin with some restrictions on the particle's wave function $\psi(x)$. Because the particle is confined to the region $0 \le x \le L$, we expect $\psi(x)$ to be zero outside that region. This agrees with the Schrödinger equation: If the term $U(x)\psi(x)$ in Eq. (40.1) is to be finite, then $\psi(x)$ must be zero where $U(x)$ is infinite.

Furthermore, $\psi(x)$ must be a *continuous* function to be a mathematically well-behaved solution to the Schrödinger equation. If it is, then $\psi(x)$ must be zero at the region's boundary, $x = 0$ and $x = L$. These two conditions serve as *boundary conditions* for the problem. They should look familiar, because they are the same conditions that we used to find the normal modes of a vibrating string in Section 15.8 (Fig. 40.3); you should review that discussion.

40.3 Normal modes of vibration for a string with length L, held at both ends.

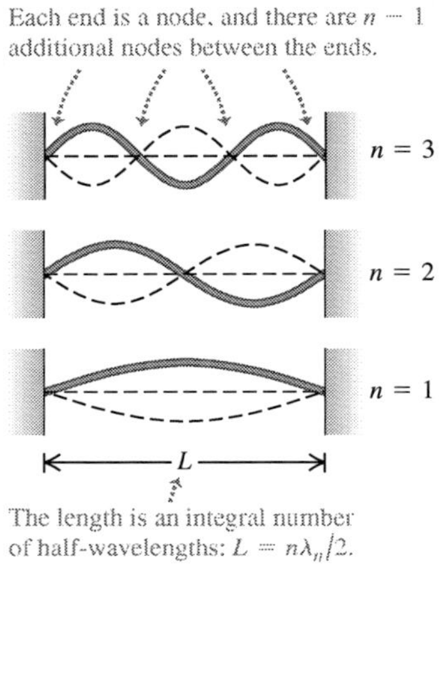

An additional condition is that to calculate the second derivative $d^2\psi(x)/dx^2$ in Eq. (40.1), the *first* derivative $d\psi(x)/dx$ must also be continuous except at points where the potential energy becomes infinite (as it does at the walls of the box). This is analogous to the requirement that a vibrating string, like those shown in Fig. 40.3, cannot have a kink in it (which would correspond to a discontinuity in the first derivative of the wave function) except at the ends of the string.

We now solve for the wave functions in the region $0 \le x \le L$ subject to the above conditions. In this region $U(x) = 0$, so the wave function in this region must satisfy

$$-\frac{\hbar^2}{2m}\frac{d^2\psi(x)}{dx^2} = E\psi(x) \qquad \text{(particle in a box)} \tag{40.3}$$

Equation (40.3) is the *same* Schrödinger equation as for a free particle, so it is tempting to conclude that the wave functions and energies are given by Eq. (40.2). It is true that $\psi(x) = Ae^{ikx}$ satisfies the Schrödinger equation with $U(x) = 0$, is continuous, and has a continuous first derivative $d\psi(x)/dx = ikAe^{ikx}$. However, this wave function does *not* satisfy the boundary conditions that $\psi(x)$ must be zero at $x = 0$ and $x = L$: At $x = 0$ the wave function in Eq. (40.2) is equal to $Ae^0 = A$, and at $x = L$ it is equal to Ae^{ikL}. (These would be equal to zero if $A = 0$, but then the wave function would be zero and there would be no particle at all!)

The way out of this dilemma is to recall Example 39.6 (Section 39.5), in which we found that a more general stationary-state solution to the Schrödinger equation with $U(x) = 0$ is

$$\psi(x) = A_1 e^{ikx} + A_2 e^{-ikx} \quad (40.4)$$

This wave function is a superposition of two waves: one traveling in the $+x$-direction of amplitude A_1, and one traveling in the $-x$-direction with the same wave number but amplitude A_2. This is analogous to the situation for standing waves on a string (Fig. 40.3), which we can regard as the superposition of two sinusoidal waves propagating in opposite directions (see Section 15.7). The energy that corresponds to Eq. (40.4) is $E = \hbar^2 k^2/2m$, just as for a single wave.

To see whether the wave function given by Eq. (40.4) can satisfy the boundary conditions, let's first rewrite it in terms of sines and cosines using Euler's formula, Eq. (39.15):

$$\begin{aligned}\psi(x) &= A_1(\cos kx + i\sin kx) + A_2[\cos(-kx) + i\sin(-kx)] \\ &= A_1(\cos kx + i\sin kx) + A_2(\cos kx - i\sin kx) \\ &= (A_1 + A_2)\cos kx + i(A_1 - A_2)\sin kx\end{aligned} \quad (40.5)$$

At $x = 0$ this is equal to $\psi(0) = A_1 + A_2$, which must equal zero if we are to satisfy the boundary condition at that point. Hence $A_2 = -A_1$, and Eq. (40.5) becomes

$$\psi(x) = 2iA_1 \sin kx = C\sin kx \quad (40.6)$$

We have simplified the expression by introducing the constant $C = 2iA_1$. (We'll come back to this constant later.) We can also satisfy the second boundary condition that $\psi = 0$ at $x = L$ by choosing values of k such that $kL = n\pi$ $(n = 1, 2, 3, \ldots)$. Hence Eq. (40.6) does give the stationary-state wave functions for a particle in a box in the region $0 \le x \le L$. (Outside this region, $\psi(x) = 0$.) The possible values of k and the wavelength $\lambda = 2\pi/k$ are

$$k = \frac{n\pi}{L} \quad \text{and} \quad \lambda = \frac{2\pi}{k} = \frac{2L}{n} \quad (n = 1, 2, 3, \ldots) \quad (40.7)$$

Just as for the string in Fig. 40.3, the length L of the region is an integral number of half-wavelengths.

Energy Levels For a Particle in a Box

The possible energy levels are given by $E = \hbar^2k^2/2m = p^2/2m$, where $p = \hbar k = (h/2\pi)(2\pi/\lambda) = h/\lambda$ is the magnitude of momentum of a free particle with wave number k and wavelength λ. This makes sense, since inside the region $0 \le x \le L$ the potential energy is zero and the energy is all kinetic. For each value of n there are corresponding values of p, λ, and E; let's call them p_n, λ_n, and E_n. Putting the pieces together, we get

$$p_n = \frac{h}{\lambda_n} = \frac{nh}{2L} \quad (40.8)$$

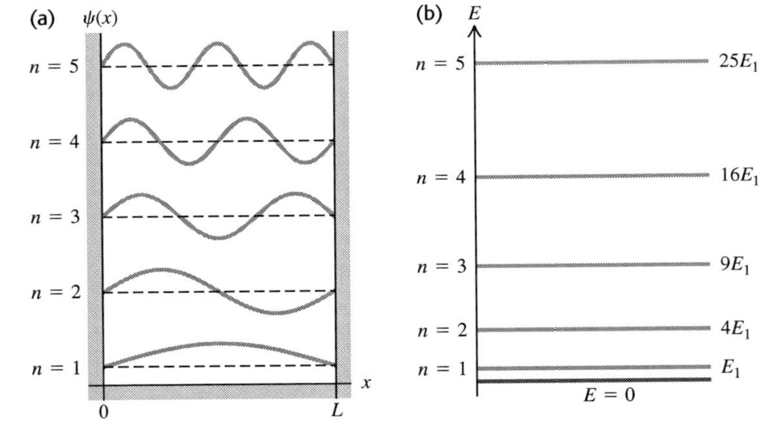

40.4 (a) Wave functions for a particle in a box, with $n = 1, 2, 3, 4$, and 5. Each of the horizontal dashed lines represents $\psi = 0$ for the respective wave function. The five graphs have been displaced vertically for clarity, as in Fig. 40.3. (b) Energy-level diagram for a particle in a box. Each energy is n^2E_1, where E_1 is the ground-level energy.

and

$$E_n = \frac{p_n^2}{2m} = \frac{n^2h^2}{8mL^2} = \frac{n^2\pi^2\hbar^2}{2mL^2} \quad (n = 1, 2, 3 \ldots) \quad \text{(energy levels, particle in a box)} \tag{40.9}$$

These are the possible energy levels for a particle in a box. Each energy level has its own value of the quantum number n and a corresponding wave function, which we denote by ψ_n. When we replace k in Eq. (40.6) by $n\pi/L$ from Eq. (40.7), we find

$$\psi_n(x) = C\sin\frac{n\pi x}{L} \quad (n = 1, 2, 3 \ldots) \tag{40.10}$$

Figure 40.4a shows graphs of the wave functions $\psi_n(x)$ for $n = 1, 2, 3, 4$, and 5. Note that these functions look identical to those for a standing wave on a string (Fig. 40.3). Figure 40.4b shows the energy-level diagram for this system. The levels of successively higher energies, proportional to n^2, are spaced farther and farther apart. There are an infinite number of levels because the walls are perfectly rigid; even a particle of infinitely great kinetic energy is confined within the box.

CAUTION **Zero energy is not permitted** Note that the energy of a particle in a box *cannot* be zero. Equation (40.9) shows that $E = 0$ would require $n = 0$, but substituting $n = 0$ into Eq. (40.10) gives a zero wave function. Since a particle is described by a nonzero wave function, this means that there cannot be a particle with $E = 0$.

Example 40.1 Electron in an atom-size box

Find the lowest energy level for a particle in a box if the particle is an electron in a box 5.0×10^{-10} m across, or a little bigger than an atom.

SOLUTION

IDENTIFY: This problem uses what we have learned in this section about a particle in a box.

SET UP: The lowest energy level (that is, the ground level) corresponds to $n = 1$ in Eq. (40.9).

EXECUTE: From Eq. (40.9),

$$E_1 = \frac{h^2}{8mL^2} = \frac{(6.626 \times 10^{-34}\ \text{J}\cdot\text{s})^2}{8(9.109 \times 10^{-31}\ \text{kg})(5.0 \times 10^{-10}\ \text{m})^2}$$
$$= 2.4 \times 10^{-19}\ \text{J} = 1.5\ \text{eV}$$

EVALUATE: A particle trapped in a box is rather different from an electron bound in an atom, but it is reassuring that this energy is of the same order of magnitude as actual atomic energy levels.

You should be able to show that replacing the electron with a proton or a neutron ($m = 1.67 \times 10^{-27}$ kg) in a box the width of a medium-sized nucleus ($L = 1.1 \times 10^{-14}$ m) gives $E_1 = 1.7$ MeV. This shows us that the energies for particles in the nucleus are about a million times greater than the energies of electrons in atoms, giving us a clue as to why each nuclear fission and fusion reaction provides so much more energy than each chemical reaction.

If you repeat this calculation for a billiard ball ($m = 0.2$ kg) bouncing back and forth between the cushions of a frictionless, perfectly elastic billiard table ($L = 1.5$ m), you'll find the separation between energy levels to be as small as 4×10^{-67} J (See Exercise 40.1). This negligible value shows that quantum effects won't have much effect on your billiard game.

Probability and Normalization

Let's look a bit more closely at the wave functions for a particle in a box, keeping in mind the *probability* interpretation of the wave function ψ that we discussed in Section 39.5. In our one-dimensional situation the quantity $|\psi(x)|^2\,dx$ is proportional to the probability that the particle will be found within a small interval dx about x. For a particle in a box,

$$|\psi(x)|^2\,dx = C^2 \sin^2\frac{n\pi x}{L}\,dx$$

Both $\psi(x)$ and $|\psi(x)|^2$ are plotted in Fig. 40.5 for $n = 1$, 2, and 3. We note that not all positions are equally likely. This is in contrast to the situation in classical mechanics, in which all positions between $x = 0$ and $x = L$ are equally likely. We see in Fig. 40.5b that $|\psi(x)|^2 = 0$ at some points, so the probability is zero of finding the particle exactly at these points. Don't let that bother you; the uncertainty principle has already shown us that we can't measure position exactly. The particle is localized only to be somewhere between $x = 0$ and $x = L$, not to be exactly at some particular value of x.

We know that the particle must be *somewhere* in the universe—that is, between $x = -\infty$ and $x = +\infty$. So the *sum* of the probabilities for all the dx's everywhere (the *total* probability of finding the particle) must equal 1. That is,

$$\int_{-\infty}^{\infty} |\psi(x)|^2\,dx = 1 \qquad \text{(normalization condition)} \qquad (40.11)$$

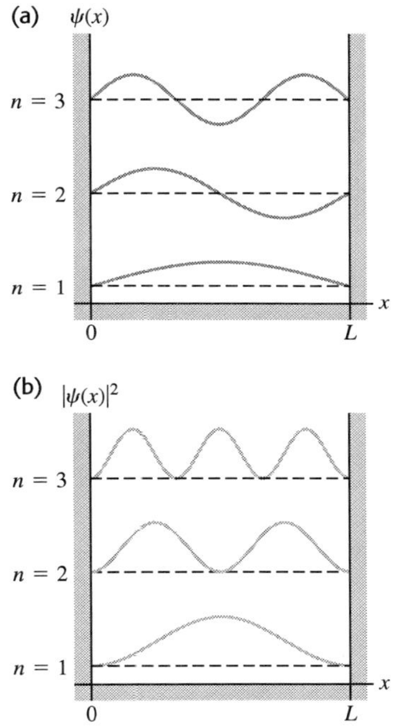

40.5 Graphs of (a) $\psi(x)$ and (b) $|\psi(x)|^2$ for the first three wave functions $(n = 1, 2, 3)$ for a particle in a box. The horizontal dashed lines represent $\psi(x) = 0$ and $|\psi(x)|^2 = 0$ for each of the three levels. The value of $|\psi(x)|^2\,dx$ at each point is the probability of finding the particle in a small interval dx about the point.

A wave function is said to be *normalized* if it has a constant such as C in Eq. (40.10) that is calculated to make the total probability equal 1 in Eq. (40.11). The process of determining the constant is called **normalization,** as we mentioned in Section 39.5. Why bother with normalization? Because for a normalized wave function, $|\psi(x)|^2\,dx$ is not merely proportional to, but *equals,* the probability of finding the particle in dx. Recall from Section 39.5 that we call $|\psi(x)|^2$ the *probability distribution function.*

Now let's normalize the wave function $\psi(x)$ given by Eq. (40.10) for the particle in a box. Since $\psi(x)$ is zero except between $x = 0$ and $x = L$, Eq. (40.11) becomes

$$\int_0^L C^2 \sin^2\frac{n\pi x}{L}\,dx = 1 \qquad (40.12)$$

You can evaluate this integral using the trigonometric identity $\sin^2\theta = \frac{1}{2}(1 - \cos 2\theta)$; the result is $C^2L/2$. Thus our probability interpretation of the wave function demands that $C^2L/2 = 1$, or $C = (2/L)^{1/2}$; the constant C is *not* arbitrary. (This is in contrast to the classical vibrating string problem, in which C represents an amplitude that depends on initial conditions.) Thus the normalized stationary-state wave functions for a particle in a box are

$$\psi_n(x) = \sqrt{\frac{2}{L}}\sin\frac{n\pi x}{L} \qquad (n = 1, 2, 3, \dots) \qquad \text{(particle in a box)} \qquad (40.13)$$

Next, let's check whether our results for the particle in a box are consistent with the uncertainty principle. We can write the position of the particle as $x = L/2 \pm L/2$, so we can estimate the uncertainty in position as $\Delta x \approx L/2$. From Eq. (40.8) the magnitude of momentum in state n is $p_n = nh/2L$. A reasonable estimate of the momentum uncertainty is the difference in momentum of two levels that differ by 1 in their n values, that is, $\Delta p_x \approx h/2L$. Then the product $\Delta x\,\Delta p_x$ is

$$\Delta x\,\Delta p_x = \frac{h}{4}$$

This is consistent with the uncertainty principle, Eq. (39.11), $\Delta x\, \Delta p_x \geq \hbar = h/2\pi$, in which the uncertainties were more precisely defined as standard deviations.

Example 40.2 A nonsinusoidal wave function?

(a) Show that $\psi(x) = Ax + B$, where A and B are constants, is a solution of the Schrödinger equation for an $E = 0$ energy level of a particle in a box. (b) What constraints do the boundary conditions at $x = 0$ and $x = L$ place on the constants A and B?

SOLUTION

IDENTIFY: This problem uses the idea that any physically reasonable wave function must satisfy both the Schrödinger equation and the appropriate boundary conditions.

SET UP: In part (a) we substitute $\psi(x)$ into the Schrödinger equation for a particle in a box, Eq. (40.3), to determine whether it is a solution. In part (b) we demand that $\psi(x)$ be equal to zero at $x = 0$ and at $x = L$.

EXECUTE: (a) From Eq. (40.3), the Schrödinger equation for this situation is $-(\hbar^2/2m)\, d^2\psi(x)/dx^2 = E\psi(x)$. Differentiating $\psi(x) = Ax + B$ twice with respect to x gives $d^2\psi(x)/dx^2 = 0$, so the left side of the equation is zero. Since we assume $E = 0$, the right side of the equation is also zero. Since $0 = 0$, $\psi(x) = Ax + B$ is a solution of this Schrödinger equation for $E = 0$. Note that both $\psi(x)$ and its derivative $d\psi(x)/dx = A$ are continuous functions, as they must be.

(b) Applying the boundary condition at $x = 0$ gives $\psi(0) = B$, so the constant B must be zero and $\psi(x) = Ax$. Applying the boundary condition at $x = L$ gives $\psi(L) = AL = 0$, so the constant A must also be zero. Thus the wave function equals zero inside the box as well as outside the box, and there is *zero* probability of finding the particle anywhere with this wave function. Hence $\psi(x) = Ax + B$ is not a physically valid wave function.

EVALUATE: The moral of this story is that there are many functions that satisfy the Schrödinger equation for a given physical situation, but most of these—including the function considered here—have to be rejected because they don't satisfy the boundary conditions.

Time Dependence

Finally, we note that the wave functions $\psi_n(x)$ in Eq. (40.13) depend only on the *spatial* coordinate x. From Section 39.5, the full wave function also depends on *time*. Equation (39.14) shows that if $\psi(x)$ is the wave function for a state of definite energy E, the full time-dependent wave function $\Psi(x, t)$ is

$$\Psi(x, t) = \psi(x)\, e^{-iEt/\hbar} \quad \text{(time-dependent wave function for a stationary state)} \tag{40.14}$$

From Eq. (40.12), it follows that the time-dependent wave functions for a particle in a box are

$$\Psi(x, t) = \sqrt{\frac{2}{L}} \sin\left(\frac{n\pi x}{L}\right) e^{-iE_n t/\hbar} \qquad (n = 1, 2, 3, \ldots) \tag{40.15}$$

In this expression the energies E_n are given by Eq. (40.9).

The absolute value of $e^{-iEt/\hbar}$ is unity: $|e^{-iEt/\hbar}|^2 = e^{+iEt/\hbar}e^{-iEt/\hbar} = e^0 = 1$. Hence $|\Psi(x, t)|^2$ is equal to $|\psi(x)|^2$, and the probability distribution function does not depend on time. We saw in Section 39.5 that this is the origin of the term *stationary state* for a state of definite energy. So in calculations with states that have definite energy, we are justified in omitting the time factor. We'll be concerned mostly with stationary states in our applications of quantum mechanics.

Test Your Understanding of Section 40.1 If a particle in a box is in the nth energy level, what is the average value of its x-component of momentum p_x? (i) $nh/2L$; (ii) $(\sqrt{2}/2)nh/L$; (iii) $(1/\sqrt{2})nh/L$; (iv) $(1/2\sqrt{2})nh/L$; (v) zero.

ActivPhysics ONLINE
20.1 Potential Energy Diagrams
20.2 Potential Wells

40.2 Potential Wells

A **potential well** is a potential-energy function $U(x)$ that has a minimum. We introduced this term in Section 7.5, and we also used it in our discussion of periodic motion in Chapter 13. In Newtonian mechanics a particle trapped in a potential well can vibrate back and forth with periodic motion. Our first application of the Schrödinger equation, the particle in a box, involved a rudimentary potential well with a function $U(x)$ that is zero within a certain interval and infinite everywhere else. This function corresponds to a few situations found in nature, as mentioned in Section 40.1, but the correspondence is only approximate.

A potential well that serves as a better approximation to several actual physical situations is a well with straight sides but *finite* height. Figure 40.6 shows a potential-energy function that is zero in the interval $0 \le x \le L$ and has the value U_0 outside this interval. This function is often called a **square-well potential.** It could serve as a simple model of an electron within a metallic sheet with thickness L, moving perpendicular to the surfaces of the sheet. The electron can move freely inside the metal but has to climb a potential-energy barrier with height U_0 to escape from either surface of the metal. The energy U_0 is related to the *work function* that we discussed in Section 38.2 in connection with the photoelectric effect. In a three-dimensional application a spherical version of a potential well can be used to approximately describe the motions of protons and neutrons within a nucleus.

40.6 A square-well potential.

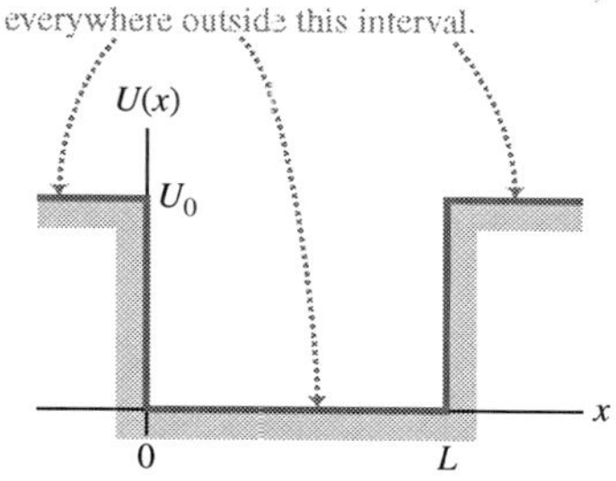

Bound States of a Square-Well Potential

In Newtonian mechanics, the particle is trapped (localized) in a well if the total energy E is less than U_0. In quantum mechanics, such a trapped state is often called a **bound state.** All states are bound for an infinitely deep well, but if E is greater than U_0 for a finite well, the particle is *not* bound.

For a finite square well, let's consider bound-state solutions of the Schrödinger equation, corresponding to $E < U_0$. The easiest approach is to consider separately the regions where $U = 0$ and where $U = U_0$. Where $U = 0$, the Schrödinger equation reduces to Eq. (40.3). Rearranging this equation, we find

$$\frac{d^2\psi(x)}{dx^2} = -\frac{2mE}{\hbar^2}\psi(x) \tag{40.16}$$

From Eq. (40.5), we can express the solutions of this equation as combinations of $\cos kx$ and $\sin kx$, where $E = \hbar^2k^2/2m$ and so $k = \sqrt{2mE}/\hbar$. Hence inside the square well $(0 \le x \le L)$ we have

$$\psi(x) = A\cos\frac{\sqrt{2mE}}{\hbar}x + B\sin\frac{\sqrt{2mE}}{\hbar}x \quad \text{(inside the well)} \tag{40.17}$$

where A and B are constants. So far, this looks a lot like the particle-in-a-box analysis in Section 40.1.

In the regions $x < 0$ and $x > L$ we use Eq. (40.1)with $U = U_0$. Rearranging, we get

$$\frac{d^2\psi(x)}{dx^2} = \frac{2m(U_0 - E)}{\hbar^2}\psi(x) \tag{40.18}$$

The quantity $U_0 - E$ is positive, so the solutions of this equation are exponential. Using κ (the Greek letter kappa) as positive in the abbreviation $\kappa = [2m(U_0 - E)]^{1/2}/\hbar$, we can write the solutions as

$$\psi(x) = Ce^{\kappa x} + De^{-\kappa x} \quad \text{(outside the well)} \tag{40.19}$$

where C and D are constants with different values in the $x < 0$ and $x > L$ regions.

We see that the bound-state wave functions for this system are sinusoidal inside the well and exponential outside it. We have to use the positive exponent in the $x < 0$ region and the negative exponent in the $x > L$ region. That is, $D = 0$ for $x < 0$ and $C = 0$ for $x > L$. Without this choice of constants, ψ would approach infinity as $|x|$ approached infinity, and the normalization condition, Eq. (40.11), could not be satisfied. We also have to match the wave functions so that they satisfy the boundary conditions that we mentioned in Section 40.1: $\psi(x)$ and $d\psi(x)/dx$ must be continuous at the boundary points ($x = 0$ and $x = L$). If the wave function $\psi(x)$ or the slope $d\psi(x)/dx$ were to change discontinuously at a point, the second derivative $d^2\psi(x)/dx^2$ would be *infinite* at that point. But that would violate the Schrödinger equation, which says that at every point, $d^2\psi(x)/dx^2$ is proportional to $U - E$. In our situation, $U - E$ is finite everywhere, so $d^2\psi(x)/dx^2$ must also be finite everywhere.

Matching the sinusoidal and exponential functions at the boundary points so that they join smoothly is possible only for certain specific values of the total energy E, so this requirement determines the possible energy levels of the finite square well. There is no simple formula for the energy levels as there was for the infinitely deep well. Finding the levels is a fairly complex mathematical problem that requires solving a transcendental equation by numerical approximation; we won't go into the details. Figure 40.7 shows the general shape of a possible wave function. The most striking features of this wave function are the "exponential tails" that extend outside the well into regions that are forbidden by Newtonian mechanics (because in those regions the particle would have negative kinetic energy). We see that there is some probability for finding the particle *outside* the potential well, despite the fact that in classical mechanics this is impossible. This penetration into classically forbidden regions is a quantum effect that has no classical analog for particles. We will discuss an amazing result of this effect in Section 40.3.

40.7 A possible wave function for a particle in a finite potential well. The function is sinusoidal inside the well ($0 \leq x \leq L$) and exponential outside it. It approaches zero asymptotically at large $|x|$. The functions must join smoothly at $x = 0$ and $x = L$; the wave function and its derivative must be continuous.

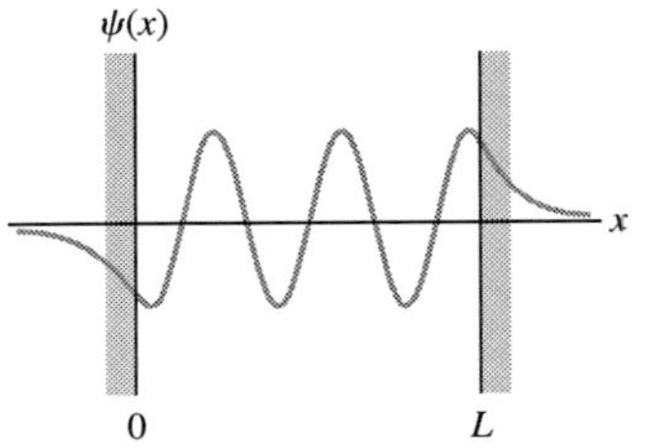

Example 40.3 Well versus box

Show that Eq. (40.19), $\psi(x) = Ce^{\kappa x} + De^{-\kappa x}$, is consistent with the corresponding particle-in-a-box wave function.

SOLUTION

IDENTIFY: Equation (40.19) gives the possible wave functions outside the potential well. Since the wave function outside the box (an infinite potential well) equals zero, we must show that $\psi(x) = Ce^{\kappa x} + De^{-\kappa x}$ approaches zero for both $x < 0$ and $x > L$ as the depth of the potential well (U_0) approaches infinity.

SET UP: We use our knowledge of the exponential function and the definition of the quantity κ in Eq. (40.19): $\kappa = [2m(U_0 - E)]^{1/2}/\hbar$.

EXECUTE: As U_0 approaches infinity, κ also approaches infinity. In the region $x < 0$, the wave function is $\psi(x) = Ce^{\kappa x}$; as $\kappa \to \infty$, $\kappa x \to -\infty$ and $e^{\kappa x} \to 0$, so the wave function approaches zero for all $x < 0$. In the region $x > L$, the wave function is $\psi(x) = De^{-\kappa x}$; as $\kappa \to \infty$, $-\kappa x \to -\infty$ and $e^{-\kappa x} \to 0$, so the wave function also approaches zero for all $x > L$.

EVALUATE: In Section 40.1 we *assumed* $\psi(x) = 0$ outside the box. This analysis validates our assumption.

In our study of mechanics, we often considered limiting cases to check our results (see Examples 5.11 and 5.13 in Section 5.2). This example shows that this technique is also useful for problems in quantum mechanics, as it is for problems in all branches of physics.

Comparing Finite and Infinite Square Wells

Let's continue the comparison of the finite-depth potential well with the infinitely deep well that we began in Example 40.3. First, because the wave functions for the finite well aren't zero at $x = 0$ and $x = L$, the wavelength of the sinusoidal part of each wave function is *longer* than it would be with an infinite well. From $p = h/\lambda$ this increase in λ corresponds to a reduced magnitude of momentum and therefore a reduced energy. Thus each energy level, including the ground level, is *lower* for a finite well than for an infinitely deep well with the same width.

Second, a well with finite depth U_0 has only a *finite* number of bound states and corresponding energy levels, compared to the *infinite* number for an infinitely deep well. How many levels there are depends on the magnitude of U_0 in comparison with the ground-level energy for the infinite well, which we call E_∞. From Eq. (40.9),

$$E_\infty = \frac{\pi^2\hbar^2}{2mL^2} \qquad (40.20)$$

40.8 (a) Wave functions for the three bound states for a particle in a finite potential well with depth U_0, for the case $U_0 = 6E_\infty$. The horizontal brown line for each wave function corresponds to $\psi = 0$; the vertical placement of these lines indicates the energy of each bound state. (b) Energy-level diagram for this system. The energies are expressed both as multiples of E_∞ and as fractions of U_0. All energies greater than U_0 are possible; states with $E > U_0$ form a continuum.

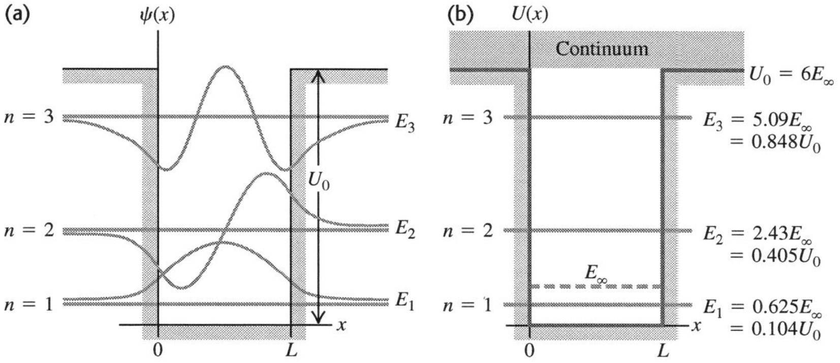

40.9 Probability distribution functions $|\psi(x)|^2$ for the square-well wave functions shown in Fig. 40.8. The horizontal brown line for each wave function corresponds to $|\psi|^2 = 0$.

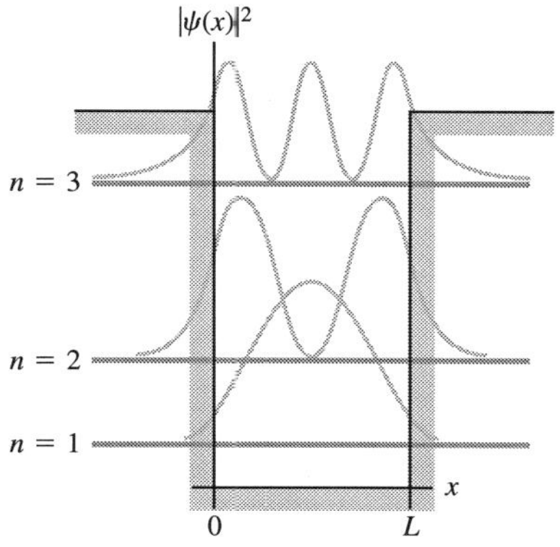

When U_0 is *much larger* than E_∞ (a very deep well), there are many bound states, and the energies of the lowest few are nearly the same as the energies for the infinitely deep well. When U_0 is only a few times as large as E_∞, there are only a few bound states. (But there is always at least one bound state, no matter how shallow the well.) As with the infinitely deep well, there is *no* state with $E = 0$; such a state would violate the uncertainty principle.

Figure 40.8 shows the particular case in which $U_0 = 6E_\infty$; for this case there are three bound states. The energy levels are expressed both as fractions of the well depth U_0 and as multiples of E_∞. Note that if the well were infinitely deep, the lowest three levels, as given by Eq. (40.9), would be E_∞, $4E_\infty$, and $9E_\infty$. The wave functions for the three bound states are also shown.

It turns out that when U_0 is less than E_∞, there is only one bound state. In the limit when U_0 is *much smaller* than E_∞ (a very shallow or very narrow well), the energy of this single state is approximately $E = 0.68U_0$.

Figure 40.9 shows graphs of the probability distributions—that is, the values of $|\psi|^2$,—for the wave functions shown in Fig. 40.8a. As with the infinite well, not all positions are equally likely. We have already commented on the possibility of finding the particle outside the well, in the classically forbidden regions.

There are also states for which E is *greater than* U_0. In this case the particle is not bound but is free to move through all values of x. Thus *any* energy E greater than U_0 is possible. These free-particle states thus form a *continuum* rather than a discrete set of states with definite energy levels. The free-particle wave functions are sinusoidal both inside and outside the well. The wavelength is shorter inside the well than outside, corresponding to greater kinetic energy in the well than outside it.

The square-well potential described in this section has a number of practical applications. We mentioned earlier the example of an electron inside a metallic sheet. A three-dimensional version, in which U is zero inside a spherical region with radius R and has the value U_0 outside, provides a simple model to represent the interaction of a neutron with a nucleus in neutron-scattering experiments. In this context the model is called the *crystal-ball model* of the nucleus, because neutrons interacting with such a potential are scattered in a way that's analogous to scattering of light by a crystal ball.

Figure 40.10 shows a graphic demonstration of particles in a *two*-dimensional finite potential well, and Example 40.4 describes another application of the square-well potential.

40.10 To make this image, 48 iron atoms (shown as yellow peaks) were placed in a circle on a copper surface. The "elevation" at each point inside the circle indicates the electron density within the circle. The standing-wave pattern is very similar to the probability distribution function for a particle in a one-dimensional finite potential well. (This image was made with a scanning tunneling microscope, to be discussed in Section 40.3.)

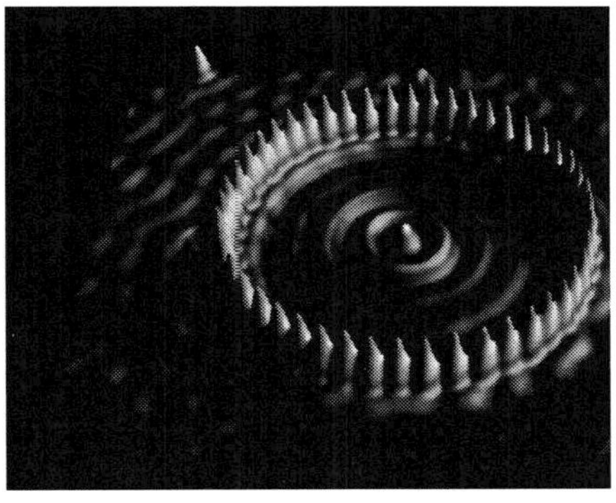

Example 40.4 Electron in a square well

An electron is trapped in a square well with a width of 0.50 nm (comparable to a few atomic diameters). (a) Find the ground-level energy if the well is infinitely deep. (b) If the actual well depth is six times the ground-level energy found in part (a), find the energy levels. (c) If the atom makes a transition from a state with energy E_2 to a state with energy E_1 by emitting a photon, find the wavelength of the photon. In what region of the electromagnetic spectrum does the photon lie? (d) If the electron is initially in its ground level and absorbs a photon, what is the minimum energy the photon must have to liberate the electron from the well? In what region of the spectrum does the photon lie?

SOLUTION

IDENTIFY: We use the ideas of a particle in a box and a square well. The energy of the photon emitted or absorbed in a transition is equal to the difference in energy between two levels involved in the transition.

SET UP: We use Eq. (40.20) to find the ground-level energy E_∞ for an infinitely deep well. The energies for a square well with $U_0 = 6E_\infty$ are shown in Fig. 40.8b. The photon wavelength is given by the relationship $E = hf = hc/\lambda$ [Eq. (38.8)].

EXECUTE: (a) From Eq. (40.20),

$$E_\infty = \frac{\pi^2\hbar^2}{2mL^2} = \frac{\pi^2(1.055 \times 10^{-34}\ \text{J}\cdot\text{s})^2}{2(9.11 \times 10^{-31}\ \text{kg})(0.50 \times 10^{-9}\ \text{m})^2}$$
$$= 2.4 \times 10^{-19}\ \text{J} = 1.5\ \text{eV}$$

(b) We are given $U_0 = 6E_\infty$, so $U_0 = 6(1.5\ \text{eV}) = 9.0\ \text{eV}$. We can simply read off the energy levels from Fig. 40.8b. Our answers are

$$E_1 = 0.104U_0 = 0.104(9.0\ \text{eV}) = 0.94\ \text{eV}$$
$$E_2 = 0.405U_0 = 0.405(9.0\ \text{eV}) = 3.6\ \text{eV}$$
$$E_3 = 0.848U_0 = 0.848(9.0\ \text{eV}) = 7.6\ \text{eV}$$

Alternatively,

$$E_1 = 0.625E_\infty = 0.625(1.5\ \text{eV}) = 0.94\ \text{eV}$$
$$E_2 = 2.43E_\infty = 2.43(1.5\ \text{eV}) = 3.6\ \text{eV}$$
$$E_3 = 5.09E_\infty = 5.09(1.5\ \text{eV}) = 7.6\ \text{eV}$$

(c) The photon energy is

$$E_2 - E_1 = 3.6\ \text{eV} - 0.94\ \text{eV} = 2.7\ \text{eV}$$

We determine the wavelength from $E = hf = hc/\lambda$:

$$\lambda = \frac{hc}{E} = \frac{(4.136 \times 10^{-15}\ \text{eV}\cdot\text{s})(3.00 \times 10^{8}\ \text{m/s})}{2.7\ \text{eV}} = 460\ \text{nm}$$

This photon is in the blue region of the visible spectrum.

(d) We see from Fig. 40.8b that the minimum energy needed to lift the electron out of the well from its $n = 1$ ground level is the well depth $(U_0 = 9.0\ \text{eV})$ minus the electron's initial energy $(E_1 = 0.94\ \text{eV})$, or 8.1 eV. Since 8.1 eV is three times the 2.7-eV photon energy of part (c), the corresponding photon wavelength is one-third of 460 nm, or 150 nm, which is in the ultraviolet region of the spectrum.

EVALUATE: To check our results for the energies of the three bound states, note that if the well had been infinitely deep, the energies would have been

$$E_1 = E_\infty = 1.5\ \text{eV}$$
$$E_2 = 4E_\infty = 6.0\ \text{eV}$$
$$E_3 = 9E_\infty = 13.5\ \text{eV}$$

The energies we found in part (b) are less than these values. This makes sense: As we mentioned earlier, the finite depth of the well lowers the energy levels compared to the levels for an infinitely deep well.

One application of these ideas is to *quantum dots,* which are nanometer-sized particles of a semiconductor such as cadmium selenide (CdSe). An electron within a quantum dot behaves much like a particle in a finite potential well with a width L equal to the size of the dot. When quantum dots are illuminated with ultraviolet light, the electrons absorb the ultraviolet photons and are excited into high energy levels like the $n = 3$ level described in this example. If the electron returns to the ground level $(n = 1)$ in two or more steps (for example, from $n = 3$ to $n = 2$ and from $n = 2$ to $n = 1$), one of the steps will involve emitting a visible-light photon as we have calculated here. (We described this process of *fluorescence* in Section 38.3.) Increasing the value of L decreases the energies of the levels and hence the spacing between them, and hence decreases the energy and increases the wavelength of the emitted photons. The photograph that opens this chapter shows quantum dots of different sizes in solution, each of which emits a characteristic wavelength depending on the dot size. Quantum dots can be injected into living tissue and their fluorescent glow used as a tracer for biological research and for medicine. They may also be the key to a new generation of lasers and ultrafast computers.

Test Your Understanding of Section 40.2 Suppose that the width of the finite potential well shown in Fig. 40.8 is reduced by one-half. How must the value of U_0 change so that there are still just three bound energy levels whose energies are the fractions of U_0 shown in Fig. 40.8b? (i) U_0 must increase by a factor of 4; (ii) U_0 must increase by a factor of 2; (iii) U_0 must remain the same; (iv) U_0 must decrease by a factor of $\frac{1}{2}$; (v) U_0 must decrease by a factor of $\frac{1}{4}$.

40.3 Potential Barriers and Tunneling

ActivPhysics Online
20.4 Potential Barriers

A **potential barrier** is the opposite of a potential well; it is a potential-energy function with a *maximum.* Figure 40.11 shows an example. In Newtonian mechanics, if the total energy is E_1, a particle that is initially to the left of the barrier must remain to the left of the point $x = a$. If it were to move to the right of this point, the potential energy U would be greater than the total energy E. Because $K = E - U$, the kinetic energy would be negative, and this is impossible since a negative value of $K = \frac{1}{2}mv^2$ would require negative mass or imaginary speed.

If the total energy is greater than E_2, the particle can get past the barrier. A roller-coaster car can surmount a hill if it has enough kinetic energy at the bottom. If it doesn't, it stops partway up and then rolls back down.

Quantum mechanics supplies us with a peculiar and very interesting phenomenon in connection with potential barriers. A particle that encounters such a barrier is not necessarily turned back; there is some probability for it to emerge on the other side, even when it doesn't have enough kinetic energy to surmount the barrier according to Newtonian mechanics. This penetration of a barrier is called **tunneling.** It's a natural name; if you dig a tunnel through a hill, you don't have to go over the top. However, in quantum-mechanical tunneling, the particle does not actually bore a hole through the barrier and it loses no energy in the tunneling process.

40.11 A potential-energy barrier. According to Newtonian mechanics, if the total energy is E_1, a particle that is on the left side of the barrier can go no farther than $x = a$. If the total energy is greater than E_2, the particle can pass the barrier.

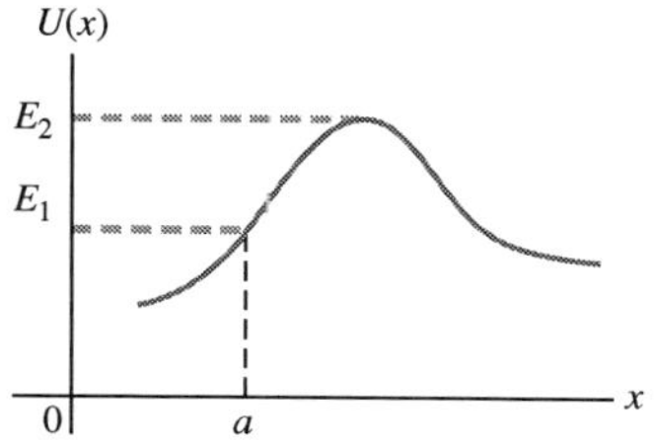

Tunneling Through a Rectangular Barrier

To understand how tunneling can occur, let's look at the potential-energy function $U(x)$ shown in Fig. 40.12. It's like Fig. 40.6 turned upside-down; the potential energy is zero everywhere except in the range $0 \le x \le L$, where it has the value U_0. This might represent a simple model for an electron and two slabs of metal separated by an air gap of thickness L. We let the potential energy be zero within either slab but equal to U_0 in the gap between them.

Let's consider solutions of the Schrödinger equation for this potential-energy function for the case in which E is less than U_0. We can use our results from Section 40.2. In the regions $x < 0$ and $x > L$ the solution is sinusoidal and given by Eq. (40.17). Within the barrier $(0 \le x \le L)$ the solution is exponential, as in Eq. (40.19). Just as with the finite potential well, the functions have to join smoothly at the boundary points $x = 0$ and $x = L$. That is, the function and its derivative have to be continuous at these points.

These requirements lead to a wave function like the one shown in Fig. 40.13. The function is *not* zero inside the barrier (the region forbidden by Newtonian mechanics). Even more remarkable, a particle that is initially to the *left* of the barrier has some probability of being found to the *right* of the barrier. How great a probability? That depends on the width L of the barrier and the particle's energy E (all kinetic) in comparison with the barrier height U_0. The probability T that the particle gets through the barrier is proportional to the square of the ratio of the amplitudes of sinusoidal wave functions on the two sides of the barrier. These amplitudes are determined by matching wave functions and their derivatives at the boundary points, a fairly involved mathematical problem. When T is much smaller than unity, it is given approximately by

$$T = Ge^{-2\kappa L} \quad \text{where} \quad G = 16\frac{E}{U_0}\left(1 - \frac{E}{U_0}\right) \quad \text{and} \quad \kappa = \frac{\sqrt{2m(U_0 - E)}}{\hbar}$$

(probability of tunneling) (40.21)

40.12 A rectangular potential-energy barrier with width L and height U_0. According to Newtonian mechanics, if the total energy E is less than U_0, a particle cannot pass over this barrier but is confined to the side where it starts.

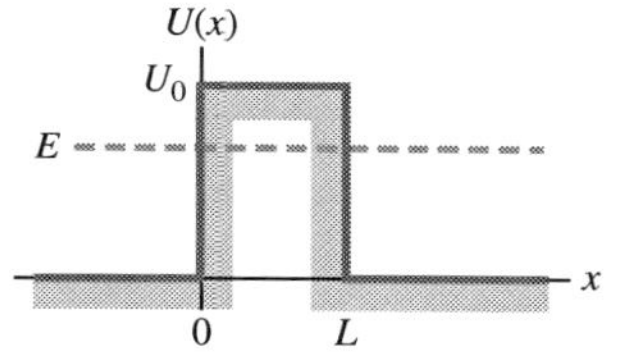

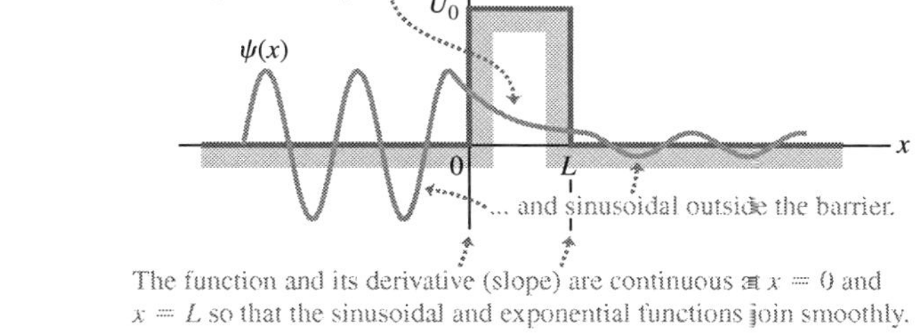

40.13 A possible wave function for a particle tunneling through the potential-energy barrier shown in Fig. 40.12.

The probability decreases rapidly with increasing barrier width L. It also depends critically on the energy difference $U_0 - E$, representing the additional kinetic energy the particle would need to be able to climb over the barrier in a Newtonian analysis.

Example 40.5 Tunneling through a barrier

A 2.0-eV electron encounters a barrier with height 5.0 eV. What is the probability that it will tunnel through the barrier if the barrier width is (a) 1.00 nm; and (b) 0.50 nm (roughly ten and five atomic diameters, respectively)?

SOLUTION

IDENTIFY: This problem uses the ideas of tunneling through a rectangular barrier like that shown in Figs. 40.12 and 40.13.

SET UP: Our target variable is the tunneling probability T in Eq. (40.21), which we evaluate for the given values $E = 2.0$ eV (electron energy), $U = 5.0$ eV (barrier height), $m = 9.11 \times 10^{-31}$ kg (mass of the electron), and $L = 1.00$ nm or 0.50 nm (barrier width).

EXECUTE: First we evaluate G and κ in Eq. (40.21), using $E = K = 2.0$ eV:

$$G = 16\frac{2.0\text{ eV}}{5.0\text{ eV}}\left(1 - \frac{2.0\text{ eV}}{5.0\text{ eV}}\right) = 3.8$$

$$U_0 - E = 5.0\text{ eV} - 2.0\text{ eV} = 3.0\text{ eV} = 4.8 \times 10^{-19}\text{ J}$$

$$\kappa = \frac{\sqrt{2(9.11 \times 10^{-31}\text{ kg})(4.8 \times 10^{-19}\text{ J})}}{1.055 \times 10^{-34}\text{ J}\cdot\text{s}} = 8.9 \times 10^{9}\text{ m}^{-1}$$

(Note that the eV energy units canceled in calculating G, but we had to convert eV to J $=$ kg $\cdot$ m^2/s^2 to find κ in m^{-1}.)

(a) When $L = 1.00\text{ nm} = 1.00 \times 10^{-9}\text{ m}$, $2\kappa L = 2(8.9 \times 10^{9}\text{ m}^{-1}) \times (1.00 \times 10^{-9}\text{ m}) = 17.8$, and

$$T = Ge^{-2\kappa L} = 3.8e^{-17.8} = 7.1 \times 10^{-8}$$

(b) When $L = 0.50$ nm, half of 1.00 nm, $2\kappa L$ is half of 17.8, or 8.9, and

$$T = 3.8e^{-8.9} = 5.2 \times 10^{-4}$$

EVALUATE: Halving the width of this barrier increases the tunneling probability by a factor of nearly ten thousand.

Applications of Tunneling

Tunneling is significant in many areas of physics, including some with considerable practical importance. For example, when you twist two copper wires together or close the contacts of a switch, current passes from one conductor to the other despite a thin layer of nonconducting copper oxide between them. The electrons tunnel through this thin insulating layer. The *tunnel diode* is a semiconductor device in which electrons tunnel through a potential barrier. The current can be switched on and off very quickly (within a few picoseconds) by varying the height of the barrier. The *Josephson junction* consists of two superconductors separated by an oxide layer a few atoms (1 to 2 nm) thick. Electron pairs in the superconductors can tunnel through the barrier layer, giving such a device unusual circuit properties. Josephson junctions are useful for establishing precise voltage standards and measuring tiny magnetic fields.

The *scanning tunneling microscope* (STM) uses electron tunneling to create images of surfaces down to the scale of individual atoms. An extremely sharp conducting needle is brought very close to the surface, within 1 nm or so (Fig. 40.14a).

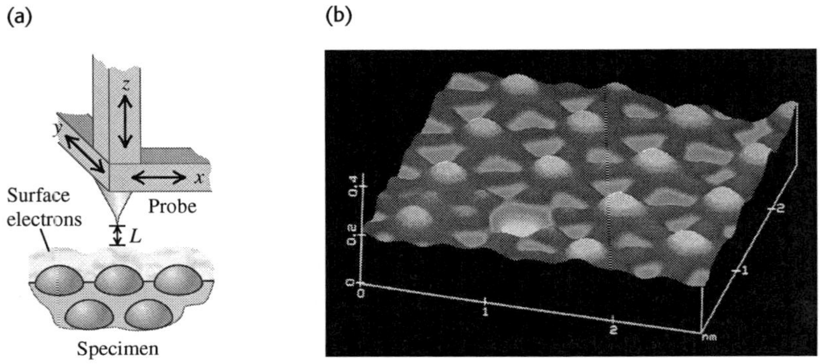

40.14 (a) Schematic diagram of the probe of a scanning tunneling microscope (STM). As the sharp conducting probe is scanned across the surface in the x- and y-directions, it is also moved in the z-direction to maintain a constant tunneling current. The changing position of the probe is recorded and used to construct an image of the surface. (b) This colored STM image shows sodium atoms (pink) adsorbed on a platinum surface. The yellow spot indicates a missing atom.

When the needle is at a positive potential with respect to the surface, electrons can tunnel through the surface potential-energy barrier and reach the needle. As Example 40.5 shows, the tunneling probability and hence the tunneling current depend strongly on the width L of the barrier (the distance between the surface and the needle tip). In one mode of operation the needle is scanned across the surface and at the same time is moved perpendicular to the surface to maintain a constant tunneling current. The needle motion is recorded, and after many parallel scans, an image of the surface can be reconstructed. Extremely precise control of needle motion, including isolation from vibration, is essential. Figure 40.14b shows an image of sodium atoms on the surface of a platinum crystal. (Figure 40.10 is also an STM image.)

Tunneling is of great importance in nuclear physics. A fusion reaction can occur when two nuclei tunnel through the barrier caused by their electrical repulsion and approach each other closely enough for the attractive nuclear force to cause them to fuse. Fusion reactions occur in the cores of stars, including the sun; without tunneling, the sun wouldn't shine. The emission of alpha particles from unstable nuclei also involves tunneling. An alpha particle at the surface of a nucleus encounters a potential barrier that results from the combined effect of the attractive nuclear force and the electrical repulsion of the remaining part of the nucleus (Fig. 40.15). The alpha particle tunnels through this barrier. Because the tunneling probability depends so critically on the barrier height and width, the lifetimes of alpha-emitting nuclei vary over an extremely wide range. We'll return to alpha decay in Chapter 43.

40.15 Approximate potential-energy function for an alpha particle interacting with a nucleus of radius R. If an alpha particle inside the nucleus has energy E greater than zero, it can tunnel through the barrier and escape from the nucleus.

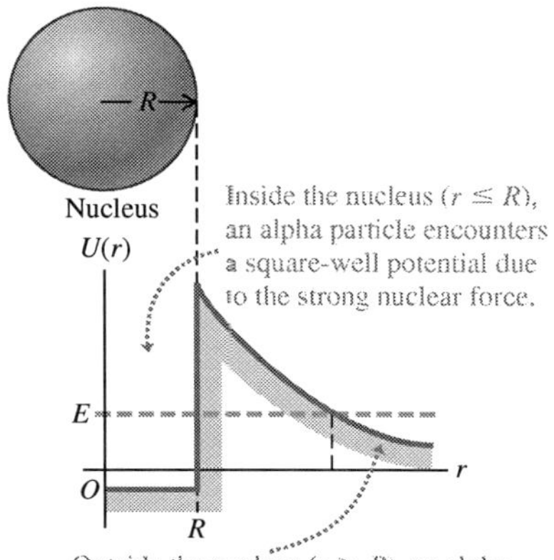

Test Your Understanding of Section 40.3 Is it possible for a particle undergoing tunneling to be found *within* the barrier rather than on either side of it?

40.4 The Harmonic Oscillator

In Newtonian mechanics a **harmonic oscillator** is a particle with mass m acted on by a conservative force component $F_x = -k'x$. (In this discussion we will use k' for the *force constant* to minimize confusion with the wave number $k = 2\pi/\lambda$.) This force component is proportional to the particle's displacement x from its equilibrium position, $x = 0$. The corresponding potential-energy function is $U = \frac{1}{2}k'x^2$ (Fig. 40.16). When the particle is displaced from equilibrium, it undergoes sinusoidal motion with angular frequency $\omega = (k'/m)^{1/2}$. We studied this system in detail in Chapter 13; you should review that discussion.

40.16 Potential-energy function for the harmonic oscillator. In Newtonian mechanics the amplitude A is related to the total energy E by $E = \frac{1}{2}k'A^2$, and the particle is restricted to the range from $x = -A$ to $x = A$.

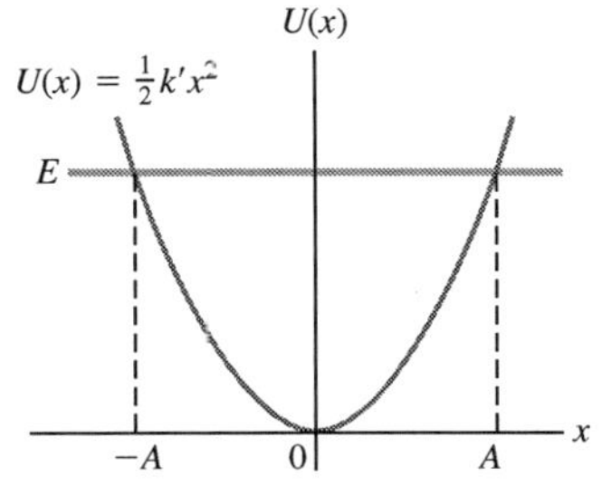

A quantum-mechanical analysis of the harmonic oscillator using the Schrödinger equation is an interesting and useful study. The solutions provide insight into vibrations of molecules, the quantum theory of heat capacities, atomic vibrations in crystalline solids, and many other situations.

Before we get into the details, let's make an enlightened guess about the energy levels. The energy E of a *photon* is related to its angular frequency ω by $E = hf = (h/2\pi)(\omega/2\pi) = \hbar\omega$. The harmonic oscillator has a characteristic angular frequency ω, at least in Newtonian mechanics. So a reasonable guess would be that in the quantum-mechanical analysis the energy levels of a harmonic oscillator would be multiples of the quantity

$$\hbar\omega = \hbar\sqrt{\frac{k'}{m}}$$

These are the energy levels that Planck assumed in deriving his radiation law (see Section 38.8). It was a good assumption; the energy levels are in fact half-integer $\left(\frac{1}{2}, \frac{3}{2}, \frac{5}{2}, \ldots\right)$ multiples of $\hbar\omega$.

For the harmonic oscillator we write $\frac{1}{2}k'x^2$ in place of U in the one-dimensional Schrödinger equation, Eq. (40.1):

$$-\frac{\hbar^2}{2m}\frac{d^2\psi}{dx^2} + \frac{1}{2}k'x^2\psi = E\psi \quad \text{(Schrödinger equation for the harmonic oscillator)} \tag{40.22}$$

The solutions of this equation are wave functions for the physically possible states of the system.

Wave Functions, Boundary Conditions, and Energy Levels

In the discussion of square-well potentials in Section 40.2 we found that the energy levels are determined by boundary conditions at the walls of the well. However, the harmonic-oscillator potential has no walls as such; what, then, are the appropriate boundary conditions? Classically, $|x|$ cannot be greater than the amplitude A, which is the maximum displacement from equilibrium. Quantum mechanics does allow some penetration into classically forbidden regions, but the probability decreases as that penetration increases. Thus the wave functions must approach zero as $|x|$ grows large.

Satisfying the requirement that $\psi(x) \rightarrow 0$ as $|x| \rightarrow \infty$ is not as trivial as it may seem. Suppose we compute solutions of Eq. (40.22) *numerically*. We start at some point x, choosing values of $\psi(x)$ and $d\psi(x)/dx$ at that point. Using the Schrödinger equation to evaluate $d^2\psi(x)/dx^2$ at the point, we can compute $\psi(x)$ and $d\psi(x)/dx$ at a neighboring point $x + \Delta x$, and so on. By using a computer to iterate this process many times with sufficiently small steps Δx, we can compute the wave function with great precision. There is no guarantee, however, that a function obtained in this way will approach zero at large x. To see what kinds of trouble we can get into, let's rewrite Eq. (40.22) in the form

$$\frac{d^2\psi(x)}{dx^2} = \frac{2m}{\hbar^2}\left(\frac{1}{2}k'x^2 - E\right)\psi(x) \tag{40.23}$$

40.17 Possible behaviors of harmonic-oscillator wave functions in the region $\frac{1}{2}k'x^2 > E$. In this region, $\psi(x)$ and $d^2\psi(x)/dx^2$ have the same sign. The curve is concave upward when $d^2\psi(x)/dx^2$ is positive and concave downward when $d^2\psi(x)/dx^2$ is negative.

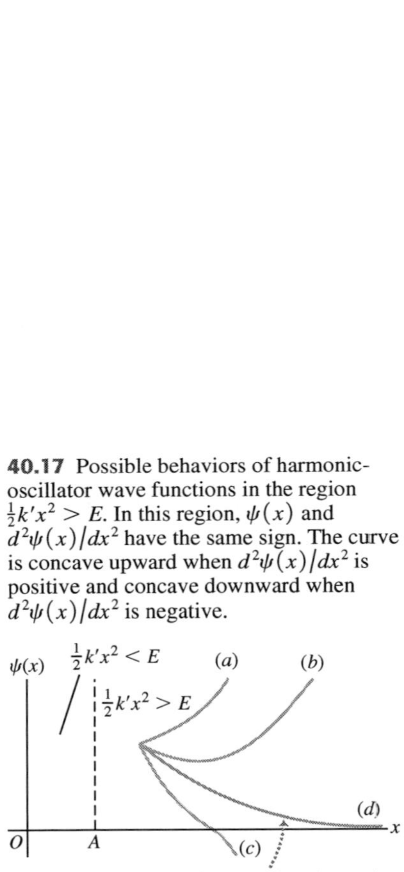

Only curve d, which approaches the x-axis asymptotically for large x, is an acceptable wave function for this system.

In this form, the equation shows that when x is large enough (either positive or negative) to make the quantity $(\frac{1}{2}k'x^2 - E)$ positive, the function $\psi(x)$ and its second derivative $d^2\psi(x)/dx^2$ have the same sign.

The second derivative of $\psi(x)$ is the rate of change of the *slope* of $\psi(x)$. Consider a point with $x > A$ for which $\frac{1}{2}k'x^2 - E > 0$. If $\psi(x)$ is positive, so is $d^2\psi(x)/dx^2$ and the function is *concave upward.* Figure 40.17 shows four possible kinds of behavior beginning at a point $x > A$. If the slope is initially positive, the function curves upward more and more steeply (curve a) and goes to infinity. If the slope is initially negative at the point, there are three possibilities. If the slope changes too quickly (curve b), the curve bends up and again goes to infinity. If the slope doesn't change quickly enough, the curve bends down and crosses the x-axis. After it has crossed, $\psi(x)$ and $d^2\psi(x)/dx^2$ are both negative (curve c) and the concave-downward curve heads for *negative* infinity. Between these infinities is the chance that the curve may bend just enough to glide in asymptotically to the x-axis (curve d). In this case, $\psi(x)$, $d\psi(x)/dx$, and $d^2\psi(x)/dx^2$ all approach zero at large x; in this possibility lies the only hope of satisfying the boundary condition that $\psi(x) \rightarrow 0$ as $|x| \rightarrow \infty$, and it occurs only for certain very special values of the constant E.

This qualitative discussion offers some insight into how the boundary conditions for this problem determine the possible energy levels. Equation (40.22) can also be solved exactly. The solutions, although not encountered in elementary calculus courses, are well known to mathematicians; they are called *Hermite functions.* Each one is an exponential function multiplied by a polynomial in x. The state with lowest energy (the ground level) has the wave function

$$\psi(x) = Ce^{-\sqrt{mk'}x^2/2\hbar} \tag{40.24}$$

The constant C is chosen to normalize the function—that is, to make $\int_{-\infty}^{\infty} |\psi|^2\, dx = 1$. You can find C using, from integral tables,

$$\int_{-\infty}^{\infty} e^{-a^2x^2}\, dx = \frac{\sqrt{\pi}}{a}$$

The corresponding energy, which we'll call E_0, is the ground-state energy; it is

$$E_0 = \frac{1}{2}\hbar\omega = \frac{1}{2}\hbar\sqrt{\frac{k'}{m}} \tag{40.25}$$

You may not believe that Eq. (40.24) really *is* a solution of Eq. (40.22) (the Schrödinger equation for the harmonic oscillator) with the energy given by Eq. (40.25). We invite you to calculate the second derivative of Eq. (40.24), substitute it into Eq. (40.22), and verify that it really is a solution if Eq. (40.25) gives E_0 (see exercise 40.28). It's a little messy, but the result is satisfying and worth the effort.

Further analysis of the Schrödinger equation for the harmonic oscillator shows that it can be written in the form

$$\frac{d^2\psi(x)}{dx^2} = \frac{2m}{\hbar^2}\left[\frac{1}{2}k'x^2 - \left(n + \frac{1}{2}\right)\hbar\sqrt{\frac{k'}{m}}\right]\psi(x)$$

Comparing this complicated expression to Eq. (40.23) gives us a pleasant surprise: The energy levels, which we'll call E_n, are given by the simple formula

$$E_n = \left(n + \frac{1}{2}\right)\hbar\sqrt{\frac{k'}{m}} = \left(n + \frac{1}{2}\right)\hbar\omega \qquad (n = 0, 1, 2, \ldots)$$
(energy levels, harmonic oscillator) (40.26)

where n is the quantum number identifying each state and energy level. Note that the ground level of energy $E_0 = \frac{1}{2}\hbar\omega$ is denoted by $n = 0$, *not* $n = 1$.

Equation (40.26) confirms our guess that the energy levels are multiples of $\hbar\omega$. Adjacent energy levels are separated by a constant interval of $\hbar\omega = hf$, as Planck assumed in 1900. There are infinitely many levels; this shouldn't be surprising because we are dealing with an infinitely deep potential well. As $|x|$ increases, $U = \frac{1}{2}k'x^2$ increases without bound.

Figure 40.18 shows the lowest six energy levels and the potential-energy function $U(x)$. For each level n, the value of $|x|$ at which the horizontal line representing the total energy E_n intersects $U(x)$ gives the amplitude A_n of the corresponding Newtonian oscillator.

40.18 Energy levels for the harmonic oscillator. The spacing between any two adjacent levels is $\Delta E = \hbar\omega$. The energy of the ground level is $E_0 = \frac{1}{2}\hbar\omega$.

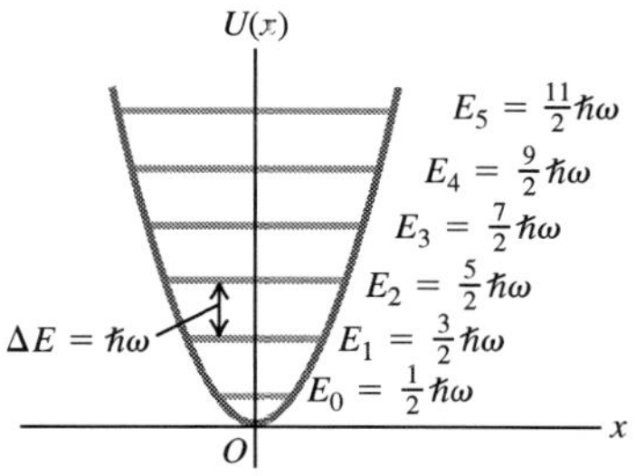

Example 40.6 Vibration in a sodium crystal

A sodium atom of mass 3.82×10^{-26} kg vibrates with simple harmonic motion in a crystal. The potential energy increases by 0.0075 eV when the atom is displaced 0.014 nm from its equilibrium position. (a) Find the angular frequency, according to Newtonian mechanics. (b) Find the spacing of adjacent energy levels in electron volts. (c) If an atom emits a photon during a transition from one vibrational level to the next lower level, what is the wavelength of the emitted photon? In what region of the electromagnetic spectrum does it lie?

SOLUTION

IDENTIFY: This problem uses both the Newtonian and quantum-mechanical descriptions of a harmonic oscillator.

SET UP: To find the angular frequency $\omega = (k'/m)^{1/2}$, we use the information given and the expression $U(x) = \frac{1}{2}k'x^2$ for the potential energy. The spacing between adjacent energy levels is given by Eq. (40.26). We calculate the wavelength of the emitted photon using the same ideas as in Example 40.4 (Section 40.2).

Continued

EXECUTE: We don't know k', but we can find it from the fact that $U = 0.0075 \text{ eV} = 1.2 \times 10^{-21}$ J when $x = 0.014 \times 10^{-9}$ m. We have

$$U = \frac{1}{2}k'x^2$$

$$k' = \frac{2U}{x^2} = \frac{2(1.2 \times 10^{-21} \text{ J})}{(0.014 \times 10^{-9} \text{ m})^2} = 12.2 \text{ N/m}$$

(a) The angular frequency is

$$\omega = \sqrt{\frac{k'}{m}} = \sqrt{\frac{12.2 \text{ N/m}}{3.82 \times 10^{-26} \text{ kg}}} = 1.79 \times 10^{13} \text{ rad/s}$$

(b) From Eq. (40.26) and Fig. 40.18 the spacing between adjacent energy levels is

$$\begin{aligned}\hbar\omega &= (1.054 \times 10^{-34} \text{ J} \cdot \text{s})(1.79 \times 10^{13} \text{ s}^{-1}) \\ &= 1.88 \times 10^{-21} \text{ J} = 0.0118 \text{ eV}\end{aligned}$$

(c) From $E = hc/\lambda$ the corresponding wavelength is

$$\begin{aligned}\lambda &= \frac{hc}{E} = \frac{(4.136 \times 10^{-15} \text{ eV} \cdot \text{s})(3.00 \times 10^8 \text{ m/s})}{0.0118 \text{ eV}} \\ &= 1.05 \times 10^{-4} \text{ m} = 105 \ \mu\text{m}\end{aligned}$$

This photon is in the infrared region of the spectrum.

EVALUATE: This example shows us that the force constants for interatomic forces are on the order of a few N/m, not very different from the values for household springs or spring-based toys such as a Slinky.™ It also suggests that physicists can learn about the vibrations of atoms by measuring the radiation that they emit as they transition from one vibrational state to a lower state. We will explore this idea further in Chapter 42.

Comparing Quantum and Newtonian Oscillators

ActivPhysics Online
20.1.6 Potential Energy Diagrams, Question 6

Figure 40.19 shows the first four harmonic-oscillator wave functions. Each graph also shows the amplitude A of a Newtonian harmonic oscillator with the same energy—that is, the value of A determined from

$$\frac{1}{2}k'A^2 = \left(n + \frac{1}{2}\right)\hbar\omega \tag{40.27}$$

In each case there is some penetration of the wave function into the regions $|x| > A$ that are forbidden by Newtonian mechanics. This is similar to the effect that we noted with the particle in a finite square well.

CAUTION **Be very careful with your algebraic symbols** This Newtonian amplitude A (in units of m) is *not* the same as the quantum-mechanical normalizing constant A (in $\text{m}^{-1/2}$) that is used in several wave functions in this chapter. Also the angular frequency ω (in rad/s) of the harmonic oscillator is *not* the same as the angular frequency ω (also in rad/s) obtained from the de Broglie relationship $E = \hbar\omega$ (which is why we expressed Eq. (40.14) in terms of $2\pi E/h = E/\hbar$ rather than in terms of ω). Finally, we have distinguished the force constant k' (in N/m or kg/s^2) from the wave number k (in rad/m). Neither of these is the same as κ (in m^{-1}) used in Eqs. (40.19) and (40.21) or the kinetic energy K (in J or eV).

Figure 40.20 shows the probability distributions $|\psi(x)|^2$ for these same states. Each graph also shows the probability distribution determined from Newtonian analysis, in which the probability of finding the particle near a randomly

40.19 The first four wave functions for the harmonic oscillator. The amplitude A of a Newtonian oscillator with the same total energy is shown for each. Each wave function penetrates somewhat into the classically forbidden regions $|x| > A$. The total number of finite maxima and minima for each function is $n + 1$, one more than the quantum number.

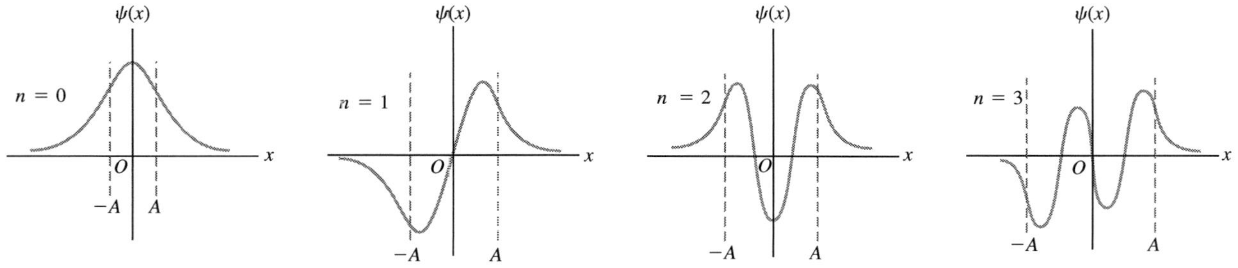

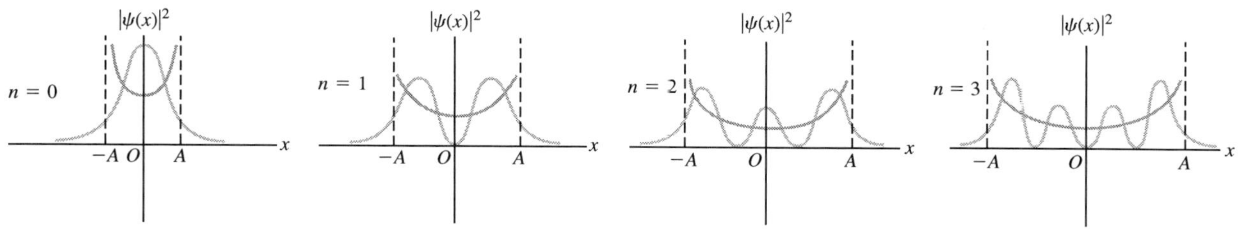

40.20 Probability distribution functions $|\psi|^2$ for the harmonic-oscillator wave functions shown in Fig. 40.19. The amplitude A of the Newtonian motion with the same energy is shown for each. The blue lines show the corresponding probability distributions for the Newtonian motion. As n increases, the averaged-out quantum-mechanical functions resemble the Newtonian curves more and more.

chosen point is inversely proportional to its speed at that point. If we average out the wiggles in the quantum-mechanical probability curves, the results for $n > 0$ resemble the Newtonian predictions. This agreement improves with increasing n; Fig. 40.21 shows the classical and quantum-mechanical probability functions for $n = 10$. Notice that the spacing between zeros of $|\psi(x)|^2$ in Fig. 40.21 increases with increasing distance from $x = 0$. This makes sense from the Newtonian perspective: As a particle moves away from $x = 0$, its kinetic energy K and the magnitude p of its momentum both decrease. Thinking quantum-mechanically, this means that the wavelength $\lambda = h/p$ increases, so the spacing between zeros of $\psi(x)$ (and hence of $|\psi(x)|^2$) also increases.

In the Newtonian analysis of the harmonic oscillator the minimum energy is zero, with the particle at rest at its equilibrium position. This is not possible in quantum mechanics; no solution of the Schrödinger equation has $E = 0$ and satisfies the boundary conditions. Furthermore, if there were such a state, it would violate the uncertainty principle because there would be no uncertainty in either position or momentum. Indeed, the energy must be at least $\frac{1}{2}\hbar\omega$ for the system to conform to the uncertainty principle. To see qualitatively why this is so, consider a Newtonian oscillator with total energy $\frac{1}{2}\hbar\omega$. We can find the amplitude A and the maximum velocity just as we did in Section 13.3. The appropriate relationships are Eqs. (13.21) for the total energy and (13.23) for the maximum speed. Setting $E = \frac{1}{2}\hbar\omega$, we find

$$E = \frac{1}{2}k'A^2 = \frac{1}{2}\hbar\omega = \frac{1}{2}\hbar\sqrt{\frac{k'}{m}} \qquad A = \sqrt{\frac{\hbar}{\sqrt{k'm}}}$$

$$p_{\text{max}} = mv_{\text{max}} = m\sqrt{\frac{k'}{m}}A = m\sqrt{\frac{k'}{m}}\sqrt{\frac{\hbar}{\sqrt{k'm}}} = \sqrt{\hbar\sqrt{k'm}}$$

Finally, if we assume that A represents the uncertainty Δx in position and p_{max} is the corresponding uncertainty Δp_x in momentum, then the product of the two uncertainties is

$$\Delta x\,\Delta p_x = \sqrt{\frac{\hbar}{\sqrt{k'm}}}\sqrt{\hbar\sqrt{k'm}} = \hbar$$

So the product equals the minimum value allowed by Eq. (39.11), $\Delta x\,\Delta p_x \geq \hbar$, and thus satisfies the uncertainty principle. If the energy had been less than $\frac{1}{2}\hbar\omega$, the product $\Delta x\,\Delta p_x$ would have been less than $\hbar$, and the uncertainty principle would have been violated.

Even when a potential-energy function isn't precisely parabolic in shape, we may be able to approximate it by the harmonic-oscillator potential for sufficiently small displacements from equilibrium. Figure 40.22 shows a typical potential-energy function for an interatomic force in a molecule. At large separations it levels off, corresponding to the absence of force at great distances. But it is approximately parabolic near the minimum point (the equilibrium position of the

40.21 Newtonian and quantum-mechanical probability distribution functions for a harmonic oscillator for the state $n = 10$. The Newtonian amplitude A is also shown.

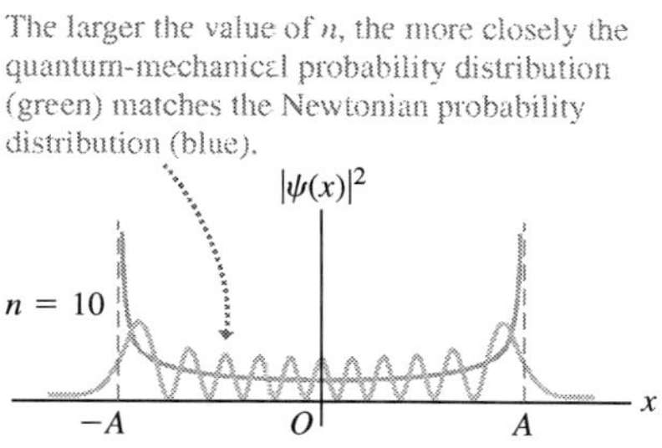

40.22 A potential-energy function describing the interaction of two atoms in a diatomic molecule. The equilibrium position is at $r = r_0$. The potential energy needed to dissociate the molecule is U_∞.

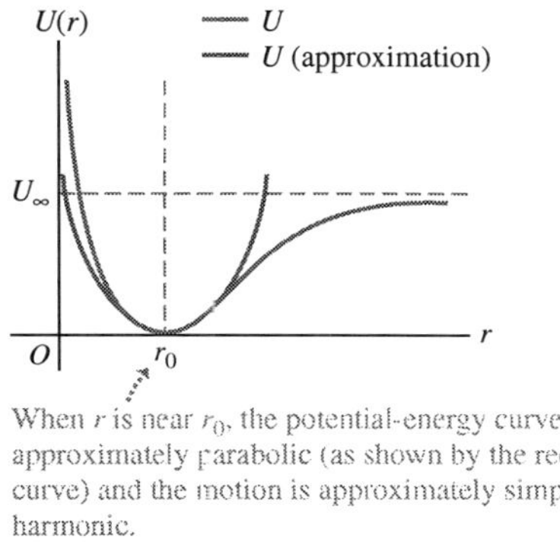

atoms). Near equilibrium the molecular vibration is approximately simple harmonic with energy levels given by Eq. (40.26), as we assumed in Example 40.6.

Test Your Understanding of Section 40.4 A quantum-mechanical system initially in its ground level absorbs a photon and ends up in the first excited state. The system then absorbs a second photon and ends up in the second excited state. For which of the following systems does the second photon have a longer wavelength than the first one? (i) a harmonic oscillator; (ii) a hydrogen atom; (iii) a particle in a box.

40.5 Three-Dimensional Problems

We have discussed the Schrödinger equation and its applications only for *one-dimensional* problems, the analog of a Newtonian particle moving along a straight line. The straight-line model is adequate for some applications, but to understand atomic structure, we need a three-dimensional generalization.

It's not difficult to guess what the three-dimensional Schrödinger equation should look like. First, the wave function ψ is a function of all three space coordinates (x, y, z). In general, the potential-energy function also depends on all three coordinates and can be written as $U(x, y, z)$. Next, note that the term $-(\hbar^2/2m)d^2\psi/dx^2$ in the one-dimensional Schrödinger equation, Eq. (40.1), is related to the kinetic energy of the particle in the state described by the wave function ψ. For example, if we insert into this term the wave function $\psi(x) = Ae^{ikx}$ for a free particle with magnitude of momentum $p = \hbar k$ and kinetic energy $K = p^2/2m$, we obtain $-(\hbar^2/2m)(ik)^2Ae^{ikx} = (\hbar^2k^2/2m)Ae^{ikx} = (p^2/2m)\psi(x) = K\psi(x)$. If the particle's momentum has three components (p_x, p_y, p_z), then the kinetic energy is

$$K = \frac{p_x^2}{2m} + \frac{p_y^2}{2m} + \frac{p_z^2}{2m} \qquad (40.28)$$

These observations, taken together, suggest that the correct generalization of the Schrödinger equation to three dimensions is

$$-\frac{\hbar^2}{2m}\left(\frac{\partial^2\psi(x, y, z)}{\partial x^2} + \frac{\partial^2\psi(x, y, z)}{\partial y^2} + \frac{\partial^2\psi(x, y, z)}{\partial z^2}\right) + U(x, y, z)\psi(x, y, z)$$
$$= E\psi(x, y, z) \quad \text{(three-dimensional Schrödinger equation)} \qquad (40.29)$$

40.23 In quantum mechanics, spherical coordinates are particularly useful in problems in which the potential energy depends only on r, the distance from the origin.

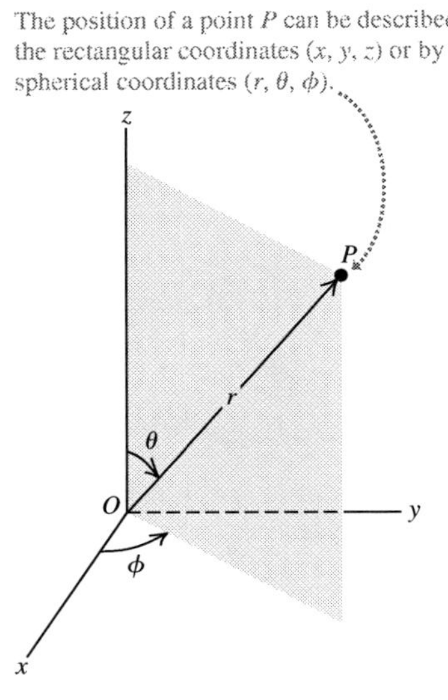

We use partial derivatives, since ψ is a function of x, y, and z. As in one dimension, E is constant for a state of particular energy.

We won't pretend that we have *derived* Eq. (40.29). Like the one-dimensional version, this equation has to be tested by comparison of its predictions with experimental results. As we will see in later chapters, Eq. (40.29) passes this test with flying colors, so we are confident that it *is* the correct equation.

In many practical problems, in atomic structure and elsewhere, the potential-energy function is *spherically symmetric;* it depends only on the distance $r = (x^2 + y^2 + z^2)^{1/2}$ from the origin of coordinates. To take advantage of this symmetry, we use *spherical coordinates* (r, θ, ϕ) (Fig. 40.23) instead of Cartesian coordinates (x, y, z). Then a spherically symmetric potential-energy function is a function only of r, not of θ or ϕ, so $U = U(r)$ This fact turns out to simplify greatly the problem of finding solutions of the Schrödinger equation, even though the derivatives in Eq. (40.29) are considerably more complex when expressed in terms of spherical coordinates. Be careful; many math texts exchange the angles θ and ϕ shown in Fig. 40.23.

For the hydrogen atom the potential-energy function $U(r)$ is the familiar Coulomb's-law function:

$$U(r) = -\frac{1}{4\pi\epsilon_0}\frac{e^2}{r} \tag{40.30}$$

We will find that for *all* spherically symmetric potential-energy functions $U(r)$, each possible wave function can be expressed as a product of three functions: one a function only of r, one only of θ, and one only of ϕ. Furthermore, the functions of θ and ϕ are *the same* for *every* spherically symmetric potential-energy function. This result is directly related to the problem of finding the possible values of *angular momentum* for the various states. We'll discuss these matters more in the next chapter.

CHAPTER 40 SUMMARY

Particle in a box: The energy levels for a particle of mass m in a box (an infinitely deep square potential well) with width L are given by Eq. (40.9). The corresponding normalized wave functions of the particle are given by Eq. (40.13). (See Examples 40.1 and 40.2.)

$$E_n = \frac{p_n^2}{2m} = \frac{n^2h^2}{8mL^2} = \frac{n^2\pi^2\hbar^2}{2mL^2} \quad (n = 1, 2, 3, \dots) \qquad (40.9)$$

$$\psi_n(x) = \sqrt{\frac{2}{L}}\sin\frac{n\pi x}{L} \quad (n = 1, 2, 3, \dots) \qquad (40.13)$$

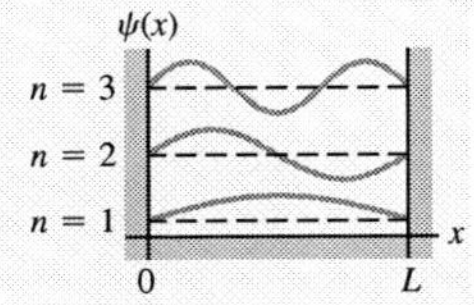

Wave functions and normalization: To be a solution of the Schrödinger equation, the wave function $\psi(x)$ and its derivative $d\psi(x)/dx$ must be continuous everywhere, except where the potential-energy function $U(x)$ has an infinite discontinuity. Wave functions are usually normalized so that the total probability for finding the particle somewhere is unity.

$$\int_{-\infty}^{\infty} |\psi(x)|^2\,dx = 1 \qquad (40.11)$$

(normalization condition)

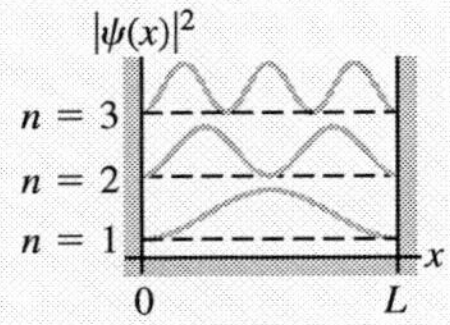

Finite potential well: In a potential well with finite depth U_0, the energy levels are lower than those for an infinitely deep well with the same width, and the number of energy levels corresponding to bound states is finite. The levels are obtained by matching wave functions at the well walls to satisfy the continuity of $\psi(x)$ and $d\psi(x)/dx$. (See Examples 40.3 and 40.4.)

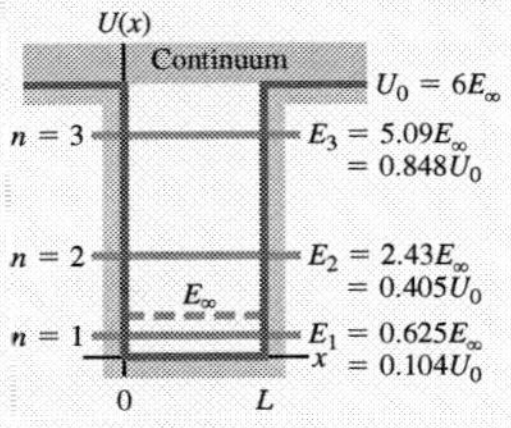

Potential barriers and tunneling: There is a certain probability that a particle will penetrate a potential-energy barrier even though its initial kinetic energy is less than the barrier height. This process is called tunneling. (See Example 40.5.)

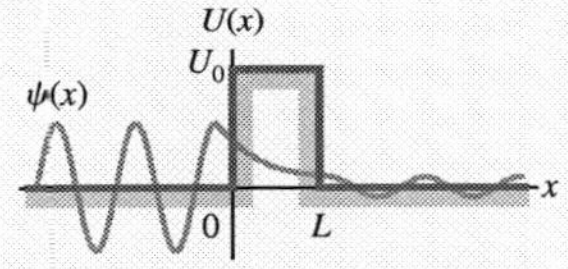

Quantum harmonic oscillator: The energy levels for the harmonic oscillator, for which $U(x) = \frac{1}{2}k'x^2$, are given by Eq. (40.26). The spacing between any two adjacent levels is $\hbar\omega$, where $\omega = \sqrt{k'/m}$ is the oscillation angular frequency of the corresponding Newtonian harmonic oscillator. (See Example 40.6.)

$$E_n = \left(n + \frac{1}{2}\right)\hbar\omega = \left(n + \frac{1}{2}\right)\hbar\sqrt{\frac{k'}{m}} \quad (n = 0, 1, 2, 3, \dots) \qquad (40.26)$$

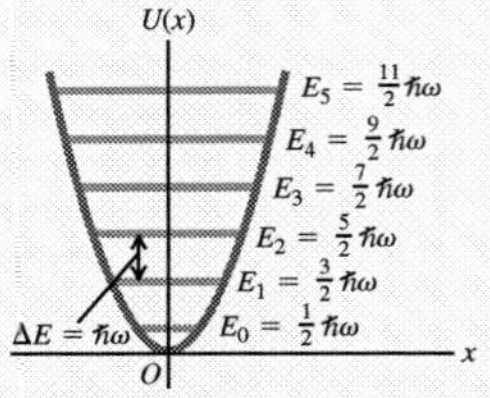

Three-dimensional problems: The Schrödinger equation for three-dimensional problems is given by Eq. (40.29).

$$-\frac{\hbar^2}{2m}\left(\frac{\partial^2\psi(x,y,z)}{\partial x^2} + \frac{\partial^2\psi(x,y,z)}{\partial y^2} + \frac{\partial^2\psi(x,y,z)}{\partial z^2}\right) + U(x,y,z)\psi(x,y,z) = E\psi(x,y,z) \qquad (40.29)$$

Key Terms

particle in a box, *1376*
normalization, *1379*
potential well, *1380*
square-well potential, *1381*
bound state, *1381*
potential barrier, *1384*
tunneling, *1385*
harmonic oscillator, *1387*

Answer to Chapter Opening Question ?

When an electron in one of these particles—called *quantum dots*—makes a transition from an excited level to a lower level, it emits a photon whose energy is equal to the difference in energy between the levels. The smaller the quantum dot, the larger the energy spacing between levels and hence the shorter (bluer) the wavelength of the emitted photons. See Example 40.4 (Section 40.2) for more details.

Answers to Test Your Understanding Questions

40.1 Answer: (v) Our derivation of the stationary-state wave functions for a particle in a box shows that they are superpositions of waves propagating in opposite directions, just like a standing wave on a string. One wave has momentum in the positive x-direction, while the other wave has an equal magnitude of momentum in the negative x-direction. The *total* x-component of momentum is zero.

40.2 Answer (i) The energy levels are arranged as shown in Fig. 40.8b if $U_0 = 6E_\infty$, where $E_\infty = \pi^2\hbar^2/2mL^2$ is the ground-level energy of an infinite well. If the well width L is reduced to $\frac{1}{2}$ of its initial value, E_∞ increases by a factor of 4 and so U_0 must also increase by a factor of 4. The energies E_1, E_2, and E_3 shown in Fig. 40.8b are all specific fractions of U_0, so they will also increase by a factor of 4.

40.3 Answer: yes Figure 40.13 shows a possible wave function $\psi(x)$ for tunneling; since $\psi(x)$ is not zero within the barrier $(0 \le x \le L)$, there is some probability that the particle can be found there.

40.4 Answer: (ii) If the second photon has a longer wavelength and hence lower energy than the first photon, the difference in energy between the first and second excited levels must be less than the difference between the ground level and the first excited level. This is the case for the hydrogen atom, for which the energy difference between levels decreases as the energy increases (see Fig. 38.9). By contrast, the energy difference between successive levels increases for a particle in a box (see Fig. 40.4b) and is constant for a harmonic oscillator (see Fig. 40.18).

PROBLEMS

For instructor-assigned homework, go to **www.masteringphysics.com**

Discussion Questions

Q40.1. For the particle in a box, we chose $k = n\pi/L$ with $n = 1, 2, 3, \ldots$ to fit the boundary condition that $\psi = 0$ at $x = L$. However, $n = 0, -1, -2, -3, \ldots$ also satisfy that boundary condition. Why didn't we also choose those values of n?

Q40.2. If ψ is normalized, what is the physical significance of the area under a graph of $|\psi|^2$ versus x between x_1 and x_2? What is the total area under the graph of $|\psi|^2$ when all x are included? Explain.

Q40.3. For a particle in a box, what would the probability distribution function $|\psi|^2$ look like if the particle behaved like a classical (Newtonian) particle? Do the actual probability distributions approach this classical form when n is very large? Explain.

Q40.4. In Chapter 15 we represented a standing wave as a superposition of two waves traveling in opposite directions. Can the wave functions for a particle in a box also be thought of as a combination of two traveling waves? Why or why not? What physical interpretation does this representation have? Explain.

Q40.5. A particle in a box is in the ground level. What is the probability of finding the particle in the right half of the box? (Refer to Fig. 40.5, but don't evaluate an integral.) Is the answer the same if the particle is in an excited level? Explain.

Q40.6. The wave functions for a particle in a box (see Fig. 40.5a) are zero at certain points. Does this mean that the particle can't move past one of these points? Explain.

Q40.7. For a particle confined to an infinite square well, is it correct to say that each state of definite energy is also a state of definite wavelength? Is it also a state of definite momentum? Explain. (*Hint:* Remember that momentum is a vector.)

Q40.8. For a particle in a finite potential well, is it correct to say that each bound state of definite energy is also a state of definite wavelength? Is it a state of definite momentum? Explain.

Q40.9. Equation 40.15 gives the time-dependent wave functions for a particle in a box. How can a wave function that has both a real part and an imaginary part describe the position (a real number) of a particle?

Q40.10. In Fig. 40.5b, the probability function is zero at the points $x = 0$ and $x = L$, the "walls" of the box. Does this mean that the particle never strikes the walls? Explain.

Q40.11. A particle is confined to a finite potential well in the region $0 < x < L$. How does the area under the graph of $|\psi|^2$ in the region $0 < x < L$ compare to the total area under the graph of $|\psi|^2$ when including all possible x?

Q40.12. Compare the wave functions for the first three energy levels for a particle in a box of width L (see Fig. 40.5a) to the corresponding wave functions for a finite potential well of the same width (see Fig. 40.8a). How does the wavelength in the interval $0 \le x \le L$ for the $n = 1$ level of the particle in a box compare to the corresponding wavelength for the $n = 1$ level of the finite potential well? Use this to explain why E_1 is less than E_∞ in the situation depicted in Fig. 40.8b.

Q40.13. It is stated in Section 40.2 that a finite potential well always has at least one bound level, no matter how shallow the well. Does this mean that as $U_0 \to 0$, $E_1 \to 0$? Does this violate the Heisenberg uncertainty principle? Explain.

Q40.14. Figure 40.8a shows that the higher the energy of a bound state for a finite potential well, the more the wave function extends outside the well (into the intervals $x < 0$ and $x > L$). Explain why this happens.

Q40.15. In classical (Newtonian) mechanics, the total energy E of a particle can never be less than the potential energy U because the kinetic energy K cannot be negative. Yet in barrier tunneling (see

Section 40.3) a particle passes through regions where E is less than U. Is this a contradiction? Explain.

Q40.16. Figure 40.10 shows the scanning tunneling microscope image of 48 iron atoms placed on a copper surface, the pattern indicating the density of electrons on the copper surface. What can you infer about the potential-energy function inside the circle of iron atoms?

Q40.17. Qualitatively, how would you expect the probability for a particle to tunnel through a potential barrier to depend on the height of the barrier? Explain.

Q40.18. The wave function shown in Fig. 40.13 is nonzero for both $x < 0$ and $x > L$. Does this mean that the particle splits into two parts when it strikes the barrier, with one part tunneling through the barrier and the other part bouncing off the barrier? Explain.

Q40.19. The probability distributions for the harmonic oscillator wave functions (see Figs. 40.20 and 40.21) begin to resemble the classical (Newtonian) probability distribution when the quantum number n becomes large. Would the distributions become the same as in the classical case in the limit of very large n? Explain.

Q40.20. In Fig. 40.21, how does the probability of finding a particle in the center half of the region $-A < x < A$ compare to the probability of finding the particle in the outer half of the region? Is this consistent with the physical interpretation of the situation?

Q40.21. Compare the allowed energy levels for the hydrogen atom, the particle in a box, and the harmonic oscillator. What are the values of the quantum number n for the ground level and the second excited level of each system?

Q40.22. In the hydrogen atom the potential-energy function depends only on distance r from the nucleus, not on direction; that is, it is spherically symmetric. Would you expect all the corresponding wave functions for the electron in the hydrogen atom to be spherically symmetric? Explain.

Q40.23. Sketch the wave function for the potential-energy well shown in Fig. 40.24 when E_1 is less than U_0 and when E_3 is greater than U_0.

Figure **40.24** Question Q40.23.

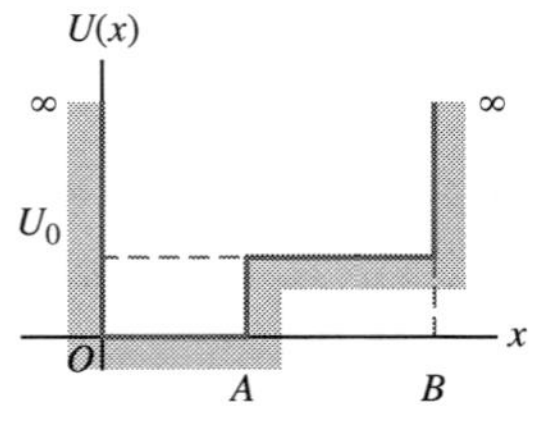

Exercises

Section 40.1 Particle in a Box

40.1. Ground-Level Billiards. (a) Find the lowest energy level for a particle in a box if the particle is a billiard ball $(m = 0.20\text{kg})$ and the box has a width of 1.5 m, the size of a billiard table. (Assume that the billiard ball slides without friction rather than rolls. That is, ignore the *rotational* kinetic energy.) (b) Since the energy in part (a) is all kinetic, to what speed does this correspond? How much time would it take at this speed for the ball to move from one side of the table to the other? (c) What is the difference in energy between the $n = 2$ and $n = 1$ levels? (d) Are quantum-mechanical effects important for the game of billiards?

40.2. A proton is in a box of width L. What must the width of the box be for the ground-level energy to be 5.0 MeV, a typical value for the energy with which the particles in a nucleus are bound? Compare your result to the size of a nucleus—that is, on the order of 10^{-14} m.

40.3. Find the width L of a one-dimensional box that would correspond to the absolute value of the ground state of a hydrogen atom.

40.4. When a hydrogen atom undergoes a transition from the $n = 2$ to the $n = 1$ level, a photon with $\lambda = 122$ nm is emitted. (a) If the atom is modeled as an electron in a one-dimensional box, what is the width of the box in order for the $n = 2$ to $n = 1$ transition to correspond to emission of a photon of this energy? (b) For a box with the width calculated in part (a), what is the ground-state energy? How does this correspond to the ground-state energy of a hydrogen atom? (c) Do you think a one-dimensional box is a good model for a hydrogen atom? Explain. (*Hint:* Compare the spacing between adjacent energy levels as a function of n.)

40.5. A certain atom requires 3.0 eV of energy to excite an electron from the ground level to the first excited level. Model the atom as an electron in a box and find the width L of the box.

40.6. Recall that $|\psi|^2\,dx$ is the probability of finding the particle that has normalized wave function $\psi(x)$ in the interval x to $x + dx$. Consider a particle in a box with rigid walls at $x = 0$ and $x = L$. Let the particle be in the ground level and use ψ_n as given in Eq. (40.13). (a) For which values of x, if any, in the range from 0 to L is the probability of finding the particle zero? (b) For which values of x is the probability highest? (c) In parts (a) and (b) are your answers consistent with Fig. 40.8? Explain.

40.7. Repeat Exercise 40.6 for the particle in the first excited level.

40.8. (a) Show that $\psi = A\sin kx$ is a solution to Eq. (40.3) if $k = \sqrt{2mE}/\hbar$. (b) Explain why this is an acceptable wave function for a particle in a box with rigid walls at $x = 0$ and $x = L$ only if k is an integer multiple of π/L.

40.9. (a) Repeat Exercise 40.8 for $\psi = A\cos kx$. (b) Explain why this cannot be an acceptable wave function for a particle in a box with rigid walls at $x = 0$ and $x = L$ no matter what the value of k.

40.10. (a) Find the excitation energy from the ground level to the third excited level for an electron confined to a box that has a width of 0.125 nm. (b) The electron makes a transition from the $n = 1$ to $n = 4$ level by absorbing a photon. Calculate the wavelength of this photon.

40.11. An electron is in a box of width 3.0×10^{-10} m. What are the de Broglie wavelength and the magnitude of the momentum of the electron if it is in (a) the $n = 1$ level; (b) the $n = 2$ level; (c) the $n = 3$ level? In each case how does the wavelength compare to the width of the box?

40.12. Show that the time-dependent wave function given by Eq. (40.15) is a solution to the one-dimensional Schrödinger equation, Eq. (40.1).

Section 40.2 Potential Wells

40.13. (a) Show that $\psi = A\sin kx$, where k is a constant, is *not* a solution of Eq. (40.1) for $U = U_0$ and $E < U_0$. 9 (b) Is this ψ a solution for $E > U_0$?

40.14. An electron is moving past the square well shown in Fig. 40.6. The electron has energy $E = 3U_0$. What is the ratio of the de Broglie wavelength of the electron in the region $x > L$ to the wavelength for $0 < x < L$?

40.15. An electron is bound in a square well of depth $U_0 = 6E_\infty$. What is the width of the well if its ground-state energy is 2.00 eV?

40.16. An electron is bound in a square well of width 1.50 nm and depth $U_0 = 6E_\infty$. If the electron is initially in the ground level and absorbs a photon, what maximum wavelength can the photon have and still liberate the electron from the well?

40.17. Calculate $d^2\psi/dx^2$ for the wave function of Eq. (40.17), and show that the function is a solution of Eq. (40.16).

40.18. Calculate $d^2\psi/dx^2$ for the wave function of Eq. (40.19), and show that for any values of C and D it is a solution of Eq. (40.1) for $x \leq 0$ and $x \geq L$ when $U = U_0$ and $E < U_0$.

40.19. A proton is bound in a square well of width 4.0 fm = 4.0×10^{-15} m. The depth of the well is six times the ground-level energy E_∞ of the corresponding infinite well. If the proton makes a transition from the level with energy E_1 to the level with energy E_3 by absorbing a photon, find the wavelength of the photon.

40.20. An electron is bound in a square well with a depth equal to six times the ground-level energy E_∞ of an infinite well of the same width. The longest-wavelength photon that is absorbed by the electron has a wavelength of 400.0 nm. Determine the width of the well.

Section 40.3 Potential Barriers and Tunneling

40.21. An electron with initial kinetic energy 6.0 eV encounters a barrier with height 11.0 eV. What is the probability of tunneling if the width of the barrier is (a) 0.80 nm and (b) 0.40 nm?

40.22. An electron with initial kinetic energy 5.0 eV encounters a barrier with height U_0 and width .060 nm. What is the transmission coefficient if (a) $U_0 = 7.0$ eV; (b) $U_0 = 9.0$ eV; (c) $U_0 = 13.0$ eV?

40.23. An electron is moving past the square barrier shown in Fig. 40.12, but the energy of the electron is *greater* than the barrier height. If $E = 2U_0$, what is the ratio of the de Broglie wavelength of the electron in the region $x > L$ to the wavelength for $0 < x < L$?

40.24. A proton with initial kinetic energy 50.0 eV encounters a barrier of height 70.0 eV. What is the width of the barrier if the probability of tunneling is 3.0×10^{-3}? How does this compare with the barrier width for an electron with the same energy tunneling through a barrier of the same height with the same probability?

40.25. (a) An electron with initial kinetic energy 32 eV encounters a square barrier with height 41 eV and width 0.25 nm. What is the probability that the electron will tunnel through the barrier? (b) A proton with the same kinetic energy encounters the same barrier. What is the probability that the proton will tunnel through the barrier?

40.26. Alpha Decay. In a simple model for a radioactive nucleus, an alpha particle ($m = 6.64 \times 10^{-27}$ kg) is trapped by a square barrier that has width 2.0 fm and height 30.0 MeV. (a) What is the tunneling probability when the alpha particle encounters the barrier if its kinetic energy is 1.0 MeV below the top of the barrier (Fig. 40.25)? (b) What is the tunneling probability if the energy of the alpha particle is 10.0 MeV below the top of the barrier?

Figure **40.25** Exercise 40.26.

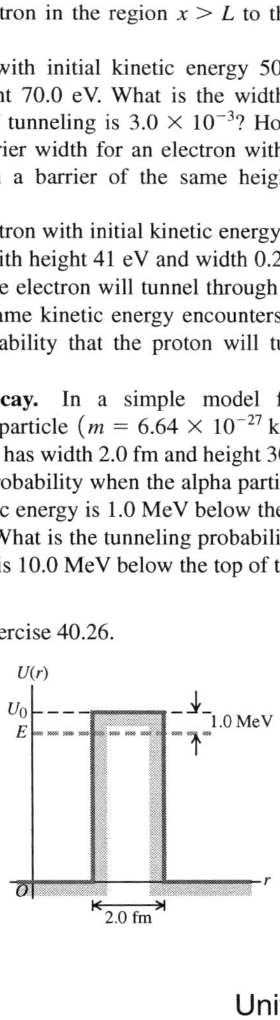

Section 40.4 The Harmonic Oscillator

40.27. A wooden block with mass 0.250 kg is oscillating on the end of a spring that has force constant 110 N/m. Calculate the ground-level energy and the energy separation between adjacent levels. Express your results in joules and in electron volts. Are quantum effects important?

40.28. Show that $\psi(x)$ given by Eq. (40.24) is a solution to Eq. (40.22) with energy $E_0 = \hbar\omega/2$.

40.29. Chemists use infrared absorption spectra to identify chemicals in a sample. In one sample, a chemist finds that light of wavelength 5.8 μm is absorbed. (a) Find the energy of this transition. (b) If the molecule has mass 5.6×10^{-26} kg, find the force constant.

40.30. The ground-state energy of a harmonic oscillator is 5.60 eV. If the oscillator undergoes a transition from its $n = 3$ to $n = 2$ level by emitting a photon, what is the wavelength of the photon?

40.31. In Section 40.4 it is shown that for the ground level of a harmonic oscillator, $\Delta x \Delta p_x = \hbar$. Do a similar analysis for an excited level that has quantum number n. How does the uncertainty product $\Delta x \Delta p_x$ depend on n?

40.32. For the ground-level harmonic oscillator wave function $\psi(x)$ given in Eq. (40.24), $|\psi|^2$ has a maximum at $x = 0$. (a) Compute the ratio of $|\psi|^2$ at $x = +A$ to $|\psi|^2$ at $x = 0$, where A is given by Eq. (40.27) with $n = 0$ for the ground level. (b) Compute the ratio of $|\psi|^2$ at $x = +2A$ to $|\psi|^2$ at $x = 0$. In each case is your result consistent with what is shown in Fig. 40.20?

40.33. For the sodium atom of Example 40.6, find (a) the ground-state energy, (b) the wavelength of a photon emitted when the $n = 4$ to $n = 3$ transition occurs; (c) the energy difference for any $\Delta n = 1$ transition.

Problems

40.34. Show that the wave function $\psi(x) = Ae^{ikx}$ is a solution of Eq. 40.1 for a particle of mass m, in a region where the potential energy is a constant $U_0 < E$. Find an expression for k, and relate it to the particle's momentum and to its de Broglie wavelength.

40.35. Wave functions like the one in Problem 40.34 can represent free particles moving with velocity $v = p/m$ in the x-direction. Consider a beam of such particles incident on a potential-energy step $U(x) = 0$, for $x < 0$, and $U(x) = U_0 < E$, for $x > 0$. The wave function for $x < 0$ is $\psi(x) = Ae^{ik_1x} + Be^{-ik_1x}$, representing incident and reflected particles, and for $x > 0$ is $\psi(x) = Ce^{ik_2x}$, representing transmitted particles. Use the conditions that both ψ and its first derivative must be continuous at $x = 0$ to find the constants B and C in terms of k_1, k_2 and A.

40.36. Let ΔE_n be the energy difference between the adjacent energy levels E_n and E_{n+1} for a particle in a box. The ratio $R_n = \Delta E_n/E_n$ compares the energy of a level to the energy separation of the next higher energy level. a) For what value of n is R_n largest, and what is this largest R_n? (b) What does R_n approach as n becomes very large? How does this result compare to the classical value for this quantity?

40.37. Photon in a Dye Laser. An electron in a long, organic molecule used in a dye laser behaves approximately like a particle in a box with width 4.18 nm. What is the wavelength of the photon emitted when the electron undergoes a transition (a) from the first excited level to the ground level and (b) from the second excited level to the first excited level?

40.38. A particle is in the ground level of a box that extends from $x = 0$ to $x = L$. (a) What is the probability of finding the particle in the region between 0 and $L/4$? Calculate this by integrating $|\psi(x)|^2\,dx$, where ψ is normalized, from $x = 0$ to $x = L/4$. (b) What is the probability of finding the particle in the region $x = L/4$ to $x = L/2$? (c) How do the results of parts (a) and (b) compare? Explain. (d) Add the probabilities calculated in parts (a) and (b). (e) Are your results in parts (a), (b), and (d) consistent with Fig. 40.5b? Explain.

40.39. What is the probability of finding a particle in a box of length L in the region between $x = L/4$ and $x = 3L/4$ when the particle is in (a) the ground level and (b) the first excited level? (*Hint:* Integrate $|\psi(x)|^2\,dx$, where ψ is normalized, between $L/4$ and $3L/4$.) (c) Are your results in parts (a) and (b) consistent with Fig. 40.5b? Explain.

40.40. Consider a particle in a box with rigid walls at $x = 0$ and $x = L$. Let the particle be in the ground level. Calculate the probability $|\psi|^2\,dx$ that the particle will be found in the interval x to $x + dx$ for (a) $x = L/4$; (b) $x = L/2$; (c) $x = 3L/4$.

40.41. Repeat Problem 40.40 for a particle in the first excited level.

40.42. A particle is confined within a box with perfectly rigid walls at $x = 0$ and $x = L$. Although the magnitude of the instantaneous force exerted on the particle by the walls is infinite and the time over which it acts is zero, the impulse (that involves a product of force and time) is both finite and quantized. Show that the impulse exerted by the wall at $x = 0$ is $(nh/L)\hat{\imath}$ and that the impulse exerted by the wall at $x = L$ is $-(nh/L)\hat{\imath}$. (*Hint:* You may wish to review Section 8.1.)

40.43. A fellow student proposes that a possible wave function for a free particle with mass m (one for which the potential-energy function $U(x)$ is zero) is

$$\psi(x) = \begin{cases} e^{+\kappa x}, & x < 0 \\ e^{-\kappa x}, & x \geq 0 \end{cases}$$

where κ is a positive constant. (a) Graph this proposed wave function. (b) Show that the proposed wave function satisfies the Schrödinger equation for $x < 0$ if the energy is $E = -\hbar^2\kappa^2/2m$—that is, if the energy of the particle is *negative*. (c) Show that the proposed wave function also satisfies the Schrödinger equation for $x \geq 0$ with the same energy as in part (b). (d) Explain why the proposed wave function is nonetheless *not* an acceptable solution of the Schrödinger equation for a free particle. (*Hint:* What is the behavior of the function at $x = 0$?) It is in fact impossible for a free particle (one for which $U(x) = 0$) to have an energy less than zero.

40.44. The penetration distance η in a finite potential well is the distance at which the wave function has decreased to $1/e$ of the wave function at the classical turning point:

$$\psi(x = L + \eta) = \frac{1}{e}\psi(L)$$

The penetration distance can be shown to be

$$\eta = \frac{\hbar}{\sqrt{2m(U_0 - E)}}$$

The probability of finding the particle beyond the penetration distance is nearly zero. (a) Find η for an electron having a kinetic energy of 13 eV in a potential well with $U_0 = 20$ ev. (b) Find η for a 20.0-MeV proton trapped in a 30.0-MeV-deep potential well.

40.45. (a) For the finite potential well of Fig. 40.6, what relationships among the constants A and B of Eq. (40.17) and C and D of Eq. (40.19) are obtained by applying the boundary condition that ψ be continuous at $x = 0$ and at $x = L$? (b) What relationships among A, B, C, and D are obtained by applying the boundary condition that $d\psi/dx$ be continuous at $x = 0$ and at $x = L$?

40.46. An electron with initial kinetic energy 5.5 eV encounters a square potential barrier with height 10.0 eV. What is the width of the barrier if the electron has a 0.10% probability of tunneling through the barrier?

40.47. A particle with mass m and total energy E tunnels through a square barrier of height U_0 and width L. When the transmission coefficient is *not* much less than unity, it is given by

$$T = \left[1 + \frac{(U_0 \sinh \kappa L)^2}{4E(U_0 - E)}\right]^{-1}$$

where $\sinh \kappa L = (e^{\kappa L} - e^{-\kappa L})/2$ is the hyperbolic sine of κL. (a) Show that if $\kappa L \gg 1$, this expression for T approaches Eq. (40.21). (b) Explain why the restriction $\kappa L \gg 1$ in part (a) implies either that the barrier is relatively wide or that the energy E is relatively low compared to U_0. (c) Show that as the particle's incident kinetic energy E approaches the barrier height U_0, T approaches $[1 + (kL/2)^2]^{-1}$, where $k = \sqrt{2mE}/\hbar$ is the wave number of the incident particle. (*Hint:* If $|z| \ll 1$, then $\sinh z \approx z$.)

40.48. A harmonic oscillator consists of a 0.020-kg mass on a spring. Its frequency is 1.50 Hz, and the mass has a speed of 0.360 m/s as it passes the equilibrium position. (a) What is the value of the quantum number n for its energy level? (b) What is the difference in energy between the levels E_n and E_{n+1}? Is this difference detectable?

40.49. For small amplitudes of oscillation the motion of a pendulum is simple harmonic. For a pendulum with a period of 0.500 s, find the ground-level energy and the energy difference between adjacent energy levels. Express your results in joules and in electron volts. Are these values detectable?

40.50. Some 164.9-nm photons are emitted in a $\Delta n = 1$ transition within a solid-state lattice. The lattice is modeled as electrons in a box having length 0.500 nm. What transition corresponds to the emitted light?

40.51. (a) Show by direct substitution in the Schrödinger equation for the one-dimensional harmonic oscillator that the wave function $\psi_0(x) = A_0 e^{-\alpha^2 x^2/2}$, where $\alpha^2 = m\omega/\hbar$, is a solution with energy corresponding to $n = 0$ in Eq. (40.22). (b) Find the normalization constant A_0. (c) Find the classical turning points and show, in contrast, that the probability density has a maximum at $x = 0$.

40.52. (a) Show by direct substitution in the Schrödinger equation for the one-dimensional harmonic oscillator that the wave function $\psi_1(x) = A_1 x e^{-\alpha^2 x^2/2}$, where $\alpha^2 = m\omega/\hbar$, is a solution with energy corresponding to $n = 1$ in Eq. (40.22). (b) Find the normalization constant A_1. (c) Show that the probability density has a minimum at $x = 0$ and maxima at $x = \pm 1/\alpha$, corresponding to the classical turning points for the ground state $n = 0$.

40.53. A Three-Dimensional Isotropic Harmonic Oscillator. An isotropic harmonic oscillator has the potential-energy function $U(x, y, z) = \frac{1}{2}k'(x^2 + y^2 + z^2)$. (*Isotropic* means that the force constant k' is the same in all three coordinate directions.) (a) Show that for this potential, a solution to Eq. (40.29) is given by $\psi = \psi_{n_x}(x)\psi_{n_y}(y)\psi_{n_z}(z)$. In this expression, $\psi_{n_x}(x)$ is a solution to the one-dimensional harmonic oscillator Schrödinger equation, Eq. (40.22), with energy $E_{n_x} = (n_x + \frac{1}{2})\hbar\omega$. The functions $\psi_{n_y}(y)$ and $\psi_{n_z}(z)$ are analogous one-dimensional wave functions for oscillations in the y- and z-directions. Find the energy associated with this ψ. (b) From your results in part (a) what are the ground-level and first-excited-level energies of the three-dimensional isotropic oscillator? (c) Show that there is only one state (one set of quantum numbers n_x, n_y, and n_z) for the ground level but three states for the first excited level.

40.54. Three-Dimensional Anisotropic Harmonic Oscillator. An oscillator has the potential-energy function $U(x, y, z) = \frac{1}{2}k_1'(x^2 + y^2) + \frac{1}{2}k_2'z^2$, where $k_1' > k_2'$. This oscillator is called *anisotropic* because the force constant is not the same in all three coordinate directions. (a) Find a general expression for the energy levels of the oscillator (see Problem 40.53). (b) From your results in part (a), what are the ground-level and first-excited-level energies of this oscillator? (c) How many states (different sets of quantum numbers n_x, n_y, and n_z) are there for the ground level and for the first excited level? Compare to part (c) of Problem 40.53.

Challenge Problems

40.55. Section 40.1 considered a box with walls at $x = 0$ and $x = L$. Now consider a box with width L but centered at $x = 0$, so that it extends from $x = -L/2$ to $x = +L/2$ (Fig. 40.26). Note that this box is symmetrical about $x = 0$. (a) Consider possible wave functions of the form $\psi(x) = A \sin kx$. Apply the boundary conditions at the wall to obtain the allowed energy levels. (b) Another set of possible wave functions are functions of the form $\psi(x) = A \cos kx$. Apply the boundary conditions at the wall to obtain the allowed energy levels. (c) Compare the energies obtained in parts (a) and (b) to the set of energies given in Eq. (40.9). (d) An odd function f satisfies the condition $f(x) = -f(-x)$, and an even function g satisfies $g(x) = g(-x)$. Of the wave functions from parts (a) and (b), which are even and which are odd?

Figure **40.26** Challenge Problem 40.55.

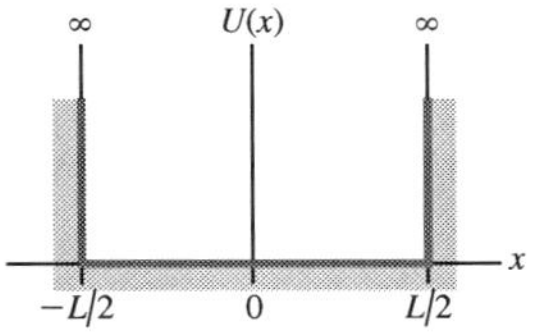

40.56. Consider a potential well defined as $U(x) = \infty$ for $x < 0$, $U(x) = 0$ for $0 < x < L$, and $U(x) = U_0 > 0$ for $x > L$ (Fig. 40.27). Consider a particle with mass m and kinetic energy $E < U_0$ that is trapped in the well. (a) The boundary condition at the infinite wall ($x = 0$) is $\psi(0) = 0$. What must the form of the function $\psi(x)$ for $0 < x < L$ be in order to satisfy both the Schrödinger equation and this boundary condition? (b) The wave function must remain finite as $x \to \infty$. What must the form of the function $\psi(x)$ for $x > L$ be in order to satisfy both the Schrödinger equation and this boundary condition at infinity? (c) Impose the boundary conditions that ψ and $d\psi/dx$ are continuous at $x = L$. Show that the energies of the allowed levels are obtained from solutions of the equation $k \cot kL = -\kappa$, where $k = \sqrt{2mE}/\hbar$ and $\kappa = \sqrt{2m(U_0 - E)}/\hbar$.

Figure **40.27** Challenge Problem 40.56.

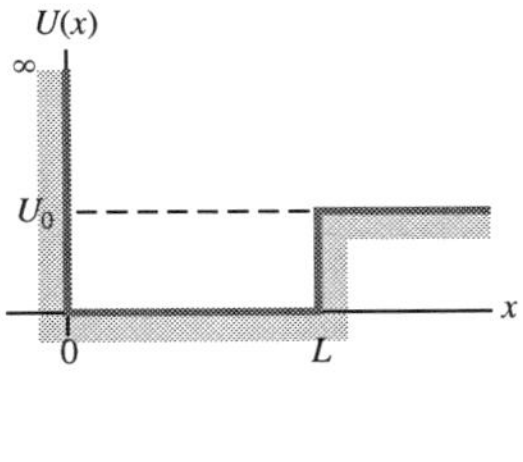

40.57. The WKB Approximation. It can be a challenge to solve the Schrödinger equation for the bound-state energy levels of an arbitrary potential well. An alternative approach that can yield good approximate results for the energy levels is the *WKB approximation* (named for the physicists Gregor Wentzel, Hendrik Kramers, and Léon Brillouin, who pioneered its application to quantum mechanics). The WKB approximation begins from three physical statements: (i) According to de Broglie, the magnitude of momentum p of a quantum-mechanical particle is $p = h/\lambda$. (ii) The magnitude of momentum is related to the kinetic energy K by the relationship $K = p^2/2m$. (iii) If there are no nonconservative forces, then in Newtonian mechanics the energy E for a particle is constant and equal at each point to the sum of the kinetic and potential energies at that point: $E = K + U(x)$, where x is the coordinate. (a) Combine these three relationships to show that the wavelength of the particle at a coordinate x can be written as

$$\lambda(x) = \frac{h}{\sqrt{2m[E - U(x)]}}$$

Thus we envision a quantum-mechanical particle in a potential well $U(x)$ as being like a free particle, but with a wavelength $\lambda(x)$ that is a function of position. (b) When the particle moves into a region of increasing potential energy, what happens to its wavelength? (c) At a point where $E = U(x)$, Newtonian mechanics says that the particle has zero kinetic energy and must be instantaneously at rest. Such a point is called a *classical turning point,* since this is where a Newtonian particle must stop its motion and reverse direction. As an example, an object oscillating in simple harmonic motion with amplitude A moves back and forth between the points $x = -A$ and $x = +A$; each of these is a classical turning point, since there the potential energy $\frac{1}{2}k'x^2$ equals the total energy $\frac{1}{2}k'A^2$. In the WKB expression for $\lambda(x)$, what is the wavelength at a classical turning point? d) For a particle in a box with length L, the walls of the box are classical turning points (see Fig. 40.1). Furthermore, the number of wavelengths that fit within the box must be a half-integer (see Fig. 40.3), so that $L = (n/2)\lambda$ and hence $L/\lambda = n/2$, where $n = 1, 2, 3, \ldots$. [Note that this is a restatement of Eq. (40.7).] The WKB scheme for finding the allowed bound-state energy levels of an *arbitrary* potential well is an extension of these observations. It demands that for an allowed energy E, there must be a half-integer number of wavelengths between the classical turning points for that energy. Since the wavelength in the WKB approximation is not a constant but depends on x, the number of wavelengths between the classical turning points a and b for a given value of the energy is the integral of $1/\lambda(x)$ between those points:

$$\int_a^b \frac{dx}{\lambda(x)} = \frac{n}{2} \qquad (n = 1, 2, 3, \ldots)$$

Using the expression for $\lambda(x)$ you found in part (a), show that the *WKB condition for an allowed bound-state energy* can be written as

$$\int_a^b \sqrt{2m[E - U(x)]}\, dx = \frac{nh}{2} \qquad (n = 1, 2, 3, \ldots)$$

(e) As a check on the expression in part (d), apply it to a particle in a box with walls at $x = 0$ and $x = L$. Evaluate the integral and show that the allowed energy levels according to the WKB approximation are the same as those given by Eq. (40.9). (*Hint:* Since the walls of the box are infinitely high, the points $x = 0$ and $x = L$ are classical turning points for *any* energy E. Inside the box, the potential energy is zero.) (f) For the finite square well shown in Fig. 40.6, show that the WKB expression given in part (d) predicts the *same* bound-state energies as for an infinite square well of the same width. (*Hint:* Assume $E < U_0$. Then the classical turning points are at $x = 0$ and $x = L$.) This shows that the WKB approximation does a poor job when the potential-energy function changes discontinuously, as for a finite potential well. In the next two problems we consider situations in which the potential-energy function changes gradually and the WKB approximation is much more useful.

40.58. The WKB approximation (see Challenge Problem 40.57) can be used to calculate the energy levels for a harmonic oscillator. In this approximation, the energy levels are the solutions to the equation

$$\int_a^b \sqrt{2m[E - U(x)]}\,dx = \frac{nh}{2} \qquad n = 1, 2, 3, \ldots$$

Here E is the energy, $U(x)$ is the potential-energy function, and $x = a$ and $x = b$ are the classical turning points (the points at which E is equal to the potential energy, so the Newtonian kinetic energy would be zero). (a) Determine the classical turning points for a harmonic oscillator with energy E and force constant k'. (b) Carry out the integral in the WKB approximation and show that the energy levels in this approximation are $E_n = \hbar\omega$, where $\omega = \sqrt{k'/m}$ and $n = 1, 2, 3, \ldots$. (*Hint:* Recall that $\hbar = h/2\pi$. A useful standard integral is

$$\int \sqrt{A^2 - x^2}\,dx = \frac{1}{2}\left[x\sqrt{A^2 - x^2} + A^2 \arcsin\left(\frac{x}{|A|}\right)\right]$$

where arcsin denotes the inverse sine function. Note that the integrand is even, so the integral from $-x$ to x is equal to twice the integral from 0 to x.) (c) How do the approximate energy levels found in part (b) compare with the true energy levels given by Eq. (40.26)? Does the WKB approximation give an underestimate or an overestimate of the energy levels?

40.59. Protons, neutrons, and many other particles are made of more fundamental particles called *quarks* and *antiquarks* (the antimatter equivalent of quarks). A quark and an antiquark can form a bound state with a variety of different energy levels, each of which corresponds to a different particle observed in the laboratory. As an example, the ψ particle is a low-energy bound state of a so-called charm quark and its antiquark, with a rest energy of 3097 MeV; the $\psi(2S)$ particle is an excited state of this same quark–antiquark combination, with a rest energy of 3686 MeV. A simplified representation of the potential energy of interaction between a quark and an antiquark is $U(x) = A|x|$ where A is a positive constant and x represents the distance between the quark and the antiquark. You can use the WKB approximation (see Challenge Problem 40.57) to determine the bound-state energy levels for this potential-energy function. In the WKB approximation, the energy levels are the solutions to the equation

$$\int_a^b \sqrt{2m[E - U(x)]}\,dx = \frac{nh}{2} \qquad (n = 1, 2, 3, \ldots)$$

Here E is the energy, $U(x)$ is the potential-energy function, and $x = a$ and $x = b$ are the classical turning points (the points at which E is equal to the potential energy, so the Newtonian kinetic energy would be zero). (a) Determine the classical turning points for the potential $U(x) = A|x|$ and for an energy E. (b) Carry out the above integral and show that the allowed energy levels in the WKB approximation are given by

$$E_n = \frac{1}{2m}\left(\frac{3mAh}{4}\right)^{2/3} n^{2/3} \qquad (n = 1, 2, 3, \ldots)$$

(*Hint:* The integrand is even, so the integral from $-x$ to x is equal to twice the integral from 0 to x.) (c) Does the difference in energy between successive levels increase, decrease, or remain the same as n increases? How does this compare to the behavior of the energy levels for the harmonic oscillator? For the particle in a box? Can you suggest a simple rule that relates the difference in energy between successive levels to the shape of the potential-energy function?

PART FOUR

PHYS 1013:

Energy and Matter

16 Temperature and Heat

■ How does this photo reveal heat loss from the house? And how can you tell that the car was recently driven?

► **To Learn**

By the end of this chapter you should be able to

- Explain the meanings of temperature and heat and the difference between them (16.1).
- Work with different temperature scales (16.1).
- Calculate the equilibrium temperature of a mixture (16.2).
- Evaluate heat loss from buildings and other systems (16.3).
- Calculate radiant-heat transfer (16.3).
- Determine a system's temperature when it's in thermal-energy balance (16.4).

◄ **To Know**

- The concept of energy is the ultimate basis for your understanding of heat and temperature (6.3, 7.1, 7.3).
- You should be comfortable expressing energies in different units (6.3).

Your own body gives you a good sense of "hot" and "cold." Questions about heat and temperature are ultimately about energy, and these concepts are crucial to an understanding of the energy flows that drive natural systems like Earth's climate and technologies such as engines, power plants, and refrigerators.

Properties like mass and kinetic energy apply equally to microscopic atoms and molecules and to cars and planets. But other properties, including temperature and pressure, apply only to macroscopic systems. It makes no sense to talk about the temperature or pressure of a single air molecule. **Thermodynamics** is the branch of physics that deals with these macroscopic properties. Ultimately, the thermodynamic behavior of matter follows from the motions of its constituent particles in response to the laws of mechanics. **Statistical mechanics** relates the macroscopic description of matter to the underlying microscopic processes. Historically, thermodynamics developed before the atomic theory of matter was fully established. The subsequent explanation of thermodynamics through statistical mechanics—the mechanics of atoms and molecules—was a triumph for physics.

16.1 Heat, Temperature, and Thermodynamic Equilibrium

Take a bottle of soda from the refrigerator, and eventually it reaches room temperature. At that point the soda and the room are in **thermodynamic equilibrium**, a state in which their macroscopic properties are no longer changing. To check for thermodynamic equilibrium we can consider any macroscopic property—length, volume, pressure, electrical resistance, whatever. If any macroscopic property changes when two systems are placed together, then they weren't originally in thermodynamic equilibrium. When changes cease, the systems have reached equilibrium.

The phrase "placed together" here has a definite meaning, stated more precisely as "placed in thermal contact." Two systems are in **thermal contact** if heating one of them results in macroscopic changes in the other. If that doesn't readily happen—for example, with a Styrofoam cup of coffee and its surroundings—then the systems are **thermally insulated**.

We're now ready to define temperature: **Two systems have the same temperature if they are in thermodynamic equilibrium**. Consider two systems *A* and *C* in thermal contact with a third system *B* but not with each other (Fig. 16.1a). Even though they're not in direct contact, *A* and *C* have the same temperature; that is, if you place *A* and *C* in thermal contact (Fig. 16.1b), no further changes occur. This fact—that two systems in equilibrium with a third system are therefore in equilibrium with each other—is so fundamental that it's called the **zeroth law of thermodynamics**.

A **thermometer** is a system with a conveniently observed macroscopic property that changes with temperature. It could be the length of a mercury column, gas pressure, electrical resistance, or the bending of a bimetal strip in a dial thermometer. Let the thermometer come to equilibrium with some system, and its temperature-dependent physical property provides a measure of temperature. The zeroth law assures consistency, in that two systems for which the thermometer gives the same reading must have the same temperature.

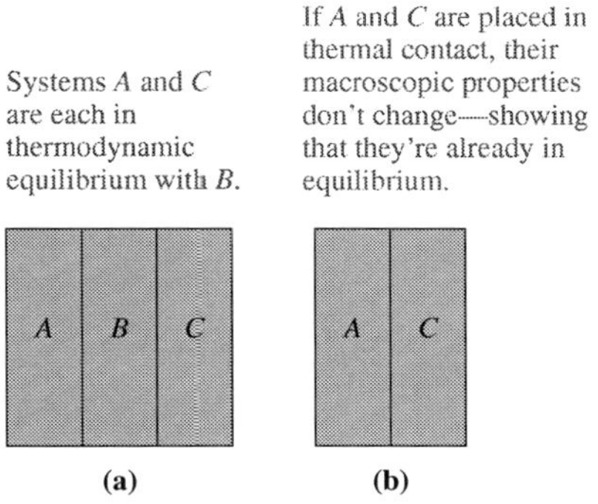

FIGURE 16.1 The zeroth law of thermodynamics.

Gas Thermometers and the Kelvin Scale

Figure 16.2 shows a **constant-volume gas thermometer**, whose temperature indication is the pressure of gas held at constant volume. We define zero temperature as that temperature at which the gas pressure would become zero. A second known point is the so-called **triple point** of water—the unique temperature at which solid, liquid, and gaseous water can coexist in equilibrium (more on this in the next chapter). The SI temperature unit, the **kelvin** (symbol K; *not* "degrees kelvin" or °K), is defined by setting the triple point at 273.16 K. Other temperatures then follow from a linear relationship as shown in Fig. 16.3.

Since a gas can't have negative pressure, the zero of the kelvin scale is an absolute lower limit on temperature and is called **absolute zero**. We'll explore the meaning of absolute zero further in Chapter 19.

Gas thermometers are useful because they work over a wide temperature range. More important, as we'll see in the next chapter, all gases behave in essentially the same way in the limit of low pressure, so gas thermometry provides a reproducible temperature standard.

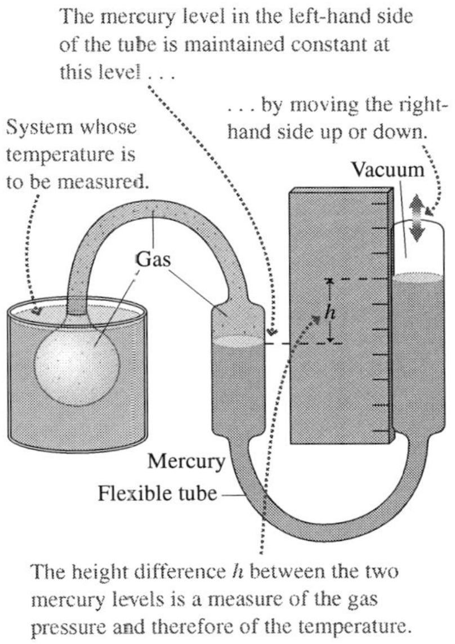

FIGURE 16.2 A constant-volume gas thermometer.

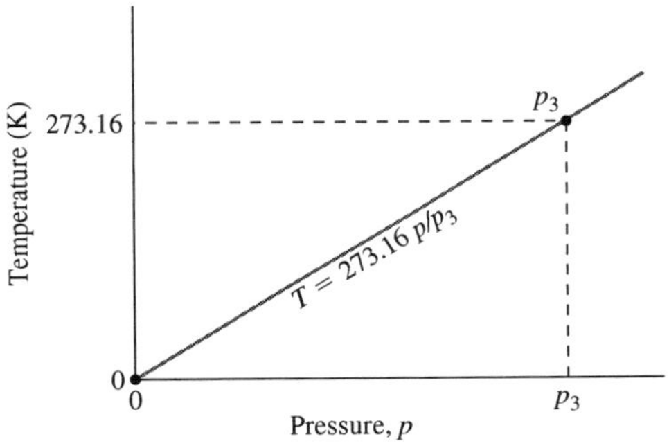

FIGURE 16.3 The Kelvin temperature scale defined using a gas thermometer; p_3 is the gas pressure at the triple point of water.

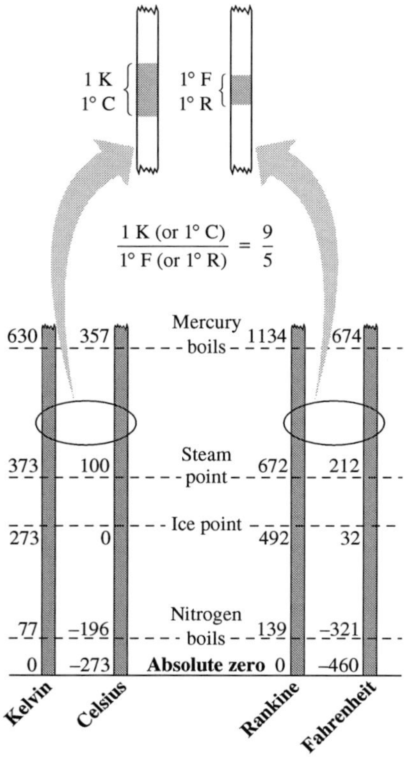

FIGURE 16.4 Relationships among four temperature scales.

Temperature Scales

Other temperature scales include Celsius (°C), Fahrenheit (°F), and Rankine (°R) (Fig. 16.4). One Celsius degree represents the same temperature difference as one kelvin, but the zero of the Celsius scale occurs at 273.15 K, so

$$T_C = T - 273.15 \tag{16.1}$$

where T is the temperature in kelvins. On the Celsius scale the melting point of ice at standard atmospheric pressure is exactly 0°C, while the boiling point is 100°C. The triple point of water occurs at 0.01°C, which accounts for the 273.15 difference between the kelvin and Celsius scales. Equation 16.1 shows that absolute zero occurs at −273.15°C.

The Fahrenheit and Rankine scales, from the British unit system, are used primarily in the United States. Fahrenheit has water melting at 32°F and boiling at 212°F, so the relation between Fahrenheit and Celsius temperatures is

$$T_F = \tfrac{9}{5}T_C + 32 \tag{16.2}$$

A Rankine degree is the same size as a Fahrenheit degree, but the zero of the Rankine scale is at absolute zero (Fig. 16.4). Engineers in the United States often use Rankine.

Heat and Temperature

A match will burn your finger, but it doesn't provide much heat. This example shows our intuitive sense of temperature and heat: Heat measures an *amount* of "something," whereas temperature is the *intensity* of that "something."

Scientists once considered heat to be a material fluid, called **caloric**, that flowed from hot bodies to colder ones. But in the late 1700s, the American-born scientist Benjamin Thompson observed essentially limitless amounts of heat being produced in the boring of cannon, and he concluded that heat could not be a conserved fluid. Instead, Thompson suggested, heat was associated with mechanical work done by the boring tool. In the next half-century, a series of experiments confirmed the association between heat and energy. These culminated in the work of the British physicist James Joule (1818–1889), who quantified the relation between heat and energy. In so doing, Joule brought thermal phenomena under the powerful conservation-of-energy principle. In recognition of this major synthesis in physics, the SI energy unit bears Joule's name.

We rarely make statements about the amount of "heat" in an object; we're more concerned that the temperature be appropriate. Rather, we think of heat as something that gets transferred from one object to another, causing a temperature change. The scientific definition reflects this sense of heat as energy in transit: **Heat is energy being transferred from one object to another because of a temperature difference alone**. Strictly speaking, **heat** refers only to energy in transit. Once heat has been transferred, we say that the **internal energy** of the object has increased, not that it contains more heat. This distinction reflects the fact that processes other than heating—such as transfer of mechanical or electrical energy—can also change an object's temperature.

16.2 Heat Capacity and Specific Heat

Experimentally, we find that the heat ΔQ transferred to an object and the resulting temperature change ΔT are directly proportional: $\Delta Q = C\,\Delta T$, where C is the **heat capacity** of the object. Since heat is a measure of energy transfer, the units of heat capacity are J/K. The heat capacity C applies to a specific object and depends on its mass and on the substance from which it's made. We characterize different substances in terms of their **specific heat** c, or heat capacity per unit mass. The heat capacity of an object is then the product of its mass and specific heat, so we can write

$$\Delta Q = mc\,\Delta T \tag{16.3}$$

The SI units of specific heat are J/kg·K. Table 16.1 lists specific heats of common materials.

TABLE 16.1 Specific Heats of Some Common Materials*

	Specific Heat, c	
Substance	**SI Units: J/kg·K**	**cal/g·°C, kcal/kg·°C, or Btu/lb·°F**
Aluminum	900	0.215
Concrete	880	0.24
Copper	386	0.0923
Iron	447	0.107
Glass	753	0.18
Mercury	140	0.033
Steel	502	0.12
Stone (granite)	840	0.20
Water:		
Liquid	4184	1.00
Ice, −10°C	2050	0.49
Wood	1400	0.33

*Temperature range 0°C to 100°C except as noted.

Scientists first studied thermodynamic phenomena before they knew the relation between heat and energy, and they used other units for heat. The **calorie** (cal) was defined as the heat needed to raise the temperature of 1 g of water from 14.5°C to 15.5°C; consequently, the specific heat of water is 1 cal/g·°C. Several different definitions of the calorie exist today, based on different methods for establishing the heat-energy equivalence. In this book we use the so-called thermochemical calorie, defined as exactly 4.184 J. The "calorie" used in describing the energy content of foods is actually a kilocalorie. In the British system, still widely used in engineering in the United States, the unit of heat is the **British thermal unit** (Btu). One Btu is the amount of heat needed to raise the temperature of 1 lb of water from 63°F to 64°F, and is equal to 1054 J.

EXAMPLE 16.1 Specific Heat: Waiting to Shower

Your whole family has showered before you, dropping the temperature in the water heater to 18°C. If the heater holds 150 kg of water, how much energy will it take to bring it up to 50°C? If the energy is supplied by a 5.0-kW electric heating element, how long will that take?

INTERPRET Here we're interested in the energy it takes to raise the water temperature, so we interpret this problem as involving specific heat. For the second part, we're given the heater's power output and asked for the time, so we need to recall (Chapter 6) that power is energy per time.

DEVELOP Equation 16.3, $\Delta Q = mc\,\Delta T$, relates energy and temperature change via specific heat, so our plan is to calculate the required energy from this equation. We'll then use the relation between power and energy to find the time.

EVALUATE Equation 16.3 gives the energy:

$$\Delta Q = mc\,\Delta T = (150\text{ kg})(4184\text{ J/kg·K})(50°\text{C} - 18°\text{C}) = 20\text{ MJ}$$

where we found the specific heat of water in Table 16.1. The heating element supplies energy at the rate of 5.0 kW or 5.0×10^3 J/s. At that rate the time needed to supply 20 MJ is

$$\Delta t = \frac{2.0 \times 10^7\text{ J}}{5.0 \times 10^3\text{ J/s}} = 4000\text{ s}$$

or a little over an hour.

ASSESS That's a long time to wait, but it's not an unreasonable answer!

✓TIP Is That C or K?

It doesn't matter when we're talking about temperature *differences*. That's why we could mix units, multiplying the specific heat in J/kg·K by the difference of Celsius temperatures.

Heat capacity and specific heat vary slightly with temperature, and they also depend on whether an object's pressure or its volume changes as it's heated. For solids and liquids, which don't expand much, that distinction is not very important. But it makes a big difference whether a gas is confined or allowed to expand when heated. Consequently, gases have two different specific heats, depending on whether volume or pressure is constant. We'll deal with that issue in Chapter 18, where we explore the thermodynamic behavior of gases.

The Equilibrium Temperature

When objects at different temperatures are in thermal contact, heat flows from the hotter object to the cooler one until they reach thermodynamic equilibrium. If the objects are thermally insulated from their surroundings, then all the energy leaving the hotter object ends up in the cooler one. Mathematically, this statement reads

$$m_1 c_1 \, \Delta T_1 + m_2 c_2 \, \Delta T_2 = 0 \tag{16.4}$$

For the hotter object, ΔT is negative, so the two terms in Equation 16.4 have opposite signs. One term represents the outflow of heat from the hotter object, the other inflow into the cooler one.

GOT IT? 16.1 A hot rock with mass 250 g is dropped into an equal mass of cool water. Which temperature changes more, that of the rock or the water? Explain.

EXAMPLE 16.2 Finding the Equilibrium Temperature: Cooling Down

An aluminum frying pan of mass 1.5 kg is at 180°C, when it's plunged into a sink containing 8.0 kg of water at 20°C. Assuming that none of the water boils and that no heat is lost to the surroundings, find the equilibrium temperature of the water and pan.

INTERPRET Here we have two objects, initially at different temperatures, that come to thermal equilibrium. So this is a problem about the equilibrium temperature, with the system of interest comprising the pan and the water.

DEVELOP Equation 16.4, $m_1 c_1 \, \Delta T_1 + m_2 c_2 \, \Delta T_2 = 0$, applies. However, we're asked for the common equilibrium temperature T, so we write the temperature differences ΔT in terms of T and the initial temperatures T_p and T_w of pan and water. Equation 16.4 then becomes $m_p c_p (T - T_p) + m_w c_w (T - T_w) = 0$.

EVALUATE We now solve for the equilibrium temperature T:

$$T = \frac{m_p c_p T_p + m_w c_w T_w}{m_p c_p + m_w c_w}$$

Using the given values of m_p, T_p, m_w, and T_w, and taking c_p and c_w from Table 16.1, we find that $T = 26$°C.

ASSESS The water has much greater mass and higher specific heat, so it makes sense that its 6°C temperature change is a lot less than the 154°C drop in the pan's temperature. ■

16.3 Heat Transfer

How is heat transferred? Engineers need to know so they can design heating and cooling systems. Scientists need to know so they can anticipate temperature changes, as in global warming. Here we'll consider three common heat-transfer mechanisms: conduction, convection, and radiation. In some situations, a single mechanism dominates; in other cases, we may need to take all three into account.

Conduction

Conduction is heat transfer through direct physical contact. It occurs as molecules in a hotter region collide with and transfer energy to those in an adjacent cooler region. **Thermal conductivity** (symbol k; SI unit W/m·K) characterizes this process. Common materials exhibit a broad range of thermal conductivities, from about 400 W/m·K for copper—a good conductor—to 0.029 W/m·K for Styrofoam, a good thermal insulator. Table 16.2 lists some thermal conductivities; they're given in both SI and British units

TABLE 16.2 Thermal Conductivities*

Material	Thermal Conductivity, k SI Units: W/m·K	British Units: Btu·in./h·ft²·°F
Air	0.026	0.18
Aluminum	237	1644
Concrete (varies with mix)	1	7
Copper	401	2780
Fiberglass	0.042	0.29
Glass	0.7–0.9	5–6
Goose down	0.043	0.30
Helium	0.14	0.97
Iron	80.4	558
Steel	46	319
Styrofoam	0.029	0.20
Water	0.61	4.2
Wood (pine)	0.11	0.78

*Temperature range 0°C to 100°C.

because the latter are widely used in heat-loss calculations for buildings. The k values in Table 16.2 reflect physical properties of the materials. Metals, for example, are good thermal conductors because they contain free electrons that move quickly. Insulators like fiberglass and Styrofoam owe their insulating properties to a physical structure that traps small volumes of air or other gas.

Figure 16.5 shows a slab of thickness Δx and area A. One side is at temperature T and the other at $T + \Delta T$. The temperature difference ΔT drives a conductive heat flow through the slab; not surprisingly, we find that the heat flow is proportional to the temperature difference, the slab area, and the thermal conductivity k. The thicker the slab, on the other hand, the more resistance to heat flow, so the flow depends inversely on thickness. Therefore

$$H = -kA\frac{\Delta T}{\Delta x} \quad \text{(conductive heat flow)} \tag{16.5}$$

where $H = -dQ/dt$ is the rate of heat flow in watts, and where the minus sign shows that the flow is opposite the direction of increasing temperature—that is, from hotter to cooler.

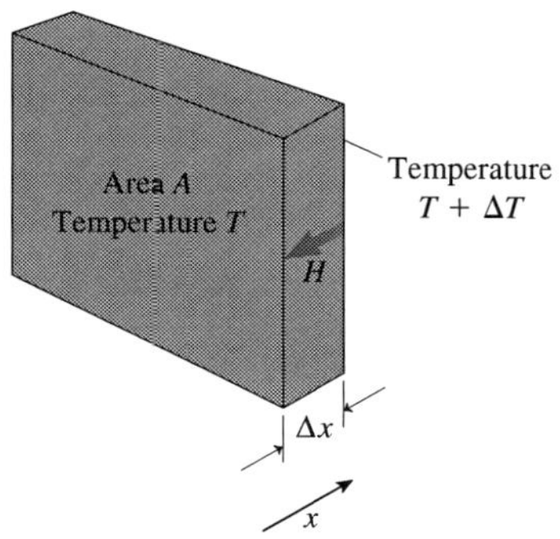

FIGURE 16.5 Heat flows from the hotter to the cooler face of the slab.

EXAMPLE 16.3 Conduction: Warming a Lake

A lake with a flat bottom and steep sides has surface area 1.5 km² and is 8.0 m deep. On a summer day, the surface water is at a temperature of 30°C and the bottom water is at 4.0°C. What is the rate of heat conduction through the lake? Assume that the temperature declines uniformly from surface to bottom.

INTERPRET This is a problem about heat conduction.

DEVELOP Our sketch, Fig. 16.6, shows that we can treat the lake like the slab shown in Fig. 16.5, provided we neglect heat flow out the sides of the lake. Then Equation 16.5, $H = -kA(\Delta T/\Delta x)$, will give the heat-flow rate.

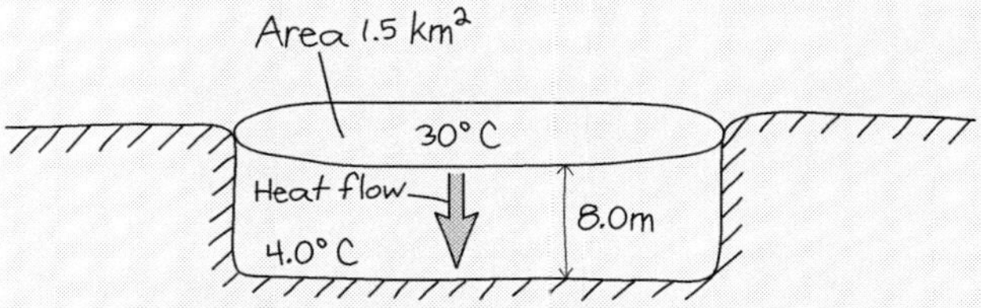

FIGURE 16.6 Our sketch for Example 16.3.

EVALUATE Substituting numerical values, including water's thermal conductivity from Table 16.2, we get

$$H = -kA\frac{\Delta T}{\Delta x}$$
$$= -(0.61\ \text{W/m}\cdot\text{K})(1.5 \times 10^6\ \text{m}^2)\frac{30°\text{C} - 4.0°\text{C}}{8.0\ \text{m}} = -3.0\ \text{MW}$$

ASSESS This is a significant flow of energy, but with direct sunlight averaging about 1 kW on every square meter, the lake's 1.5-km² surface area absorbs plenty of solar energy, and that's what maintains the temperature difference that drives the conductive heat flow. Figure 16.5 shows x increasing in the direction of increasing temperature, so negative sign in our answer indicates that the flow is downward. ■

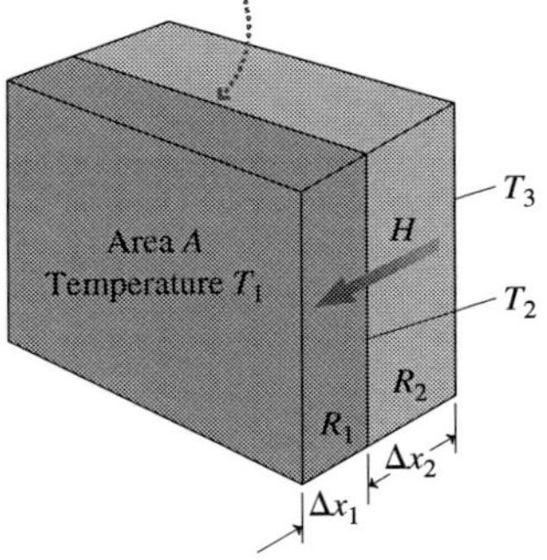

FIGURE 16.7 A composite slab.

Equation 16.5 is strictly correct only when the temperature varies uniformly from one surface to the other. That's the case when two surfaces at different temperatures have the same area. With other geometries—as in the insulation surrounding a cylindrical pipe—we need to write $\Delta T/\Delta x$ as the derivative dT/dx and integrate to find the heat flow. Problems 72 and 78 explore this situation.

Often heat flows through several different materials. A building wall, for example, may contain wood, plaster, and fiberglass insulation. Figure 16.7 shows such a composite structure, with temperature T_1 on one side and T_3 on the other. The heat-flow rate H must be the same through both slabs so energy doesn't accumulate at the interface between the two. Then Equation 16.5 gives

$$H = -k_1 A \frac{T_2 - T_1}{\Delta x_1} = -k_2 A \frac{T_3 - T_2}{\Delta x_2}$$

where k_1 and k_2 are the thermal conductivities of the two materials, and T_2 is the temperature at the interface. We can express the heat-flow rate in terms of the surface temperatures T_1 and T_3 alone if we define the **thermal resistance** R of each slab:

$$R = \frac{\Delta x}{kA} \tag{16.6}$$

The SI units of R are K/W. Unlike the thermal conductivity k, which is a property of a *material*, R is a property of a *particular piece* of material, reflecting both its conductivity and its geometry. In terms of thermal resistance, our heat-flow equation becomes

$$H = -\frac{T_2 - T_1}{R_1} = -\frac{T_3 - T_2}{R_2}$$

so $R_1 H = T_1 - T_2$ and $R_2 H = T_2 - T_3$. Adding these two equations gives

$$(R_1 + R_2)H = T_1 - T_2 + T_2 - T_3 = T_1 - T_3$$

or

$$H = \frac{T_1 - T_3}{R_1 + R_2} \tag{16.7}$$

Equation 16.7 shows that the composite slab acts like a single slab whose thermal resistance is the sum of the resistances of the two slabs that compose it. We could easily extend this treatment to show that the thermal resistances of three or more slabs add when the slabs are arranged so the same heat flows through all of them.

GOT IT? 16.2 The figure shows three slabs with the same thickness but different thermal conductivities: k, $3k$, and $2k$; the left side is hotter, as shown. Rank order the three temperature differences ΔT.

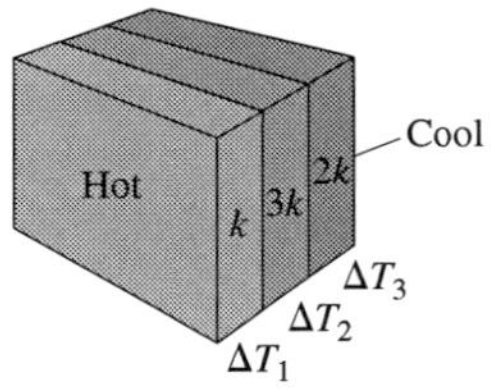

Insulating properties of building materials are described by the **$\mathcal{R}$-factor**, which is the thermal resistance for a slab of unit area:

$$\mathcal{R} = RA = \frac{\Delta x}{k} \quad (16.8)$$

The SI units of $\mathcal{R}$ are $m^2 \cdot K/W$, and that's how you'll find it listed if you buy insulation in Europe or other SI-based regions. In the United States, $\mathcal{R}$ is in $ft^2 \cdot °F \cdot h/Btu$, although the units are almost never stated. This means that $\mathcal{R}$-19 fiberglass insulation loses $\frac{1}{19}$ Btu per hour for each square foot of insulation for each degree Fahrenheit temperature difference across the insulation (Fig. 16.8).

FIGURE 16.8 Each square foot of this $\mathcal{R}$-19 fiberglass insulation loses $\frac{1}{19}$ Btu per hour for every °F of temperature difference ΔT.

EXAMPLE 16.4 Calculating Heat Loss: The Cost of Oil

Figure 16.9 shows a house whose walls consist of plaster ($\mathcal{R} = 0.17$), $\mathcal{R}$-11 fiberglass insulation, plywood ($\mathcal{R} = 0.65$), and cedar shingles ($\mathcal{R} = 0.55$). The roof has the same construction except it uses $\mathcal{R}$-30 fiberglass insulation. The average outdoor temperature in winter is 20°F, and the house is maintained at 70°F. The house's oil furnace produces 100,000 Btu for every gallon of oil, and oil costs \$2.20 per gallon. How much does it cost to heat the house for a month?

INTERPRET Although the problem asks for the monthly cost of oil, this isn't economics! We interpret this as a problem about heat loss and identify the walls and roof as systems for which we need to know the heat flow. This is a rare case of a problem stated in English units.

DEVELOP We're given the drawing in Fig. 16.9. We have the $\mathcal{R}$-factors; in English units, their inverses give the heat-loss rate on a square-foot basis. So our plan is to find the square footage of the walls and roof separately, calculate the total heat-loss rate, and then find the amount and cost of oil to compensate for a month's heat loss.

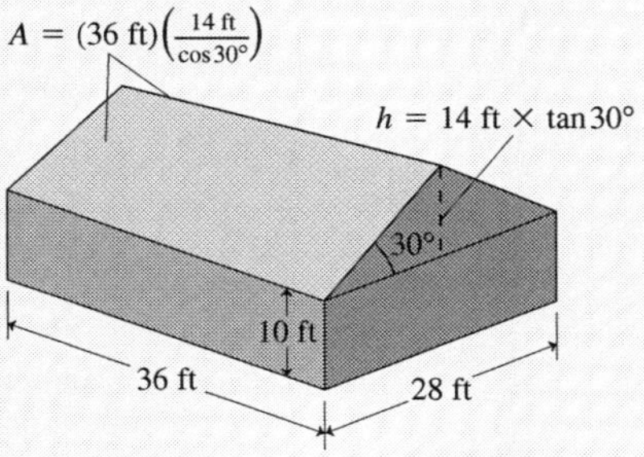

FIGURE 16.9 House for Example 16.4.

EVALUATE The $\mathcal{R}$-factors for the wall materials sum to give $\mathcal{R}_{wall} = 12.4$; similarly, $\mathcal{R}_{roof} = 31.4$. The perimeter of the house measures $2 \times 28\text{ ft} + 2 \times 36\text{ ft} = 128\text{ ft}$, so the 10-ft vertical walls have area 1280 ft². There are also the triangular gables. Since there are two of them, each with area $\frac{1}{2}bh$, they give another bh or $(28\text{ ft})(14\text{ ft}\tan 30°) = 226\text{ ft}^2$, so $A_{wall} = 1506\text{ ft}^2$. These $\mathcal{R}$-12.4 walls lose 1/12.4 Btu/h/ft²/°F. With 1506 ft² and a temperature difference of 50°F, the total heat-loss rate through the walls is

$$H_{wall} = \left(\tfrac{1}{12.4}\text{Btu/h/ft}^2\text{/°F}\right)(1506\text{ ft}^2)(50\text{°F}) = 6073\text{ Btu/h}$$

The area of the pitched roof is larger than that of a flat roof by the factor 1/cos 30°, so the heat-loss rate through the roof is

$$H_{roof} = \left(\tfrac{1}{31.4}\text{Btu/h/ft}^2\text{/°F}\right)\frac{(36\text{ ft})(28\text{ ft})}{\cos 30°}(50\text{°F}) = 1853\text{ Btu/h}$$

The total heat-loss rate is then 7926 Btu/h. In a month, this results in a heat loss of $Q = (7926\text{ Btu/h})(30\text{ days/month})(24\text{ h/day}) = 5.7\text{ MBtu}$.

Now for the oil: With 10^5 Btu (0.1 MBtu) per gallon, we'll burn 57 gallons per month to produce that 5.7 MBtu. At \$2.20/gal, that will cost \$126.

ASSESS If you've paid for heat in a northern climate, you know that this figure is, if anything, low. That's because we neglected heat losses through windows, doors, and the floor, as well as cold-air infiltration. On the other hand, we also left out any solar energy gained through the windows on sunny days. Problem 69 provides a more realistic look at this house. ■

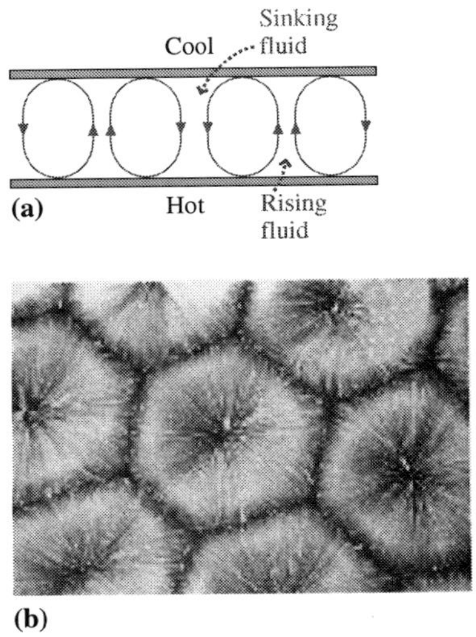

FIGURE 16.10 (a) Convection between two plates at different temperatures. (b) Top view of convection cells in a laboratory experiment. Fluid rises at the centers and sinks at the edges of the hexagonal cells.

Convection

Convection is heat transfer by fluid motion. It occurs as heated fluid becomes less dense and therefore rises. Figure 16.10a shows two plates at different temperatures, with fluid between them. Fluid heated by the lower plate rises and transfers heat to the upper plate. The cooled fluid sinks, and the process repeats. The pattern of rising and sinking fluid often acquires a striking regularity, as shown in Fig. 16.10b.

Convection is important in many technological and natural environments. When you heat water on a stove, convection carries heat through the water. Houses usually rely on convection from heat sources near floor level to circulate warm air throughout a room. Insulating materials trap air and thereby inhibit convection that would otherwise cause excessive heat loss. Convection associated with solar heating of Earth's surface drives the vast air movements that establish our overall climate. Violent convection, as in thunderstorms, is associated with localized temperature differences. On a much longer time scale, convection in Earth's mantle drives continental drift. Convection plays a crucial role in many astrophysical processes, including the generation of magnetic fields in stars and planets.

As with conduction, the convective heat-loss rate often is approximately proportional to the temperature difference. But the calculation of convective heat loss is complicated because of the associated fluid motion. The study of convection processes is an important research area in many fields of contemporary science and engineering.

Radiation

Turn a stove burner to "high" and it glows brightly; turn it to "low" and you can still sense its heat although it doesn't glow visibly. Either way, the burner loses energy by emitting electromagnetic waves, or **radiation**. The radiated power P increases rapidly with temperature, as described by the **Stefan-Boltzmann law**:

$$P = e\sigma AT^4 \quad \left(\begin{array}{c}\text{Stefan-Boltzmann law;}\\ \text{radiated power}\end{array}\right) \tag{16.9}$$

where A is the area of the emitting surface, T the temperature in kelvins, and σ the **Stefan-Boltzmann constant**, approximately $5.67 \times 10^{-8}\ \text{W/m}^2\cdot\text{K}^4$. The quantity e is the **emissivity**, a number from 0 to 1 that measures the material's effectiveness in emitting radiation. For radiation of a given wavelength, a material is equally good at emitting and absorbing radiation. A perfect emitter has $e = 1$ and is also a perfect absorber. Such an object would appear black at room temperature and is therefore called a **blackbody**. A shiny object, in contrast, reflects most of the radiation that hits it and is therefore also a poor emitter. Wood stoves are usually painted black to increase their emissivity; Thermos bottles, on the other hand, have a shiny coating to reduce radiation.

Because of the strong T^4 dependence on temperature, radiation tends to be less important at low temperatures but dominant at high temperatures. And it's not just the amount of radiation that changes with temperature; as our stove burner example suggests, it's also the wavelength. Objects near room temperature, for example, emit mostly invisible infrared radiation; very hot objects like the Sun or a lightbulb filament emit more visible light. We'll take a quantitative look at this relation in Chapter 34.

GOT IT? 16.3 Name the dominant form of heat transfer from (a) a red-hot stove burner with nothing on it, (b) a burner in direct contact with a pan of water; and (c) the bottom to the top of the water in the pan once it's begun to boil.

EXAMPLE 16.5 Calculating Radiation: The Sun's Temperature

The Sun radiates energy at the rate $P = 3.9 \times 10^{26}$ W, and its radius is 7.0×10^8 m. Treating the Sun as a blackbody $(e = 1)$, find its surface temperature.

INTERPRET This is a problem about the radiation from a hot object.

DEVELOP The Stefan-Boltzmann law, Equation 16.9, gives the radiated power in terms of the temperature, emissivity, and surface area: $P = e\sigma AT^4$. Our plan is to solve this equation for T. For the Sun, radiation comes from the entire spherical surface of area $4\pi R^2$, as our sketch shows (Fig. 16.11).

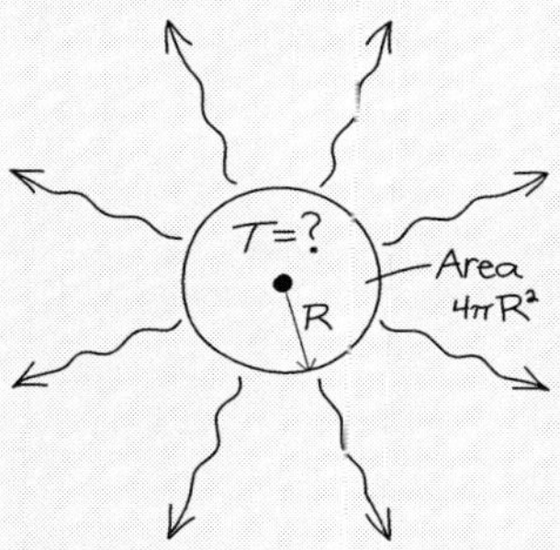

FIGURE 16.11 The Sun radiates from its spherical surface area $4\pi R^2$.

EVALUATE Using the Sun's spherical surface area and solving for T give

$$T = \left(\frac{P}{4\pi R^2 \sigma}\right)^{1/4}$$

$$= \left[\frac{3.9 \times 10^{26}\ \text{W}}{4\pi(7.0 \times 10^8\ \text{m})^2(5.7 \times 10^{-8}\ \text{W/m}^2 \cdot \text{K}^4)}\right]^{1/4} = 5.8 \times 10^3\ \text{K}$$

ASSESS Make sense? Yes: Our answer has the unit of temperature and agrees with observational measurements. ■

16.4 Thermal-Energy Balance

You keep your house at a comfortable temperature in winter by balancing heat loss with energy from your heating system (Fig. 16.12). This state of **thermal-energy balance** occurs throughout science and engineering. Understanding thermal-energy balance enables engineers to specify a building's heat sources, and helps scientists predict Earth's future climate.

Engineered systems actively control the thermal-energy balance to achieve a desired temperature. But even without active control, systems with a fixed rate of energy input naturally tend toward energy balance. That's because all heat-loss mechanisms give increased loss with increasing temperature. If the rate of energy input to a system is greater than the loss rate, then the system gains energy and its temperature increases—and so, therefore, does the loss rate. Eventually the two come to balance at some fixed temperature. If the loss exceeds the gain, the system cools until again it's in balance. Problems involving thermal-energy balance are similar regardless of the energy-loss mechanism or whether the application is to a technological or a natural system.

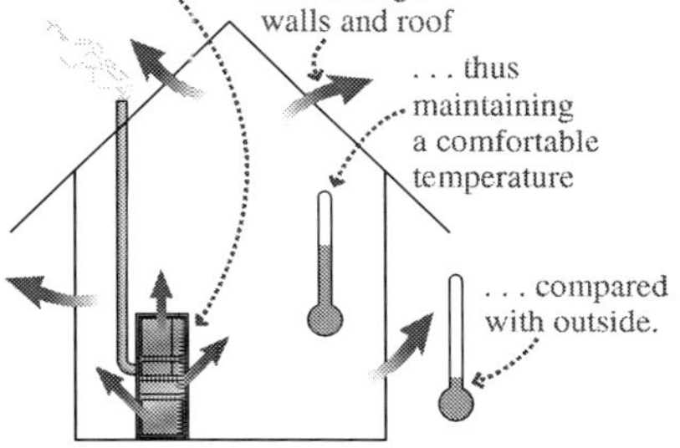

FIGURE 16.12 A house in thermal-energy balance.

PROBLEM SOLVING STRATEGY 16.1 Thermal-Energy Balance

INTERPRET Interpret the problem to be sure it deals with heat gains and losses. Identify the system of interest, the source(s) of energy input to the system, and the significant heat-loss mechanism(s).

DEVELOP Determine which equation(s) govern the heat loss; these will necessarily involve the system's temperature. Your plan is then to equate the rate of energy loss with the rate of energy input.

EVALUATE Write an equation that expresses equality of energy loss and input. Then evaluate by solving for the quantity the problem asks for—often the system's temperature.

ASSESS If your answer is a temperature, does it seem reasonable? Is the temperature of a heated system higher than that of its surroundings?

EXAMPLE 16.6 Thermal-Energy Balance: Hot Water

A poorly insulated electric water heater loses heat by conduction at the rate of 120 W for each Celsius degree difference between the water and its surroundings. It's heated by a 2.5-kW electric heating element and is located in a basement kept at 15°C. What's the water temperature if the heating element operates continuously?

INTERPRET The concept here is energy balance, and we identify the system of interest as the water. Its energy input comes from the heating element at the rate of 2.5 kW. The heat loss is by conduction.

DEVELOP Figure 16.13 is a sketch suggesting energy balance in the heater. We're given the conductive heat loss of 120 W/°C, meaning that the total heat-loss rate is $H = (120\ \text{W/°C})(\Delta T)$. We then equate the heat-loss rate to the energy-input rate: $(120\ \text{W/°C})(\Delta T) = 2.5\ \text{kW}$.

EVALUATE Solving for (ΔT) gives

$$\Delta T = \frac{2.5\ \text{kW}}{120\ \text{W/°C}} = 21°\text{C}$$

With the basement at 15°C, the water temperature is then 36°C.

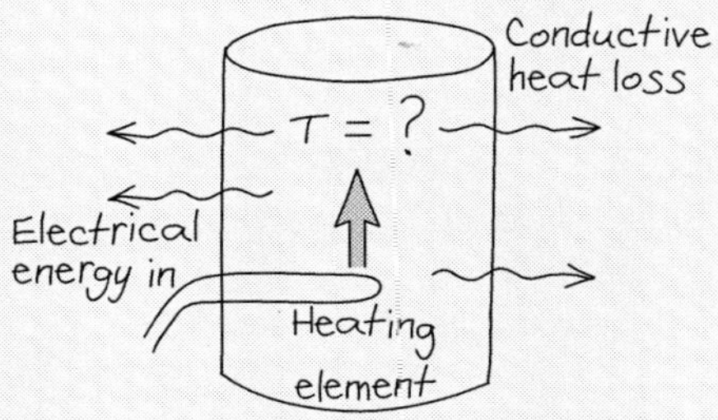

FIGURE 16.13 Balance between the heat supplied by the electric element and the conductive loss determines the water temperature.

ASSESS Is this answer reasonable? Not if you want a hot shower; our answer is 1°C below body temperature! But we're told the insulation is bad, so it's time for a new water heater! ■

EXAMPLE 16.7 Thermal-Energy Balance: A Solar Greenhouse

A solar greenhouse has 300 ft² of opaque $\mathcal{R}$-30 walls, and 250 ft² of $\mathcal{R}$-1.8 double-pane glass that admits solar energy at the average rate of 40 Btu/h/ft². Find the greenhouse temperature on a day when the outdoor temperature is 15°F.

INTERPRET Again the concept is energy balance, now with the greenhouse as the system of interest. We're given $\mathcal{R}$-factors, suggesting that the energy loss is by conduction through walls and glazing. The energy input is sunlight.

DEVELOP As we saw in Example 16.4, the $\mathcal{R}$-factor determines a heat-loss rate that is related directly to area and temperature difference and inversely to the $\mathcal{R}$-factor. So we have

$$H_w = \frac{A_w\,\Delta T}{\mathcal{R}_w} = \left(\frac{300}{30}\right)\Delta T = (10\ \text{Btu/h/°F})\ \Delta T$$

for the heat loss through the walls and

$$H_g = \frac{A_g\,\Delta T}{\mathcal{R}_g} = \left(\frac{250}{1.8}\right)\Delta T = (139\ \text{Btu/h/°F})\ \Delta T$$

for the heat loss through the glass, giving a total heat loss $H = (149\ \text{Btu/h/°F})\ \Delta T$. Meanwhile, the energy input through the entire 250-ft² of glass is $(40\ \text{Btu/h/ft}^2)(250\ \text{ft}^2) = 1.0 \times 10^4\ \text{Btu/h}$. Our plan is to equate energy input and loss and then solve for ΔT.

EVALUATE Equating loss and gain gives

$$(149\ \text{Btu/h/°F})\ \Delta T = 1.0 \times 10^4\ \text{Btu/h}.$$

We then solve for ΔT:

$$\Delta T = \frac{1.0 \times 10^4\ \text{Btu/h}}{149\ \text{Btu/h/°F}} = 67°\text{F}$$

So when it's 15°F outside, the greenhouse is at a tropical 82°F.

ASSESS This seems a reasonable greenhouse temperature. Our calculation assumes that solar input remains constant; in a real greenhouse the temperature would fluctuate as the Sun's angle changes and clouds pass over. We could minimize these fluctuations by giving the greenhouse a large heat capacity, perhaps by incorporating a massive concrete slab or concrete walls. ■

APPLICATION The Greenhouse Effect and Global Warming

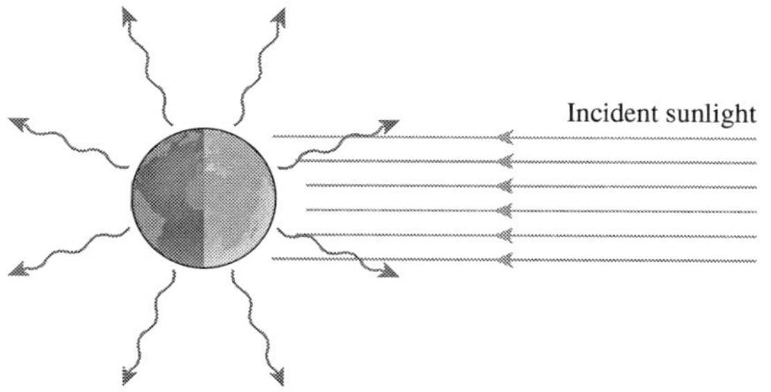

Earth's energy balance.

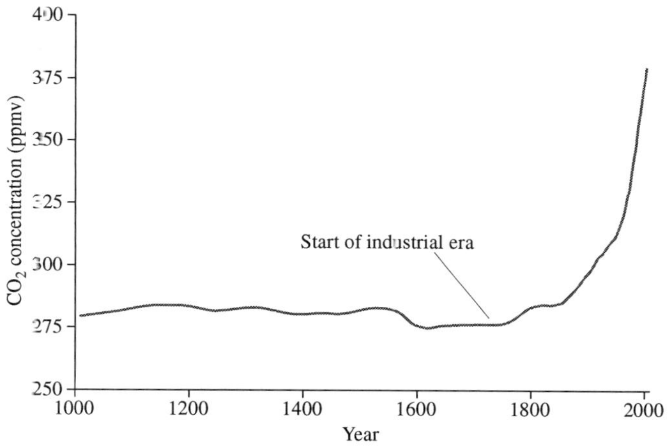

Increase in atmospheric CO_2 over the past 1000 years, in parts per million by volume (ppmv).

Earth absorbs energy from the Sun at a rate $S = 960\ \text{W/m}^2$ averaged over the planet's cross-sectional area πR_E^2. It therefore warms and, in thermal-energy balance, radiates energy at the same rate. Earth is much cooler than the Sun, so this outgoing radiation is invisible infrared; furthermore, it's radiated from the planet's entire surface area, $4\pi R_E^2$. Assuming emissivity $e = 1$ in the infrared, energy balance using Equation 16.9 gives $\pi R_E^2 S = \sigma 4\pi R_E^2 T^4$. Solving yields $T = 255\ \text{K} = -18°\text{C}$ or 0°F. Is this reasonable? It's certainly in the right ballpark—not so hot as to boil the oceans or freeze the atmosphere. But 0°F seems a bit cold for a global average temperature. And it is: Earth's average temperature is around 15°C or 59°F. Why the discrepancy?

The answer lies with Earth's atmosphere. The dominant atmospheric gases, nitrogen and oxygen, are largely transparent to both incoming sunlight and outgoing infrared. But others—the so-called **greenhouse gases**, especially water vapor and carbon dioxide—let sunlight pass through but impede outgoing infrared. As a result, Earth's surface temperature has to be higher to get the same total radiation to space. This is the **natural greenhouse effect**, and it explains the 33°C temperature difference between our naïve calculation and Earth's actual surface temperature. Neighbor planets confirm this reasoning. Mars, with very little atmosphere, exhibits almost no greenhouse warming. Venus, whose atmosphere is 100 times denser than Earth's and largely CO_2, has a "runaway" greenhouse effect that keeps its surface hotter than an oven.

As the CO_2 graph shows, we humans have increased atmospheric carbon dioxide some 36% since the start of the industrial era, to levels the planet has not seen for millions of years. Combustion of fossil fuels is the dominant source of this CO_2, although processes like deforestation also contribute, as do other greenhouse gases such as methane. Basic physics then dictates that Earth's surface temperature should rise. How much and how fast depend on complex interactions among atmosphere, surface, oceans, and life, and on future greenhouse emissions. Nevertheless, a consensus among climate scientists suggests that Earth warmed by some 0.6°C during the 20th century, mostly attributable to human activities (see the temperature graph). Further warming in the range of 1.5°C–6°C is projected by 2100. Although this may seem modest, the rate of increase is far greater than most natural climate change. And the increase is expected to be greater over land and at high latitudes. Even a few degrees' increase in the global average will result in significant climate change and a rise in sea level.

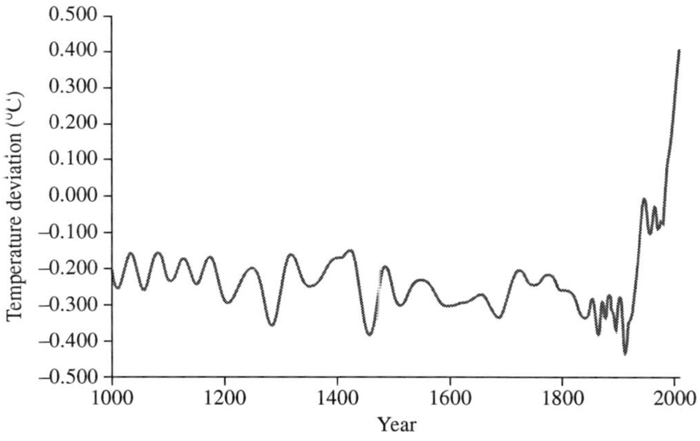

Global temperature over the past 1000 years, given as deviations from the average temperature for 1961–1990. Data before 1856 are a reconstruction based on tree rings and other proxies for temperature; data from 1856 on are from thermometer records.

CHAPTER 16 SUMMARY

Big Picture

The big ideas here are **temperature** and **heat**. **Temperature** is a property common to systems in **thermodynamic equilibrium**. Temperature is quantified in SI units using the **kelvin scale**, defined in terms of gas-based thermometers.

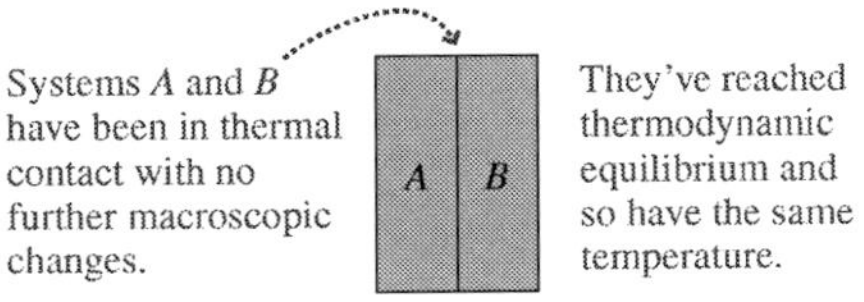

Heat is energy in transit as a result of a temperature difference.

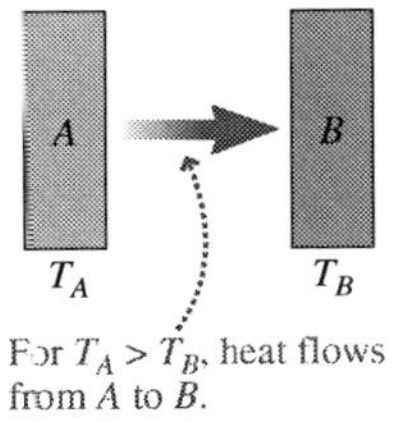

Key Concepts and Equations

Heat capacity and **specific heat** quantify the energy ΔQ required to raise an object's temperature by ΔT:

$$\Delta Q = mc\,\Delta T$$

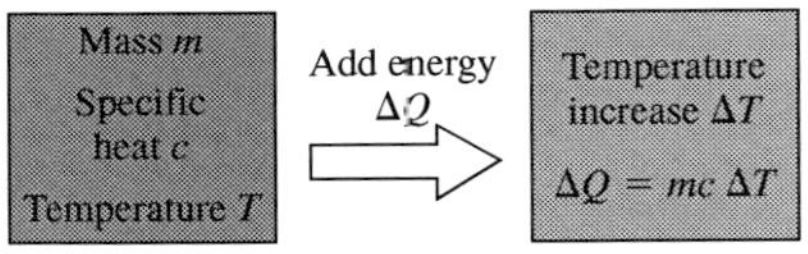

Three important heat-transfer mechanisms are:

Conduction

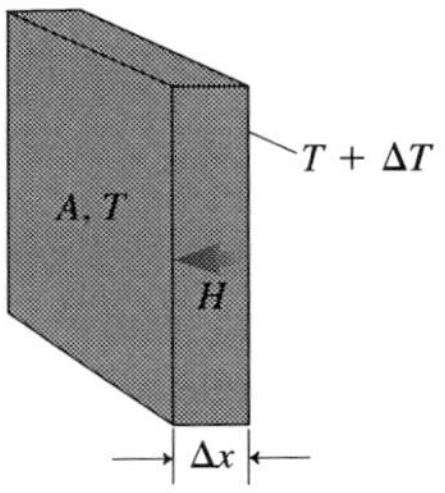

$$H = -kA\frac{\Delta T}{\Delta x} \quad \text{(conductive heat flow)}$$

Convection

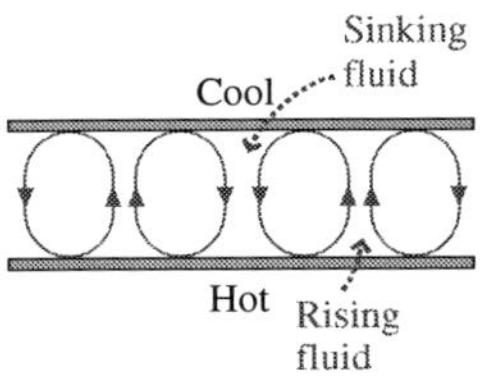

Radiation

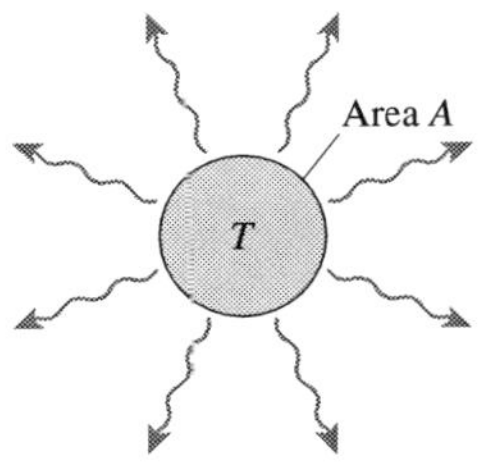

$$P = e\sigma AT^4 \quad \left(\begin{array}{c}\text{Stefan-Boltzmann law;}\\ \text{radiated power}\end{array}\right)$$

Cases and Uses

Temperature scales include Kelvin (K), Celsius (°C), Fahrenheit (°F), and Rankine (°R).

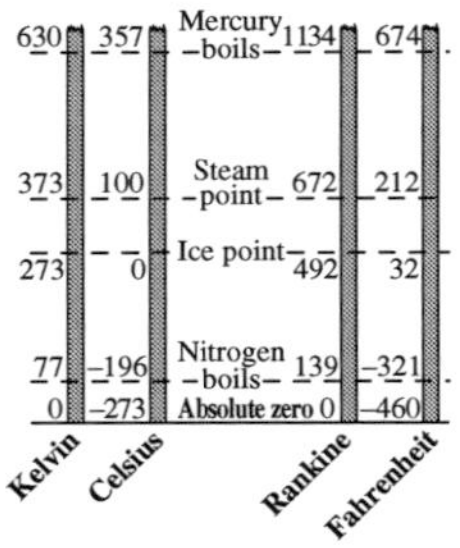

The Kelvin and Celsius scales are related by $T_C = T - 273.15$. The relation between Fahrenheit and Celsius scales is $T_F = \frac{9}{5}T_C + 32$.

Equilibrium temperature: Combining two systems at different temperatures results in a common equilibrium temperature given by $m_1c_1\,\Delta T_1 + m_2c_2\,\Delta T_2 = 0$.

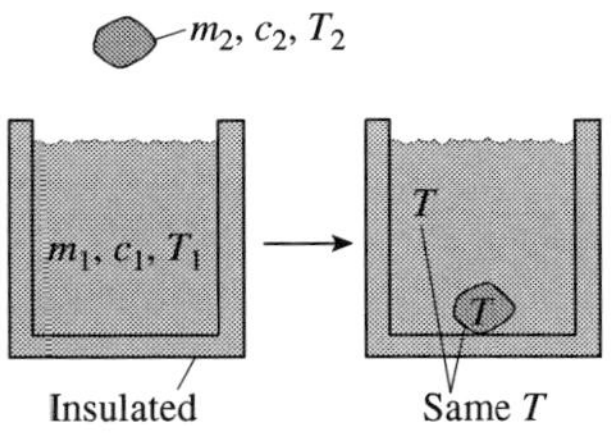

Energy balance: A system experiencing both energy input and energy loss comes to energy balance at the temperature for which the energy-loss rate equals the rate of energy input.

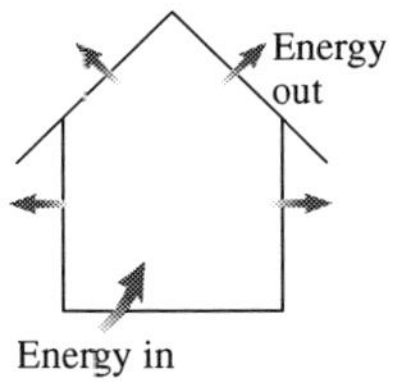

For Thought and Discussion

1. If system A is not in thermodynamic equilibrium with system B, and B is not in equilibrium with C, can you draw any conclusions about the temperatures of the three systems?
2. Does a thermometer measure its own temperature or the temperature of its surroundings? Explain.
3. Compare the relative sizes of the kelvin, the degree Celsius, the degree Fahrenheit, and the degree Rankine.
4. If you put a thermometer in direct sunlight, what do you measure: the air temperature, the temperature of the Sun, or some other temperature?
5. Why does the temperature in a stone building usually vary less than in a wooden building?
6. Why do large bodies of water exert a temperature-moderating effect on their surroundings?
7. A Thermos bottle consists of an evacuated, double-wall glass liner. The glass is coated with a thin layer of aluminum. How does a Thermos bottle work?
8. Stainless-steel cookware often has a layer of aluminum or copper embedded in the bottom. Why?
9. What method of energy transfer dominates in baking? In broiling?
10. After a calm, cold night, the temperature a few feet above ground often drops just as the Sun comes up. Explain in terms of convection.
11. Glass and fiberglass are made from the same material, yet have dramatically different thermal conductivities. Why?
12. To keep your hands warm while skiing, you should wear mittens instead of gloves. Why?
13. Since Earth is exposed to solar radiation, why doesn't Earth have the same temperature as the Sun?

Exercises and Problems

Exercises

Section 16.1 Heat, Temperature, and Thermodynamic Equilibrium

14. A Canadian meteorologist predicts an overnight low of −15°C. How would a U.S. meteorologist express that same prediction?
15. Normal room temperature is 68°F. What is this in Celsius?
16. The outdoor temperature rises by 10°C. What is that rise in Fahrenheit?
17. At what temperature do the Fahrenheit and Celsius scales coincide?
18. The normal boiling point of nitrogen is 77.3 K. Express this in Celsius and Fahrenheit.
19. A sick child's temperature reads 39.1 on a Celsius thermometer. What's the child's temperature on the Fahrenheit scale?

Section 16.2 Heat Capacity and Specific Heat

20. Find the heat capacity of a 55-tonne slab of concrete (1 tonne = 1000 kg).
21. Find the energy needed to raise a 2.0-kg chunk of aluminum by 18°C.
22. It takes 7.5 kJ to increase the temperature of a 1-kg block of material by 3.0°C. What's the material's specific heat?
23. The average human diet contains about 2000 kcal per day. If all this food energy is released rather than stored as fat, what is the approximate average power output of the human body?
24. Walking at 3 km/h requires an energy expenditure rate of about 200 W. How far would you have to walk to "burn off" a 300-kcal hamburger?
25. You bring a 350-g wrench into the house from your car. The house is 15°C warmer than the car, and it takes 2.52 kJ of energy to warm the wrench by this amount. Find (a) the heat capacity of the wrench and (b) the specific heat of the metal it's made from.
26. (a) How much heat does it take to bring a 3.4-kg iron skillet from 20°C to 130°C? (b) If the heat is supplied by a stove burner at the rate of 2.0 kW, how long will it take to heat the pan?

Section 16.3 Heat Transfer

27. Building heat loss in the United States is usually expressed in Btu/h. What is 1 Btu/h in SI units?
28. Find the heat-loss rate through a 1.0-m^2 slab of (a) wood and (b) Styrofoam, each 2.0 cm thick, if one surface is at 20°C and the other at 0°C.
29. The top of a steel wood stove measures 90 cm by 40 cm and is 0.45 cm thick. The fire maintains the inside surface of the stovetop at 310°C, while the outside surface is at 295°C. Find the rate of heat conduction through the stovetop.
30. How thick a concrete wall would be needed to give the same insulating value as $3\frac{1}{2}$ in. of fiberglass?
31. An 8.0 m by 12 m house is built on a concrete slab 23 cm thick. What is the heat-loss rate through the floor if the interior is at 20°C while the ground is at 10°C?
32. What is the $\mathcal{R}$-factor of a wall that loses 0.040 Btu each hour through each square foot for each °F temperature difference?
33. Compute the $\mathcal{R}$-factors for 1-in. thicknesses of air, concrete, fiberglass, glass, Styrofoam, and wood.
34. A horseshoe has a surface area of 50 cm^2, and a blacksmith heats it to a red-hot 810°C. At what rate does it emit energy by radiation?

Section 16.4 Thermal-Energy Balance

35. An oven loses energy at the rate of 14 W per °C temperature difference between its interior and the 20°C temperature of the kitchen. What average power must be supplied to maintain the oven at 180°C?
36. A home heating system can supply energy at the maximum rate of 40 kW. If the house loses energy at the rate of 8.0 kW per °C temperature difference between interior and exterior, what's the minimum outdoor temperature at which the heating system can maintain 20°C (68°F) indoors?
37. The filament of a 100-W lightbulb is at 3.0 kK. What's the surface area of the filament?
38. The solar input at Venus amounts to about 1.9 kW/m^2. Use an energy-balance calculation like that done for Earth in the greenhouse effect application to estimate Venus's surface temperature in the absence of a greenhouse effect, taking $e = 1$. The actual surface temperature is close to 500°C.

Problems

39. A constant-volume gas thermometer is filled with air whose pressure is 101 kPa at the normal melting point of ice. What would its pressure be at (a) the normal boiling point of water, (b) the normal boiling point of oxygen (90.2 K), and (c) the normal boiling point of mercury (630 K)?

40. A constant-volume gas thermometer is at 55-kPa pressure at the triple point of water. By how much does its pressure change for each kelvin temperature change?
41. In the gas thermometer of Fig. 16.14, the height h is 60.0 mm at the triple point of water. When the thermometer is immersed in boiling sulfur dioxide, the height drops to 57.8 mm. What is the boiling point of SO_2 in kelvins and in degrees Celsius?

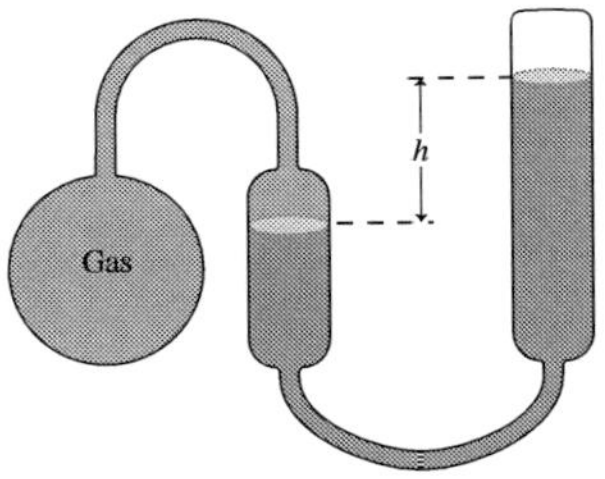

FIGURE 16.14 Problem 41

42. If your mass is 60 kg, what is the minimum number of calories you would "burn off" climbing a 1700-m-high mountain? *Note:* The actual metabolic energy used would be much greater.
43. Typical fats contain about 9 kcal per gram. If the energy in body fat could be utilized with 100% efficiency, how much mass could a runner lose in a 26.2-mile marathon? The runner's energy expenditure rate is 125 kcal/mile.
44. A circular lake 1.0 km in diameter is 10 m deep (Fig. 16.15). Solar energy is incident on the lake at an average rate of 200 W/m^2. If the lake absorbs all this energy and does not exchange heat with its surroundings, how long will it take to warm from 10°C to 20°C?

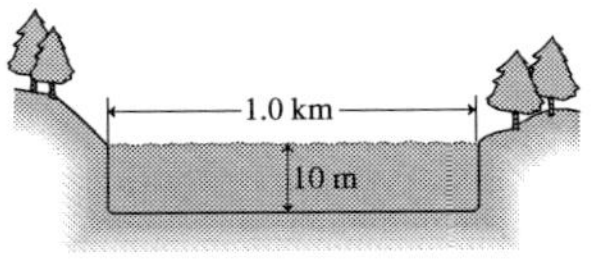

FIGURE 16.15 Problem 44

45. How much heat is required to raise an 800-g copper pan from 15°C to 90°C if (a) the pan is empty; (b) the pan contains 1.0 kg of water; and (c) the pan contains 4.0 kg of mercury?
46. Initially, 100 g of water and 100 g of another substance listed in Table 16.1 are at 20°C. Heat is then transferred to each substance at the same rate for 1.0 min. At the end of that time, the water is at 32°C and the other substance at 76°C. (a) What is the other substance? (b) What is the heating rate?
47. How long will it take a 625-W microwave oven to bring 250 mL of water from 10°C to the boiling point?
48. Two neighbors return from Florida to find their houses at a frigid 35°F. Each house has a furnace with heat output of 100,000 Btu/h. One house is made of stone and weighs 75 tons. The other is made of wood and weighs 15 tons. How long does it take each house to reach 65°F? Neglect heat loss, and assume the entire house mass reaches a uniform 65°F temperature.
49. A stove burner supplies heat at the rate of 1.0 kW, a microwave oven at 625 W. You can heat water in the microwave in a paper cup of negligible heat capacity, but the stove requires a pan whose heat capacity is 1.4 kJ/K. (a) How much water do you need before it becomes quicker to heat on the stovetop? (b) What will be the rate at which the temperature of this much water rises?
50. When a nuclear power plant's reactor is shut down, radioactive decay continues to produce heat at about 10% of the reactor's normal power level of 3.0 GW. In a major accident, a pipe breaks and all the reactor cooling water is lost. The reactor is immediately shut down, the break sealed, and 420 m^3 of 20°C water injected into the reactor. If the water were not actively cooled, how long would it take to reach its normal boiling point?
51. A 1.2-kg iron tea kettle sits on a 2.0-kW stove burner. If it takes 5.4 min to bring the kettle and the water in it from 20°C to the boiling point, how much water is in the kettle?
52. Two cars collide head-on at 90 km/h. If all their kinetic energy ended up as heat, what would be the temperature increase of the wrecks? The specific heat of the cars is essentially that of iron.
53. A 1500-kg car moving at 40 km/h is brought to a sudden stop. If all the car's energy is dissipated in heating its four 5.0-kg steel brake disks, by how much do the disk temperatures increase?
54. A child complains that her cocoa is too hot. The cocoa is at 90°C. Her father pours 2 oz of milk at 3°C into the 6 oz of cocoa. Assuming milk and cocoa have the same specific heat as water, what is the new temperature of the cocoa?
55. A piece of copper at 300°C is dropped into 1.0 kg of water at 20°C. If the equilibrium temperature is 25°C, what is the mass of the copper?
56. A biology lab's walk-in cooler measures 3.0 m by 2.0 m by 2.3 m and is insulated with 8.0-cm-thick Styrofoam. If the surrounding building is at 20°C, at what average rate must the cooler's refrigeration unit remove heat in order to maintain 4.0°C in the cooler?
57. One end of an iron rod 40 cm long and 3.0 cm in diameter is in ice water, the other in boiling water (Fig. 16.16). The rod is well insulated so no heat is lost out the sides. What is the heat-flow rate along the rod?

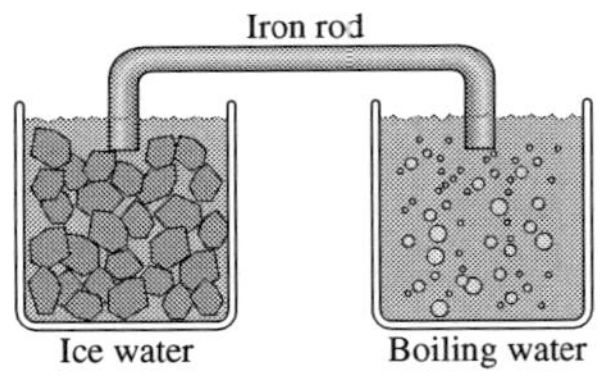

FIGURE 16.16 Problem 57

58. (a) What is the $\mathcal{R}$-factor for a wall consisting of $\frac{1}{4}$-in. pine paneling, $\mathcal{R}$-11 fiberglass insulation, $\frac{3}{4}$-in. pine sheathing, and 2.0-mm aluminum siding? (b) What is the heat-loss rate through a 20 ft by 8 ft section of wall when the temperature difference across the wall is 55°F?
59. You're considering installing a 5 ft by 8 ft picture window in a north-facing wall that now has $\mathcal{R}$-19 insulation. The window has $\mathcal{R} = 2.1$ ft^2·°F·h/Btu. If you install the window, how much more oil, at 100,000 Btu per gallon, will you have to burn in a winter month when the outdoor temperature averages 15°F and the indoor temperature is 68°F?
60. Repeat Problem 59 for a south-facing window where the average sunlight intensity is 180 W/m^2.

61. A house is insulated so its total heat loss is 370 W/°C. On a night when the outdoor temperature is 12°C, the owner throws a party and 40 people come. The average power output of the human body is 100 W. If there are no other heat sources in the house, what will be the house temperature during the party?
62. An electric stove burner has surface area 325 cm^2 and emissivity $e = 1.0$. The burner is at 900 K and the electric power input to the burner is 1500 W. If room temperature is 300 K, what fraction of the burner's heat loss is by radiation?
63. An electric current passes through a metal strip 0.50 cm by 5.0 cm by 0.10 mm, heating it at the rate of 50 W. The strip has emissivity $e = 1.0$ and its surroundings are at 300 K. What will be the temperature of the strip if it's enclosed in (a) a vacuum bottle transparent to all radiation and (b) an insulating box with thermal resistance $R = 8.0$ K/W that blocks all radiation?
64. The average human body produces heat at the rate of 100 W and has a total surface area of about 1.5 m^2. What is the coldest outdoor temperature in which a down sleeping bag with 4.0-cm loft (thickness) can be used without the body temperature dropping below 37°C? Consider only conductive heat loss.
65. A blacksmith heats a 1.1-kg iron horseshoe to 550°C, then plunges it into a bucket containing 15 kg of water at 20°C. What is the final temperature?
66. What is the power output of a microwave oven that can heat 430 g of water from 20°C to the boiling point in 5.0 min? Neglect the heat capacity of the container.
67. A cylindrical log 15 cm in diameter and 65 cm long is glowing red hot in a fireplace. If it's emitting radiation at the rate of 34 kW, what is its temperature? The log's emissivity is essentially 1.
68. A star whose surface temperature is 50 kK radiates 4.0×10^{27} W. If the star behaves like a blackbody, what is its radius?
69. Rework Example 16.4, now assuming that the house has ten single-glazed windows, each measuring 2.5 ft by 5.0 ft. Four of the windows are on the south and admit solar energy at the average rate of 30 Btu/h·ft^2. *All* the windows lose heat; their $\mathcal{R}$-factor is 0.90. (a) What is the total heating cost for the month? (b) How much is the solar gain worth?
70. A black wood stove with surface area 4.6 m^2 is made from cast iron 4.0 mm thick. The interior wall of the stove is at 650°C, while the exterior is at 647°C. (a) What is the rate of heat conduction through the stove wall? (b) What is the rate of heat loss by radiation from the stove? (c) Use the results of parts (a) and (b) to find how much heat the stove loses by a combination of conduction and convection in the surrounding air.
71. What should be the average temperature on Pluto, at its 5.9×10^{12}-m distance from the Sun? Treat Pluto as a blackbody.
72. In a cylindrical pipe where area isn't constant, Equation 16.5 takes the form $H = -kA(dT/dr)$. Use this equation to show that the heat-loss rate from a cylindrical pipe of radius R_1 and length L is

$$H = \frac{2\pi kL(T_1 - T_2)}{\ln(R_2/R_1)}$$

where the pipe is surrounded by insulation of outer radius R_2 and thermal conductivity k and where T_1 and T_2 are the temperatures at the pipe surface and at the outer surface of the insulation, respectively. *Hint:* Consider the heat flow through a thin section of pipe, with thickness dr, as shown in Fig. 16.17. Then integrate.

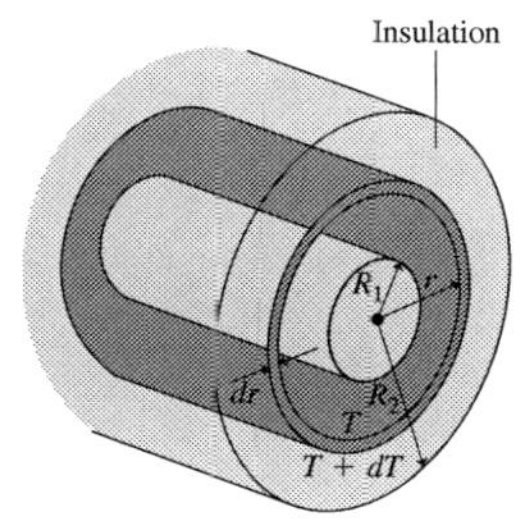

FIGURE 16.17 Problem 72

73. A friend of yours is writing a novel about the impact of a small asteroid on Earth. Such an event has been credited with causing a "nuclear winter" that may have killed the dinosaurs. Dust from the impact would darken the atmosphere and cause a reduction in the solar radiation received at the surface and thus lower the average surface temperature. Your friend would like to know the change in Earth's average temperature of 287 K if the impact caused a 10% reduction in solar intensity.
74. Your professor, who is building his own cabin, is examining a data sheet for some European-made insulation. According to the design specs for his cabin, he needs at least $\mathcal{R}$-19 insulation in the walls. The data sheet gives the $\mathcal{R}$-factor of the European insulation in SI units—namely, 3.5 m^2·K/W. Will the European insulation work?
75. You're the public affairs manager for an electric power utility. A nuclear power plant has been refueled and needs to power up again. The news media want to know how long it will take before the reactor is online again. The reactor needs to heat 5.4×10^6 kg of cooling water from 10°C to 350°C (it's a pressurized-water reactor, so the water remains liquid). The reactor's thermal power output is 1.42 GW. Ignoring the heat capacity of the reactor vessel and plumbing, how long will that take?
76. A young relative gets a pet rabbit that will live in a backyard hutch. The hutch is enclosed and has a thermal resistance of 0.25 K/W. A rabbit-care book says the hutch should be kept at a temperature of 0°C or higher so the rabbit's water won't freeze. On a day when the outside temperature is −15°C, you put a 50-W heat lamp in the hutch. Will the bunny be able to take a drink of water? *Note:* Neglect the heat due to the animal's metabolism.
77. At low temperatures the specific heat of a solid is approximately proportional to the cube of the absolute temperature; for copper $c = 31(T/343\ \text{K})^3$ J/g·K. Integrate Equation 16.3 in differential form, $dQ = mc\,dT$, to find the heat required to bring a 40-g sample of copper from 10 K to 25 K.
78. Use the method outlined in Problem 72 to show that the steady heat flow rate in the direction of the axis of a truncated cone with conductivity k, faces of radii R_1 and R_2, and length L is $H = \pi k R_1 R_2 (T_1 - T_2)/L$, where T_1 and T_2 are the temperatures on the faces and insulation prevents any heat flow out the sides (see Figure 16.18).

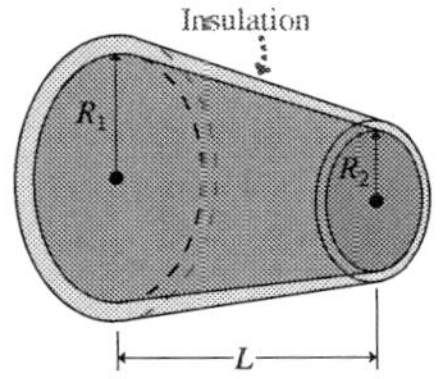

FIGURE 16.18 Problem 78

79. A house is at 20°C on a winter day when the outside temperature is a steady −15°C. The heat capacity of the house is 6.5 MJ/K and its thermal resistance is 6.67 mK/W. If the furnace suddenly fails (and there are no other sources of heat) how long will it take for the house temperature to reach the freezing point? (*Hint:* Combine the differential forms of Equation 16.3 for heat capacity and Equation 16.5, in terms of the thermal resistance, to show that the rate of temperature change is proportional to the temperature difference between the house and its surroundings. This relation is known as Newton's law of cooling.)

80. A 1000-W electric clothes iron has a surface area 300 cm^2 and emissivity $e = 0.97$. If the surface temperature of the iron is 500 K and its surroundings are at 300 K, (a) what is the *net* rate of energy transfer by radiation from the iron? (b) What must be the rate of energy transfer by conduction and convection?

Answers to Chapter Questions

Answer to Chapter Opening Question

The photo is taken in infrared light, and the amount of infrared radiation increases rapidly with increasing temperature. The car's wheels are glowing with infrared, a result of frictional heating when the brakes were recently applied.

Answers to GOT IT? Questions

16.1 The rock's temperature changes more because its specific heat is lower.

16.2 $\Delta T_2 > \Delta T_3 > \Delta T_1$; since H and Δx are the same for each slab, the product $k\,\Delta T$ must be constant, so a higher conductivity means a lower ΔT.

16.3 (a) Radiation; (b) conduction; (c) convection.

17 The Thermal Behavior of Matter

■ What unusual property of water is evident in this photo?

▶ To Learn

By the end of this chapter you should be able to

- Explain the relation between the statistical mechanics of molecules and the corresponding thermodynamic properties, for simple gases. This includes especially the relation between molecular energy and temperature (17.1).
- Describe the three principal phases of matter, and calculate the energies required to change phase (17.2).
- Interpret phase diagrams to determine conditions under which the various phases exist (17.2).
- Calculate thermal-expansion effects in solids, liquids, and gases (17.3).
- Describe the unusual thermal-expansion properties of water (17.3).

◀ To Know

- This chapter links thermodynamics with the statistical mechanics of atoms and molecules. Key concepts in this linkage are conservation of energy and conservation of momentum (7.3, 9.2).

Matter responds to heating in several ways. It may get hotter or it may melt. It may change size, shape, or pressure. This chapter explores the thermal behavior of matter. We start with a simple gaseous state, whose behavior follows from Newtonian mechanics at the molecular level. We then move to liquids and solids, whose behavior is still grounded in the molecular properties of matter, but whose description is more empirical.

17.1 Gases

Gases are simple because their molecules are far apart and only rarely interact. That makes gas behavior and its physical explanation particularly straightforward. Developing that explanation will clarify the relation between macroscopic properties—such as temperature and pressure—and the underlying microscopic properties of gas molecules.

The Ideal-Gas Law

The macroscopic state of a gas in thermodynamic equilibrium is determined by its temperature, pressure, and volume. Moreover, it turns out that all gases exhibit, to a very good approximation, the same relation among these three quantities.

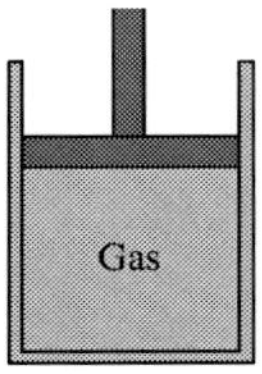

FIGURE 17.1 A piston-cylinder system.

A simple system for studying gas behavior consists of a gas-filled cylinder sealed by a movable piston (Fig. 17.1). This is not just a pedagogical abstraction: Practical devices including engines, pumps, and air compressors contain piston-cylinder systems, while lungs, balloons, gas bubbles, and many other natural systems are analogous to our piston-cylinder system.

If we maintain the system of Fig. 17.1 at constant temperature and move the piston to vary the gas volume, we find that the pressure varies inversely with the volume. If we increase the temperature while holding the volume fixed, the pressure rises in direct proportion to the temperature. If we double the amount of gas while holding temperature and volume constant, the pressure doubles. Putting all these results together, we can write

$$pV = NkT \quad \text{(ideal-gas law)} \tag{17.1}$$

with p, V, and T the pressure, volume, and temperature, respectively, and N the number of molecules in the gas. The constant $k = 1.38 \times 10^{-23}$ J/K is **Boltzmann's constant**, named for the Austrian physicist Ludwig Boltzmann (1844–1906), who was instrumental in developing the microscopic description of thermal phenomena. Equation 17.1 is the **ideal-gas law**. Most real gases obey this law to a very good approximation.

Because the number of molecules N in a typical gas sample is astronomically large, we often express the ideal-gas law in terms of the number of **moles** (mol) of gas molecules. One mole is an SI unit equal to Avogadro's number, $N_A = 6.022 \times 10^{23}$, of atoms or molecules. Formally, Avogadro's number is defined as the number of carbon-12 atoms in 12 grams of carbon-12.

If we have n moles of a gas, then $N = nN_A$ is the number of molecules, so the ideal-gas law becomes

$$pV = nN_AkT = nRT \tag{17.2}$$

where $R = N_Ak = 8.314$ J/K·mol is called the **universal gas constant**.

EXAMPLE 17.1 The Ideal Gas Law: STP

What volume is occupied by 1.00 mol of an ideal gas at standard temperature and pressure (STP), where $T = 0°\text{C}$ and $p = 101.3$ kPa?

INTERPRET We're dealing with an ideal gas, and we're given the amount of gas, the temperature, and the pressure.

DEVELOP Because we're given the number of moles n (and not the number of molecules), we'll use the ideal-gas law in the form of Equation 17.2, $pV = nRT$, to find the volume.

EVALUATE Solving for V gives

$$V = \frac{nRT}{p} = \frac{(1.00\text{ mol})(8.314\text{ J/K}\cdot\text{mol})(273\text{ K})}{1.01 \times 10^5\text{ Pa}} = 22.4 \times 10^{-3}\text{ m}^3 = 22.4\text{ L}$$

where we expressed $T = 0°\text{C}$ as 273 K.

ASSESS This result may be familiar to you from earlier chemistry or physics courses: 1 mole of any ideal gas—no matter what its chemical composition—occupies 22.4 L at standard temperature and pressure. Recall that 1 liter (L) is 1000 cm^3 or 10^{-3} cm^3. ■

The ideal-gas law is remarkably simple. Neither its form nor the constants k and R depend on the substance making up the gas or on the mass of the gas molecules. Yet most real gases follow the ideal-gas law very closely over a wide range of pressures. This nearly ideal behavior is what gives gas thermometers their high precision over a wide temperature range.

Kinetic Theory of the Ideal Gas

Why do gases obey such a simple relation among temperature, pressure, and volume? Here we answer that question with an analysis based ultimately on Newtonian mechanics.

We start with some simplifying assumptions:

1. The gas consists of many identical molecules, each with mass m but negligible size and no internal structure. This assumption is approximately true for real gases when the distance between molecules is large compared with their size. This allows us to

neglect intermolecular collisions, an assumption that simplifies our analysis but isn't crucial to the ideal gas.

2. The molecules don't exert action-at-a-distance forces on each other. Thus there's no intermolecular potential energy, and therefore the molecules have only kinetic energy. This assumption is fundamental to an ideal gas.
3. The molecules move in random directions with a distribution of speeds that's independent of direction.
4. Collisions with the container walls are elastic, conserving the molecules' energy and momentum. Here's where we tie our gas model to Newtonian mechanics.

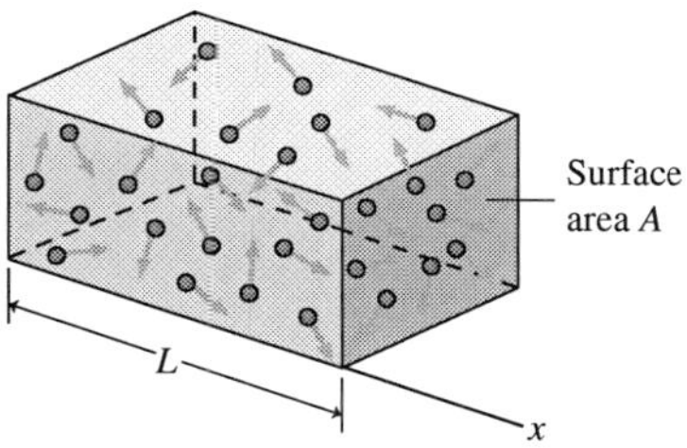

FIGURE 17.2 Gas molecules confined to a rectangular box.

Consider N molecules confined to a rectangular box with length L (Fig. 17.2). Each molecule that collides with a wall exerts a force. There are so many molecules that individual collisions aren't evident; instead the wall experiences an essentially constant average force. The gas pressure p is a measure of this force on a unit area. We're going to find an expression for p and show that it takes in the form of the ideal-gas law.

Figure 17.3 shows one molecule colliding with the right-hand wall. Since the collision is elastic, the y component of the molecule's velocity is unchanged, while the x component reverses sign. Thus the molecule undergoes a momentum change of magnitude $2mv_{xi}$, where i labels this particular molecule. After the molecule collides with the right-hand wall, nothing will change its x velocity until it hits the left-hand wall and its x velocity again reverses. So it will be back at the right-hand wall in the time $\Delta t = 2L/v_{xi}$ that it takes to go back and forth along the container.

Now each time our molecule collides with the right-hand wall, it delivers momentum $2mv_{xi}$ to the wall. Newton's second law says that force is the rate of change of momentum. So we can calculate the average force $\overline{F_i}$ due to one molecule by dividing the momentum delivered, $2mv_{xi}$, by the time, $2L/v_{xi}$ between collisions:

$$\overline{F_i} = \frac{2mv_{xi}}{2L/v_{xi}} = \frac{mv_{xi}^2}{L}$$

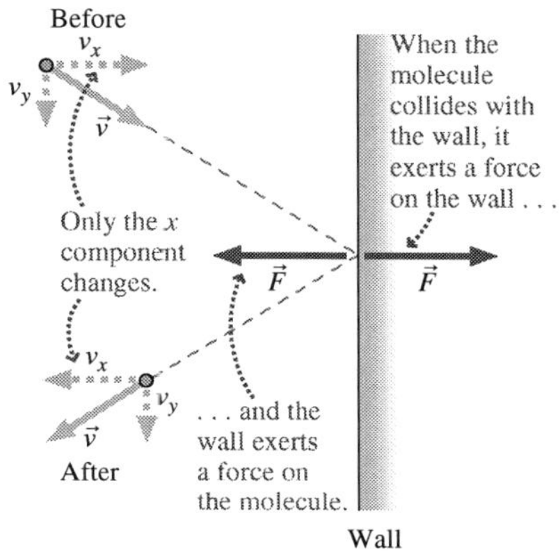

FIGURE 17.3 A molecule undergoes an elastic collision, reversing its x component and transferring momentum $2mv_x$ to the wall.

To get the total force on the wall, we sum over all N molecules with their different x velocities. Dividing by the wall area A then gives the pressure:

$$p = \frac{\overline{F}}{A} = \frac{\sum \overline{F_i}}{A} = \frac{\sum mv_{xi}^2/L}{A} = \frac{m\sum v_{xi}^2}{AL}$$

The last step follows because the box length L and molecular mass m are the same for all molecules, so they factor out of the sum. We can simplify by noting that the denominator AL is just the volume V. Let's also multiply by 1 in the form N/N, with N the number of molecules. Then we have

$$p = \frac{m\sum v_{xi}^2}{AL} = \frac{mN}{V}\frac{\sum v_{xi}^2}{N}$$

In the final expression here, the term $\sum v_{xi}^2/N$ is the average of the squares of all the x-velocity components of all the molecules; we designate this quantity $\overline{v_x^2}$. So the pressure becomes

$$p = \frac{mN}{V}\overline{v_x^2}$$

We still haven't used assumption 3—that the molecules move in random directions with speeds independent of direction. If we grab a molecule at random, that means we're just as likely to find it moving in the x direction, the y direction, the z direction, or any direction in between—and its speed, on average, won't depend on its direction of motion. So the average quantities $\overline{v_x^2}$, $\overline{v_y^2}$, and $\overline{v_z^2}$ must be equal. Since the three directions x, y, and z are perpendicular, the average of the molecular speeds squared is $\overline{v^2} = \overline{v_x^2} + \overline{v_y^2} + \overline{v_z^2}$. We've just argued that all three terms on the right are equal, so we can write $\overline{v^2} = 3\overline{v_x^2}$, or $\overline{v_x^2} = \frac{1}{3}\overline{v^2}$. Then our expression for pressure becomes

$$p = \frac{mN}{3V}\overline{v^2}$$

Multiplying through by V and by 1 in the form 2/2, we have

$$pV = \tfrac{2}{3}N\left(\tfrac{1}{2}m\overline{v^2}\right)$$

This looks a lot like the ideal-gas law (Equation 17.1), except that instead of kT we have $\frac{2}{3}\left(\frac{1}{2}m\overline{v^2}\right)$. Take a good look at the quantity in parentheses: You'll see that it's just the average kinetic energy of a gas molecule.

Think about what we've done here. We applied the fundamental laws of mechanics to an ideal gas and came up with an equation that looks like the experimentally verified ideal-gas law, except that it's expressed in terms of a microscopic quantity—molecular kinetic energy—rather than the macroscopic quantity temperature. Since our equation describes the behavior of an ideal gas, it *must be* the ideal-gas law. Comparing with the ideal-gas law in the form 17.1, we must therefore have

$$\tfrac{1}{2}m\overline{v^2} = \tfrac{3}{2}kT \quad \text{(temperature and molecular energy)} \qquad (17.3)$$

Our derivation shows why, in terms of Newtonian mechanics, a gas obeying our four assumptions should obey the ideal-gas law. In Equation 17.3 we get an added bonus—a microscopic understanding of the meaning of temperature: **Temperature measures the average kinetic energy associated with random translational motion of the molecules.**

EXAMPLE 17.2 Molecular Energy and Speed: An Air Molecule

Find the average kinetic energy of a molecule in air at room temperature (20°C or 293 K), and determine the speed of a nitrogen molecule (N_2) with this energy.

INTERPRET This problem asks about the linkage between thermodynamic quantities and molecular energy. We just found that linkage: The temperature of a gas is a measure of the average kinetic energy of its molecules.

DEVELOP Equation 17.3, $\frac{1}{2}m\overline{v^2} = \frac{3}{2}kT$, quantifies the relation between temperature and molecular kinetic energy. Once we find the molecular kinetic energy, we'll need the molecular mass to determine the speed. We can get that using the atomic weight of nitrogen and the fact that an N_2 molecule contains two atoms.

EVALUATE We first evaluate the average molecular kinetic energy: $\overline{K} = \frac{1}{2}m\overline{v^2} = \frac{3}{2}kT = \frac{3}{2}(1.38 \times 10^{-23}\ \text{J/K})(293\ \text{K}) = 6.07 \times 10^{-21}\ \text{J}$

We can solve for the corresponding speed if we know the molecular mass m. A nitrogen molecule consists of two atoms each with mass 14 u (see Appendix D), so its mass is

$$m = 2(14\ \text{u})(1.66 \times 10^{-27}\ \text{kg/u}) = 4.65 \times 10^{-26}\ \text{kg}$$

Since $\overline{K} = \frac{1}{2}m\overline{v^2}$, the speed corresponding to this kinetic energy is

$$v = \sqrt{\frac{2K}{m}} = \sqrt{\frac{2(6.07 \times 10^{-21}\ \text{J})}{4.65 \times 10^{-26}\ \text{kg}}} = 511\ \text{m/s}$$

ASSESS Make sense? Not surprisingly, the answer is the same order of magnitude as the speed of sound ($\sim$340 m/s) in air at room temperature. At the microscopic level, the speed of the individual molecules limits the rate at which information can be transmitted by disturbances—sound waves—propagating through the gas. ■

We call the speed calculated in Example 17.2 the **thermal speed**. In terms of temperature, Equation 17.3 shows

$$v_{\text{th}} = \sqrt{\frac{3kT}{m}} \qquad (17.4)$$

GOT IT? 17.1 If you double the kelvin temperature of a gas, what happens to the thermal speed of the gas molecules?

The Distribution of Molecular Speeds

The thermal speed v_{th} is a typical molecular speed, but it doesn't tell us much about the distribution of speeds. Are molecular speeds limited to a narrow band about v_{th}? Or are lots of molecules moving much faster or much slower?

In the 1860s, the Scottish physicist James Clerk Maxwell showed that elastic collisions among molecules result in a speed distribution that peaks near the thermal speed but may extend considerably higher. Figure 17.4 plots this **Maxwell-Boltzmann distribution** for two different temperatures. Note that increasing temperature results in a higher thermal speed, as expected, but that it also broadens the distribution so there are more molecules at lower and higher speeds. The high-speed "tail" of the distribution is especially important to chemists because high-energy molecules participate most readily in chemical reactions. The rapid extension of the high-energy tail with increasing temperature shows why reaction rates are strongly temperature sensitive, and explains why foods keep much longer with even modest refrigeration. High-energy molecules are also the first to evaporate from a liquid, leaving slower, cooler molecules behind and thus explaining evaporative cooling. Without this effect, Earth's atmosphere would be much drier and it would rain far less frequently.

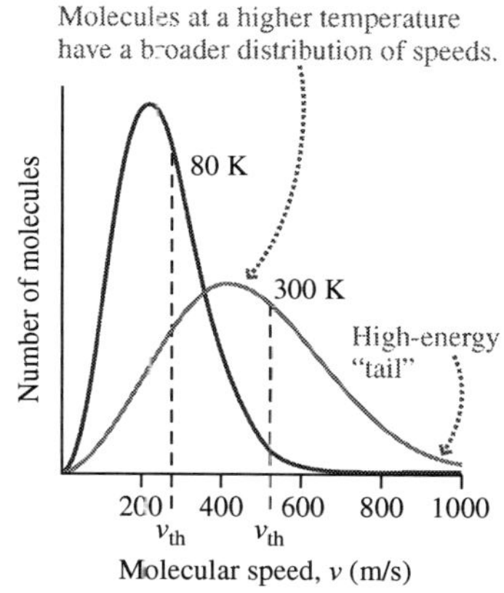

FIGURE 17.4 Maxwell-Boltzmann distribution of molecular speeds for nitrogen (N_2) at temperatures of 80 K and 300 K.

Real Gases

The ideal-gas law is a good approximation to the behavior of most real gases, but it's not perfect because our assumptions aren't entirely realistic. Two factors are especially important. First, real molecules take up space. This reduces the available volume, altering the ideal-gas law. Second, electrical effects that we'll explore in Chapter 20 result in a weak attractive force between nearby molecules. As they move apart, molecules do work against this **van der Waals force**, and their kinetic energy drops. Again, the effect is a deviation from ideal-gas behavior. That deviation is small for diffuse gases with widely spaced molecules, but at high density it becomes significant.

17.2 Phase Changes

Step out of a steamy shower, and you'll find the mirror fogged with water condensed on the cool glass. Climb a mountain in winter, and you'll be treated to the lovely spectacle of every branch and pine needle covered with a delicate coating of frost that's formed right from the air. Burn a rewritable CD or DVD, and you've stored information with a laser that melts tiny spots on the spinning disc. These examples involve **phase changes** between gas and liquid, gas and solid, and solid and liquid.

Heat and Phase Changes

Drop ice cubes into a drink and stir. What's the temperature of the drink? It's 0°C, and it stays at 0°C as long as any ice remains. The melting of a pure solid occurs at a fixed temperature. During the process, energy goes into breaking the molecular bonds that hold the material in its solid form. This increases the potential energy of the molecules but not their kinetic energy. Since temperature is a measure of molecular kinetic energy, that means the temperature doesn't change either.

Figure 17.5 shows the temperature history for a block of ice as it's warmed to the melting point and then beyond, all with a fixed rate of energy input. Note that the temperature rises steadily while the water is in the solid, liquid, or gas phase, but remains constant during the solid-liquid and liquid-gas transitions.

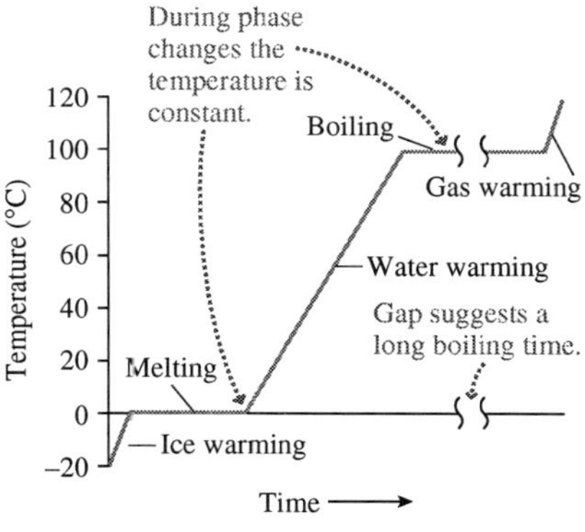

FIGURE 17.5 Temperature versus time for what's initially a block of ice at −20°C, supplied with energy at a constant rate. The process takes place at atmospheric pressure.

The energy per unit mass required to change phase is called a **heat of transformation** L; for the solid-liquid change it's the **heat of fusion** L_f, and for liquid-gas it's the **heat of vaporization** L_v. Less familiar is the **heat of sublimation** for the transition from solid directly to gas. These quantities have units of J/kg, so the energy required to change the phase of a mass m is

$$Q = Lm \quad \text{(heat of transformation)} \qquad (17.5)$$

TABLE 17.1 Heats of Transformation (at Atmospheric Pressure)

Substance	Melting Point (K)	L_f (kJ/kg)	Boiling Point (K)	L_v (kJ/kg)
Alcohol, ethyl	159	109	351	879
Copper	1357	205	2840	4726
Lead	601	24.7	2013	858
Mercury	234	11.3	630	296
Oxygen	54.8	13.8	90.2	213
Sulfur	388	38.5	718	287
Water	273	334	373	2257
Uranium	1406	82.8	4091	1875

To reverse the change requires removing the same energy. Table 17.1 lists heats of transformation for some common materials. These quantities are typically quite large; water's heat of fusion, for example, is 334 kJ/kg or 80 cal/g—meaning it takes as much energy to melt 1 gram of ice as to heat the resulting water to 80°C.

GOT IT? 17.2 You bring a pot of water to boil and then forget about it. Ten minutes later you come back to the kitchen to find the water still boiling. Is its temperature (a) less than, (b) greater than, or (c) equal to 100°C?

EXAMPLE 17.3 The Heat of Fusion: Meltdown!

A nuclear power plant's reactor vessel cracks, and all the cooling water drains out. Although nuclear fission stops, radioactive decay continues to heat the reactor's 2.5×10^5 kg uranium core at the rate of 120 MW. Once the melting point is reached, how much energy will it take to melt the core? How long will the melting take?

INTERPRET Since this problem is about melting, it must involve the heat of fusion. We identify the material in question as uranium.

DEVELOP Our plan is to find uranium's heat of fusion in Table 17.1 and then use Equation 17.5, $Q = Lm$, to calculate the energy required for melting. We're given the rate of energy generation by radioactive decay, and from that we'll be able to get the time.

EVALUATE Using uranium's L_f value from Table 17.1 in Equation 17.5, we have

$$Q = L_f m = (2.5 \times 10^5\ \text{kg})(82.8\ \text{kJ/kg}) = 20.7\ \text{GJ}$$

With a heating rate of 120 MW or 0.12 GJ/s, the time to melt the uranium is $(20.7\ \text{GJ})/(0.12\ \text{GJ/s}) = 173$ s.

ASSESS The time to meltdown is just under 3 minutes! Failsafe emergency cooling systems are essential to prevent nuclear meltdowns. ■

Often we're interested in the total energy needed to bring a material to its transition point and then to make the phase transition. Then we need to combine specific-heat considerations of Chapter 16 with the heats of transformation introduced here.

EXAMPLE 17.4 Heating and Phase Change: Enough Ice?

When 200 g of ice at −10°C are added to 1.0 kg of water at 15°C, is there enough ice to cool the water to 0°C? If so, how much ice is left in the mixture?

INTERPRET This problem involves both a temperature rise and a phase change. We identify water as the substance involved.

DEVELOP Equation 16.3, $\Delta Q = mc\,\Delta T$, determines the energy for the temperature rise, and Equation 17.5, $Q = Lm$, determines the phase-change energy. But we don't know whether all the ice melts. So our plan is to find the energy that it *would* take to heat the ice to 0°C and then melt all of it; if *more* than that much is available in cooling the water to 0°C, we'll know that we end up with all water at $T > 0$°C. But if there isn't sufficient energy, then we'll have a mixture with both ice and water at 0°C, and we can use the energy extracted in cooling the water to find out how much ice melts.

EVALUATE We begin by evaluating the energy Q_1 to heat the ice and then melt it all, adding the energies from Equations 16.3 and 17.5 and then getting the specific heat and heat of fusion from Tables 16.1 and 17.1, respectively:

$$\begin{aligned} Q_1 &= m_{\text{ice}} c_{\text{ice}}\,\Delta T_{\text{ice}} + m_{\text{ice}} L_f \\ &= (0.20\ \text{kg})(2.05\ \text{kJ/kg·K})(10\ \text{K}) + (0.20\ \text{kg})(334\ \text{kJ/kg}) \\ &= 4.1\ \text{kJ} + 66.8\ \text{kJ} = 70.9\ \text{kJ} \end{aligned}$$

Cooling the water to 0°C would extract energy Q_2 given by Equation 16.3:

$$Q_2 = m_{\text{water}} c_{\text{water}}\,\Delta T_{\text{water}} = (1.0\ \text{kg})(4.184\ \text{kJ/kg·K})(15\ \text{K}) = 62.8\ \text{kJ}$$

This is far more than the 4.1 kJ needed to bring the ice to 0°C, but not quite the 70.9 kJ needed to leave it all melted. So there's enough ice to

cont'd.

cool the water to 0°C, with some left over. How much? Our calculation of Q_1 shows that 4.1 kJ go into raising the ice temperature. Of the 62.8 kJ extracted from the water, the remaining 58.7 kJ go to melting ice. From Equation 17.5, the amount of ice melted is then

$$m_{\text{melted}} = \frac{Q}{L_f} = \frac{58.7\text{ kJ}}{334\text{ kJ/kg}} = 0.176\text{ kg} = 176\text{ g}$$

So we're left with 24 g of ice in 1176 g of water, all at 0°C.

ASSESS Make sense? Our 62.8 kJ was nearly enough to bring all the ice to the liquid phase, so it makes sense that only a small fraction of the ice remains. ■

Phase Diagrams

Why can't mountaineers enjoy piping hot coffee? Because water's boiling point drops with the decreasing pressure at high altitudes. In general, the temperatures at which phase changes occur depend on pressure. A **phase diagram** shows the different phases on a plot of pressure versus temperature. Figure 17.6 is a phase diagram for a typical substance. Most phase diagrams are similar, although water's is slightly unusual for reasons we'll discuss in the next section.

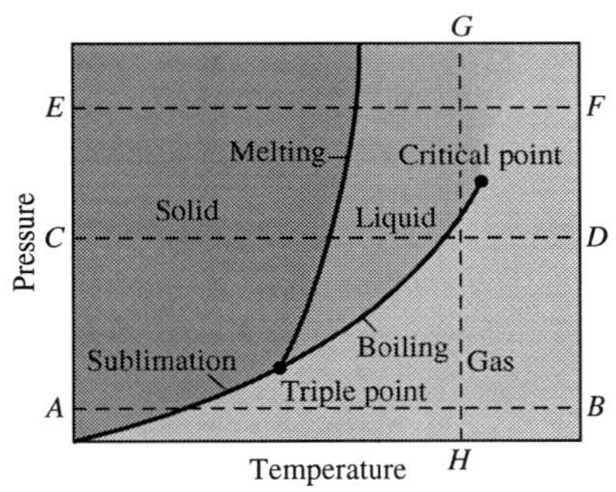

FIGURE 17.6 A phase diagram showing solid, liquid, and gas phases on a plot of pressure versus temperature.

The phase diagram divides pressure-temperature space into regions corresponding to solid, liquid, and gas phases. Lines separating these regions mark the phase transitions. Everyday experience suggests that heating takes a substance from solid, to liquid, to gas—as with water in Fig. 17.5. But Fig. 17.6 shows that this sequence doesn't always occur. At low pressure (line *AB* in Fig. 17.6) the substance goes directly from solid to gas. This is **sublimation**. We don't see this with water because normal atmospheric pressure is too high. For carbon dioxide, though, atmospheric pressure is low in the phase diagram, which is why "dry ice" turns directly into gaseous CO_2 without becoming liquid. At higher pressures (line *CD*) we get the familiar solid-liquid-gas sequence. Higher still (line *EF*), we're above the **critical point**, where the abrupt distinction between liquid and gas disappears. Instead, the substance starts out as a thick fluid whose properties change gradually from liquidlike to gaslike as it's heated.

We think of changing phase by applying heat, but Fig. 17.6 shows we can also change phase by changing pressure. Lowering pressure along line *GH*, for example, takes the substance from liquid to gas without any heat input. You may have seen a demonstration of water boiling vigorously at room temperature in a closed container pumped down to low pressure.

Don't let Fig. 17.6 fool you into thinking that phase transitions occur instantaneously. Those heats of transformation are large, and a substance moving, say, along line *CD* in response to heating will linger at each phase transition until all of it has changed phase; that's what the level portions of Fig. 17.5 showed.

The dividing curves in Fig. 17.6 show where two phases can coexist simultaneously, like ice floating in water at 0°C and atmospheric pressure. It's because phase changes occur along curves that terms like "melting point" and "boiling point" are meaningless unless pressure is specified. But there's one unique **triple point** where solid, liquid, and gas all coexist in equilibrium. Here temperature and pressure have unique, unambiguous values—which is why the 273.16-K triple point of water is used to define the kelvin scale.

17.3 Thermal Expansion

FIGURE 17.7 Thermal expansion distorted these tracks, causing a derailment. Expansion of long structures like this is best described using the coefficient of linear expansion.

We've seen how heating causes changes in temperature and phase. But heating also results in pressure or volume changes. For a gas at constant pressure, for example, the ideal-gas law shows that volume increases in direct proportion to temperature. The volume and pressure relations for liquids and solids aren't so simple. Because their molecules are closely spaced, liquids and solids aren't very compressible, so thermal expansion is less pronounced.

We characterize the change in the volume with temperature using the **coefficient of volume expansion** β, defined as the fractional change in volume when a substance undergoes a small temperature change ΔT:

$$\beta = \frac{\Delta V/V}{\Delta T} \tag{17.6}$$

This equation assumes that β is independent of temperature; if it varies significantly, then we would need to define β in terms of the derivative dV/dT (Problem 69). Our definition of β also assumes constant pressure; we could entirely inhibit thermal expansion with appropriate pressure increases.

Often we want to know how one linear dimension of a solid changes with temperature. This is especially true with long structures, where the absolute change is greatest along the long dimension (Fig. 17.7). We then speak of the **coefficient of linear expansion** α, defined by

$$\alpha = \frac{\Delta L/L}{\Delta T} \tag{17.7}$$

The volume- and linear-expansion coefficients are related in a simple way: $\beta = 3\alpha$, as you can show in Problem 72. However, the linear-expansion coefficient α is really meaningful only with solids, because liquids and gases deform and don't expand proportionately in all directions. Table 17.2 lists the expansion coefficients for some common substances.

TABLE 17.2 Expansion Coefficients*

Solids	α (K^{-1})	Liquids and Gases	β (K^{-1})
Aluminum	24×10^{-6}	Air	3.7×10^{-3}
Brass	19×10^{-6}	Alcohol, ethyl	75×10^{-5}
Copper	17×10^{-6}	Gasoline	95×10^{-5}
Glass (Pyrex)	3.2×10^{-6}	Mercury	18×10^{-5}
Ice	51×10^{-6}	Water, 1°C	-4.8×10^{-5}
Invar†	0.9×10^{-6}	Water, 20°C	20×10^{-5}
Steel	12×10^{-6}	Water, 50°C	50×10^{-5}

*At approximately room temperature unless noted.

†Invar, consisting of 64% iron and 36% nickel, is an alloy designed to minimize thermal expansion.

GOT IT? 17.3 The figure shows a donut-shaped object. If it's heated, will the hole get (a) larger or (b) smaller?

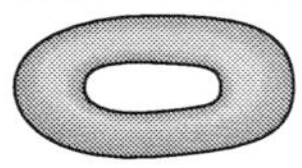

EXAMPLE 17.5 Thermal Expansion: Spilled Gasoline

A steel gas can holds 20 L at 10°C. It's filled to the brim with gas at 10°C. If the temperature now increases to 25°C, by how much does the can's volume increase? How much gas spills out?

INTERPRET This is a problem about thermal expansion. Since it involves volume, we identify the relevant quantity as the coefficient of volume expansion β.

DEVELOP Equation 17.6, $\beta = (\Delta V/V)/\Delta T$, determines the volume change. Our plan is to calculate the expanded volume of the tank and then of the gasoline. The difference will be the amount that spills out. Table 17.2 lists β for gasoline but α for steel; therefore we'll use the equation $\beta = 3\alpha$ for the steel.

EVALUATE First we use Equation 17.6 to evaluate the volume change ΔV of the can. Using $\beta = 3\alpha$, we have

$$\Delta V_{can} = \beta V \Delta T = (3)(12 \times 10^{-6}\,K^{-1})(20\,L)(15\,K) = 0.0108\,L$$

Similarly, for the gasoline,

$$\Delta V_{gas} = \beta V \Delta T = (95 \times 10^{-5}\,K^{-1})(20\,L)(15\,K) = 0.285\,L$$

We therefore lose 0.275 L.

ASSESS Make sense? The thermal-expansion coefficient for gasoline is so much greater than for steel that the can's expansion is negligible and the gas has nowhere to go. By the way, that spill wastes nearly 10 MJ of energy! ■

Thermal Expansion of Water

The entry for water at 1°C in Table 17.2 is remarkable, the negative expansion coefficient showing that water at this temperature actually *contracts* on heating. This unusual behavior occurs because ice has a relatively open crystal structure (Fig. 17.8) and therefore is less dense than liquid water. That's why ice floats. Immediately above the melting point, the intermolecular forces that bond H_2O molecules in ice still exert an influence, giving cold liquid water a lower density than at slightly higher temperatures. At 4°C water reaches its maximum density, and above this temperature the effect of molecular kinetic energy in keeping molecules apart wins out over intermolecular forces. From there on, water exhibits the more normal behavior of expansion with increasing temperature.

This unusual property of water near its melting point is reflected in its phase diagram, shown in Fig. 17.9. Note that the solid-liquid boundary extends leftward from the triple point, in contrast to the more typical behavior in Fig. 17.6. That means that ice at a fixed temperature will melt if the pressure is *increased*—an unusual property known as pressure melting.

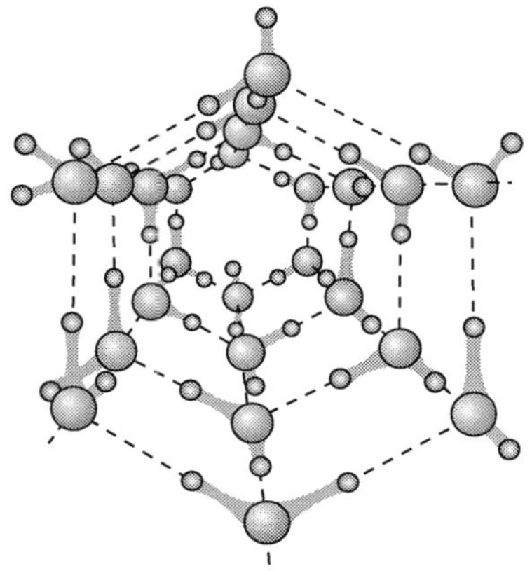

FIGURE 17.8 Water molecules in an ice crystal form an open structure, giving solid water a lower density than the liquid.

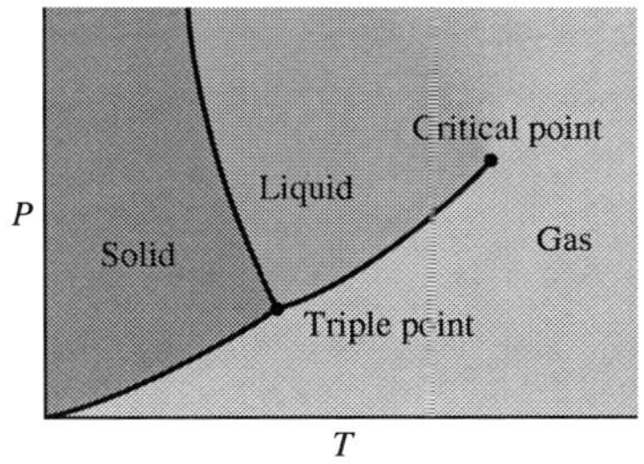

FIGURE 17.9 Phase diagram for water. Compare the solid-liquid boundary with that of Fig. 17.6.

APPLICATION Aquatic Life and Lake Turnover

The anomalous behavior of water has important consequences for life. If ice didn't float, then ponds, lakes, and even oceans would freeze solid from the bottom up, making aquatic life impossible. What actually happens, instead, is that a thin layer of ice forms on the surface, insulating the water below and keeping it liquid; as a result, ice cover in temperate climates rarely exceeds a meter or so. Because water has the greatest density at 4°C, water at this temperature sinks to the bottom. At lake depths greater than a few meters, sunlight is inadequate to raise the temperature, which therefore remains year-round at 4°C.

Water's unusual density behavior also causes the twice-yearly turnover of lakes in temperate climates. In the summer, a lake's surface water is warm, but deep water remains at 4°C. In the winter, water just beneath the ice is at 0°C, while the bottom water is still at 4°C. Both situations are stable, with less dense and therefore more buoyant water at the surface. But in the spring, ice melts and the surface water warms. When that water reaches 4°C, there's no density variation and the lake water mixes freely. This is the spring overturning. A similar overturning occurs in the fall, as the surface water cools through 4°C. Turnover is important to aquatic life because it brings up nutrients that would otherwise be trapped in the deep water.

CHAPTER 17 SUMMARY

Big Picture

The big idea here is that matter responds to heating in a variety of ways in addition to changing temperature. Other responses include changes of phase and of volume and/or pressure. The ideal gas provides a particularly simple system for understanding volume and pressure changes. Analyzing ideal-gas behavior provides a link between the Newtonian mechanics of molecules and macroscopic thermodynamics, showing that temperature is a measure of the average molecular kinetic energy.

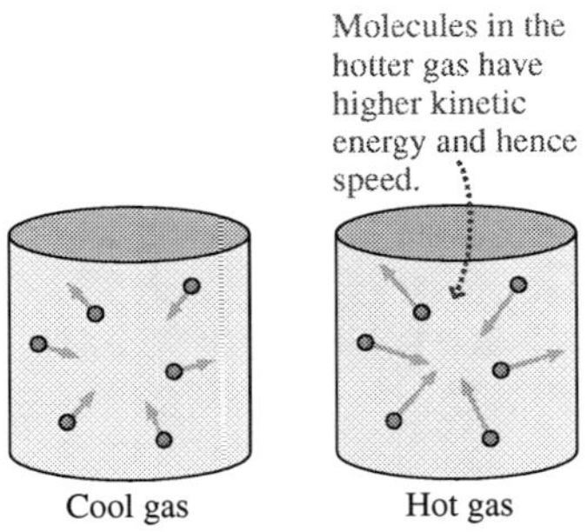

Key Concepts and Equations

The **ideal-gas law** relates pressure, volume, temperature, and the number of molecules in a gas:

$$pV = NkT \qquad \text{(ideal-gas law)}$$

where **Boltzmann's constant** is $k = 1.38 \times 10^{-23}$ J/K.

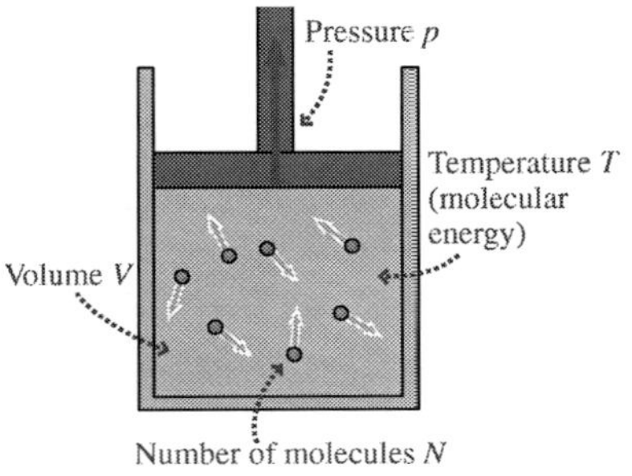

In terms of the number of moles n, the ideal-gas law is

$$pV = nN_AkT = nRT$$

where the **universal gas constant** $R = N_Ak = 8.314$ J/K·mol.

The temperature of an ideal gas is a measure of the gas molecules' average kinetic energy:

$$\tfrac{1}{2}m\overline{v^2} = \tfrac{3}{2}kT \qquad \text{(temperature and molecular energy)}$$

Heats of transformation L describe the energy per unit mass needed to effect phase changes. The total energy required to change the phase of a mass m is given by

$$Q = Lm \qquad \text{(heat of transformation)}$$

Phase diagrams plot solid, liquid, and gas phases against temperature and pressure, and reveal the **triple point**, where all three phases can coexist, and the **critical point**, where the liquid-gas distinction disappears.

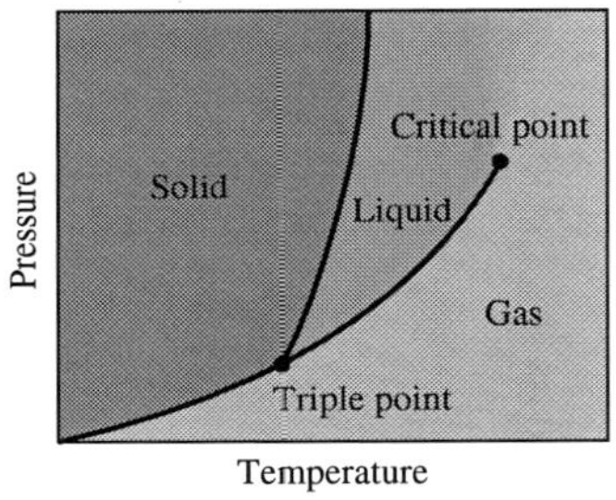

Cases and Use

Thermal expansion results when pressure is held constant. It's described by the **coefficient of volume expansion** and its linear counterpart. The volume-expansion coefficient relates the fractional volume change $\Delta V/V$ to the temperature change ΔT:

$$\beta = \frac{\Delta V/V}{\Delta T} \qquad \text{(volume-expansion coefficient)}$$

while the **coefficient of linear expansion** relates the fractional length $\Delta L/L$ change to ΔT:

$$\alpha = \frac{\Delta L/L}{\Delta T} \qquad \text{(linear-expansion coefficient)}$$

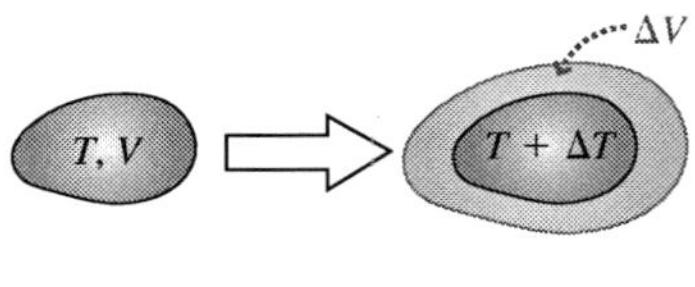

For Thought and Discussion

1. If the volume of an ideal gas is increased, must the pressure drop proportionately? Explain.
2. According to the ideal-gas law, what should the volume of a gas be at absolute zero? Why is this result absurd?
3. Why are you supposed to check the pressure in a tire when the tire is cold?
4. The average *speed* of the molecules in a gas increases with increasing temperature. What about the average *velocity?*
5. Suppose you start running while holding a jar of air. Do you change the average speed of the air molecules? The average velocity? The temperature?
6. Do all molecules in a gas have the same speed? If not, how can the notion of temperature be meaningful?
7. Two different gases are at the same temperature, and both are at low enough densities that they behave like ideal gases. Do their molecules have the same thermal speeds? Explain.
8. Some people think that ice and snow must be at 0°C. Is this always true?
9. What is the temperature of water just under the ice layer of a frozen lake? At the bottom of a deep lake?
10. Some ice and water have been together in a glass for a long time. Is the water hotter than the ice?
11. Does it take more heat to melt a gram of ice at 0°C than to bring the resulting water to the boiling point? Once at the boiling point, does it take more heat to boil all the water than it did to bring it from the melting point to the boiling point?
12. Why does removing the plastic wrap from a package of frozen hamburger help it to thaw faster?
13. Why do we use the triple point of water for thermometer calibration? Why not just use the melting point or boiling point?
14. How is it possible to have boiling water at a temperature other than 100°C?
15. How does a pressure cooker work?
16. Suppose mercury and glass had the same coefficient of volume expansion. Could you build a mercury thermometer?
17. A bimetallic strip consists of thin pieces of brass and steel bonded together (Fig. 17.10). Such strips are often used in thermostats. What will happen when the strip is heated? *Hint:* Consult Table 17.2.

FIGURE 17.10 For Thought and Discussion 17

Exercises and Problems

Exercises

Section 17.1 Gases

18. Mars's atmospheric pressure is about 1% that of Earth, and its average temperature is around 215 K. Find the volume of 1 mol of the Martian atmosphere.
19. How many molecules are in an ideal-gas sample at 350 K that occupies 8.5 L when the pressure is 180 kPa?
20. What is the pressure of an ideal gas if 3.5 mol occupy 2.0 L at a temperature of −150°C?
21. An ideal gas occupies a volume V at 100°C. If the gas pressure is held constant, by what factor does the volume change (a) if the Celsius temperature is doubled and (b) if the kelvin temperature is doubled?
22. (a) If 2.0 mol of an ideal gas are at an initial temperature of 250 K and a pressure of 1.5 atm, what is the gas volume? (b) The pressure is now increased to 4.0 atm, and the gas volume drops to half its initial value. What is the new temperature?
23. A pressure of 1.0×10^{-10} Pa is readily achievable with laboratory vacuum apparatus. If the residual air in this "vacuum" is at 0°C, how many air molecules are in 1 L?
24. What is the thermal speed of hydrogen (H_2) molecules at 800 K?
25. In which gas are the molecules moving faster: hydrogen (H_2) at 75 K or sulfur dioxide (SO_2) at 350 K?

Section 17.2 Phase Changes

26. How much energy does it take to melt a 65-g ice cube?
27. It takes 200 J to melt an 8.0-g sample of one of the substances in Table 17.1. What is the substance?
28. If it takes 840 kJ to vaporize a sample of liquid oxygen, how large is the sample?
29. Carbon dioxide sublimes (changes from solid to gas) at 195 K. The heat of sublimation is 573 kJ/kg. How much heat must be extracted from 250 g of CO_2 gas at 195 K in order to solidify it?
30. Find the energy needed to convert 28 kg of liquid oxygen at its boiling point into gas.
31. What is the density, in moles per cubic meter, of air in a tire whose absolute pressure is 300 kPa at 34°C?

Section 17.3 Thermal Expansion

32. A copper wire is 20 m long on a winter day when the temperature is −12°C. By how much does its length increase on a 26°C summer day?
33. You have exactly 1 L of ethyl alcohol at room temperature (20°C). You put it in a refrigerator at 2°C. What's its new volume?
34. A Pyrex glass marble is 1.00000 cm in diameter at 20°C. What will its diameter be at 85°C?
35. At 0°C, the hole in a steel washer is 9.52 mm in diameter. To what temperature must it be heated in order to fit over a 9.55-mm-diameter bolt?
36. Suppose a single piece of welded steel railroad track stretched 5000 km across the continental United States. If the track were free to expand, by how much would its length change if the entire track went from a cold winter temperature of −25°C to a hot summer day at 40°C?

Problems

37. The solar corona is an extended atmosphere of hot $(2\times10^6\ \text{K})$ gas surrounding the cooler visible surface of the Sun. The gas pressure in the solar corona is about 0.03 Pa. What is the coronal density in particles per cubic meter? Compare with Earth's atmosphere.
38. A helium balloon occupies 8.0 L at 20°C and 1.0-atm pressure. The balloon rises to an altitude where the air pressure is 0.65 atm and the temperature is −10°C. What is its volume when it reaches equilibrium at the new altitude? *Note:* Neglect tension forces in the material of the balloon.

39. A compressed air cylinder stands 100 cm tall and has an internal diameter of 20.0 cm. At room temperature, the pressure is 180 atm. (a) How many moles of air are in the cylinder? (b) What volume would this air occupy at 1.0 atm and room temperature?
40. An aerosol can of whipped cream is pressurized at 440 kPa when it's refrigerated at 3°C. The can warns against temperatures in excess of 50°C. What is the maximum safe pressure for the can?
41. A 3000-mL flask is initially open while in a room containing air at 1.00 atm and 20°C. The flask is then closed and immersed in a bath of boiling water. When the air in the flask has reached thermodynamic equilibrium, the flask is opened and air is allowed to escape. The flask is then closed and cooled back to 20°C. (a) What is the maximum pressure reached in the flask? (b) How many moles escape when air is released from the flask? (c) What is the final pressure in the flask?
42. A stove burner supplies heat to a pan at the rate of 1500 W. How long will it take to boil away 1.1 kg of water, from the time the water is at its boiling point?
43. If a 1-megaton nuclear bomb were exploded deep in the Greenland ice cap, how much ice would it melt? Assume the ice is initially at about its freezing point, and consult Appendix C for the appropriate energy conversion.
44. How much ice can a 625-W microwave oven melt in 1.0 min if the ice is initially at 0°C?
45. At winter's end, Lake Superior's 82,000-km^2 surface is frozen to a depth of 1.3 m. The density of ice is 917 kg/m^3. (a) How much energy does it take to melt the ice? (b) If the ice disappears in 3 weeks, what is the average power supplied to melt it?
46. A refrigerator extracts energy from its contents at the rate of 95 W. How long will it take to freeze 750 g of water already at 0°C?
47. An ice layer 50 cm thick covers a lake; soot from air pollution covers the ice and absorbs 75% of the incident sunlight, whose intensity averages 200 W/m^2. How long will it take to melt the ice, assuming an initial temperature of about 0°C? The density of ice is 917 kg/m^3.
48. Repeat Example 17.4 if the initial mass of ice is 50 g.
49. How much energy does it take to melt 10 kg of ice initially at -10°C?
50. Water is brought to its boiling point and then allowed to boil away completely. If the energy needed to raise the water to the boiling point is one-tenth of that needed to boil it away, what was the initial temperature?
51. During a nuclear accident, 420 m^3 of emergency cooling water at 20°C are injected into a reactor vessel where the reactor core is producing heat at the rate of 200 MW. If the water is allowed to boil at normal atmospheric pressure, how long will it take to boil the reactor dry?
52. What is the minimum amount of ice in Example 17.4 that will ensure a final temperature of 0°C?
53. A bowl contains 16 kg of punch (essentially water) at a warm 25°C. What is the minimum amount of ice at 0°C that will cool the punch to 0°C?
54. A 50-g ice cube at -10°C is placed in an equal mass of water. What must the initial water temperature be if the final mixture still contains equal amounts of ice and water?
55. A 40-kg block of aluminum is initially at 50°C. A jet of steam at 100°C hits the block and condenses, and the resulting water drops off before its temperature changes. How much steam must hit the block to raise its temperature to 100°C?
56. What power is needed to melt 20 kg of ice in 6.0 min?
57. You put 300 g of water into a 500-W microwave oven and accidentally set the time for 20 min instead of 2.0 min. If the water is initially at 20°C, how much is left at the end of 20 min?
58. If 4.5×10^5 kg of emergency cooling water at 10°C are dumped into a malfunctioning nuclear reactor whose core is producing energy at the rate of 200 MW, and if no circulation or cooling of the water is provided, how long will it be before half the water has boiled away?
59. Describe the composition and temperature of the equilibrium mixture after 1.0 kg of ice at -40°C is added to 1.0 kg of water at 5.0°C.
60. A glass marble 1.000 cm in diameter is to be dropped through a hole in a steel plate. At room temperature the hole diameter is 0.997 cm. By how much must the steel temperature be raised so that the marble will fit through the hole?
61. A 2000-mL graduated cylinder is filled with liquid at 350 K. When the liquid is cooled to 300 K, the cylinder is full to only the 1925-mL mark. Use Table 17.1 to identify the liquid.
62. A steel ball bearing is encased in a Pyrex glass cube 1.0 cm on a side. At 330 K, the ball bearing fits tightly inside the cube. At what temperature will it have a clearance of 1.0 μm all around?
63. Gasoline comes from its underground tank at 10°C. On a summer day, how much gas can you put in your car's 60-L tank if the tank is not to overflow when the gas reaches the ambient temperature of 25°C?
64. A rod of length L_0 is clamped rigidly at both ends. Its temperature increases by an amount ΔT, and in the ensuing expansion it cracks to form two straight pieces, as shown in Fig. 17.11. Find an expression for the distance d shown in the figure, in terms of L_0, ΔT, and the coefficient of linear expansion α.

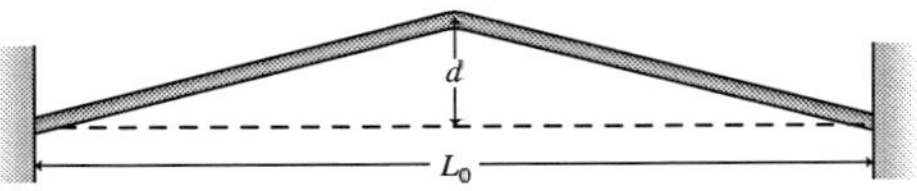

FIGURE 17.11 Problem 64

65. The thermal resistance of a refrigerator's walls is 0.12 K/W. During a power failure you put a 15-kg block of ice at 0°C in the refrigerator, bringing the interior temperature to 0°C. If room temperature is 20°C, how long will the ice last?
66. A solar-heated house stores energy in 5.0 tons of Glauber salt ($Na_2SO_4 \cdot 10\,H_2O$), a substance that melts at 90°F. The heat of fusion of Glauber salt is 104 Btu/lb, and the specific heats of the solid and liquid are, respectively, 0.46 Btu/lb·°F and 0.68 Btu/lb·°F. After a week of sunny weather, the storage medium is all liquid at 95°F. Then a cool, cloudy period sets in during which the house loses heat at an average rate of 20,000 Btu/h. (a) How long is it before the temperature of the storage medium drops below 60°F? (b) How much of this time is spent at 90°F?
67. Show that the coefficient of volume expansion of an ideal gas at constant pressure is the reciprocal of its kelvin temperature.
68. Water's coefficient of volume expansion in the temperature range from 0°C to about 20°C is given approximately by $\beta = a + bT + cT^2$, where T is in Celsius and $a = -6.43 \times 10^{-5}$ °C^{-1}, $b = 1.70 \times 10^{-5}$ °C^{-2}, and $c = -2.02 \times 10^{-7}$ °C^{-3}. Show that water has its greatest density at approximately 4.0°C.
69. When the expansion coefficient varies with temperature, Equation 17.6 should be written $\beta = (1/V)(dV/dT)$. If a sample of water occupies 1.00000 L at 0°C, find its volume at 12°C. Use the information from Problem 68, and integrate the equation above.

70. Ignoring air resistance, find the height from which you must drop an ice cube at 0°C so it melts completely on impact. Assume no heat exchange with the environment.
71. The timekeeping of an old clock is regulated by a brass pendulum 20.0 cm long. If the clock is accurate at 20°C but is in a room at 18°C, how long will it be before the clock is in error by 1 minute? Will it be too fast or too slow?
72. Prove the equation $\beta = 3\alpha$ (Section 17.3) by considering a cube of side s and therefore volume $V = s^3$ that undergoes a small temperature change dT and corresponding length and volume changes ds and dV.
73. Venus's atmospheric pressure is 90 times that of Earth, and its average temperature is 730 K. Suppose you're on the team planning a mission to Venus to collect samples of the atmosphere and soil for analysis. The design engineers are planning a container for the atmospheric sample that can hold 1 L of gas. The atmospheric scientists need at least 1 mol of gas in order to do their experiments. Will the design work?
74. Your chemistry-major roommate is in the midst of an experiment. A 140-W heat source is being applied to 250 g of ethyl alcohol in order to evaporate it completely. Rather than sit around and wait for the whole process, your roommate, who has not had lunch, wants to know if there's at least 30 minutes of time before the alcohol entirely evaporates. In addition to a lecture on not leaving boiling alcohol unattended, how do you respond?
75. For your first case out of law school you're representing a microwave oven manufacturer. The plaintiff complains the oven was defective and caused a fire. The argument claims the 500-W oven can completely vaporize 500 g of ice at 0°C in 10 minutes. Is that correct?
76. Perhaps for a literature class you have read the book *Fahrenheit 451* or seen the film version. The setting (in England) is an Orwellian society where books are banned and all information is disseminated by a large TV screen in each home. Fire departments respond not to put out fires but to burn books, which combust at a temperature of 451°F—hence the title. Suppose the fire department decides to use the burning books to heat water for afternoon tea. This is done with a burner that transfers 80% of the heat from the burning book to 500 mL of water on top of the burner. The water is initially at 25°C and needs to be raised to 100°C. How many copies of a 500-page book would need to burn if the energy released in combustion is 5 J/page?

Answers to Chapter Questions

Answer to Chapter Opening Question

Water's solid phase is less dense than the liquid, which causes ice to float. Our world would be a very different place if ice were denser than water.

Answers to GOT IT? Questions

17.1 It increases by a factor of $\sqrt{2}$.
17.2 (c) Equal to 100°C.
17.3 (a) Larger, since all linear dimensions of the object expand equally.

18 Heat, Work, and the First Law of Thermodynamics

■ A jet aircraft engine converts the energy of burning fuel into mechanical energy. How does energy conservation apply in this process?

► To Learn

By the end of this chapter you should be able to

- Explain how the first law of thermodynamics extends the conservation-of-energy principle to include thermal energy (18.1).
- Describe quantitatively the effects of basic thermodynamic processes—isothermal, adiabatic, isobaric, and constant-volume—on an ideal gas (18.2).
- Compute the specific heat of an ideal gas based on its molecular structure (18.3).

◄ To Know

- This chapter joins the concept of work, introduced in Chapter 6 (6.1, 6.2), and the behavior of ideal gases as described in Chapter 17 (17.1).
- You should be ready to extend the conservation-of-energy principle introduced in Chapter 7 (7.3).

In Chapter 7 we introduced the powerful idea of energy conservation, limited then to mechanical energy and conservative forces. Now we've learned that thermal processes involve energy—and that sets the stage for a broadening of the conservation-of-energy principle. Here we explore this broader principle and see how it describes energy interchanges in thermodynamic systems ranging from engines to Earth's atmosphere.

18.1 The First Law of Thermodynamics

Figure 18.1 shows two ways to raise the temperature in a beaker of water: by heating with a flame and by stirring vigorously with a spoon. Using the flame involves heat—energy in transit because of the temperature difference between flame and water. But there's no temperature difference between spoon and water; here the energy transfer occurs because the spoon does mechanical work on the water. We already know that doing work can increase the kinetic or potential energy of a macroscopic object; here we see it, instead, changing the **internal energy** associated ultimately with individual molecules. The point is that both processes—heating and mechanical work—result in exactly the same final state—

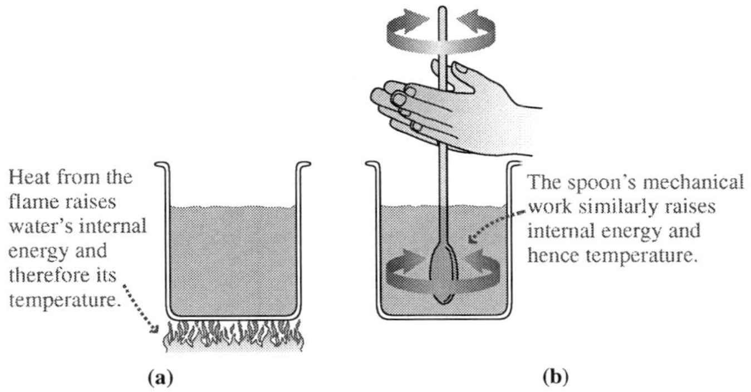

FIGURE 18.1 Two ways to raise temperature.

namely, water with a higher temperature and therefore greater internal energy. It's this common result that made possible Joule's quantitative identification of heat as a form of energy (Fig. 18.2).

Keep track of all the energy entering and leaving a system—both heat and work—and you'll find that the change in the system's internal energy depends only on the net energy transferred. In one sense this is hardly surprising; it just extends the idea of energy conservation to include heat. But in another way it's remarkable; it doesn't matter at all *how* the energy gets into the system—heat, work, or some combination of the two. This statement constitutes the **first law of thermodynamics**:

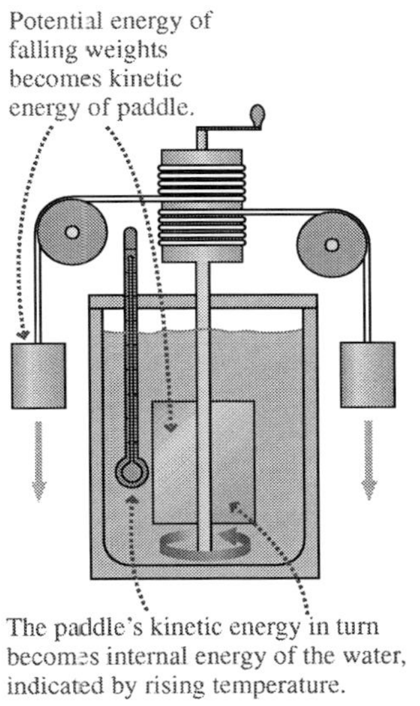

FIGURE 18.2 Joule's apparatus for determining what he called "the mechanical equivalent of heat."

> **First law of thermodynamics** The change in the internal energy of a system depends only on the net heat transferred to the system and the net work done by the system, independent of the particular processes involved.

Mathematically, the first law is

$$\Delta U = Q - W \quad \text{(first law of thermodynamics)} \tag{18.1}$$

where ΔU is the change in a system's internal energy, Q the heat transferred *to* the system, and W the work done *by* the system. A positive Q represents an energy gain, while a positive W represents an energy loss; that's the minus sign in Equation 18.1. This sign convention is used because the first law was developed in connection with engines, which take heat from a source and give out mechanical work.

The first law says that the change in a system's internal energy doesn't depend on how the energy gets transferred, but only on the net energy. Internal energy is therefore a **thermodynamic state variable**, meaning a quantity whose value doesn't depend on how a system got into its particular state. Temperature and pressure are also thermodynamic state variables; heat and work are not.

We're frequently concerned with *rates* of energy flow. Differentiating the first law with respect to time gives a statement about rates:

$$\frac{dU}{dt} = \frac{dQ}{dt} - \frac{dW}{dt} \tag{18.2}$$

where dU/dt is the rate of change of a system's internal energy, dQ/dt the rate of heat transfer to the system, and dW/dt the rate at which the system does work.

EXAMPLE 18.1 The First Law of Thermodynamics: Thermal Pollution

The reactor in a nuclear power plant supplies energy at the rate of 3.0 GW, boiling water to produce steam that turns a turbine-generator. The spent steam is then condensed through thermal contact with water taken from a river. If the power plant produces electrical energy at the rate of 1.0 GW, at what rate is heat transferred to the river?

INTERPRET This problem is about heat and mechanical energy, which are related by the first law of thermodynamics. We identify the system as the entire power plant, comprising the nuclear reactor, including its fuel, and the turbine-generator. We identify U as the internal energy stored in the fuel, W as the mechanical work that ends up as electrical energy, and Q as the heat transferred to the river.

DEVELOP Since we're dealing here with *rates*, Equation 18.2, $dU/dt = dQ/dt - dW/dt$, applies. The reactor extracts internal energy from its fuel, so the rate dU/dt is negative; the power plant delivers electrical energy to the outside world, so dW/dt is positive. Our plan is then to solve for dQ/dt, the rate of energy transfer to the river.

EVALUATE Solving, we have

$$\frac{dQ}{dt} = \frac{dU}{dt} + \frac{dW}{dt} = -3.0\text{ GW} + 1.0\text{ GW} = -2.0\text{ GW}$$

ASSESS Make sense? Since positive Q represents heat transferred *to* the system, the minus sign shows that heat is transferred *from* the power plant to the river at the rate of 2 GW. The numbers here are typical for large nuclear and coal-burning power plants, and show that about two-thirds of the energy extracted from the fuel is wasted in heating the environment. We'll see in the next chapter just why this waste occurs.

✓TIP Identify the System

The first law of thermodynamics deals with energy flows into and out of a system. It's up to you to define the system, and how you do so affects the meanings of the terms in the first law. In this Example we included the nuclear reactor, with the internal energy of its fuel, as part of the system. If we had considered only the turbine-generator, then we would have had 3 GW of heat coming in from the reactor and no change in internal energy. But the result would be the same: 1 GW going out as electricity and 2 GW of heat dumped into the river. ■

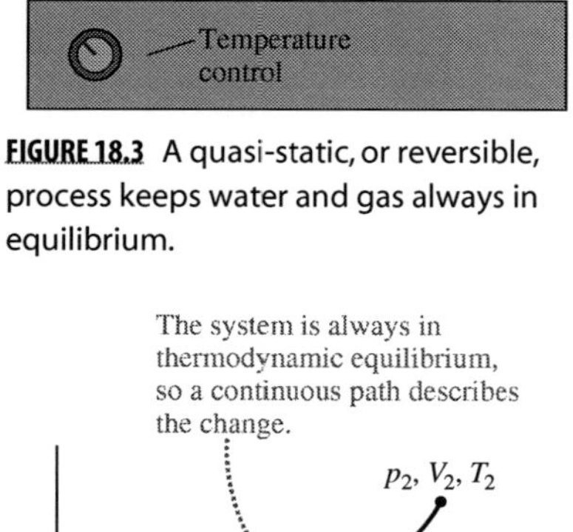

FIGURE 18.3 A quasi-static, or reversible, process keeps water and gas always in equilibrium.

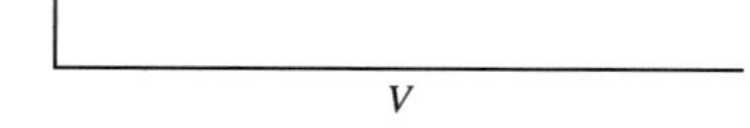

FIGURE 18.4 The pV diagram of a system undergoing quasi-static change.

18.2 Thermodynamic Processes

Although the first law applies to *any* system, it's easiest to understand when applied to an ideal gas. The ideal-gas law relates the temperature, pressure, and volume of a given gas sample: $pV = nRT$. The thermodynamic state is completely determined by any two of the quantities p, V, or T. We'll find it convenient to represent different states as points on a ***pV* diagram**—a graph whose vertical and horizontal axes represent pressure and volume, respectively.

Reversible and Irreversible Processes

Imagine a gas sample immersed in a large reservoir of water and allowed to come to equilibrium (Fig. 18.3). If we then raise the reservoir temperature very slowly, both water and gas temperatures will rise essentially in unison, and the gas will remain in equilibrium. Such a slow change is called a **quasi-static process**. Because a system undergoing a quasi-static process is always in thermodynamic equilibrium, its evolution from one state to another is described by a continuous sequence of points—a curve—in its pV diagram (Fig. 18.4).

We could reverse this heating process by slowly lowering the reservoir temperature; the gas would cool, reversing its path in the pV diagram. For that reason, a quasi-static process is also called a **reversible process**. A process like suddenly plunging a cool gas sample into hot water is, in contrast, **irreversible**. During an irreversible process the system is not in equilibrium, and thermodynamic variables like temperature and pressure don't have well-defined values. It therefore makes no sense to think of a path in the pV diagram. A process may be irreversible even though it returns a system to its original state. The distinction lies not in the end states but in the *process* that takes the system between states.

There are many ways to change the thermodynamic state of a system. Here we consider several important special cases involving an ideal gas. These illustrate the physical principles behind a myriad of technological devices and natural phenomena, from the operation of a gasoline engine to the propagation of a sound wave to the oscillations of a star.

Our system consists of an ideal gas confined to a cylinder sealed with a movable piston (Fig. 18.5). The piston and cylinder walls are perfectly insulating—they block all heat transfer—and the bottom is a perfect conductor of heat. We can change the thermodynamic state of the gas mechanically by moving the piston, or thermally by transferring heat through the bottom. We'll consider only reversible processes, which we can describe by paths in the pV diagram for the gas.

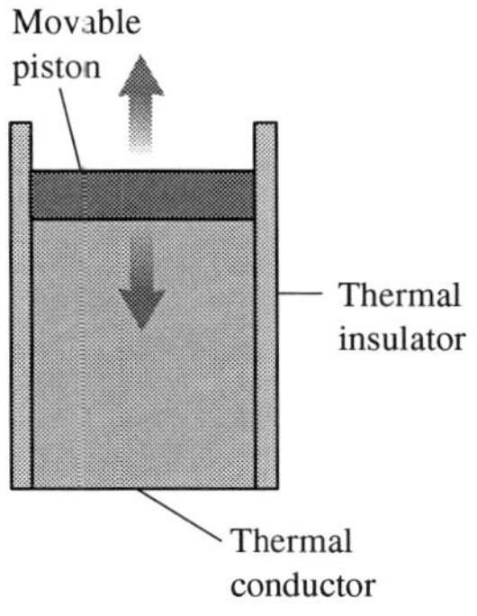

FIGURE 18.5 A gas-cylinder system with insulating walls and a conducting bottom.

Work and Volume Changes

We begin by developing a relation between volume change and work that holds for all processes. If A is the cross-sectional area of our piston and cylinder and p the gas pressure, then $F = pA$ is the force the gas exerts on the piston. If the piston moves a small distance Δx, the work done by the gas is $\Delta W = F\,\Delta x = pA\,\Delta x = p\Delta V$, where $\Delta V = A\,\Delta x$ is the change in gas volume (Fig. 18.6a). The work ΔW is positive when the volume increases, indicating that the gas does work on the piston, and negative when ΔV is negative, indicating that work is done on the gas. Pressure may vary with volume, in which case our result is strictly valid only in the limit of very small volume changes. To find the work associated with a larger change, we take the limit $\Delta V \to 0$ and integrate between the initial volume V_1 and final volume V_2:

$$W = \int dW = \int_{V_1}^{V_2} p\,dV \quad \text{(work done during volume change)} \tag{18.3}$$

Figure 18.6b shows that the work done by the gas is the area under the pV curve.

We'll now explore several basic thermodynamic processes, in each case holding one thermodynamic variable constant.

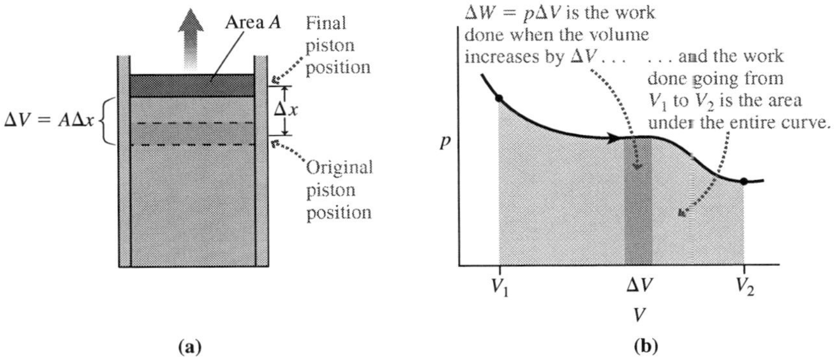

FIGURE 18.6 Work done as the piston rises is the area under the pV curve.

GOT IT? 18.1 Two identical gas-cylinder systems are taken from the same initial state to the same final state, but by different processes. Which are the same in both cases: (a) the work done on or by the gas; (b) the heat added or removed; or (c) the change in internal energy?

Isothermal Processes

An **isothermal process** occurs at constant temperature. Figure 18.7 shows one way to effect an isothermal process: We place our gas cylinder in thermal contact with a heat reservoir whose temperature is constant. We then move the piston to change the gas volume, slowly enough that the gas remains in equilibrium with the heat reservoir. The system

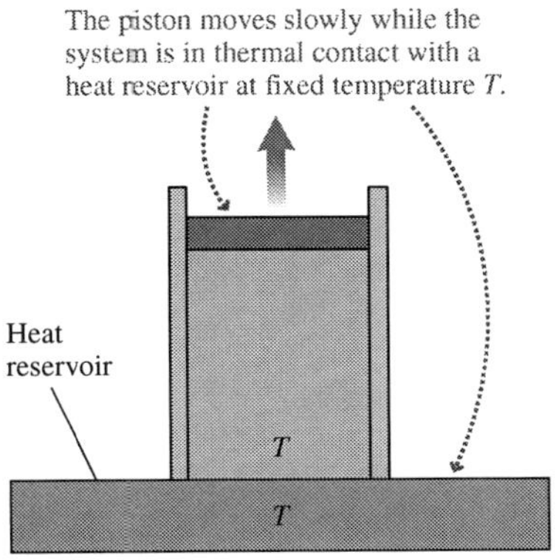

FIGURE 18.7 An isothermal process.

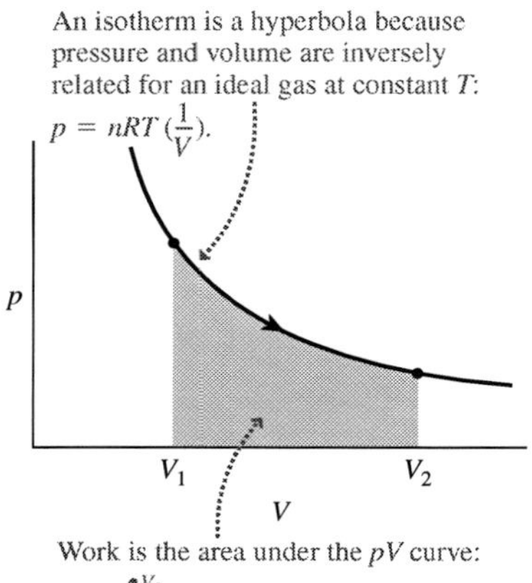

FIGURE 18.8 A pV diagram for an isothermal process.

moves from its initial state to its final state along a curve of constant temperature—an **isotherm**—in the pV diagram (Fig. 18.8). The work done in the process is given by Equation 18.3 and is equal to the area under the isotherm.

To find that work, we relate pressure and volume through the ideal-gas law: $p=(nRT)/V$. Then Equation 18.3 becomes

$$W = \int_{V_1}^{V_2} \frac{nRT}{V}\,dV$$

For an isothermal process, the temperature T is constant, giving

$$W = nRT\int_{V_1}^{V_2} \frac{dV}{V} = nRT\ln V\Big|_{V_1}^{V_2} = nRT\ln\left(\frac{V_2}{V_1}\right)$$

The internal energy of an ideal gas consists only of the kinetic energy of its molecules, which, in turn, depends only on temperature. Thus, there's no change in the internal energy of an ideal gas during an isothermal process. The first law of thermodynamics then gives $\Delta U = 0 = Q - W$, so

$$Q = W = nRT\ln\left(\frac{V_2}{V_1}\right) \quad \text{(isothermal process)} \qquad (18.4)$$

Does this result $Q=W$ make sense? Recall that Q is the heat transferred to the gas and W is the work done by the gas. Therefore, our result says that when an ideal gas does work W on the external world with no change in temperature, it must absorb an equal amount of heat from outside. Similarly, if work is done on the gas, then the gas must transfer an equal amount of heat to the outside if its temperature is not to change.

EXAMPLE 18.2 An Isothermal Process: Bubbles!

A scuba diver is 25 m down, where the pressure is 3.5 atm or about 350 kPa. The air she exhales forms bubbles 8.0 mm in radius. How much work does each bubble do as it rises to the surface, assuming the bubbles remain at the uniform 300 K temperature of the water?

INTERPRET The constant 300 K temperature tells us we're dealing with an isothermal process.

DEVELOP Equation 18.4, $W = nRT\ln(V_2/V_1)$, determines the work. To use this equation, we need the quantity nRT and the volume ratio V_2/V_1. We know p and V (actually the radius, from which we can get V) at the 25-m depth, so we can use the ideal-gas law $pV = nRT$ to get nRT and also the bubble volume just before it reaches the surface. Then we'll have everything we need to apply Equation 18.4.

EVALUATE The ideal-gas law gives $nRT = pV = \frac{4}{3}\pi r^3 p$. The number of moles n doesn't change and R is a constant, so pV is itself constant in the isothermal process. That means $p_1V_1 = p_2V_2$, showing that the volume expands by a factor of 3.5 as the pressure drops from 3.5 atm to 1 atm at the surface—so $V_2/V_1 = 3.5$. Then Equation 18.4 gives

$$W = nRT\ln\left(\frac{V_2}{V_1}\right) = \tfrac{4}{3}\pi r^3 p\ln 3.5$$

Using the 8-mm bubble radius and the 350-kPa pressure gives 0.94 J for the work. Note that we needed to use pressure in SI units here; to find the volume ratio, any units would do because V_2/V_1 followed from the pressure *ratio* p_1/p_2.

ASSESS Make sense? The work is positive because an expanding bubble pushes water outward and ultimately upward. It therefore raises the ocean's gravitational potential energy. When the bubble breaks, this excess potential energy becomes kinetic energy, appearing as small waves on the water surface. The bubble, in turn, gets its energy from heat that flows in to keep it at constant temperature. Energy is conserved! ■

Constant-Volume Processes and Specific Heat

A **constant-volume process** (also called isometric, isochoric, or isovolumic) occurs in a rigid closed container whose volume can't change. We could tightly clamp the piston in Fig. 18.5 for a constant-volume process. Because the piston doesn't move, the gas does no work, and the first law becomes simply $\Delta U = Q$. To express this result in terms of a temperature change ΔT, we introduce the **molar specific heat at constant volume** C_V, defined by

$$Q = nC_V\,\Delta T \quad \text{(constant-volume process)} \qquad (18.5)$$

where n is the number of moles. This molar specific heat is like the specific heat defined in Chapter 16, except it's per mole rather than per unit mass. Using Equation 18.5 for Q in the statement $\Delta U = Q$ gives

$$\Delta U = nC_V\,\Delta T \quad \text{(any process)} \tag{18.6}$$

For an ideal gas, the internal energy is a function of temperature alone, so $\Delta U/\Delta T$ has the same value no matter what process the gas undergoes. Therefore Equation 18.6, relating the temperature change ΔT and internal-energy change ΔU, applies not only to a constant-volume process but to *any* ideal-gas process. Why, then, have we been so careful to label C_V the specific heat *at constant volume?* Although Equation 18.6, $\Delta U = nC_V\,\Delta T$, holds for any process, it's only when there's no work that the first law lets us write $Q = \Delta U$, and therefore only for a constant-volume process that Equation 18.5 holds.

Isobaric Processes and Specific Heat

Isobaric means constant pressure. Processes occurring in systems exposed to the atmosphere are essentially isobaric. In a reversible isobaric process, a system moves along an isobar, or curve of constant pressure, in its pV diagram (Fig. 18.9). The work done as the volume changes from V_1 to V_2 is the area under the isobar, or

$$W = p(V_2 - V_1) = p\,\Delta V \tag{18.7}$$

a result we could obtain formally by integrating Equation 18.3.

Solving the first law (Equation 18.1) for Q and using our expression for work give $Q = \Delta U + W = \Delta U + p\,\Delta V$. For an ideal gas, we've just found that the change in internal energy is given by $\Delta U = nC_V\,\Delta T$ for *any* process. Therefore $Q = nC_V\,\Delta T + p\,\Delta V$ for an ideal gas undergoing an isobaric process. We define the **molar specific heat at constant pressure** C_p as the heat required to raise 1 mol of gas by 1 K at constant pressure, or $Q = nC_p\,\Delta T$. Equating our two expressions for Q gives

$$nC_p\,\Delta T = nC_V\,\Delta T + p\,\Delta V \quad \text{(isobaric process)} \tag{18.8}$$

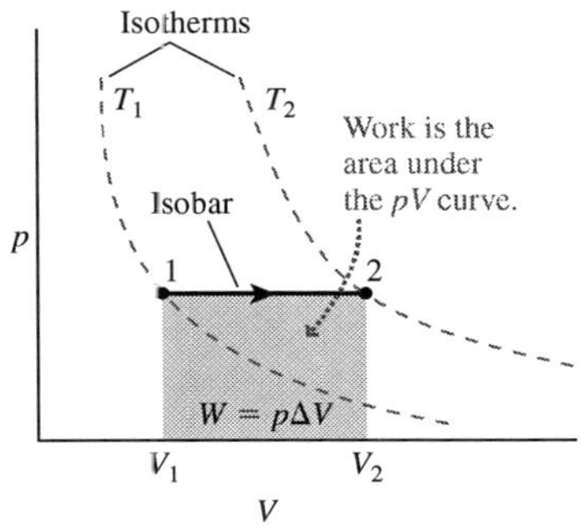

FIGURE 18.9 A pV diagram for an isobaric process; also shown are isotherms for the initial and final temperatures.

This is a useful form for calculating temperature changes in an isobaric process if we know both specific heats C_p and C_V. However, we really need only one of these specific heats because a simple relation holds between the two. The ideal-gas law, $pV = nRT$, allows us to write $p\,\Delta V = nR\,\Delta T$ for an isobaric process. Using this expression in Equation 18.8 gives $nC_p\,\Delta T = nC_V\,\Delta T + nR\,\Delta T$, so

$$C_p = C_V + R \quad \text{(molar specific heats)} \tag{18.9}$$

Does this make sense? Specific heat measures the heat needed to cause a given temperature change. In a constant-volume process, no work is done and all the heat goes into raising the internal energy and thus the temperature of an ideal gas. In a constant-pressure process, work *is* done and some of the added heat ends up as mechanical energy, leaving less available for raising the temperature. Therefore a constant-pressure process requires *more* heat for a given temperature change. Thus the specific heat at constant pressure is greater than at constant volume, as reflected in Equation 18.9.

Why didn't we distinguish specific heats at constant volume and constant pressure earlier? Because we were concerned mostly with solids and liquids, whose coefficients of expansion are far lower than those of gases. As a result, much less work is done by a solid or liquid than by a gas. Since work is what gives rise to the difference between C_V and C_p, the distinction is less significant for solids and liquids. As a practical matter, measured specific heats are usually at constant pressure.

Adiabatic Processes

In an **adiabatic process**, no heat flows between a system and its environment. The way to achieve this is to surround the system with perfect thermal insulation. Even without insulation, processes that occur quickly are often approximately adiabatic because they're over before significant heat transfer has had time to occur. In a gasoline engine, for example,

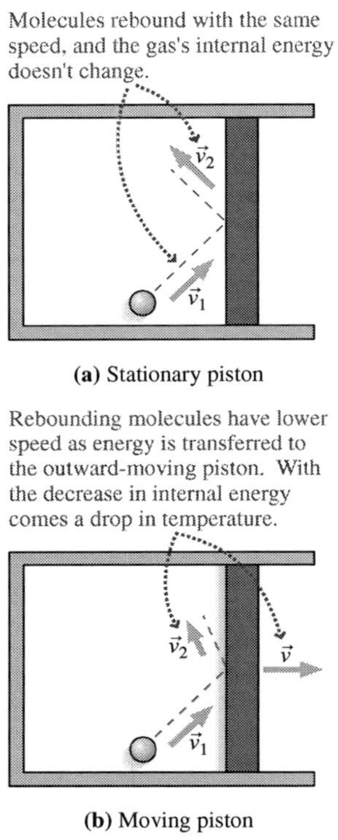

FIGURE 18.10 Internal energy decreases during an adiabatic expansion, for the reason shown in (b).

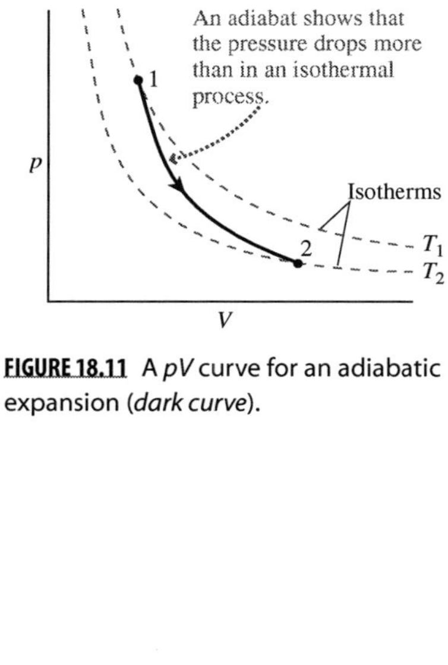

FIGURE 18.11 A pV curve for an adiabatic expansion (*dark curve*).

compression of the gasoline-air mixture and expansion of the combustion products are nearly adiabatic because they occur so rapidly that little heat flows through the cylinder walls.

Since the heat Q is zero in an adiabatic process, the first law becomes simply

$$\Delta U = -W \qquad \text{(adiabatic process)} \tag{18.10}$$

This says that if a system does work on its surroundings, then—in the absence of heat transfer—its internal energy must decrease by the same amount. Microscopically, this energy loss occurs in a gas-cylinder system when molecules give up energy as they collide with the moving piston (Fig. 18.10).

As a gas expands adiabatically, its volume increases while its internal energy and temperature decrease. The ideal-gas law, $pV = nRT$, then requires that the pressure decrease as well—and by more than it would in an isothermal process where T remains constant. In a pV diagram, the path of an adiabatic process—called an **adiabat**—is therefore steeper than the isotherms (Fig. 18.11).

Tactics 18.1 details the math involved in finding the adiabatic path; the result is

$$pV^{\gamma} = \text{constant} \qquad \text{(adiabatic process)} \tag{18.11a}$$

where $\gamma = C_p/C_V$ is the ratio of the specific heats. Because $C_p = C_V + R$, the ratio $\gamma = C_p/C_V$ is always greater than 1. As expected, an adiabatic process therefore results in a greater pressure change than would a comparable isothermal process, as reflected in the steeper adiabatic path in Fig. 18.11. Physically, the adiabatic path is steeper because the gas loses internal energy as it does work, so its temperature drops. Problem 66 shows how to rewrite Equation 18.11a in terms of temperature:

$$TV^{\gamma-1} = \text{constant} \qquad \text{(adiabatic process)} \tag{18.11b}$$

GOT IT? 18.2 The ideal-gas law says $pV = nRT$, but Equation 18.11a says $pV^{\gamma} = \text{constant}$. Which is true: (a) the ideal-gas law, (b) Equation 18.11a, or (c) both? Explain.

TACTICS 18.1 Deriving the Adiabatic Equation

Equation 18.6 gives the infinitesimal change in internal energy for *any* process: $dU = nC_V\,dT$. The corresponding work is $dW = p\,dV$ so, with $Q = 0$ in an adiabatic process, the first law becomes $nC_V\,dT = -p\,dV$. We can eliminate dT by differentiating the ideal-gas law, now letting *both* p and V change: $nR\,dT = d(pV) = p\,dV + V\,dp$. Solving for dT, substituting in our first-law statement, and multiplying through by R lead to $C_V V\,dp + (C_V + R)p\,dV = 0$. But $C_V + R = C_p$; substituting this and dividing through by $C_V pV$ give

$$\frac{dp}{p} + \frac{C_p}{C_V}\frac{dV}{V} = 0$$

Defining $\gamma \equiv C_p/C_V$ and integrating give

$$\ln p + \gamma \ln V = \ln(\text{constant})$$

where we've chosen to call the constant of integration ln(constant). Since $\gamma \ln V = \ln V^{\gamma}$, it follows by exponentiation that

$$pV^{\gamma} = \text{constant}$$

It's another exercise in calculus to integrate Equation 18.3 for the work done in an adiabatic process. You can do this yourself in Problem 62; the result is

$$W = \frac{p_1V_1 - p_2V_2}{\gamma - 1} \tag{18.12}$$

EXAMPLE 18.3 An Adiabatic Process: Diesel Power

Fuel ignites in a diesel engine from the heat of compression as the piston moves toward the top of the cylinder; there's no spark plug as in a gasoline engine. Compression is fast enough that the process is essentially adiabatic. If the ignition temperature is 500°C, what compression ratio V_{max}/V_{min} is needed (Fig. 18.12)? Air's specific-heat ratio is $\gamma = 1.4$, and before compression the air is at 20°C.

INTERPRET We identify the thermodynamic process here as adiabatic compression.

DEVELOP The problem involves temperature and volume, so Equation 18.11b applies, giving $T_{min}V_{min}^{\gamma-1} = T_{max}V_{max}^{\gamma-1}$.

EVALUATE Solving for the compression ratio V_{max}/V_{min} gives

$$\frac{V_{max}}{V_{min}} = \left(\frac{T_{min}}{T_{max}}\right)^{1/(\gamma-1)} = \left(\frac{773\text{ K}}{293\text{ K}}\right)^{1/0.4} = 11$$

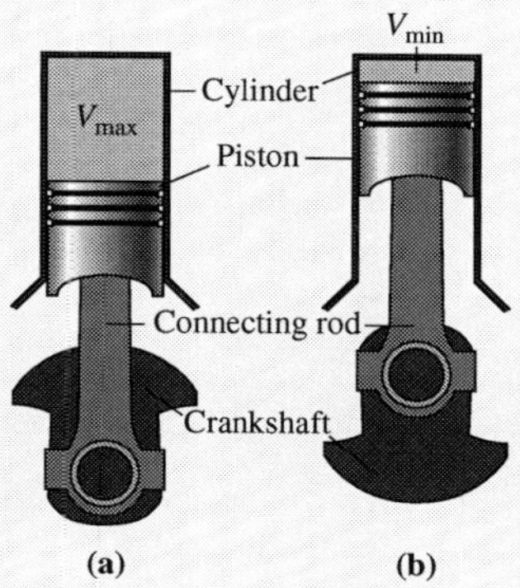

FIGURE 18.12 One cylinder of a diesel engine, shown with the piston (a) at the bottom of its stroke and (b) at the top. The compression ratio is V_{max}/V_{min}.

ASSESS Practical diesel engines have higher ratios to ensure reliable ignition. Their high compression makes diesels heavier than their gasoline counterparts, but also more fuel efficient. Problem 67 explores the diesel engine further. ■

APPLICATION Smog Alert!

The smog that blankets urban areas is an unfortunate manifestation of our prolific fossil-fueled energy consumption. Adiabatic processes in the atmosphere determine whether or not smog lingers over a city. Consider a volume of air that's heated, perhaps because it's over hot pavement that absorbs solar energy. The air becomes less dense, and its buoyancy makes it rise. As it ascends into regions of lower pressure, it expands, doing work against the surrounding atmosphere. Air is a poor heat conductor, so the process is essentially adiabatic. Therefore the gas cools as it does work.

Now, temperature in the atmosphere normally decreases with altitude. So here's the crucial question: Does the rising air cool faster or slower than the surrounding atmosphere? If it cools more slowly, then it continues to be warmer, and it continues to rise. Any pollution is carried high into the atmosphere where it's dispersed. But if the decrease in air temperature with altitude isn't great, or in an **inversion** where it's actually warmer aloft, the rising air will soon reach equilibrium with its surroundings and won't rise any higher. The effect is to trap air and its entrained pollutants near the surface, as shown in this photo of Los Angeles. Smog alert!

GOT IT? 18.3 Name the basic thermodynamic process involved when each of the following is done to a piston-cylinder system containing ideal gas, and tell also whether temperature, pressure, volume, and internal energy increase or decrease: (a) The piston is locked in place and a flame is applied to the bottom of the cylinder; (b) the cylinder is completely insulated and the piston is pushed downward; (c) the piston is exposed to atmospheric pressure and is free to move, while the cylinder is cooled by placing it on a block of ice.

Cyclic Processes

Many natural and technological systems undergo **cyclic processes**, in which the system returns periodically to the same thermodynamic state. Engineering examples include engines and refrigerators whose mechanical construction ensures cyclic behavior. Many natural oscillations, like those of a sound wave or a pulsating star, are essentially cyclic.

Cyclic processes often involve the four basic processes we've just explored, as summarized in Table 18.1. We've seen that the work done in any reversible process is just the area under the pV curve. A cyclic process returns to the same point in the pV diagram, so it involves both expansion and compression (Fig. 18.13). During expansion the gas does work on its surroundings; during compression work gets done on the gas. The net work is the difference between the two Figure 18.13c shows that this work is the area enclosed by the cyclic path in the pV diagram.

TABLE 18.1 Ideal-Gas Processes

	ISOTHERMAL	CONSTANT-VOLUME	ISOBARIC	ADIABATIC
pV diagram	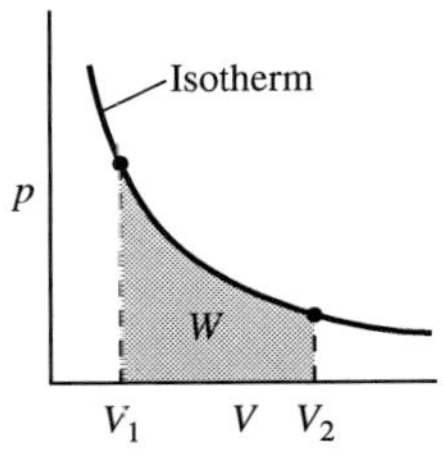	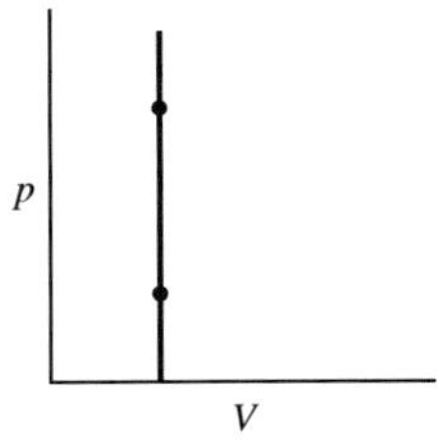	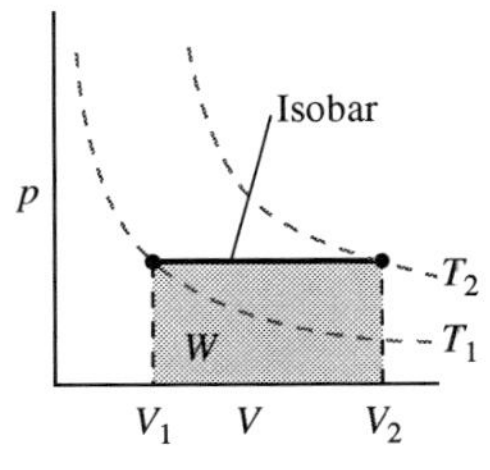	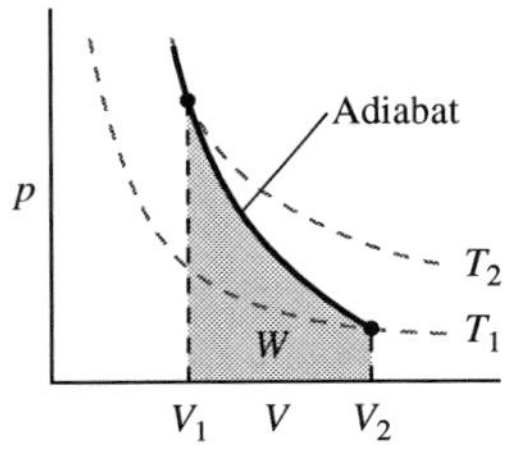
Defining characteristic	$T =$ constant	$V =$ constant	$p =$ constant	$Q = 0$
First law	$Q = W$	$Q = \Delta U$	$Q = \Delta U + W$	$\Delta U = -W$
Work done by gas	$W = nRT \ln\left(\frac{V_2}{V_1}\right)$	$W = 0$	$W = p(V_2 - V_1)$	$W = \frac{p_1V_1 - p_2V_2}{\gamma - 1}$
Other relationships	$pV =$ constant	$Q = nC_V\Delta T$	$Q = nC_p\Delta T$ $C_p = C_V + R$	$pV^{\gamma} =$ constant $TV^{\gamma-1} =$ constant

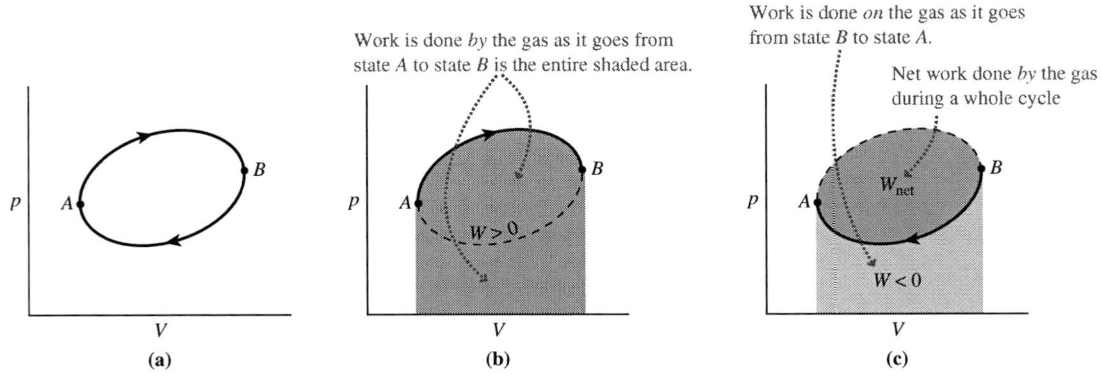

FIGURE 18.13 (a) A pV diagram for a cyclic process. (b), (c) Work done in one cycle is the area inside the closed path.

EXAMPLE 18.4 A Cyclic Process: Finding the Work

An ideal gas with $\gamma = 1.4$ occupies 4.0 L at 300 K and 100 kPa pressure. It's compressed adiabatically to one-fourth of its original volume, then cooled at constant volume back to 300 K, and finally allowed to expand isothermally to its original volume. How much work is done on the gas?

INTERPRET This problem involves a cyclic process, and we identify three separate thermodynamic processes that make up the cycle: adiabatic, constant-volume, and isothermal.

DEVELOP Here it helps to draw a pV diagram, shown in Fig. 18.14. Our plan is to use equations in Table 18.1 to determine the work for each of the basic processes and then combine them to get the net work. For the adiabatic process AB, Table 18.1 gives $W_{AB} = (p_A V_A - p_B V_B)/(\gamma - 1)$; for the constant-volume process BC, $W_{BC} = 0$; and for the isothermal process CA, the work is $W_{CA} = nRT \ln(V_A/V_C)$.

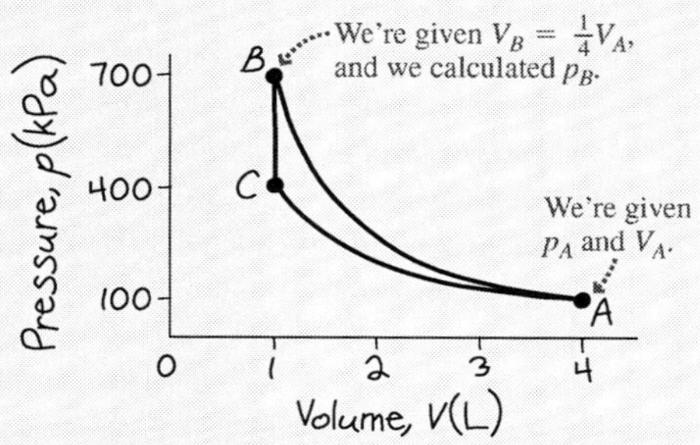

FIGURE 18.14 The cyclic process $ABCA$ of Example 18.4 includes adiabatic (AB), constant-volume (BC), and isothermal (CA) sections.

EVALUATE For the adiabatic process AB we're given all quantities except p_B. This we can get from the adiabatic equation $pV^\gamma = \text{constant}$, or $p_B V_B^\gamma = p_A V_A^\gamma$. Solving gives $p_B = p_A(V_A/V_B)^\gamma = 696.4$ kPa, where we used the given information $p_A = 100$ kPa, $\gamma = 1.4$, and a compression to one-fourth the original volume $(V_A/V_B = 4)$. We now have enough information to find the work done over the adiabatic path:

$$W_{AB} = \frac{p_A V_A - p_B V_B}{\gamma - 1} = -741 \text{ J}$$

where, with pressures in kPa $(= 10^3 \text{ Pa})$ and volumes in L $(= 10^{-3} \text{ m}^3)$, the factors $10^{\pm 3}$ cancel and there's no need to convert. The work W_{AB} is *negative* because work is done *on* the gas when it's compressed.

In the expression $W_{CA} = nRT \ln(V_A/V_C)$ for the isothermal work, we can evaluate the quantity nRT at *any* point on the isothermal curve because T is constant. The ideal-gas law says that $nRT = pV$, and we know both p and V at point A. So $nRT = p_A V_A = 400$ J, where again we could multiply $p_A = 100$ kPa by $V_A = 4.0$ L to get an answer in SI units. The isothermal work is then

$$W_{CA} = nRT \ln\left(\frac{V_A}{V_C}\right) = (400 \text{ J})(\ln 4) = 555 \text{ J}$$

This is positive because the gas does work in expanding from C to A.

Combining our results for all three segments gives the net work:

$$W_{ABCA} = W_{AB} + W_{BC} + W_{CA} = -741 \text{ J} + 0 \text{ J} + 555 \text{ J} = -186 \text{ J}$$

ASSESS Make sense? The final answer is negative because we've done net work *on* the gas; that's always the case in going counterclockwise around a cyclic path in a pV diagram. Since the system returns to its original state, its internal energy undergoes no net change. That means all the work that's done on it must be transferred to its surroundings as heat. Since no heat flows during the adiabatic process AB, and since the gas *absorbs* heat during the isothermal expansion CA, the only time it transfers heat to its surroundings is during the constant-volume cooling process BC. ■

18.3 Specific Heats of an Ideal Gas

We've found that the thermodynamic behavior of an ideal gas depends on the specific heats C_V and C_p. What are the values of those quantities?

Our ideal-gas model of Chapter 17 assumed the gas molecules were structureless point particles with only translational kinetic energy. The internal energy U of the gas is the sum of all those molecular kinetic energies. But the average kinetic energy is directly proportional to the temperature: $\frac{1}{2}m\overline{v^2} = \frac{3}{2}kT$. If we have n moles of gas, the internal energy is then $U = nN_A\left(\frac{1}{2}m\overline{v^2}\right) = \frac{3}{2}nN_A kT$, where N_A is Avogadro's number. But $N_A k = R$, the gas constant, so $U = \frac{3}{2}nRT$. Solving Equation 18.6 for the molar specific heat then gives

$$C_V = \frac{1}{n}\frac{\Delta U}{\Delta T} = \frac{3}{2}R \qquad (18.13)$$

For this gas of structureless particles, the adiabatic exponent γ is therefore

$$\gamma = \frac{C_p}{C_V} = \frac{C_V + R}{C_V} = \frac{\frac{5}{2}R}{\frac{3}{2}R} = \frac{5}{3} = 1.67$$

Some gases, notably the inert gases helium (He), neon (Ne), argon (Ar), and others in the last column of the periodic table, have adiabatic exponents and specific heats given by these equations. But others do not. At room temperature, for example, hydrogen (H_2), oxygen (O_2), and nitrogen (N_2) obey adiabatic laws with γ very nearly $\frac{7}{5}(= 1.4)$ and, correspondingly, specific heat $C_V = \frac{5}{2}R$. On the other hand, sulfur dioxide (SO_2) and nitrogen dioxide (NO_2) have specific-heat ratios close to 1.3 and therefore C_V of about $3.4R$.

What's going on here? A clue lies in the structure of individual gas molecules, reflected in their chemical formulas. The inert-gas molecules are **monatomic**, consisting of single atoms. To the extent that these atoms behave like structureless mass points, the only energy they can have is kinetic energy of translational motion. We can think of that kinetic energy as being a sum of *three* terms, each associated with motion in one of the three mutually perpendicular directions. We call each separate term in the energy of a system a **degree of freedom**, meaning a way that system can take on energy. So a monatomic molecule has three degrees of freedom.

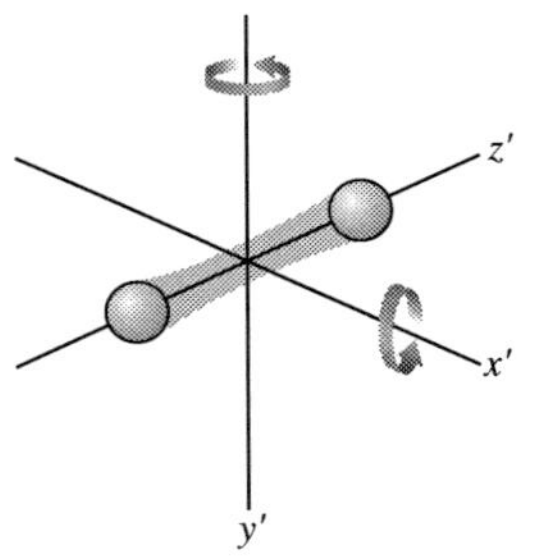

FIGURE 18.15 A diatomic molecule can have significant rotation about two perpendicular axes.

In contrast, hydrogen, oxygen, and nitrogen molecules are **diatomic**, as shown in Fig. 18.15. Although a gas of such molecules should still obey the ideal-gas law $PV = nRT$, these molecules can have rotational as well as translational kinetic energy. Then the kinetic energy of a diatomic molecule consists of *five* terms, three for the three directions of translational motion and two for rotational motions about the two mutually perpendicular axes shown in Fig. 18.15. So a diatomic molecule has five degrees of freedom. We'll now see how this difference between *three* degrees of freedom for monatomic molecules and *five* for diatomic molecules accounts for the difference between their specific heats.

The Equipartition Theorem

We showed in Chapter 17 that the average kinetic energy associated with a gas molecule's motion in one direction is $\frac{1}{2}kT$. We then argued that all three directions are equally probable, making the molecular kinetic energy, on average, $\frac{3}{2}kT$. The argument from one direction to three is based on the assumption that random collisions will share energy equally among the possible motions. When a molecule can rotate as well as translate, energy should be shared also among possible rotational motions. The 19th-century Scottish physicist James Clerk Maxwell first proved this fact, which is known as the **equipartition theorem**:

Equipartition theorem When a system is in thermodynamic equilibrium, the average energy per molecule is $\frac{1}{2}kT$ for each degree of freedom.

We've just seen that a diatomic molecule has five degrees of freedom: three translational and two rotational. The average energy of such a molecule is then $5\left(\frac{1}{2}kT\right) = \frac{5}{2}kT$, so the total internal energy in n moles of a diatomic gas is $U = nN_A\left(\frac{5}{2}kT\right) = \frac{5}{2}nRT$. Equation 18.6 then gives the molar specific heat at constant volume:

$$C_V = \frac{1}{n}\frac{\Delta U}{\Delta T} = \frac{5}{2}R \qquad \text{(diatomic molecule)}$$

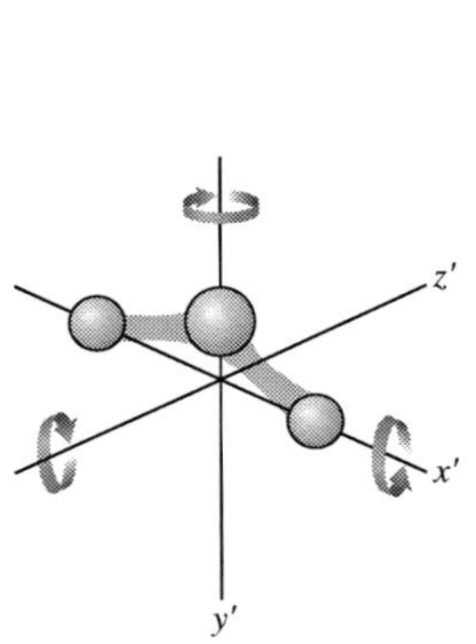

FIGURE 18.16 A triatomic molecule like NO_2 has three rotational degrees of freedom.

Our result $C_p = C_V + R$ still holds, since it was derived from the first law of thermodynamics without regard to molecular structure, so $C_p = \frac{7}{2}R$ and $\gamma = C_p/C_V = \frac{7}{5} = 1.4$. These results describe the observed behavior of diatomic gases like hydrogen, oxygen, and nitrogen at room temperature.

A polyatomic molecule like NO_2 can rotate about any of three perpendicular axes (Fig. 18.16). It then has a total of six degrees of freedom, giving $U = 3nRT$ and corresponding specific heats $C_V = 3R$ and $C_p = C_V + R = 4R$. The adiabatic exponent is then $\gamma = \frac{4}{3} \simeq 1.33$, reasonably close to the experimental value $\gamma = 1.29$ for NO_2.

EXAMPLE 18.5 Specific Heat: A Gas Mixture

A gas mixture consists of 2.0 mol of oxygen (O_2) and 1.0 mol of argon (Ar). Find the volume specific heat of the mixture.

INTERPRET This problem is about specific heat and molecular structure. We identify the molecules involved as diatomic O_2 and monatomic Ar.

DEVELOP Equation 18.6, $\Delta U = nC_V\,\Delta T$, determines the volume specific heat, so we need to find how the internal energy U depends on temperature. Our plan is to use the equipartition theorem to get the energy per molecule for each gas, then find the total energy as a function of temperature, and from that the specific heat.

EVALUATE Being diatomic, O_2 has five degrees of freedom, so the equipartition theorem gives the average energy per molecule as $\frac{5}{2}kT$. Then the total energy in $n=2$ moles of oxygen is $U_{O_2} = nN_A\left(\frac{5}{2}kT\right) = \frac{5}{2}nRT = 5.0RT$, where we used $N_Ak = R$. Monotonic Ar has three degrees of freedom, so the internal energy in our 1 mole of argon is, similarly, $U_{Ar} = \frac{3}{2}nRT = 1.5RT$. The total internal energy is then $U = 6.5RT$, so Equation 18.6 gives

$$C_V = \frac{1}{n}\frac{\Delta U}{\Delta T} = \frac{6.5R}{3.0\text{ mol}} = 2.2R$$

ASSESS Make sense? Our answer lies between the values $1.5R$ and $2.5R$ that we found for monatomic and diatomic gases, respectively. It's closer to $2.5R$ because there's more oxygen in the mixture. ■

Quantum Effects

Relating molecular structure and gas behavior is a remarkable triumph for Newtonian physics. But hidden in our analysis is an assumption that Newtonian physics can't justify. Real atoms have size, so even monatomic molecules should rotate. Why not more degrees of freedom? The answer lies in quantum physics, which requires a certain minimum energy for a periodic motion such as rotation. At normal temperatures, the average thermal energy is too low to excite rotation of monatomic molecules, or of diatomic molecules about their long axis. So these molecules exhibit three and five degrees of freedom, respectively. That results in the volume specific heats $\frac{3}{2}R$ and $\frac{5}{2}R$ that we've seen. For diatomic molecules at higher temperatures, still another motion comes into play—the simple harmonic oscillation of the two atoms due to the springlike bond between them. That adds two more degrees of freedom, corresponding to the kinetic and potential energies of this oscillation, and the specific heat increases correspondingly. At very low temperatures, in contrast, there isn't enough thermal energy to excite any rotation in a diatomic gas, and it then exhibits the specific heat $C_V = \frac{3}{2}R$ that we normally associate with a monatomic gas. Figure 18.17 shows these effects for diatomic hydrogen (H_2).

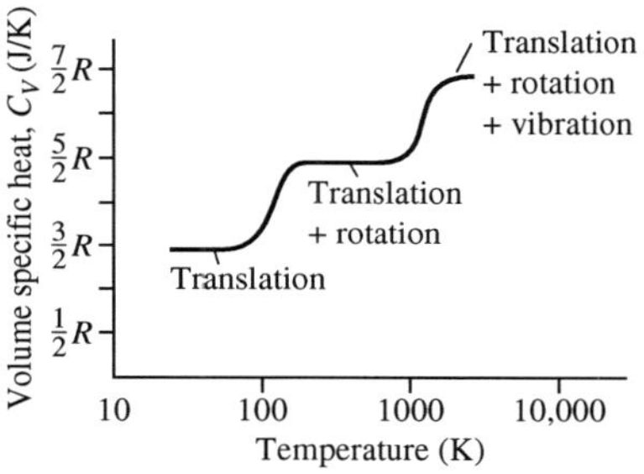

FIGURE 18.17 Volume specific heat of H_2 gas as a function of temperature. Below 20 K hydrogen is liquid, and above 3200 K it dissociates into individual atoms.

Are you bothered by the strange restrictions quantum mechanics imposes on molecular rotation and vibration? You should be! Nothing in your experience suggests that a rotating object can't have any amount of energy you care to give it. But quantum mechanics deals with a realm much smaller than that of our daily experience. The quantization of energy is only one of many unusual things that occur in the quantum realm. We'll explore more quantum phenomena in Part 6.

CHAPTER 18 SUMMARY

Big Picture

The big idea here is conservation of energy, now expanded to include heat. The expanded statement of energy conservation is the **first law of thermodynamics**, which relates the change in a system's internal energy to the heat flowing into the system and the work done by the system. The first law combines with the ideal-gas law to give a quantitative description of basic thermodynamic processes applied to ideal gases; these are described graphically using ***pV* diagrams**. The **equipartition theorem** states that in thermodynamic equilibrium, energy is shared equally among the possible energy modes of a system.

Key Concepts and Equations

Quantitatively, the first law of thermodynamics states

$$\Delta U = Q - W$$

Meaning of terms in the first law:

- ΔU is the change in a system's internal energy.
- Q is the heat transferred *to* the system.
  - Positive Q means a net heat input to the system.
  - Negative Q means heat leaves the system.
- W is the work done *by* the system.
  - Positive W means the system does work on its surroundings.
  - Negative W means work is done on the system.

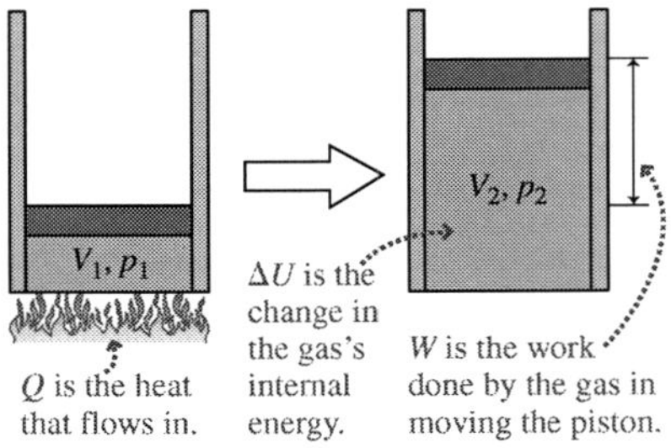

In general, the work done by a system is related to the changes in pressure and volume:

$$W = \int_{V_1}^{V_2} p\, dV$$

Cases and Uses

Ideal-gas processes:

ISOTHERMAL

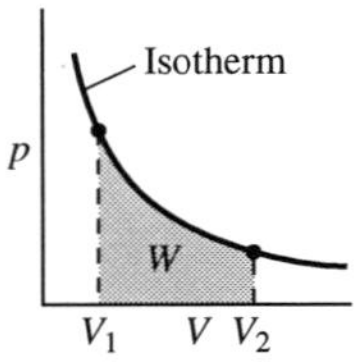

$T = \text{constant}$
$Q = W$
$W = nRT \ln\left(\frac{V_2}{V_1}\right)$
$pV = \text{constant}$

CONSTANT-VOLUME

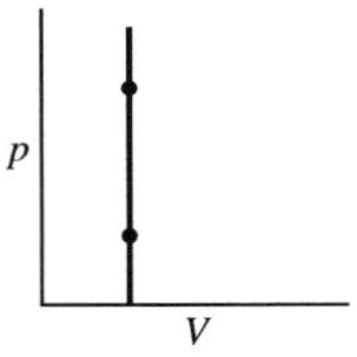

$V = \text{constant}$
$Q = \Delta U$
$W = 0$
$Q = nC_V\Delta T$

ISOBARIC

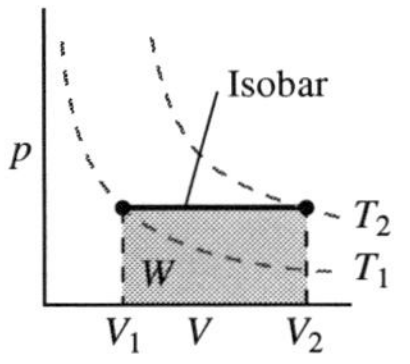

$p = \text{constant}$
$Q = \Delta U + W$
$W = p(V_2 - V_1)$
$Q = nC_p\Delta T$
$C_p = C_V + R$

ADIABATIC

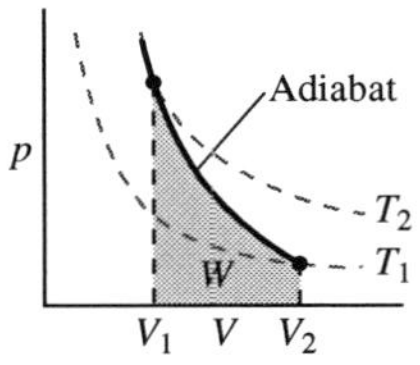

$Q = 0$
$\Delta U = -W$
$W = \frac{p_1V_1 - p_2V_2}{\gamma - 1}$
$pV^\gamma = \text{constant}$
$TV^{\gamma-1} = \text{constant}$

The specific heats of an ideal gas follow from the **degrees of freedom** of each molecule:

Monatomic
3 degrees of freedom
$C_V = \frac{3}{2}R$

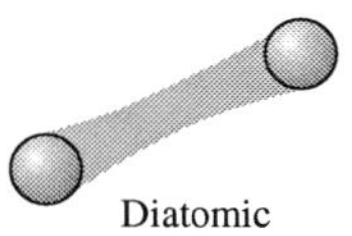

Diatomic
5 degrees of freedom
$C_V = \frac{5}{2}R$

For Thought and Discussion

1. The temperature of the water in a jar is raised by violently shaking the jar. Which of the terms Q and W in the first law of thermodynamics is involved in this case?
2. What is the difference between heat and internal energy?
3. Some water is tightly sealed in a perfectly insulated container. Is it possible to change the water temperature? Explain.
4. Are the initial and final equilibrium states of an irreversible process describable by points in a pV diagram? Explain.
5. Why can't an irreversible process be described by a path in a pV diagram?
6. Does the first law of thermodynamics apply to an irreversible process?
7. A quasi-static process begins and ends at the same temperature. Is the process necessarily isothermal?
8. Figure 18.18 shows two processes, A and B, that connect the same initial and final states, 1 and 2. For which process is more heat added to the system?

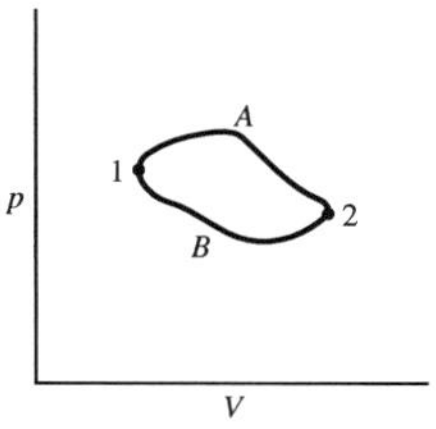

FIGURE 18.18 For Thought and Discussion 8

9. When you let air out of a tire, the air seems cool. Why? What kind of process is occurring?
10. Blow on the back of your hand with your mouth wide open. Your breath will feel hot. Now tighten your lips into a small opening and blow again. Now your breath feels cool. Why?
11. Water is boiled in an open pan. Of which of the four basic processes we considered is this an example?
12. Three identical gas-cylinder systems expand from the same initial state to final states that have the same volume. One system expands isothermally, one adiabatically, and one isobarically. Which does the most work? Which does the least work?
13. Why is the specific heat at constant pressure greater than at constant volume?
14. In what sense can a gas of diatomic molecules be considered an ideal gas, given that its molecules aren't point particles?

Exercises and Problems

Exercises

Section 18.1 The First Law of Thermodynamics

15. In a perfectly insulated container, 1.0 kg of water is stirred vigorously until its temperature rises by 7.0°C. How much work was done on the water?
16. In a closed but uninsulated container, 500 g of water are shaken violently until the temperature rises by 3.0°C. The mechanical work required in the process is 9.0 kJ. (a) How much heat is transferred during the shaking? (b) How much mechanical energy would have been required if the container had been perfectly insulated?
17. A 40-W heat source is applied to a gas sample for 25 s, during which time the gas expands and does 750 J of work on its surroundings. By how much does the internal energy of the gas change?
18. What is the rate of heat flow into a system whose internal energy is increasing at the rate of 45 W, given that the system is doing work at the rate of 165 W?
19. In a certain automobile engine, 17% of the total energy released in burning gasoline ends up as mechanical work. What is the engine's mechanical power output if its heat output is 68 kW?

Section 18.2 Thermodynamic Processes

20. An ideal gas expands from the state (p_1, V_1) to the state (p_2, V_2), where $p_2 = 2p_1$ and $V_2 = 2V_1$. The expansion proceeds along the straight diagonal path AB shown in Fig. 18.19. Find an expression for the work done by the gas during this process.
21. Repeat Problem 20 for a process that follows the path ACB in Fig. 18.19.

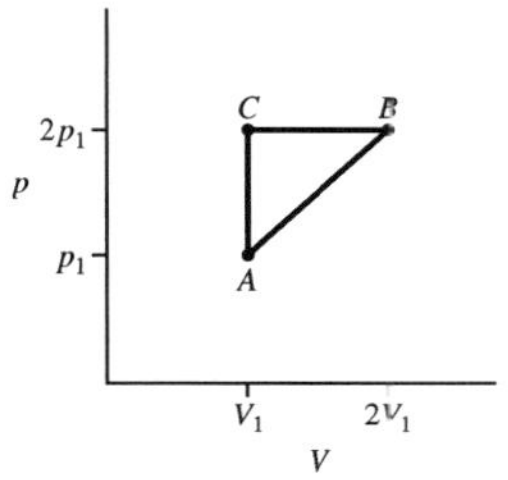

FIGURE 18.19 Exercises 20, 21 and Problem 70.

22. A balloon contains 0.30 mol of helium. It rises, while maintaining a constant 300 K temperature, to an altitude where its volume has expanded five times. How much work is done by the gas in the balloon during this isothermal expansion? Neglect tension forces in the balloon.
23. The balloon of Problem 22 starts at a pressure of 100 kPa and rises to an altitude where the pressure is 75 kPa, maintaining a constant 300 K temperature. (a) By what factor does its volume increase? (b) How much work does the gas in the balloon do?
24. How much work does it take to compress 2.5 mol of an ideal gas to half its original volume while maintaining a constant 300 K temperature?
25. By what factor must the volume of a gas with $\gamma = 1.4$ be changed in an adiabatic process if the kelvin temperature is to double?

Section 18.3 Specific Heats of an Ideal Gas

26. A gas mixture contains 2.5 mol of O_2 and 3.0 mol of Ar. What are the molar specific heats at constant volume and pressure for this mixture?
27. A mixture of monatomic and diatomic gases has specific-heat ratio $\gamma = 1.52$. What fraction of the molecules are monatomic?
28. What should be the approximate specific-heat ratio of a gas consisting of 50% $NO_2\,(\gamma = 1.29)$, 30% $O_2\,(\gamma = 1.4)$, and 20% $Ar\,(\gamma = 1.67)$?
29. By how much does the temperature of (a) an ideal monatomic gas and (b) an ideal diatomic gas (with molecular rotation but no vibration) change in an adiabatic process in which 2.5 kJ of work are done on each mole of gas?

Problems

30. Water flows over Niagara Falls (height 50 m) at the rate of about 10^6 kg/s. Suppose that all the water passes through a turbine connected to an electric generator producing 400 MW of electric power. If the water has negligible kinetic energy after leaving the turbine, by how much has its temperature increased between the top of the falls and the outlet of the turbine?
31. An ideal gas expands to 10 times its original volume, maintaining a constant 440 K temperature. If the gas does 3.3 kJ of work on its surroundings, (a) how much heat does it absorb, and (b) how many moles of gas are there?
32. A 0.25-mol sample of an ideal gas initially occupies 3.5 L. If it takes 61 J of work to compress the gas isothermally to 3.0 L, what is the temperature?
33. As the heart beats, blood pressure in an artery varies from a high of 125 mm of mercury to a low of 80 mm of mercury. These values are gauge pressures—that is, excesses over atmospheric pressure. An air bubble trapped in the artery has diameter 1.52 mm when the blood pressure is at its minimum. (a) What will its diameter be at maximum pressure? (b) How much work does the blood (and ultimately the heart) do in compressing this bubble, assuming the air remains at the same 37.0°C temperature as the blood?
34. It takes 600 J to compress a gas isothermally to half its original volume. How much work would it take to compress it by a factor of 10 starting from its original volume?
35. A gas undergoes an adiabatic compression during which its volume drops to half its original value. If the gas pressure increases by a factor of 2.55, what is its specific-heat ratio γ?
36. A gas with $\gamma = 1.4$ is at 100 kPa pressure and occupies 5.00 L. (a) How much work does it take to compress the gas adiabatically to 2.50 L? (b) What is its final pressure?
37. A gas sample undergoes the cyclic process *ABCA* shown in Fig. 18.20, where *AB* lies on an isotherm. The pressure at point *A* is 60 kPa. Find (a) the pressure at *B* and (b) the net work done on the gas.

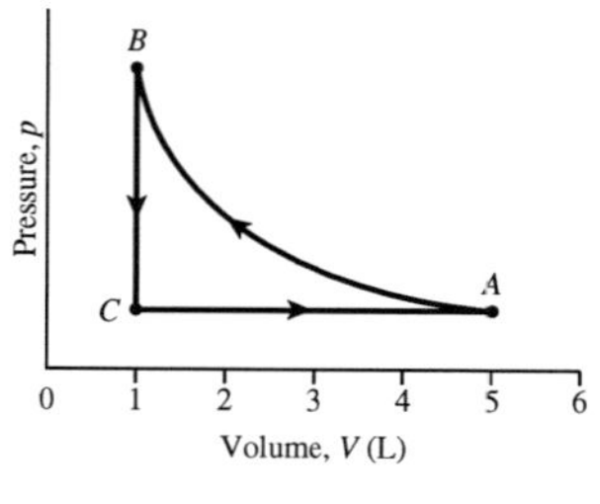

FIGURE 18.20 Problems 37 and 38

38. Repeat Problem 37 taking *AB* to be on an adiabat and using a specific-heat ratio of $\gamma = 1.4$.
39. A gasoline engine has a compression ratio of 8.5 (see Example 18.3 for the meaning of this term). If the fuel-air mixture enters the engine at 30°C, what will its temperature be at maximum compression? Assume the compression is adiabatic and the mixture has $\gamma = 1.4$.
40. By how much must the volume of a gas with $\gamma = 1.4$ be changed in an adiabatic process if the pressure is to double?
41. Volvo's B5340 engine, used in the V70 series cars, has a compression ratio of 10.2. If air at 320 K and atmospheric pressure fills an engine cylinder at its maximum volume, (a) what will be the air temperature at the point of maximum compression? (b) What will be the pressure at this point? Assume the compression is adiabatic, with $\gamma = 1.4$.
42. A gas expands isothermally from state *A* to state *B*, in the process absorbing 35 J of heat. It is then compressed isobarically to state *C*, where its volume equals that of state *A*. During the compression, 22 J of work are done on the gas. The gas is then heated at constant volume until it returns to state *A*. (a) Draw a *pV* diagram for this process. (b) How much work is done on or by the gas during the complete cycle? (c) How much heat is transferred to or from the gas as it goes from *B* to *C* to *A*?
43. A 2.0-mol sample of ideal gas with molar specific heat $C_V = \frac{5}{2}R$ is initially at 300 K and 100 kPa pressure. Determine the final temperature and the work done by the gas when 1.5 kJ of heat are added to the gas (a) isothermally, (b) at constant volume, and (c) isobarically.
44. Prove that the slope of an adiabat at a given point in a *pV* diagram is γ times the slope of the isotherm passing through the same point.
45. An ideal gas with $\gamma = 1.67$ starts at point *A* in Fig. 18.21, where its volume and pressure are 1.00 m^3 and 250 kPa, respectively. It then undergoes an adiabatic expansion that triples its volume, ending at point *B*. It's then heated at constant volume to point *C*, and then compressed isothermally back to *A*. Find (a) the pressure at *B*, (b) the pressure at *C*, and (c) the net work done on the gas.

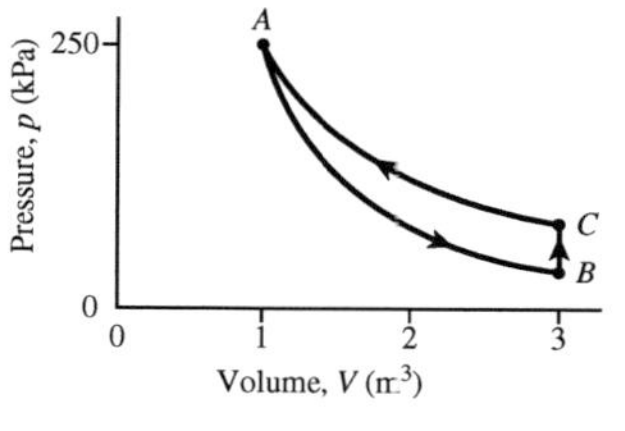

FIGURE 18.21 Problem 45

46. The gas of Example 18.4 starts at state *A* in Fig. 18.14 and is compressed adiabatically until its volume is 2.0 L. It's then cooled at constant pressure until it reaches 300 K, then allowed to expand isothermally back to state *A*. Find (a) the net work done on the gas and (b) the minimum volume reached.
47. The gas of Example 18.4 starts at state *A* in Fig. 18.14 and is heated at constant volume until its pressure has doubled. It's then compressed adiabatically until its volume is one-fourth its original value, then cooled at constant volume to 300 K, and finally allowed to expand isothermally to its original state. Find the net work done on the gas.

48. A 25-L sample of an ideal gas with $\gamma = 1.67$ is at 250 K and 50 kPa. The gas is compressed isothermally to one-third of its original volume, then heated at constant volume until its state lies on the adiabatic curve that passes through its original state, and then allowed to expand adiabatically to that original state. Find the net work involved. Is work done on or by the gas?
49. A 25-L sample of an ideal gas with $\gamma = 1.67$ is at 250 K and 50 kPa. The gas is compressed adiabatically until its pressure triples, then cooled at constant volume back to 250 K, and finally allowed to expand isothermally to its original state. (a) How much work is done on the gas? (b) What is the minimum volume reached? (c) Sketch this cyclic process in a pV diagram.
50. A bicycle pump consists of a cylinder 30 cm long when the pump handle is all the way out. The pump contains air $(\gamma = 1.4)$ at 20°C. If the pump outlet is blocked and the handle pushed until the internal length of the pump cylinder is 17 cm, by how much does the air temperature rise? Assume that no heat is lost.
51. A tightly sealed flask contains 5.0 L of air at 0°C and 100 kPa pressure. How much heat is required to raise the air temperature to 20°C? The molar specific heat of air at constant volume is $2.5R$.
52. External forces compress 21 mol of ideal monatomic gas; during the process the gas transfers 15 kJ of heat to its surroundings, yet its temperature rises by 160 K. How much work was done on the gas?
53. A gas with $\gamma = \frac{7}{5}$ is at 273 K when it's compressed isothermally to one-third of its original volume and then further compressed adiabatically to one-fifth of its original volume. What is its final temperature?
54. An ideal gas with $\gamma = 1.3$ is initially at 273 K and 100 kPa. The gas is compressed adiabatically to 240 kPa pressure. What is its final temperature?
55. The curved path in Fig. 18.22 lies on the 350 K isotherm for an ideal gas with $\gamma = 1.4$. (a) Calculate the net work done on the gas as it goes around the cyclic path $ABCA$. (b) How much heat flows into or out of the gas on the segment AB?

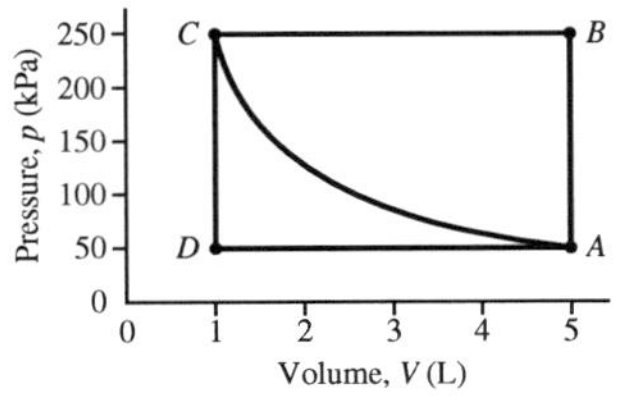

FIGURE 18.22 Problems 55 and 56

56. Repeat part (a) of Problem 55 for the path $ACDA$ in Fig. 18.22. (b) How much heat flows into or out of the gas on the segment CD?
57. A gas mixture contains monatomic argon and diatomic oxygen. An adiabatic expansion that doubles its volume results in the pressure dropping to one-third of its original value. What fraction of the molecules are argon?
58. How much of a triatomic gas with $C_V = 3R$ would you have to add to 10 mol of monatomic gas to get a mixture whose thermodynamic behavior was like that of a diatomic gas?
59. An 8.5-kg rock at 0°C is dropped into a well-insulated vat containing a mixture of ice and water at 0°C. When equilibrium is reached, there are 6.3 g less ice. From what height was the rock dropped?
60. A piston-cylinder arrangement containing 0.30 mol of nitrogen at high pressure is in thermal equilibrium with an ice-water bath containing 200 g of ice. The pressure of the ambient air is 1.0 atm. The gas is allowed to expand isothermally until it's in pressure balance with its surroundings. After the process is complete, the bath contains 210 g of ice. What was the original gas pressure?
61. A piston-cylinder arrangement contains 1.0 g of water at 50°C. The piston, which has negligible mass, is free to move and is exposed to atmospheric pressure (1.0 atm). The system is heated to 200°C. Assuming that liquid water is incompressible and that steam is an ideal gas with volume specific heat $4.3R$, determine (a) the work done by the system and (b) the heat added to it.
62. Show that the application of Equation 18.3 to an adiabatic process results in Equation 18.12.
63. A horizontal piston-cylinder system containing n mol of ideal gas is surrounded by air at temperature T_0 and pressure P_0. If the piston is displaced slightly from equilibrium, show that it executes simple harmonic motion with angular frequency $\omega = Ap_0/\sqrt{MnRT_0}$, where A and M are the piston area and mass, respectively. Assume the gas temperature remains constant.
64. An ideal gas is taken clockwise around the circular path shown in Fig. 18.23. (a) How much work does the gas do? (b) If there are 1.3 mol of gas, what is the maximum temperature reached?

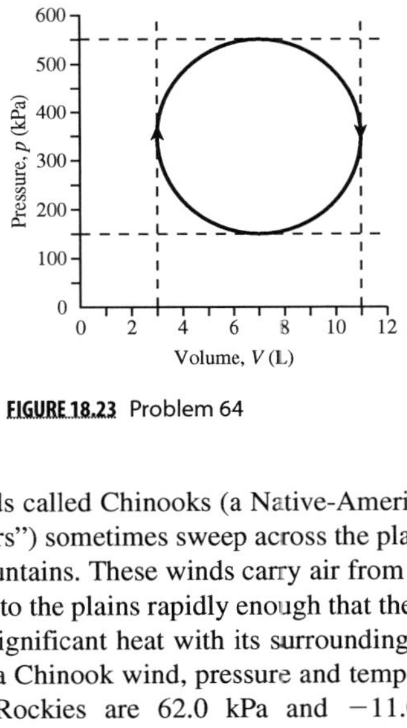

FIGURE 18.23 Problem 64

65. Warm winds called Chinooks (a Native-American term meaning "snow eaters") sometimes sweep across the plains just east of the Rocky Mountains. These winds carry air from high in the mountains down to the plains rapidly enough that the air has no time to exchange significant heat with its surroundings (Fig. 18.24). On the day of a Chinook wind, pressure and temperature high in the Colorado Rockies are 62.0 kPa and −11.0°C, respectively. (a) What will the air temperature be when the Chinook has carried it to the plain where pressure is 86.5 kPa? (b) How much work is done on what was initially a cubic meter of air as it descends to the plain?

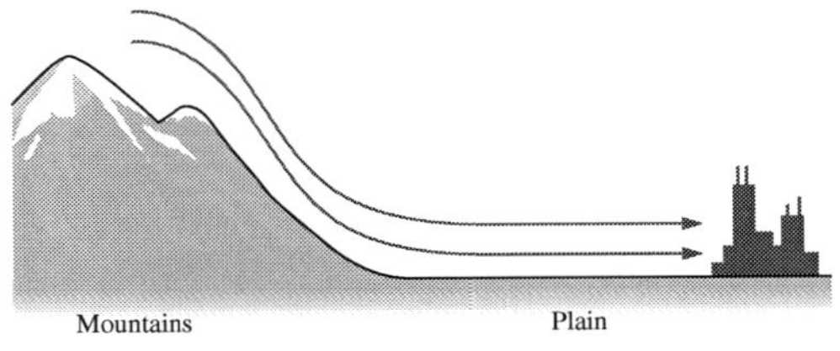

FIGURE 18.24 Problem 65

66. Use the ideal-gas law to eliminate pressure in Equation 18.11a, and show that the result can be written as Equation 18.11b.
67. Figure 18.25 shows the thermodynamic cycle of a diesel engine. Note that this cycle differs from that of a gasoline engine in that combustion takes place isobarically. As in Example 18.3, the compression ratio r is the ratio of maximum to minimum volume: $r = V_1/V_2$. In addition, the so-called *cutoff ratio* is defined by $r_c = V_3/V_2$. Find an expression for the engine's efficiency, in terms of the ratios r and r_c and the specific-heat ratio γ. Although your expression suggests that the diesel engine might be less efficient than the gasoline engine (see Problem 54 of Chapter 19), the diesel's higher compression ratio more than compensates, giving it a higher efficiency (the ratio of the work delivered to the heat extracted during the combustion phase).

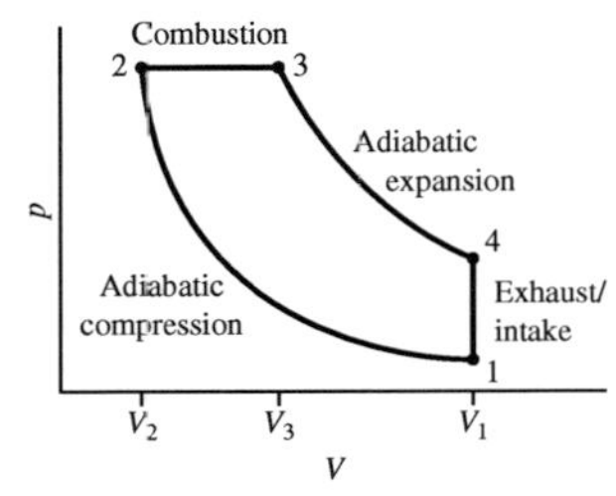

FIGURE 18.25 Problem 67

68. In a reversible process, a volume of air $V_0 = 17\ \text{m}^3$ at pressure $p_0 = 1$ atm is compressed such that the pressure and volume are related by $(p/p_0)^2 = V/V_0$. How much work is done by the gas in reaching a final pressure of 1.4 atm? (*Hint:* You may need to do an integral.)
69. A real gas is more accurately described using the Van der Waals equation: $[p + a(n/V)^2](V - nb) = nRT$, where a and b are constants. Find the expression, corresponding to Equation 18.4, for the work done by a Van der Waals gas undergoing an isothermal expansion from V_1 to V_2.
70. Repeat Exercise 20 for an expansion along the path $p = p_1[1 + (V - V_1)^2/V_1^2]$ in the pV diagram.
71. When air rises in the atmosphere, it expands adiabatically and its temperature falls with altitude. (This is called the adiabatic lapse rate.) Express dT in terms of dp for an adiabatic process, and use the hydrostatic equation (Equation 15.2) to express dp in terms of dy. Then, calculate this lapse rate dT/dy. Take the average molecular weight of air to be 29 g, $\gamma = 1.4$, and remember that the altitude y is the negative of the depth h in Equation 15.2.
72. The nuclear power plant at which you're the public affairs manager has a backup high-temperature gas-turbine system. It's a so-called combined-cycle system that maximizes the conversion of thermal energy into useful work. One such turbine produces electrical energy at the rate of 360 MW, while extracting energy from natural gas at the rate of 670 MW. The local town council has raised concern over waste thermal energy dumped into the environment. The council adopted a standard that the thermal waste power must not exceed 400 MW and all power generation must be at least 50% efficient. Does the turbine meet this standard, which was set after minimal technical research by the council?
73. Your class on alternative habitats has an assignment to design an underwater habitat. A small diving bell will be lowered to the habitat. A hatch at the bottom of the bell is open, so water can enter to compress the air and thus keep the air pressure inside equal to the pressure of the surrounding water. The bell is lowered slowly enough so that the inside air remains at the same temperature as the water. But the water temperature increases with depth in such a way that the air pressure and volume are related by $p = p_0\sqrt{V_0/V}$, where $V_0 = 17\ \text{m}^3$ and $p_0 = 1$ atm are the surface values. Suppose the diving bell's air volume cannot be less than 8.66 m^3 and the pressure must not exceed 1.5 atm when submerged. Are these criteria met?
74. The phone rings. It's your roommate calling from the chemistry building. "I'm not sure I figured out the right equation for γ given the volume and temperature change of a gas in an adiabatic process. The final volume was 0.177 of the initial volume, and the temperature changed by a factor of 2. I get $\gamma = -0.6$. I don't think it's correct. Help. What is γ?"
75. A magician friend asks for your help in designing an illusion involving a "frozen" balloon. He will take the balloon containing 5.0 L of air at 0 °C out of a freezer and heat it with a small explosive charge. He needs to know how much heat energy will be required to raise the temperature to 20°C. The molar specific heat of air at constant volume is $2.5R$. Neglecting the tension in the balloon's latex, what do you tell him?

Answers to Chapter Questions

Answer to Chapter Opening Question

Energy is conserved, provided thermal energy is included. The engine produces both mechanical energy and thermal energy of its exhaust gases; together, they sum to the energy released in combustion.

Answers to GOT IT? Questions

18.1 (c). Only the internal energy is the same, since it's a thermodynamic state variable unique to a point in the pV diagram.

18.2 (c). Superficially they seem to be contradictory, but that's only if you read the ideal-gas law as stating that pV remains constant. That's true in an isothermal process, but not in an adiabatic process to which Equation 18.11a applies. The ideal-gas law is true in both cases, and in an adiabatic process T changes in such a way that pV^γ, rather than pV, remains constant.

18.3 (a) Constant-volume; T and p increase, V doesn't change, U increases as heat flows into the gas. (b) Adiabatic; T and p increase, V decreases, U increases as work is done on the gas. (c) Isobaric; T decreases, p doesn't change, V decreases, U decreases as heat flows out of the gas.

19 The Second Law of Thermodynamics

■ Most of the energy extracted from the fuel in power plants is dumped to the environment as waste heat, here using a large cooling tower. Why is so much energy wasted?

► To Learn

By the end of this chapter you should be able to

- Explain the second law of thermodynamics and the limitations it places on our ability to extract useful work from thermal-energy sources (19.2).
- Calculate the efficiencies of heat engines and refrigerators (19.2, 19.3).
- Describe the concepts of energy quality and entropy (19.4).
- Compute entropy changes for basic thermodynamic processes (19.4).

◄ To Know

- This chapter builds on the first law of thermodynamics and the use of *pV* diagrams (18.1, 18.2).
- You should be familiar with the basic thermodynamic processes, particularly isothermal and adiabatic processes (18.2).

The first law of thermodynamics relates heat and other forms of energy. Much of our world depends on this relationship. Cars extract energy from the heat of burning gasoline. Electricity that powers lights and motors and computers originates in heat released by burning fuels or fissioning uranium. Our own bodies run on energy that begins as heat in the Sun's core. But the first law doesn't tell the whole story. Heat and mechanical energy aren't the same, and the difference makes the conversion of heat to work a more subtle task than the first law would imply.

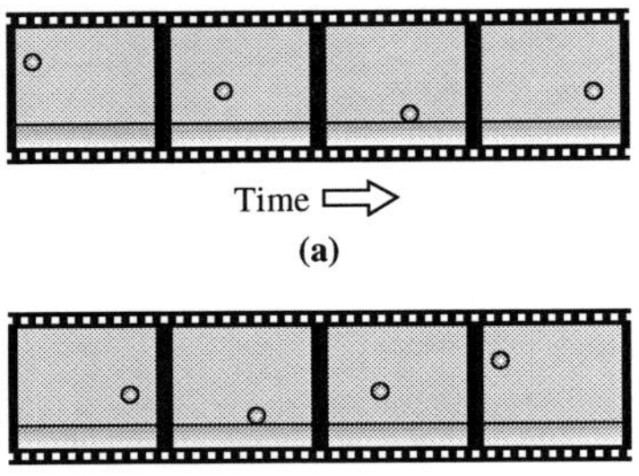

FIGURE 19.1 A movie of a bouncing ball makes sense whether it's shown forward or backward.

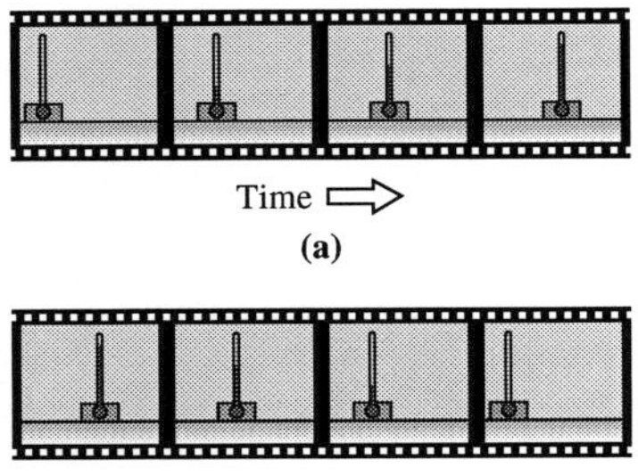

FIGURE 19.2 (a) A block warming (note thermometer) as friction dissipates its kinetic energy. (b) The reverse sequence would never happen, even though it doesn't violate energy conservation.

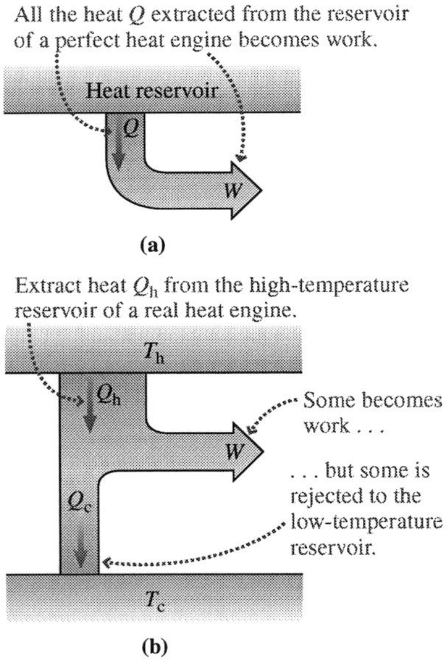

FIGURE 19.3 (a) Energy-flow diagram for a perfect heat engine. (b) A real engine delivers as work only a fraction of the energy extracted from the high-temperature reservoir.

19.1 Reversibility and Irreversibility

Figure 19.1 shows a movie of a bouncing ball. Run it backward and it still makes sense. Figure 19.2 shows a block sliding along a table, slowing to a stop because of friction—and warming in the process. Play this film backward and it makes no sense. You'll never see a block at rest suddenly start to move, cooling as it goes. Yet energy would be conserved if it did, so the first law of thermodynamics would be satisfied. Beat an egg, blending yolk and white. Reverse the beater, and you'll never see them separate again. Put cups of cold and hot water in contact; the hot water gets colder and the cold gets hotter. The opposite never occurs—although energy would still be conserved.

Why are these events **irreversible**? In each case we start with matter in an organized state. The molecules of the sliding block share a common motion. The yolk molecules are all in one place. The hot water has more energetic molecules. Of all the possible states, these *organized* ones are relatively rare. There are many more *disorganized* states—for example, all the possible arrangements of molecules in a scrambled egg. As a system evolves, chances are it will end up less organized, simply because there are far more such states available to it. It's very unlikely to assume spontaneously a more organized state.

A key word here is "spontaneous." We could restore organization—for example, by putting one cup of water in the refrigerator and the other in the microwave—but that requires a rather deliberate and energy-consuming process.

Irreversibility is a probabilistic notion. Events that *could* occur without violating the principles of Newtonian physics nevertheless *don't* occur because they're too improbable. As a practical consequence, harnessing the internal energy associated with random molecular motions is difficult because those motions won't spontaneously become organized. That makes much of the world's energy unavailable for doing useful work.

GOT IT? 19.1 Which of these processes is irreversible: (a) stirring sugar into coffee, (b) building a house, (c) demolishing a house with a wrecking ball, (d) demolishing a house by taking it apart piece by piece, (e) harnessing the energy of falling water to drive machinery, (f) harnessing the energy of falling water to heat a house?

19.2 The Second Law of Thermodynamics

Heat Engines

It's impossible to convert *all* the internal energy of a system to useful work. But devices called **heat engines** can extract *some* of that internal energy. Examples include gasoline and diesel engines, fossil-fueled and nuclear power plants, and jet aircraft engines.

Figure 19.3a is an energy-flow diagram for a "perfect" heat engine—one that extracts heat from a heat reservoir and converts it all to work. Such an engine would do exactly what we argued against in the preceding section: It would convert the random energy of thermal motion entirely to the ordered motion associated with mechanical work. In fact a perfect heat engine is impossible, for the same reason that we can't unscramble an egg or make a block accelerate spontaneously at the expense of its internal energy. This fact is one statement of the **second law of thermodynamics**:

Second law of thermodynamics (Kelvin-Planck statement) It is impossible to construct a heat engine operating in a cycle that extracts heat from a reservoir and delivers an equal amount of work.

The phrase "in a cycle" means that a practical engine must go through a repeated sequence of steps, as in the back-and-forth motions of the pistons in a gasoline engine.

A simple heat engine consists of a gas-cylinder system and a heat reservoir, the latter kept hot, perhaps, by burning a fuel. With the gas initially at high pressure, we place the cylinder in contact with the heat reservoir. The gas expands and does work W on the piston. In this isothermal process, the gas extracts heat $Q = W$ from the reservoir. Eventually the gas reaches pressure equilibrium and stops expanding. The piston must then be returned to its original position if it's to do more work.

If we just push the piston back, we'll have to do as much work as we got during the expansion, and our engine won't produce any net work. Instead we can cool the gas to reduce its volume, through thermal contact with a cool reservoir. But then some energy leaves the system as heat rather than work, as shown conceptually in Fig. 19.3b. Our engine extracts heat from a source and delivers mechanical work, but over a full cycle work delivered is less than the heat extracted. The remaining energy is rejected to the lower-temperature reservoir, usually the environment. In practical terms, much of the energy released from fuels in car engines and power plants ends up as waste heat.

The second law of thermodynamics says we can't build a perfect heat engine. But how close can we come? We define the **efficiency** e of an engine as the ratio of what we want from the engine—namely, the work W—to what we have to supply (and usually pay for)—namely, the heat energy Q_h: $e = W/Q_h$. Since the process is cyclic, there's no net change in internal energy over one cycle. Then the first law of thermodynamics, $\Delta U = Q - W$, ensures that the work done is the difference between the heat Q_h extracted from the high-temperature reservoir and the heat Q_c rejected to the cool reservoir. So the efficiency is

$$e = \frac{W}{Q_h} = \frac{Q_h - Q_c}{Q_h} = 1 - \frac{Q_c}{Q_h} \qquad (19.1)$$

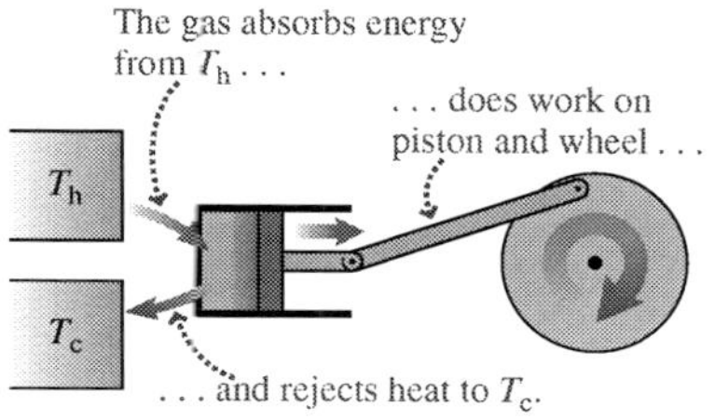

FIGURE 19.4 A simple heat engine.

Figure 19.4 shows a heat engine whose efficiency we can calculate. The engine consists of a cylinder containing an ideal gas, sealed by a movable piston. The piston is connected to a rod that turns a wheel. The engine gets its energy from a heat reservoir at a high temperature T_h, and it rejects heat to a cooler reservoir at temperature T_c. Figure 19.5 shows how the engine works in a cycle of four steps, starting with the piston in its leftmost position (state A in Fig. 19.5), where the gas volume is a minimum:

1. Isothermal expansion: The high-temperature reservoir is placed in thermal contact with the cylinder. The gas absorbs heat Q_h from the hot reservoir and expands isothermally along path AB. Since temperature remains constant, so does internal energy. The first law then shows that the engine does work $W = Q$ on the piston and wheel.
2. Adiabatic expansion: At B we remove the hot reservoir. The expansion becomes adiabatic and follows path BC. We design the engine so the gas has cooled to T_c when the piston reaches its rightmost position (state C), the point of maximum gas volume.
3. Isothermal compression: At C we bring the cool reservoir into contact with the cylinder. The wheel's inertia keeps it turning, so the piston does work on the gas, compressing it isothermally from state C to D. This work ends up as heat rejected to the cool reservoir.
4. Adiabatic compression: At D we remove the cool reservoir and the compression continues adiabatically until the gas temperature is once again at T_h and the engine is back at state A.

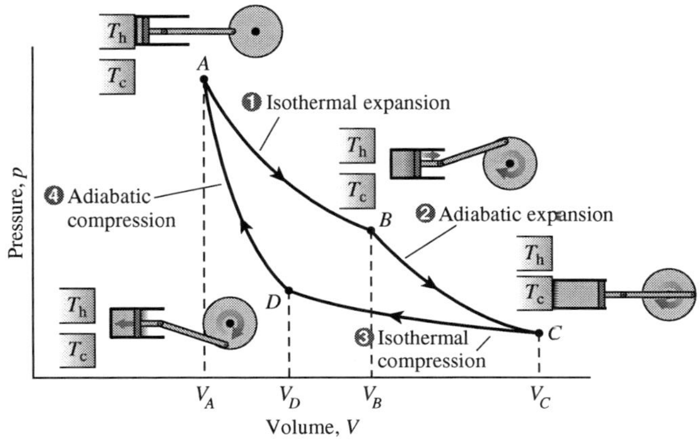

FIGURE 19.5 A pV diagram for the Carnot engine.

This cyclic process of two isothermal and two adiabatic steps is called a **Carnot cycle** and the engine a **Carnot engine**, after the French engineer Sadi Carnot (1796–1832). The particular configuration of the engine is not important, nor is the choice of an ideal gas as the engine's **working fluid**. What distinguishes the Carnot cycle from others is the sequence of thermodynamic processes and the fact that these processes are reversible. The Carnot engine is an example of a **reversible engine**—one in which thermodynamic equilibrium is maintained so that all steps could, in principle, be reversed.

What's the efficiency of a Carnot engine? To find out, we need the heats Q_h and Q_c absorbed and rejected during the isothermal parts of the cycle shown in Fig. 19.5. Equation 18.4 gives the heat Q_h absorbed during the isothermal expansion AB:

$$Q_h = nRT_h \ln\left(\frac{V_B}{V_A}\right)$$

and the heat Q_c rejected during the isothermal compression CD:

$$Q_c = -nRT_c \ln\left(\frac{V_D}{V_C}\right) = nRT_c \ln\left(\frac{V_C}{V_D}\right)$$

We put the minus sign here because our statement of the first law describes Q as the heat *absorbed*, while Equation 19.1 for the engine efficiency requires that Q_c be the heat *rejected.* To calculate engine efficiency according to Equation 19.1, we need the ratio Q_c/Q_h:

$$\frac{Q_c}{Q_h} = \frac{T_c \ln(V_C/V_D)}{T_h \ln(V_B/V_A)} \tag{19.2}$$

This expression can be simplified by applying Equation 18.11b to the adiabatic processes BC and DA in the Carnot cycle: $T_h V_B^{\gamma-1} = T_c V_C^{\gamma-1}$ and $T_h V_A^{\gamma-1} = T_c V_D^{\gamma-1}$. Dividing these two equations gives

$$\left(\frac{V_B}{V_A}\right)^{\gamma-1} = \left(\frac{V_C}{V_D}\right)^{\gamma-1} \quad \text{or} \quad \frac{V_B}{V_A} = \frac{V_C}{V_D}$$

so Equation 19.2 becomes simply $Q_c/Q_h = T_c/T_h$. Using this result in Equation 19.1 then gives the efficiency of the Carnot engine:

$$e_{\text{Carnot}} = 1 - \frac{T_c}{T_h} \quad \text{(Carnot engine efficiency)} \tag{19.3}$$

where the temperatures are measured on an absolute scale (Kelvin or Rankine). Equation 19.3 shows that the Carnot engine's efficiency depends only on the highest and lowest temperatures of its working fluid. In practice, the low temperature is usually that of the environment; then maximizing efficiency requires making the high temperature as high as possible. Real engines trade off efficiency with the ability of materials to withstand high temperature and pressure.

EXAMPLE 19.1 Calculating Efficiency: A Carnot Engine

A Carnot engine extracts 240 J from its high-temperature reservoir during each cycle, and rejects 100 J to the environment at 15°C. How much work does the engine do in one cycle? What's its efficiency? What's the temperature of the hot reservoir?

INTERPRET This problem is about a Carnot engine, which operates via the Carnot cycle.

DEVELOP Equation 19.3, $e_{\text{Carnot}} = 1 - T_c/T_h$, relates the two temperatures and the efficiency. Here $Q_h = 240$ J, $Q_c = 100$ J, and $T_c = 15$°C or 288 K. The first law of thermodynamics (Equation 18.1, $\Delta U = Q - W$) relates work and heat flows. So our plan is to use the first law to find the work, then find the efficiency, and then use Equation 19.3 to find T_h.

EVALUATE Since there's no change in internal energy over one cycle, the first law reads $Q = W$ and thus requires that the net heat, 240 J − 100 J, be equal to the work done. So $W = 140$ J. The efficiency is the ratio of work delivered to heat extracted, so $e = W/Q_h = 140\text{ J}/240\text{ J} = 58.3\%$. Knowing the efficiency, we solve Equation 19.3 for T_h:

$$T_h = \frac{T_c}{1 - e} = \frac{288\text{ K}}{1 - 0.583} = 691\text{ K} = 418°\text{C}$$

ASSESS Make sense? Given that the engine rejects somewhat less than half the 240 J as waste heat, we should expect efficiency somewhat over 50%. And of course T_h must be greater than T_c, as our calculation confirms. ■

Engines, Refrigerators, and the Second Law

Why so much emphasis on the Carnot engine? Because understanding this device will help answer the broader question of how much work we can ever hope to extract from thermal energy. That, in turn, will help us understand practical limitations on humankind's attempts to harness ever more energy, and will ultimately bring a deeper understanding of the second law of thermodynamics.

Why is Carnot's engine special? Couldn't we build a better engine with greater efficiency? The answer is no. The special role of the Carnot cycle is embodied in **Carnot's theorem**:

Carnot's theorem All Carnot engines operating between temperatures T_h and T_c have the same efficiency (given by Equation 19.3), and no other engine operating between the same two temperatures can have a greater efficiency.

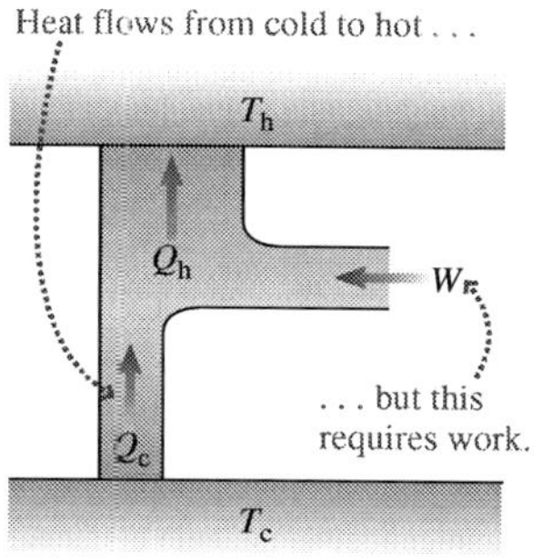

FIGURE 19.6 Energy-flow diagram for a real refrigerator.

To prove Carnot's theorem, we introduce the **refrigerator**. A refrigerator is the opposite of an engine: It extracts heat from a cool reservoir and rejects it to a hotter one, taking in work in the process (Fig. 19.6). A refrigerator forces heat to flow from cold to hot, but to do so it requires work. A household refrigerator cools its contents and warms the house (you can feel the heat coming out the back), but it uses electricity. That heat doesn't flow spontaneously from cold to hot constitutes another statement of the second law of thermodynamics:

Second law of thermodynamics (Clausius statement) It is impossible to construct a refrigerator operating in a cycle whose sole effect is to transfer heat from a cooler object to a hotter one.

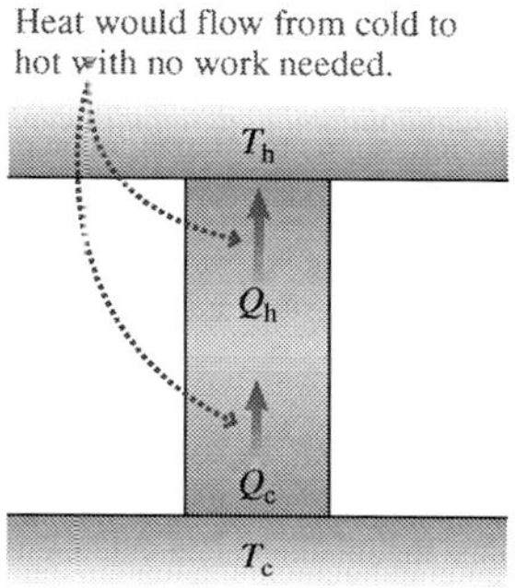

FIGURE 19.7 A perfect refrigerator is impossible.

The Clausius statement rules out a perfect refrigerator (Fig. 19.7).

Suppose the Clausius statement were false. Then we could build the device of Fig. 19.8a, consisting of a reversible Carnot engine and a perfect refrigerator. In each cycle the engine would extract, say, 100 J from the hot reservoir, put out 60 J of useful work, and reject 40 J to the cool reservoir. The perfect refrigerator could transfer the 40 J back to the hot reservoir. The net effect would be to extract 60 J from the hot reservoir and convert it entirely to work (Fig. 19.8b)—and we would have a perfect heat engine, in violation of the Kelvin-Planck statement of the second law. A similar argument (Problem 38) shows that if a perfect heat engine is possible, then so is a perfect refrigerator. So the Clausius and Kelvin-Planck statements of the second law are equivalent, in that if one is false, then so is the other.

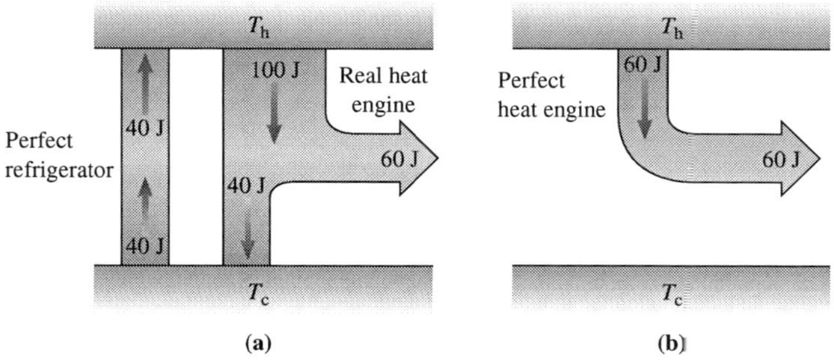

FIGURE 19.8 (a) A real heat engine combined with a perfect refrigerator is equivalent to (b) a perfect heat engine.

Because the Carnot engine is reversible, we could run it backward and reverse its path in Fig. 19.5. The engine would extract heat from the cool reservoir, take in work, and reject heat to the hot reservoir. It would be a refrigerator. Although real refrigerators aren't designed exactly like engines, the two are, in principle, interchangeable.

We're now ready to prove Carnot's assertion that Equation 19.3 gives the maximum engine efficiency. Consider again the Carnot engine in Fig. 19.8a. It extracts 100 J of heat and delivers 60 J of work, so it's 60% efficient. Suppose we had another engine operating

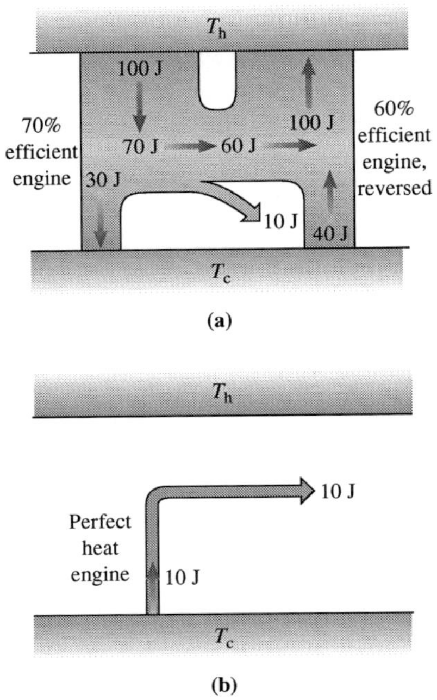

FIGURE 19.9 (a) A 60% efficient reversible engine run as a refrigerator, along with a hypothetical engine with 70% efficiency. (b) The combination is equivalent to a perfect heat engine.

between the same two reservoirs, but with 70% efficiency. Since the Carnot engine is reversible, we can run it as a refrigerator. If we put the two together, we get the device of Fig. 19.9a. Its net effect is to extract 10 J from the cool reservoir and deliver 10 J of work—so it's a perfect heat engine, in violation of the second law (Fig. 19.9b). It's therefore impossible to make an engine that's more efficient than a Carnot engine, and thus Equation 19.3 gives the maximum possible efficiency for *any* heat engine operating between the same two fixed temperatures. For that reason the Carnot efficiency of Equation 19.3 is also called the **thermodynamic efficiency**.

Irreversible engines, because they involve processes that dissipate organized motion, are necessarily *less* efficient. So are reversible engines, if their heat exchange doesn't take place solely at the highest and lowest temperatures. The ordinary gasoline engine is a case in point; even if it could be made perfectly reversible, its efficiency would be less than that of a comparable Carnot engine (see Problem 54).

19.3 Applications of the Second Law

The world abounds with thermal energy, but the second law of thermodynamics limits our ability to turn that energy to our own uses. Any device we construct that involves the interchange of heat and work is a heat engine or refrigerator, subject to the second law.

Limitations on Heat Engines

Most of our electricity is produced in large power plants that are heat engines powered by the fossil fuels coal, oil, or natural gas, or by nuclear fission. Figure 19.10 diagrams such a power plant. The working fluid is water, heated in a boiler and converted to steam at high pressure. The steam expands adiabatically to spin a fanlike turbine. The turbine turns a generator that converts mechanical work to electrical energy.

Steam leaving the turbine is still gaseous and is hotter than the water supplied to the boiler. Here's where the second law applies! Had the water returned from the turbine in its original state, we would have extracted as work all the energy acquired in the boiler, in violation of the second law. So we run the steam through a **condenser**, where it contacts pipes carrying cool water, typically from a river, lake, or ocean. The condensed steam, now cool water, is fed back into the boiler to repeat the cycle.

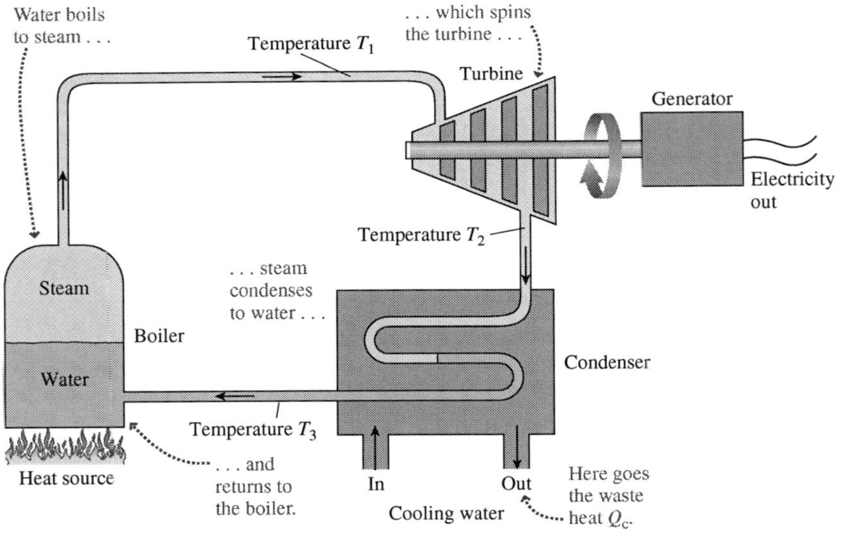

FIGURE 19.10 Schematic diagram of an electric power plant.

The maximum steam temperature in a power plant is limited by the materials used in its construction. For a conventional fossil-fueled plant, current technology permits high temperatures around 650 K. Potential damage to nuclear fuel rods limits the temperature in a nuclear plant to around 570 K. Cooling-water temperature averages about 40°C (310 K), so the maximum possible efficiencies for these power plants, given by Equa-tion 19.3, are

$$e_{\text{fossil}} = 1 - \frac{310\text{ K}}{650\text{ K}} = 52\% \quad \text{and} \quad e_{\text{nuclear}} = 1 - \frac{310\text{ K}}{570\text{ K}} = 46\%$$

Temperature differences between the exhausted steam and the cooling water, mechanical friction, and the need to divert energy for driving pumps and pollution-control devices all reduce efficiency further, to about 40% for fossil-fueled plants and 34% for nuclear plants. This means that, when we make electricity, roughly two-thirds of the fuel energy ends up as waste heat.

A typical large power plant produces 1 GW of electricity, so another 2 GW of waste heat goes into the cooling water. The resulting temperature rise can cause serious ecological problems. The huge cooling towers you see at power plants reduce such "thermal pollution" by transferring much of the waste heat to the atmosphere (see this chapter's opening photo). The need for power-plant cooling water nevertheless remains so great that a substantial fraction of all rainwater falling on the United States eventually finds its way through the condensers of power plants (see Problem 29).

EXAMPLE 19.2 Improving Efficiency: A Combined-Cycle Power Plant

The gas turbine in a combined-cycle power plant (see the Application below) operates at 1450°C. Its waste heat at 500°C is the input for a conventional steam cycle, with its condenser fed by river water at 8°C. Find the thermodynamic efficiency of the combined cycle, and compare with the efficiencies of the individual components if they were operated independently.

INTERPRET This problem is about the thermodynamic efficiency of a combined-cycle power plant. As described in the Application, that means a plant using a high-temperature gas turbine whose waste heat becomes the energy input to a conventional steam turbine.

DEVELOP Figure 19.11 is a conceptual diagram of the combined-cycle plant, based on the Application. Equation 19.3, $e = 1 - T_c/T_h$, gives the thermodynamic efficiencies of each cycle and of the combination. We identify the 1450°C = 1723 K temperature as T_h in Equation 19.3 for the gas turbine. The intermediate temperature 500°C = 773 K serves as T_c for the gas turbine but as T_h for the steam cycle. Finally, the 8°C or 281-K cooling-water temperature is T_c for the steam cycle.

EVALUATE To treat the entire plant as a single heat engine in Equation 19.3, we use the highest and lowest temperatures:

$$e_{\text{combined}} = 1 - \frac{T_c}{T_h} = 1 - \frac{281\text{ K}}{1723\text{ K}} = 0.84 = 84\%$$

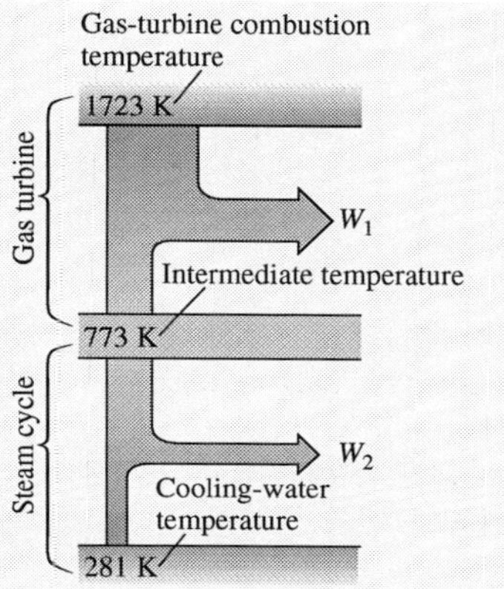

FIGURE 19.11 Conceptual diagram of a combined-cycle power plant.

Friction and other losses would reduce this figure substantially, but a combined-cycle plant operating at these temperatures could have a practical efficiency near 60%. The efficiencies of the individual components also follow from Equation 19.3:

$$e_{\text{gas turbine}} = 1 - \frac{773\text{ K}}{1723\text{ K}} = 55\% \quad \text{and} \quad e_{\text{steam}} = 1 - \frac{281\text{ K}}{773\text{ K}} = 64\%$$

ASSESS Make sense? Because of its extreme temperatures, the combined cycle gives an efficiency that's better than either of its parts! You can learn more about combined-cycle power plants in the Application below, and by working Problem 30. ■

APPLICATION Combined-Cycle Power Plants

Improving power-plant efficiency helps reduce air pollution and greenhouse-gas emissions, not to mention the cost of electricity. Modern *combined-cycle* power plants achieve efficiencies approaching 60% by combining a conventional steam system like that of Fig. 19.10 with a *gas turbine* similar to a jet aircraft engine. Gas turbines operate at high temperatures—between 1000 K and 2000 K—but they aren't very efficient because their exhaust temperature (T_c in Equation 19.3) is also high. In a combined-cycle plant, the hot exhaust from a gas turbine drives a conventional steam cycle whose condenser operates at near the ambient temperature. The overall effect is the same as a single heat engine operating between the gas turbine's high combustion temperature and the low temperature of the environment (see Problem 30). The second law still limits the efficiency, but with high T_h and low T_c, the efficiency can be much greater than in a conventional plant. Combined-cycle plants burning natural gas are becoming more widespread, but their expense relative to coal plants makes them less popular in two of the world's most coal-rich and polluting countries: the United States and China. The photo shows a gas-fired combined-cycle plant.

Gasoline and diesel engines provide another pervasive example of heat engines. A typical automobile engine has a theoretical maximum efficiency around 50%, but irreversible thermodynamic processes make the actual efficiency much lower. Mechanical friction dissipates additional energy, with the end result that less than 20% of the fuel energy reaches the driving wheels. Problems 54 and 55 explore the thermodynamics of the gasoline engine.

We wouldn't be so concerned with efficiency if we didn't have to pay for fuel. Engines with "free" fuel include solar-thermal power plants that concentrate sunlight to boil a fluid that drives a turbine, and ocean thermal-energy conversion (OTEC) schemes that extract useful work from the modest temperature difference between tropical surface waters and the deep ocean. Although pilot projects of both types have been constructed, neither has yet proven technologically or economically practical.

Refrigerators and Heat Pumps

Figure 19.6 showed that a refrigerator works like an engine in reverse: It takes in mechanical work and transfers heat from cooler to hotter. The heat you feel at the back of your kitchen refrigerator includes both energy removed from the refrigerator's contents and energy supplied as electricity to run the appliance. An efficient refrigerator should maximize what we want—namely, cooling, as measured by the heat Q_c removed—compared with what we have to supply—namely, mechanical work W or its equivalent in electrical energy. The **coefficient of performance** (COP) quantifies this ratio:

$$\text{COP} = \frac{Q_c}{W} = \frac{Q_c}{Q_h - Q_c} = \frac{T_c}{T_h - T_c} \quad \left(\begin{array}{c}\text{refrigerator;}\\ \text{for heat pump c} \rightarrow \text{h in numerator}\end{array}\right) \tag{19.4}$$

The second equality here follows from the first law of thermodynamics, and the third expresses the equality $Q_c/Q_h = T_c/T_h$ that we found in deriving Equation 19.3. The first equality in Equation 19.4 is a general definition of the COP; the final equality gives a theoretical maximum that real refrigerators can approach but not exceed. When the temperatures T_h and T_c are close, Equation 19.4 gives a high COP—meaning that the refrigerator takes relatively little work to do its job. But as the difference increases, the COP drops and we have to supply more work.

EXAMPLE 19.3 The COP: A Home Freezer

A typical home freezer operates between a low of 0°F (−18°C or 255 K) and a high of 86°F (30°C or 303 K). What's its maximum possible COP? With this COP, how much electrical energy would it take to freeze 500 g of water initially at 0°C?

INTERPRET This problem is about a refrigerator—in this case a freezer. We identify T_h and T_c with the values 303 K and 255 K, respectively.

DEVELOP Equation 19.4, $\text{COP} = T_c/(T_h - T_c)$, will determine the COP. Then we'll use Equation 17.5, $Q = Lm$, to find the heat Q_c that the freezer must extract to freeze the water. From there we'll be able to use the first equality in Equation 19.4 to find the work—equivalently, the electrical energy—required.

EVALUATE Equation 19.4 gives

$$\text{COP} = \frac{T_c}{T_h - T_c} = \frac{255\ \text{K}}{303\ \text{K} - 255\ \text{K}} = 5.31$$

From Equation 17.5, we find the heat that needs to be removed in freezing 500 g of ice: $Q_c = Lm = (0.50\ \text{kg})(334\ \text{kJ/kg}) = 167\ \text{kJ}$, where we found the heat of fusion in Table 17.1. The COP is the ratio of the heat removed to the work or electrical energy required, so we have $W = Q_c/\text{COP} = 167\ \text{kJ}/5.31 = 31\ \text{kJ}$.

ASSESS Make sense? A COP of 5.3 means that each unit of work can transfer 5.3 units of heat from inside the freezer—so the electrical-energy requirement here is modest. A practical freezer operating between these temperatures would have a lower COP and require correspondingly more electrical energy. ■

Refrigerators aren't confined to kitchens. An air conditioner is a refrigerator that cools a room or building. A **heat pump** is a refrigerator that cools in summer and heats in winter (Fig. 19.12). In warmer climates, heat pumps exchange energy between a building and the outside air; in cooler climates, they exchange energy through pipes buried in the ground where the temperature is typically around 10°C year-round. Heat pumps require electricity, but like any refrigerator they transfer more heat energy than they use in electricity—although that gain is somewhat offset by the inefficiency of the power plant producing the electricity. Because the desired output is heat at the higher temperature, the COP of a heat pump is often defined as Q_h/W rather than Q_c/W as for a refrigerator; that means Q_c and T_c become Q_h and T_h, respectively, in the numerator of Equation 19.4 for a heat pump. Several end-of-chapter problems explore the physics and economics of heat pumps.

In summer the heat pump cools the house by extracting energy and rejecting it to the outdoor environment.

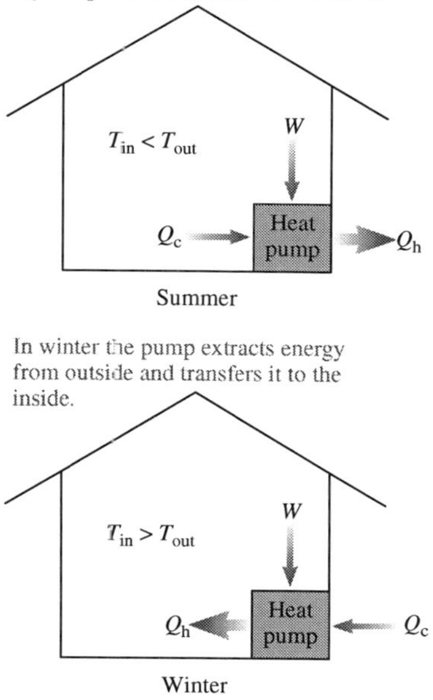

FIGURE 19.12 A heat pump.

GOT IT? 19.2 A clever engineer decides to increase the efficiency of a Carnot engine by cooling the low-temperature reservoir using a refrigerator with the maximum possible COP. Will the overall efficiency of this system (a) exceed, (b) be less than, or (c) equal that of the original engine alone?

19.4 Entropy and Energy Quality

If offered a joule of energy, would you rather have it delivered in the form of mechanical work, heat at 1000 K, or heat at 300 K? Your answer might depend on what you want to do. To lift or accelerate a mass, you'd be smart to take your energy as work. But if you want to keep warm, heat at 300 K would be perfectly acceptable.

If, on the other hand, you're not sure what you want to do with the energy, then which should you take? The second law of thermodynamics makes the answer clear: You should take the work. Why? Because you could use it directly as mechanical energy, or you could, through friction or other irreversible processes, use it to raise the temperature of something.

If you chose 300 K heat for your joule of energy, then you could supply a full joule only to objects cooler than 300 K. You couldn't do mechanical work unless you ran a heat engine. With its maximum temperature only a little above ambient, your engine would be very inefficient, and you could extract only a small fraction of a joule of mechanical energy. You would be better off with 1000-K heat since you could transfer it to anything cooler than 1000 K, or you could run a heat engine to produce up to 0.70 joule of mechanical energy.

Taking your energy in the form of work gives you the most options. Anything you can do with a joule of energy, you can do with the work. Heat is less versatile, with 300-K heat

APPLICATION **Energy Quality, End Use, and Cogeneration**

Smart use of energy means matching the quality of energy sources to our energy needs. It makes little sense to produce high-quality electrical energy and then use it to heat water for a shower. You'd be much better off burning fuel in a water heater where nearly all its energy gets transferred to the water than burning it in a 40% efficient power plant to make electricity, in the process discharging 60% of the fuel energy to the environment as waste heat. Yet electricity is such a convenient energy source that we tend to use it for everything, including low-grade heat. Is there a way to avoid this thermodynamic folly? There is, and it involves using the waste heat produced in generating electricity to supply low-quality energy needs—a process known as **cogeneration**. In parts of Europe, whole cities are heated with the waste heat from electric power plants. In the United States, industrial and institutional energy users are beginning to combine their heating plants with smaller electric generators to extract the most possible high-quality energy before using the lower-quality remainder for industrial-process heat or space heating. The photo shows a 600-kW cogenerating unit in the heating plant at Middlebury College.

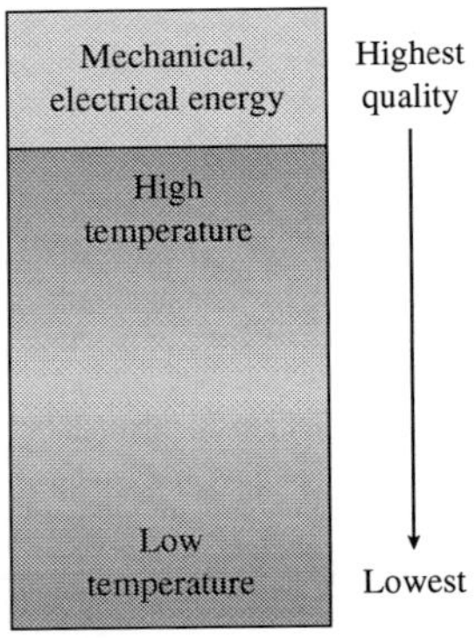

FIGURE 19.13 Energy quality measures the versatility of different energy forms.

the least useful of the three. We're not talking here about the quantity of energy—we have exactly 1 joule in each case—but about **energy quality**, indicated by the ability to do a variety of useful tasks (Fig. 19.13). We can readily convert an entire amount of energy from higher to lower quality, but the second law of thermodynamics prevents us from going in the opposite direction with 100% efficiency.

Entropy

Can we quantify the notion of energy quality? Imagine mixing glasses of hot and cold water (Fig. 19.14). Soon they come to equilibrium at a lukewarm temperature, with no overall energy loss. But something has been lost—namely, the ability to do useful work. In the initial state we could have run a heat engine between the hot and cold water. In the final state there's no temperature difference and therefore no way to run a heat engine. The quantity of energy hasn't changed, but its quality has decreased. Can we find some property whose value reflects this change?

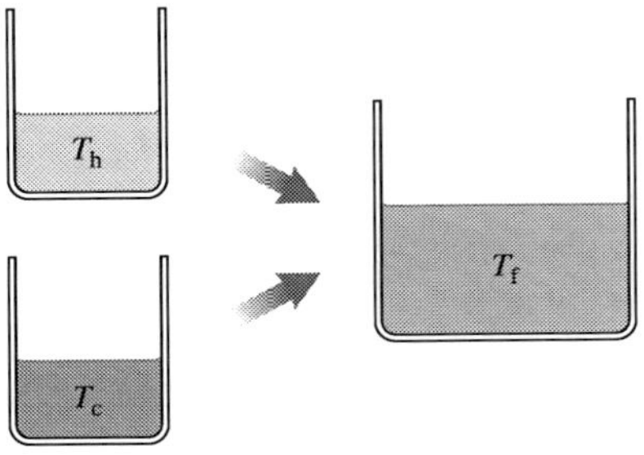

FIGURE 19.14 Mixing hot and cold water gives lukewarm water at final temperature T_f. There's no energy loss, but the system loses the ability to do work.

Consider an ideal gas undergoing a Carnot cycle of two isothermal and two adiabatic processes (recall Fig. 19.5). In deriving Equation 19.3 for the efficiency of this cycle, we found that $Q_c/Q_h = T_c/T_h$, where Q_c was the heat *rejected* from the system to the low-temperature reservoir at T_c, and Q_h the heat *added* from the reservoir at T_h.

We now change the definition of Q_c so it also means heat *added* to the system. This redefinition just changes the sign of Q_c, so now our equality can be written

$$\frac{Q_c}{T_c} + \frac{Q_h}{T_h} = 0$$

We can generalize this result to any reversible cycle by approximating the cycle as a sequence of adiabatic and isothermal step (Fig. 19.15). Figure 19.15 shows how summing over the individual isothermal segments then gives $\Sigma Q/T = 0$. We can approximate the closed cycle ever closer by more and more adiabatic and isothermal segments. In the limit the approximation becomes exact and the sum becomes an integral:

$$\oint \frac{dQ}{T} = 0 \qquad (19.5)$$

where the circle indicates that we integrate over a *closed* path.

Equation 19.5 describes a quantity that doesn't change when we take a system around a cyclic path. This quantity is called **entropy**, S. If we take a system around a path that isn't closed, then its entropy will, in general, change. That change is given by integrating over the path in question:

$$\Delta S = \int_1^2 \frac{dQ}{T} \quad \text{(change in entropy)} \qquad (19.6)$$

where ΔS is the change in entropy between the two thermodynamic states 1 and 2. Note that entropy has the units J/K.

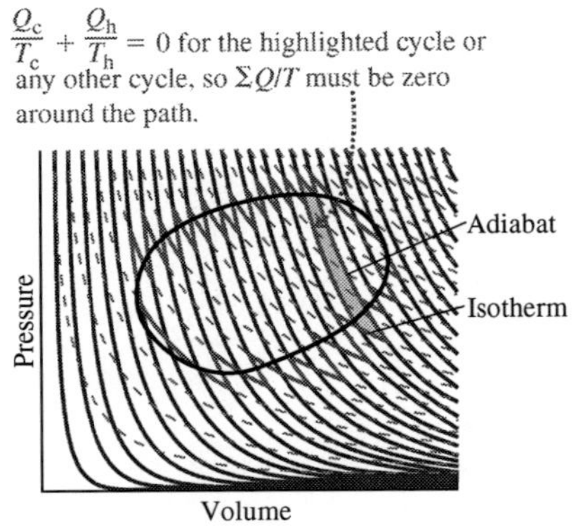

FIGURE 19.15 An arbitrary cycle approximated by isothermal (dashed curves) and adiabatic (solid curves) steps. Heat transfer occurs only during the isothermal steps.

Suppose we take a system part way around a closed path, from state 1 to state 2 in Fig. 19.16; Equation 19.6 gives the corresponding entropy change ΔS_{12}. If we go back to state 1 by *any* path, the resulting entropy change ΔS_{21} must be $-\Delta S_{12}$, so that there's no entropy change around the closed path. Thus the entropy change of Equation 19.6 is independent of path; it depends only on the initial and final states. Like pressure and temperature, entropy is a thermodynamic state variable—a quantity that characterizes a given state independently of how the system got into that state.

Now, Equation 19.6 applies only to reversible processes, since irreversible processes take a system out of thermodynamic equilibrium and therefore aren't described by paths in the pV diagram. But because entropy depends only on the initial and final states, we can calculate the entropy change in an *irreversible* process by using Equation 19.6 for a *reversible* process that goes between the same two states. Doing so for a couple of examples will help you understand the meaning of entropy.

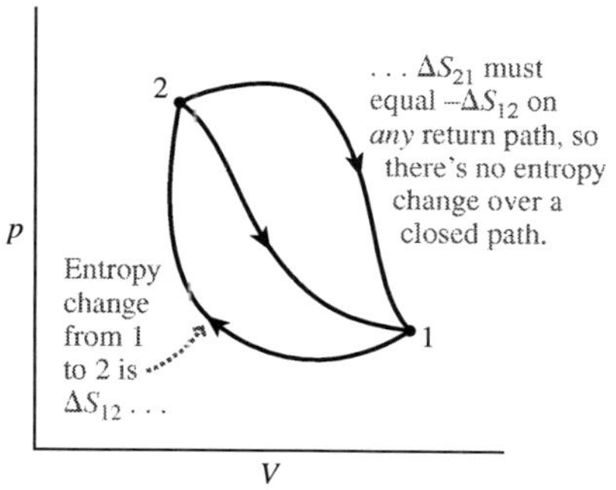

FIGURE 19.16 Entropy change is path-independent.

Irreversible Heat Transfer

Mixing hot and cold water in Fig. 19.14 is an irreversible process, as we described in Section 19.1. But we could produce the same change reversibly, by putting each glass in contact with a separate heat reservoir, then cooling or heating until each is at the final temperature T_f, and then mixing them. The entropy change for the cold water would then be $\Delta S_c = \int_{T_c}^{T_f} dQ/T$. Rather than evaluate this integral, note that there must be some intermediate temperature T_1 that lies between T_c and T_f and for which $\Delta S_c = Q_c/T_1$, where Q_c is the heat absorbed as the cold water warms to T_f. We've marked T_1 on Fig. 19.17. Similarly, the entropy change for the hot water is $\Delta S_h = Q_h/T_2$, where T_2 lies between T_f and T_h. Then the entropy change for the whole system is $\Delta S = Q_c/T_1 + Q_h/T_2$. But the heat lost from the hot water in the actual irreversible process is equal to the heat gained by the cold water, so $Q_h = -Q_c$, and the entropy change is

$$\Delta S = \frac{Q_c}{T_1} - \frac{Q_c}{T_2} = Q_c\left(\frac{1}{T_1} - \frac{1}{T_2}\right)$$

Now $T_1 < T_2$ (see Fig. 19.17), and that makes ΔS a *positive* quantity. Therefore entropy has *increased* during the irreversible mixing process.

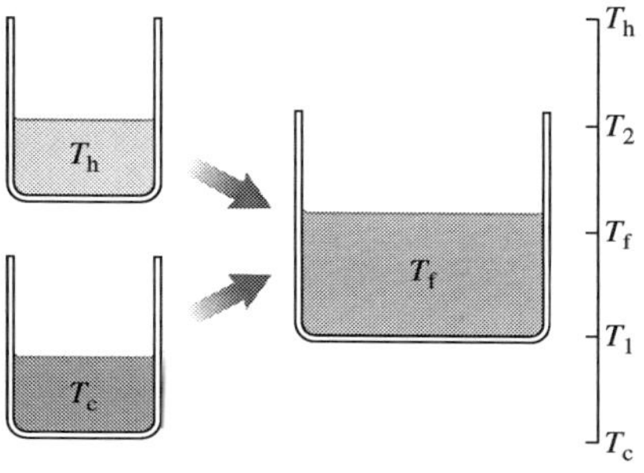

FIGURE 19.17 Repeat of Fig. 19.14, showing intermediate temperatures T_1 and T_2 used in calculating entropy change.

Adiabatic Free Expansion

Figure 19.18 shows another irreversible process that results in a loss of energy quality. In Fig. 19.18a, a partition confines an ideal gas to one side of an insulated box; on the other side is vacuum. Since the box is insulated, no heat can flow into or out of the gas, so the situation is adiabatic. Remove the partition, and the gas expands freely to fill the box (Fig. 19.18b). The process is therefore an **adiabatic free expansion**. In the vacuum there's no pressure to oppose the gas, so it does no work and therefore its internal energy doesn't change. Figure 19.18c shows how we could instead have used the expanding gas to turn a paddle wheel, extracting useful work. We can't do that with the uniform-pressure gas of Fig. 19.18b, so the free expansion results in the system's losing its ability to do work.

Now let's find the entropy change for the irreversible process of Fig. 19.18a → b. Since the internal energy of the gas doesn't change, neither does the temperature. But we've seen that we can calculate the entropy change by considering a *reversible* process that takes the gas between the same two states. An isothermal expansion is such a process, for which Equation 18.4 gives the heat that would be added: $Q = nRT\ \ln(V_2/V_1)$. Since the temperature is constant, the entropy change of Equation 19.6 becomes

$$\Delta S = \int \frac{dQ}{T} = \frac{1}{T}\int dQ = \frac{Q}{T} = nR\ \ln\left(\frac{V_2}{V_1}\right)$$

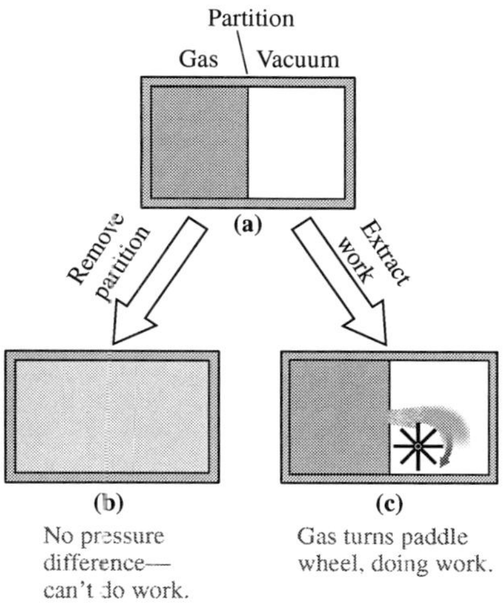

FIGURE 19.18 Two ways for a gas to expand into a vacuum.

The final volume V_2 is larger than V_1, so entropy has *increased.* Although we computed this result for the reversible process, we emphasize that it holds for *any* process that takes the system between the same initial and final states—including our irreversible adiabatic free expansion.

Entropy and the Availability of Work

Entropy increases during both the irreversible processes we just considered. Energy quality deteriorates in both processes, in that the systems lose ability to do work. Suppose we had let the gas in Fig. 19.18 undergo a reversible isothermal expansion instead of the adiabatic free expansion. Then it would have done work equal to the heat gained, which we just calculated:

$$W = Q = nRT \ln\left(\frac{V_2}{V_1}\right).$$

After the irreversible free expansion, the gas can no longer do this work, even though its energy is unchanged. Comparing with the entropy change we calculated above, we see that the energy that becomes unavailable to do work is $E_{\text{unavailable}} = T\,\Delta S$. This is an example of a more general relation between entropy and the quality of energy:

> During an irreversible process in which the entropy of a system increases by ΔS, energy $E = T_{\text{min}}\,\Delta S$ becomes unavailable to do work, where T_{min} is the coolest temperature available to the system.

This statement shows that entropy provides our measure of energy quality. Given two systems with identical energy content, the one with the lower entropy contains the higher-quality energy. An entropy increase corresponds to a degradation in energy quality, as energy becomes unavailable to do work.

EXAMPLE 19.4 Increasing Entropy: The Loss of Energy Quality

A 2.0-L cylinder contains 5.0 mol of compressed gas at 300 K. If the cylinder is discharged into a 150-L vacuum chamber and its temperature remains 300 K, how much energy has become unavailable to do work?

INTERPRET This problem asks about the loss of energy quality during an irreversible and therefore entropy-increasing process, namely an adiabatic free expansion.

DEVELOP Figure 19.19 is a before-and-after sketch of the situation, which is similar to that of Fig. 19.18 except that here the gas volume changes more dramatically. In analyzing the adiabatic free expansion of Fig. 19.18, we found $\Delta S = nR\ln(V_2/V_1)$. Our statement relating entropy and energy quality says that the energy made unavailable to do work is $T_{\text{min}}\,\Delta S$. So our plan is to calculate ΔS and multiply by T_{min} to find that unavailable energy.

EVALUATE Because the temperature doesn't change, T_{min} is the 300-K temperature we're given, and we have

$$E_{\text{unavailable}} = T\,\Delta S = nRT\ln\left(\frac{V_2}{V_1}\right)$$

$$= (5.0\text{ mol})(8.314\text{ J/K}\cdot\text{mol})(300\text{ K})\ln\left(\frac{152\text{ L}}{2.0\text{ L}}\right) = 54\text{ kJ}$$

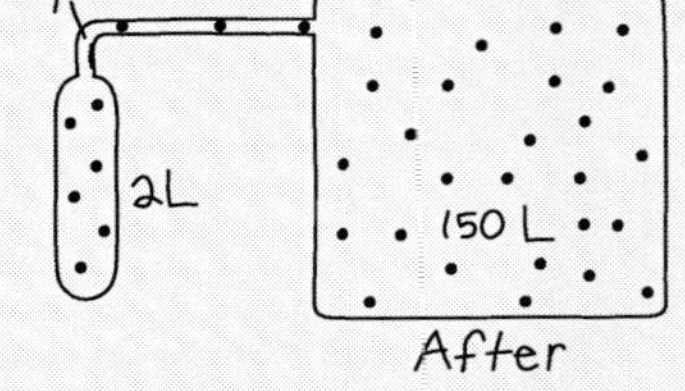

FIGURE 19.19 Our sketch for Example 19.4.

ASSESS Make sense? Yes: This is the work we could have gained from a reversible isothermal expansion. By letting the gas undergo an irreversible process, we gave up the possibility of extracting this work. ■

Entropy and the Second Law of Thermodynamics

We started this chapter arguing that natural processes are generally irreversible, going from ordered states to disordered states. It's this loss of order that makes energy unavailable to do work. Entropy is a measure of this disorder. Given the tendency of systems to evolve toward disordered states, we can make a statement about entropy that is, in fact, a general statement of the second law of thermodynamics:

Second law of thermodynamics The entropy of a closed system can never decrease.

At best, the entropy of a closed system remains constant—and this happens only in an ideal, reversible process. If anything irreversible occurs—a slight amount of friction or a deviation from exact thermodynamic equilibrium—then entropy increases. There's no going back. As entropy increases, energy becomes unavailable to do work, and nothing within the closed system can restore that energy to its original quality. This statement of the second law in terms of entropy is equivalent to our previous statements about the impossibility of perfect heat engines and refrigerators, for the operation of either device would require a decrease in entropy.

What about a system that isn't closed? Can't we decrease its entropy? Yes—but only by supplying high-quality energy from outside. Running a refrigerator decreases the entropy of its contents, but this requires electrical energy to make heat flow in the direction it doesn't normally go. That high-quality electrical energy deteriorates into additional heat that's rejected to the refrigerator's environment. If we consider the entire system, not just the refrigerator's contents, this means that overall entropy has increased.

Any system whose entropy seems to decrease—that gets more rather than less organized—cannot be a closed system. If we enlarge a system's boundaries to encompass the entire universe, then we definitely have a closed system. Therefore the ultimate statement of the second law becomes

Second law of thermodynamics The entropy of the universe can never decrease.

As examples of this broad statement, consider the growth of a living thing from the random mix of molecules in its environment, or the construction of a skyscraper from materials that were originally dispersed about Earth, or the appearance of ordered symbols on a printed page from a bottle of ink. All these are processes in which matter goes from near chaos to a highly organized state—akin to separating yolk and white from a scrambled egg. They are certainly processes in which entropy decreases. But Earth isn't a closed system. It gets high-quality energy from the Sun, energy that is ultimately responsible for life and all its actions. If we consider the Earth-Sun system, then the entropy decrease associated with life and civilization is more than balanced by the entropy increase associated with the degradation of high-quality solar energy. We living things represent a remarkable phenomenon—the organization of matter in a universe governed by a tendency toward disorder. But we can't escape the second law of thermodynamics. Our highly organized selves and society, and the entropy decreases they represent, come into being only at the expense of greater entropy increases elsewhere.

GOT IT? 19.3 In each of the following processes, does the entropy of the named system alone increase, decrease, or stay the same? (a) A balloon deflates; (b) cells differentiate in a growing embryo, forming different physiological structures; (c) an animal dies, and its remains gradually decay; (d) an earthquake demolishes a building; (e) a plant utilizes sunlight, carbon dioxide, and water to manufacture sugar; (f) a power plant burns coal and produces electrical energy; (g) a car's friction-based brakes stop the car.

CHAPTER 19 SUMMARY

Big Picture

The big idea behind this chapter is the **second law of thermodynamics**—ultimately, the statement that systems tend naturally toward disorder, or states of higher **entropy**. The second law is manifest in our practical world in forbidding the construction of perfect heat engines and perfect refrigerators—and therefore preventing us from extracting as useful work all the energy that's contained in random thermal motions. Ultimately, the second law says that the entropy of any closed system, including the entire universe, cannot decrease.

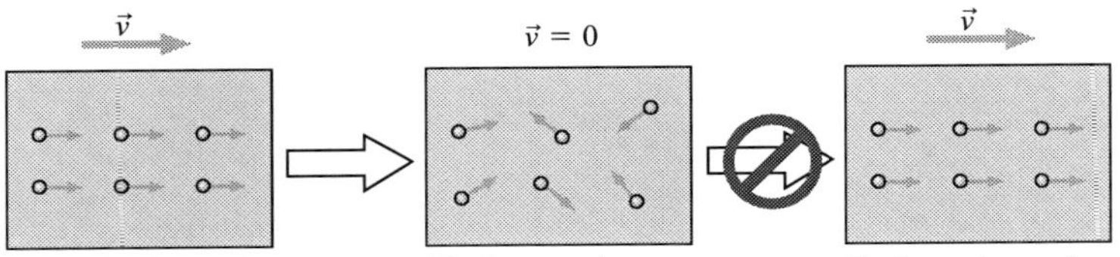

Block moving; molecular motion ordered.

Block stopped; molecular motion disordered.

Can't spontaneously restore order!

Key Concepts and Equations

Entropy is a quantitative measure of energy quality and of disorder; the higher the entropy, the lower the energy quality and the greater the disorder. The highest-quality energy is mechanical or electrical energy, followed by the internal energy of systems at high temperature, and finally low-temperature internal energy. Whenever entropy increases, energy becomes unavailable to do work.

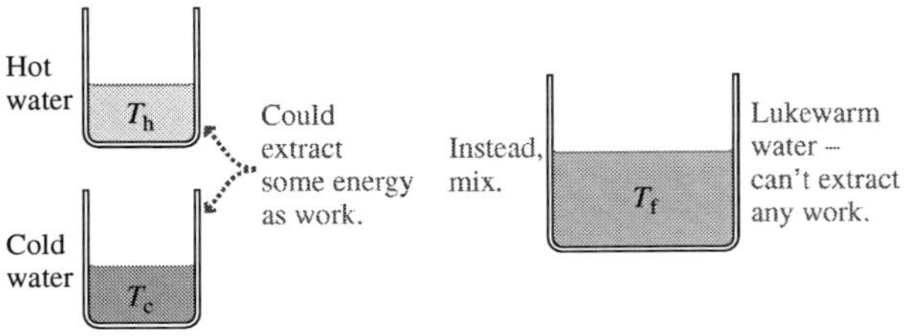

- $\Delta S = \int_1^2 \frac{dQ}{T}$ gives the entropy change as a system goes from state 1 to state 2.
- $E_{\text{unavailable}} = T_{\min} \Delta S$ is the energy that becomes unavailable as a result of entropy increase ΔS.

Cases and Uses

The second law sets the maximum possible efficiency of any heat engine as that of the **Carnot engine**, an engine that combines adiabatic and isothermal processes.

$$e = \frac{W}{Q_h} \leq e_{\max} = 1 - \frac{T_c}{T_h}$$

This defines an engine's efficiency. This is the maximum possible efficiency.

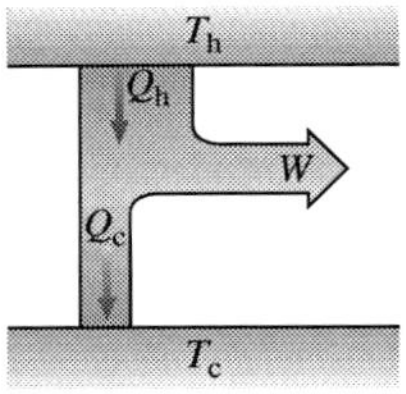

Energy-flow diagram for an engine

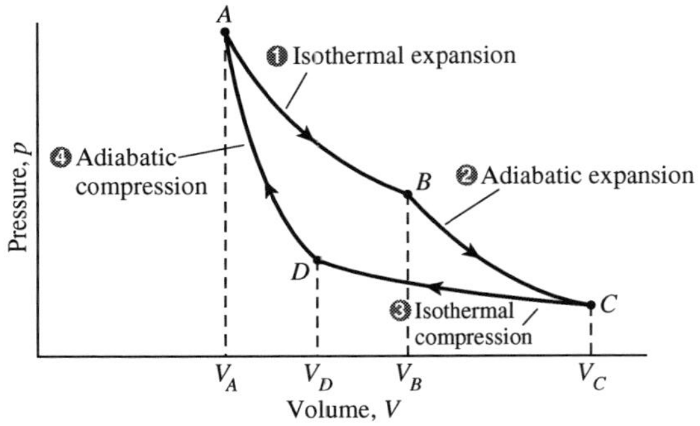

pV diagram for Carnot engine

Similarly, the second law limits the **coefficient of performance** of a refrigerator to

$$\text{COP} = \frac{Q_c}{W} \leq \text{COP}_{\max} = \frac{T_c}{T_h - T_c}$$

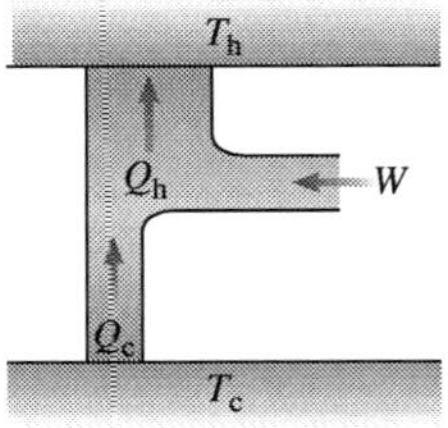

For Thought and Discussion

1. Could you cool the kitchen by leaving the refrigerator door open? Explain.
2. Could you heat the kitchen by leaving the oven open? Explain.
3. Should a car get better mileage in the summer or the winter? Explain.
4. Is there a limit to the maximum temperature that can be achieved by focusing sunlight with a lens? If so, what is it?
5. Name some irreversible processes that occur in a real engine.
6. A power company claims that electric heat is 100% efficient. Discuss this claim.
7. A hydroelectric power plant, using the energy of falling water, can operate with an efficiency arbitrarily close to 100%. Why?
8. To maximize the COP of a refrigerator, should you strive for a large or a small temperature difference? Explain.
9. The manufacturer of a heat pump claims that the device will heat your home using only energy already available in the ground. Is this true?
10. Does sunlight represent high- or low-quality energy? Explain. *Note:* See also For Thought and Discussion 4.
11. The heat Q added during adiabatic free expansion is zero. Why can't we then argue from Equation 19.6 that the entropy change is zero?
12. Energy is conserved, so why can't we recycle it as we do materials?
13. Why doesn't the evolution of human civilization violate the second law of thermodynamics?

Exercises and Problems

Exercises

Sections 19.2 and 19.3 The Second Law of Thermodynamics and Its Applications

14. What are the efficiencies of reversible heat engines operating between (a) the normal freezing and boiling points of water, (b) the 25°C temperature at the surface of a tropical ocean and deep water at 4°C, and (c) a 1000°C candle flame and room temperature?
15. A cosmic heat engine might operate between the Sun's 5600 K surface and the 2.7 K temperature of intergalactic space. What would be its efficiency?
16. A reversible Carnot engine operating between helium's melting point and its 4.25K boiling point has an efficiency of 77.7%. What is the melting point?
17. A Carnot engine absorbs 900 J of heat each cycle and provides 350 J of work. (a) What is its efficiency? (b) How much heat is rejected each cycle? (c) If the engine rejects heat at 10°C, what is its maximum temperature?
18. What is the COP of a reversible refrigerator operating between 0°C and 30°C?
19. How much work does a refrigerator with a COP of 4.2 require to freeze 670 g of water already at its freezing point?

Section 19.4 Entropy and Energy Quality

20. Calculate the entropy change associated with melting 1.0 kg of ice at 0°C.
21. You heat 250 g of water from 10°C to 95°C. By how much does the entropy of the water increase?
22. A 2.0-kg sample of water is heated to 35°C. If the entropy change is 740 J/K, what was the initial temperature?
23. Melting a block of lead already at its melting point results in an entropy increase of 900 J/K. What is the mass of the lead? *Hint:* Consult Table 17.1.
24. How much energy becomes unavailable for work in an isothermal process at 440 K, if the entropy increase is 25 J/K?

Problems

25. A Carnot engine extracts 890 J from a 550 K reservoir during each cycle and rejects 470 J to a cooler reservoir. (a) How much work does it do during each cycle? (b) What is its efficiency? (c) What is the temperature of the cool reservoir? (d) If the engine undergoes 22 cycles per second, what is its mechanical power output?
26. The maximum temperature in a nuclear power plant is 570 K. The plant rejects heat to a river whose temperature is 0°C in the winter and 25°C in the summer. What are the maximum possible efficiencies for the plant in these seasons?
27. A power plant's electrical output is 750 MW. Cooling water at 15°C flows through the plant at 2.8×10^4 kg/s, and its temperature rises by 8.5°C. Assuming that the plant's only energy loss is to the cooling water and that the cooling water is effectively the low-temperature reservoir, find (a) the rate of energy extraction from the fuel, (b) the plant's efficiency, and (c) its highest temperature.
28. A power plant extracts energy from steam at 250°C and delivers 800 MW of electric power. It discharges waste heat to a river at 30°C. The overall efficiency of the plant is 28%. (a) How does this efficiency compare with the maximum possible at these temperatures? (b) What is the rate of waste-heat discharge to the river? (c) How many houses, each requiring 18 kW of heating power, could be heated with the waste heat from this plant?
29. The electric power output of all the thermal electric power plants in the United States is about 2×10^{11} W, and these plants operate at an average efficiency around 33%. What is the rate at which all these plants use cooling water, assuming an average 5°C rise in cooling-water temperature? Compare with the 1.8×10^7 kg/s average flow at the mouth of the Mississippi River.
30. Consider a Carnot engine operating between temperatures T_h and T_i, where T_i is intermediate between T_h and the ambient temperature T_c (Fig. 19.20). It should be possible to operate a second engine between T_i and T_c. Show that the maximum overall efficiency of such a two-stage engine is the same as that of a single engine operating between T_h and T_c. *Note:* This problem shows why combined-cycle power plants achieve their high efficiencies.

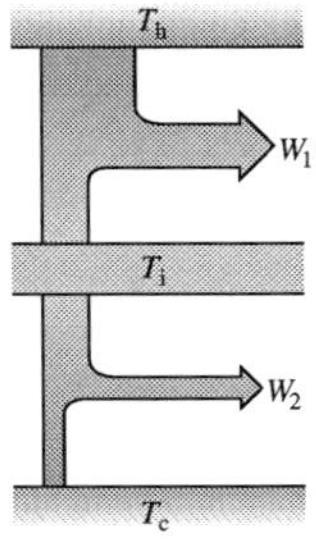

FIGURE 19.20 Problem 30

31. An industrial freezer operates between 0°C and 32°C, consuming electrical energy at the rate of 12 kW. Assuming the freezer is perfectly reversible, (a) what is its COP? (b) How much water at 0°C can it freeze in 1 hour?

32. Use appropriate energy-flow diagrams to analyze the situation in Got It? 19.2; that is, show that using a refrigerator to cool the low-temperature reservoir can't increase the overall efficiency of a Carnot engine when the work input to the refrigerator is included.
33. Cooling water circulates through a reversible Carnot engine at 3.2 kg/s. The water enters at 23°C and leaves at 28°C; the average temperature is essentially that of the engine's cool reservoir. If the engine's mechanical power output is 150 kW, what are (a) its efficiency and (b) its highest temperature?
34. It costs \$180 to heat a house with electricity in a typical winter month. (Electric heat simply converts all the incoming electrical energy to heat.) What would the monthly heating bill be following conversion to an electrically powered heat-pump system with COP = 3.1?
35. A 4.0-L sample of water at 9.0°C is put into a refrigerator. The refrigerator's 130-W motor then runs for 4.0 min to cool the water to the refrigerator's low temperature of 1.0°C. (a) What is the COP of the refrigerator? (b) How does this compare with the maximum possible COP if the refrigerator exhausts heat at 25°C?
36. A refrigerator maintains an interior temperature of 4°C while the temperature near its heat exhaust is 30°C. The refrigerator's insulation is imperfect, and heat leaks into the refrigerator at the rate of 340 W. Assuming the refrigerator is reversible, at what rate must it consume electrical energy to maintain a constant 4°C interior?
37. A store is heated by an oil furnace that supplies 30 kWh of heat from each gallon. The store's owners are considering switching to a heat-pump system. Oil costs \$1.75/gallon, and electricity costs 16.5¢/kWh. What is the minimum heat-pump COP that will result in a savings in heating costs?
38. Use appropriate energy-flow diagrams to show that the existence of a perfect heat engine would permit the construction of a perfect refrigerator, thus violating the Clausius statement of the second law.
39. A heat pump extracts energy from the ground at 10°C and transfers it to water at 70°C to heat a building. Find (a) its COP and (b) its electric power consumption if it supplies heat at the rate of 20 kW. (c) Compare the hourly operating cost of the pump with an oil furnace if electricity costs 15.5¢/kWh and oil costs 1.95/gallon and releases about 30 kWh/gal when burned.
40. A reversible engine contains 0.20 mol of ideal monatomic gas, initially at 600 K and confined to 2.0 L. The gas undergoes the following cycle:
   - Isothermal expansion to 4.0 L.
   - Isovolumic cooling to 300 K.
   - Isothermal compression to 2.0 L.
   - Isovolumic heating to 600 K.

   (a) Calculate the net heat added during the cycle and the net work done. (b) Determine the engine's efficiency, defined as the ratio of the work done to only the heat *absorbed* during the cycle.
41. (a) Determine the efficiency for the cycle shown in Fig. 19.21, using the definition given in Problem 40. (b) Compare with the efficiency of a Carnot engine operating between the same temperature extremes. Why are the two efficiencies different?

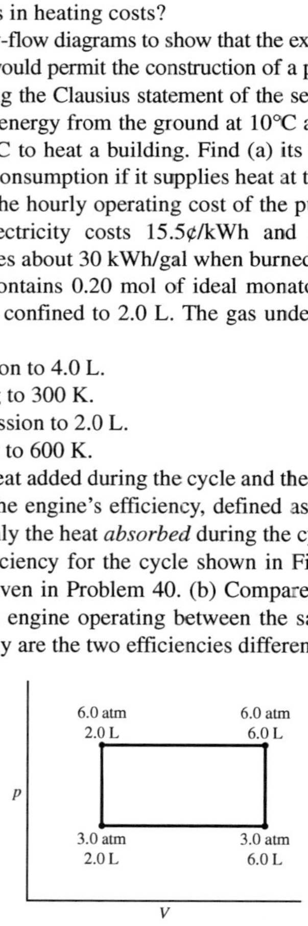

FIGURE 19.21 Problem 41

42. A 0.20-mol sample of an ideal gas goes through the Carnot cycle of Fig. 19.22. Calculate (a) the heat Q_h absorbed, (b) the heat Q_c rejected, and (c) the work done. (d) Use these quantities to determine the efficiency. (e) Find the maximum and minimum temperatures, and show explicitly that the efficiency as defined in Equation 19.1 is equal to the Carnot efficiency of Equation 19.3.

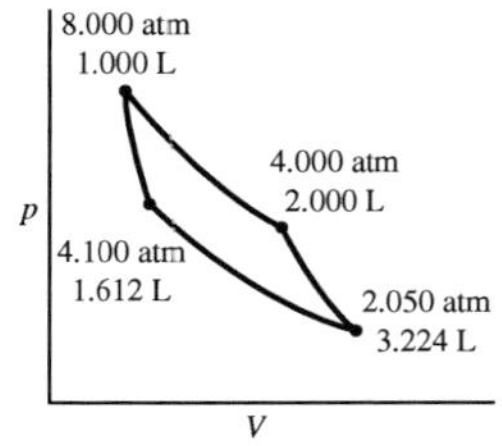

FIGURE 19.22 Problem 42

43. A shallow pond contains 94,000 kg of water. In winter it's entirely frozen. By how much does the entropy of the pond increase when the ice, already at 0°C, melts and then heats to its summer temperature of 15°C?
44. The temperature of n moles of ideal gas is changed from T_1 to T_2 while the gas volume is held constant. Show that the corresponding entropy change is $\Delta S = nC_V \ln(T_2/T_1)$.
45. The temperature of n moles of ideal gas is changed from T_1 to T_2 while the gas pressure is held constant. Show that the corresponding entropy change is $\Delta S = nC_p \ln(T_2/T_1)$.
46. A 5.0-mol sample of an ideal diatomic gas $\left(C_V = \frac{5}{2}R\right)$ is initially at 1.0 atm pressure and 300 K. What is the entropy change if the gas is heated to 500 K (a) at constant volume, (b) at constant pressure, and (c) adiabatically?
47. The interior of a house is maintained at 20°C while the outdoor temperature is −10°C. The house loses heat at the rate of 30 kW. At what rate does the entropy of the universe increase because of this irreversible heat flow?
48. A 250-g sample of water at 80°C is mixed with 250 g of water at 10°C. Find the entropy changes for (a) the hot water, (b) the cool water, and (c) the system.
49. In an adiabatic free expansion, 8.7 mol of ideal gas at 450 K expand 10-fold in volume. How much energy becomes unavailable to do work?
50. Ideal gas occupying 1.0 cm^3 is placed in a 1.0-m^3 vacuum chamber, where it expands adiabatically. If 6.5 J of energy become unavailable to do work, what was the initial gas pressure?
51. Find the entropy change when a 2.4-kg aluminum pan at 155°C is plunged into 3.5 kg of water at 15°C.
52. You're lying in a bathtub of water at 42°C. Suppose, in violation of the second law of thermodynamics, that the water spontaneously cooled to room temperature (20°C) and that the energy so released was transformed into your gravitational potential energy. Estimate the height to which you would rise above the bathtub.
53. An engine with mechanical power output 8.5 kW extracts heat from a source at 420 K and rejects it to a 1000-kg block of ice at its melting point. (a) What is its efficiency? (b) How long can it maintain this efficiency if the ice is not replenished?
54. Gasoline engines operate approximately on the **Otto cycle**, consisting of two adiabatic and two constant-volume segments. The Otto cycle for a particular engine is shown in Fig. 19.23. (a) If the gas in the engine has specific-heat ratio γ, find the engine's efficiency, assuming all processes are reversible. (b) Find the maximum temperature in terms of the minimum temperature T_{min}.

(c) How does the efficiency compare with that of a Carnot engine operating between the same two temperature extremes? *Note:* Figure 19.23 neglects the intake of fuel-air and the exhaust of combustion products, which together involve essentially no net work.

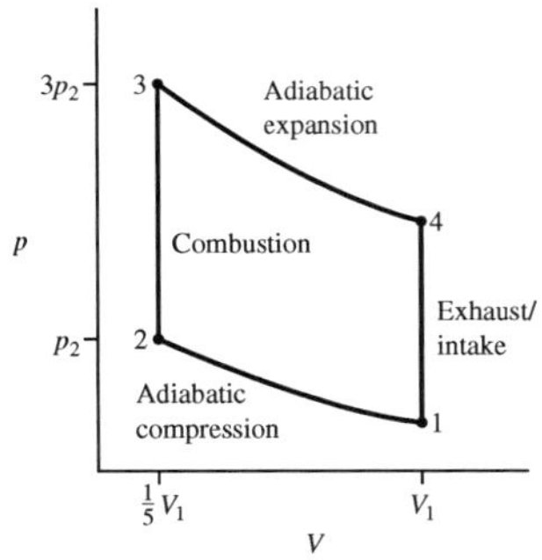

FIGURE 19.23 Problems 54 and 55

55. The compression ratio r of an engine is the ratio of maximum to minimum gas volume. For the engine of Problem 54, Fig. 19.23 shows that the compression ratio is 5. (a) Find an expression for the engine efficiency as a function of the compression ratio. Assume that pressure continues to triple during the combustion phase, as shown in Fig. 19.23. (b) Make a graph of efficiency versus r, and on the same plot show also the efficiency of a Carnot engine operating between the same temperature extremes.
56. A 5.0-mol sample of ideal monatomic gas undergoes the cycle shown in Fig. 19.24, in which the process BC is isothermal. Calculate the entropy change associated with each of the three steps, and show explicitly that there is zero net entropy change over the full cycle.

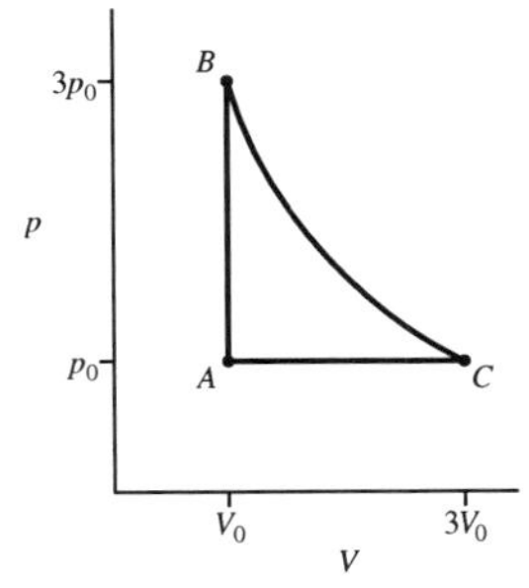

FIGURE 19.24 Problem 56

57. The McNeil Generating Station in Burlington, Vermont, is one of the world's largest wood-fired electric power plants. McNeil produces steam at 950°F to drive its turbines, and condensed steam returns to the boiler as water at 90°F. (Note the temperatures in Fahrenheit, used in U.S. engineering situations.) Find McNeil's maximum thermodynamic efficiency, and compare with its actual efficiency of 25%. *Note:* Some of the difference comes from having to evaporate moisture out of the wood-chip fuel.
58. A 500-g copper block at 80°C is dropped into 1.0 kg of water at 10°C. Find (a) the final temperature and (b) the entropy change of the system.
59. A Carnot engine extracts heat from a block of mass m and specific heat c that is initially at temperature T_{h0} but that has no heat source to maintain that temperature. The engine rejects heat to a reservoir at a constant temperature T_c. The engine is operated so its mechanical power output is proportional to the temperature difference $T_h - T_c$:

$$P = P_0 \frac{T_h - T_c}{T_{h0} - T_c}$$

where T_h is the instantaneous temperature of the hot block and P_0 is the initial power output. (a) Find an expression for T_h as a function of time, and (b) determine how long it takes for the engine's power output to reach zero.
60. Find an expression for the entropy change of the whole system in Fig. 19.14, in terms of T_c and T_h, by evaluating the integrals in the paragraph under the heading on Irreversible Heat Transfer on page 319. Assume both glasses contain the same mass, m, of water, whose specific heat c is constant, and neglect the heat capacity of the glasses. Prove that your expression is positive.
61. Problem 77 of Chapter 16 provided an approximate expression for the specific heat of copper at low absolute temperatures; $c = 31(T/343\text{ K})^3$ J/g·K. Use this to find the entropy change in a 40-g sample of copper which is cooled from 25 K to 10 K. Why is the change negative?
62. The molar specific heat at constant pressure for a certain gas is given by $c_p = a + bT + cT^2$, where $a = 33.6$ J/mol·K, $b = 2.93 \times 10^{-3}$ J/mol·K^2, and $c = 2.13 \times 10^{-5}$ J/mol·K^3. What is the change in entropy of 2 moles of this gas when heated from 20°C to 200°C?
63. You have 50 kg of steam at 100°C, but no heat source to maintain it in that condition. You also have a heat reservoir at 0°C. Suppose you operate a reversible heat engine with this system, so the steam gradually condenses and cools until it reaches 0°C. (a) Calculate the total entropy change of the steam and subsequent water. (b) Calculate the total entropy change of the reservoir. (c) Find the total amount of work that the engine can do. *Hint:* Consider the entropy change of the entire system. This change occurs as the result of a reversible process that supplies useful work. But if it had occurred irreversibly, how much energy would have become unavailable to do work?
64. A city council member contacts you, the public-affairs manager for a nuclear power plant, concerning the environmental effects of discharging hot water into the river that flows by the plant. The flow rate of the river is 3.4×10^4 kg/s. The thermal-energy input rate to the river is 100 MW. The council member says a 5°C increase in temperature is acceptable. Does your plant meet this standard?
65. A friend has a scheme to run an engine that will deposit its heat into a block of ice. The engine has a power output of 8.5 kW and extracts heat from a source at 420 K and rejects heat to a 1000-kg block of ice at its melting point. Your friend claims his gadget will run for an 8-hour workday. Is he right? *Hint:* Calculate the efficiency and then how long the ice will last without replenishment.

Answers to Chapter Questions

Answer to Chapter Opening Question

The second law of thermodynamics prevents us from converting thermal energy to mechanical energy with 100% efficiency, and practical limits on temperature make it hard to achieve 50% efficiency in conventional power plants.

Answers to GOT IT? Questions

19.1 (a), (c), and (f).
19.2 (c); see Problem 32 for a proof.
19.3 (a) increase; (b) decrease; (c) increase; (d) increase; (e) decrease; (f) increase; (g) increase.

APPENDICES

Appendix One

Mathematics

A-1 Algebra and Trigonometry

Quadratic Formula

If $ax^2 + bx + c = 0$, then $x = \dfrac{-b \pm \sqrt{b^2 - 4ac}}{2a}$.

Circumference, Area, Volume

Where $\pi \simeq 3.14159\ldots$

circumference of circle	$2\pi r$
area of circle	πr^2
surface area of sphere	$4\pi r^2$
volume of sphere	$\frac{4}{3}\pi r^3$
area of triangle	$\frac{1}{2}bh$
volume of cylinder	$\pi r^2 l$

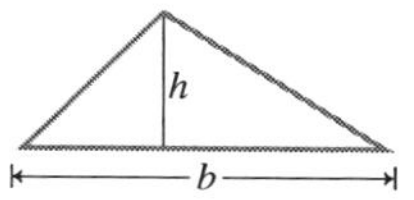

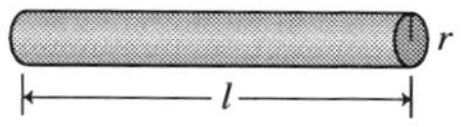

Trigonometry

definition of angle (in radians): $\theta = \dfrac{s}{r}$

2π radians in complete circle

1 radian $\simeq 57.3°$

Trigonometric Functions

$$\sin\theta = \frac{y}{r}$$

$$\cos\theta = \frac{x}{r}$$

$$\tan\theta = \frac{\sin\theta}{\cos\theta} = \frac{y}{x}$$

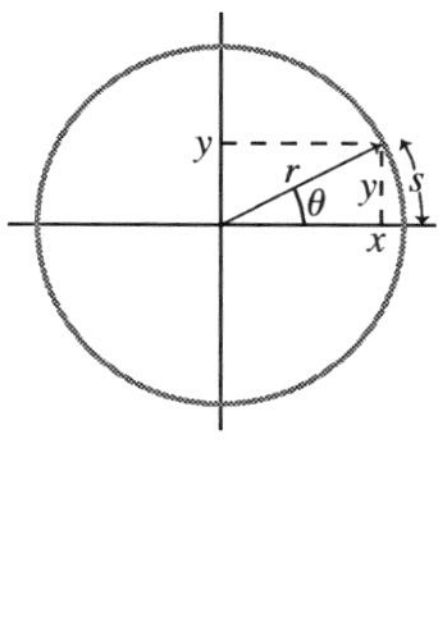

Values at Selected Angles

$\theta \rightarrow$	**0**	$\frac{\pi}{6}$ **(30°)**	$\frac{\pi}{4}$ **(45°)**	$\frac{\pi}{3}$ **(60°)**	$\frac{\pi}{2}$ **(90°)**
$\sin\theta$	0	$\frac{1}{2}$	$\frac{\sqrt{2}}{2}$	$\frac{\sqrt{3}}{2}$	1
$\cos\theta$	1	$\frac{\sqrt{3}}{2}$	$\frac{\sqrt{2}}{2}$	$\frac{1}{2}$	0
$\tan\theta$	0	$\frac{\sqrt{3}}{3}$	1	$\sqrt{3}$	∞

Graphs of Trigonometric Functions

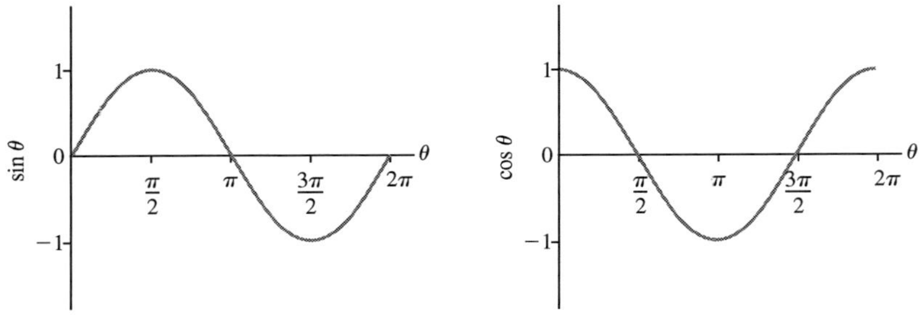

Trigonometric Identities

$$\sin(-\theta) = -\sin\theta$$

$$\cos(-\theta) = \cos\theta$$

$$\sin\left(\theta \pm \frac{\pi}{2}\right) = \pm\cos\theta$$

$$\cos\left(\theta \pm \frac{\pi}{2}\right) = \mp\sin\theta$$

$$\sin^2\theta + \cos^2\theta = 1$$

$$\sin 2\theta = 2\sin\theta\cos\theta$$

$$\cos 2\theta = \cos^2\theta - \sin^2\theta = 1 - 2\sin^2\theta = 2\cos^2\theta - 1$$

$$\sin(\alpha \pm \beta) = \sin\alpha\cos\beta \pm \cos\alpha\sin\beta$$

$$\cos(\alpha \pm \beta) = \cos\alpha\cos\beta \mp \sin\alpha\sin\beta$$

$$\sin\alpha \pm \sin\beta = 2\sin\left[\tfrac{1}{2}(\alpha \pm \beta)\right]\cos\left[\tfrac{1}{2}(\alpha \mp \beta)\right]$$

$$\cos\alpha + \cos\beta = 2\cos\left[\tfrac{1}{2}(\alpha + \beta)\right]\cos\left[\tfrac{1}{2}(\alpha - \beta)\right]$$

$$\cos\alpha - \cos\beta = -2\sin\left[\tfrac{1}{2}(\alpha + \beta)\right]\sin\left[\tfrac{1}{2}(\alpha - \beta)\right]$$

Laws of Cosines and Sines

Where A, B, C are the sides of an arbitrary triangle and α, β, γ the angles opposite those sides:

Law of cosines

$C^2 = A^2 + B^2 - 2AB\cos\gamma$

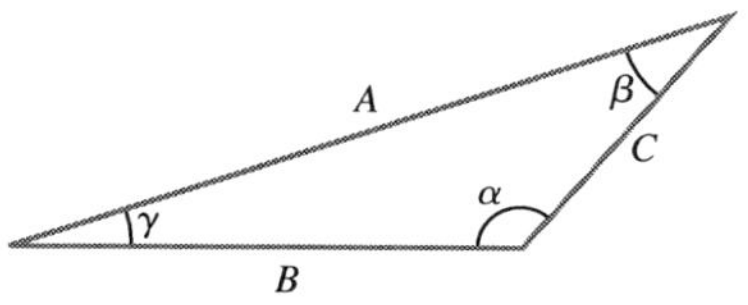

Law of sines

$$\frac{\sin\alpha}{A} = \frac{\sin\beta}{B} = \frac{\sin\gamma}{C}$$

Exponentials and Logarithms

$e^{\ln x} = x, \quad \ln e^x = x \quad e = 2.71828\ldots$

$a^x = e^{x\ln a} \qquad \ln(xy) = \ln x + \ln y$

$a^x a^y = a^{x+y} \qquad \ln\left(\frac{x}{y}\right) = \ln x - \ln y$

$(a^x)^y = a^{xy} \qquad \ln\left(\frac{1}{x}\right) = -\ln x$

$\log x \equiv \log_{10} x = \ln(10)\ln x \simeq 2.3\ln x$

Approximations

For $|x| \ll 1$, the following expressions provide good approximations to common functions:

$e^x \simeq 1 + x$

$\sin x \simeq x$

$\cos x \simeq 1 - \frac{1}{2}x^2$

$\ln(1 + x) \simeq x$

$(1 + x)^p \simeq 1 + px$ (binomial approximation)

Expressions that don't have the forms shown may often be put in the appropriate form. For example:

$$\frac{1}{\sqrt{a^2 + y^2}} = \frac{1}{a\sqrt{1 + \frac{y^2}{a^2}}} = \frac{1}{a}\left(1 + \frac{y^2}{a^2}\right)^{-1/2} \simeq \frac{1}{a}\left(1 - \frac{y^2}{2a}\right) \quad \text{for} \quad y^2/a^2 \ll 1, \text{ or } y^2 \ll a^2.$$

Vector Algebra

Vector Products

$\vec{A} \cdot \vec{B} = AB\cos\theta$

$|\vec{A} \times \vec{B}| = AB\sin\theta$, with direction of $\vec{A} \times \vec{B}$ given by the right-hand rule:

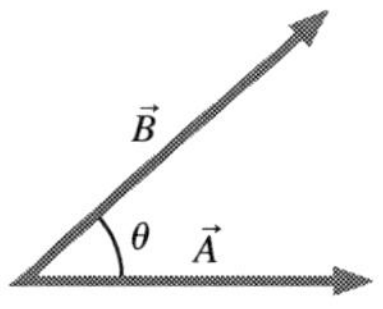

Unit Vector Notation

An arbitrary vector $\vec{A}$ may be written in terms of its components A_x, A_y, A_z and the unit vectors $\hat{\imath}, \hat{\jmath}, \hat{k}$ that have length 1 and lie along the x, y, z axes:

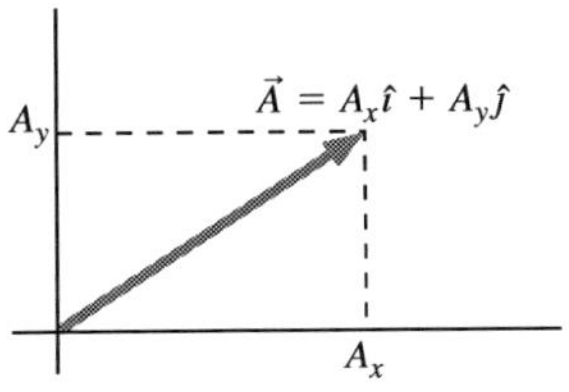

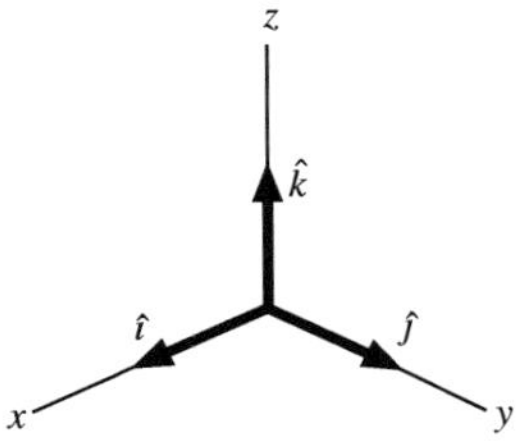

In unit vector notation, vector products become

$$\vec{A} \cdot \vec{B} = A_xB_x + A_yB_y + A_zB_z$$

$$\vec{A} \times \vec{B} = (A_yB_z - A_zB_y)\hat{\imath} + (A_zB_x - A_xB_z)\hat{\jmath} + (A_xB_y - A_yB_x)\hat{k}$$

Vector Identities

$$\vec{A} \cdot \vec{B} = \vec{B} \cdot \vec{A}$$

$$\vec{A} \times \vec{B} = -\vec{B} \times \vec{A}$$

$$\vec{A} \cdot (\vec{B} \times \vec{C}) = \vec{B} \cdot (\vec{C} \times \vec{A}) = \vec{C} \cdot (\vec{A} \times \vec{B})$$

$$\vec{A} \times (\vec{B} \times \vec{C}) = (\vec{A} \cdot \vec{C})\vec{B} - (\vec{A} \cdot \vec{B})\vec{C}$$

A-2 Calculus

Derivatives

Definition of the Derivative

If y is a function of x, then the **derivative of y with respect to x** is the ratio of the change Δy in y to the corresponding change Δx in x, in the limit of arbitrarily small Δx:

$$\frac{dy}{dx} = \lim_{\Delta x \to 0} \frac{\Delta y}{\Delta x}$$

Algebraically, the derivative is the rate of change of y with respect to x; geometrically, it is the slope of the y versus x graph—that is, of the tangent line to the graph at a given point:

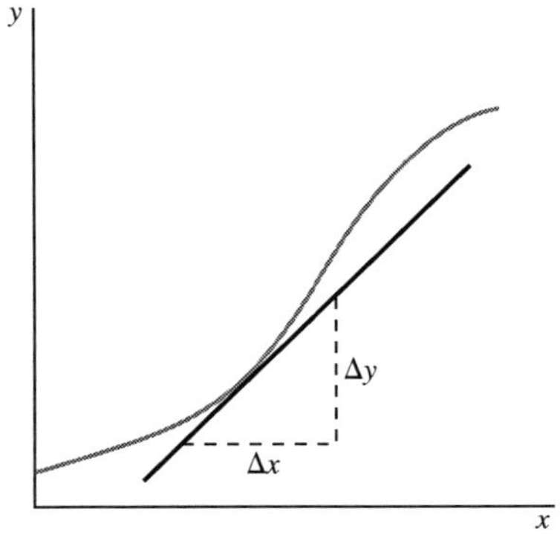

Derivatives of Common Functions

$$\frac{da}{dx} = 0 \quad (a \text{ is a constant})$$

$$\frac{dx^n}{dx} = nx^{n-1} \quad (n \text{ need not be an integer})$$

$$\frac{d}{dx}\sin x = \cos x$$

$$\frac{d}{dx}\cos x = -\sin x$$

$$\frac{d}{dx}\tan x = \frac{1}{\cos^2 x}$$

$$\frac{de^x}{dx} = e^x$$

$$\frac{d}{dx}\ln x = \frac{1}{x}$$

Derivatives of Sums, Products, and Functions of Functions

1. Derivative of a constant times a function

$$\frac{d}{dx}[af(x)] = a\frac{df}{dx} \qquad (a \text{ is a constant})$$

2. Derivative of a sum

$$\frac{d}{dx}[f(x) + g(x)] = \frac{df}{dx} + \frac{dg}{dx}$$

3. Derivative of a product

$$\frac{d}{dx}[f(x)g(x)] = g\frac{df}{dx} + f\frac{dg}{dx}$$

Examples

$$\frac{d}{dx}(x^2\cos x) = \cos x\frac{dx^2}{dx} + x^2\frac{d}{dx}\cos x = 2x\cos x - x^2\sin x$$

$$\frac{d}{dx}(x\ln x) = \ln x\frac{dx}{dx} + x\frac{d}{dx}\ln x = (\ln x)(1) + x\left(\frac{1}{x}\right) = \ln x + 1$$

4. Derivative of a quotient

$$\frac{d}{dx}\left[\frac{f(x)}{g(x)}\right] = \frac{1}{g^2}\left(g\frac{df}{dx} - f\frac{dg}{dx}\right)$$

Example

$$\frac{d}{dx}\left(\frac{\sin x}{x^2}\right) = \frac{1}{x^4}\left(x^2\frac{d}{dx}\sin x - \sin x\frac{dx^2}{dx}\right) = \frac{\cos x}{x^2} - \frac{2\sin x}{x^3}$$

5. Chain rule for derivatives

If f is a function of u and u is a function of x, then

$$\frac{df}{dx} = \frac{df}{du}\frac{du}{dx}$$

Examples

a. Evaluate $\frac{d}{dx}\sin(x^2)$. Here $u = x^2$ and $f(u) = \sin u$, so

$$\frac{d}{dx}\sin(x^2) = \frac{d}{du}\sin u\frac{du}{dx} = (\cos u)\frac{dx^2}{dx} = 2x\cos(x^2)$$

b. $\frac{d}{dt}\sin\omega t = \frac{d}{d\,\omega t}\sin\omega t\frac{d}{dt}\omega t = \omega\cos\omega t \qquad (\omega \text{ is a constant})$

c. Evaluate $\frac{d}{dx}\sin^2 5x$. Here $u = \sin 5x$ and $f(u) = u^2$, so

$$\frac{d}{dx}\sin^2 5x = \frac{d}{du}u^2\frac{du}{dx} = 2u\frac{du}{dx} = 2\sin 5x\frac{d}{dx}\sin 5x$$
$$= (2)(\sin 5x)(5)(\cos 5x) = 10\sin 5x\cos 5x = 5\sin 2x$$

Second Derivative

The second derivative of y with respect to x is defined as the derivative of the derivative:

$$\frac{d^2y}{dx^2} = \frac{d}{dx}\left(\frac{dy}{dx}\right)$$

Example

If $y = ax^3$, then $dy/dx = 3ax^2$, so

$$\frac{d^2y}{dx^2} = \frac{d}{dx}3ax^2 = 6ax$$

Partial Derivatives

When a function depends on more than one variable, then the partial derivatives of that function are the derivatives with respect to each variable, taken with all other variables held constant. If f is a function of x and y, then the partial derivatives are written

$$\frac{\partial f}{\partial x} \quad \text{and} \quad \frac{\partial f}{\partial y}$$

Example

If $f(x, y) = x^3 \sin y$, then

$$\frac{\partial f}{\partial x} = 3x^2 \sin y \quad \text{and} \quad \frac{\partial f}{\partial y} = x^3 \cos y$$

Integrals

Indefinite Integrals

Integration is the inverse of differentiation. The **indefinite integral**, $\int f(x)\,dx$, is defined as a function whose derivative is $f(x)$:

$$\frac{d}{dx}\left[\int f(x)\,dx\right] = f(x)$$

If $A(x)$ is an indefinite integral of $f(x)$, then because the derivative of a constant is zero, the function $A(x) + C$ is also an indefinite integral of $f(x)$, where C is any constant. Inverting the derivatives of common functions listed in the preceding section gives the integrals that follow (a more extensive table appears at the end of this appendix).

$$\int a\,dx = ax + C \qquad \int \cos x\,dx = \sin x + C$$

$$\int x^n\,dx = \frac{x^{n+1}}{n+1} + C, \quad n \neq -1 \qquad \int e^x\,dx = e^x + C$$

$$\int \sin x\,dx = -\cos x + C \qquad \int x^{-1}\,dx = \ln x + C$$

Definite Integrals

In physics we're most often interested in the **definite integral**, defined as the sum of a large number of very small quantities, in the limit as the number of quantities grows arbitrarily large and the size of each arbitrarily small:

$$\int_{x_1}^{x_2} f(x)\,dx \equiv \lim_{\substack{\Delta x \to 0 \\ N \to \infty}} \sum_{i=1}^{N} f(x_i)\,\Delta x$$

where the terms in the sum are evaluated at values x_i between the limits of integration x_1 and x_2; in the limit $\Delta x \to 0$, the sum is over all values of x in the interval.

The key to evaluating the definite integral is provided by the **fundamental theorem of calculus**. The theorem states that, if $A(x)$ is an *indefinite* integral of $f(x)$, then the *definite integral* is given by

$$\int_{x_1}^{x_2} f(x)\,dx = A(x_2) - A(x_1) \equiv A(x)\Big|_{x_1}^{x_2}$$

Geometrically, the definite integral is the area under the graph of $f(x)$ between the limits x_1 and x_2.

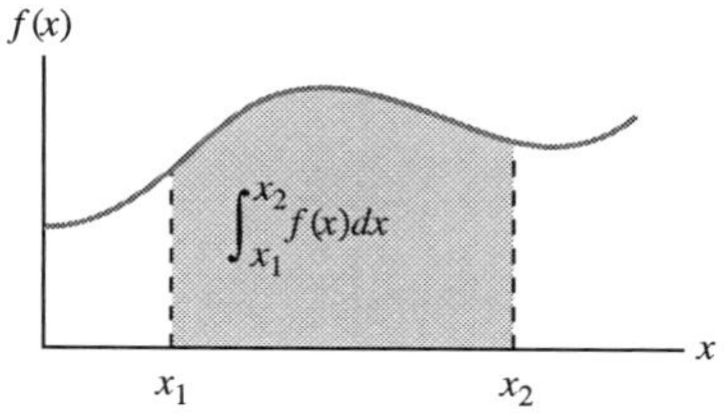

Evaluating Integrals

The first step in evaluating an integral is to express all varying quantities within the integral in terms of a single variable; Chapter 9 outlines a general strategy for setting up an integral. Once you've set up an integral, you can evaluate it yourself or look it up in tables. Two common techniques can help you evaluate integrals or convert them to forms listed in tables:

1. **Change of variables**
   An unfamiliar integral can often be put into familiar form by defining a new variable. For example, it is not obvious how to integrate the expression

$$\int \frac{x\,dx}{\sqrt{a^2 + x^2}}$$

   where a is a constant. But let $z = a^2 + x^2$. Then

$$\frac{dz}{dx} = \frac{da^2}{dx} + \frac{dx^2}{dx} = 0 + 2x = 2x$$

   so $dz = 2x\,dx$. Then the quantity $x\,dx$ in our unfamiliar integral is just $\frac{1}{2}dz$, while the quantity $\sqrt{a^2 + x^2}$ is just $z^{1/2}$. So the integral becomes

$$\int \frac{1}{2} z^{-1/2}\,dz = \frac{\frac{1}{2}z^{1/2}}{1/2} = \sqrt{z}$$

   where we have used the standard form for the integral of a power of the independent variable. Substituting back $z = a^2 + x^2$ gives

$$\int \frac{x\,dx}{\sqrt{a^2 + x^2}} = \sqrt{a^2 + x^2}$$

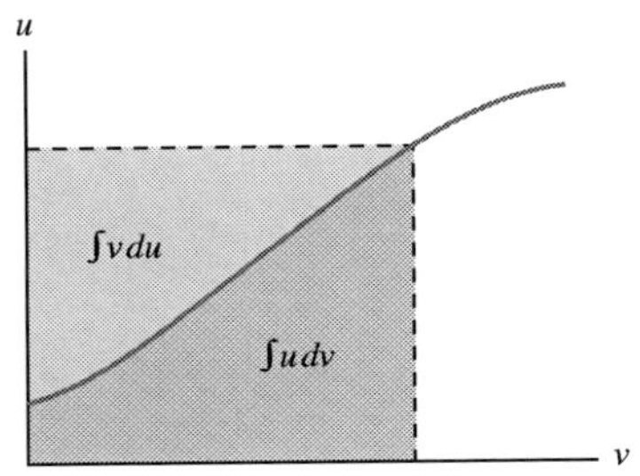

2. Integration by parts

The quantity $\int u\,dv$ is the area under the curve of u as a function of v between specified limits. In the figure, that area can also be expressed as the area of the rectangle shown minus the area under the curve of v as a function of u. Mathematically, this relation among areas may be expressed as a relation among integrals:

$$\int u\,dv = uv - \int v\,du \qquad \text{(integration by parts)}$$

This expression may often be used to transform complicated integrals into simpler ones.

Example

Evaluate $\int x \cos x\,dx$. Here let $u = x$, so $du = dx$. Then $dv = \cos x\,dx$, so we have $v = \int dv = \int \cos x\,dx = \sin x$. Integrating by parts then gives

$$\int x \cos x\,dx = (x)(\sin x) - \int \sin x\,dx = x \sin x + \cos x$$

where the + sign arises because $\int \sin x\,dx = -\cos x$.

Table of Integrals

More extensive tables are available in many mathematical and scientific handbooks; see, for example, ***Handbook of Chemistry and Physics*** (Chemical Rubber Co.) or Dwight, ***Tables of Integrals and Other Mathematical Data*** (Macmillan). Some math software, including *Mathematica* and *Maple*, can also evaluate integrals symbolically. Wolfram Research provides *Mathematica*-based integration at *http://integrals.wolfram.com.*

In the expressions below, a and b are constants. An arbitrary constant of integration may be added to the right-hand side.

$$\int e^{ax}\,dx = \frac{e^{ax}}{a}$$

$$\int \sin ax\,dx = -\frac{\cos ax}{a}$$

$$\int \cos ax\,dx = \frac{\sin ax}{a}$$

$$\int \tan ax\,dx = -\frac{1}{a}\ln(\cos ax)$$

$$\int \sin^2 ax\,dx = \frac{x}{2} - \frac{\sin 2ax}{4a}$$

$$\int \cos^2 ax\,dx = \frac{x}{2} + \frac{\sin 2ax}{4a}$$

$$\int x \sin ax\,dx = \frac{1}{a^2}\sin ax - \frac{1}{a}x\cos ax$$

$$\int x \cos ax\,dx = \frac{1}{a^2}\cos ax + \frac{1}{a}x\sin ax$$

$$\int \frac{dx}{\sqrt{a^2 - x^2}} = \sin^{-1}\left(\frac{x}{a}\right)$$

$$\int \frac{dx}{\sqrt{x^2 \pm a^2}} = \ln\left(x + \sqrt{x^2 \pm a^2}\right)$$

$$\int \frac{dx}{x^2 + a^2} = \frac{1}{a}\tan^{-1}\left(\frac{x}{a}\right)$$

$$\int \frac{x\,dx}{\sqrt{a^2 - x^2}} = -\sqrt{a^2 - x^2}$$

$$\int \frac{x\,dx}{\sqrt{x^2 \pm a^2}} = \sqrt{x^2 \pm a^2}$$

$$\int \frac{dx}{(x^2 \pm a^2)^{3/2}} = \frac{\pm x}{a^2\sqrt{x^2 \pm a^2}}$$

$$\int x e^{ax}\,dx = \frac{e^{ax}}{a^2}(ax - 1)$$

$$\int x^2 e^{ax}\,dx = \frac{x^2 e^{ax}}{a} - \frac{2}{a}\left[\frac{e^{ax}}{a^2}(ax - 1)\right]$$

$$\int \frac{dx}{a + bx} = \frac{1}{b}\ln(a + bx)$$

$$\int \frac{dx}{(a + bx)^2} = -\frac{1}{b(a + bx)}$$

$$\int \ln ax\,dx = x\ln ax - x$$

The International System of Units (SI)

APPENDIX B

This material is from the U.S. edition of the English translation of the seventh edition of "Le Système International d'Unités (SI)," the definitive publication in the French language issued in 1991 by the International Bureau of Weights and Measures (BIPM). The year the definition was adopted is given in parentheses.

length (meter): The meter is the length of the path traveled by light in vacuum during a time interval of 1/299 792 458 of a second. (1983)

mass (kilogram): The kilogram is equal to the mass of the international prototype of the kilogram. (1889)

time (second): The second is the duration of 9 192 631 770 periods of the radiation corresponding to the transition between the two hyperfine levels of the ground state of the cesium-133 atom. (1967)

electric current (ampere): The ampere is that constant current which, if maintained in two straight parallel conductors of infinite length, of negligible circular cross section, and placed 1 meter apart in vacuum, would produce between these conductors a force equal to 2×10^{-7} newton per meter of length. (1948)

temperature (kelvin): The kelvin, unit of thermodynamic temperature, is the fraction 1/273.16 of the thermodynamic temperature of the triple point of water. (1967)

amount of substance (mole): The mole is the amount of substance of a system that contains as many elementary entities as there are atoms in 0.012 kilogram of carbon-12. (1971)

luminous intensity (candela): The candela is the luminous intensity, in a given direction, of a source that emits monochromatic radiation of frequency 540×10^{12} hertz and that has a radiant intensity in that direction of $(1/683)$ watt per steradian. (1979)

SI Base and Supplementary Units

	SI Unit	
Quantity	**Name**	**Symbol**
Base Unit		
Length	meter	m
Mass	kilogram	kg
Time	second	s
Electric current	ampere	A
Thermodynamic temperature	kelvin	K
Amount of substance	mole	mol
Luminous intensity	candela	cd
Supplementary Units		
Plane angle	radian	rad
Solid angle	steradian	sr

SI Prefixes

Factor	Prefix	Symbol
10^{24}	yotta	Y
10^{21}	zetta	Z
10^{18}	exa	E
10^{15}	peta	P
10^{12}	tera	T
10^{9}	giga	G
10^{6}	mega	M
10^{3}	kilo	k
10^{2}	hecto	h
10^{1}	deka	da
10^{0}	—	—
10^{-1}	deci	d
10^{-2}	centi	c
10^{-3}	milli	m
10^{-6}	micro	μ
10^{-9}	nano	n
10^{-12}	pico	p
10^{-15}	femto	f
10^{-18}	atto	a
10^{-21}	zepto	z
10^{-24}	yocto	y

Some SI Derived Units with Special Names

	SI Unit			
Quantity	**Name**	**Symbol**	**Expression in Terms of Other Units**	**Expression in Terms of SI Base Units**
Frequency	hertz	Hz		s^{-1}
Force	newton	N		$m \cdot kg \cdot s^{-2}$
Pressure, stress	pascal	Pa	N/m^2	$m^{-1} \cdot kg \cdot s^{-2}$
Energy, work, heat	joule	J	$N \cdot m$	$m^2 \cdot kg \cdot s^{-2}$
Power	watt	W	J/s	$m^2 \cdot kg \cdot s^{-3}$
Electric charge	coulomb	C		$s \cdot A$
Electric potential, potential difference, electromotive force	volt	V	J/C	$m^2 \cdot kg \cdot s^{-3} \cdot A^{-1}$
Capacitance	farad	F	C/V	$m^{-2} \cdot kg^{-1} \cdot s^4 \cdot A^2$
Electric resistance	ohm	Ω	V/A	$m^2 \cdot kg \cdot s^{-3} \cdot A^{-2}$
Magnetic flux	weber	Wb	$V \cdot s$	$m^2 \cdot kg \cdot s^{-2} \cdot A^{-1}$
Magnetic field	tesla	T	Wb/m^2	$kg \cdot s^{-2} \cdot A^{-1}$
Inductance	henry	H	Wb/A	$m^2 \cdot kg \cdot s^{-2} \cdot A^{-2}$
Radioactivity	becquerel	Bq	1 decay/s	s^{-1}
Absorbed radiation dose	gray	Gy	J/kg, 100 rad	$m^2 \cdot s^{-2}$
Radiation dose equivalent	sievert	Sv	J/kg, 100 rem	$m^2 \cdot s^{-2}$

Conversion Factors

APPENDIX C

The listings below give the SI equivalents of non-SI units. To convert from the units shown to SI, multiply by the factor given; to convert the other way, divide. For conversions within the SI system, see the table of SI prefixes in Appendix B, Chapter 1, or the inside front cover. Conversions that are not exact by definition are given to, at most, four significant figures.

Length

1 inch (in) = 0.0254 m
1 foot (ft) = 0.3048 m
1 yard (yd) = 0.9144 m
1 mile (mi) = 1609 m
1 nautical mile = 1852 m
1 angstrom (Å) = 10^{-10} m
1 light year (ly) = 9.46×10^{15} m
1 astronomical unit (AU) = 1.5×10^{11} m
1 parsec = 3.09×10^{16} m
1 fermi = 10^{-15} m = 1 fm

Mass

1 slug = 14.59 kg
1 metric ton (tonne; t) = 1000 kg
1 unified mass unit (u) = 1.661×10^{-27} kg

Force units in the English system are sometimes used (incorrectly) for mass. The units given below are actually equal to the number of kilograms multiplied by *g*, the acceleration of gravity.

1 pound (lb) = weight of 0.454 kg
1 ton = 2000 lb = weight of 908 kg
1 ounce (oz) = weight of 0.02835 kg

Time

1 minute (min) = 60 s
1 hour (h) = 60 min = 3600 s
1 day (d) = 24 h = 86 400 s
1 year (y) = 365.2422 d* = 3.156×10^7 s

Area

1 hectare (ha) = 10^4 m^2
1 square inch (in^2) = 6.452×10^{-4} m^2
1 square foot (ft^2) = 9.290×10^{-2} m^2
1 acre = 4047 m^2
1 barn = 10^{-28} m^2
1 shed = 10^{-30} m^2

Volume

1 liter (L) = 1000 cm^3 = 10^{-3} m^3
1 cubic foot (ft^3) = 2.832×10^{-2} m^3
1 cubic inch (in^3) = 1.639×10^{-5} m^3
1 fluid ounce = 1/128 gal = 2.957×10^{-5} m^3
1 barrel = 42 gal = 0.1590 m^3
1 gallon (U.S.; gal) = 3.785×10^{-3} m^3
1 gallon (British) = 4.546×10^{-3} m^3

Angle, Phase

1 degree (°) = π/180 rad = 1.745×10^{-2} rad
1 revolution (rev) = 360° = 2π rad
1 cycle = 360° = 2π rad

* The length of the year changes very slowly with changes in Earth's orbital period.

Speed, Velocity

1 km/h = (1/3.6) m/s = 0.2278 m/s

1 mi/h (mph) = 0.4470 m/s

1 ft/s = 0.3048 m/s

1 ly/y = 3.00×10^8 m/s

Angular Speed, Angular Velocity, Frequency, and Angular Frequency

1 rev/s = 2π rad/s = 6.283 rad/s (s^{-1})

1 Hz = 1 cycle/s = 2π s^{-1}

1 rev/min (rpm) = 0.1047 rad/s (s^{-1})

Force

1 dyne = 10^{-5} N

1 pound (lb) = 4.448 N

Pressure

1 dyne/cm^2 = 0.10 Pa

1 atmosphere (atm) = 1.013×10^5 Pa

1 torr = 1 mm Hg at 0°C = 133.3 Pa

1 bar = 10^5 Pa = 0.987 atm

1 lb/in^2 (psi) = 6.895×10^3 Pa

1 in H_2O (60°F) = 248.8 Pa

1 in Hg (60°F) = 3.377×10^3 Pa

Energy, Work, Heat

1 erg = 10^{-7} J

1 calorie* (cal) = 4.184 J

1 electronvolt (eV) = 1.602×10^{-19} J

1 foot-pound (ft · lb) = 1.356 J

1 Btu* = 1.054×10^3 J

1 kWh = 3.6×10^6 J

1 megaton (explosive yield; Mt) = 4.18×10^{15} J

Power

1 erg/s = 10^{-7} W

1 horsepower (hp) = 746 W

1 Btu/h (Btuh) = 0.293 W

1 ft · lb/s = 1.356 W

Magnetic Field

1 gauss (G) = 10^{-4} T

1 gamma (γ) = 10^{-9} T

Radiation

1 curie (ci) = 3.7×10^{10} Bq

1 rad = 10^{-2} Gy

1 rem = 10^{-2} Sv

Energy Content of Fuels

Energy Source	Energy Content
Coal	29 MJ/kg = 7300 kWh/ton = 25×10^6 Btu/ton
Oil	43 MJ/kg = 39 kWh/gal = 1.3×10^5 Btu/gal
Gasoline	44 MJ/kg = 36 kWh/gal = 1.2×10^5 Btu/gal
Natural gas	55 MJ/kg = 30 kWh/100 ft^3 = 1000 Btu/ft^3
Uranium (fission)	
Normal abundance	5.8×10^{11} J/kg = 1.6×10^5 kWh/kg
Pure U-235	8.2×10^{13} J/kg = 2.3×10^7 kWh/kg
Hydrogen (fusion)	
Normal abundance	7×10^{11} J/kg = 3.0×10^4 kWh/kg
Pure deuterium	3.3×10^{14} J/kg = 9.2×10^7 kWh/kg
Water	1.2×10^{10} J/kg = 1.3×10^4 kWh/gal = 340 gal gasoline/gal
H_2O	
100% conversion, matter to energy	9.0×10^{16} J/kg = 931 MeV/u = 2.5×10^{10} kWh/kg

* Values based on the thermochemical calorie; other definitions vary slightly.

The Elements

The atomic weights of stable elements reflect the abundances of different isotopes; values given here apply to elements as they exist naturally on Earth. For stable elements, parentheses express uncertainties in the last decimal place given. For elements with no stable isotopes (indicated in **boldface**), at most three isotopes are given; for elements 99 and beyond, only the longest-lived isotope is given. (Exceptions are the unstable elements thorium, protactinium, and uranium, for which atomic weights reflect natural abundances of long-lived isotopes.) See also the periodic table inside the back cover.

Atomic Number	Names	Symbol	Atomic Weight
1	Hydrogen	H	1.00794 (7)
2	Helium	He	4.002602 (2)
3	Lithium	Li	6.941 (2)
4	Beryllium	Be	9.012182 (3)
5	Boron	B	10.811 (5)
6	Carbon	C	12.011 (1)
7	Nitrogen	N	14.00674 (7)
8	Oxygen	O	15.9994 (3)
9	Fluorine	F	18.9984032 (9)
10	Neon	Ne	20.1797 (6)
11	Sodium (Natrium)	Na	22.989768 (6)
12	Magnesium	Mg	24.3050 (6)
13	Aluminum	Al	26.981539 (5)
14	Silicon	Si	28.0855 (3)
15	Phosphorus	P	30.973762 (4)
16	Sulfur	S	32.066 (6)
17	Chlorine	Cl	35.4527 (9)
18	Argon	Ar	39.948 (1)
19	Potassium (Kalium)	K	39.0983 (1)
20	Calcium	Ca	40.078 (4)
21	Scandium	Sc	44.955910 (9)
22	Titanium	Ti	47.88 (3)
23	Vanadium	V	50.9415 (1)
24	Chromium	Cr	51.9961 (6)
25	Manganese	Mn	54.93805 (1)
26	Iron	Fe	55.847 (3)
27	Cobalt	Co	58.93320 (1)
28	Nickel	Ni	58.69 (1)
29	Copper	Cu	63.546 (3)
30	Zinc	Zn	65.39 (2)
31	Gallium	Ga	69.723 (1)
32	Germanium	Ge	72.61 (2)
33	Arsenic	As	74.92159 (2)

cont'd.

Atomic Number	Names	Symbol	Atomic Weight
34	Selenium	Se	78.96 (3)
35	Bromine	Br	79.904 (1)
36	Krypton	Kr	83.80 (1)
37	Rubidium	Rb	85.4678 (3)
38	Strontium	Sr	87.62 (1)
39	Yttrium	Y	88.90585 (2)
40	Zirconium	Zr	91.224 (2)
41	Niobium	Nb	92.90638 (2)
42	Molybdenum	Mo	95.94 (1)
43	**Technetium**	**Tc**	**97, 98, 99**
44	Ruthenium	Ru	101.07 (2)
45	Rhodium	Rh	102.90550 (3)
46	Palladium	Pd	106.42 (1)
47	Silver	Ag	107.8682 (2)
48	Cadmium	Cd	112.411 (8)
49	Indium	In	114.82 (1)
50	Tin	Sn	118.710 (7)
51	Antimony (Stibium)	Sb	121.75 (3)
52	Tellurium	Te	127.60 (3)
53	Iodine	I	126.90447 (3)
54	Xenon	Xe	131.29 (2)
55	Cesium	Cs	132.90543 (5)
56	Barium	Ba	137.327 (7)
57	Lanthanum	La	138.9055 (2)
58	Cerium	Ce	140.115 (4)
59	Praseodymium	Pr	140.90765 (3)
60	Neodymium	Nd	144.24 (3)
61	**Promethium**	**Pm**	**145, 147**
62	Samarium	Sm	150.36 (3)
63	Europium	Eu	151.965 (9)
64	Gadolinium	Gd	157.25 (3)
65	Terbium	Tb	158.92534 (3)
66	Dysprosium	Dy	162.50 (3)
67	Holmium	Ho	164.93032 (3)
68	Erbium	Er	167.26 (3)
69	Thulium	Tm	168.93421 (3)
70	Ytterbium	Yb	173.04 (3)
71	Lutetium	Lu	174.967 (1)
72	Hafnium	Hf	178.49 (2)
73	Tantalum	Ta	180.9479 (1)
74	Tungsten (Wolfram)	W	183.85 (3)
75	Rhenium	Re	186.207 (1)
76	Osmium	Os	190.2 (1)
77	Iridium	Ir	192.22 (3)
78	Platinum	Pt	195.08 (3)
79	Gold	Au	196.96654 (3)
80	Mercury	Hg	200.59 (3)
81	Thallium	Tl	204.3833 (2)
82	Lead	Pb	207.2 (1)
83	Bismuth	Bi	208.98037 (3)

cont'd.

Atomic Number	Names	Symbol	Atomic Weight
84	Polonium	Po	209, 210
85	Astatine	At	210, 211
86	Radon	Rn	211, 220, 222
87	Francium	Fr	223
88	Radium	Ra	223, 224, 226
89	Actinium	Ac	227
90	Thorium	Th	232.0381 (1)
91	Protactinium	Pa	231.03588 (2)
92	Uranium	U	238.0289 (1)
93	Neptunium	Np	237, 239
94	Plutonium	Pu	239, 242, 244
95	Americium	Am	241, 243
96	Curium	Cm	245, 247, 248
97	Berkelium	Bk	247, 249
98	Californium	Cf	249, 250, 251
99	Einsteinium	Es	252
100	Fermium	Fm	257
101	Mendelevium	Md	258
102	Nobelium	No	259
103	Lawrencium	Lr	262
104	Rutherfordium	Rf	263
105	Dubnium	Db	268
106	Seaborgium	Sg	266
107	Bohrium	Bh	272
108	Hassium	Hs	277
109	Meitnerium	Mt	276
110	Darmstadtium	Ds	281
111	Roentgenium	Rg	280
112	—	—	285
113	—	—	284
114	—	—	289
115	—	—	288
116	—	—	292

Appendix E Astrophysical Data

Sun, Planets, Principal Satellites

Body	Mass (10^{24} kg)	Mean Radius (10^6 m Except as Noted)	Surface Gravity (m/s^2)	Escape Speed (km/s)	Sidereal Rotation Period* (days)	Mean Distance from Central Body† (10^6 km)	Orbital Period	Orbital Speed (km/s)
Sun	1.99×10^6	696	274	618	36 at poles 27 at equator	2.6×10^{11}	200 My	250
Mercury	0.330	2.44	3.70	4.25	58.6	57.6	88.0 d	48
Venus	4.87	6.05	8.87	10.4	−243	108	225 d	35
Earth	5.97	6.37	9.81	11.2	0.997	150	365.3 d	30
Moon	0.0735	1.74	1.62	2.38	27.3	0.385	27.3 d	1.0
Mars	0.642	3.38	3.74	5.03	1.03	228	1.88 y	24.1
Phobos	9.6×10^{-9}	9–13 km	0.001	0.008	0.32	9.4×10^{-3}	0.32 d	2.1
Deimos	2×10^{-9}	5–8 km	0.001	0.005	1.3	23×10^{-3}	1.3 d	1.3
Jupiter	1.90×10^3	69.1	26.5	60.6	0.414	778	11.9 y	13.0
Io	0.0888	1.82	1.8	2.6	1.77	0.422	1.77 d	17
Europa	0.479	1.57	1.3	2.0	3.55	0.671	3.55 d	14
Ganymede	0.148	2.63	1.4	2.7	7.15	1.07	7.15 d	11
Callisto	0.107	2.40	1.2	2.4	16.7	1.88	16.7 d	8.2
and 13 smaller satellites								
Saturn	569	56.8	11.8	36.6	0.438	1.43×10^3	29.5 y	9.65
Tethys	0.0007	0.53	0.2	0.4	1.89	0.294	1.89 d	11.3
Dione	0.00015	0.56	0.3	0.6	2.74	0.377	2.74 d	10.0
Rhea	0.0025	0.77	0.3	0.5	4.52	0.527	4.52 d	8.5
Titan	0.135	2.58	1.4	2.6	15.9	1.22	15.9 d	5.6
and 12 smaller satellites								
Uranus	86.6	25.0	9.23	21.5	−0.65	2.87×10^3	84.1 y	6.79
Ariel	0.0013	0.58	0.3	0.4	2.52	0.19	2.52 d	5.5
Umbriel	0.0013	0.59	0.3	0.4	4.14	0.27	4.14 d	4.7
Titania	0.0018	0.81	0.2	0.5	8.70	0.44	8.70 d	3.7
Oberon	0.0017	0.78	0.2	0.5	13.5	0.58	13.5 d	3.1
and 11 smaller satellites								
Neptune	103	24.0	11.9	23.9	0.768	4.50×10^3	165 y	5.43
Triton	0.134	1.9	2.5	3.1	5.88	0.354	5.88 d	4.4
and 7 smaller satellites								
Pluto	0.015	1.2	0.4	1.2	−6.39	5.91×10^3	249 y	4.7
Charon	0.001	0.6			−6.39	0.02	6.39 d	0.2
and 2 smaller satellites								

*Negative rotation period indicates retrograde motion, in opposite sense from orbital motion. Periods are sidereal, meaning the time for the body to return to the same orientation relative to the distant stars rather than the Sun.

†Central body is galactic center for Sun, Sun for planets, and planet for satellites.

Appendix Two

APPENDIX F

NUMERICAL CONSTANTS

Fundamental Physical Constants*

Name	Symbol	Value
Speed of light	c	2.99792458×10^8 m/s
Magnitude of charge of electron	e	$1.60217653(14) \times 10^{-19}$ C
Gravitational constant	G	$6.6742(10) \times 10^{-11}$ N · m²/kg²
Planck's constant	h	$6.6260693(11) \times 10^{-34}$ J · s
Boltzmann constant	k	$1.3806505(24) \times 10^{-23}$ J/K
Avogadro's number	N_A	$6.0221415(10) \times 10^{23}$ molecules/mol
Gas constant	R	8.314472(15) J/mol · K
Mass of electron	m_e	$9.1093826(16) \times 10^{-31}$ kg
Mass of proton	m_p	$1.67262171(29) \times 10^{-27}$ kg
Mass of neutron	m_n	$1.67492728(29) \times 10^{-27}$ kg
Permeability of free space	μ_0	$4\pi \times 10^{-7}$ Wb/A · m
Permittivity of free space	$\epsilon_0 = 1/\mu_0 c^2$	$8.854187817 \ldots \times 10^{-12}$ C²/N · m²
	$1/4\pi\epsilon_0$	$8.987551787 \ldots \times 10^9$ N · m²/C²

Other Useful Constants*

Mechanical equivalent of heat		4.186 J/cal (15° calorie)
Standard atmospheric pressure	1 atm	1.01325×10^5 Pa
Absolute zero	0 K	−273.15°C
Electron volt	1 eV	$1.60217653(14) \times 10^{-19}$ J
Atomic mass unit	1 u	$1.66053886(28) \times 10^{-27}$ kg
Electron rest energy	$m_e c^2$	0.510998918(44) MeV
Volume of ideal gas (0°C and 1 atm)		22.413996(39) liter/mol
Acceleration due to gravity (standard)	g	9.80665 m/s²

*Source: National Institute of Standards and Technology (**http://physics.nist.gov/cuu**). Numbers in parentheses show the uncertainty in the final digits of the main number; for example, the number 1.6454(21) means 1.6454 ± 0.0021. Values shown without uncertainties are exact.

Appendix Three

THE GREEK ALPHABET

	UPPERCASE	LOWERCASE
Alpha	A	α
Beta	B	β
Gamma	Γ	γ
Delta	Δ	δ
Epsilon	E	ϵ
Zeta	Z	ζ
Eta	H	η
Theta	Θ	θ
Iota	I	ι
Kappa	K	κ
Lambda	Λ	λ
Mu	M	μ
Nu	N	ν
Xi	Ξ	ξ
Omicron	O	o
Pi	Π	π
Rho	P	ρ
Sigma	Σ	σ
Tau	T	τ
Upsilon	Υ	υ
Phi	Φ	ϕ
Chi	X	χ
Psi	Ψ	ψ
Omega	Ω	ω

Appendix Four

ANSWERS TO ODD-NUMBERED PROBLEMS

Chapter 13

13.1 a) 4.54×10^{-3} s, 1.38×10^{3} rad/s
b) 2.27×10^{-3} s, 2.76×10^{3} rad/s
13.3 5.53×10^{3} rad/s, 1.14×10^{-3} s
13.5 0.0500 s
13.7 a) 0.167 s b) 37.7 rad/s c) 8.44×10^{-2} kg
13.9 a) 0.375 s b) 2.66 Hz c) 16.7 rad/s
13.11 a) 0.98 m b) $\pi/2$ rad
c) $x(t) = (-0.98\text{ m})\sin([12.2\text{ rad/s}]t)$
13.13 a) -2.71 m/s^2
b) $x(t) = (1.46\text{ cm})\cos([15.7\text{ rad/s}]t + 0.715\text{ rad})$,
$v_x(t) = (-22.9\text{ cm/s})\sin([15.7\text{ rad/s}]t + 0.715\text{ rad})$,
$a_x(t) = (-359\text{ cm/s}^2)\cos([15.7\text{ rad/s}]t + 0.715\text{ rad})$
13.15 120 kg
13.17 a) 0.253 kg b) 1.22 cm c) 3.05 N
13.19 a) 1.51 s b) 26.0 N/m c) 0.308 m/s
d) 1.92 N e) −0.0125 m; 0.303 m/s; 0.216 m/s^2
13.21 a) 1.48 m/s b) 2.96×10^{-5} J
13.23 a) 1.20 m/s b) 1.11 m/s c) 36 m/s^2
d) 13.5 m/s^2 e) 0.36 J
13.25 $m = 3M$; $\frac{3}{4}$
13.27 0.240 m
13.29 a) 0.0778 m b) 1.28 Hz c) 0.624 m/s
13.31 a) 4.06 cm b) 1.21 m/s c) 29.8 rad/s
13.33 b) 23.9 cm; 1.45 Hz
13.35 a) $2.7 \times 10^{-8}\text{ kg}\cdot\text{m}^2$
b) $4.3 \times 10^{-6}\text{ N}\cdot\text{m/rad}$
13.37 $5.12 \times 10^{-2}\text{ kg}\cdot\text{m}^2$
13.41 a) 0.25 s b) 0.25 s
13.43 0.407 swings/s
13.45 2.00 m
13.47 10.7 m/s^2
13.49 $0.129\text{ kg}\cdot\text{m}^2$
13.53 A: $2\pi\sqrt{L/g}$; B: $(4\pi\sqrt{2}/3)\sqrt{L/g} = 0.943T_A$; pendulum A
13.55 A: $2\pi\sqrt{L/g}$; B: $2\pi\sqrt{\dfrac{11L}{10g}} = 1.05T_A$; pendulum B
13.57 a) 0.393 Hz b) 1.73 kg/s
13.59 a) A b) magnitude $= bA/2m$, in $-x$-direction; slope is negative
c) $a_x(0) = A\left(\dfrac{b^2}{2m^2} - \dfrac{k}{m}\right)$; if $b < \sqrt{2mk}$, $a(0) < 0$; if $b = \sqrt{2mk}$, $a(0) = 0$; if $b > \sqrt{2mk}$, $a(0) > 0$
13.61 a) kg/s b) (i) $5.0F_{\max}/k$ (ii) $2.5F_{\max}/k$
13.63 a) $6.72 \times 10^{3}\text{ m/s}^2$ b) 3.02 kN
c) 18.3 m/s, 75.6 J d) 17.6 kW e) 12.1 kN, 36.7 m/s, 302 J, 141 kW
13.65 a) all unchanged b) 1/4 as large c) halved
d) $1/\sqrt{5}$ as great e) U: unchanged; K: 1/5 as large
13.67 a) 24.4 cm b) 0.220 s c) 1.19 m/s
13.69 a) 0.318 Hz, 0.500 m, 3.14 s b) 1.57 s
13.71 $\dfrac{1}{2\pi}\sqrt{\dfrac{3\sqrt{2}}{5}}\sqrt{\dfrac{g}{L}} = 0.921\left(\dfrac{1}{2\pi}\sqrt{\dfrac{g}{L}}\right)$
13.73 a) 1.49 s b) -2.12×10^{-4} s per s; shorter
c) 0.795 s
13.75 a) 0.150 m/s b) 0.112 m/s^2, downward
c) 0.700 s d) 4.38 m
13.77 a) 2.6 m/s b) 0.21 m c) 0.49 s
13.79 9.08×10^{24} kg
13.81 1.17 s
13.83 a) yes c) 2.40×10^{3} s d) no
13.87 c) -7.57×10^{-19} J e) 8.39×10^{12} Hz
13.89 0.705 Hz; 14.5°
13.91 $2\pi\sqrt{M/3k}$
13.93 $\dfrac{1}{4\pi}\sqrt{\dfrac{3g}{\sqrt{2}L}}$
13.95 c) 0.430 m
13.97 a) $k_{\text{eff}} = k_1 + k_2$ b) $k_{\text{eff}} = k_1 + k_2$
c) $k_{\text{eff}} = \dfrac{k_1k_2}{k_1 + k_2}$ d) $\sqrt{2}$
13.99 a) $Mv^2/6$ c) $\omega = \sqrt{3k/M}$; $M' = M/3$
13.101 579 N/m

Chapter 15

15.1 a) 0.439 m; 1.28 ms b) 0.219 m
15.3 220 m/s = 800 km/h
15.5 a) 4.3×10^{14} Hz to 7.5×10^{14} Hz; 1.3×10^{-15} s to 2.3×10^{-15} s b) no
15.7 a) $f = 25.0$ Hz, $T = 0.0400$ s, $k = 19.6$ rad/m
b) $y(x, t) = (0.0700\text{ m})\cos 2\pi\left(\dfrac{x}{0.320\text{ m}} + \dfrac{t}{0.0400\text{ s}}\right)$ c) +0.0495 m d) 0.0050 s
15.9 c) $-x$-direction for both
d) $v_y(x, t) = \omega A\cos(kx + \omega t)$,
$a_y(x, t) = -\omega^2 A\sin(kx + \omega t)$
15.11 a) 4.0 mm b) 0.040 s c) 0.14 m, 3.6 m/s
d) 0.24 m, 6.0 m/s e) no
15.13 b) $+x$-direction
15.15 a) 16.3 m/s b) 0.136 m c) both increase by factor of $\sqrt{2}$
15.17 0.390 s
15.19 a) 10.0 m/s b) 0.250 m
c) $y(x, t) = (3.00\text{ cm})\cos[(8.00\pi\text{ rad/s})x - (80.0\pi\text{ rad/s})t]$ d) 1890 m/s^2 e) yes
15.21 a) 95 km b) $2.5 \times 10^{-7}\text{ W/m}^2$
c) 1.1×10^{5} W
15.23 a) 0.050 W/m^2 b) 2.2×10^{4} J
15.25 707 W
15.33 a) $(1.33\text{ m})n$, $n = 0, 1, 2, 3, \ldots$
b) $(1.33\text{ m})(n + \frac{1}{2})$, $n = 0, 1, 2, 3, \ldots$
15.39 a) 96.0 m/s b) 461 N c) 1.13 m/s, 426 m/s^2
15.41 a) 2.80 cm c) 277 cm d) 185 cm, 0.126 s, 1470 cm/s e) 280 cm/s
f) $y(x, t) = (5.60\text{ cm})\sin[(0.0907\text{ rad/cm})x]\sin([133\text{ rad/s}]t)$
15.43 a) $y(x,t) = (4.60\text{ mm})\sin[(6.98\text{ rad/m})x]\sin([742\text{ rad/s}]t)$ b) 3rd c) 39.4 Hz
15.45 a) 45.0 cm b) no
15.47 a) 311 m/s b) 246 Hz c) 245 Hz, 1.40 m
15.49 a) 20.0 Hz, 126 rad/s, 3.49 rad/m
b) $y(x,t) = (2.50\text{ mm})\cos[(3.49\text{ rad/m})x - (126\text{ rad/s})t]$
c) $y(t) = (2.50\text{ mm})\cos[(126\text{ rad/s})t]$
d) $y(t) = (2.50\text{ mm})\cos[3\pi/2 - (126\text{ rad/s})t]$ e) 0.314 m/s
f) −2.50 mm, 0
15.51 a) $(7L/2)\sqrt{\mu_1/F}$ b) no
15.53 a) $(2\pi A/\lambda)\sqrt{FL/M}$ b) increase by a factor of 4
15.55 a) $4\pi^2 F\Delta x/\lambda^2$
15.57 a) 1, 0; 2, +; 3, −; 4, 0; 5, −; 6, + b) 1, −; 2, +; 3, −; 4, +; 5, −; 6, 0 c) (a): answers would reverse sign; (b): no change
15.61 c) C/B
15.63 b) k decreases by a factor of $2\sqrt{2}$; ω decreases by a factor of $\sqrt{2}$
15.65 a) 7.07 cm b) 400.0 W
15.67 $\alpha = (v_1^2 - v_2^2)\rho/Y\Delta T$
15.69 $n(0.800\text{ Hz})$, $n = 1, 2, 3, \ldots$
15.71 c) yes
15.73 c) $2A$, $2A\omega$, $2A\omega^2$
15.75 230 N
15.77 a) 0, L b) 0, $L/2$, L d) no
15.79 a) 148 N b) 26%
15.81 b) $\frac{1}{2}\mu\omega^2A^2\sin^2(kx - \omega t)$
e) $\frac{1}{2}Fk^2A^2\sin^2(kx - \omega t)$
15.83 $\pi/\omega\sqrt{2}$
15.85 a) 99.4 N c) −4.25 Hz, falls

Chapter 16

16.1 a) 0.344 m b) 1.2×10^{-5} m
c) 6.9 m, 50 Hz
16.3 a) 7.78 Pa b) 77.8 Pa c) 778 Pa
16.5 a) 1.33×10^{10} Pa b) 9.47×10^{10} Pa
16.7 90.8 m
16.9 81.5°C
16.11 0.208 s
16.13 $Y/900$
16.15 a) 9.44×10^{-11} m; 0.434 m
b) 5.66×10^{-9} m; 0.101 m
c) air; $A_{\text{air}}/A_{\text{water}} = 60.0$
16.17 a) 1.94 Pa b) $4.58 \times 10^{-3}\text{ W/m}^2$
c) 96.6 dB
16.19 a) $4.4 \times 10^{-12}\text{ W/m}^2$ b) 6.39 dB
c) 5.8×10^{-11} m
16.21 14.0 dB
16.23 a) 20.0
16.25 a) *fundamental:* displacement node at 0.60 m, pressure nodes at 0 and 1.20 m; *first overtone:* displacement nodes at 0.30 m and 0.90 m, pressure nodes at 0. 0.60 m, and 1.20 m; *second overtone:* displacement nodes at 0.20 m, 0.60 m, and 1.00 m, pressure nodes at 0, 0.40 m, 0.80 m, and 1.20 m b) *fundamental:* displacement node at 0, pressure node at 1.20 m; *first overtone:* displacement nodes at 0 and 0.80 m, pressure nodes at 0.40m and 1.20 m, and; *second overtone:* displacement nodes at 0, 0.48 m, and 0.96 m, pressure nodes at 0.24 m, 0.72 m, 1.20 m
16.27 506 Hz, 1520 Hz, 2530 Hz
16.29 a) 267 Hz b) no
16.31 a) 614 Hz b) 1.23 kHz
16.33 a) 172 Hz b) 86 Hz
16.35 0.125 m
16.37 a) 4 beats/s b) 3.0×10^{-8} m, 0
16.39 1.3 Hz
16.41 780 m/s
16.43 a) 375 Hz b) 371 Hz c) 4 Hz
16.45 a) 0.25 m/s b) 0.91 m
16.47 19.8 m/s
16.49 26.8 Hz
16.51 $0.095c$; toward us
16.53 a) 36.0° b) 2.23 s
16.55 a) 1.00 b) 8.00 c) 47.3 nm
16.57 b) $3f_0$ c) $v = 4Lf_0$
16.59 flute harmonic $3N$ ($N = 1, 3, \ldots$) resonates with string harmonic $4N$
16.61 a) stopped b) $n = 7$, $n = 9$ c) 43.9 cm
16.63 a) $v/(2L)$, v/L, $3v/(2L)$
16.65 a) 375 m/s b) 1.39 c) 0.8 cm
16.67 a) $n(77.3\text{ Hz})$, $n = 1, 2, 3, \ldots$
16.69 1.27
16.71 a) 548 Hz b) 652 Hz
16.73 a) $I = 2\pi^2\sqrt{\rho B}f^2(\Delta R)^2$
b) $P = 8\pi^3\sqrt{\rho B}f^2R^2(\Delta R)^2$
c) $A = (R/d)\Delta R$, $p_{\max} = 2\pi\sqrt{\rho B}(Rf/d)\Delta R$, $I = 2\pi^2\sqrt{\rho B}(fR/d)^2(\Delta R)^2$
16.75 a) 6.74 cm b) 147 Hz
16.77 b) 2.0 m/s
16.79 a) 1.2×10^{6} m/s b) 3.6×10^{16} m = 3.8 ly
c) 5.2×10^{3} ly; 4100 BCE
16.81 a) $f_0\left(\dfrac{2v_W}{v - v_W}\right)$ b) $f_0\left(\dfrac{2v_W}{v + v_W}\right)$
16.83 d) 9.69 cm/s; $6.67 \times 10^{2}\text{ m/s}^2$

Chapter 33

33.1 39.4°
33.3 a) 1.55 b) 549 nm
33.5 a) 5.17×10^{-7} m b) 3.40×10^{-7} m
33.7 a) 47.5° b) 66.0°
33.9 2.51×10^{8} m/s
33.13 a) frequency = f; wavelength = $n\lambda$; speed = $nf\lambda = nv$ b) frequency = f; wavelength = $\left(\dfrac{n}{n'}\right)\lambda$; speed = $\left(\dfrac{n}{n'}\right)f\lambda = \left(\dfrac{n}{n'}\right)v$
33.15 71.8°
33.17 a) 51.3° b) 33.8°
33.19 a) 58.1° b) 22.8°
33.21 1.77
33.23 24.4°
33.25 a) A: $I_0/2$ B: $I_0/8$ C: $3I_0/32$ b) 0
33.27 a) 1.40 b) 35.5°

33.29 $\alpha = \arccos\left(\frac{\cos\theta}{\sqrt{2}}\right) = \cos^{-1}\left(\frac{\cos\theta}{\sqrt{2}}\right)$
33.31 $6.38\ \text{W/m}^2$
33.33 a) first: $I = I_0/2$, second: $I = 0.25I_0$, third: $I = 0.125I_0$ all linearly polarized along the axis of their respective filters.
33.35 a) $I_R = 0.374I$ b) $I_V = 2.35I$
33.39 a) $\sin\theta_3 = (n_1 \sin\theta_1)/n_3$ c) yes
33.41 72.0°
33.45 1.53
33.47 1.8
33.49 a) 48.6° b) 48.6°
33.51 39.1°
33.53 a) $n = 1.11$ b) i) 9.75 ns ii) 4.07 ns; total = 8.95 ns
33.55 b) 0.22°
33.61 b) 38.9° c) 5.0°
33.63 a) 35° b) $10.1\ \text{W/m}^2$, $19.9\ \text{W/m}^2$
33.67 a) $\Delta = 2\theta_a^A - 6\sin^{-1}\left(\frac{1}{n}\sin\theta_a^A\right) + 2\pi$
b) $\cos^2\theta_2 = (n^2 - 1)/8$
c) red: $\theta_2 = 71.9°$; $\Delta = 230.1°$; violet: $\theta_2 = 71.6°$, $\Delta = 233.2°$; violet

Chapter 34

34.1 39.2 cm to right of mirror; 4.85 cm
34.3 image at (x_0, y_0)
34.5 b) 33.0 cm to left of vertex, 1.20 cm tall, inverted, real
34.7 0.213 mm
34.9 18.0 m from convex side of glass shell, 0.50 cm tall, erect, virtual
34.11 a) $m = \frac{f}{(f - s)}$ c) $s > f$ d) $s < f$ e) $-\infty$ f) $s = f$ g) $s' = 0$ i) $s < f$ j) $s > f$ k) $s > 2f$ l) it becomes infinite
34.13 a) concave b) $f = 2.50$ cm, $R = 5.00$ cm
34.15 2.67 cm
34.17 a) at the center of the ball, $m = +1.33$ b) no
34.19 $s = 0.395$ m
34.21 8.35 cm to left of vertex, 0.326 mm, erect
34.23 a) 1.06 m to right of lens, 17.7 mm tall, real, inverted b) all same as (a)
34.25 71.2 cm to right of lens, $m = -2.97$
34.27 $f = 3.69$ cm, object is 2.82 cm to left of lens
34.29 $n = 1.67$
34.33 Object is 26.3 cm from lens with height 1.24 cm; image is erect; same side
34.35 10.2 m
34.37 a) 1.4×10^{-4} b) 5.25×10^{-4} c) 1.50×10^{-3}
34.39 a) 85 mm b) 135 mm
34.41 a) 11 b) 2.160×10^{-3} s
34.43 a) convex b) 50 mm to 56 mm
34.45 a) 80.0 cm b) 76.9 cm
34.47 a) +2.33 diopters b) −1.67 diopters
34.49 a) 6.06 cm b) 4.12 mm
34.51 4.17 cm from lens; image is located on same side as ant
34.53 a) 8.37 mm b) 21.4 c) 297
34.55 19.4 m
34.57 a) −6.33 b) 1.90 cm c) 0.126 rad = 7.22°
34.59 a) 66.1 cm b) −59.1
34.61 4.80 m/s
34.63 $n/2$
34.65 a) 13.3 cm b) 26.2 cm
34.67 a) 46.2 cm from mirror, on opposite side of mirror; virtual b) 2.88 cm, erect c) no
34.69 a) $-12.0\ \text{cm} < s < 0$ b) erect
34.71 $f = \pm 4.4$ cm, ± 13.3 cm
34.73 $v = 31$ m/s
34.75 b) i) 120.00 cm from mirror, 119.96 cm from mirror ii) $m = -0.600$, $m' = -0.360$
c) faces perpendicular to axis: squares with side 0.600 mm: faces parallel to axis: rectangles with sides of length 0.360 mm (parallel to axis) and 0.600 mm (perpendicular to axis)
34.77 b) image = 2.4 cm high; $m = -0.13$
34.79 a) −3.3 cm b) virtual c) 1.9 cm to right of vertex at right end of rod d) real, inverted e) 105 mm
34.81 a) $f = 58.7$ cm, converging b) $h = 4.47$ mm, virtual
34.83 a) 2.53 mm
34.85 a) $R = 8.8$ mm b) no. behind the retina c) $s' = 14$ mm from the cornea. In front of the retina. Yes. The lens needs to complete the focusing.
34.87 2.00
34.89 a) 3.75 cm to left of first lens b) 332 cm c) real d) $h = 60.0$ mm. inverted.
34.91 10.6 cm
34.93 a) 0.24 m b) 0.24 m
34.95 Inside the glass, 72.1 cm from the spherical surface
34.97 0.80 cm
34.99 −26.7 cm
34.101 1.24 cm above page
34.103 a) 46.7 m b) 35.0 m
34.105 134 cm to left of object

Chapter 35

35.1 a) 2.50 m b) 1.00 m, 4.00 m
35.3 0.75 m, 2.00 m, 3.25 m, 4.50 m, 5.75 m, 7.00 m, 8.25 m
35.5 a) 2.0 m b) constructively c) 1.0 m; destructively
35.9 0.83 mm
35.11 590 nm
35.13 12.6 cm
35.15 1200 nm
35.17 a) $m = 19$, 39 bright fringes b) $m = \pm 19$, $\theta = \pm 73.3°$
35.19 a) $0.750I_0$ b) 80 nm
35.21 1670 rad
35.23 a) 0.888 mm b) 0.444 mm
35.25 71.4 m
35.27 114 nm
35.29 0.0235°
35.31 a) $\Delta T = 56$ nm b) i) 2180 nm ii) 198.5 nm; 11.0 wavelengths
35.33 a) 514 nm; green b) 603 nm; orange
35.35 0.11 μm
35.37 0.570 mm
35.39 1.82 mm
35.41 $n = 1.730$
35.43 27.3°, 66.5°
35.45 $n = 1.57$
35.47 b) constructive: $r_2 - r_1 = (m + \phi/2\pi)\lambda$, $m = 0, \pm 1, \pm 2, \pm 3, \ldots$; destructive: $r_2 - r_1 = \left(m + \frac{1}{2} + \phi/2\pi\right)\lambda$, $m = 0, \pm 1, \pm 2, \pm 3, \ldots$
35.49 a) $\sqrt{x^2 + (y + d)^2} - \sqrt{x^2 + (y - d)^2} = m\lambda$ c) $\sqrt{x^2 + (y + d)^2} - \sqrt{x^2 + (y - d)^2} = \left(m + \frac{1}{2}\right)\lambda$
35.51 $6.8 \times 10^{-5}\ (\text{C°})^{-1}$
35.53 $\lambda/2d$, independent of m
35.55 b) 72 cm
35.57 $n = 1.42$
35.59 a) pattern moves down the screen b) $I = I_0\cos^2[(\pi/\lambda)(d\sin\theta + (n - 1)L)]$ c) $d\sin\theta = m\lambda - (n - 1)L$
35.61 14.0

Chapter 36

36.1 506 nm
36.3 $m_{max} = 113$; 226 dark fringes
36.5 ± 45.4 cm
36.9 $\pm 16.0°$, $\pm 33.4°$, $\pm 55.6°$
36.11 0.920 μm
36.13 a) 10.8 mm b) 5.4 mW
36.15 a) 6.75 mm b) $2.43 \times 10^{-6}\ \text{W/m}^2$
36.17 a) 668 nm b) $9.36 \times 10^{-5}I_0$
36.19 a) $\pm 13.0°$, $\pm 26.7°$, $\pm 42.4°$, $\pm 64.1°$ b) $I = 2.08\ \text{W/m}^2$
36.21 a) 3 b) 2
36.23 a) $\pm 0.0627°$ b) $0.249I_0$ c) $0.0256I_0$
36.25 cases (i), (iii): slits 1 and 3 and slits 2 and 4; case (ii): slits 1 and 2 and slits 3 and 4
36.27 $a = 1.50 \times 10^4$ nm in width; $d = 4.50 \times 10^4$ nm in separation
36.29 a) 4790 b) 19.0°, 40.7° c) no
36.31 a) yes b) 13.3 nm
36.33 23.3°, 52.3°
36.35 10.5°, 21.3°, 33.1°
36.37 a) $R = 17{,}500$ b) yes c) i) 587.8170 nm ii) 587.7834 nm iii) $587.7834\ \text{nm} < \lambda < 587.8170\ \text{nm}$
36.39 0.232 nm
36.41 a) 0.461 m
36.43 1.9 m
36.45 92 cm
36.47 1.45 m
36.49 a) Hubble: 77 m; Arecibo: 1.1×10^6 m b) 1500 km
36.51 no
36.53 a) i) 25.6° ii) 10.2° iii) 5.1° b) i) 60.0° ii) 23.1° iii) 11.5°
36.55 2.07
36.57 a) 1.80 mm b) 0.798 mm
36.59 $\Delta\theta_{\pm} = \frac{2\lambda}{dN}$
36.61 b) for $3\pi/2$: any two slits separated by one other slit; for the other cases: any two slits separated by three other slits
36.65 513 nm
36.67 second order
36.69 c) ± 2.6 rad
36.71 492 km

Chapter 38

38.1 a) about $8.3 \times 10^{-19}\ \text{J} = 5.2\ \text{eV}$ b) about $6.1 \times 10^{-34}\ \text{J} \cdot \text{s} = 3.8 \times 10^{-15}\ \text{eV} \cdot \text{s}$
38.3 a) 5.77×10^{14} Hz, $1.27 \times 10^{-27}\ \text{kg} \cdot \text{m/s}$, $3.82 \times 10^{-19}\ \text{J} = 2.38\ \text{eV}$
38.5 a) 5.92×10^{20} Hz b) 5.06×10^{-13} m
38.7 2.5×10^5 m/s
38.9 a) 5.0×10^{14} Hz b) 2.3×10^{20} photons/s c) no
38.11 a) $K_2 = 4K_1$ b) $E_2 = 2E_1$
38.13 a) 264 nm b) 4.70 eV
38.15 a) 434.1 nm b) 6.906×10^{14} Hz c) $4.576 \times 10^{-19}\ \text{J} = 2.856\ \text{eV}$
38.17 a) − 5.08 eV b) − 5.63 eV
38.19 a) $E_1 = -17.50$ eV, $E_1 = -4.38$ eV, $E_1 = -1.95$ eV, $E_1 = -1.10$ eV $E_1 = -0.71$ eV
38.21 a) $5.82 \times 10^{-13}\ \text{J} = 3.63\ \text{MeV}$ b) $5.82 \times 10^{-13}\ \text{J} = 3.63\ \text{MeV}$ c) 1.32×10^7 m/s
38.23 $3\hbar = 3.16 \times 10^{-34}\ \text{kg} \cdot \text{m}^2/\text{s}$
38.25 a) −218 eV, 16 times greater b) 218 eV, 16 times greater c) 7.63 nm d) 1/4 as large
38.27 a) $v_1 = 2.19 \times 10^6$ m/s, $v_2 = 1.09 \times 10^6$ m/s, $v_3 = 7.29 \times 10^5$ m/s b) $T_1 = 1.52 \times 10^{-16}$ s, $T_2 = 1.22 \times 10^{-15}$ s, $T_3 = 4.10 \times 10^{-15}$ s c) 8.22×10^6
38.29 4.00×10^{17}
38.31 a) 1.2×10^{-33} b) 3.4×10^{-17} c) 5.9×10^{-9}
38.33 0.310 nm, the same
38.35 a) $f_{min} = 6.04 \times 10^{18}$ Hz b) $\lambda_{min} = 0.0497$ nm
38.37 0.0714 nm, 180°
38.39 a) 4.39×10^{-4} nm b) 0.04294 nm c) $\Delta E_r = -300$ eV d) $\Delta E_e = +300$ eV
38.43 a) 2.06×10^4 K b) 141 nm
38.45 1.06 mm, microwave
38.47 a) $\lambda = \frac{hc}{4.965kT}$ b) $\lambda_m T = 2.90 \times 10^{-3}\ \text{m} \cdot \text{K}$
38.51 a) 1.04 eV b) 1.20 μm c) 2.51×10^{14} Hz d) 4.14×10^{-7} eV
38.53 a) 4.59×10^{14} Hz b) 653 nm c) 1.89 eV d) $6.59 \times 10^{-34}\ \text{J} \cdot \text{s}$
38.55 a) $hc(\lambda_1 - \lambda_2)/e\lambda_1\lambda_2$ b) 0.476 V
38.57 a) 5.1×10^{-3} J b) 11 W c) $N = 1.5 \times 10^{16}$
38.59 a) 1.69×10^{-28} kg b) − 2.53 keV c) 0.653 nm
38.61 a) $6.99 \times 10^{-24}\ \text{kg} \cdot \text{m/s}$ b) 705 eV
38.63 a) 12.09 eV b) 3 possibilities: $3 \rightarrow 2$ (656 nm), $3 \rightarrow 1$ (103 nm), $2 \rightarrow 1$ (122 nm)
38.65 a) 0.90 eV
38.67 a) 5×10^{49} photons/s b) 3×10^4
38.69 2.98×10^4 K

38.71 a) $\Delta\lambda = \frac{h}{2mc}$ b) 6.61×10^{-16} m for any n
38.73 a) 5×10^{-33} m b) $(4 \times 10^{-9})°$ c) 0.1 mm
38.75 a) 5.10×10^{-17} J = 319 eV, 1.06×10^{7} m/s
b) 3.89 nm
38.77 5.67×10^{-8} W/(m$^2 \cdot$ K^4)
38.79 $f \approx \frac{me^4}{4\epsilon_0^2 n^3 h^3}$
38.81 b) 7.08×10^{-15} m c) gamma rays

Chapter 39

39.1 a) 1.55×10^{-10} m b) 8.44×10^{-14} m
39.3 a) 2.37×10^{-24} kg·m/s
b) 3.67×10^{-18} J = 19.2 eV
39.5 a) 3.32×10^{-10} m b) 1.33×10^{-9} m
39.7 a) $m = 75$ kg, $v = 1.0$ m/s: $\lambda = 8.8 \times 10^{-36}$ m
39.9 a) photon: 620 nm; electron: 0.274 nm
b) photon: 4.96 eV; electron: 2.41×10^{-5} eV
c) 250 nm; electron
39.11 3.90×10^{-34} m; no
39.13 a) 7.3×10^{6} m/s b) 150 eV c) 12 keV
d) electrons
39.15 0.431 eV
39.17 a) $m = 1$: 2.07°; $m = 2$: 4.14°
b) 1.81 cm, 3.61 cm, 1.81 cm
39.19 a) 8.79×10^{-32} m/s; no
39.21 the claim is not valid
39.23 2.03×10^{-32} J = 1.27×10^{-13} eV
39.25 0.087 MeV = $2.8 \times 10^{-5} E$
39.27 a) 8.25×10^{3} b) electrons
39.29 a) $x = (2n + 1)(\lambda/4)$, n an integer
b) $x = n\lambda/2$, n an integer
39.31 a) $\int_{-\infty}^{\infty} |\Psi|^2 dV = 1$
integrating the probability density function between infinite limits must equal unity
b) no. no. not a valid wave function.
c) $A = \sqrt{2b}$ m$^{-1/2}$
39.33 1
39.35 a) 1.414 m$^{-1/2}$ c) i) $P = 0.865$ ii) $P = \frac{1}{2}$
iii) $P = 0.0585$
39.37 ψ cannot be a solution
39.39 a) 12 eV b) 1.5×10^{-4} V, 7.3×10^{3} m/s,
c) 8.2×10^{-8} V; 4.0 m/s
39.41 a) 9.42×10^{5} m/s; nonrelativistic
b) $V_{acc} = 2.53$ V
39.43 a) $E = c\sqrt{2mK} = 4.05 \times 10^{-7}\sqrt{K}$
b) $E_{photon} > E_{electron}$
39.45 1.66×10^{-17} m; no
39.47 a) 1.10×10^{-10} m b) 9.09×10^{-13} m
39.49 a) $(1/\sqrt{15})(h/mc)$
b) i) 1.53 MeV, 6.26×10^{-13} m
ii) 2.81×10^{3} MeV, 3.41×10^{-16} m
39.51 a) 2.1×10^{-20} kg·m/s b) 39 MeV
c) 0.29 MeV; no
39.53 1.4×10^{-35} kg = $(5.8 \times 10^{-8}\, m_{pion})$
39.55 a) 1.1×10^{-35} m/s b) 2.3×10^{27} y, no
39.57 a) 7.27×10^{5} m/s, nonrelativistic
b) $V = 1.50$ V
39.59 a) $2d\sin\theta = m\lambda$, $m = 1, 2, 3 \ldots$ b) 53.1°
c) less
39.61 a) $A|x|/x$, $x \neq 0$ b) $(3/2)(h^2A^2/m)^{1/3}$
39.63 $\omega = E/\hbar$
39.65 a) 0.21 kg·m/s b) 1.7 m
39.67 $P = 4\pi|A|^2 r^2 e^{-2\alpha r^2} dr$, r = $1/\sqrt{2\alpha}$; no
39.69 a) $\psi(x) = (\sin k_0 x)/k_0 x$ b) L c) $2L$
d) h, h
39.71 2.2×10^{-16} m

Chapter 40

40.1 a) 1.2×10^{-67} J b) 1.1×10^{-33} m/s
c) 1.4×10^{33} s d) 3.7×10^{-67} J
40.3 $L = 1.66 \times 10^{-10}$ m
40.5 0.61 nm
40.7 a) 0, $L/2$, L b) $L/4$, $3L/4$ c) yes
40.11 a) 6.0×10^{-10} m, 1.1×10^{-24} kg·m/s
b) 3.0×10^{-10} m, 2.2×10^{-24} kg·m/s
c) 2.0×10^{-10} m, 3.3×10^{-24} kg·m/s
40.15 3.43×10^{-10} m
40.19 2.2×10^{-14} m
40.21 a) 4.3×10^{-8} b) 4.2×10^{-4}
40.23 $1/\sqrt{2}$
40.25 a) 1.3×10^{-3} b) 10^{-148}
40.27 a) 1.11×10^{-33} J = 6.90×10^{-15} eV;
2.21×10^{-33} J = 1.38×10^{-14} eV
40.29 a) 0.21 eV b) 5900 N/m
40.31 $\Delta x \Delta p = (2n + 1)\hbar$
40.33 a) 5.9×10^{-3} eV b) 106 μm
c) 0.0118 eV = 1.18×10^{-2} eV
40.35 $B = \frac{K_1 - K_2}{K_1 + K_2}A$; $C = \frac{2K_1}{K_1 + K_2}A$
40.37 a) 19.2 μm b) 1[illegible].5 μm
40.39 a) $\left(\frac{1}{2} + 1/\pi\right)$ b) $\frac{1}{2}$ c) yes
40.41 a) $2dx/L$ b) 0 c) $2dx/L$
40.45 a) $B = C$, $A\sin\frac{\sqrt{2mE}}{\hbar}L + B\cos\frac{\sqrt{2mE}}{\hbar}L = De^{-\kappa L}$
b) $\frac{\sqrt{2mE}}{\hbar}A = \kappa C$, $\frac{\sqrt{2mE}}{\hbar}\left(A\cos\frac{\sqrt{2mE}}{\hbar}L - B\sin\frac{\sqrt{2mE}}{\hbar}L\right) = -\kappa De^{-\kappa L}$
40.49 6.63×10^{-34} J = 4.14×10^{-15} eV,
1.33×10^{-33} J = 8.27×10^{-15} eV, no
40.51 b) $A_0 = \left(\frac{m\omega}{\hbar\pi}\right)^{1/4}$
c) classical turning points $A = \pm\sqrt{\frac{\hbar}{\omega m}}$
40.53 a) $(n_x + n_y + n_z + (3/2))\hbar\omega$
b) $(3/2)\hbar\omega$, $(5/2)\hbar\omega$
40.55 a) $E_n = n^2h^2/8mL^2$, $n = 2, 4, 6 \ldots$
b) $E_n = n^2h^2/8mL^2$, $n = 1, 3, 5 \ldots$ c) same
d) odd in part (a), even in part (b)
40.57 b) increases c) infinite
40.59 a) $-E/A$, $+E/A$ c) decrease

Appendix Five

Answers to Odd-Numbered Exercises and Problems

Chapter 16

15. 20° C
17. −40
19. 102.4°F
21. 32 kJ
23. 96.9 W
25. (a) 168 J/K (b) 480 J/kg·K
27. 0.293 W
29. 55 kW
31. −4.17 kW
33. $R_{air} = 5.56$ per in; $R_{conc} = 0.143$ per in; $R_{fg} = 3.45$ per in; $R_{glass} = 0.167$ to 0.2 per in; $R_{sf} = 5$ per in; $R_{pine} = 1.28$ per in
35. 2240 W
37. 22 mm^2
39. (a) 138 kPa (b) 33.4 kPa (c) 233 kPa
41. −10.0°C
43. 0.36 kg
45. (a) 23 kJ (b) 337 kJ (c) 65 kJ
47. 2.51 min
49. (a) 560 g (b) 0.27 K/s
51. 1.8 kg
53. 9.22 K
55. 197 g
57. −14.2 W
59. 6.47 gal
61. 23°C
63. (a) 1.15×10^3 K (b) 700 K
65. 24°C
67. 1151 K
69. (a) $87 (b) $10
71. 44.5 K
73. 7.5 K
75. 3.3 h
77. 2.9 J

Chapter 17

19. 3.17×10^{23}
21. (a) $1.27V_2$ (b) $2V_2$
23. 2.7×10^7
25. H_2
27. Lead
29. 143 kJ
31. 120 mol/m^3
33. 0.985 L
35. 263°C.
37. 2.69×10^{25} m^{-3}, about 2.5×10^{10} times greater
39. (a) 235 mol (b) 5.65 m^3
41. (a) 1.27 atm (b) 0.0268 mol (c) 0.786 atm
43. 1.3×10^{10} kg
45. (a) 3.26×10^{19} J to melt at 0°C (b) 1.80×10^{13} W
47. 11.8 d
49. 3.55 MJ
51. 1.51 h
53. 5.0 kg
55. 0.798 kg
57. 79 g
59. 1.2 kg ice, 0.80 kg water, all at 0°C
61. ethyl alcohol
63. 59.2 L
65. 8.35 h
69. 1.00034 L
71. 36.5 d, fast
73. Yes. On Venus, one mole would occupy 0.67 L, so a 1 L container is big enough.
75. No; it would take 50 minutes.

Chapter 18

15. 29 kJ
17. Increase by 250 J
19. 13.9 kW
21. $2P_1V_1$
23. (a) 1.33 (b) 215 J
25. $V = 0.18V_0$
27. 57.7%
29. (a) 200 K (b) 120 K
31. (a) 3.3 kJ (b) 0.392 mol
33. (a) 1.49 mm (b) 10.7 μJ
35. 1.35
37. (a) 300 kPa (b) −243 J
39. 440°C
41. (a) 810 K (b) 25.8 atm
43. (a) 300 K, 1.5 kJ (b) 336 K 0 J (c) 326 K, 429 J
45. (a) 39.9 kPa (b) 83.3 kPa (c) 80.2 kJ
47. 928 J
49. (a) 211 J (b) 12.9 L
51. 91.6 J
53. 335 K
55. (a) 598 J (b) 2500 J flows in
57. 79%
59. 25 m
61. (a) 218 J (b) 2.71 kJ
65. (a) 15.2°C (b) 15.5 kJ
67. $e = 1 - r_1 - y\left[\dfrac{r\dfrac{\gamma}{c} - 1}{\gamma(r_c - 1)}\right]$
69. $W = nRT \ln\left[\dfrac{V_2 - nb}{V_1 - nb}\right] + an^2\left[\dfrac{1}{V_2} - \dfrac{1}{V_1}\right]$
71. −9.8 K/km
73. 1.4 atm, $V = 8.67$ L
75. 130 J

Chapter 19

15. 99.95%
17. (a) 39% (b) 550 J (c) 190°C
19. 53.3 kJ
21. 275 J/K
23. 21.9 kg
25. (a) 420 J (b) 47.2% (c) 290 K (d) 9.24 kW
27. (a) 1.75GW (b) 43% (c) 505 K
29. 2×10^7 kg/s, slightly more than the Mississippi's flow
31. (a) 8.53 (b) 1.10×10^3 kg
33. (a) 69% (b) 967 K
35. (a) 4.3 (b) maximum COP = 11
37. 2.38
39. (a) 5.7 (b) 3.5 kW (c) heat pump 53¢/h, oil $1.30/h,
41. (a) 17.4% (b) 83.3%.
43. 1.35×10^8 J/K
47. 11.7 J/K
49. 74.9 kJ
51. 163 J/K
53. (a) 35% (b) 5.88 h
55. (a) $e_{otto} = 1 - r^{1-\gamma}$, $e_{carnot} = 1 - (1/3)r^{1-\gamma}$
57. 61%
59. (a) $Th = T_{h0}e^{-P0t/mc}(T_{h0} - T_c)$ (b) $b = 0$ at $t = \dfrac{mc(T_{h0} - T_c)}{P_0}\ln\left(\dfrac{Th^0}{Tc}\right)$
61. −0.150 J/k; cooling produces a more ordered state
63. (a) −368 kJ/K (b) 490 kJ/K (c) −65.3 kJ/K
65. No, 5.9 h